The Most Successful Way of JEE Preparation

DPP

Daily Practice Problems

JEE Main & Advanced

Laws of Motion, Work, Power & Energy

with Friction & Circular Motion

PHYSICS Vol-2

The Most Successful Way of JEE Preparation

DPP

Daily Practice Problems

JEE Main & Advanced

Laws of Motion, Work, Power & Energy

with Friction & Circular Motion

PHYSICS Vol-2

DB Singh

arihant

ARIHANT PRAKASHAN (SERIES), MEERUT

DPP
Daily Practice Problems

arihant
ARIHANT PRAKASHAN (SERIES), MEERUT

卐 Administrative & Production Offices

Corporate Office 'Ramchhaya' 4577/15, Agarwal Road, Darya Ganj, New Delhi -110002
Tele: 011- 47630600, 43518550; Fax: 011- 23280316

Head Office Kalindi, TP Nagar, Meerut (UP) - 250002
Tel: 0121-2401479, 2512970, 4004199; Fax: 0121-2401648

卐 Sales & Support Offices

Agra, Ahmedabad, Bengaluru, Bareilly, Chennai, Delhi, Guwahati, Hyderabad, Jaipur, Jhansi, Kolkata, Lucknow, Meerut, Nagpur & Pune

卐 ISBN : 978-93-13193-32-6

Published by Arihant Publications (I) Ltd.

Production Team

Publishing Managers Mahendra Singh Rawat & Keshav Mohan
Project Head Sorabh Chaudhary
Cover Designer Shanu Mahad
Inner Designer Mazher Chaudhary
Page Layouting Sunil Sharma, Hemant
Proof Reading Ashish

For further information about the products from Arihant, log on to www.arihantbooks.com or email to info@arihantbooks.com

/arihantpub /@arihantpub Arihant Publications /arihantpub

ABOUT DPP

Daily Practice Problems

Indian Institutes of Technology better known as IITs have always been the dream destination of each science group students who want to pursue their career as Engineers. These are the institutes which have got very high value not even in India but in the whole world. To get the admission in such institutes one need to face entrance Examinations; JEE Main & JEE Advanced, out of which JEE Main is considered as the preliminary exam & JEE Advanced is the final exam to get into the IITs. Approximately per year 12 lakh students enroll themselves for JEE Main and only the top 2.25 lakh in JEE Main merit can attempt for JEE Advanced, the fact clearly express the degree of toughness of JEE Main & JEE Advanced.

To have success in such an exam one needs to be very planned & systematic in its preparation. A very popular wise saying 'Rome was not built in a day' suggests that to achieve a big target like IITs we need to work continuously. For continuous work, one need to have proper planning that how much work he/she has to do on regular basis, i.e., on daily basis. Each student preparing for JEE has many books but unfortunately none of them gives the students a plan for daily studies, which is the most effective success key for JEE.

Keeping the daily practice needs of the students in mind we have come up with DPP Series (Daily Practice Problems) having approximately 26 books. The primary focus of this series is on achieving success through practice & proper planning. For that we have divided the whole syllabus of JEE Main & JEE Advanced; Physics, Chemistry & Mathematics, into different volumes (around 8 volumes of Mathematics & 9 volumes each of Physics & Chemistry), each volume has Daily Practice Problems Sheets having specific questions on various Topics of the individual chapter of the volume ensuring the complete Practice & Assessment of the topic.

DPP

Daily Practice Problems

In this way we have tried to give you an effective daily planning for JEE keeping in mind the daily hours that a student can devote to his practice, you are just required to practice one sheet of each subject on daily basis. This habit of daily practice will keep you away from exam stress and thoughts like; how much I have prepared/am I preparing properly or not/ do I need to study more on daily basis or what I am doing is ok, etc., which regularly tease your mind and can give you panic at any stage of your preparation.

Some special features that we have tried to incorporate in our Daily Practice Problems (DPP) series are discussed below

1. Micro Level coverage of all the topics of a Chapter/ Unit given in JEE Main & Advanced Syllabus.
2. Each DPP has various types of questions; Subjective Questions, Single/ Multiple Correct Options, Statements type, Comprehension based, Integer Value Answer type, Matching, etc., making you practice & ready for all the question formats of JEE.
3. New & innovative questions in each DPP challenging & improving your problem solving skills.
4. Along with Topical Coverage, Revisal DPPs for JEE Main & JEE Advanced with each Chapter, covering multidisciplinary questions proving comprehensive practice.
5. JEE Main & JEE Advanced Archive (collection of last 13 years' exams questions) with each chapter.
6. Complete Solutions for each DPP.

We hope that DPP series will prove to be very beneficial for the students preparing for JEE, by giving them daily practice in a systematic & planned manner. The suggestions/criticism from the students, teachers & parents about DPP series will be welcomed whole heartedly and we promise to update this series accordingly from time to time.

DPP
Daily Practice Problems

DAILY TASKS...

1 LAWS OF MOTION

2 FRICTION

DPP
Daily Practice Problems

③ CIRCULAR MOTION

④ WORK, POWER and ENERGY

1. Laws of Motion

DPP-1 Normal Reaction

Subjective Questions

Direction (Q. Nos. 1-2) *These questions are subjective in nature, need to be solved completely on notebook.*

Working Space

1 Find normal reaction exerted by the surface on the block in the following cases.

i. 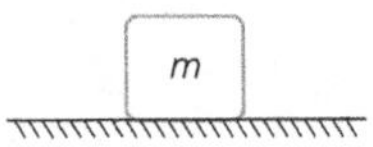

ii. 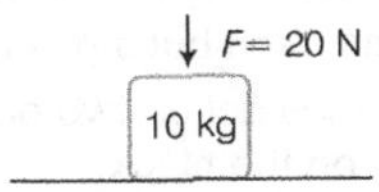

iii. 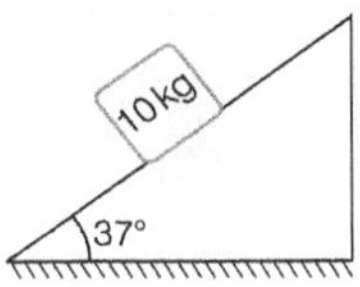

iv. 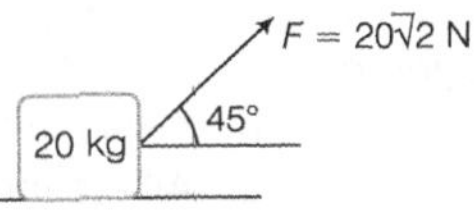

2 Find normal contact force between smooth stationary blocks in the following cases.

i. 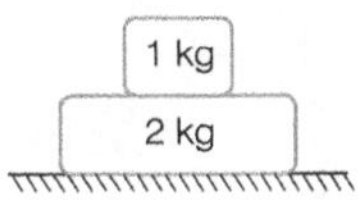

ii. 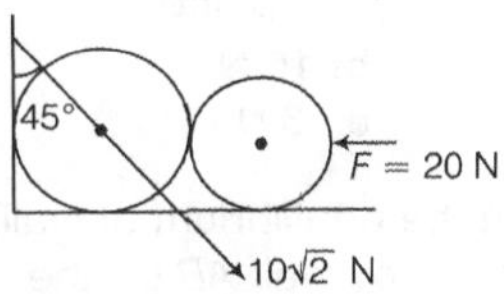

iii. 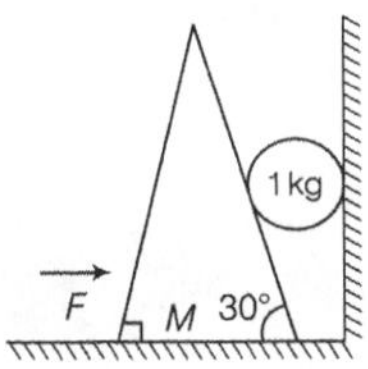

iv. 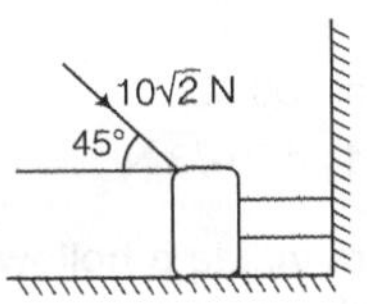

Only One Option Correct Type

Direction (Q. Nos. 3-15) *This section contains 13 multiple choice questions. Each question has four choices (a), (b), (c) and (d), out of which ONLY ONE option is correct.*

3 If action force is gravitational force, then reaction force

a. may be electromagnetic force
b. may be electric force
c. may be spring force
d. must be gravitational force

4 The magnitude of action force is

a. greater than reaction force
b. equal to reaction force
c. lesser than reaction force
d. None of these

5 If you push a wall by a force of 10 N. Then, the force by wall on you is

a. 10 N
b. 20 N
c. zero
d. $10 \le F \le 10$

6 If you push a book by applying a force $\mathbf{F} = 3\hat{i} + 4\hat{j}$, the force which is exerted by the book on you is

a. $3\hat{i} + 4\hat{j}$
b. $3\hat{i} - 4\hat{j}$
c. $-3\hat{i} - 4\hat{j}$
d. $\hat{i} + \hat{j}$

Working Space

7 The nature of normal reaction force is

a. gravitational force
b. electromagnetic force
c. nuclear force
d. weak force

8 Generally it is said, "Normal reaction always acts perpendicular to the surface of contact."

a. The statement is always correct
b. The statement is sometimes correct
c. The statement is incorrect
d. We cannot comment

9 In a general debate on topic "normal reaction and *mg* never form action and reaction pair" a number of students of JEE Mentors institute took part.

i. **Agam Goel said** Action and reaction forces should be of same nature. Here, normal reaction is electromagnetic force but *mg* is the gravitational force.
ii. **Aditi Mehta said** Action and reaction forces act on two different bodies. Here, normal reaction and *mg* both are acting on the block.
iii. **Asharya said** Action reaction forces balance each other. So, normal reaction and *mg* should be action and reaction pair.

Mark correct option as your judgement,

a. All are correct
b. Only (iii) is correct
c. Both (i) and (ii) are correct
d. All are incorrect

10 A block of mass 1 kg is moving on a horizontal *XY*-plane in the influence of force $\mathbf{F} = 3\hat{i} + 4\hat{j} - 5\hat{k}$. The acceleration due to gravity is directed in negative *z*-direction. Find the normal reaction on the block by the plane.

a. 10 N
b. 15 N
c. 20 N
d. 5 N

11 The smooth rod of mass 10 kg is in the equilibrium of walls which are fixed. Find the force exerted by wall *AB* on the rod.

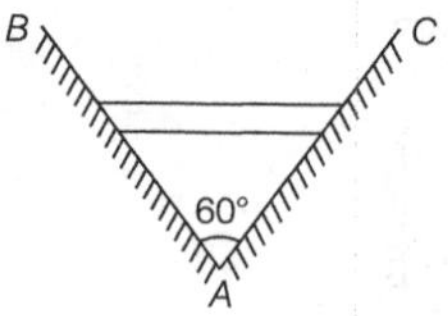

a. 100 N
b. 50 N
c. $50\sqrt{3}$ N
d. $100\sqrt{3}$ N

12 A smooth rod of mass 1 kg is in equilibrium inside a hollow fixed sphere such that $\angle ACB = 120°$. Find the normal reaction on the rod at end *A*.

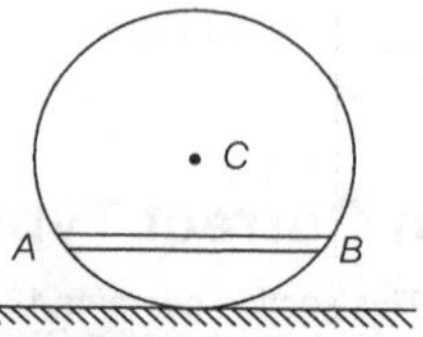

a. 20 N
b. $10\sqrt{3}$ N
c. 10 N
d. $5\sqrt{3}$ N

13 A smooth light rod not reaching the rough floor is inserted between two identical blocks. A horizontal force *F* is applied to upper end of rod. The contact force between block *A* and rod is F_1 and contact force between block *B* and rod is F_2. Then,

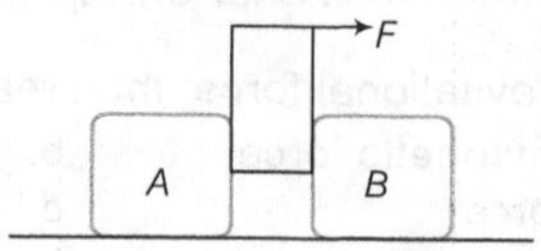

a. $F_1 = F_2$
b. $F_1 > F_2$
c. $F_1 < F_2$
d. $F_1 = 0$ and $F_2 > 0$

14 Three identical smooth cylinders each of mass $\sqrt{3}$ kg are placed as shown in the figure. Find normal reaction between cylinder *A* and cylinder *B*.

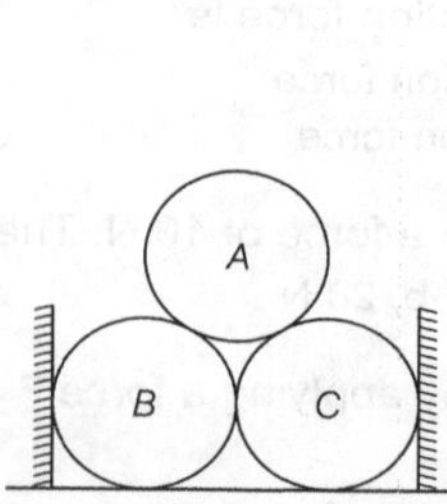

a. 10 N
b. 20 N
c. $10\sqrt{3}$ N
d. Cannot be calculated

Working Space

15 A force of $5t$ N is applied on a block of mass 1 kg. The magnitude of normal reaction on the block by horizontal ground is N_1 and magnitude of normal reaction on block due to vertical wall is N_2. N_1 *versus* N_2 graph is

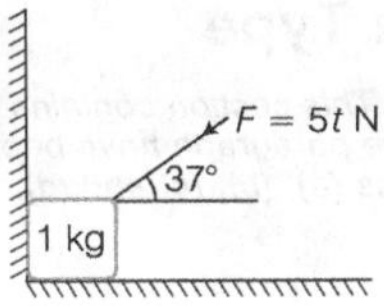

a. 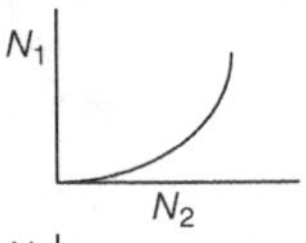

b. 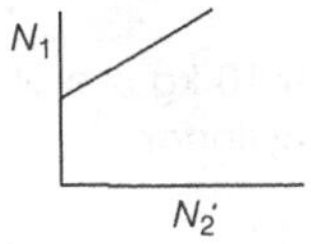

c. 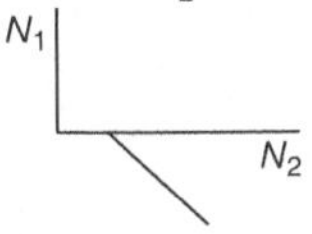

d. None of these

One or More than One Options Correct Type

Direction (Q. Nos. 16-19) *This section contains 4 multiple choice questions. Each question has four choices (a), (b), (c) and (d), out of which ONE or MORE THAN ONE are correct.*

16 A smooth block of mass 1 kg is placed on the smooth surface as shown in the figure. A variable force $F = \sqrt{2}t$ N is acting on block at an angle 45° with horizontal. Normal reaction on the block due to ground is N_1 and normal reaction due to vertical wall is N_2. Mark correct option(s).

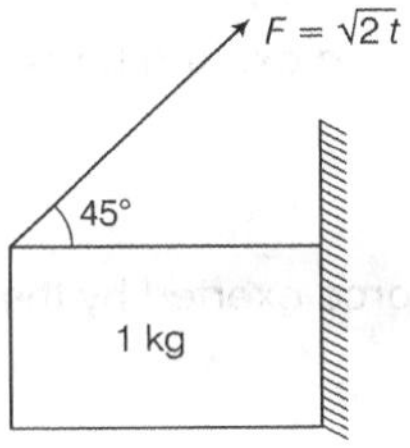

a. $N_1 = 9$ N, $N_2 = 1$ N at $t = 1$ s
b. $N_1 = 0$, $N_2 = 10$ N at $t = 10$ s
c. $N_1 = 0$, $N_2 = 5$ N at $t = 5$ s
d. None of the above

17 A smooth block A rests on a horizontal plank. The angle of inclination θ of the plank with horizontal is gradually increased from 0° to 90°. Then,

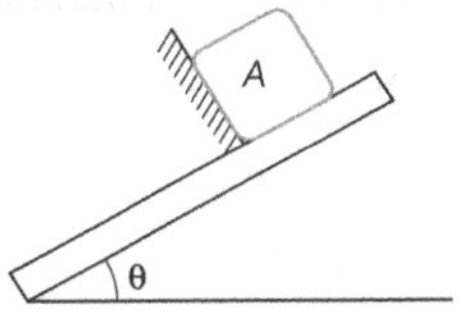

a. the normal reaction between plank and block gradually decreases up to zero
b. the normal contact force between wall and the block gradually increases up to mg
c. the normal contact force between the wall and block is always zero
d. the normal contact force between the plank and block gradually increases up to mg

18 For the situation shown in figure, the rod of mass m is stationary and vertical wall is smooth. Then,

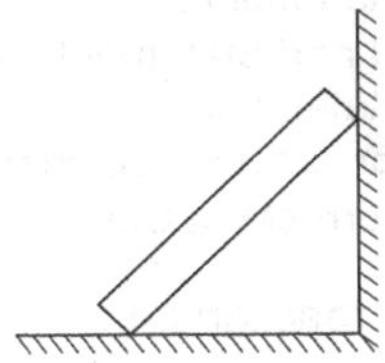

a. the normal contact force by vertical wall is $\frac{mg}{2}$
b. the normal contact force between the rod and horizontal surface is mg
c. the force exerted by horizontal surface is greater than mg
d. the normal contact force by vertical wall is zero

19 A uniform rope of linear mass density λ is placed inside a fixed curved tube in vertical plane as shown in the figure. The rope is stationary. Find normal reaction on the rope by the tube.

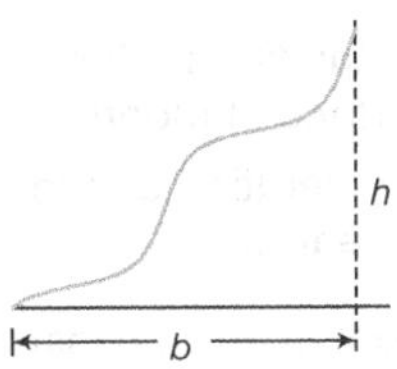

a. λgh
b. $\lambda g b$
c. $\lambda g\sqrt{b^2 + h^2}$
d. Non-zero

Comprehension Type

Working Space

Direction (Q. Nos. 20-23) *This section contains a paragraph, describing theory, experiments, data, etc. Four questions related to the paragraph have been given. Each question has only one correct answer among the four given options (a), (b), (c) and (d).*

Passage

Two identical smooth cylinders each of mass 10 kg are placed on two racks so that each rack supports half the weight of a cylinder.

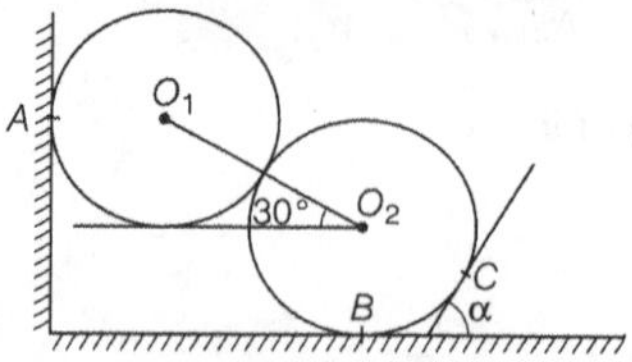

20 Find the reaction force exerted by the rack at point *A*.

a. $200\sqrt{3}$ N
b. $100\sqrt{3}$ N
c. $100\sqrt{6}$ N
d. $50\sqrt{3}$ N

21 Find the reaction force exerted by rack at point *C*.

a. 100 N
b. $100\sqrt{3}$ N
c. $100\sqrt{6}$ N
d. $50\sqrt{3}$ N

22 Find the reaction force exerted by the rack at the point *B* if $\alpha = \frac{\pi}{4}$.

a. $100(2-\sqrt{3})$ N
b. $100\sqrt{6}$ N
c. $100(2+\sqrt{3})$ N
d. $100\sqrt{3}$ N

23 Find the least value of $\tan\alpha$ for which equilibrium is possible.

a. $\sqrt{3}$
b. 1
c. $\frac{1}{\sqrt{3}}$
d. $\frac{\sqrt{3}}{2}$

Statement Type

Direction (Q. Nos. 24-26) *This section is based on Statement I and Statement II. Select the correct answer from the codes given below.*

a. Both Statement I and Statement II are correct and Statement II is the correct explanation of Statement I
b. Both Statement I and Statement II are correct and Statement II is not the correct explanation of Statement I
c. Statement I is correct but Statement II is incorrect
d. Statement II is correct but Statement I is incorrect

24 **Statement I** Normal reaction on a block does not accelerate the block.

Statement II Normal reaction always acts perpendicular to the surface of contact, no matter what the angle of that surface.

25 **Statement I** A man standing on the surface of earth feels more normal reaction than that on the surface of moon.

Statement II The acceleration due to gravity is smaller on the moon than acceleration on the earth.

26 **Statement I** Assume that earth consists of only neutrons. The normal reaction on a man standing on the surface of earth is zero.

Statement II Normal reaction is a electromagnetic force.

Matching List Type

Direction (Q. No. 27) Choices for the correct combination of elements from Column I and Column II are given as options (a), (b), (c) and (d), out of which one is correct.

27 Three smooth cylinders *A* of mass 3 kg, *B* of mass 8 kg and *C* of mass 4 kg are piled so that $\cos\phi = \frac{4}{5}$ and $\cos\theta = \frac{7}{11}$. Match Column I with Column II, with respect to the given information.

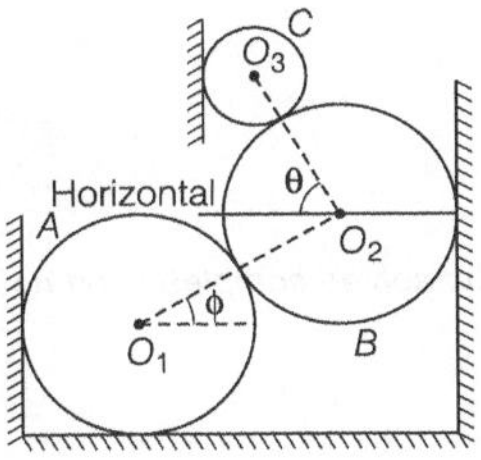

	Column I		Column II
i.	The normal force exerted by the wall on the cylinder *C*	p.	193 N
ii.	The normal force exerted by the wall on the cylinder *B*	q.	160 N
iii.	The normal reaction exerted by the cylinder *B* on cylinder *A*	r.	33 N
iv.	Normal reaction by vertical wall on the cylinder *A*	s.	200 N

Codes

	i	ii	iii	iv		i	ii	iii	iv
a.	p	q	r	s	b.	r	p	s	q
c.	q	p	r	s	d.	p	r	s	q

One Integer Value Correct Type

Direction (Q. Nos. 28 and 29) This section contains 2 questions. When worked out will result in an integer from 0 to 9 (both inclusive).

28 Two heavy smooth identical spherical balls, each of mass *m* rests in a V-shaped trough whose sides are inclined at angles α and β from the horizontal. The normal reaction exerted on the lower ball at the point *A* is $\frac{nmg\sin\alpha}{\sin(\alpha+\beta)}$. Find the value of *n*.

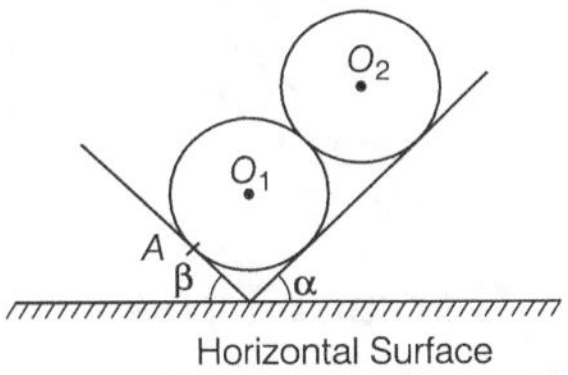

29 Three smooth identical cylinders *A*, *B* and *C* are placed on smooth inclined plane shown in the figure. The minimum value of $\tan\theta$ that prevent the arrangement from collapse is $\frac{1}{3\sqrt{n}}$. Find the value of *n*.

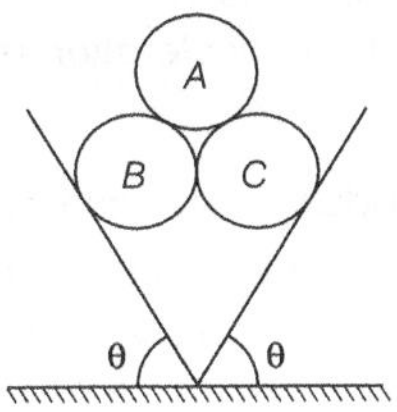

Working Space

DPP-2 Relation between Force and Acceleration

Subjective Question

Direction *(Q. No. 1) This question is subjective in nature, need to be solved completely on notebook.*

Working Space

1 i. The acceleration of the block is ... .

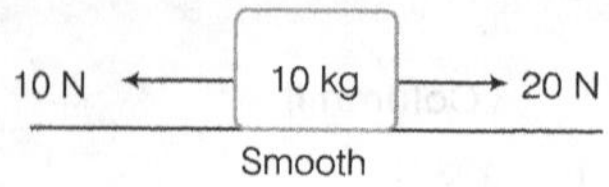

ii. The acceleration of the block is ... .

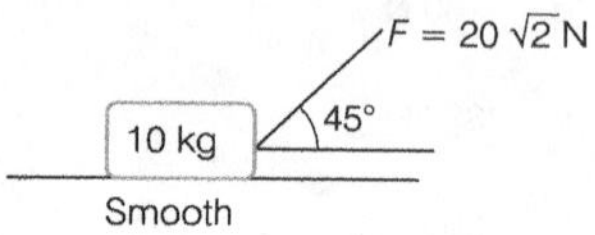

iii. If inclined plane is fixed and block is smooth. The acceleration of block is ....

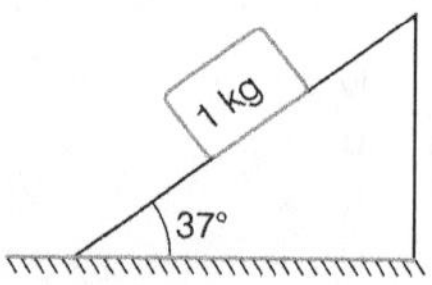

iv. A horizontal force F of 20 N is applied on the smooth block of mass 1 kg. The acceleration of block is ... .

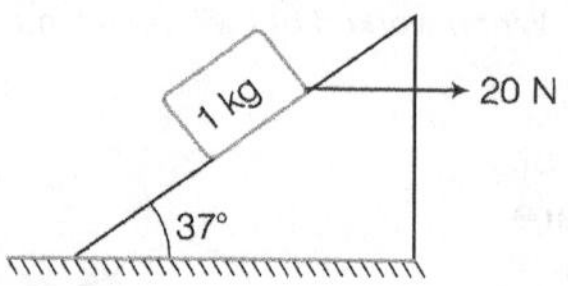

Only One Option Correct Type

Direction *(Q. Nos. 2-13) This section contains 12 multiple choice questions. Each question has four choices (a), (b), (c) and (d) out of which ONLY ONE option is correct.*

2 Mark the correct option.

a. For finding resultant force on a block, the law of vector addition is applicable
b. If resultant force on a body is zero, the acceleration of body is zero
c. Component of resultant force in perpendicular direction of acceleration is zero
d. All of the above

3 If a force F is applied on a block of mass m_1, then it produces an acceleration a_1. The same force is applied to a second block of mass m_2, then it produces an acceleration a_2. If the same force is applied on a block of mass $(m_1 + m_2)$, then acceleration is

a. $a_1 + a_2$ b. $\frac{a_1 + a_2}{2}$ c. $a_1 - a_2$ d. $\frac{a_1 a_2}{a_1 + a_2}$

Working Space

4 A block of mass 1 kg is moving along X-axis. A force of 1 N acts on the block. The force F always points towards the origin O. The magnitude of F is constant except at point O, where it is zero. At $t = 0$, the block is released from rest at $x = 2$ m. The period of motion of the block is

a. 8 s b. 4 s c. 2 s d. 1 s

5 A force $F = 5\sqrt{2}\,t$ N is applied on the block as shown in the figure. The block starts from rest at $t = 0$, then

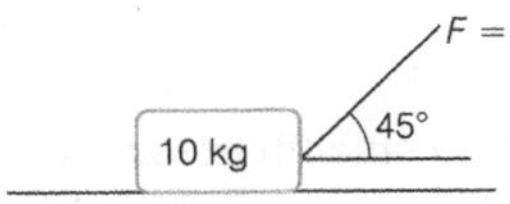

a. the block leaves the horizontal surface at $t = 20$ s
b. the block never leaves the horizontal surface
c. the velocity of block at $t = 20$ s is 100 m/s
d. Both (a) and (c) are correct

6 The inclined plane is fixed. A small smooth block is released from rest from point A. The ball periodically slides on the inclined plane under gravity. If the linear dimensions of the inclined plane ACB are increased by sixteen times.
How many times will the time period of this motion changes?

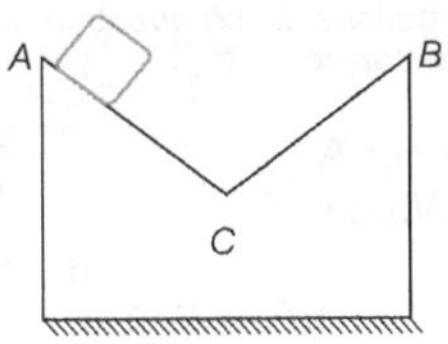

a. 2 times b. 4 times c. 8 times d. 16 times

7 A block of mass 1 kg is placed on the smooth inclined plane of inclination 37°. A force of 8 N is applied along the plane of inclined plane and perpendicular to greatest slope. The acceleration of the block is

a. 10 m/s^2 b. 8 m/s^2 c. 6 m/s^2 d. 2m/s^2

8 A block of mass 1 kg is held against a smooth wall by applying a horizontal force F of 5 N on the block and another force F_2 parallel to plane of the wall. The value of F_2 is

a. 5 N b. 10 N c. $5\sqrt{5}$ N d. 15 N

9 A box of mass 1 kg starts to move up on a frictionless inclined plane of inclination 30° with an acceleration $a_0 = 3$ m/s^2 by applying a constant force F along the inclined plane. When a block of mass m_0 is put inside the box, the box slides down with same acceleration a_0. The value of m_0 is

a. 1 kg b. 2 kg c. 3 kg d. 4 kg

10 The force on an object of mass 1 kg is $F = 5(1 - t)$ N along X-axis, where t is in second. The object starts from rest from origin at $t = 0$. Find the maximum distance, object has covered on $+x$-coordinate.

a. 10 m b. $\frac{20}{3}$ m c. $\frac{10}{3}$ m d. 2 m

11 The position of a particle of mass 10 kg is (0, 100 m) at $t = 0$. The equation of motion of the particle is $v_x = 50 - 16t$ and $y = 100 - 4t^2$. Find the net force on particle at $y = 0$. (where v_x is in m/s, t is in second and y is in metre.

a. 80 N b. 160 N c. 240 N d. 179 N

Working Space

12 The force F which is applied to 1 kg block initially at rest varies linearly with time as shown in the figure. Find velocity of the block at $t = 4$ s.

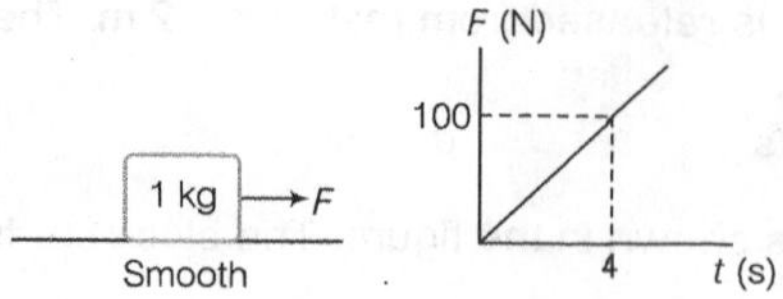

a. 100 m/s
b. 200 m/s
c. 20 m/s
d. 4 m/s

13 The figure shows three blocks in contact and placed on a smooth horizontal fixed table. The ratio of force exerted by block A to B and the force by block B to C is.

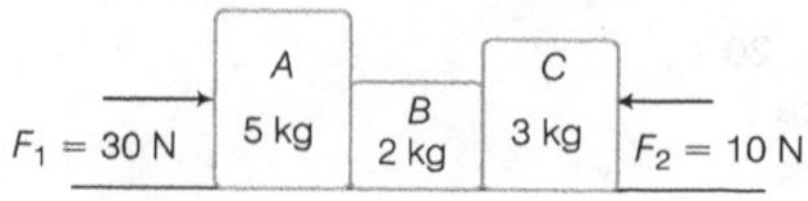

a. 1 : 2
b. 2 : 3
c. 3 : 2
d. 5 : 4

One or More than One Options Correct Type

Direction (Q. Nos. 14-16) *This section contains 3 multiple choice questions. Each question has four choices (a), (b), (c) and (d) out of which ONE or MORE THAN ONE are correct.*

14 A block of mass 1 kg is placed on the horizontal surface. A variable force of $2t$ N is applied in upward direction. Which of the given statement(s) is/are correct?

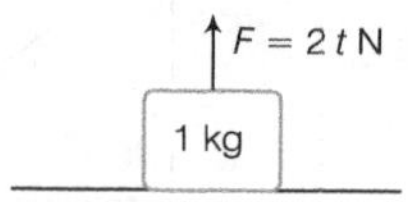

a. At $t = 2$ s, normal reaction on the block is 6 N
b. At $t = 10$ s, acceleration of block is 10 m/s^2
c. At $t = 10$ s, velocity of block is 25 m/s
d. None of the above

15 Two blocks are in contact on a smooth horizontal surface. In situation (i), the contact force between blocks is F_1 and contact force in situation (ii) is F_2. Mark the correct option(s).

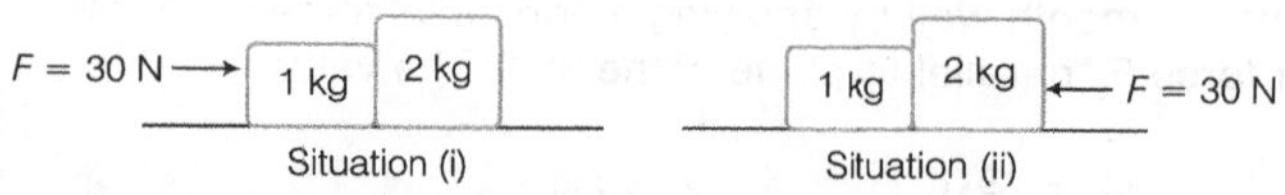

a. $F_1 = F_2$
b. $F_1 > F_2$
c. $F_1 < F_2$
d. $F_1 = 20$ N and $F_2 = 10$ N

16 A smooth plank of mass 2 kg is placed on frictionless horizontal surface. A smooth block of mass 1 kg and negligible width is placed on right end of the plank. The length of plank is 2.5 m. Which of the given statement(s) is/are correct?

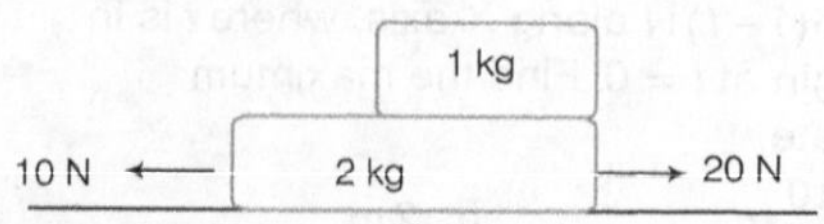

a. The acceleration of block is zero
b. The acceleration of plank is $\frac{10}{3}$ m/s^2
c. The acceleration of plank is 5 m/s^2
d. The block will drop from plank after 1 s

Comprehension Type

Working Space

Direction (Q. Nos. 17-19) This section contains a paragraph, each describing theory, experiments, data, etc. Three questions related to the paragraph have been given. Each question has only one correct answer among the four given options (a), (b), (c) and (d).

Passage

Two forces $\mathbf{F}_1 = (-\hat{i} - \hat{j})$ N and $\mathbf{F}_2 = (-8\hat{i} + 4\hat{j})$ N act on an object of mass 2 kg, that is initially at rest at coordinates (–2m, 4m).

17 Find velocity of the object at $t = 10$ s.

a. $-45\hat{i} + 15\hat{j}$ b. $15\hat{i} - 45\hat{j}$

c. $45\hat{i} - 15\hat{j}$ d. $-10\hat{i} - 6\hat{j}$

18 At what angle with positive X-axis is the object moving at $t = 10$ s?

a. $\tan^{-1}\left(\frac{1}{3}\right)$ b. 30°

c. $\tan^{-1}\left(-\frac{1}{3}\right)$ d. 60°

19 What is the position vector of the object at $t = 10$ s?

a. $6\hat{i} + 10\hat{j}$ b. $-227\hat{i} + 79\hat{j}$

c. $-225\hat{i} + 75\hat{j}$ d. $-6\hat{i} - 8\hat{j}$

Statement Type

Direction (Q. Nos. 20-24) This section is based on Statement I and Statement II. Select the correct answer from the codes given below.

a. Both Statement I and Statement II are correct and Statement II is the correct explanation of Statement I
b. Both Statement I and Statement II are correct and Statement II is not the correct explanation of Statement I
c. Statement I is correct but Statement II is incorrect
d. Statement II is correct but Statement I is incorrect

20 **Statement I** Internal forces of a body may cause of acceleration of the body.

Statement II $\mathbf{F} = m\mathbf{a}$

21 **Statement I** Action and reaction forces act at the same time.

Statement II Action and reaction forces balance each other.

22 **Statement I** Mass is a measure of inertia.

Statement II The change in momentum is the combined effect of force and time.

23 **Statement I** A moving car decelerates to stop. You are sitting in the car. You will fall forward.

Statement II Due to inertia of motion, the seat covers less distance than you are covering.

24 **Statement I** A thief jumped from a building with a box of mass 20 kg on his head. During the fall, the weight of box felt by the thief is zero.

Statement II The true weight of box during falling is zero.

One Integer Value Correct Type

Working Space

Direction (Q. Nos. 25-27) *This section contains 3 questions. When worked out will result in an integer from 0 to 9 (both inclusive).*

25 A block is at rest in gravity free space. It is acted by four forces, 3 N in East direction, 1 N in South direction, 1 N in downward direction and an unknown force $\mathbf{F}_0$. Due to these forces, the block starts to move in South-East direction. The minimum possible value of $\mathbf{F}_0$ is $\sqrt{n}$ newton. Find the value of n.

26 A force $\mathbf{F} = bt - cs$, (here $b = 1$ N/s and $c = 1$ N/m) is acting on a block of mass 1 kg placed on a smooth horizontal surface. During motion, the acceleration of the block is $a = a_0 \sin \omega t$. Find the value of ω.

27 It is found that when 0.5 m was cut-off from the muzzle of a gun firing block of 50 kg, the velocity of the block was changed from 700 m/s to 600 m/s. The force exerted on the block by the powder of gas at the muzzle when expanded in the bore was 6.5×10^n N. Find the value of n.

Matching List Type

Direction (Q. No. 28) *Choices for the correct combination of elements from Column I and Column II are given as options (a), (b), (c) and (d), out of which one is correct.*

28 A man of mass 80 kg is standing on a light weighing machine placed in an elevator. ($g = 10$ m/s^2)

Match Column I with Column II and choose the correct option from codes given below.

	Column I (Cases of elevator motion)		Column II (Reading of weighing machine or apparent weight of man)
i.	If elevator is moving upward with a constant velocity 10 m/s.	p.	Same as true weight
ii.	If elevator is moving upward with an acceleration 2 m/s^2.	q.	800 N
iii.	If elevator is moving downward with an acceleration 2 m/s^2.	r.	960 N
iv.	If elevator is moving downward with an acceleration 10 m/s^2.	s.	640 N
		t.	Lesser than true weight
		u.	Weightlessness

Codes

	i	ii	iii	iv
a.	p,q	r	s,t	t,u
b.	p,q,r	r	s	t
c.	u	s,t	p,q	r
d.	r,s	t,u	r,q	p

DPP-3 Tension in Strings

Subjective Questions

Working Space

Direction (*Q. Nos. 1-5*) *These questions are subjective in nature, need to be solved completely on notebook.*

1 Find the tension in light string in the following cases.

i. A block of mass 10 kg is suspended from a ceiling through a light string.

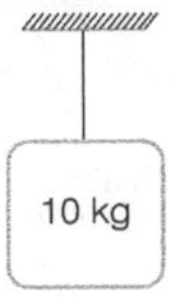

ii. Two blocks *A* and *B* are connected by light string.

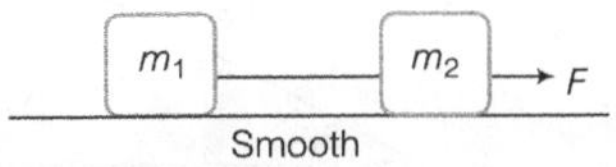

iii. Two blocks *A* and *B* are connected by a light string passing through a smooth pulley.

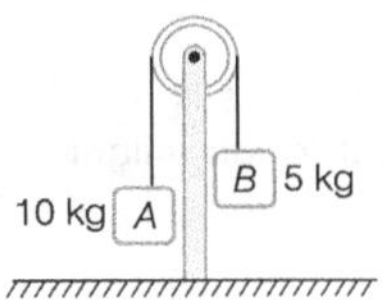

iv. A block of mass 1 kg is connected with a light string passing over a fixed smooth pulley.

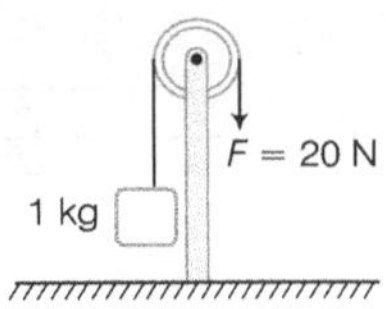

A force F of 20 N is applied at other end of string.

v. A block of mass 1 kg placed on the smooth inclined plane is connected through a string from a fixed support.

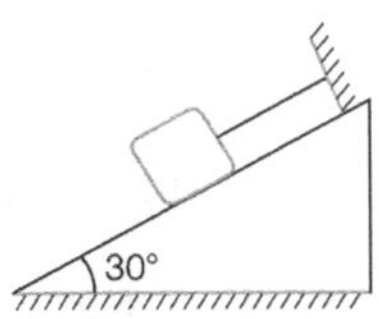

vi. Block *A* placed on the smooth fixed inclined plane is connected to block *B* through a light string passing over smooth fixed pulley.

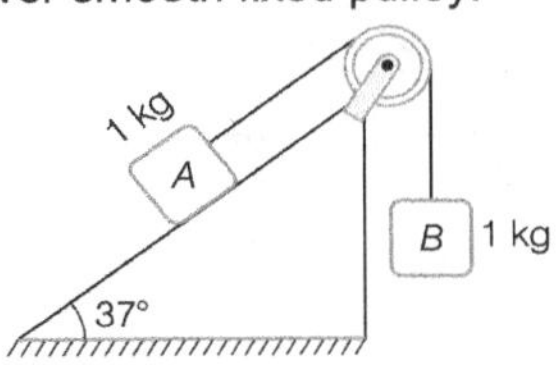

Working Space

2 Find the clamp force in pulley in the following cases.

i. The pulley is fixed and smooth. The mass of pulley is 1 kg.

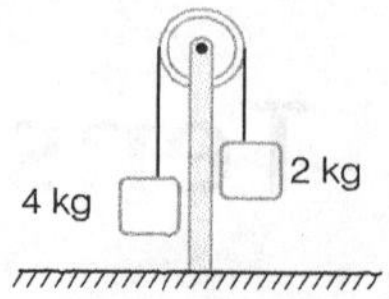

ii. The smooth pulley of mass $m_1 = 1$ kg is fixed. A block of mass (m_0) 3 kg is pulled slowly through a string by applying a force F at the other end of the light string.

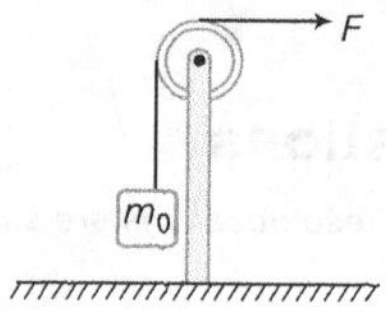

iii. The pulley is fixed, smooth and light.

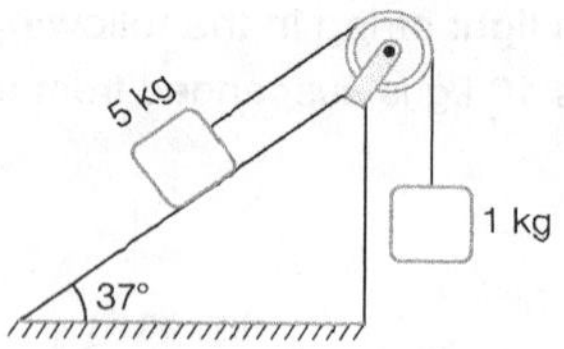

iv. Block A of mass $\sqrt{3}$ kg is connected with block B of mass 1 kg through a light string passing over a fixed smooth pulley of mass $\sqrt{3}$ kg. The inclined plane is smooth.

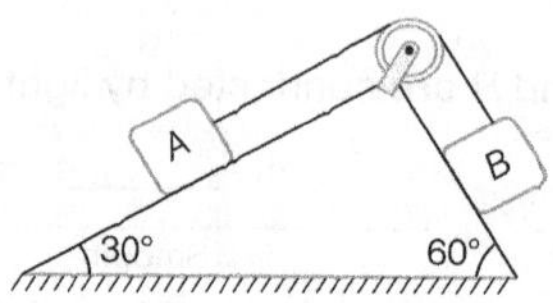

3 Find the ratio of magnitudes of tension in light strings A and B in the following cases.

i. For the situation shown in figure, blocks are connected with a light string B. The upper block is attached to a fixed ceiling through a light string A.

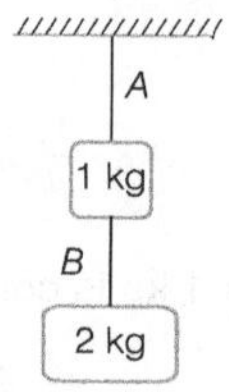

ii. The blocks of masses 1 kg and 2 kg are connected through a light string B. The hanging block is connected with 2 kg block kept on a smooth horizontal surface through a light string A passing over a smooth fixed pulley.

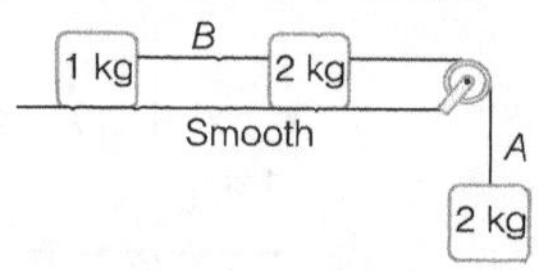

iii. For the situation shown in figure, strings A and B are light and inclined plane is smooth. The pulley is light and smooth.

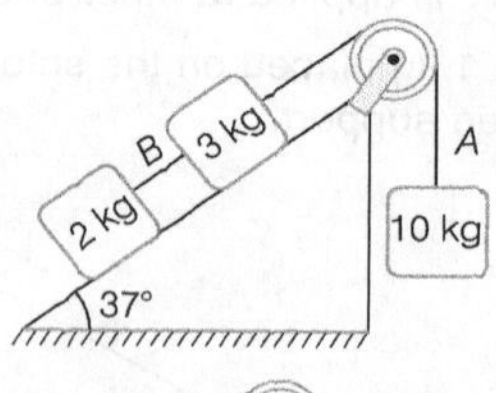

iv. For the situation shown in figure, all pulleys are smooth and fixed. The inclined plane is also smooth. A constant force F of 38 N is applied at the end of the string.

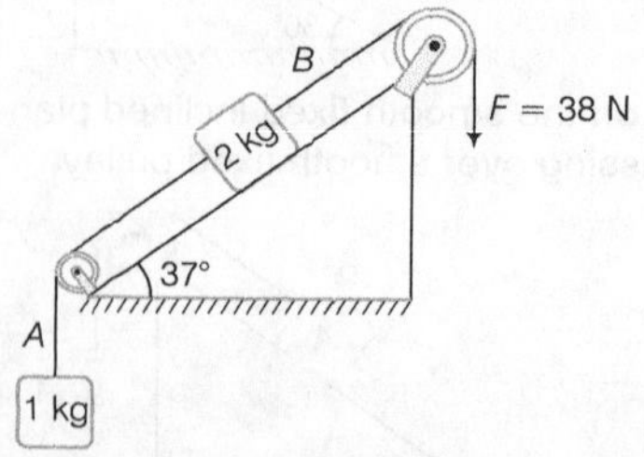

Working Space

4 Find the ratio of tensions in light strings A, B and C in the following cases.

i. Three light strings A, B and C are connected at point O.

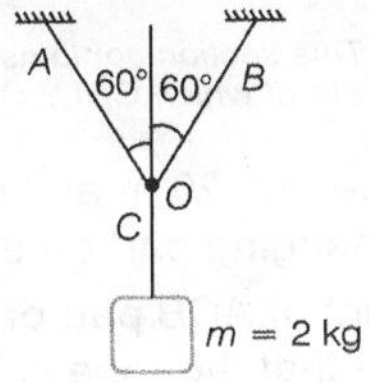

ii. A box of mass m is shifted upward slowly by applying the forces on strings.

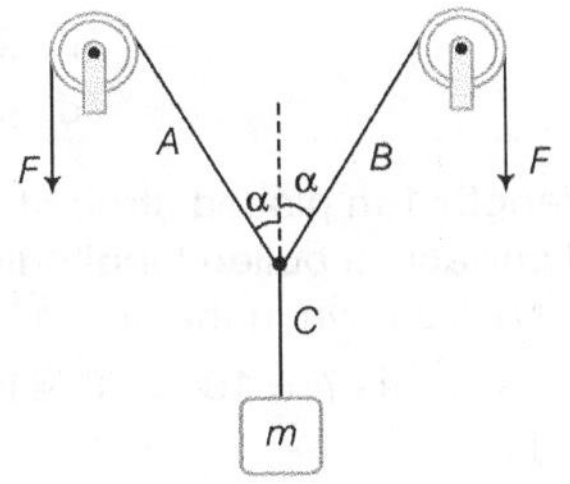

iii. The one end of a light string $O_1 O_2$ of 0.5 m long is fixed to a point O_2 and the other end is fastened to a small block of mass $m_2 = 0.8$ kg. The horizontal distance between O_1 and O_3 is 0.3 m.

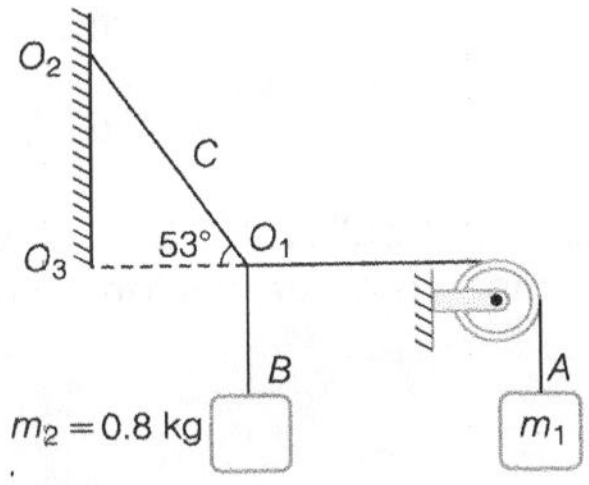

iv. Two light strings are knotted at point P, is pulled downward through a string by applying a force F of 25 N.

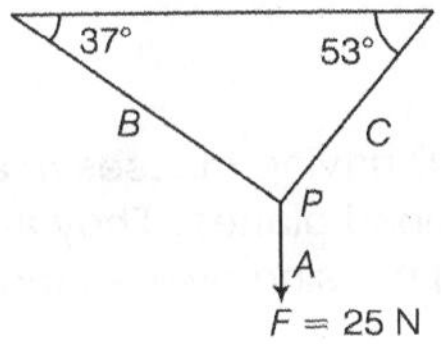

5 Find the tension at point A in a uniform massive rope in the following cases.

i. The massive rope is placed on smooth horizontal surface.

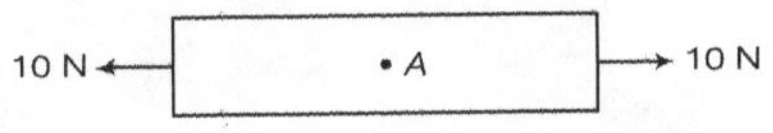

ii. A massive rope of mass 3 kg is placed on smooth horizontal surface. The length of the rope is 3 m and the distance of A from end B is 1 m.

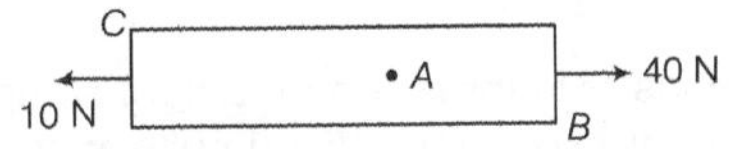

iii. A massive rope of mass 3 kg is pulled upward by applying a force F of 60 N. A is the middle point of the rope.

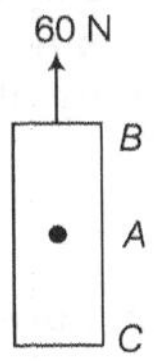

iv. A rope of mass $m = 10$ kg hangs between two fixed points B and C which are at the same horizontal level. The inclination of the rope with vertical is 45°. The point A is mid-point of the rope.

v. Two smooth balls each of mass 1 kg are connected by a massive rope of mass 3 kg and length 3 m. The point A is at distance 1 m from rear ball.

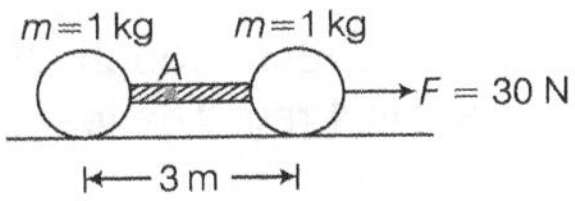

Only One Option Correct Type

Working Space

Direction *(Q. Nos. 6-13) This section contains 8 multiple choice questions. Each question has four choices (a), (b), (c) and (d) out of which ONLY ONE option is correct.*

6 A uniform rope of length 20 m and mass 20 kg is shown in the figure. The hanging part on each side is 6 m. C is the middle point of ACB part of rope. Find the tension at point C. Assume that the rope is in equilibrium.

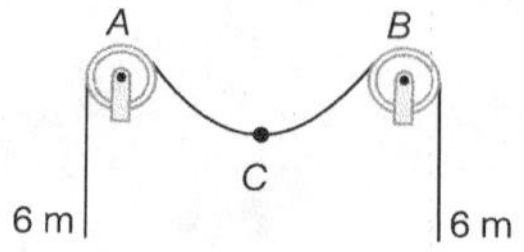

a. 20 N
b. $20\sqrt{5}$ N
c. 60 N
d. 40 N

7 A massive rod of length 1 m placed straight on a smooth horizontal surface is pulled longitudinally by a force F of 10 N as shown in the figure. The tension in rod varies as $T = 10\sqrt{x}$. The linear mass density of rod is

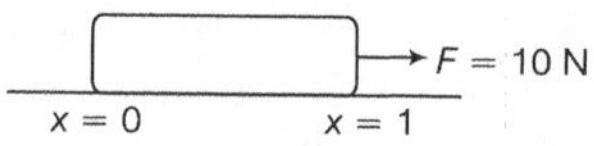

a. $\lambda \propto x$
b. $\lambda \propto \frac{1}{\sqrt{x}}$
c. $\lambda \propto \sqrt{x}$
d. $\lambda \propto x^2$

8 The radius of smooth light pulley is 0.2 m. The mass of each block is 10 kg. Find the normal force per unit length of its length, the string applies the pulley.

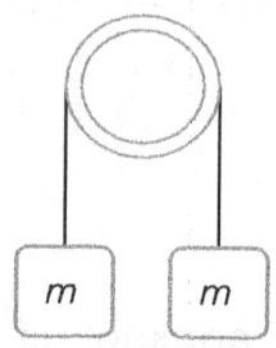

a. 500 N/m
b. 100 N/m
c. 200 N/m
d. 250 N/m

9 Two blocks A and B having masses m and M $(m << M)$ are on smooth inclined planes. They are connected through light string passing over a smooth pulley. The tension in string is

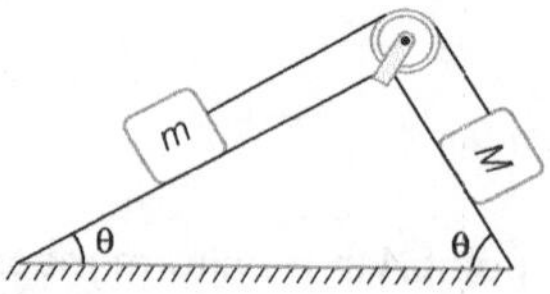

a. $(M - m)g\sin\theta$
b. $2Mg\sin\theta$
c. $2mg\sin\theta$
d. $(M + m)g\sin\theta$

10 A smooth ring of mass M is threaded on a light string whose ends are then threaded over two smooth fixed pulleys with masses $m_1 = 3$ kg and $m_2 = 6$ kg tied on to them respectively. The various portions of the strings are vertical. The system is free to move. If ring remains stationary, then the mass M of the ring is

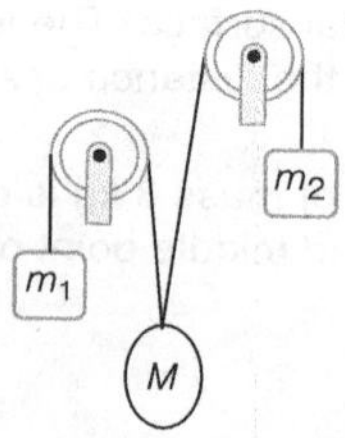

a. 2 kg
b. 6 kg
c. 8 kg
d. 4 kg

11 For the given figure, upper pulley is smooth and fixed. The lower light pulley is free to move. A horizontal velocity is imparted to block A. Then,

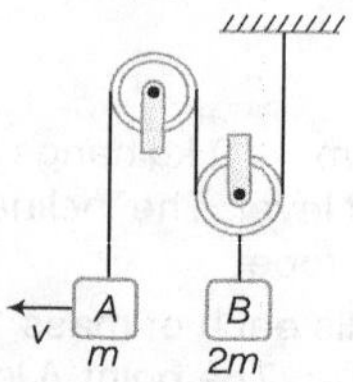

a. both blocks always remain in same level
b. after sometime, block B is lower than block A
c. after sometime, block A is lower than the block B
d. no sufficient information

12 A block of mass m is to be raised by a light rope from rest to rest through a height 60 m, the maximum tension of which the rope safely bear is 4 mg. The least time in which the ascent can be made is

a. 1 s
b. 2 s
c. 3 s
d. 4 s

Working Space

13 A block A of mass 1 kg is connected with a block B of mass 4 kg through a light string passing over a smooth fixed pulley. After striking with ground, the block B abruptly comes in rest. Blocks are released from rest. Find the maximum height attained by the block A.

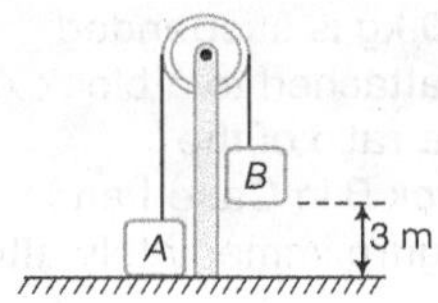

a. 3 m
b. 6 m
c. 4.8 m
d. 3.6 m

One or More than One Options Correct Type

Direction (Q. Nos. 14 and 15) *This section contains 2 multiple choice questions. Each question has four choices (a), (b), (c) and (d) out of which ONE or MORE THAN ONE are correct.*

14 For the situation shown in the figure, a man of mass 50 kg is standing on a light weighing machine placed inside a cabin of mass 10 kg. The cabin is joined by a light string with a block of mass 60 kg placed on a smooth inclined plane. Then, choose the correct options.

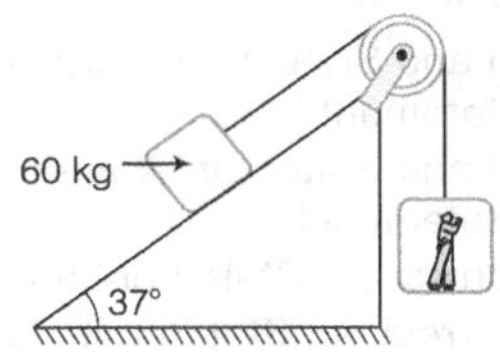

a. The acceleration of the block is 2 m/s^2
b. The reading of weighing machine is 40 kg
c. True weight of a man is 50 kg
d. The true weight of a man is 40 kg

15 Pulleys P_2 and P_3 are smooth and fixed. The light pulley P_1 is moving upward by applying a force F of 120 N. The pulley P_4 is also light and movable. Blocks A, B and C of their respective masses 1 kg, 2 kg and 3 kg are initially placed at the smooth horizontal surface. Then, choose the correct options.

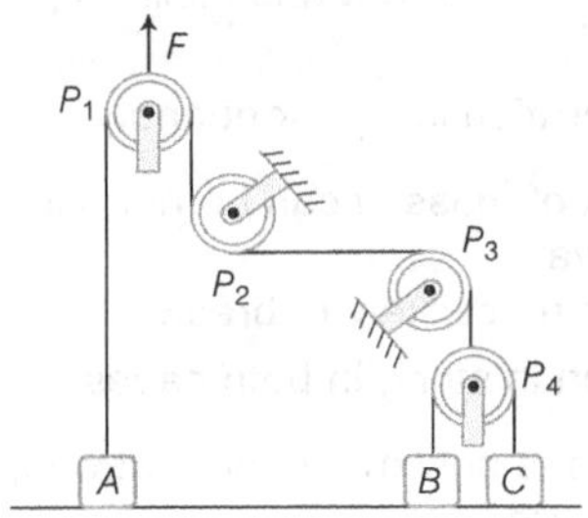

a. The acceleration of block A is 50 m/s^2
b. The acceleration of block B is 5 m/s^2
c. The acceleration of block C is zero
d. The pulley P_1 has non-zero acceleration

One Integer Value Correct Type

Direction (Q. Nos. 19-21) *This section contains 3 questions. When worked out will result in an integer from 0 to 9 (both inclusive).*

16 A block of mass m is connected to a peg at point D through a light and inextensible string of length 18 m. The length of section AB is 3 m. The point C of string is pulled down by applying a force F of 128 N. The block A remains stationary. If the mass of block A is $\frac{16}{\sqrt{n}}$ kg, find the value of n.

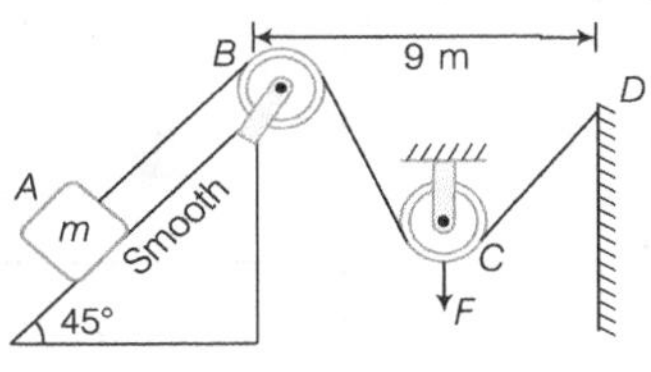

17 A block B of mass 1kg is placed on a light plank shown in figure. A force F of 10 N is applied on the plank horizontally. Find the acceleration (in m/s^2) of block B.

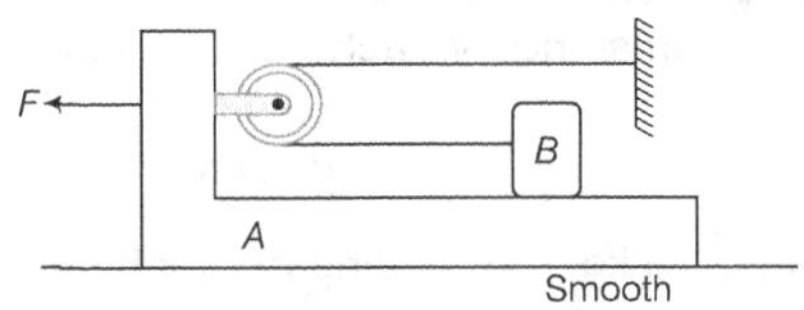

Working Space

18 A block of mass 20 kg is suspended from a light string attached to a block A of mass 40 kg. The ratio of the acceleration of block B in Case I and case II shown in figure immediately after the system is released from rest is $\frac{6n}{4\sqrt{2}}$. Find the value of n (Neglect friction).

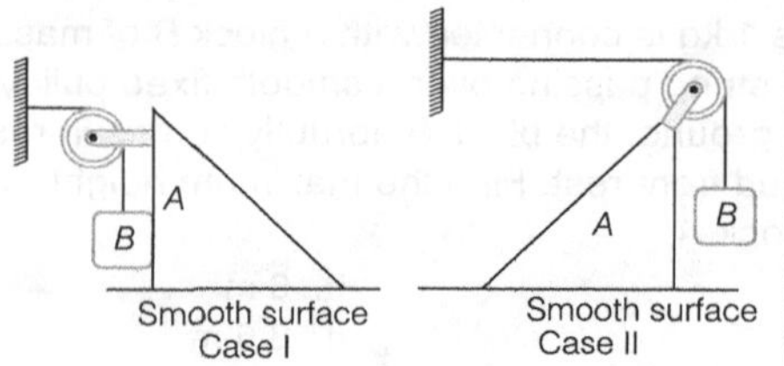

Statement Type

Direction (Q. Nos. 19-21) *This section is based on Statement I and Statement II. Select the correct answer from the codes given below.*

a. Both Statement I and Statement II are correct and Statement II is the correct explanation of Statement I
b. Both Statement I and Statement II are correct and Statement II is not the correct explanation of Statement I
c. Statement I is correct but Statement II is incorrect
d. Statement II is correct but Statement I is incorrect

19 **Statement I** The magnitude of acceleration of block B calculated by a student is $2\ m/s^2$. For this calculation, pulley is smooth, string is light and inclined plane is smooth and fixed.

Statement II Acceleration is vector quantity.

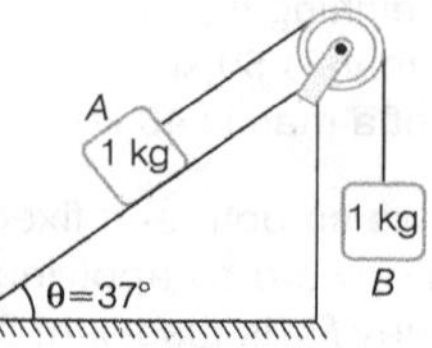

20 **Statement I** A block of mass m can be hung in any of the following two ways.
In Case II, string is most likely to break.

Statement II Tension in string in both cases is same.

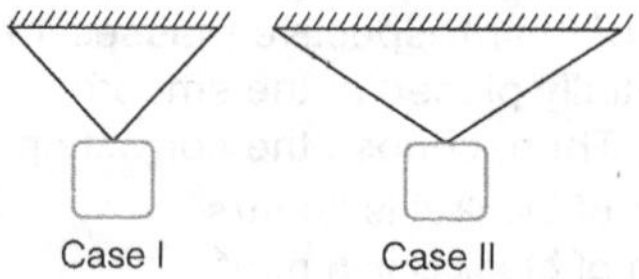

21 **Statement I** If a rope is only made by neutrons, the tension in rope is zero.

Statement II Tension is an electromagnetic force.

Comprehension Type

Direction (Q. Nos. 22-27) *This section contains a paragraph, each describing theory, experiments, data, etc. Six questions related to the paragraph have been given. Each question has only one correct answer among the four given options (a), (b), (c) and (d).*

Passage I

A weighing machine of mass 20 kg is fixed in a lift of mass 10 kg. A man of mass 60 kg is standing on the weighing machine and is pulling the lift through light strings passing over light fixed smooth pulleys.

22 If lift is stationary, then reading of weighing machine is

a. 60 kg b. 120 kg
c. 80 kg d. 15 kg

23 If reading of weighing machine is equal to true weight of man, the force exerted on string by each hand of man is

a. 1800 N b. 900 N
c. 450 N d. zero

24 If the mass of a man is 20 kg, the reading of weighing machine is

a. zero b. 90 kg c. 15 kg d. 20 kg

Working Space

Passage II

Two smooth blocks A of mass 1 kg and B of mass 2 kg are connected by a light string passing over a smooth pulley as shown. The block B is sliding down with a velocity 2 m/s. A force F is applied on the block A so that the block B will reverse its direction of motion after 3 s.

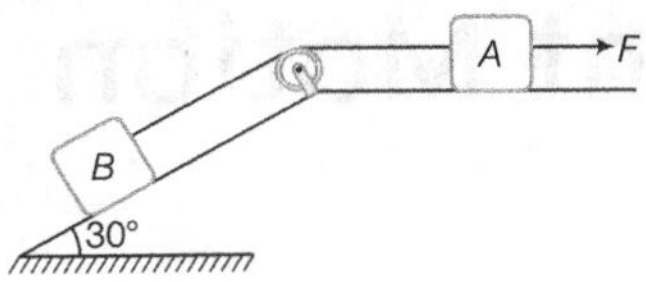

25 Find the acceleration of block A.

a. 1 m/s^2　　b. $\frac{2}{3}$ m/s^2　　c. 3 m/s^2　　d. $\frac{4}{9}$ m/s^2

26 Find the tension in the string.

a. 10 N　　b. 11.33 N　　c. 13.67 N　　d. 40 N

27 The value of F is

a. 9 N　　b. 10 N　　c. 11 N　　d. 12 N

Matching List Type

Direction (Q. No. 28) Choices for the correct combination of elements from Column I and Column II are given as options (a), (b), (c) and (d), out of which one is correct.

28 The masses of blocks as shown in the figure are $m_A = 1$ kg, $m_B = 2$ kg and $m_C = 3$ kg. Light and smooth pulleys P_2 and P_3 are fixed. But, light and smooth pulleys P_1 and P_4 are movable.

Match the Column I with Column II and choose the correct options from the codes given below.

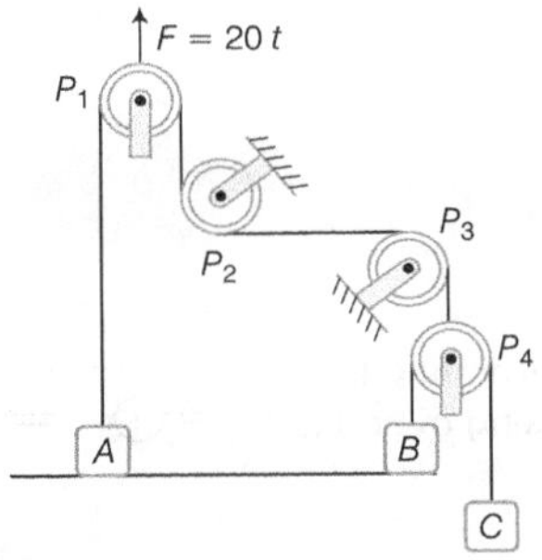

	Column I		Column II
i.	Block A	p.	$a = 0$ at $t = 2$ s
ii.	Block B	q.	$a \neq 0$ at $t = 2$ s
iii.	Block C	r.	$v = 5$ m/s, $t = 2$ s
		s.	$v = 50$ m/s, $t = 2$ s

Codes

	i	ii	iii			i	ii	iii
a.	q,r	p	s		b.	p	q	s
c.	p,s	q,r	q		d.	s	p	q

DPP-4 Constraint Motion or Dependent Motion

Subjective Questions

Direction (Q. Nos. 1-7) *These questions are subjective in nature, need to be solved completely on notebook.*

Working Space

1 Find the relation between u and v in following cases.

i. Two balls are connected with a string and moving on a horizontal surface. During motion, the string is taut.

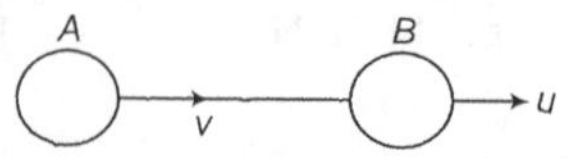

ii. Two balls connected with a taut string are moving on a smooth horizontal surface.

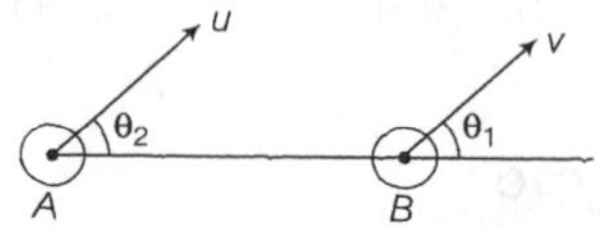

iii. Two blocks are connected through a light string passing over a fixed pulley.

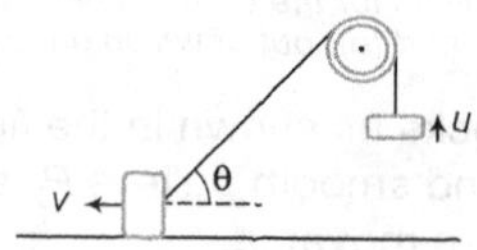

iv. For the situation shown in figure, the blocks A and B are moving downward with constant speed v while block C is moving with speed u.

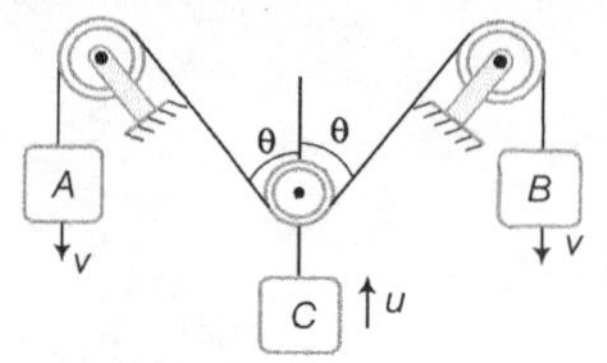

v. A smooth ring A slides on a horizontal fixed rod. The ring is connected to a block B through a light string passing over fixed pulley.

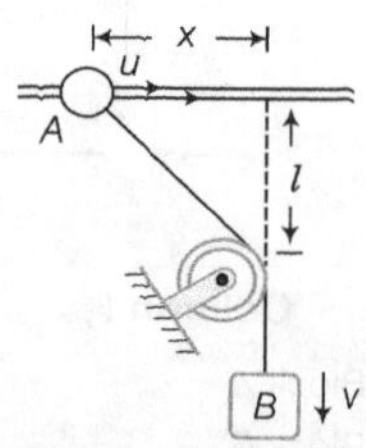

vi. A sleeve C can move freely on a smooth fixed horizontal rod. It is connected by blocks A and B by a light string passing over fixed smooth pulleys as shown in the figure. The velocity of A is u and velocity of B is v.

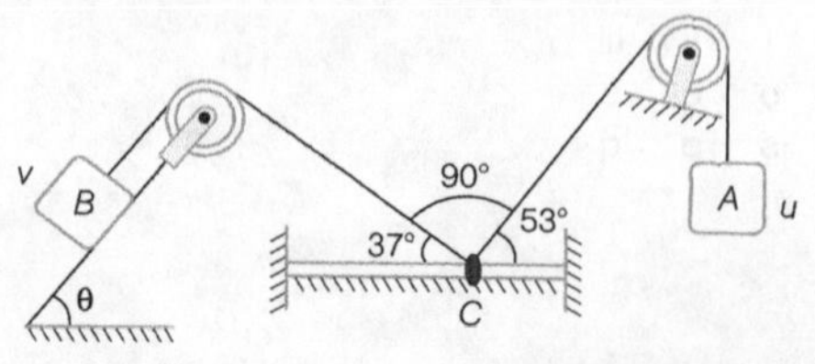

Working Space

2 Find the relation between u and v in following cases.

i. The pulley A is smooth and fixed. The pulley B is free to move. The velocity of block m_1 is u in downward direction and velocity of m_2 is v.

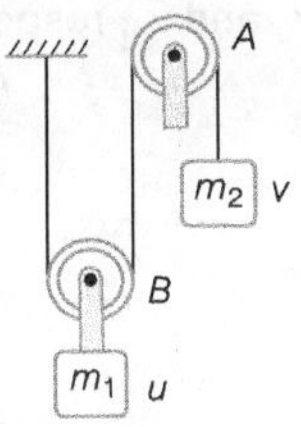

ii. Movable pulley P is smooth and light.

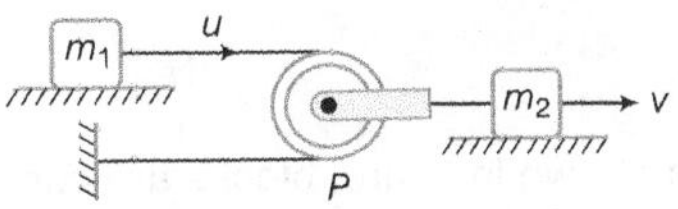

iii. Pulleys A and B are movable. The velocity of pulley B is v and velocity of A is u.

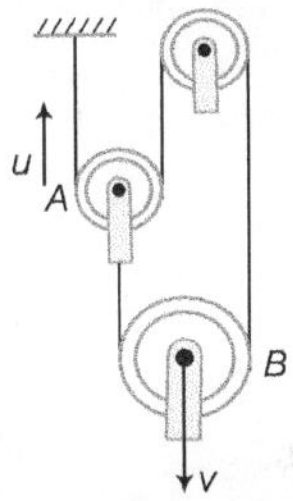

iv. Pulleys A and B are light and movable.

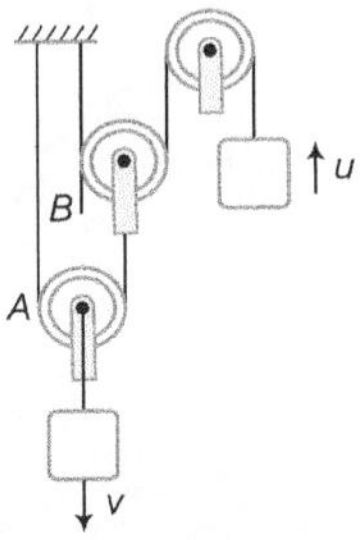

3 Find the relation among v_1, v_2 and v_3 in the following cases.

i. Blocks A and B are placed on the smooth horizontal table. A smooth light pulley P_1 is movable but smooth pulley P_2 is fixed. The velocities of blocks A, B and C are v_1, v_2 and v_3, respectively.

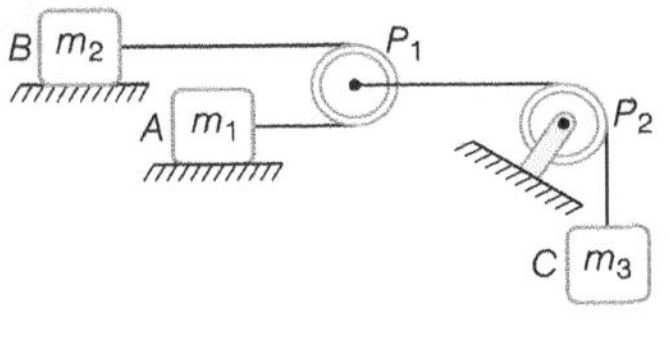

ii. All ideal pulleys are fixed except pulley P_3. The velocities of A, B and C are v_1, v_2 and v_3, respectively.

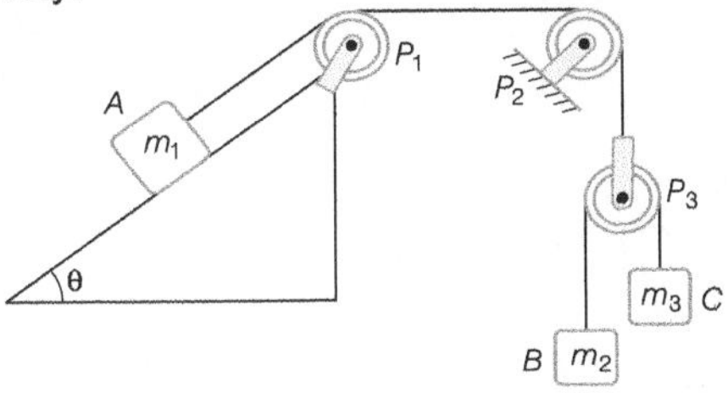

Working Space

iii. All ideal pulleys are fixed except pulleys P_1 and P_2. The velocities of masses m_1, m_2 and m_3 are v_1, v_2 and v_3, respectively.

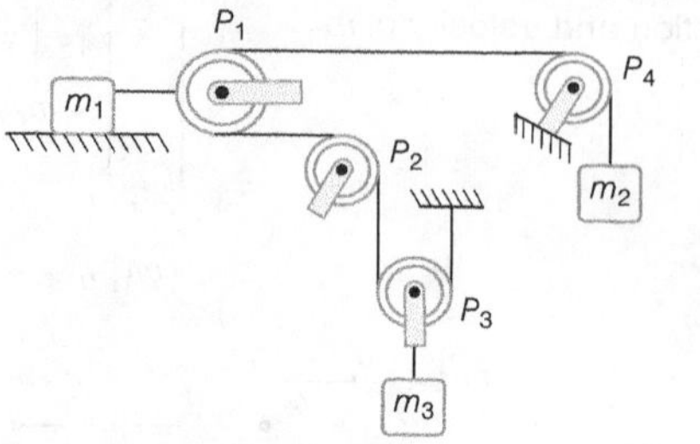

iv. For the situation shown in figure, blocks are connected through light strings. Light pulley P_1 is movable while other smooth pulleys are fixed.

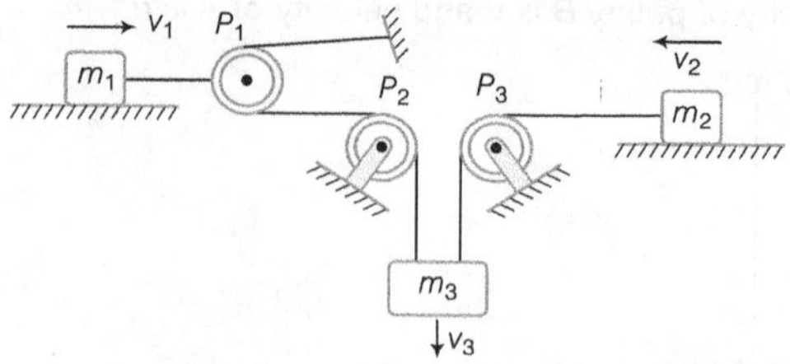

4 Find the relation between a_1 and a_2 in the following cases.

i. The smooth and light pulley P is movable. The accelerations of blocks A and B are a_1 and a_2, respectively.

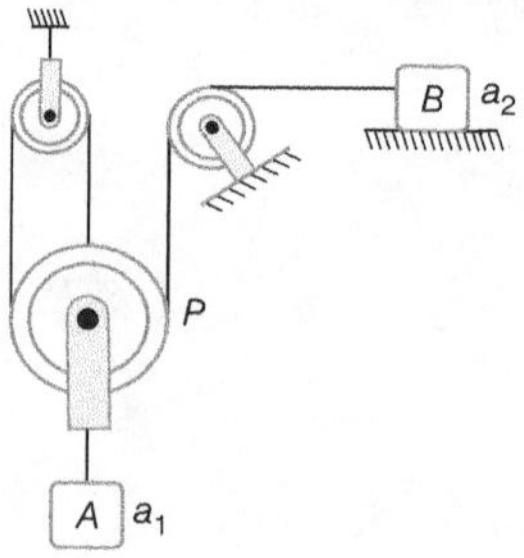

ii. All pulleys are smooth and light. The pulleys P_2 and P_3 are fixed, but P_1 and P_4 are movable.

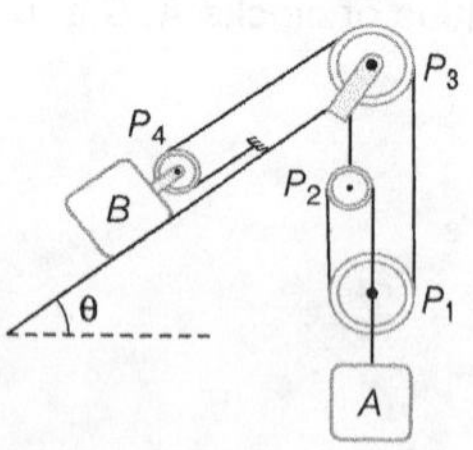

iii. Blocks A and B are connected through light string passing through light smooth pulleys as shown in figure.

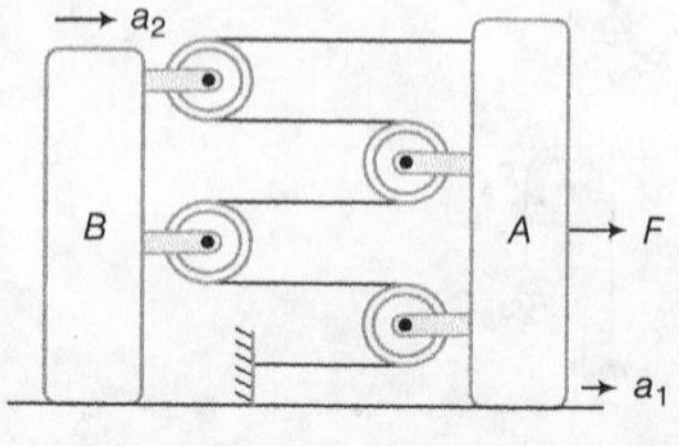

Working Space

iv. All pulleys are smooth and light, only pulleys P_1 and P_2 are movable.

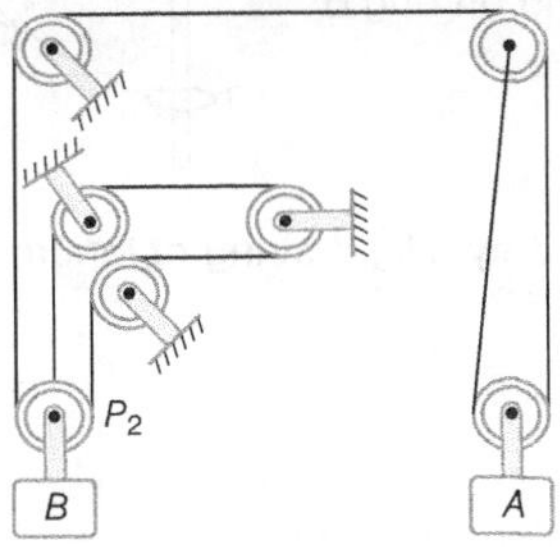

v. Point P is moving with an acceleration a_0.

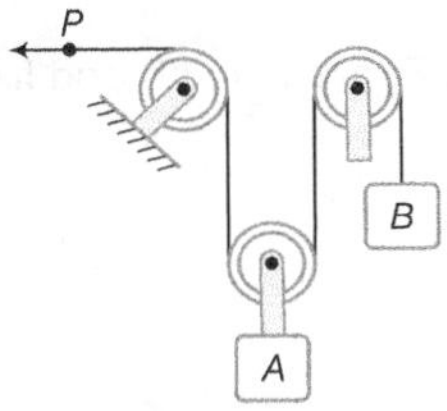

vi. All pulleys are smooth and light. Only pulleys P_1, P_2 and P_3 are movable, but other pulleys are fixed.

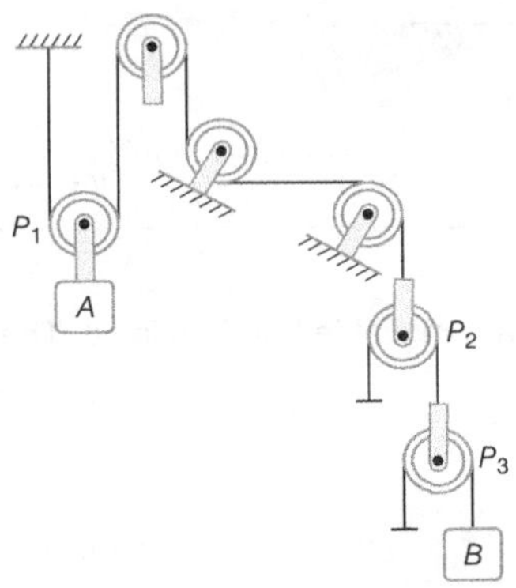

5 Find the relation between velocities of A and B.

i. For the arrangement shown in figure, block B is moving downward with a uniform speed while block A is moving upward.

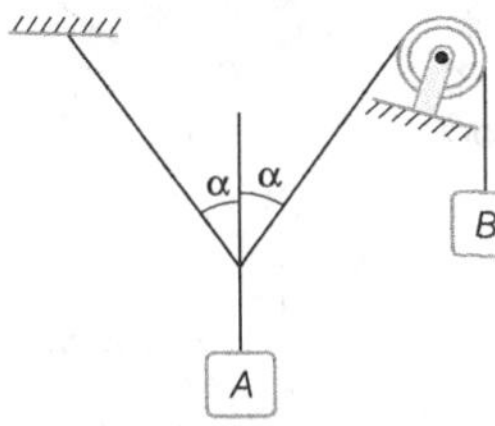

ii. A hinged construction consists of two rhombuses with the ratio of sides 3 : 5. The vertex A_2 moves in the horizontal direction with velocity v_0.

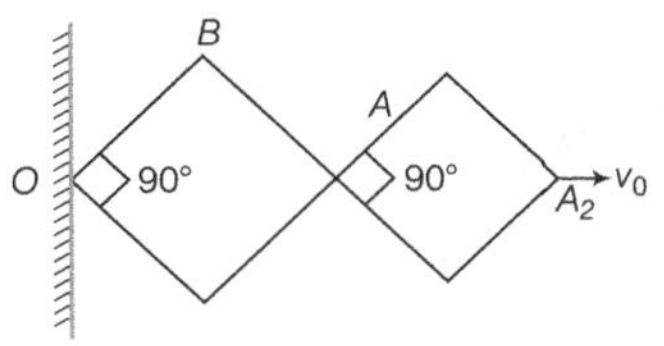

Working Space

iii. Two light rings of negligible radius are put on two vertical fixed rods CD and EF. A light string is fixed at point C and on ring B and is passed through ring A.

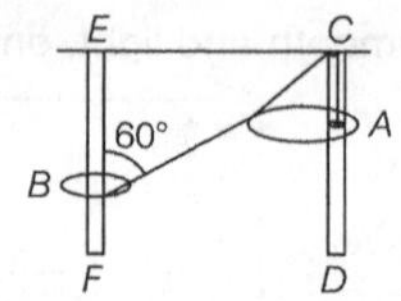

iv. Rings A and B slide along the fixed rods, connected by a light string of length L.

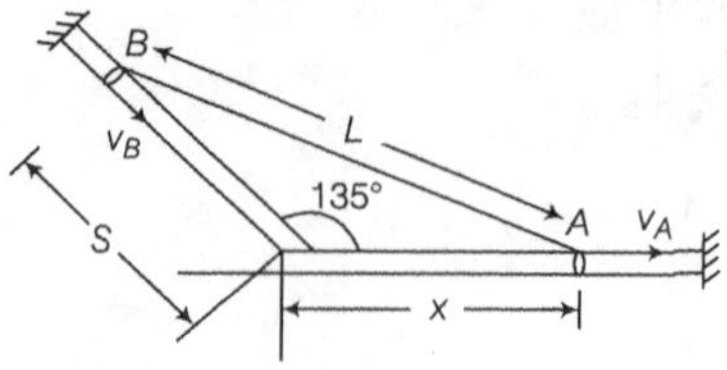

v. For the arrangement shown in figure, pulleys P_1 and P_3 are smooth and fixed while pulley P_2 is movable.

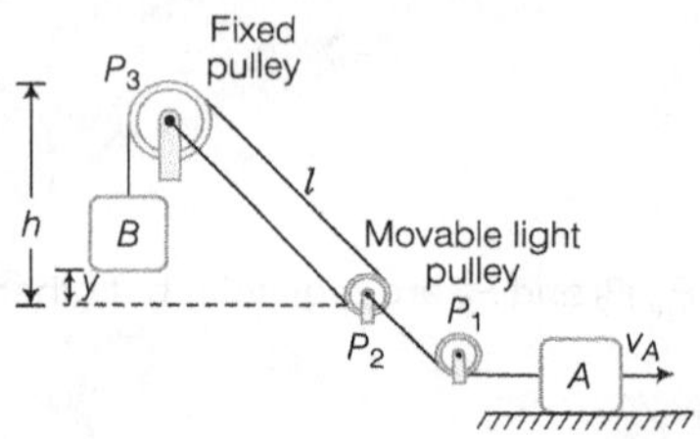

6 Find the acceleration of blocks in the following cases.

i. Blocks of masses m_1 and m_2 are connected through a light string passing over a smooth, light and movable pulley. The horizontal surface is frictionless.

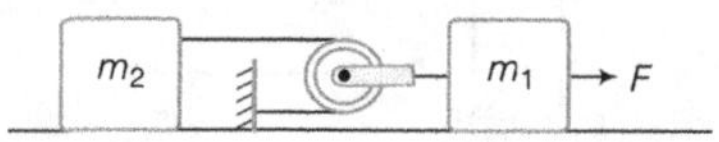

ii. Pulleys are light and smooth. Pulleys P_1 and P_2 are movable but P_3 is fixed. The horizontal surface is smooth.

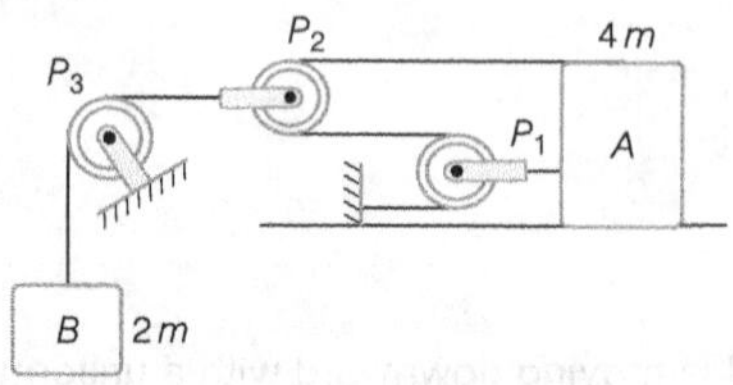

iii. For the arrangement shown in the figure, block A of mass $m_1 = 2$ kg is placed on smooth inclined plane of inclination 37°. Pulleys are light and smooth. Pulley P_1 is movable but pulley P_2 is fixed. The block A and block B of mass $m_2 = 2$ kg, are connected with light string.

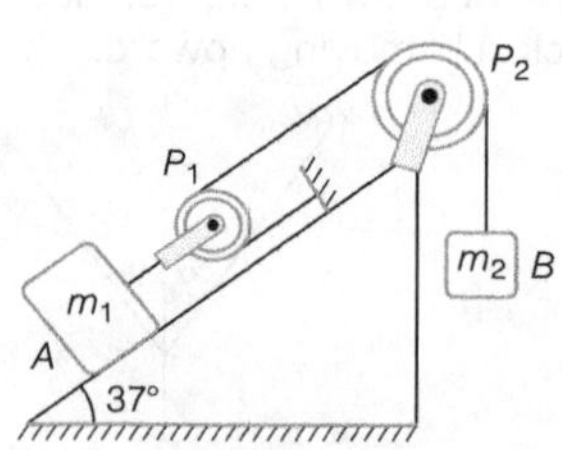

iv. Three identical blocks A, B and C each of mass 2 kg are connected through light string passing over smooth and light pulleys.

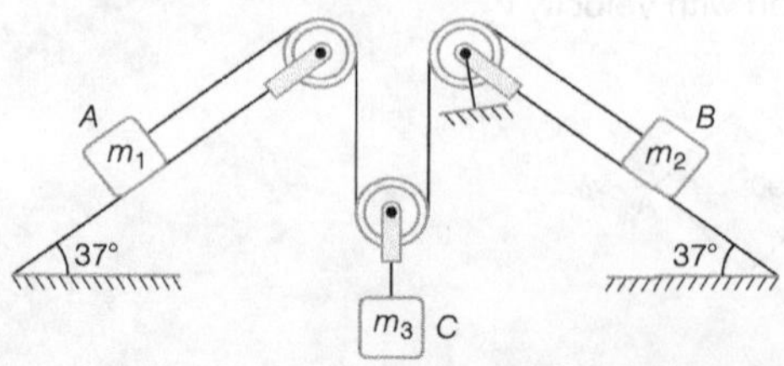

Working Space

v. In the situation shown in the figure, pulley P_1 is fixed but pulley P_2 is free to move. Both pulleys are light and smooth. Find the acceleration of block A.

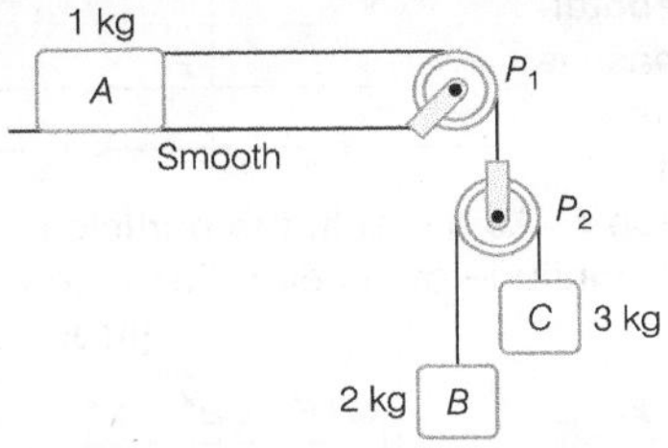

7 Find constraint relation in relative motion in the following cases.

i. For the situation shown in figure, find the relation between accelerations of blocks A and B. All surfaces of contact are frictionless.

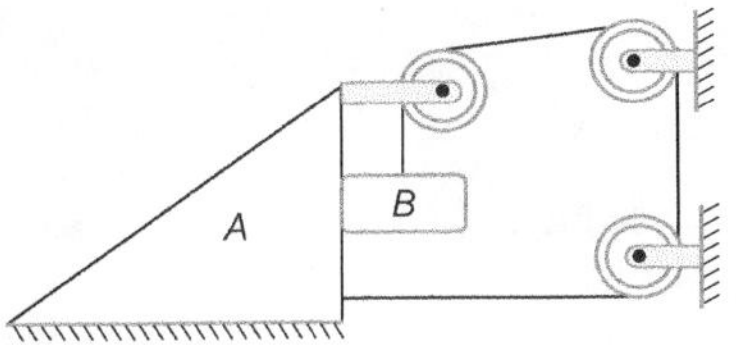

ii. Pulley is light and of negligible radius. The string section between B and pulley always remains vertical. Determine the relation between velocity of blocks A and B.

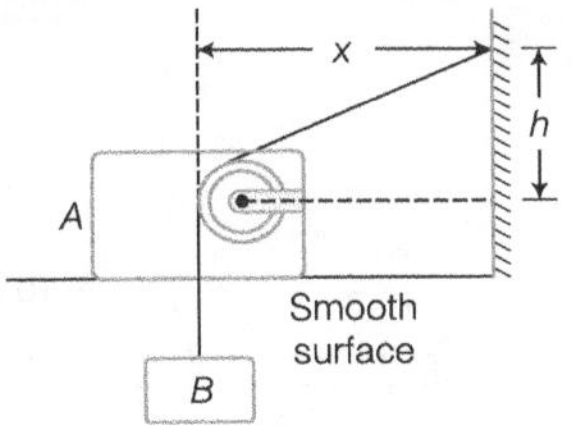

iii. For the arrangement shown in the figure, the acceleration of mass m_2 with respect to mass m_1 is $a_{rel} = 2\ \text{m/s}^2$ upward along the inclination of triangular block. The block of mass m_3 is moving downward with an acceleration $a_0 = 5\ \text{m/s}^2$. Determine the acceleration of the triangular block.

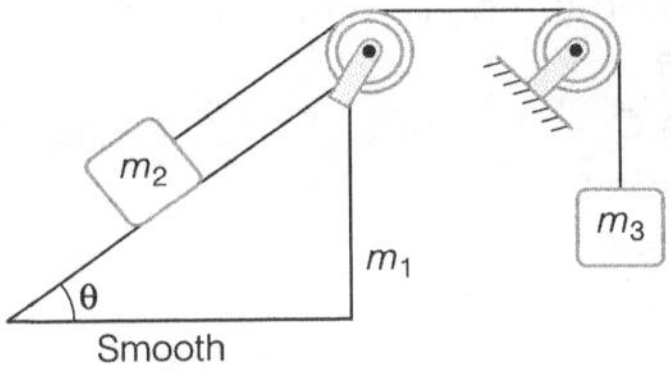

Only One Option Correct Type

Direction (Q. Nos. 8-16) This section contains 9 multiple choice questions. Each question has four choices (a), (b), (c) and (d) out of which ONLY ONE option is correct.

8 Two blocks A and B of masses 10 kg and 20 kg, respectively, connected by a light string are placed on the smooth horizontal table. The middle point of string is pulled through another light string by applying a force F. Due to application of force, both blocks begin to move with same magnitude of acceleration $\frac{100}{\sqrt{3}}\ \text{m/s}^2$.

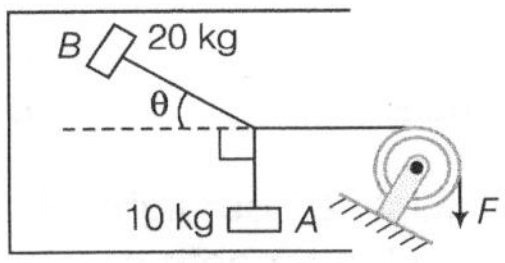

Find the value of F.

a. 2000 N
b. $1000\sqrt{3}$ N
c. 1000 N
d. 500 N

9 Two particles of mass m each, are tied at the ends of a light string of length $2a$.
The whole system is kept on a frictionless horizontal surface with the string held tight so that each mass is at a distance a from the centre P as shown in the figure. Now, the mid-point of the string is pulled vertically upwards with a small, but constant force F. As a result, the particles move towards each other on the surface. The magnitude of acceleration when separation between them becomes $2x$, is **[IIT JEE 2007]**

Working Space

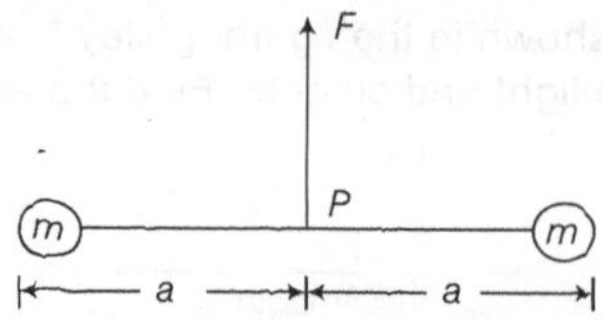

a. $\frac{F}{2m}\frac{a}{\sqrt{a^2-x^2}}$ b. $\frac{F}{2m}\frac{x}{\sqrt{a^2-x^2}}$ c. $\frac{F}{2m}\frac{x}{a}$ d. $\frac{F}{2m}\frac{\sqrt{a^2-x^2}}{x}$

10 All pulleys are smooth and light. Pulley E is fixed but pulleys C and D are free to move. Find the acceleration of blocks A and B.

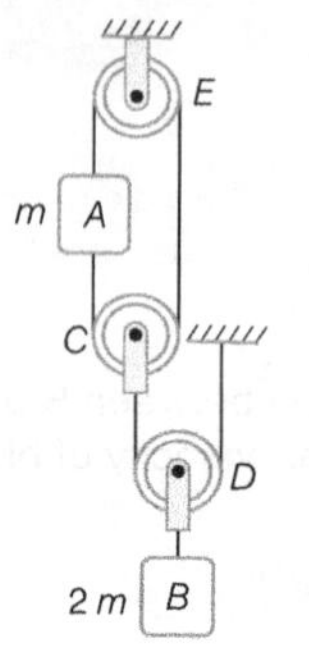

a. $a_A = g, a_B = 0$
b. $a_A = 0, a_B = 0$
c. $a_A = 0, a_B = g$
d. $a_A = g, a_B = g$

11 OO' and CC' are two fixed rods. Rings A and B have negligible radius. A light string is fixed at O and is connecting the ring B after passing through ring A. Find the speed of ring B at the instant as shown in the figure.

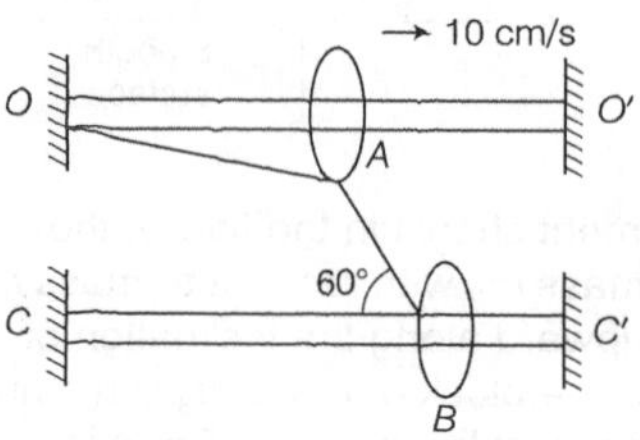

a. 10 cm/s towards C
b. $10\sqrt{3}$ cm/s towards C
c. $5\sqrt{3}$ cm/s towards C
d. None of these

12 Blocks are placed on the smooth horizontal surface. Each block has mass m and charge q. A uniform electric field E is present in the region as shown in the figure. The block B is projected with velocity v_0 as shown in the figure. After sometime, which block is closer to pulley?

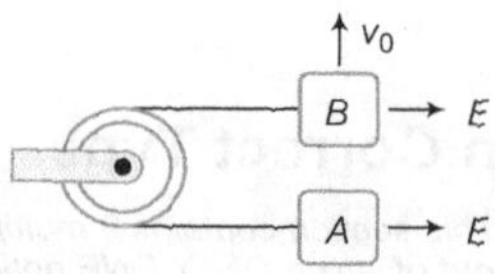

[**Hint** *The force on a charge q by the electric field E is F = qE in the direction of E.*]

a. A
b. B
c. May be A or may be B
d. None of these

13 The construction consists of four light wires of each length l. It is placed on the smooth horizontal surface. Find the speed of hinged point D at the instant as shown in the figure given below.

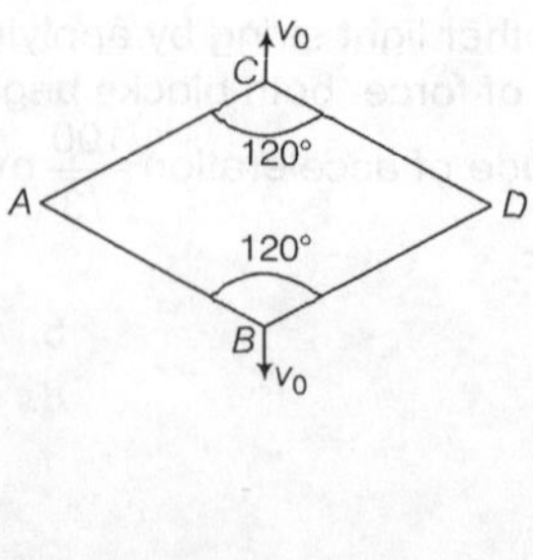

a. v_0
b. $2v_0$
c. $\frac{v_0}{2}$
d. $\frac{v_0}{\sqrt{3}}$

Working Space

14 Find the distance moved by the block A in 2 s, if hoisting drum fixed in the plank wraps string 1 cm/s. The plank is moving downward with a speed of 2 cm/s.

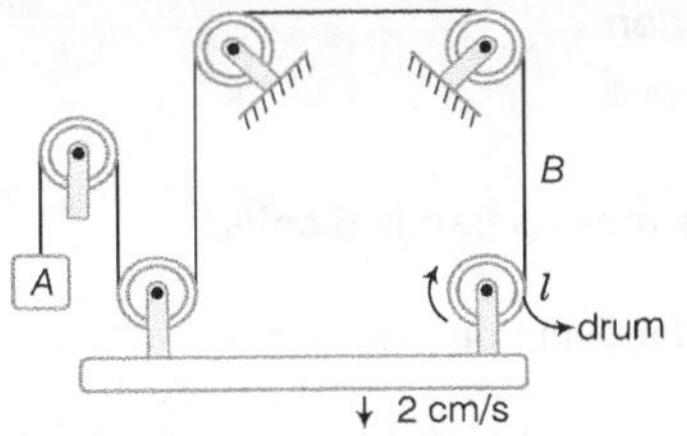

a. 10 cm b. 7 cm c. 14 cm d. 12 cm

15 In the situation shown in the figure, pulleys are smooth and light. The blocks are connected through light string. Blocks A and C are moving rightward with velocities 5 m/s and 2 m/s, respectively. The velocity of block B is

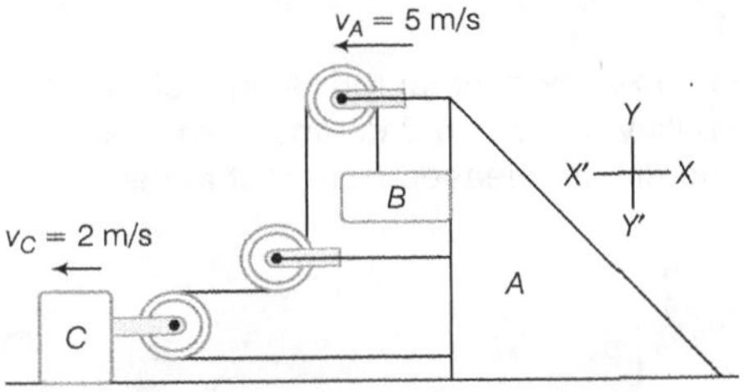

a. $-\hat{i}-\hat{j}$ b. $-5\hat{i}-6\hat{j}$ c. $-3\hat{i}-4\hat{j}$ d. $-5\hat{i}+6\hat{j}$

16 For the arrangement shown in the figure, blocks A and B are placed on the smooth horizontal surface. The free end P of the thread is being pulled by a constant force $F = 1$ N. Find the acceleration of free end P.

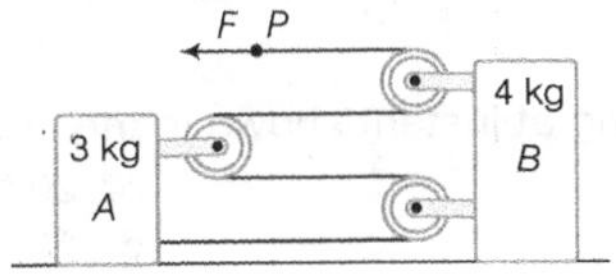

a. $6\,m/s^2$ b. $2\,m/s^2$ c. $7\,m/s^2$ d. $1\,m/s^2$

One or More than One Options Correct Type

Direction (Q. Nos. 17 and 18) *This section contains 2 multiple choice questions. Each question has four choices (a), (b), (c) and (d) out of which ONE or MORE THAN ONE are correct.*

17 The mass of pulley P_1 is m. The light pulley P_2 is placed on the smooth horizontal surface. Both pulleys are connected by a light string passing over fixed pulleys as shown in the figure.

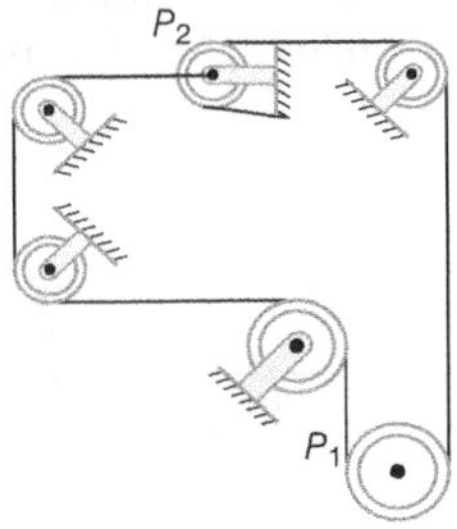

a. The acceleration of pulley P_1 is g in downward direction
b. Pulley P_2 is moving rightward with an acceleration $2g$
c. Both pulleys P_1 and P_2 are in rest
d. Pulley P_2 is stationary

Working Space

18 A long thin movable carpet is laid on the floor. One end of the carpet is bent back and then pulled backwards with constant velocity $v_0 = 8$ cm/s just above the part of the carpet which is still at rest on the floor. Then,

a. the speed of geometrical centre of mass of the moving part is 6 cm/s
b. the speed of point A is 4 cm/s
c. the speed of the geometrical centre of mass of moving part is 8 cm/s
d. the speed of the point A is 8 cm/s

Comprehension Type

Direction (Q. Nos. 19-23) This section contains 2 paragraphs, each describing theory, experiments, data etc. Five questions related to the paragraphs have been given. Each question has only one correct answer among the four given options (a), (b), (c) and (d).

Passage I

A smooth ring of mass $m = 1$ kg can slide on a fixed horizontal rod. A light string tied to the ring passes over a fixed smooth pulley of mass 0.2 kg and carries a block of mass $2m$ as shown in the figure. The ring is released from rest at the instant shown in figure.

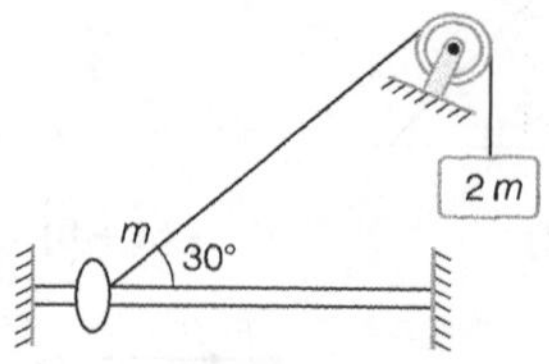

19 The tension in string at instant shown in the figure, is
a. 8 N
b. 20 N
c. 10 N
d. 15 N

20 The normal force on rod due to ring is
a. 8 N
b. 6 N
c. 10 N
d. 4 N

21 Reaction force (i.e. clamp force) on the pulley due to clamp is
a. $8\sqrt{3}$ N
b. 8 N
c. $\sqrt{244}$ N
d. None of these

Passage II

A man is standing on light weighing machine fixed inside a box of mass $m_1 = 20$ kg and holding a light string passing over ideal pulleys. Another mass $m_2 = 40$ kg is hanging as shown in the figure.

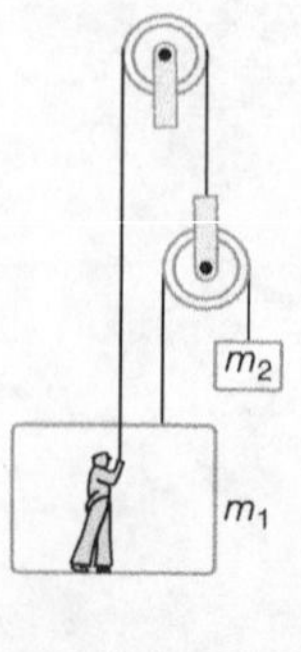

22 If the man manage to keep the box at rest, the find the reading of weighing machine.

a. 40 kg b. 20 kg
c. 10 kg d. 50 kg

23 If hanging mass is moving downward with an acceleration $5 \, m/s^2$, then find the reading of weighing machine.

a. $\frac{110}{3}$ kg-wt b. 20 kg-wt
c. 40 kg-wt d. $\frac{200}{3}$ kg-wt

One Integer Value Correct Type

Direction (Q. Nos. 24 and 25) This section contains 2 questions. When worked out will result in an integer from 0 to 9 (both inclusive).

24 For the situation shown in the figure, pulleys are light and smooth. Pulleys C and D are fixed, but A and B are movable. Find the value of m (in kg) if m is in equilibrium. (take, $m_1 = 2$ kg and $m_2 = 1$ kg)

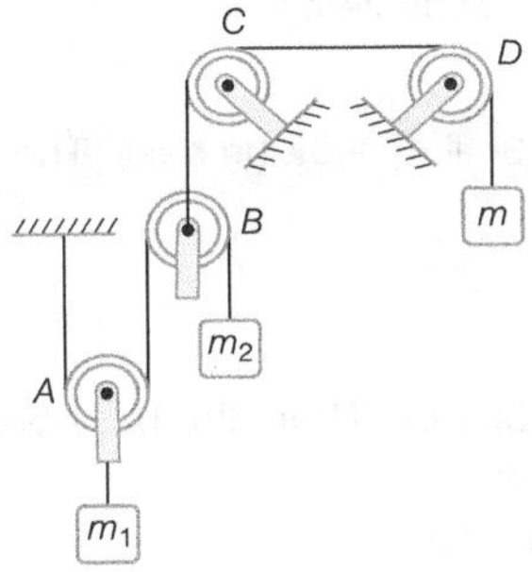

25 Consider the infinite number of arrangement of pulleys as shown in the figure. A light string passes over each light and smooth pulley with one end to a block of mass m and connected to another pulley. The system is released from rest. The acceleration of the left most block is $\frac{g}{x}$. Find the value of x.

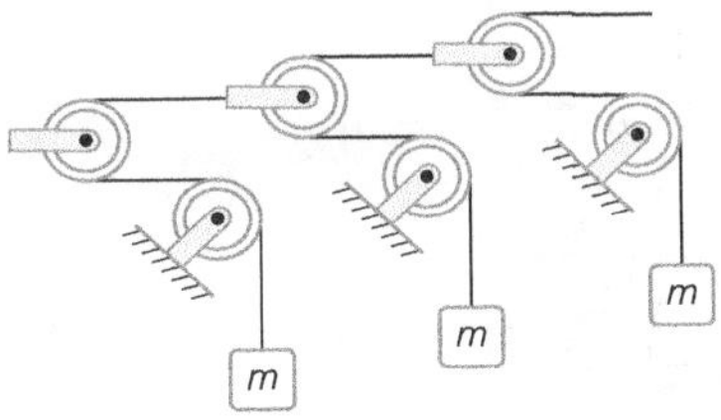

DPP-5 Wedge Constraints

Comprehension Type

Working Space

Direction (*Q. Nos. 1-5) This section contains 2 paragraphs, each describing theory, experiments, data, etc. Five questions related to the paragraphs have been given. Each question has only one correct answer among the four given options (a), (b), (c) and (d).*

Passage I

A block of mass m is placed on a horizontal xy-plane, i.e. suppose the block is placed on a table. The block is constrained with the table.

1 Assuming that the table is fixed the required condition for constraint's relation is

a. the block always remains in contact with the table
b. the z-component of velocity of block must be zero
c. the x-component and y-component of velocities must be zero
d. None of the above

2 Suppose that the table is moving with velocity $\mathbf{v} = 3\hat{\mathbf{i}} + 4\hat{\mathbf{j}} + 5\hat{\mathbf{k}}$ (in m/s). Then, the z-component of velocity of the block is

a. zero
b. 5 m/s
c. 3 m/s
d. 4 m/s

3 In previous problem, if the velocity of block is $a\hat{\mathbf{i}} + b\hat{\mathbf{j}} + c\hat{\mathbf{k}}$. Then, the possible values of a, b and c are

a. $a = 3, b = 4, c = 0$
b. $a = 0, b = 10, c = 5$
c. $a = 20, b = 30, c = 5$
d. Both (b) and (c) are correct

Passage II

A smooth rod AB is subjected to vertical up and down movement system, while resting on a hemisphere of radius R. The hemisphere moves to the right with a velocity v_0 and an acceleration of a_0.

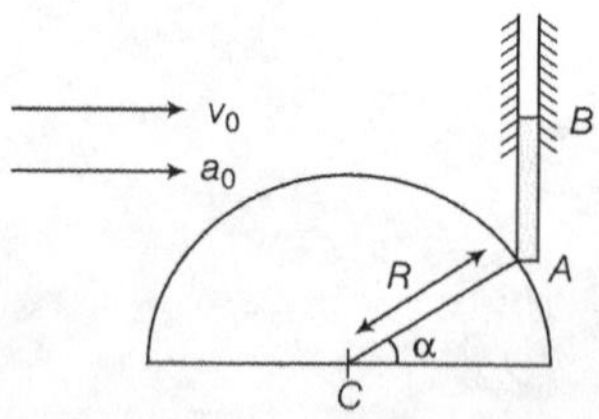

4 The velocity of the point B is

a. $v_0 \tan\alpha$
b. $v_0 \sin\alpha$
c. $v_0 \cot\alpha$
d. $v_0 \cos\alpha$

5 The acceleration of the point B is

a. $a_0 \cot\alpha$
b. $\dfrac{a_0 R\sin\alpha + v_0^2 + v_0^2 \tan^2\alpha}{R\cos\alpha}$
c. $\dfrac{a_0 R\cos\alpha + v_0^2 + \tan^2\alpha}{R\cos\alpha}$
d. None of these

Only One Option Correct Type

Working Space

Direction (Q. Nos. 6 -9) *This section contains 4 multiple choice questions. Each question has four choices (a), (b), (c) and (d) out of which ONLY ONE is correct.*

6 If the acceleration of block A is $5\ m/s^2$, then, acceleration of block B is

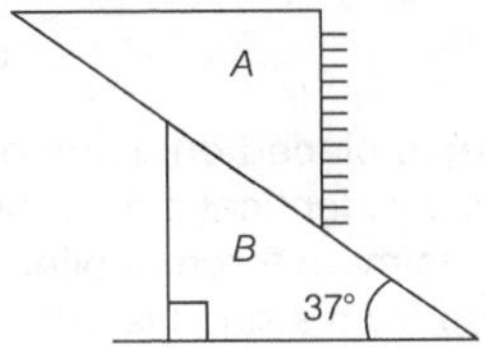

a. $5\ m/s^2$
b. $10\ m/s^2$
c. zero
d. $6.67\ m/s^2$

7 The end A of rod OA is in touch of triangular block B. Its other end O is hinged at the point O. The speed of end A of rod at the instant shown in the figure is

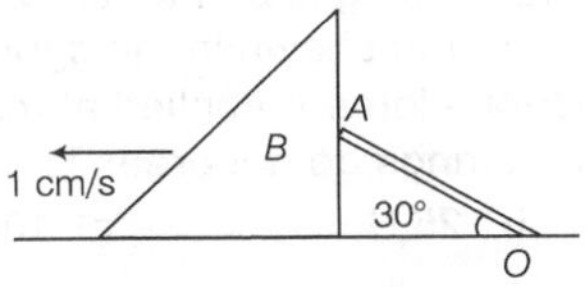

a. 1 cm/s
b. 2 cm/s
c. $\sqrt{3}$ cm/s
d. $2\sqrt{3}$ cm/s

8 Two smooth surfaced identical cylinders A and B of radius R and mass m each, are placed touching along their lengths on a smooth horizontal surface. A third identical cylinder C of the same mass and same radius is placed lengthwise on them. If initial acceleration of cylinder C is $\sqrt{3}\ m/s^2$, then initial acceleration of the cylinder A is

a. $\sqrt{3}\ m/s^2$
b. $3\ m/s^2$
c. $\frac{1}{\sqrt{3}}\ m/s^2$
d. $2\ m/s^2$

9 For the situation shown in the figure, find the speed of cylinder at the instant shown in figure.

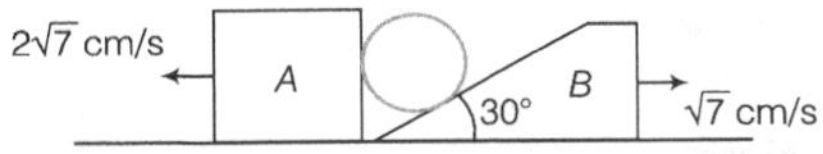

a. $\sqrt{21}$ cm/s
b. $\sqrt{17}$ cm/s
c. 7 cm/s
d. $2\sqrt{7}$ cm/s

One or More than One Options Correct Type

Direction (Q. Nos. 10-14) *This section contains 5 multiple choice questions. Each question has four choices (a), (b), (c) and (d) out of which ONE or MORE THAN ONE are correct.*

10 Two wedges ABC and DEF are rigidly attached to each other as shown in the figure. The composite wedge is placed on a smooth horizontal surface. When the smooth rod is released, then

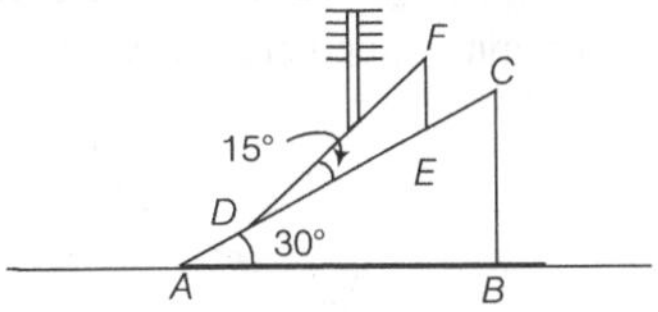

a. the rod is moving downward
b. the magnitude of acceleration of rod and the magnitude of acceleration of composite wedge are same
c. the composite wedge moves horizontally rightward
d. the composite wedge remains stationary

11 A circular wire is attached with a thin wire DE. O is the centre of the circular wire. Beads A and B moving along the circle freely is connected with a sleeve C through light string. At the instant shown in the figure, $\angle ACD = \frac{\pi}{4} = \angle OCB$, $\angle AOD = \frac{5\pi}{12}$ and $\angle OBC = \frac{\pi}{3}$. The speed of the sleeve C is $\sqrt{2}$ cm/s in downward direction at the instant shown in figure.

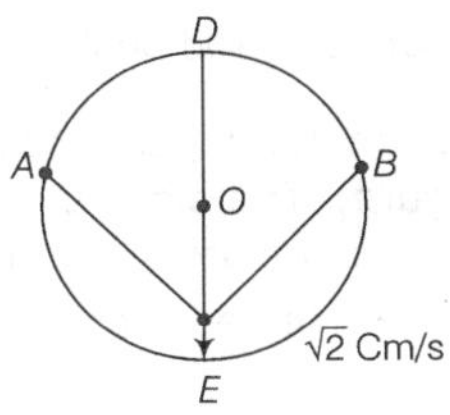

a. The speed of bead A is 2 cm/s
b. The speed of bead B is $\frac{2}{\sqrt{3}}$ cm/s
c. Both beads move with same speed
d. The speed of bead B is zero

Working Space

12 A light rod AB is fixed on an inclined plane. A sleeve starts from A slides along the smooth rod to reach on the inclined plane in minimum time. The inclination of rod with vertical is

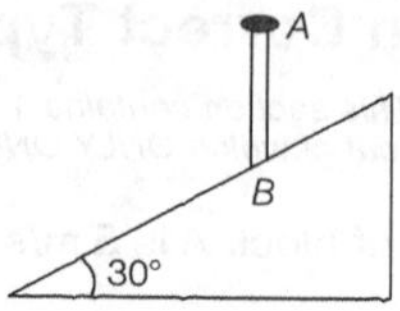

a. 30°
b. 0°
c. 15°
d. non-zero

13 A rod of mass 10 kg is placed on a smooth horizontal surface and is fixed by identical strings as shown in the figure (i). The minimum force applied at mid-point of rod to free the rod from strings is 100 N. Now, the same rod is placed on smooth horizontal surface and a smooth block of mass 10 kg is connected with the same strings to ends of rod shown in the figure (ii). The maximum horizontal force F applied at mid-point of the rod so that the strings do not break is

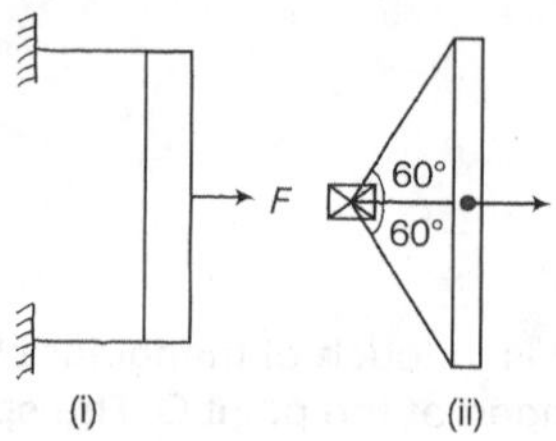

a. zero
b. 25 N
c. 100 N
d. non-zero

14 A particle P is sliding down a frictionless hemispherical bowl, it passes through the point A at $t = 0$. At this instant of time, the horizontal component of its velocity is v. A bead Q of the same mass as P is ejected from A at $t = 0$ along the horizontal string AB with the speed v. Friction between the bead and the string may be neglected. Let t_P and t_Q be the respective times taken by P and Q to reach the point B. Then,

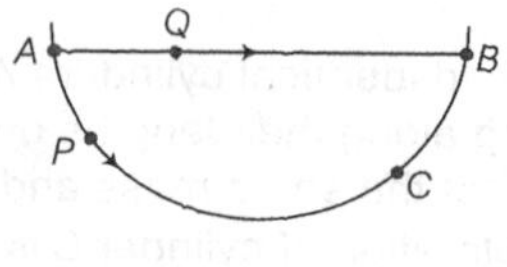

a. $t_P < t_Q$
b. $t_P = t_Q$
c. $t_P > t_Q$
d. $\frac{t_P}{t_Q} < 1$

Subjective Questions

Direction (Q. Nos. 15 -21) *These questions are subjective in nature, need to be solved completely on notebook.*

15 A block of mass 80/3 kg is resting over a smooth horizontal plane. The distance of block A from the wall is 0.4 cm and it is held at rest by a light thread AC. A block B of mass10 kg is resting against the vertical smooth wall. The blocks A and B are hinged at the ends of a light rigid rod. Now, the string AC is burnt.

i. Find relation between initial acceleration a_1 of the block A and initial acceleration a_2 of the block B.

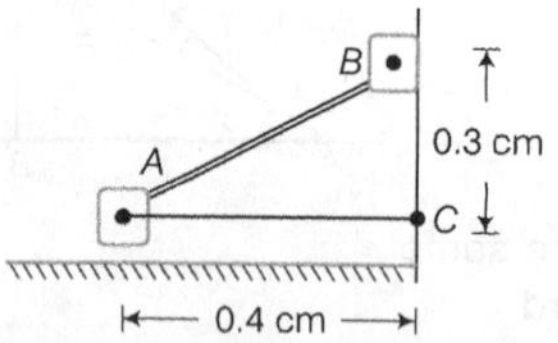

ii. Find initial acceleration of the block B.

iii. Find the force exerted by rod on the block A, initially.

16 A 1 kg block A rests on the upper surface of block B of mass 1 kg (shown in the figure). The block B is on an inclined plane of inclination 37°. All surfaces are smooth.

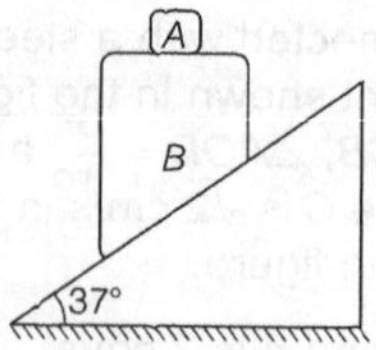

i. Find the relation between acceleration of the block A and the block B.

ii. Find the acceleration of the block A.

iii. Find the normal force exerted by block A on block B.

Working Space

17 In the situation shown in the figure, all surfaces are smooth. The block B of mass 1 kg is moving rightward with an acceleration 3 m/s^2.

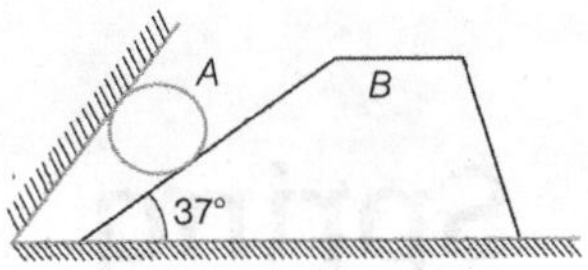

i. Find the acceleration of sphere A.
ii. Find the normal reaction exerted by the sphere on the block B.
iii. Find the normal reaction exerted by the vertical wall on the sphere.
iv. Find the mass of the sphere A.

18 Two identical triangular blocks m_1 and m_2 of mass 1 kg each, rest on a smooth horizontal surface, m_2 being held against a vertical wall as shown in the figure. A horizontal force F is applied to m_1 so that the two triangular blocks support to keep a symmetric smooth surface mass $m_3 = 6$ kg at rest at height of 10 m from the floor. At $t = 0$, the force F is withdrawn.

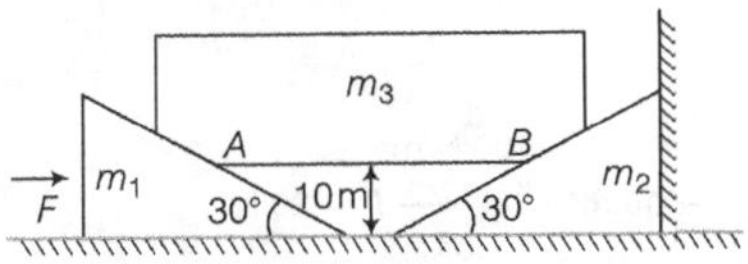

i. Find the relation between acceleration of m_1 and m_3, when F is withdrawn.
ii. Find the time taken by m_3 to strike with ground.
iii. Find the velocity of m_1 at the instant when m_3 strikes with ground.

19 The arrangement shown in the figure, the masses of modified wedge A and the rod B are $m_1 = m_2 = 1$kg respectively, all surfaces are frictionless.

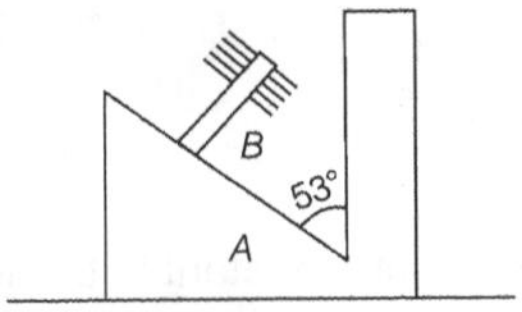

i. If the acceleration of rod is a_2 and acceleration of wedge is a_1. Find the relation between a_1 and a_2.
ii. Find the acceleration of the rod.
iii. Find the force exerted by rod on wedge.

20 A ball of mass $m_1 = 4$ kg is suspended from the roof of a box of mass $m_2 = 5$ kg through a light string. The box and ball are released from rest. Find the acceleration of ball just after the system is released.

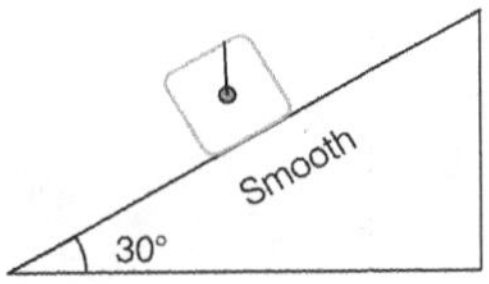

21 Two smooth balls, A of mass 4 kg and radius 1 cm and B of mass 3 kg and radius 4 cm are placed at a smooth corner. A horizontal force F is acting on the ball A horizontally.

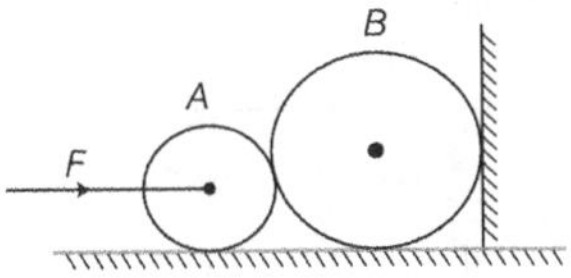

i. The minimum value of F to lift the ball B from the ground is $10F_{min}$. Find the value of F_{min}.
ii. Force $F = 44$ N, find the initial acceleration of ball.

DPP-6 Spring

Subjective Questions

Direction (Q. Nos. 1-8) *These questions are subjective in nature, need to be solved completely on notebook.*

Working Space

1 Find elongation or compression in spring at equilibrium of the block(s) in the following cases.

i.

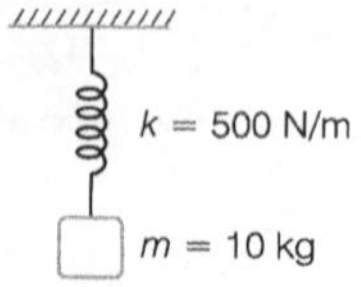

ii.

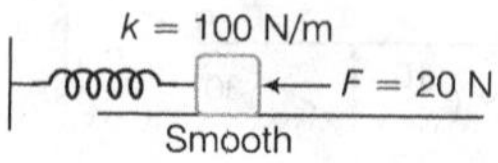

iii.

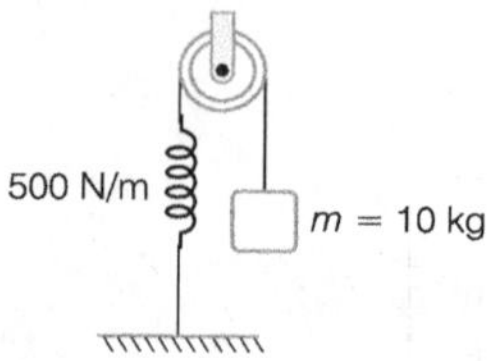

iv.

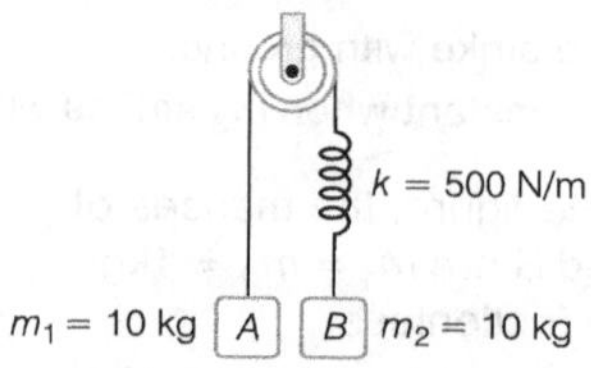

2 Find the equivalent spring constant in the following cases.

i.

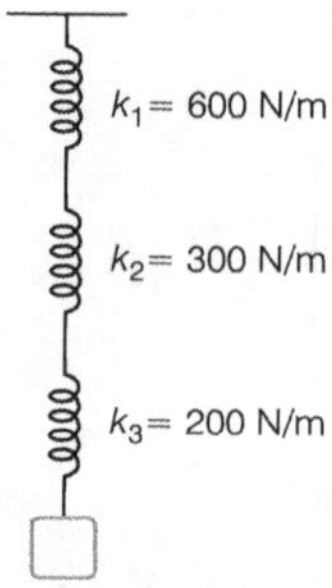

ii.

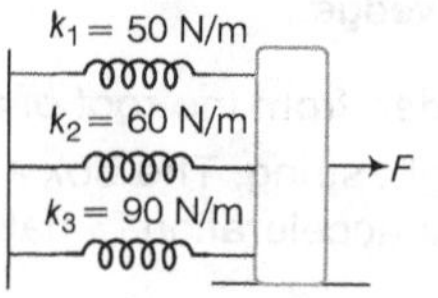

iii.

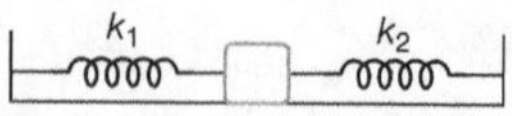

iv.

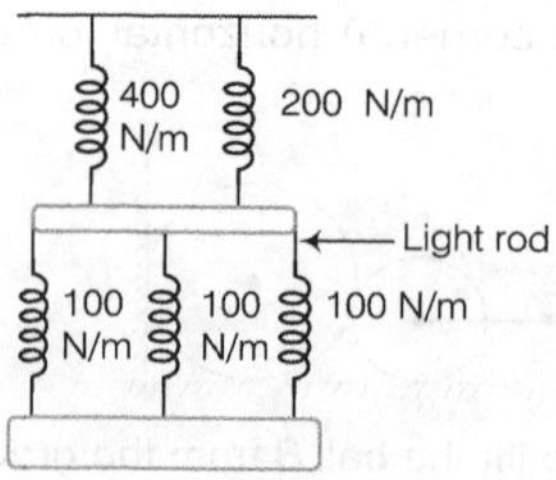

v. Infinite number of identical springs, each of spring constant $k = \dfrac{20}{(\sqrt{5}+1)}$ are combined as shown in the figure.

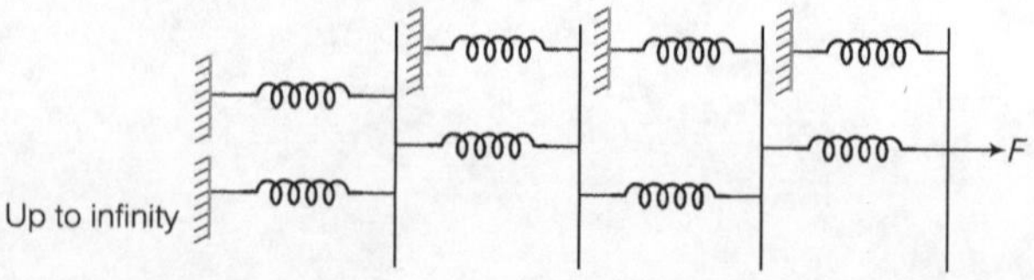

Working Space

3 For the following cases, block is initially in equilibrium, when gravity is switch OFF. Find the distance moved by the block A, when springs come in natural lengths.

i. 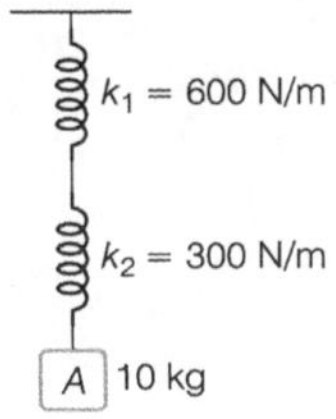

ii. 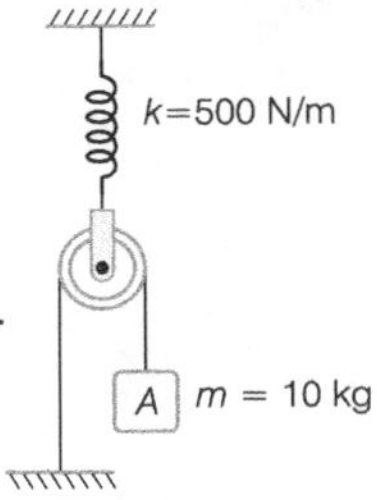

iii. 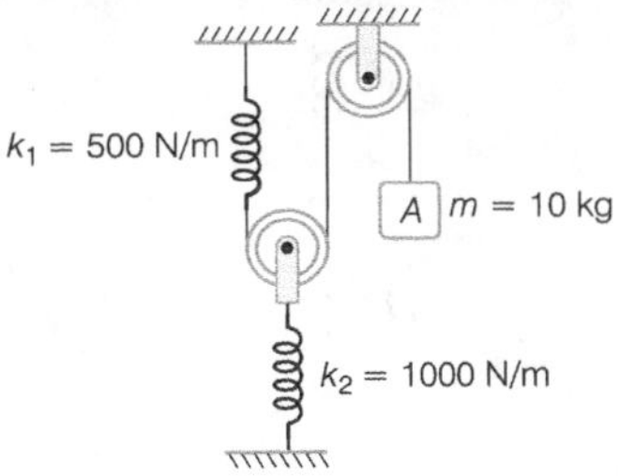

4 i. A spring of spring constant 100 N/m is cut into two equal parts, then connected in parallel. Find the equivalent spring constant.

ii. A spring of spring constant $k = 100$ N/m is cut into three parts, whose length ratio is 1 : 2 : 3. Find spring constant of each part.

iii. A spring of force constant k is cut into three equal parts. What is force constant of each part?

iv. If a spring of stiffness k is cut into two parts A and B of length ratio $l_A : l_B = 2 : 3$. Find the stiffness of spring A. **[AIEEE 2011]**

5 Systems shown in the following cases are in equilibrium. The springs and strings are light. The pulleys are light and smooth. Find acceleration of block just after cutting the string or spring A.

i. The tension in string A is 15 N.

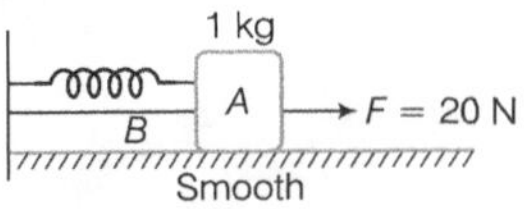

ii. The tension in string A is 20 N.

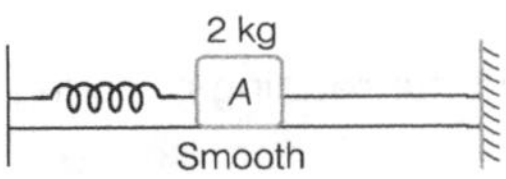

iii. $k_A = k_B = 100$ N/m

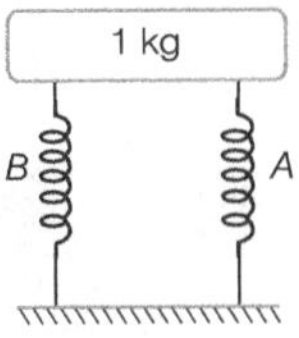

iv. A ball of mass 2 kg is hanging in equilibrium as shown in the figure.

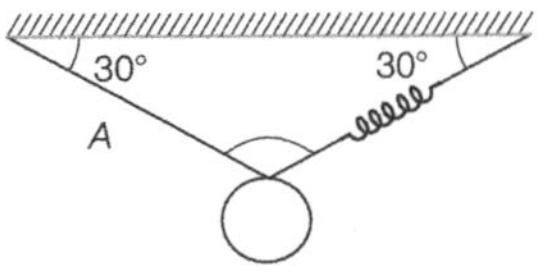

6 The systems shown in the following cases are in equilibrium. Find acceleration of each block just after cutting the string or spring or light rod A.

i. 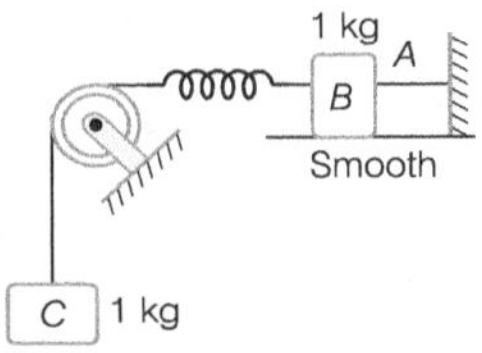

ii. 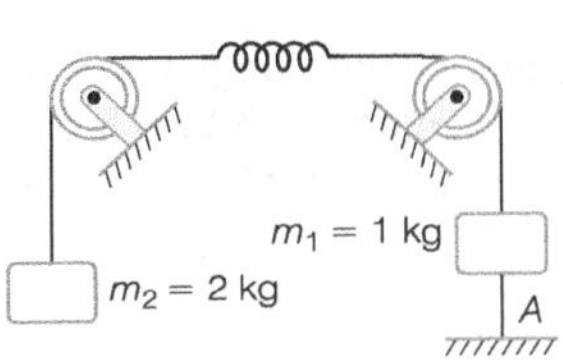

iii. 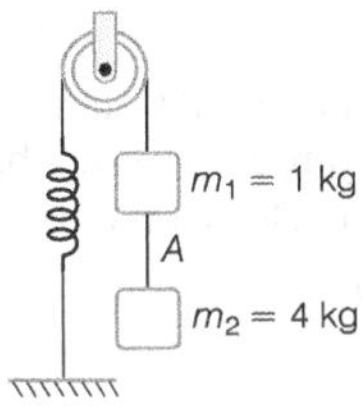

Working Space

iv. System shown in the figure is in equilibrium and at rest. The spring and string are massless. Now, the string is cut. Find the acceleration of mass $2m$ and m just after the string is cut. **[IIT JEE 2006]**

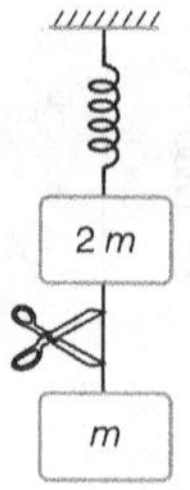

v.

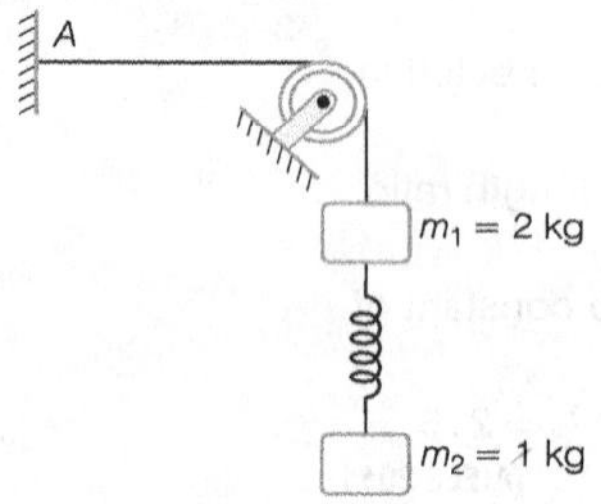

vi.

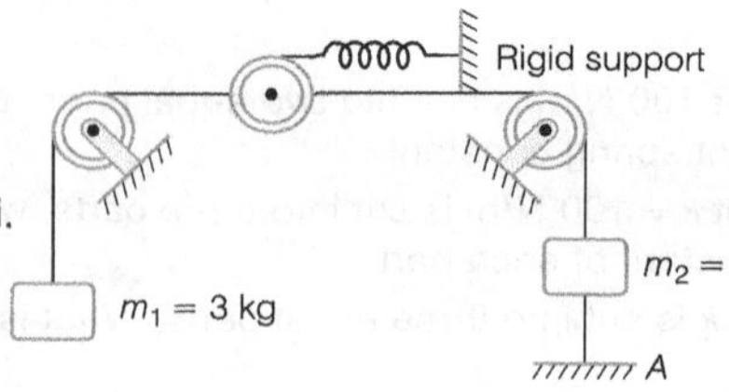

vii.

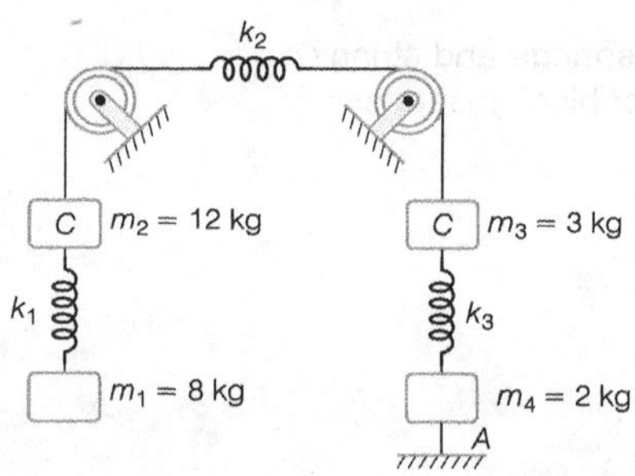

7 All spring balances are light and in steady state. Find the reading of spring balances (S, S_1 and S_2) in the following cases.

i.

ii.

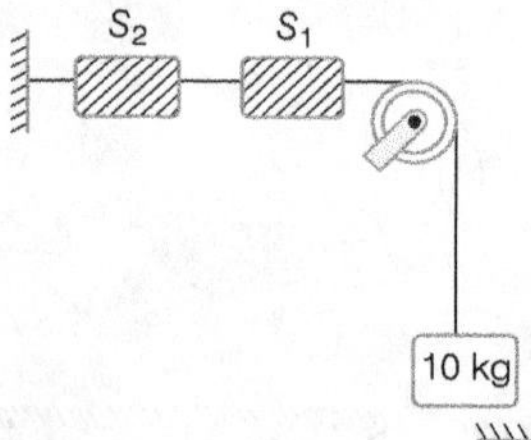

iii.

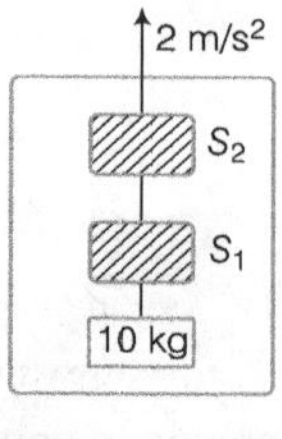

iv.

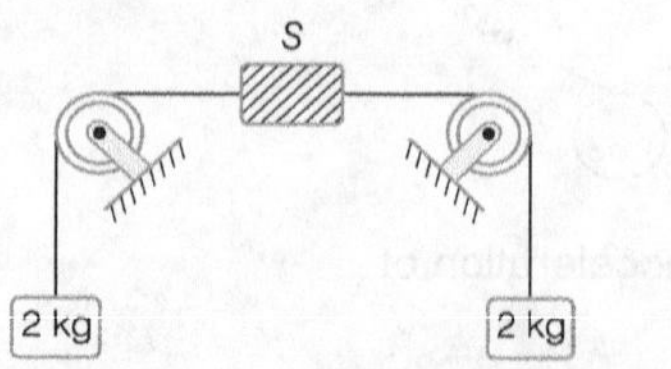

v.

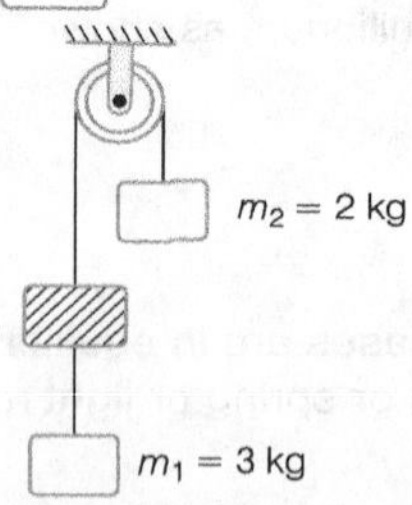

vi. A spring balance is attached to the ceiling of a lift. A man hangs his bag on the spring and the spring reads 49 N, when the lift is stationary. If the lift moves downward with an acceleration of 5 m/s^2. Find the reading of the spring balance. (take, $g = 9.8$ m/s^2) **[AIEEE 2003]**

Working Space

8 Initially, light spring A is in its natural length. Find maximum speed of block B and maximum change in length of spring A in the following cases.

i. 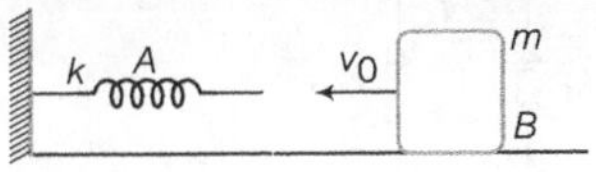

ii. 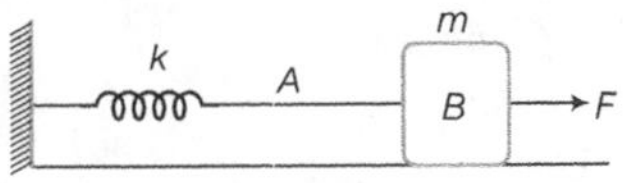

iii. An ideal spring with spring constant k is hung from the ceiling and a block of mass M is attached to its lower end. The mass is released with the spring initially unstretched. What is the maximum extension in the spring? **[IIT JEE 2002]**

Only One Option Correct Type

Direction (Q. Nos. 9-17) This section contains 9 multiple choice questions. Each question has four choices (a), (b), (c) and (d), out of which ONLY ONE option is correct.

9 For the given figure, three light springs are shown. Their spring constants are k, $2k$ and $3k$. The respective compressions springs are x_1, x_2 and x_3. The value of $x_1 : x_2 : x_3$ is

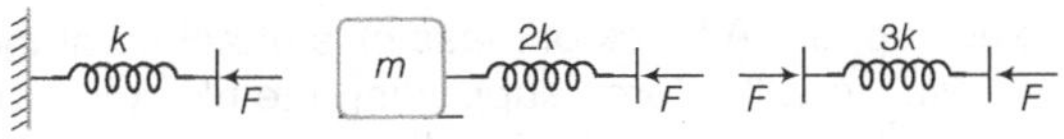

a. 1 : 1 : 1 b. 1 : 2 : 3 c. 6 : 3 : 2 d. None of these

10 At an instant, $t = t_0$, the acceleration of block A is 8 m/s^2. The acceleration of the block B at $t = t_0$ is

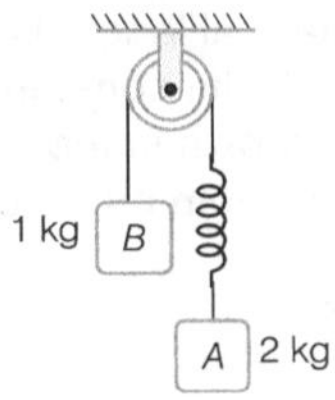

a. 8 m/s^2 upward

b. 6 m/s^2 downward direction

c. zero

d. 3.33 m/s^2 in upward direction

11 A spring of length L is shown in the figure. Find the velocity of a turn at distance x from end A.

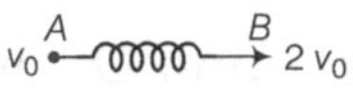

a. v_0 b. $\frac{v_0}{L}x$ c. $\frac{3v_0}{L}x - v_0$ d. $\frac{2v_0}{L}x$

12 A block of mass 10 kg is hanging with three identical massless springs (as shown in the figure). Find the magnitude of acceleration of the block at the moment, the right spring CP breaks. Assume that elongation in spring BP is 5 cm and its spring constant is 1000 N/m ($g = 10$ m/s^2).

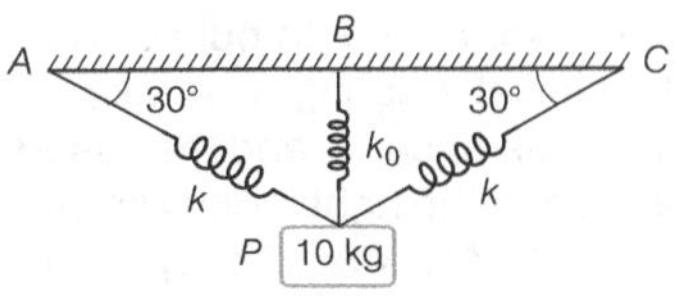

a. zero b. 5 m/s^2 c. 4.33 m/s^2 d. 2.5 m/s^2

13 Two springs are light but the spring balances have mass 2 kg each. They are used as shown in the diagram. If the extension is stationary, the readings in S_1 and S_2 are

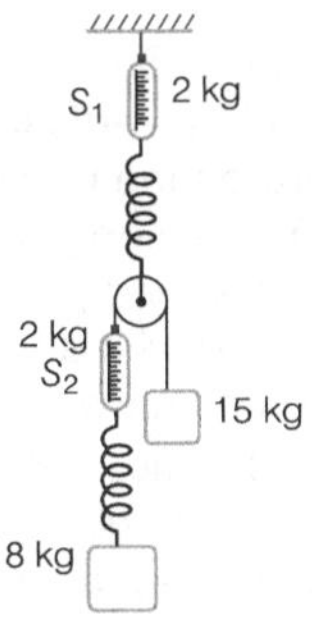

a. $\frac{108}{7}$ kg, $\frac{240}{7}$ kg

b. $\left(\frac{108}{7} + 2\right)$ kg, $\left(\frac{240}{7} + 2\right)$ kg

c. 80 kg, 35 kg

d. None of these

14 For the arrangement shown in the figure, the springs are light and have stiffness k_1 and k_2. The light pulley is smooth and fixed. The block of mass m is initially held at rest maintaining springs vertically straight and springs relaxed. Now, the force supporting the block is gradually reduced and eventually removed. Now, the block hangs motionless in equilibrium. How far does the block descend during the process of reduction of the force supporting it?

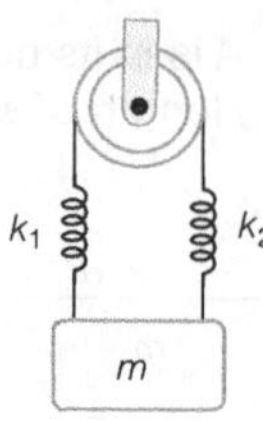

a. $\frac{mg}{k_1+k_2}$ b. $\frac{mg(k_1+k_2)}{k_1k_2}$ c. $\frac{mg(k_1+k_2)}{4k_1k_2}$ d. $\frac{mg(k_1+k_2)}{2k_1k_2}$

15 A light spring of spring force constant 100 N/m is compressed between two supports (shown in the figure), when a force of $F = 5$ N is exerted at point A in downward direction along the spring, what will be the displacement of point A?

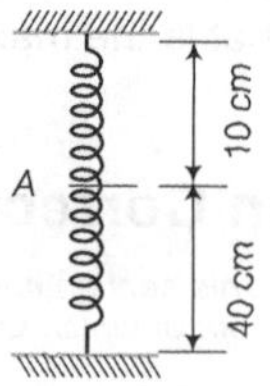

a. 10^{-3} m b. 8×10^{-3} m c. 4×10^{-3} m d. 5×10^{-3} m

16 A light spring of length 0.3 m and spring constant 600 N/m is suspended from a fixed support. A block of mass m is attached at the lower end and is initially held at rest. Now, the force supporting the block is gradually reduced and eventually removed. Now, the block hangs motionless in equilibrium. During this 6th coil of the spring from fixed end comes at the point where 8th coil of relaxed spring was. Find the mass of hanging block.

a. 1 kg b. 2 kg c. 3 kg d. 6 kg

17 A linear spring of natural length is placed on smooth horizontal table. Now, end A and end B are pulled with constant speed 1 cm/s and 5 cm/s, respectively in opposite direction (shown in the figure). Find the shift in a point at distance 2 m from end B of the relaxed spring in 5 s.

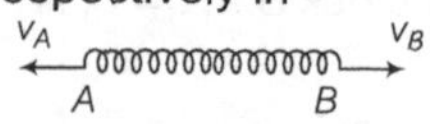

a. 5 cm towards A b. 5 cm towards B
c. 10 cm towards B d. zero

One or More than One Options Correct Type

Direction (Q. Nos. 18 and 19) *This section contains 2 multiple choice questions. Each question has four choices (a), (b), (c) and (d) out of which ONE or MORE THAN ONE option are correct.*

18 For the given figure, two blocks of equal mass m are connected by a light string that passes over a smooth pulley. The block A is connected by spring constant k. Initially spring is relaxed. Then, the block A is slowly pulled down a distance x and released. The magnitude of accelerations of the left and the right blocks immediately after releasing are a_1 and a_2, respectively. Mark correct option (s).

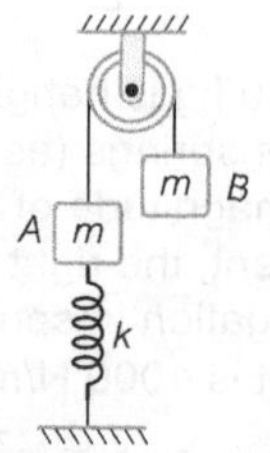

a. $a_1 = a_2$ b. $a_1 = a_2$, if $k \le \frac{2mg}{x}$

c. If $a_1 = \frac{k}{m}x$, $a_2 = 0$ d. $a_1 < a_2$, if $k > \frac{2mg}{x}$

19 A spider has fastened one end of a super elastic silk thread of length 1 m to a vertical wall. A small caterpillar is sitting somewhere on the thread. The hungry spider whilst not moving from its original position, starts pulling in the other end of the thread with a uniform speed, $v_0 = 1\,\text{cms}^{-1}$. Meanwhile, the caterpillar starts fleeing towards the wall with a uniform speed of 1 mm s^{-1} with respect to the moving thread. The caterpillar starts from $x = x_0$.

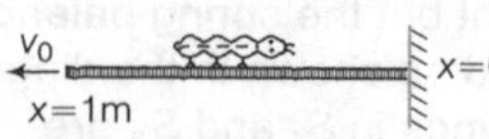

a. If $x_0 = 10$ cm, the distance between caterpillar and the wall does not change
b. If $x_0 < 10$ cm, caterpillar will reach at the wall
c. If $x_0 > 10$ cm, caterpillar may reach at the wall
d. If $x_0 > 10$ cm, distance between caterpillar and the wall increases

Working Space

Comprehension Type

Direction (Q. Nos. 20-22) *This section contains a paragraph, each describing theory, experiments, data, etc. Three questions related to the paragraph have been given. Each question has only one correct answer among the four given options (a), (b), (c) and (d).*

Passage

For the situation shown in the figure, horizontal spring and the spring connecting blocks B and C are light. The blocks A, B and C all are of same mass m. The pulley is smooth and fixed.

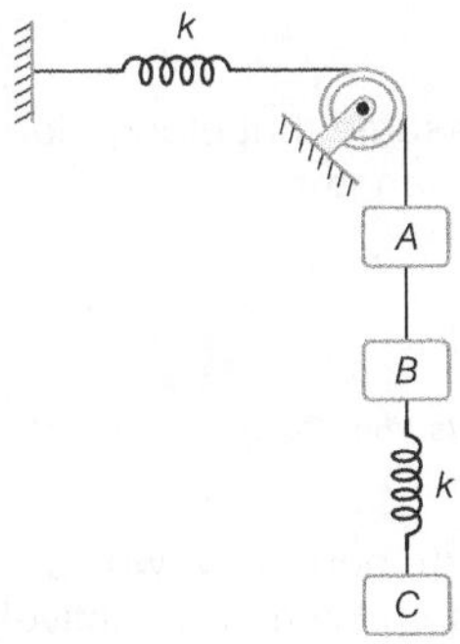

20 The accelerations of A, B and C just after cutting the horizontal spring are

a. $a_A = a_B = 15\text{ m/s}^2$ and $a_C = 0$
b. $a_A = a_B = a_C = 0$
c. $a_A = a_B = 10\text{ m/s}^2$ and $a_C = 0$
d. $a_C = a_B = 0$ and $a_C = 10\text{ m/s}^2$

21 The accelerations of the blocks just after cutting the string between A and B are

a. $a_A = a_B = a_C = 0$
b. $a_A = 20\text{ m/s}^2$, $a_B = 20\text{ m/s}^2$, $a_C = 0$
c. $a_A = 0$, $a_B = 10\text{ m/s}^2$, $a_C = 20\text{ m/s}^2$
d. None of these

22 The accelerations of A, B and C just cutting spring between B and C are

a. $a_A = a_B = 0$, $a_C = g$
b. $a_A = a_B = 5\text{ m/s}^2$, $a_C = 10\text{ m/s}^2$
c. $a_A = a_B = a_C = 0$
d. None of these

One Integer Value Correct Type

Direction (Q. Nos. 23-25) *This section contains 3 questions when worked out will result in an integer from 0 to 9 (both inclusive).*

23 For the system shown in the figure, pulley is smooth and light. Each spring has spring constant $k = 100$ N/m.

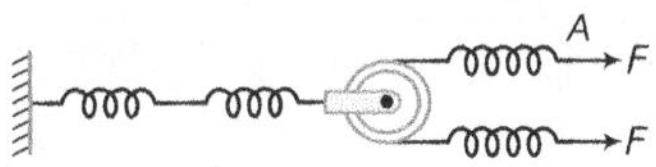

The least force $F = 100$ N is required to shift the point A through a distance x_0. Find the value of x_0.

24 Two light springs BP and AP of negligible natural lengths are knotted to point P. The knot point is connected to a block of mass $m = 100$ kg through a light string passing over a fixed pulley. In the given figure, $OA = OB$ and the coordinate of point P for equilibrium of the block is $(3m, 4m)$. Find the spring constant. If it is 1×10^x, then what is the value of x?

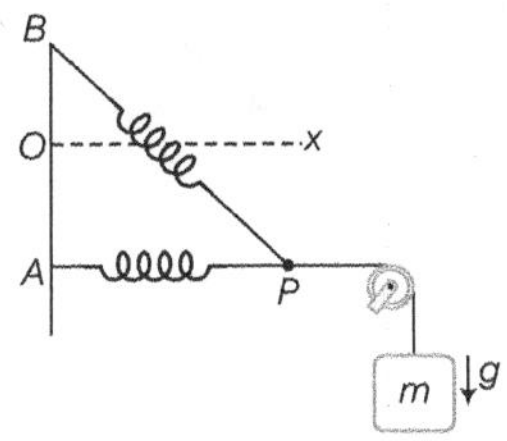

Working Space

25 Three segments cut from a long light spring are attached at point *A* (shown in the figure). The point *A* is pulled horizontally on a smooth horizontal table by a force $F = 10$ N.

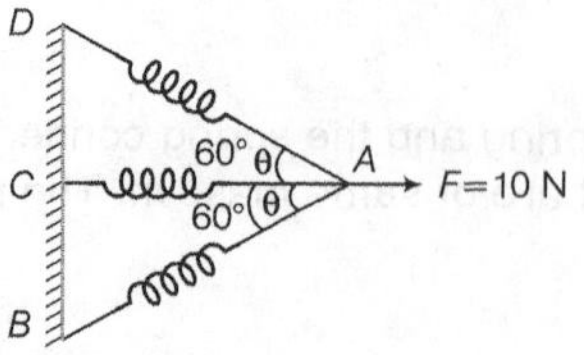

At steady state, find the tension in middle section. Assume that elongation in spring segments are much smaller than their relaxed lengths.

Matching List Type

Direction (Q. No. 26) *Choices for the correct combination of elements from Column I and Column II are given as options (a), (b), (c) and (d), out of which one is correct.*

26 The rigid light horizontal bar (shown in the figure) is supported by two light springs. *A* and *B* having spring constant 20 kN/m and 10 kN/m, respectively. A 150 kg block is then placed upon the bar above the spring *A*. *A* sudden blow projects the block towards *B* with an initial velocity of 2 m/s. (neglect friction).

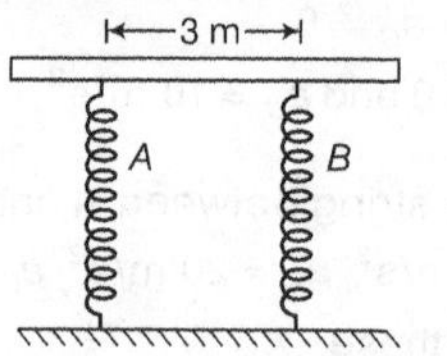

Match the Column I and Column II and mark the correct option from the given codes.

	Column I		Column II
i.	Compression (in mm) in spring *A* when block is above spring *A*	p.	2.34
ii.	Compression (in mm) in spring *B* when block is above spring *A*	q.	zero
iii.	Compression (in mm) in spring *B* when block is above spring *B*	r.	75
iv.	The speed of block (in m/s) when block is above spring *B*.	s.	150

Codes

	i	ii	iii	iv		i	ii	iii	iv
a.	s	q	s	p	b.	p	s	r	q
c.	r	p	q	s	d.	p	q	r	s

DPP-7 Pseudo Force

Subjective Questions

Working Space

Direction *(Q. Nos. 1-4) These questions are subjective in nature, need to be solved completely on notebook.*

1 i. A car A is moving with a velocity $3t\hat{i} + 4t^2\hat{j}$. Another car B is moving with a velocity of $t^2\hat{i} + 2t^2\hat{j}$. The driver of mass 50 kg of car B is observed in the frame of car A. At $t = 1$s, find pseudo force on the driver of car B.

ii. A particle of mass 1 kg is moving horizontally with an acceleration 2 m/s^2. The particle is observed by a frame moving with a constant velocity of 10 m/s. Find the magnitude of pseudo force on the particle.

iii. A block is on smooth surface. The net force on it is zero. An observer measures non-zero acceleration of the block. The observer is in .... reference frame.

iv. If car A and car B are accelerating with respect to each other. The pseudo force acting on a particle in the frame of car A is F_A and in the frame of car B is F_B. Then, $F_A = F_B = 0$ (true or false).

v. A reference frame attached to the earth is .... reference frame.

2 i. A block is moving with an acceleration 2 m/s^2. The smooth block B is stationary with respect to block A. Find tension in the string as shown in the figure.

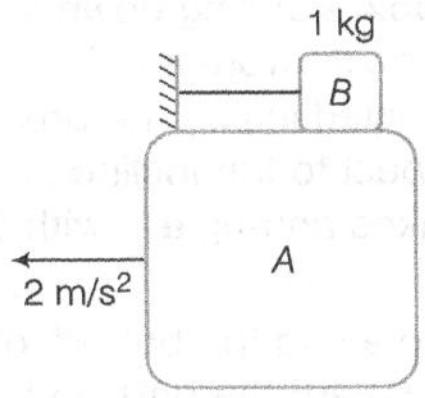

ii. A block B of mass 5 kg is attached with smooth triangular block moving horizontally with an acceleration a_0. The block B is stationary with respect to the triangular block. Find minimum acceleration of the triangular block so that string slacks.

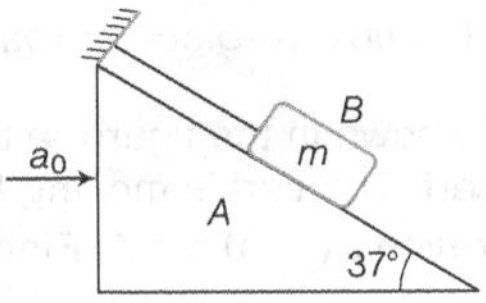

iii. A block B of mass 1 kg is kept on a triangular block moving horizontally with an acceleration a_0 (shown in the figure). Find maximum value of a_0 so that block B does not leave the surface of the triangular block.

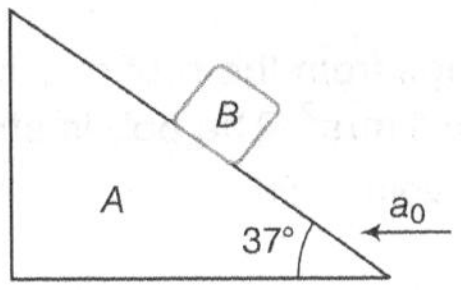

iv. A block is kept on a frictionless inclined surface with an angle of inclination α. The incline is given an acceleration a to keep the block stationary. What is the value of a?
[AIEEE 2005]

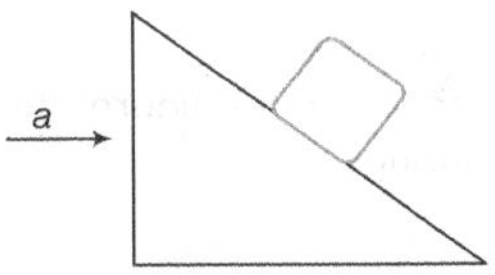

v. A smooth block A of mass 10 kg is connected with a block B with a light spring of spring constant 100 N/m. The block B is moving with an acceleration 5 m/s^2. Block A is stationary with respect to block B. Find the compression in the spring as shown in the figure.

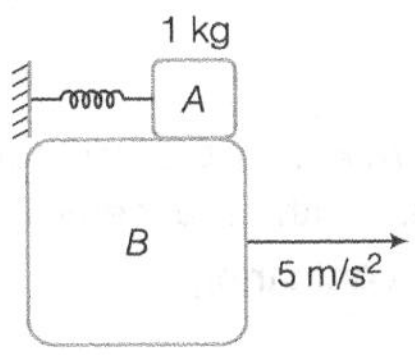

3 i. A smooth inclined plane of angle 37° and length 20 m is moving horizontally with an acceleration $a_0 = 5$ m/s^2. Find the time taken by the block to slide down the length of the inclined plane if it is released from the top of the inclined plane.

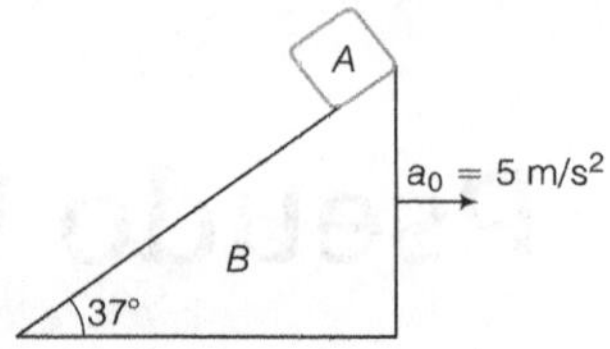

Working Space

ii. A block *B* is moving upward with an acceleration 5 m/s^2. Find the time taken by block *A* to slide down the length 18 m of the incline if it is released from the top of the incline.

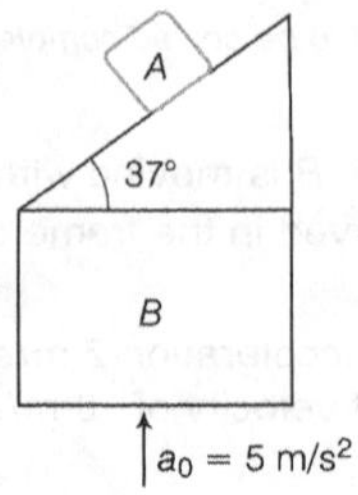

iii. A cubical box of side length 16 cm is sliding on smooth inclined plane of inclination 37°. A bolt starts to fall from point *P*. Find the time after which bolt strikes the bottom surface of box.

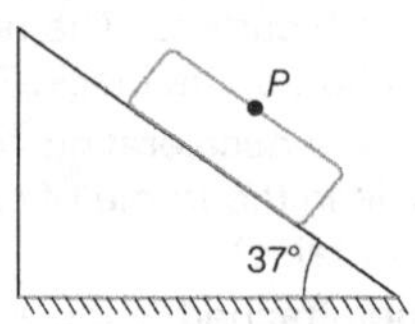

iv. A large, heavy box is sliding down without friction a smooth plane of inclination θ. From a point *P* on the bottom of the box, a particle is projected inside the box. The initial speed of the particle with respect to the incline plane is *u* and the direction of projection makes an angle α with the bottom as shown in the figure.

Find the distance along the bottom of the box between the point of projection *P* and the point *Q*, where the particle lands. assume that the particle does not hit any other surface of the box. Neglect air resistance.

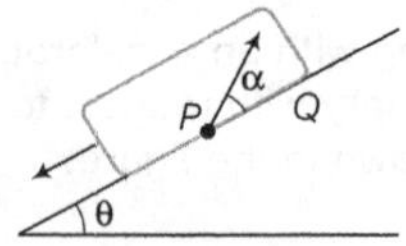

4 i. For the situation shown in the figure, a ball is stationary with respect to cart. The cart is moving horizontally with constant acceleration $a_0 = 10$ m/s^2. Find the value of θ.

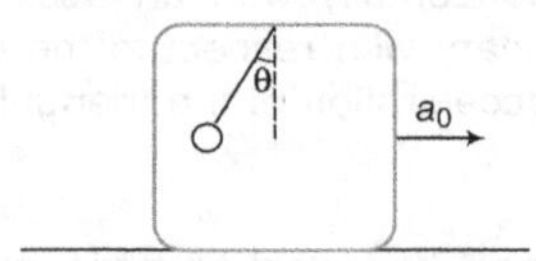

ii. A pendulum hangs from the roof of a cart moving with an acceleration $a_0 = 1$ m/s^2. The bob is stationary with respect to cart. Find the value of θ.

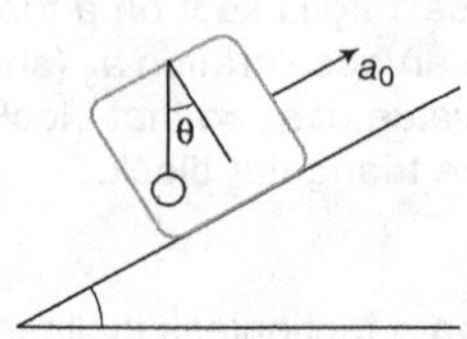

iii. For the situation shown in the figure, find the ratio of tension in lower and upper strings.

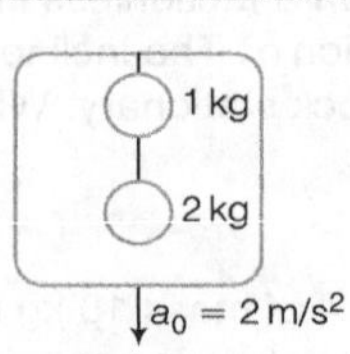

iv. A block of mass $m = 1$ kg is suspended inside an elevator moving downward with an acceleration 2 m/s^2. Find the tension in upper light string.

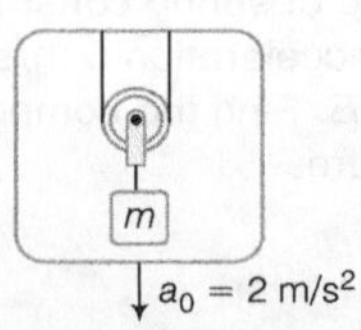

Working Space

v. Two blocks of mass 2.9 kg and 1.9 kg are suspended from a rigid support S by two inextensible wires each of length 1 m, shown in the figure. The upper wire has negligible mass and the lower wire has a uniform mass of 0.2 kg/m. The whole system of blocks wires and support have an upward acceleration of 0.2 m/s^2. Acceleration due to gravity is 9.8 m/s^2.
a. Find the tension at the mid-point of the lower wire.
b. Find the tension at the mid-point of the upper wire.

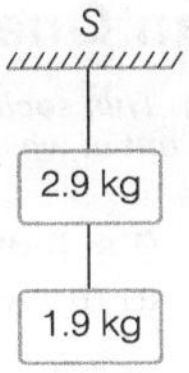

Only One Option Correct Type

Direction (Q. Nos. 5-9) *This section contains 5 multiple choice questions. Each question has four choices (a), (b), (c) and (d), out of which ONLY ONE is correct.*

5 A lift with block system is moving upward with an acceleration 2 m/s^2. The spring balance is light. The mass of each block is 1 kg. The reading of spring balance A is

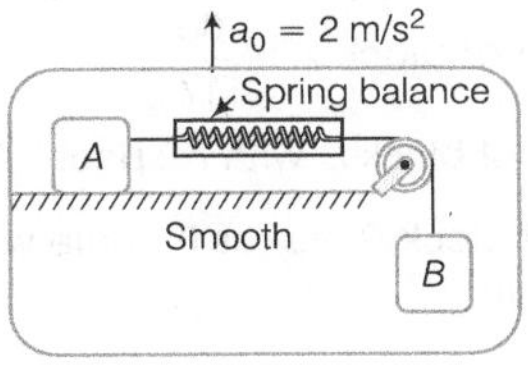

a. 1 N b. 10 N
c. 6 N d. zero

6 A wedge of mass M is free to move on smooth horizontal surface. Two blocks of mass m_1 and m_2 are free to move on smooth inclined surfaces of the wedge.

If $\frac{m_1}{m_2} = \frac{\sin 2\theta_2}{\sin 2\theta_1}$, the wedge

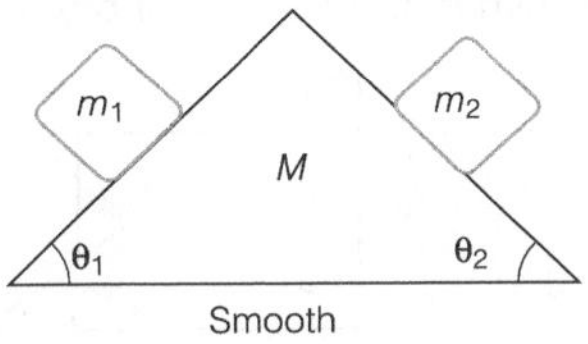

a. remains stationary b. may accelerate leftward
c. may accelerate rightward d. None of these

7 A ball of mass m is hanging over a smooth pulley inside a cart through a light string. Other end of the string connected with a drum which wraps string on at the rate of $v = a_0 t$. The cart is moving with constant acceleration a_0 directed horizontally (shown in the figure). The tension in string is

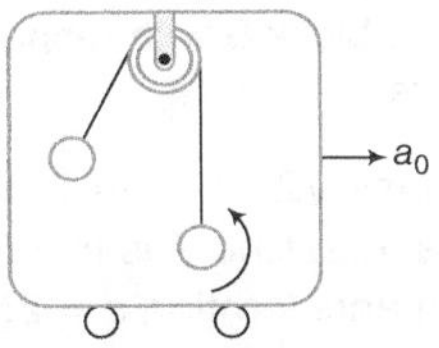

a. mg b. $m\sqrt{g^2 + 4a_0^2}$
c. $m\sqrt{g^2 + a_0^2} - ma_0$ d. $m\sqrt{g^2 + a_0^2} + ma_0$

8 For the given arrangement, pulleys and string are light and smooth. All surfaces are smooth. Find the magnitude of acceleration of block A of mass M.

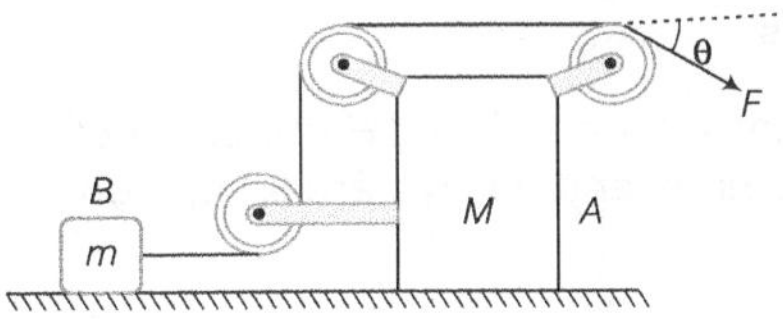

a. $\frac{F(1-\cos\theta)}{M}$ b. $\frac{F\cos\theta}{m+M}$ c. $\frac{F\cos\theta}{m}$ d. None of these

9 A small sphere of mass 1 kg rests at one corner of a cubical box. The box is moving as such its position is $\mathbf{r} = 4t^2\hat{\mathbf{i}} - \frac{2}{3}t^3\hat{\mathbf{j}}$. The force exerted by box on the ball at t = 1s is (assume that figure shows vertical plane of the cube)

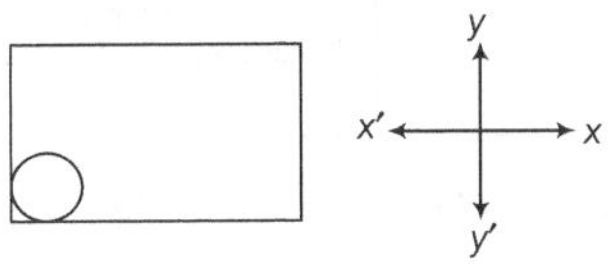

a. 6 N b. 10 N c. 15 N d. 20 N

One or More than One Options Correct Type

Working Space

Direction (Q. Nos. 10 -13) *This section contains 4 multiple choice questions. Each question has four choices (a), (b), (c) and (d), out of which ONE or MORE THAN ONE are correct.*

10 Block A of mass 1 kg is placed on a smooth horizontal surface. A smooth block B of mass 1 kg starts to move on the inclined surface of block A. Then,

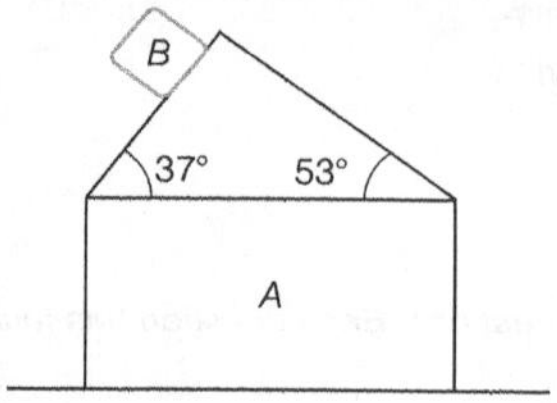

a. the acceleration of block A is $\frac{6g}{17}$

b. the acceleration of block B with respect to block A is $\frac{15g}{17}$

c. if acceleration of block B makes θ angle with horizontal, then $|\tan\theta| = 1.5$

d. None of the above

11 For the situation shown in the figure, all blocks are free to move. Then,

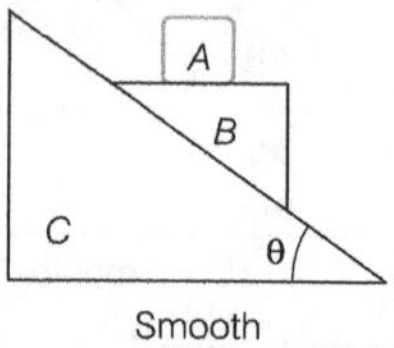

a. the block A is accelerated downward

b. the acceleration of block C is horizontally rightward

c. the acceleration of block B with respect to block A is leftward

d. None of the above

12 A circular ring of mass $\sqrt{2}$ kg rests in a smooth vertical right angle isosceles triangle. End A is fixed in a toy car accelerating horizontally with an acceleration $a = 2t\,(\text{m/s}^2)$ as shown in the figure. Then,

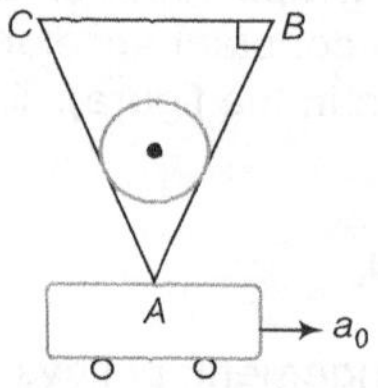

a. the normal reaction exerted by AC on the ring at an instant t is 20 N

b. the normal reaction on the ring due to AB decreases with time

c. the normal reaction on the ring due to AB increases with time

d. the ring starts to rise up the wire AC at $t = 5$ s

13 For the given figure, all blocks are of same mass. Blocks E and F are connected by light rod. All surfaces are frictionless and strings and pulleys are light, then

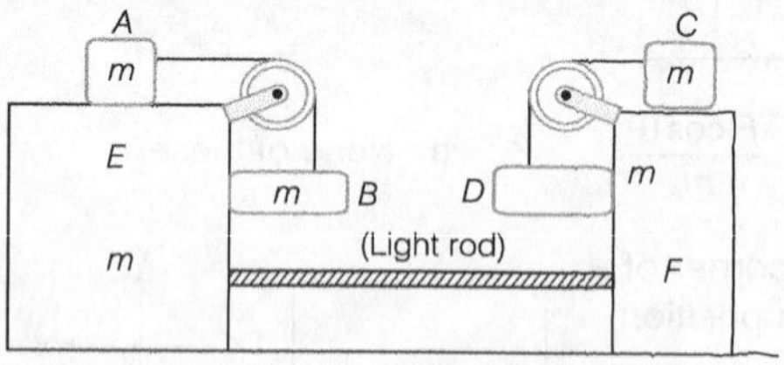

a. acceleration of block B is zero

b. acceleration of block B is $\frac{g}{2}$

c. acceleration of block E is zero

d. acceleration of block E is $\frac{g}{2}$

Working Space

Comprehension Type

Direction (Q. Nos. 14-18) *This section contains a paragraph, each describing theory, experiments, data, etc. Five questions related to the paragraph have been given. Each question has only one correct answer among the four given options (a), (b), (c) and (d).*

Passage

A block resting over a horizontal floor has a symmetric track ABC as shown in the figure. The mass of the block is $M = 3.12$ kg and length $AB = BC = 1$ m. A block of mass $m = 2$ kg is put on the track at A and the system is released from rest. (neglecting friction and impact at B).

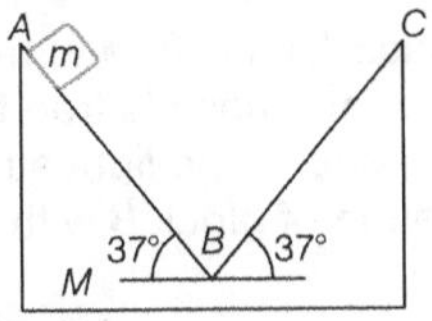

14 During motion, the centre of mass of the system moves

a. centre of mass of system remains at same position
b. centre of mass of system moves periodically in the horizontal direction
c. centre of mass of system moves periodically in vertical direction
d. centre of mass of system moves in an elliptical path

15 When smaller block is on AB part of track, then acceleration of bigger block is

a. 8 m/s^2 b. 2.5 m/s^2
c. 10 m/s^2 d. 4.5 m/s^2

16 The acceleration of smaller block with respect to bigger block is

a. 8 m/s^2 b. 2.5 m/s^2
c. 10 m/s^2 d. 6.18 m/s^2

17 When the smaller block is at point B, the speed of bigger block is

a. minimum b. maximum
c. same as initial d. zero

18 When the smaller block is at point C, then bigger block is

a. in rest b. in motion in leftward
c. in motion in rightward d. insufficient information

One Integer Value Correct Type

Direction (Q. Nos. 19-23) *This section contains 5 questions. Each question, when worked out will result in an integer from 0 to 9 (both inclusive).*

19 For the arrangement shown in the figure. The magnitude of the force F exerted on a large block C of mass $m_3 = 1.9$ kg is such that block A of mass $m_1 = 1.1$ kg and block B of mass $m_2 = 1$ kg remain stationary with respect to the block C. All surfaces are smooth. The value of F is $\frac{16\sqrt{21}}{n}$ N. Find the value of n.

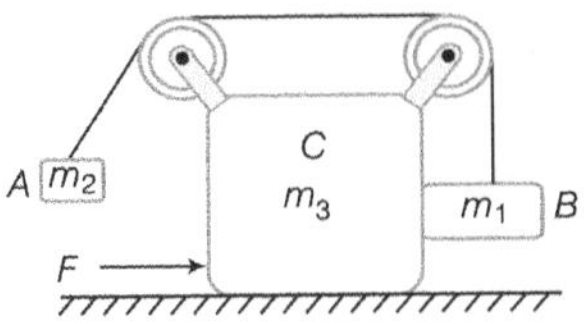

Working Space

20 The masses of blocks *A* and *B* are same. The mass of each block is 1 kg. All surfaces are smooth. A force $F = 20$ N is applied on the light string at point *P*. The acceleration of point *P* is $5x$ m/s^2. Find the value of *x*.

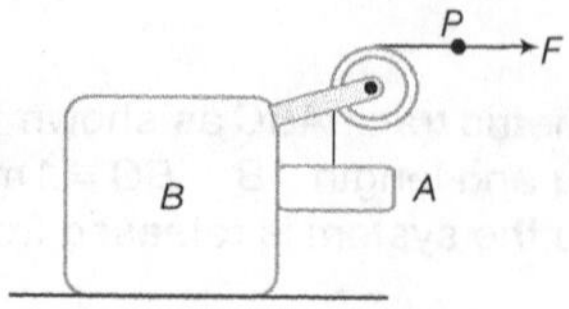

21 For the situation shown in the figure, the block *A* of mass 1 kg is free to move on the horizontal smooth surface. Another block *B* of the same mass is free to move on the smooth inclined surface of the block *A*. The free end of the light string is pulled by a constant force $F = 10$ N. Find the acceleration of block *B* with respect to the block *A*.

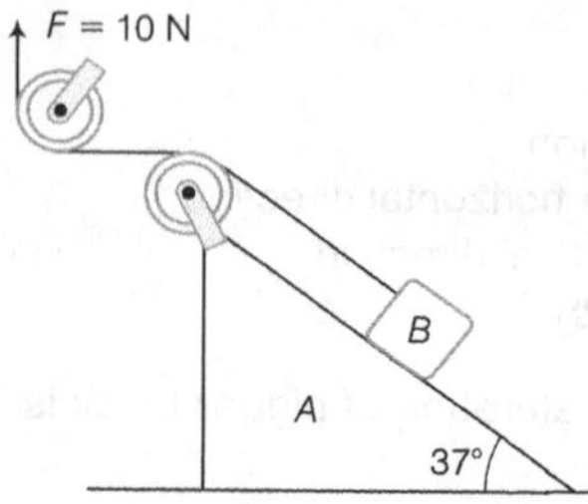

22 For the situation shown in the figure, blocks *A* and *B* each of mass 1 kg are free to move. Find the acceleration of block *B* (all surfaces are smooth). If it is $\sqrt{x}\,g$, then find the value of *x*.

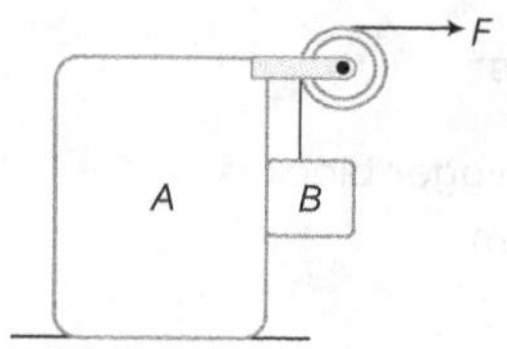

23 The box shown in the figure can slide without friction on a smooth surface. A collar *A* is free to slide along the smooth light rod mounted in the box. If the path of collar is along dotted line shown in the figure. Find the value of $\left(\frac{\text{mass of block}}{\text{mass of box}}\right)$.

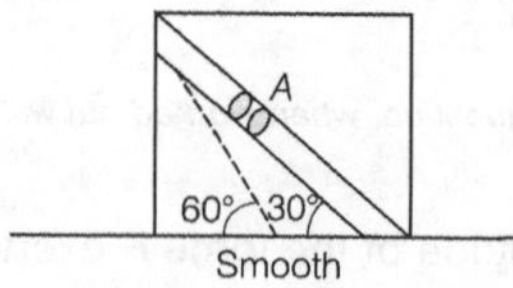

DPP-8 Problems Related to Climb on a String

Subjective Questions

Working Space

Direction *(Q. Nos. 1-4) These questions are subjective in nature, need to be solved completely on notebook.*

1 Find the tension in light string in the following cases.

i. A monkey of mass 10 kg is moving upward with a constant velocity.

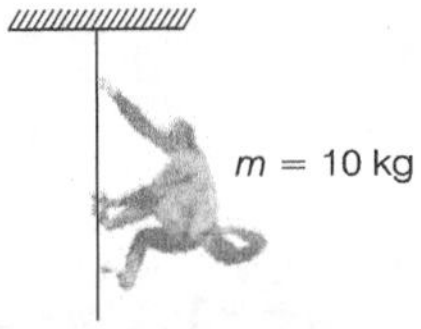

ii. A monkey of mass 20 kg holds the string and is stationary.

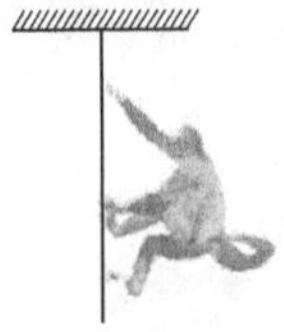

iii. A monkey of mass 5 kg is moving upward with an acceleration $2\,m/s^2$.

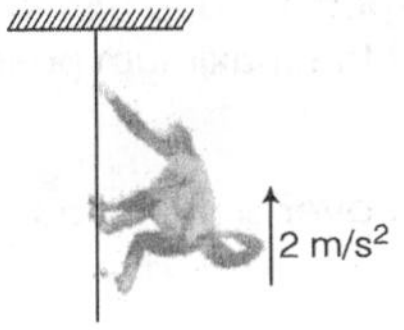

iv. A monkey of mass 8 kg is moving downward with an acceleration of $3\,m/s^2$.

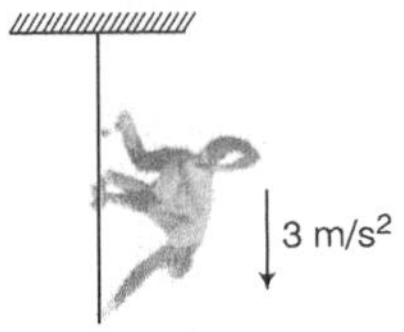

2 If maximum tension in a light string A is 200 N. Find the maximum acceleration of monkey in the following cases.

i. A monkey of mass 10 kg climbs up with an acceleration a.

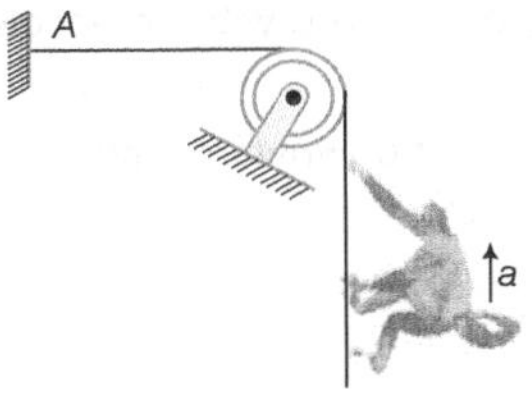

ii. For the situation shown in the figure, a monkey of mass 5 kg climbs up with an acceleration *a*.

Working Space

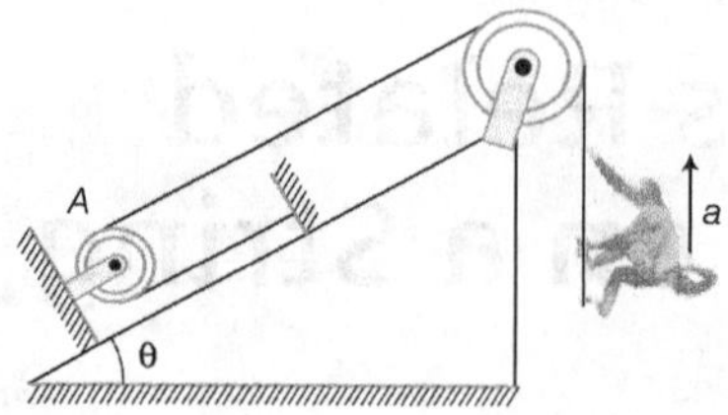

iii. One end of massless rope which passes over a massless and frictionless pulley *P* is tied to block *C* while the other end is free. Maximum tension that rope can bear is 720 N with what value of maximum safe acceleration (in ms^{-2}) can a monkey of 60 kg climb on the rope? **[AIEEE 2002]**

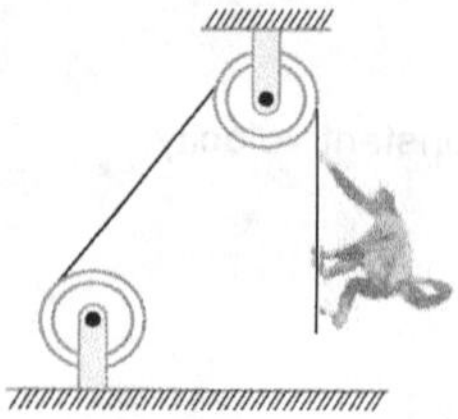

3 For the given figure, both monkey are climbing up with same acceleration of 2 m/s^2.

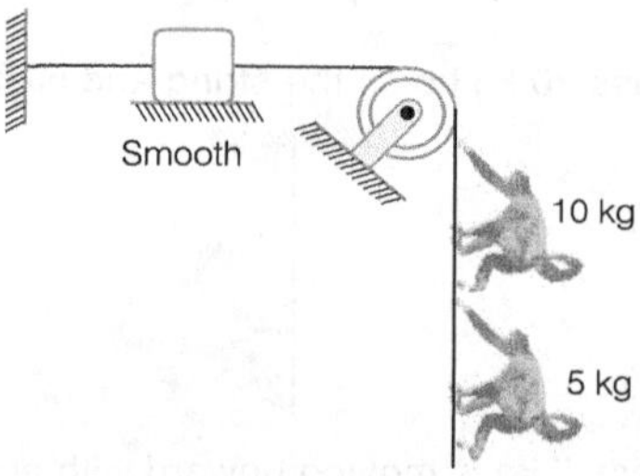

i. Find the tension in left most string.
ii. If maximum permissible tension in left most string is 300 N. The monkey shown below is stationary and is holding the string. Find the maximum possible acceleration of upper monkey.
iii. The maximum permissible tension in left most string is 300 N. If upper monkey is stationary with respect to rope and lower monkey is climbing upward with an acceleration. Find the maximum possible acceleration of the lower monkey as shown in figure.

4 A monkey of mass 4 kg climbs upon a light string hung over a pulley as shown in the figure.

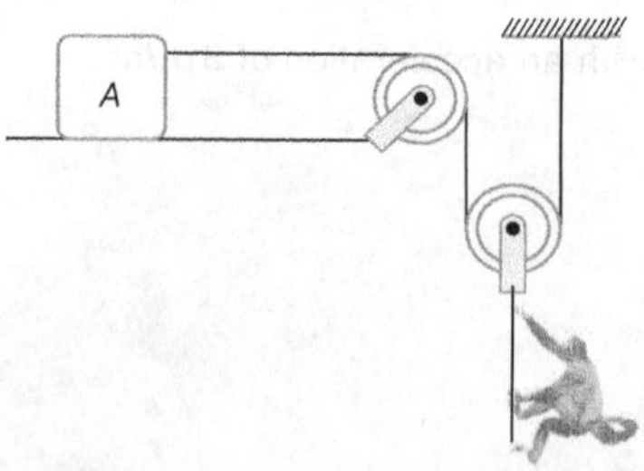

i. Find the acceleration of the block *A* of mass $m = 1$ kg if the monkey is stationary with respect to the string.
ii. Find acceleration of the block if the monkey is moving upward with an acceleration 2 m/s^2 with respect to the string.
iii. Find the acceleration of the block *A* if monkey is moving downward with acceleration 2 m/s^2 with respect to the string.

Only One Option Correct Type

Working Space

Direction *(Q. Nos. 5-12) This section contains 8 multiple choice questions. Each question has four choices (a), (b), (c) and (d) out of which ONLY ONE option is correct.*

5 A monkey's head is bent over her baby. The mass of her head is 1.75 kg and is supported by the muscle force $\mathbf{F}_m$ exerted by the neck extensor muscles and by the contact force $\mathbf{F}_C$ exerted at the atlanto occipital joint. If $\mathbf{F}_m$ is 37.5 N and is directed 37° below the horizontal. Find the magnitude of $\mathbf{F}_C$.

a. 30 N b. 40 N c. 50 N d. 60 N

6 A dead monkey has fallen in a ditch of depth 44 m. Mass of monkey is 10 kg. Two men are pulling the monkey using light rope as shown in the figure. If the maximum tension in the string is 200 N. Find the minimum time in which monkey comes out from ditch.

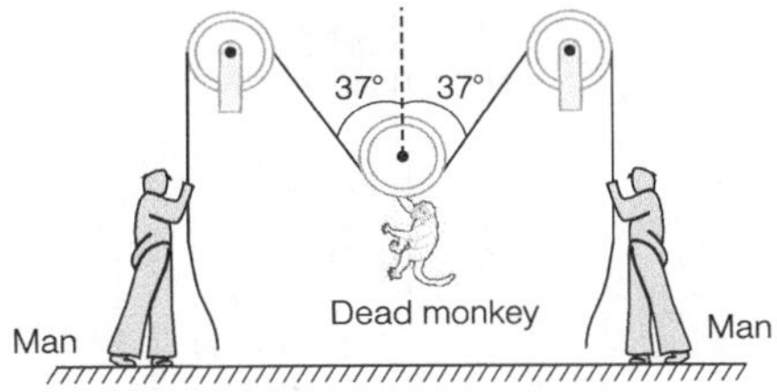

a. 1 s
b. 2 s
c. 3 s
d. 4 s

7 A monkey is coming down with an acceleration a_0. The maximum possible acceleration of the monkey is

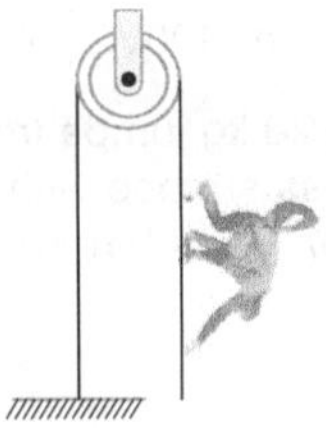

a. $\frac{g}{2}$
b. g
c. $2g$
d. $3g$

8 For the situation shown in the figure, the block of mass $m_2 = 20$ kg is on the horizontal floor. What is the maximum acceleration with which monkey of mass 5 kg climb up on a light rope so that block does not leave the horizontal floor?

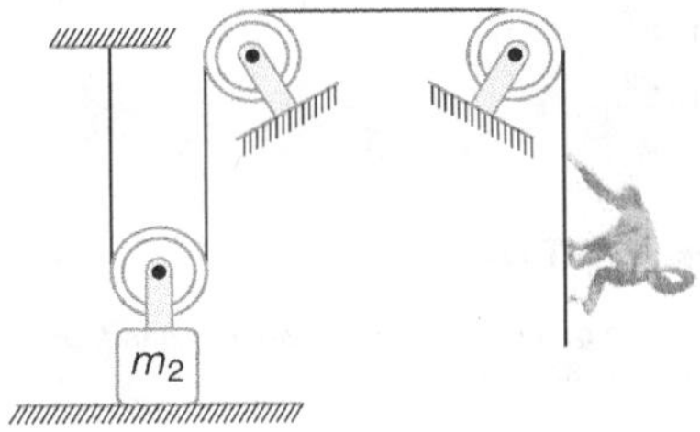

a. Zero b. $30\,m/s^2$ c. $10\,m/s^2$ d. $5\,m/s^2$

9 A boy of mass 10 kg rests on a light weighing machine fixed in box of mass 2 kg. A stationary monkey on the branch of a tree is pulling the box with a force of 156 N. The reading of balance is

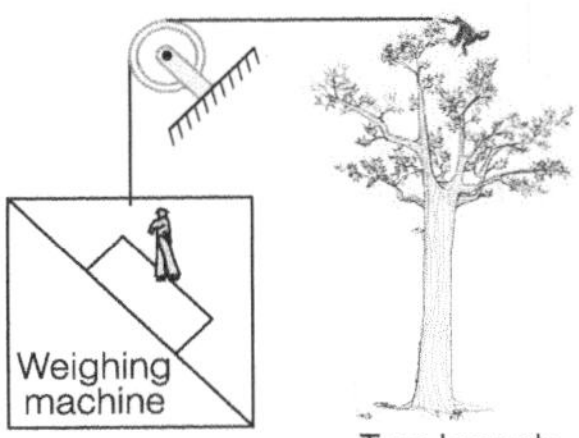

a. 100 N b. 80 N
c. 56 N d. 104 N

Working Space

10 Three monkeys *A*, *B* and *C* of masses 5 kg, 3 kg and 2 kg, respectively are moving along a vertical rope of mass 10 kg as shown in the figure. The monkey *A* is moving upward with an acceleration of 2 m/s^2, the monkey *B* is moving downward with an acceleration of 1 m/s^2 while the monkey *C* is moving downward with constant velocity of 4 m/s. The force exerted by the rope on the ceiling is

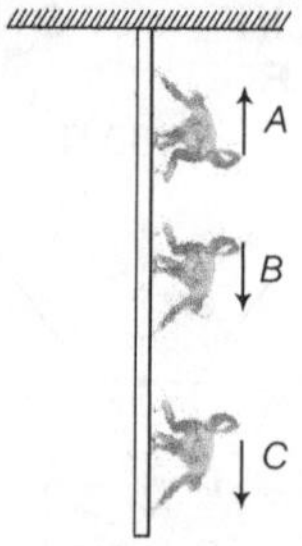

a. 107 N
b. 113 N
c. 207 N
d. 13 N

11 In a circus, a clever monkey of mass 50 kg is driving a bike with an acceleration 2 m/s^2 on a straight road. The horizontal component of force exerted by the bike on the monkey is

a. 500 N b. $100\sqrt{26}$ N c. 100 N d. 600 N

12 A monkey of mass 20 kg jumps from a plank of mass 60 kg placed on the horizontal frictionless surface with an acceleration 5 m/s^2 with respect to the plank at an angle of 37° with horizontal. The initial acceleration of plank is

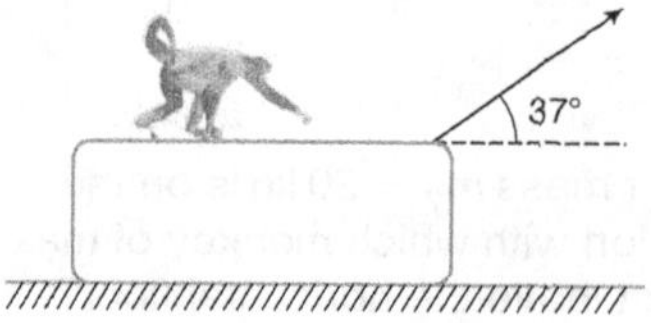

a. 1 m/s^2
b. 2 m/s^2
c. 3 m/s^2
d. 4 m/s^2

One or More than One Options Correct Type

Direction (Q. Nos. 13-16) *This section contains 4 multiple choice questions. Each question has four choices (a), (b), (c) and (d), out of which ONE or MORE THAN ONE are correct.*

13 A monkey of mass 10 kg is moving on a plank of 30 kg with an acceleration 1 m/s^2 in rightward direction.

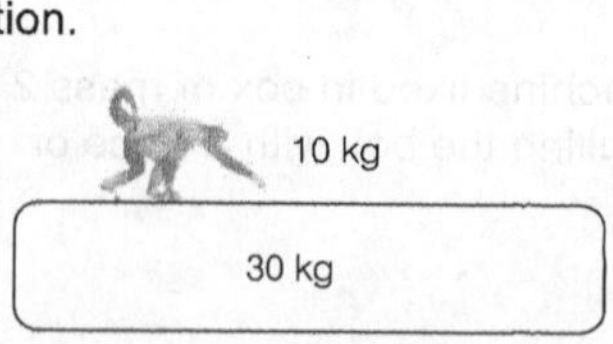

a. The horizontal force exerted by plank on monkey is 10 N
b. The acceleration of plank is 0.25 m/s^2
c. The acceleration of plank is $\frac{1}{3}$ m/s^2
d. The acceleration of plank is zero

Working Space

14 A bomboo piece of mass 10 kg is held on an angle 37° to the horizontal. A monkey of mass 30 kg try to go from roof *B* to roof *A* through the bomboo piece. Roofs are frictionless. If bomboo piece is stationary. Then,

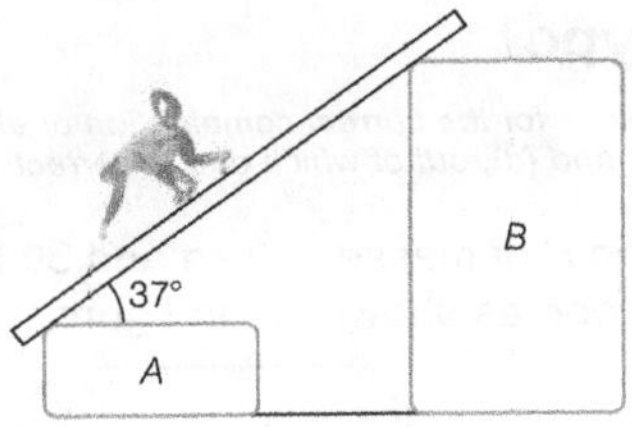

a. monkey should move slowly
b. monkey should accelerate with $8\,m/s^2$
c. monkey should decelerate with $8\,m/s^2$
d. monkey will complete the course

15 An arrow of mass 0.01 kg is shot towards a monkey from a 100 N draw bow, at a full draw of 0.4 m. Assume a linear relationship between draw and force (just like spring force). The speed with which the arrow leaves the bow is

a. 90 m/s
b. $20\sqrt{10}$ m/s
c. $50\sqrt{2}$ m/s
d. non-zero

16 For the situation shown in figure, the inclined surfaces of both identical wedges are smooth. Both monkey start from rest from top of their respective wedges. Which monkey comes on horizontal surface first?

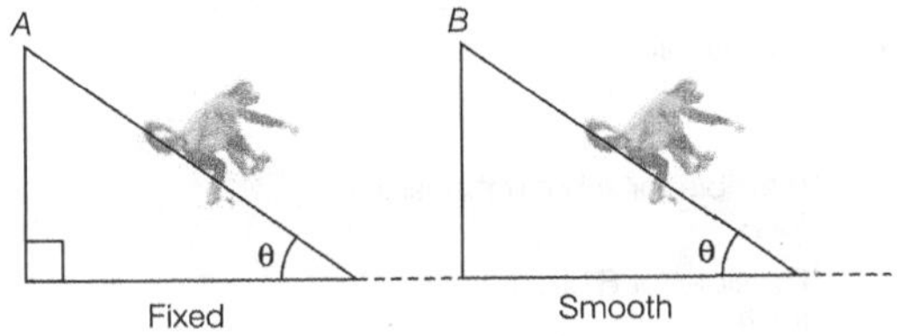

a. *A*
b. *B*
c. Both at same time
d. Both at different times

Comprehension Type

Direction (Q. Nos. 17 and 18) This section contains a paragraph, describing theory, experiments, data, etc. Two questions related to the paragraph have been given. Each question has only one correct answer among the four given options (a), (b), (c) and (d).

Passage

For the situation shown in figure, the monkey *A* of mass 20 kg climbs up the light rope with an acceleration $1\,m/s^2$ relative to the rope. The monkey *B* of mass 20 kg climbs up to light rope with an acceleration of $2\,m/s^2$ relative to rope. The mass of block *C* is 10 kg.

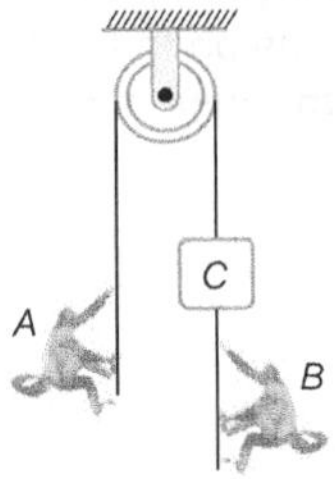

17 The acceleration of the block *C* is

a. $1\,m/s^2$
b. $2\,m/s^2$
c. $0.6\,m/s^2$
d. $\sqrt{5}\,m/s^2$

18 Initially, the monkey *A* is 0.8 m below the fixed pulley. The system is released from rest. The time taken by the monkey *A* to reach the pulley is

a. 0.6 s
b. 1 s
c. 0.3 s
d. 2 s

Matching List Type

Direction (Q. No. 19) *Choices for the correct combination of elements from Column I and Column II are given as options (a), (b), (c) and (d), out of which one is correct.*

19 Two monkeys A and B of masses 10 kg and 30 kg, respectively hold on to the two sections of a light rope as shown in the figure.

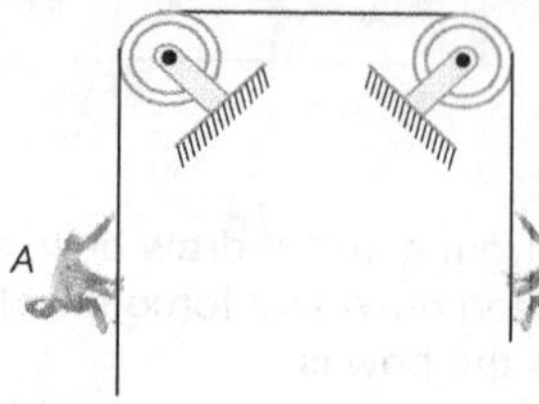

Macth the Column I with Column II. Select the correct option from the codes given below.

	Column I		Column II
i.	When A is stationary while B slides down with some acceleration.	p.	Possible
ii.	When monkey B is stationary while the monkey A climbs up with some acceleration.	q.	Impossible
iii.	Both A and B move with same acceleration in same direction.	r.	Possible for A but not possible for B
iv.	Both move with same magnitude of acceleration but opposite in direction.	s.	Possible for B but not possible for A

Codes

	i	ii	iii	iv			i	ii	iii	iv
a.	p	p	q	p		b.	p	q	r	s
c.	s	r	q	p		d.	q	s	r	p

One Integer Value Correct Type

Direction (Q. No. 20) *This section contains a question. When worked out will result in an integer from 0 to 9 (both inclusive).*

20 The baby monkey C shown in figure in holding on to the tail of the monkey B which is climbing up a light rope. The masses of the monkeys B, C and the block D are 10 kg, 5 kg and 1 kg, respectively. If the maximum tension in the tail of monkey is 80 N, then the ratio of minimum and maximum tension in the string A is $\frac{7x}{23}$. Find the value of x.

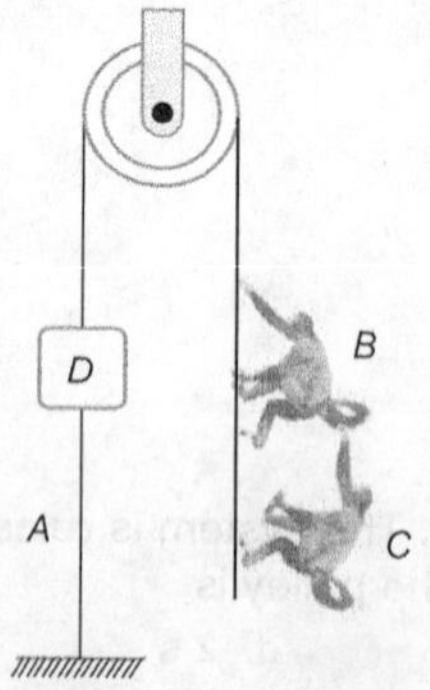

Working Space

Revisal Problems
for JEE Main

Only One Option Correct Type

Working Space

Direction (Q. Nos. 1-20) *This section contains 20 multiple choice questions. Each question has four choices (a), (b), (c) and (d), out of which ONLY ONE option is correct.*

1 In the situation shown in the figure, block A is stopped for a moment, after 3 system is set into motion. Find the time elapsed before the string is tight again.

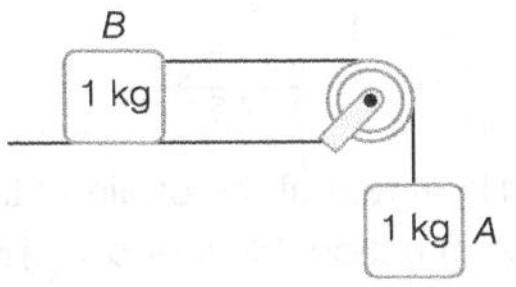

a. 1 s
b. 2 s
c. 3 s
d. 4 s

2 In the situation shown in the figure, both pulleys A and B are smooth and light. The radius of pulley A is 10 cm. Find normal reaction per unit length on string due to the pulley A.

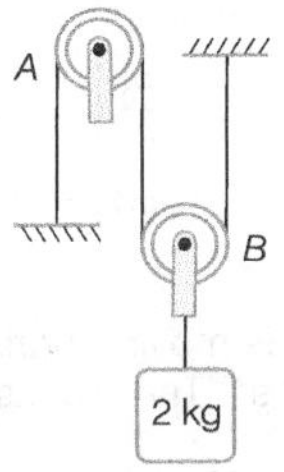

a. 1 N/cm
b. 2 N/cm
c. $\frac{2}{\pi}$ N/cm
d. $\frac{1}{\pi}$ N/cm

3 The structure of NH_3 is triangular pyramid. In an excited NH_3 molecule, the separation between nitrogen atom and hydrogen atom is l and that of between two any hydrogen atoms is $\frac{3}{2}l$ (only hypothetical concept). The force of interaction between nitrogen and hydrogen atoms is F_0. Find the force applied on nitrogen to keep it stationary (assume that H-atoms are fixed).

a. F_0
b. $3F_0$
c. $\frac{3}{2}F_0$
d. $\frac{\sqrt{3}}{2}F_0$

4 The pulleys and strings shown in the figure, are smooth and of negligible mass for the system to remain in equilibrium, the angle θ should be **[IIT JEE 2002]**

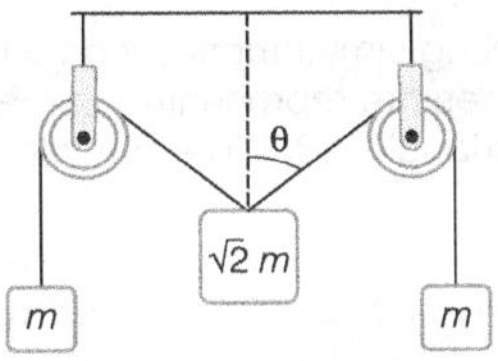

a. 0°
b. 30°
c. 45°
d. 60°

5 A smooth plank of mass 1 kg and length 8 m is on smooth horizontal surface. The gap between fixed pulley P and plank is larger than the length of plank. A smooth block D is placed on the plank. The masses of blocks A and B are 1 kg and 3 kg, respectively. Find the time elapsed before the block D separates from the plank.

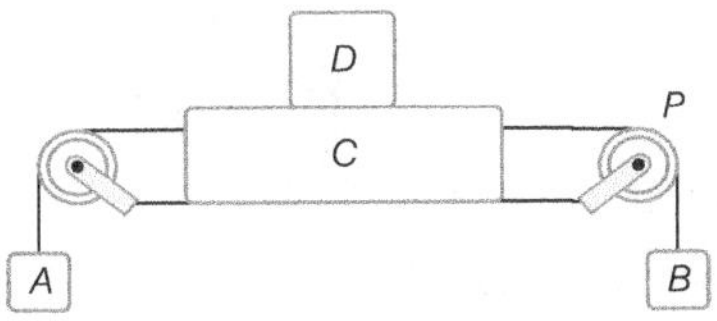

a. 1 s
b. 2 s
c. 3 s
d. Insufficient information

6 A billiard ball of mass $\sqrt{7}$ kg is moving in horizontal direction with a speed of 2 m/s towards a wall. After collision with the wall, the ball moves with a speed of 1 m/s in a direction making an angle 120° with its initial direction. The duration of collision between wall and ball is 0.01 s. The magnitude of average force applied by the wall on the ball is

a. $10\sqrt{7}$ N
b. 700 N
c. 70 N
d. $100\sqrt{21}$ N

Working Space

7 A player caught a cricket ball of mass 150 g moving at a rate of 20 m/s. If the catching process is completed in 0.1 s. The force of blow exerted by the ball on the hand of the player is equal to **[AIEEE 2006]**

a. 150 N   b. 3 N   c. 30 N   d. 300 N

8 The wedge is moving toward s right with an acceleration $a = a_0 t$ (where, $a_0 = 1\ m/s^2$). A block of mass 1 kg rests on the inclined smooth surface of the wedge is connected to wedge through a light string. The tension *versus* time graph is

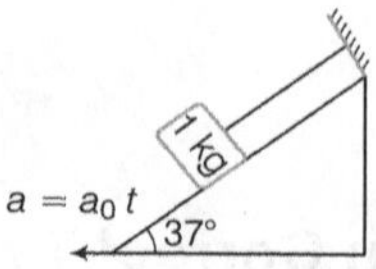

a. 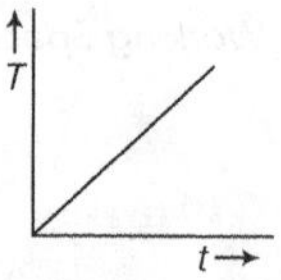
b. 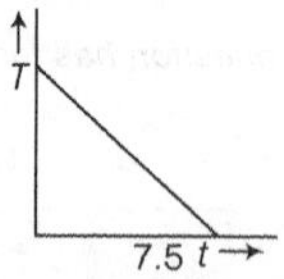
c. 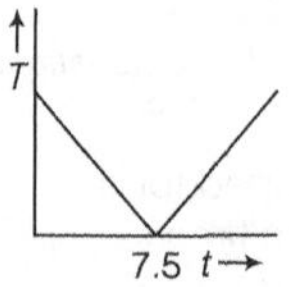
d. 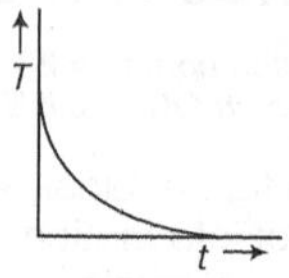

9 A particle of mass m is at rest at the origin at time $t = 0$. It is subjected to a force $F(t) = F_0 e^{-bt}$ in the X-direction. Its speed $v(t)$ is depicted by which of the following curves? **[AIEEE 2012]**

a. 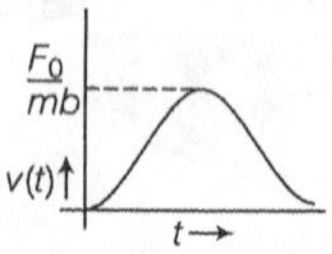
b. 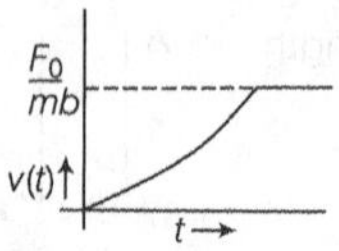
c. 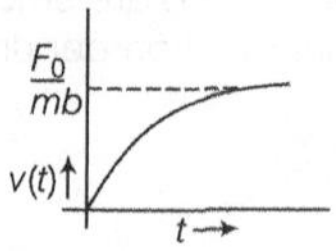
d. 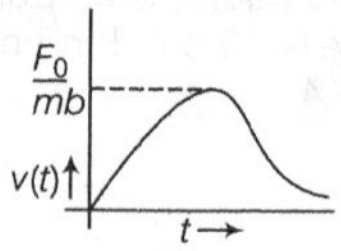

10 The triangular block is moving horizontally with an acceleration of $5\ m/s^2$. The blocks A and B have same mass of 1 kg each.

The reading of light spring balance S is

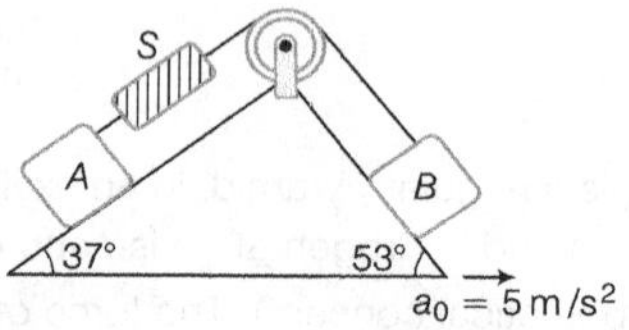

a. 2 kg   b. 0.2 kg
c. 0.85 kg   d. 1 kg

11 A particle is moving with constant speed on a complicated curved path, whose some part is parabolic, some part is circular and some part is sinusoidal curve. The resultant force on the particle is

a. always perpendicular to its velocity
b. always along the path of the particle
c. always constant in magnitude
d. always directed towards a fixed point

12 Three forces start acting simultaneously on a particle moving with velocity **v**. These forces are represented in magnitude and direction by the three sides of a triangle ABC. The particle will now move with velocity **[AIEEE 2003]**

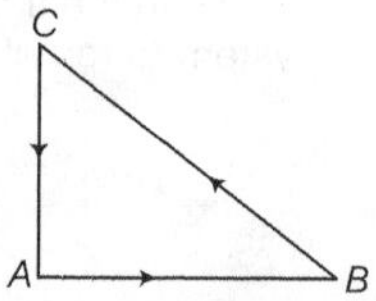

a. less than **v**
b. greater than **v**
c. |**v**| in direction of largest force BC
d. **v** remaining unchanged

13 A small block of mass m is in connected through light springs each of spring constant k. Initially both springs are relaxed and the block is $l_0 = 20$ cm below the pulley. Now block brings slowly at equilibrium position. At what distance from the pulley the block is in equilibrium?

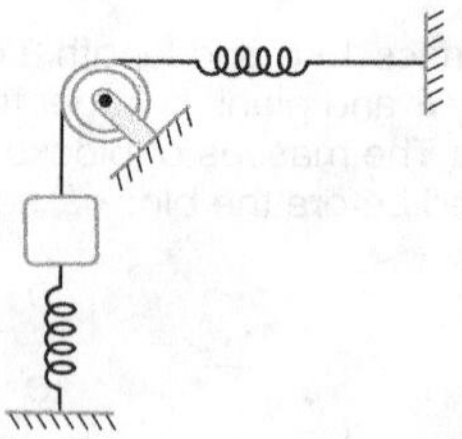

a. 20 cm
b. 25 cm
c. 30 cm
d. 5 cm

14 In the situation shown in the figure, the mass of blocks A and B are 1 kg and 2 kg, respectively. The blocks are stationary and in equilibrium. The acceleration of the block A just after cutting right spring is

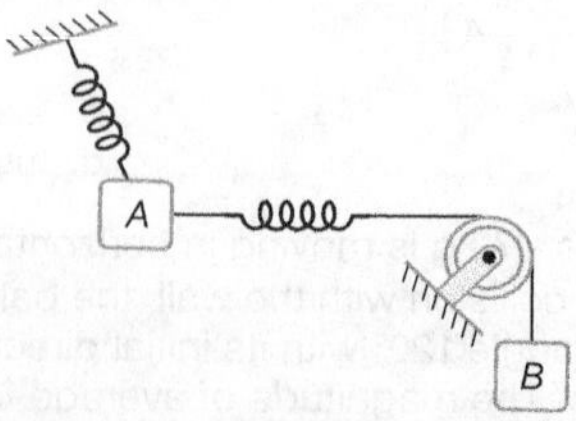

a. g in downward direction
b. $2g$ in rightward direction
c. $2g$ in leftward direction
d. g in upward direction

Working Space

15 A block of mass 10 kg splits in two blocks of masses m_1 and m_2. Now they are connected through a massless string as shown in figure. For the maximum tension in the string, the ratio of m_1 and m_2 is

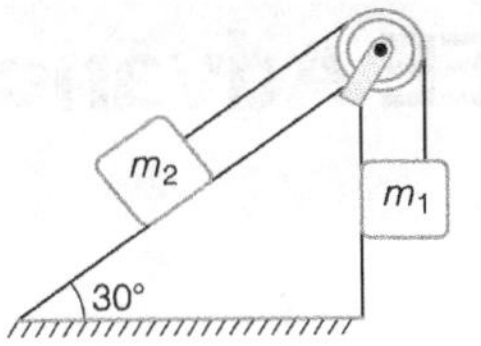

a. 1 : 1 b. 3 : 2 c. 2 : 3 d. 1 : 3

16 Two masses $m_1 = 5$ kg and $m_2 = 4.8$ kg tied to a string are hanging over a light frictionless pulley. What is the acceleration of the masses when they are free to move ($g = 9.8 \text{ m/s}^2$)? **[AIEEE 2004]**

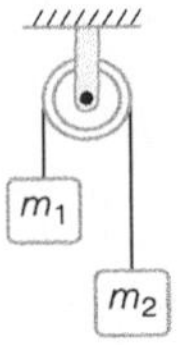

a. 0.2 ms^{-2} b. 9.8 ms^{-2} c. 5 ms^{-2} d. 4.8 ms^{-2}

17 A light string passing over a smooth light pulley connects two blocks of masses m_1 and m_2. If the acceleration of the system is $\frac{g}{8}$, then the ratio of the masses is **[AIEEE 2002]**

a. 8 : 1 b. 9 : 7 c. 4 : 3 d. 5 : 3

18 Block *B* is moving downward with a velocity of 12 m/s. The speed of point *D* on the string is

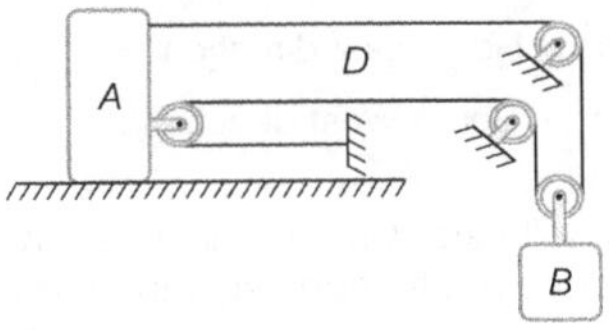

a. 12 m/s b. 8 m/s c. 16 m/s d. 24 m/s

19 In the situation shown in the figure, all pulleys are fixed except pulley *C*. The pulley *C* is moving horizontally with a speed of $v_p = 10 \text{ m/s}$. A motor *A* is moving with a speed of 7 m/s. It winds the string at constant rate of 3 m/s. The speed of the block *B* is

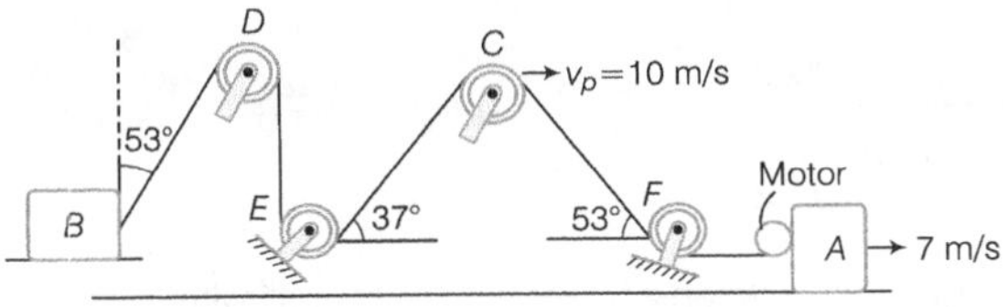

a. 5 m/s b. 10 m/s c. 15 m/s d. 12 m/s

20 Two fixed frictionless inclined planes making angles of 30° and 60° with the vertical are shown in the figure. Two blocks *A* and *B* are placed on the two planes. What is the relative vertical acceleration of *A* with respect to *B*? **[AIEEE 2010]**

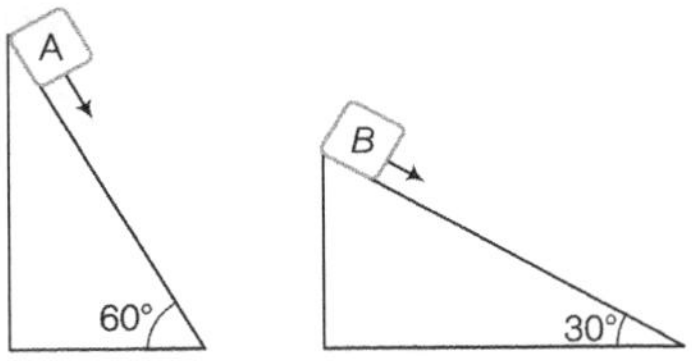

a. 4.9 ms^{-2} in horizontal direction b. 9.8 ms^{-2} in vertical direction
c. Zero d. 4.9 ms^{-2} in vertical direction

SET - 2

Revisal Problems
for JEE Advanced

Only One Option Correct Type

Direction (Q. Nos. 1-8) *This section contains 8 multiple choice questions. Each question has four choices (a), (b), (c) and (d), out of which ONLY ONE option is correct.*

Working Space

1 The acceleration of the block of mass 2 kg is along negative X-axis. The magnitude of acceleration of the block is

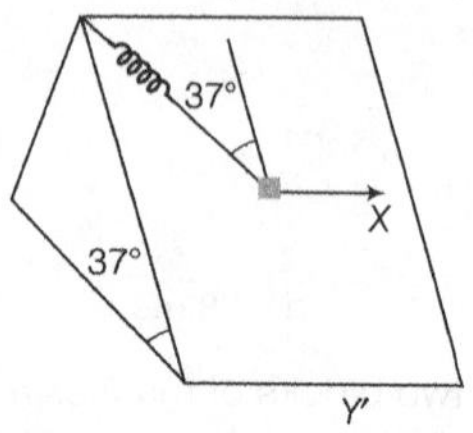

a. $1\ m/s^2$ b. $9\ m/s^2$
c. $4.5\ m/s^2$ d. $7.5\ m/s^2$

2 Two men each of mass 50 kg are standing on a horizontal plank of mass 100 kg. Both men jumps from plank simultaneously with acceleration $2\ m/s^2$, one in Eastward and other in Northward. The initial acceleration of plank is

a. $\sqrt{2}\ m/s^2$ in South-West direction b. $\sqrt{2}\ m/s^2$ in North-East direction
c. $2\ m/s^2$ in South-West direction d. $\frac{1}{\sqrt{2}}\ m/s^2$ in South-West direction

3 The situation shown in the figure, the block C is moving leftward with an acceleration a_0. The mass of block B is twice of that of block A. All surfaces are frictionless. Find the value of a_0, if blocks A and B do not slide on the block C.

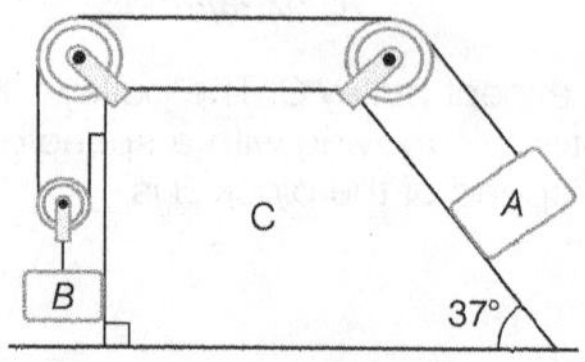

a. Zero b. $5\ m/s^2$
c. $10\ m/s^2$ d. $0.6\ m/s^2$

4 A monkey of mass 40 kg slides down on a light rope with an acceleration of $5\ m/s^2$ with respect to rope. The lift is going up with an acceleration of $a_0 = 5\ m/s^2$.
A light spring balance S is connected with a light pulley as shown in the figure. The reading of spring balance S is

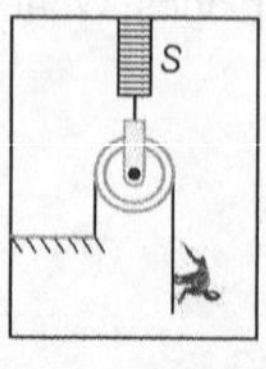

a. 40 kg b. 80 kg
c. 120 kg d. zero

Working Space

5 A piece of wire is bent in the shape of parabola $y = kx^2$ (Y-axis vertical) with a bead of mass m on it. The bead can slide on the wire without friction. It stays at the lowest point of the parabola when the wire is at rest. The wire is now accelerated parallel to the X-axis with a constant acceleration a. The distance of the new equilibrium position of the bead, where the bead can stay at rest with respect to the wire, from the Y-axis is **[IIT JEE 2009]**

a. $\frac{a}{gk}$ b. $\frac{a}{2gk}$ c. $\frac{2a}{gk}$ d. $\frac{a}{4gk}$

6 Two identical smooth cylinders each of mass $8\sqrt{3}$ kg and radius 5 cm lie in lateral contact in a horizontal hollow fixed cylinder of radius 0.4 m. The contact force exerted by one cylinder to another cylinder is

a. 5 N b. $80\sqrt{3}$ N c. 17.5 N d. 30 N

7 A trolley C can run on a smooth horizontal table. Two, much smaller but equal masses A and B are hung by strings which pass over smooth pulleys. The strings are long enough such that when C is in equilibrium, A and B both just rest on the ground. The trolley is pulled slowly to one side and released as shown in the figure. The graph of its velocity v against time t will be as

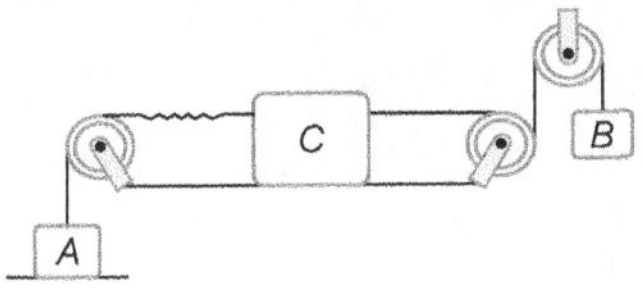

a. 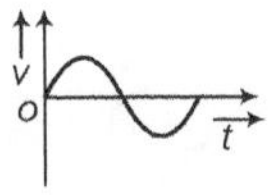

b. 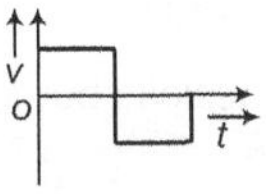

c. 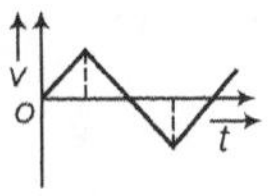

d. 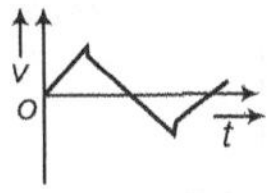

8 A hoisting drum hoists one of two blocks of equal mass. The hoisting drum wraps the string with constant speed v_0. When the blocks are at the same height, the upper pulley is released.

Which load will touch the floor first?

a. A
b. B
c. Both at same time
d. Either A or B

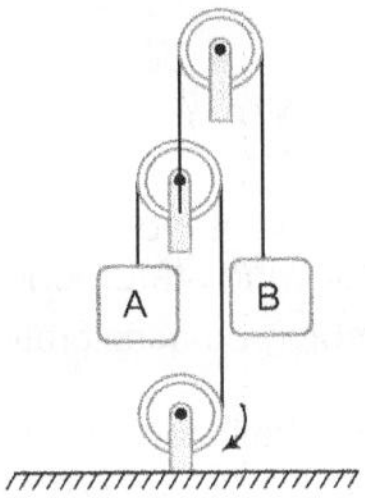

One or More than One Options Correct Type

Direction (Q. Nos. 9-13) ***This section contains 5 multiple choice questions. Each question has four choices (a), (b), (c) and (d), out of which ONE or MORE THAN ONE option are correct.***

9 Two smooth blocks A and B of equal mass, $m = 1$ kg are connected by light string as shown in the figure. Mark the correct options.

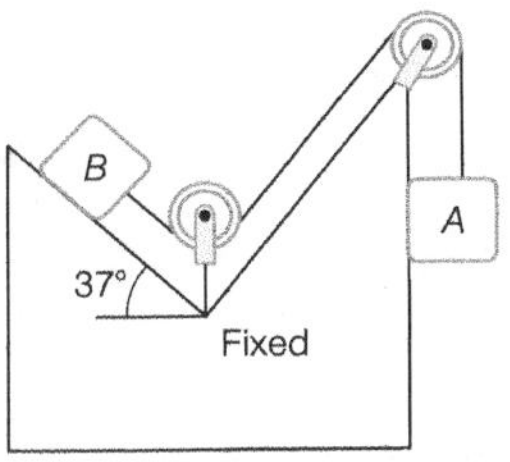

a. Tension in string is 8 N
b. The acceleration of block B is 2 m/s^2
c. The magnitude of acceleration of block B with respect to block A is $\frac{4}{\sqrt{5}}$ m/s^2
d. The acceleration of block A is 2 m/s^2

10 The situation shown in the figure, all surfaces are frictionless. Blocks A and C each of mass 5 kg are connected through a light string passing over smooth pulleys fixed in a truncated triangular block B of mass 5 kg. The system is released from rest. Choose the correct options.

Working Space

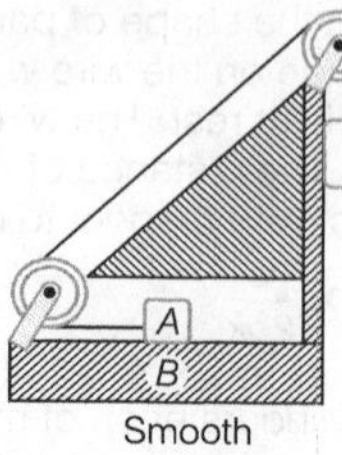

a. The acceleration of block B is 2 m/s^2
b. The tension in string is 20 N
c. The normal reaction on the block B from ground is 120 N
d. The acceleration of block A with respect to block B is 6 m/s^2

11 The situation shown in the figure, $m_A = m_B = m_C = 1$ kg and $m_D = 2$ kg. All surfaces are smooth. Strings are light and pulleys are smooth. Choose the correct options.

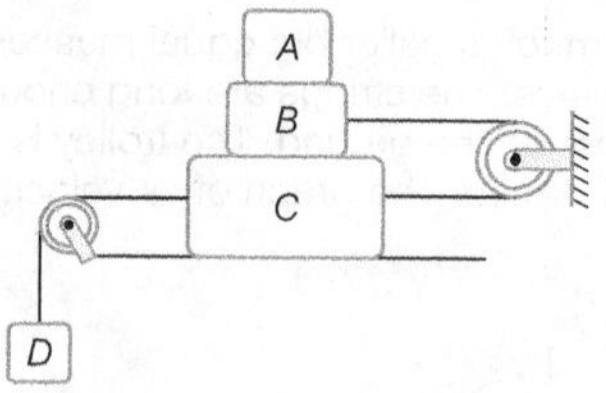

a. The initial acceleration of block A is zero
b. The initial acceleration of block B is 5 m/s^2
c. The tension in the string connecting blocks B and C is 5 N
d. The tension in the string connecting blocks C and D is 10 N

12 The situation shown in the figure, left ladder C is massless but right ladder is of mass 5 kg. A man A of mass 30 kg is standing on the ladder C.

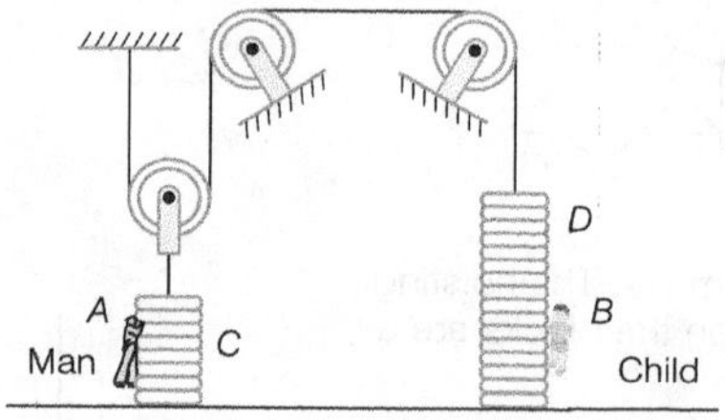

A child B of mass 10 kg is standing on ladder D. Initially both A and B are at the same level. Which of the following statements is/are correct?

a. The maximum climbing up acceleration of the child is zero to maintain stationary condition of ladder C
b. If child B starts up on the ladder D with acceleration 4 m/s^2 with respect to ground, vertical distance between A and B in 1 s is 0.4 m
c. If both A and B are climbing up with constant acceleration 4 m/s^2 with respect to ground, the acceleration of D is 4 m/s^2 in upward direction
d. None of the above

13 The situation shown in the figure, all surfaces are smooth. m_1, m_2, m_3 and m_4 are blocks of variable mass. The sum of masses of blocks is constant. A constant horizontal force of non-zero magnitude is applied on the mass m_1.

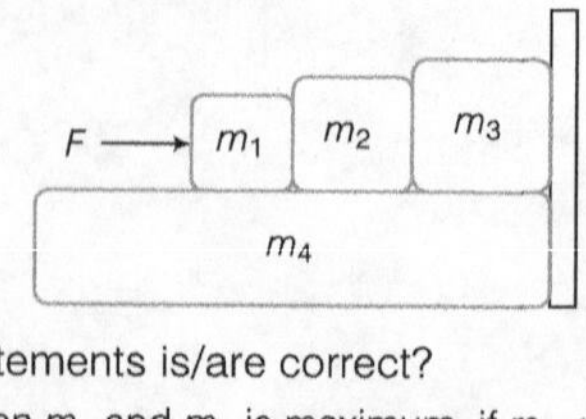

Which of the following statements is/are correct?

a. The contact force between m_2 and m_3 is maximum, if $m_1 < m_2 < m_3 < m_4$
b. The contact force between m_1 and m_2 is maximum, if $m_1 < m_2 < m_3$
c. The contact force between m_1 and m_2 is maximum, if $m_1 = m_2$ and $2m_1 < m_3 + m_4$
d. The normal reaction exerted by ground on m_4 is constant

Working Space

Comprehension Type

Direction (Q. Nos. 14-17) *This section contains 2 paragraphs, each describing theory, experiments, data, etc. Four questions related to the paragraphs have been given. Each question has only one correct answer among the four given options (a), (b), (c) and (d).*

Passage I

When Robert Goddard proposed the possibility of rocket propelled vehicles, the New York times agreed that such vehicles would be useful and successful within the Earth's atmosphere ("Topics of the Times," New York times, January 13, 1920 p.12). But the times balked at the idea of using such a rocket in the vacuum of space, noting that, "...its flight would be neither accelerated nor maintained by the explosion of the charges it then might have left. To claim that it would be is to deny a fundamental law of dynamics, and only Dr. Einstein and his chosen dozen, so few and fit, are licensed to do that...That Professor Goddard, with his chair in clark college and the contenancing of the Smithsonian Institution, does not know the relation of the action to reaction and of the neet to have something better than a vacuum against which is react to say that would be absurd. Of course, he only seems to lack the knowledge ladled out daily in high schools. What did the writer of this passage overlook?

14 Which of the following common mistakes made by writer of the passage?

a. Gases do not need to push on anything
b. A rocket works by expelling gases which push on something, propelling the rocket forward
c. In absence of gravity, the momentum of gas + rocket system remains conserved
d. None of the above

15 According to passage writer

a. it is impossible to see how a rocket fired in empty space would work
b. gases would not need to push on anything
c. a rocket can function in a vacuum as well as in an atmosphere
d. None of the above

Passage II

A light string passing over a fixed smooth pulley carries two varing variable unequal masses placed on smooth inclined planes as shown in the figure.

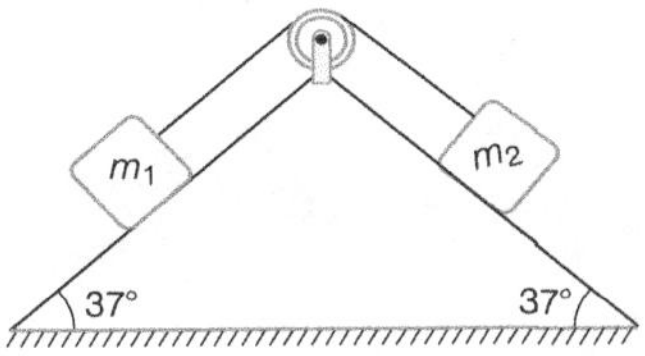

The sum of masses is 8 kg. The maximum tension in the string is 22.5 N.
The blocks are released from rest.

16 Ratio of masses of smaller and larger block is

a. $\frac{4}{5}$ b. $\frac{3}{5}$ c. $\frac{2}{5}$ d. $\frac{8}{5}$

17 The minimum possible value of acceleration is

a. 1 m/s^2 b. 6 m/s^2 c. 8 m/s^2 d. 1.5 m/s^2

One Integer Value Correct Type

Direction (Q. Nos. 18-22) *This section contains 5 questions. Each question, when worked out will result in one integer from 0 to 9 (both inclusive).*

18 The situation shown in the figure, all surfaces are frictionless. Masses of blocks *B* and *C* are 1 kg and 2 kg, respectively. Find the mass of the blocks *A* for which the block *B* remains stationary with respect to block *C*.

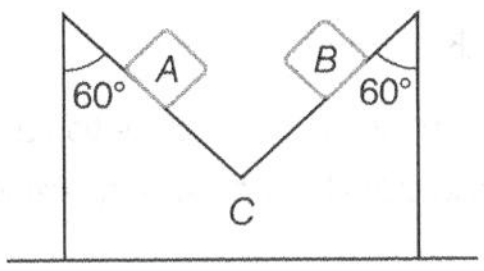

Working Space

19 The situation shown in the figure, all surfaces are smooth. The mass of sphere is 7 kg.

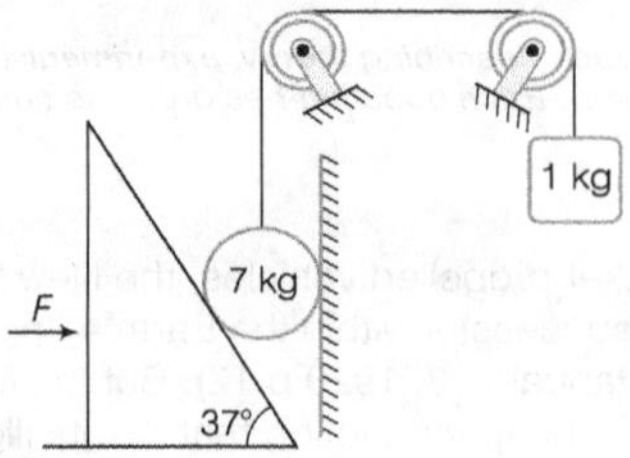

If the force F required to keep the system in equilibrium is $9n$ N.
Find the value of n.

20 The situation shown in the figure, all surfaces are smooth. Blocks A and B each have same mass 1 kg. A light rod is fixed in the block A. The maximum force F which can be applied to the rod, so that B will not slip up on the block A is $10n$ N. Find the value of n.

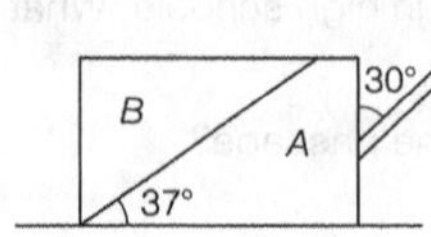

21 Two laddu (spherical Indian sweet likes by Lord Ganesh) having radius R each are put in a cylinder of diameter d. The normal force exerted by vertical side wall on lower laddu is $mg\dfrac{(d-2R)}{\sqrt{nRd-d^2}}$, where m is mass of each laddu. Find the value of n.

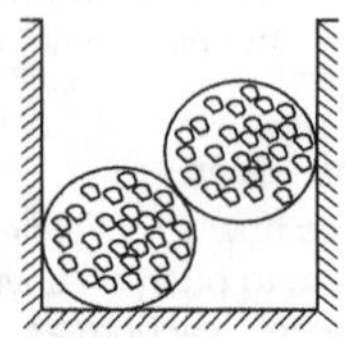

22 All surfaces are smooth. Find acceleration of triangular block in m/s^2. (take, $g = 9\ m/s^2$)

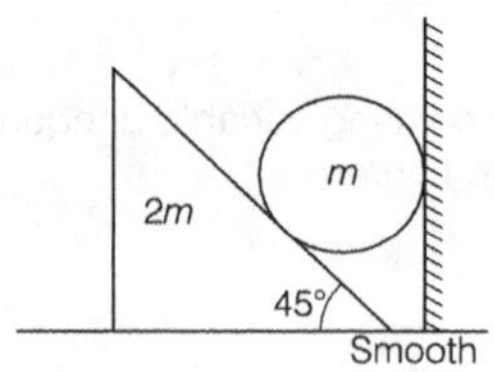

Statement Type

Direction *(Q. Nos. 23-27)* ***This section is based on Statement I and Statement II. Select the correct answer from the codes given below.***

a. Both Statement I and Statement II are correct and Statement II is the correct explanation of Statement I
b. Both Statement I and Statement II are correct and Statement II is not the correct explanation of Statement I
c. Statement I is correct but Statement II is incorrect
d. Statement II is correct but Statement I is incorrect

23 Two blocks A and B are on two frictionless inclined planes as shown in the figure.

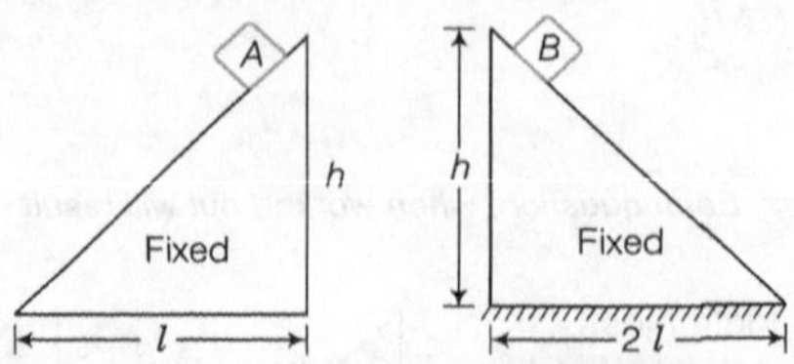

Statement I B takes more time than A to reach at the bottom.

Statement II The acceleration of A is greater than the acceleration of B.

24 **Statement I** If net force on a particle is constant in magnitude and direction, the path of particle is either straight line or parabola.

Statement II If a particle moves on a parabolic path, acceleration may be perpendicular to velocity at every instant.

25 **Statement I** If a particle is in rest, the particle must be in equilibrium.

Statement II For being a particle in equilibrium, net force on particle should be zero.

26 **Statement I** If the net force on a particle of mass 2 kg is measured by an observer inside a car moving with a velocity $\mathbf{v} = 3t\hat{\mathbf{i}}$ is $(20\,\text{N})\,\hat{\mathbf{i}}$. The acceleration of the particle is $\mathbf{a} = (10\ \text{m/s}^2)\hat{\mathbf{i}}$.

Statement II According to Newton's second law, $\mathbf{F} = m\mathbf{a}$.

27 **Statement I** A man runs for a certain distance after getting out from a moving train.

Statement II Man runs for certain distance in the direction of motion to maintain the same velocity of all parts of his body relative to the train, so that he may not fall forward due to the inertia of rest of the lower part of his body and inertia of motion of the upper part of his body.

Working Space

Matching List Type

Direction (Q. No. 28) Choices for the correct combination of elements from Column I and Column II are given as options (a), (b), (c) and (d), out of which one is correct.

28 The situation shown in the figure, a boy of mass m_3 is standing on a platform of mass 10 kg and in holding a string passing over smooth and light pulleys. The mass of hanging block is $m_2 = 20$ kg. Match the Column I with Column II.

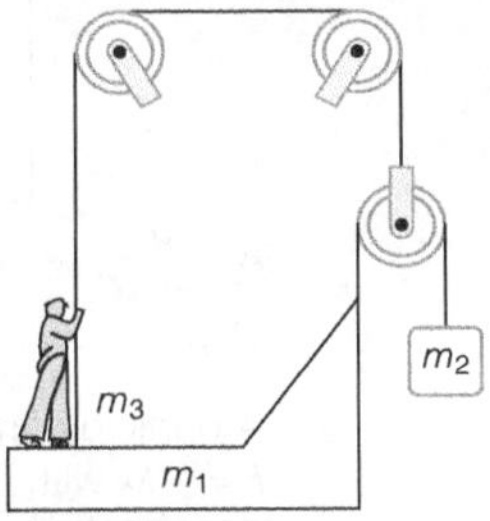

	Column I		Column II
i.	Value of m_3g for equilibrium	p.	600 N
ii.	Force exerted by the boy on string to accelerate upward	q.	500 N
iii.	Force exerted by boy on string to accelerate downward	r.	150 N
iv.	Normal reaction of platform on man in equilibrium	s.	100 N

Codes

	i	ii	iii	iv		i	ii	iii	iv
a.	q	p,q	r,s	s	b.	p	q	r	s
c.	q	p,s	r	s	d.	p,q	q	r,s	r

JEE Main & AIEEE Archive

(Compilation of Last 13 Years Questions)

1 The figure shows the position-time (x-t) graph of one-dimensional motion of a body of mass 0.4 kg. The magnitude of each impulse is **(AIEEE 2010)**

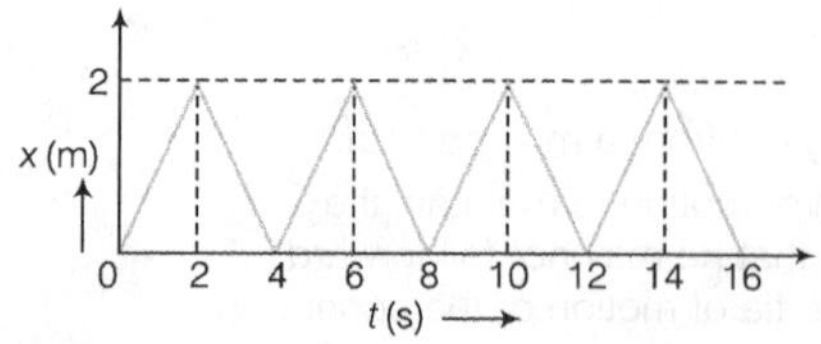

a. 0.4 N- s
b. 0.8 N-s
c. 1.6 N-s
d. 0.2 N-s

2 Two fixed frictionless inclined plane making the angles 30° and 60° with the vertical are shown in the figure. Two blocks A and B are placed on the two planes. What is the relative vertical acceleration of A with respect to B? **(AIEEE 2010)**

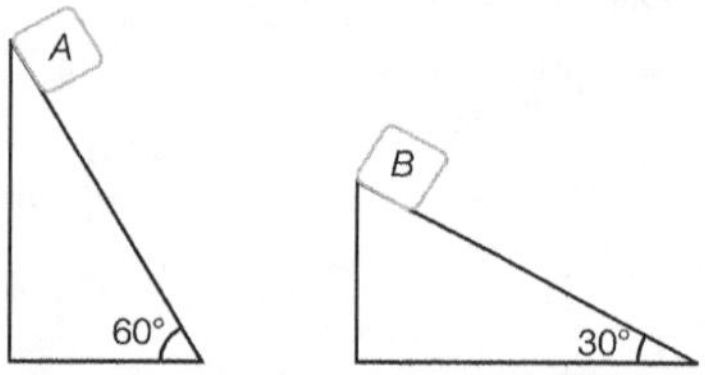

a. 4.9 ms^{-2} in horizontal direction
b. 9.8 ms^{-2} in vertical direction
c. zero
d. 4.9 ms^{-2} in vertical direction

3 A block of mass m is connected to another block of mass M by a spring (massless) of spring constant k. The blocks are kept on a smooth horizontal plane. Initially the blocks are at rest and the spring is unstretched. Then, a constant force F starts acting on the block of mass M to pull it. Find the force on the block of mass m. **(AIEEE 2007)**

a. $\frac{mF}{M}$
b. $\frac{(M+m)F}{m}$
c. $\frac{mF}{(m+M)}$
d. $\frac{MF}{(m+M)}$

4 A mass of M kg is suspended by a weightless string. The horizontal force that is required to displace it until the string makes an angle of 45° with the initial vertical direction is **(AIEEE 2006)**

a. $Mg(\sqrt{2}+1)$
b. $Mg\sqrt{2}$
c. $\frac{Mg}{\sqrt{2}}$
d. $Mg(\sqrt{2}-1)$

5 A ball of mass 0.2 kg is thrown vertically upwards by applying a force by hand. If the hand moves 0.2 m while applying the force and the ball goes upto 2 m height further, find the magnitude of the force. (take, $g = 10\ m/s^2$). **(AIEEE 2006)**

a. 4 N
b. 16 N
c. 20 N
d. 22 N

6 A player caught a cricket ball of mass 150 g moving at a rate of 20 m/s. If the catching process is completed in 0.1 s, the force of the blow exerted by the ball on the hand of the player is equal to **(AIEEE 2006)**

a. 150 N
b. 3 N
c. 30 N
d. 300 N

7 A block is kept on a frictionless inclined surface with angle of inclination α. The incline is given an acceleration a to keep the block stationary. Then, a is equal to **(AIEEE 2005)**

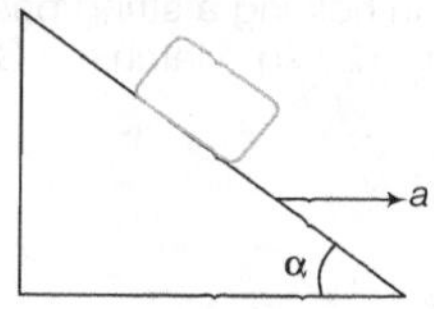

a. $\frac{g}{\tan\alpha}$
b. $g\ \mathrm{cosec}\ \alpha$
c. g
d. $g\tan\alpha$

8 A particle of mass 0.3 kg is subjected to a force $F = -kx$ with $k = 15\ Nm^{-1}$. What will be its initial acceleration, if it is released from a point 20 cm away from the origin? **(AIEEE 2005)**

a. 3 ms^{-2}
b. 15 ms^{-2}
c. 5 ms^{-2}
d. 10 ms^{-2}

9 A machine gun fires a bullet of mass 40 g with a velocity 1200 ms^{-1}. The man holding it, can exert a maximum force of 144 N on the gun. How many bullets can he fire per second at the most? **(AIEEE 2004)**

a. 1
b. 4
c. 2
d. 3

10 Two masses $m_1 = 5$ kg and $m_2 = 4.8$ kg tied to a string are hanging over a light frictionless pulley. What is the acceleration of the masses when lift is free to move? (take, $g = 9.8\ ms^{-2}$) **(AIEEE 2004)**

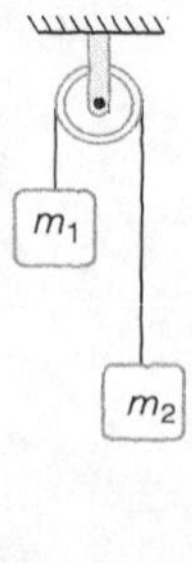

a. 0.2 ms^{-2}
b. 9.8 ms^{-2}
c. 5 ms^{-2}
d. 4.8 ms^{-2}

11 A spring balance is attached to the ceiling of a lift. A man hangs his bag on the spring and the spring reads 49 N, when the lift is stationary. If the lift moves downward with an acceleration of 5 ms^{-2}, the reading of the spring balance will be

a. 24 N b. 74 N **(AIEEE 2003)**
c. 15 N d. 49 N

12 Three forces start acting simultaneously on a particle moving with velocity v. These forces are represented in magnitude and direction by the three sides of a ΔABC (as shown). The particle will now move with velocity **(AIEEE 2003)**

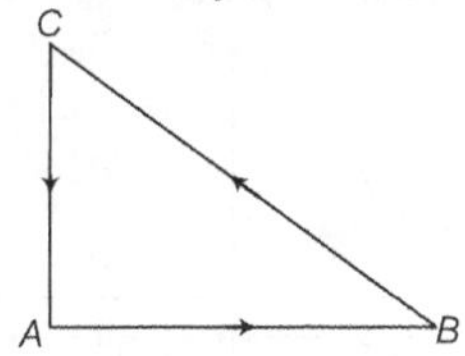

a. less than **v**
b. greater than **v**
c. $|v|$ in the direction of largest force *BC*
d. **v**, remaining unchanged

13 Consider the following two statements.

I. Linear momentum of a system of particles is zero.

II. Kinetic energy of a system of particles is zero. Then, **(AIEEE 2003)**

a. I does not imply II and II does not imply I
b. I implies II but II does not imply I
c. I does not imply II but II implies I
d. I implies II and II implies I

14 A block of mass *M* is pulled along a horizontal frictionless surface by a rope of mass *m*. If a force *P* is applied at the free end of the rope, the force exerted by the rope on the block is

a. $\frac{Pm}{M+m}$ b. $\frac{Pm}{M-m}$ **(AIEEE 2003)**
c. P d. $\frac{PM}{M+m}$

15 A light spring balance hangs from the hook of the other light spring balance and a block of mass *M* kg hangs from the former one. Then, the true statement about the scale reading is **(AIEEE 2003)**

a. both the scales read *M* kg each
b. the scale of the lower one reads *M* kg and of the upper one zero
c. the reading of the two scales can be anything but the sum of the readings will be *M* kg
d. both the scales read *M*/2 kg

16 A rocket with a lift-off mass 3.5×10^4 kg is blasted upwards with an initial acceleration of 10 ms^{-2}. Then, the initial thrust of the blast is **(AIEEE 2003)**

a. 3.5×10^5 N b. 7.0×10^5 N
c. 14.0×10^5 N d. 1.75×10^5 N

17 A lift is moving down with acceleration a. A man in the lift drops a ball inside the lift. The acceleration of the ball as observed by the man in the lift and a man standing stationary on the ground are respectively **(AIEEE 2002)**

a. g, g b. $g-a, g-a$
c. $g-a, g$ d. a, g

18 Two forces are such that the sum of their magnitudes is 18 N and their resultant which has magnitude 12 N, perpendicular to the smaller force. Then, the magnitudes of the forces are

a. 12 N, 6 N b. 13 N, 5 N **(AIEEE 2002)**
c. 10 N, 8 N d. 16 N, 2 N

19 When forces F_1, F_2, F_3 are acting on a particle of mass *m* such that F_2 and F_3 are mutually perpendicular, then the particle remains stationary. If the force F_1 is now removed, then the acceleration of the particle is **(AIEEE 2002)**

a. $\frac{F_1}{m}$ b. $\frac{F_2 F_3}{mF_1}$
c. $\frac{(F_2 - F_3)}{m}$ d. $\frac{F_2}{m}$

20 A light string passing over a smooth light pulley connects two blocks of masses m_1 and m_2 (vertically). If the acceleration of the system is $g/8$, then the ratio of the masses is **(AIEEE 2002)**

a. 8 : 1 b. 9 : 7 c. 4 : 3 d. 5 : 3

21 One end of massless rope, which passes over a massless and frictionless pulley *P* is tied to a hook *C* while the other end is free. Maximum tension that the rope can bear is 360 N. With what value of maximum safe acceleration (in ms^{-2}) can a man of 60 kg climb on the rope? **(AIEEE 2002)**

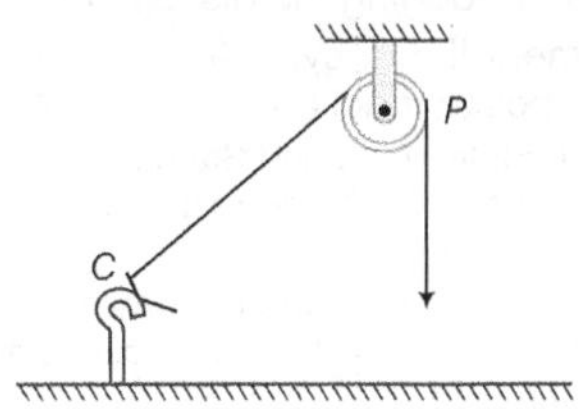

a. 16 b. 6 c. 4 d. 80

JEE Archive

JEE Advanced & IIT JEE Archive

(Compilation of Last 10 Years Questions)

1 A small block of mass 0.1 kg lies on a fixed inclined plane PQ which makes an angle θ with the horizontal. A horizontal force of 1 N acts on the block through its centre of mass as shown in the figure. The block remains stationary, if (take $g = 10 \text{ m/s}^2$)

(2012, One or More than One Options Correct Type)

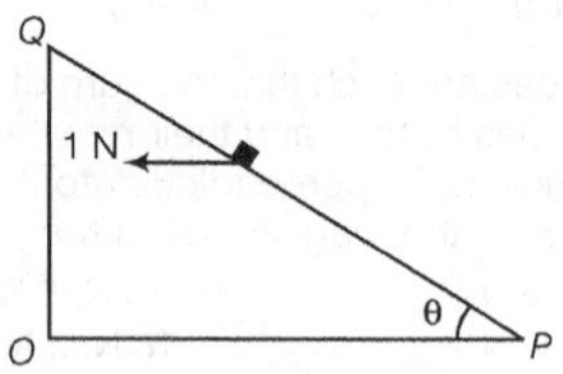

a. $\theta = 45°$

b. $\theta > 45°$ and a frictional force acts on the block towards P

c. $\theta > 45°$ and a frictional force acts on the block towards Q

d. $\theta < 45°$ and a frictional force acts on the block towards Q

2 A piece of wire is bent in the shape of a parabola $y = kx^2$ (Y-axis vertical) with a bead of mass m on it. The bead can slide on the wire without friction. It stays at the lowest point of the parabola when the wire is at rest. The wire is now accelerated parallel to the X-axis with a constant acceleration a. The distance of the new equilibrium position of the bead, where the bead can stay at rest with respect to the wire, from the Y-axis is **(2009, Only One Option Correct Type)**

a. $\frac{a}{gk}$ b. $\frac{a}{2gk}$ c. $\frac{2a}{gk}$ d. $\frac{a}{4gk}$

3 **Statement I** A cloth covers a table. Some dishes are kept on it. The cloth can be pulled out without dislodging the dishes from the table.

Statement II For every action there is an equal and opposite reaction. **(2007, Statement Type)**

a. Both Statement I and Statement II are correct and Statement II is the correct explanation of Statement I

b. Both Statement I and Statement II are correct and Statement II is not the correct explanation of Statement I

c. Statement I is correct but Statement II is incorrect

d. Statement II is correct but Statement I is incorrect

4 Two particles of mass m each are tied at the ends of a light string of length $2a$. The whole system is kept on a frictionless horizontal surface with the string held tight, so that each mass is at a distance a from the centre P (as shown in the figure).

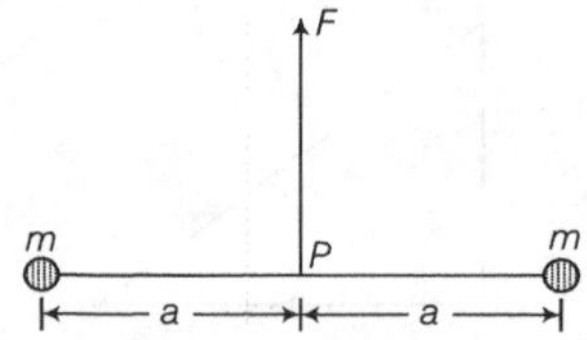

Now, the mid-point of the string is pulled vertically upwards with a small but constant force F. As a result, the particles move towards each other on the surface. The magnitude of acceleration, when the separation between them becomes $2x$, is **(2007, Only One Option Correct Type)**

a. $\frac{F}{2m}\frac{a}{\sqrt{a^2 - x^2}}$

b. $\frac{F}{2m}\frac{x}{\sqrt{a^2 - x^2}}$

c. $\frac{F}{2m}\frac{x}{a}$

d. $\frac{F}{2m}\frac{\sqrt{a^2 - x^2}}{x}$

5 System shown in figure is in equilibrium and at rest. The spring and string are massless, now the string is cut. The acceleration of mass $2m$ and m just after the string is cut will be

(2006, Only One Option Correct Type)

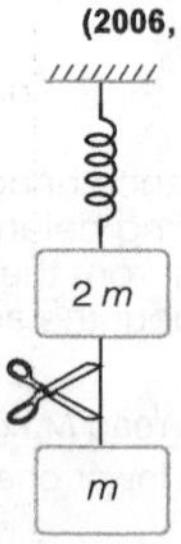

a. $g/2$ upwards, g downwards

b. g upwards, $g/2$ downwards

c. g upwards, $2g$ downwards

d. $2g$ upwards, g downwards

2. Friction

DPP-1 Basic Concepts of Static Friction

Subjective Questions

Working Space

Direction (*Q. Nos.* 1-5) *These questions are subjective in nature, need to be solved completely on notebook.*

1 State True or False.

i. The static frictional force opposes the tendency of relative motion between contact surfaces.

ii. The static frictional forces oppose the tendency of sliding of contact surfaces over each other.

iii. Static frictional force is real force in nature.

iv. Static frictional force obeys Newton's laws of motion.

v. If frictional force on a book is in backward direction, then equal magnitude of frictional force acting on table is in forward direction.

vi. Static frictional force has no certain formula. Its value is $0 \le f \le \mu_S N$.

vii. Static friction is self adjusting in nature.

viii. Static friction opposes the relative motion between contact surfaces.

ix. In the case of static friction, relative velocity between contact surfaces is zero.

x. In the case of static friction, relative acceleration between contact surfaces is zero.

xi. Static friction is just like cold weilding between contact surfaces.

2 In the following cases, blocks are stationary. Find magnitude and direction of friction on the blocks. (Given, μ_s = coefficient of static friction, μ_k = coefficient of kinetic friction). Also, calculate the contact force.

i. 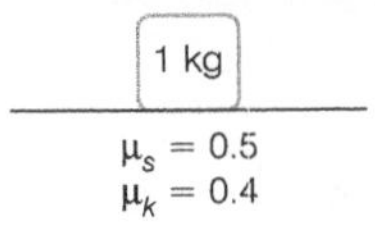

ii. 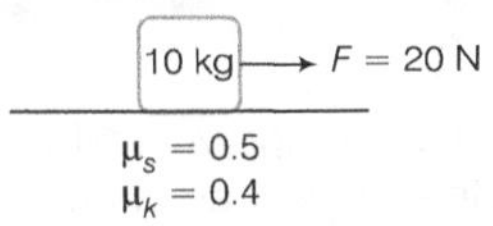

iii. 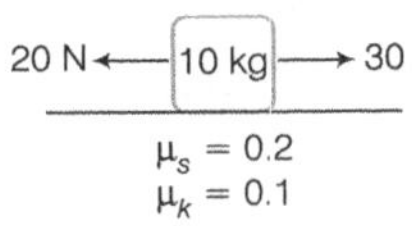

iv. 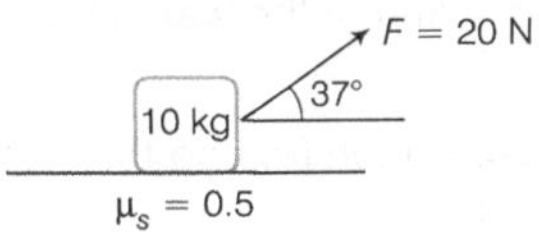

v. 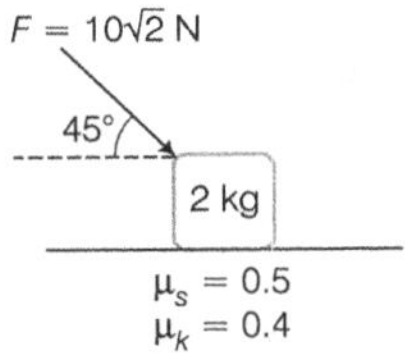

vi. 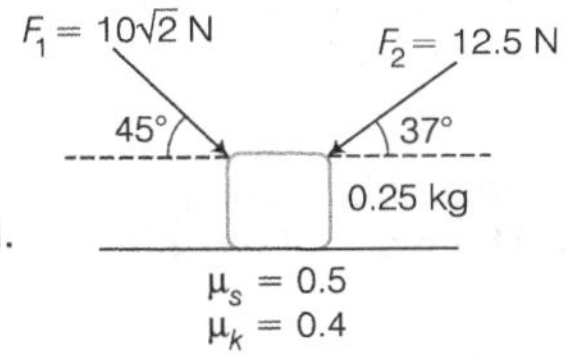

vii. A portion of a chain of mass 9 kg and length 9 m is placed on a rough horizontal surface. The length of the hanging portion is 3 m.

Working Space

3 In the following cases, blocks are stationary. Find the magnitude and direction of friction. Also, calculate contact force.

i. A block of mass $m = 1$ kg is placed on a rough inclined plane of inclination of 37°.

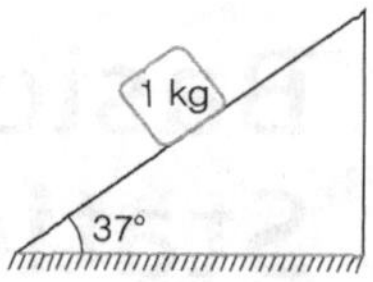

ii. A force of 10 N is applied on the block of mass 1 kg placed on rough inclined surface.

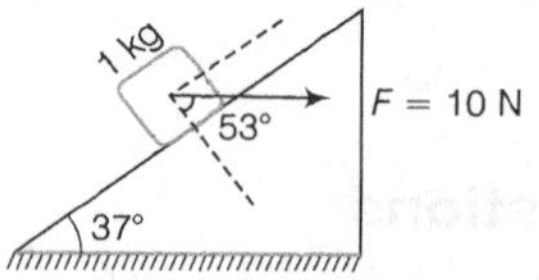

iii. Two forces $F_1 = 18$ N and $F_2 = 20$ N are acted on a block parallel to rough inclined surface as shown in the figure. Another force $F_3 = 10$ N is acting on the block making an angle of 37° with the inclined plane.

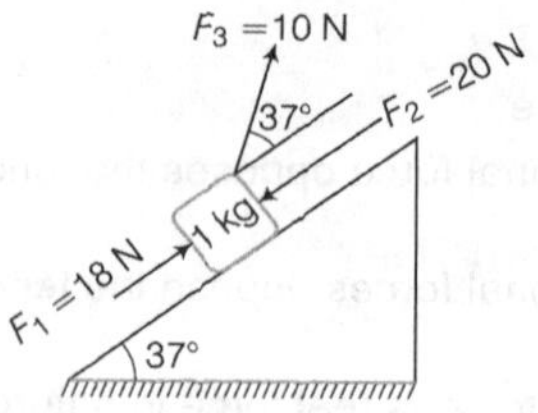

iv. A block of mass 1 kg is placed on a rough inclined plane of inclination 30°. A horizontal force $F = 12$ N is applied on the block parallel to the plane.

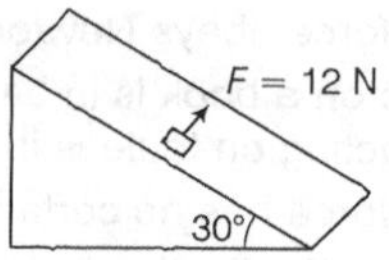

4 Calculate maximum force F required to hold a block stationary in the following cases.

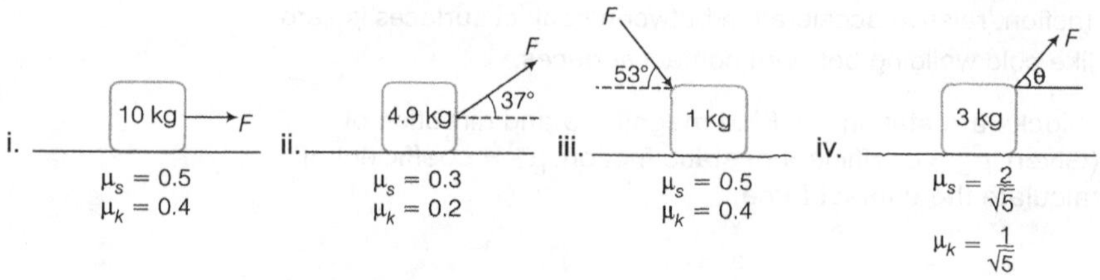

v. A block of mass 0.5 kg is pushed against a rough vertical wall with a force of $6\sqrt{5}$ N. Another force F is applied on the block at an angle of 37° with the direction of acceleration due to gravity in the plane of the wall. The coefficient of static friction is 0.5.

5 Calculate the maximum value of acceleration (a_0) so that there is no relative motion between the blocks in following cases.

i. A block B of mass 10 kg is placed on a block A, moving with acceleration a_0. The block B does not slide on the block A. The coefficient of static friction between blocks is 0.5.

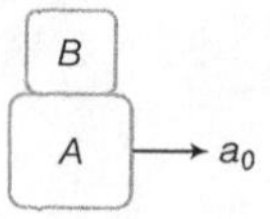

ii. In the situation shown in the figure, block B is moving horizontally with an acceleration a_0. The block of mass 1 kg does not slide on the block B.

The coefficient of static friction between blocks is $\mu_s = \dfrac{34}{37}$.

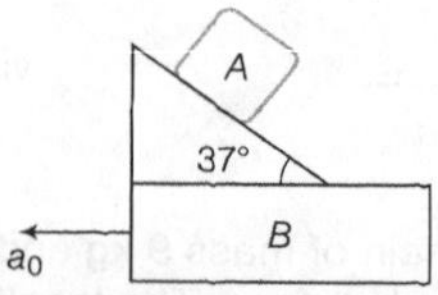

iii. A block of mass 1 kg lies on a horizontal surface in a truck. The coefficient of static friction between the block and surface is 0.6. If the acceleration of the truck is 5 m/s^2, find the frictional force acting on the block.

Working Space

Only One Option Correct Type

Direction *(Q. Nos. 6-22) This section contains 17 multiple choice questions. Each question has four choices (a), (b), (c) and (d), out of which ONLY ONE option is correct.*

6 A block B of mass 2 kg rests on a rough inclined plane making an angle of 30° with the horizontal surface. The coefficient of static friction between the block and the plane is 0.7. The frictional force on the block is

a. 9.8 N b. $0.7 \times 9.8\sqrt{3}$ N c. 9.8×7 N d. 0.8×9.8 N

7 A block of mass 0.1 kg is held against a wall by applying a horizontal force of 5 N on the block. If the coefficient of friction between the block and the wall is 0.5, the magnitude of the friction force acting on the block is

a. 2.5 N b. 0.98 N c. 4.9 N d. 0.49 N

8 A horizontal force of 10 N is necessary to hold a block stationary against a wall. The coefficient of friction between the block and the wall is 0.2. The weight of the block is **[AIEEE 2003]**

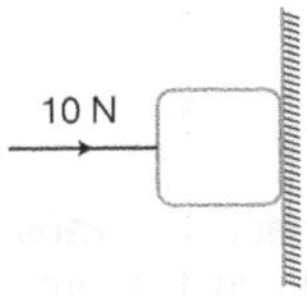

a. 20 N b. 50 N
c. 100 N d. 2 N

9 A block rests on a rough inclined plane making an angle of 30° with the horizontal surface. The coefficient of static friction between the block and the plane is 0.8. If the frictional force on the block is 10 N, the mass of the block (in kg) is (take, $g = 10 \text{ m/s}^2$) **[AIEEE 2004]**

a. 2.0 b. 4.0 c. 1.6 d. 2.5

10 What is the maximum value of the force F such that the block shown in the arrangement, does not move? **[IIT JEE 2003]**

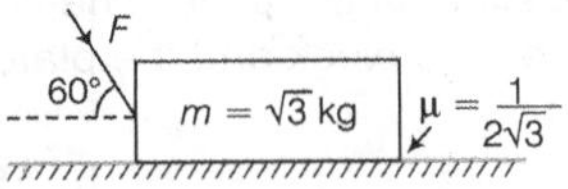

a. 20 N b. 10 N
c. 12 N d. 15 N

11 A block A of mass m is kept on a rough inclined plane of inclination θ. The coefficient of static friction between the block and the surface is $\mu_s = 0.75$. If the angle θ is slowly increased from 0° to 37°, then the friction *versus* θ graph is

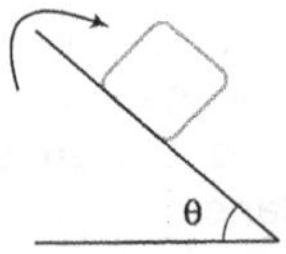

a. 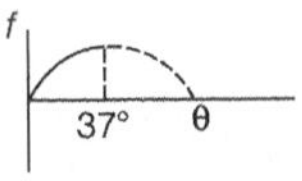

b. 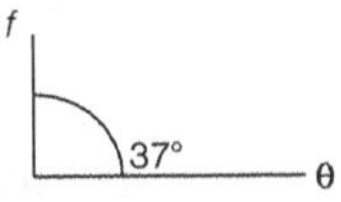

c. 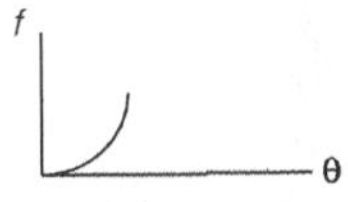

d. 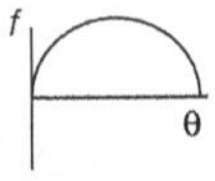

12 A bottle of circular shaped mouth of radius 5 cm is closed by a tightly fitted light cork. A light hook is attached to the cork (shown in figure). The coefficient of friction between cork and bottle is $\frac{7}{22}$. The normal contact force exerted by bottle on cork per unit circumference of the cork is 100 N/m. Find the maximum force F applied on the hook, so that the cork does not pop out.

a. 10 N b. 100 N c. 50 N d. Zero

13 A block of mass $\frac{\sqrt{2}}{10}$ kg is kept on an inclined plane of inclination of 45°. The coefficient of static friction between block and inclined plane is $\mu_s = \frac{1}{\sqrt{3}}$. Find the minimum force applied on block, up the inclined plane in order to support it.

a. $\left(\frac{\sqrt{3}-1}{\sqrt{3}}\right)$ N b. 1 N c. $\frac{1}{\sqrt{3}}$ N d. $\sqrt{3}$ N

Working Space

14 A block of mass m is placed on a rough fixed sphere. The coefficient of static friction between the sphere and the block is $\mu_s = \frac{1}{\sqrt{3}}$. Find the maximum value of θ, so that the block remains stationary.

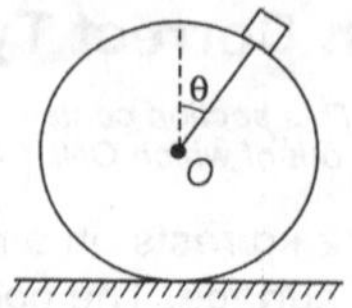

a. 30° b. 45° c. 60° d. 90°

15 A uniform rod is placed on an inclined plane of inclination of θ. The coefficients of static friction for sections AB, BC and CD are $\frac{1}{2}, \frac{5}{8}$ and $\frac{3}{8}$, respectively. Find the maximum value of $\tan\theta$, so that the rod remains stationary.

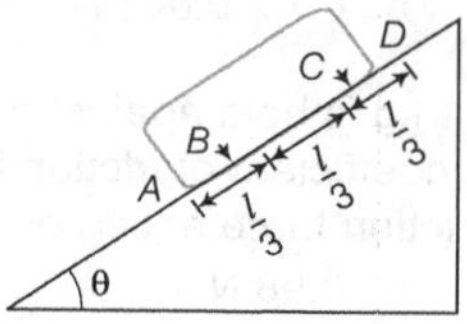

a. $\frac{1}{2}$ b. $\frac{5}{8}$

c. $\frac{3}{8}$ d. $\frac{13}{16}$

16 If the force which is acting parallel to an inclined plane of angle 45° is sufficient to pull a block up to be n times, the force which will let it be on the point of sliding down. Find the value of coefficient of static friction between block and inclined plane.

a. $\frac{n+1}{n-1}$ b. $\frac{n-1}{n+1}$ c. $\frac{2n+1}{2n+3}$ d. $\frac{n}{n+1}$

17 A block A of mass 1 kg is in contact with a vertical plate. The plate is moving horizontally with an acceleration a_0. The block does not slide on the plate. The coefficient of static friction between the block and the plate is 0.5. Find minimum value of a_0.

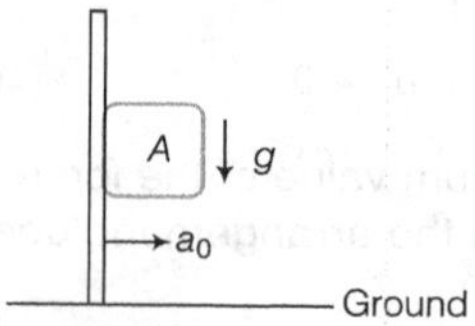

a. 10 m/s^2 b. 15 m/s^2

c. 20 m/s^2 d. Zero

18 A force F of $\sqrt{13}$ N along X-axis on a block of mass 1 kg is placed on a rough inclined plane as shown in the figure. Find minimum coefficient of friction, such that the block does not slide on inclined plane.

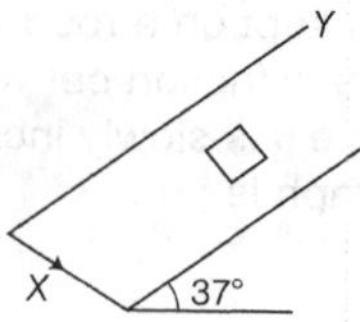

a. 1 b. $\frac{1}{2}$

c. $\frac{7}{8}$ d. $\frac{7}{16}$

19 A rod of mass m and length l is placed on a rough horizontal surface. The coefficient of static friction is $\mu = bx$, where b is a positive constant. Find maximum force F applied on the rod, so that rod does not move.

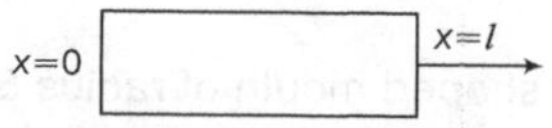

a. $mgbl$ b. $\frac{mgbl}{2}$

c. $\frac{mgbl}{3}$ d. Zero

20 Priyansh's mummy cook 'AMLET' in her fancy kitchen. The coefficient of static friction between teflon and scrambled eggs is 0.4. What is the smallest angle from the horizontal surface that will cause the eggs to slide across the bottom of a teflon-coated skillet? (use the approximation $\sin\theta \approx \theta$ and $\tan\theta \approx \theta$)

a. 0.04° b. 2.23°

c. 3° d. Cannot be calculated

Working Space

21 A uniform massive rope ABC has length of 6 m. The rope is hold by applying a force F at an angle of $\alpha = 37°$ with horizontal surface. If hanging part is less than 2 m, then the rope does not slip on the horizontal rough surface. The value of coefficient of static friction is

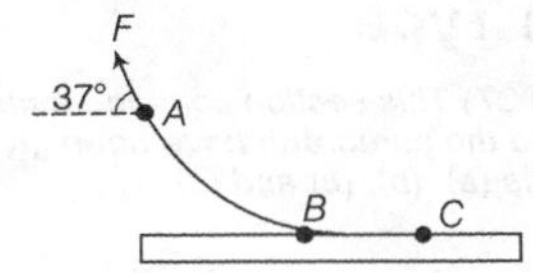

a. 0.5 b. $\frac{3}{5}$ c. $\frac{5}{6}$ d. $\frac{2}{3}$

22 Three identical billiard balls each of mass 1 kg are placed as shown in the figure. The upper ball is smooth. Find minimum coefficient of static friction between lower balls and ground, so that balls remain stationary.

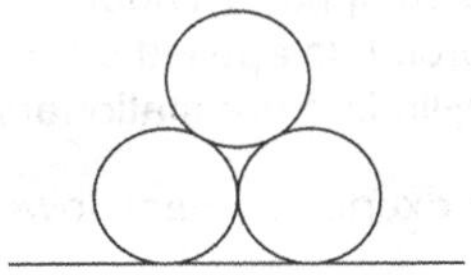

a. $\frac{1}{\sqrt{3}}$ b. $\frac{1}{3\sqrt{3}}$

c. $\frac{2}{\sqrt{3}}$ d. 0.5

One or More than One Options Correct Type

Direction (Q. Nos. 23-25) *This section contains 3 multiple choice questions. Each question has four choices (a), (b), (c) and (d), out of which ONE or MORE THAN ONE are correct.*

23 In the figure, a block of mass 2 kg rests in an inclined right angled trough, then

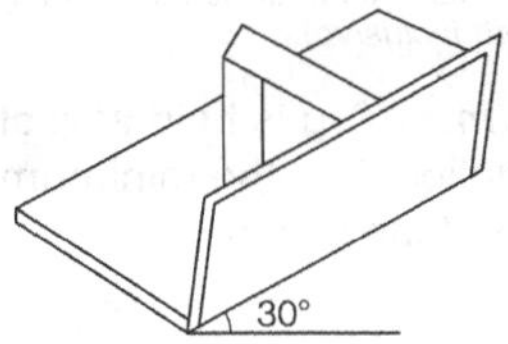

a. frictional force on the block is 10 N
b. contact force on the block is 20 N
c. contact force on the block is 10 N
d. frictional force on the block is 20 N

24 A force $\mathbf{F} = \hat{\mathbf{i}} + 4\hat{\mathbf{j}}$ acts on a block as shown. Ther,

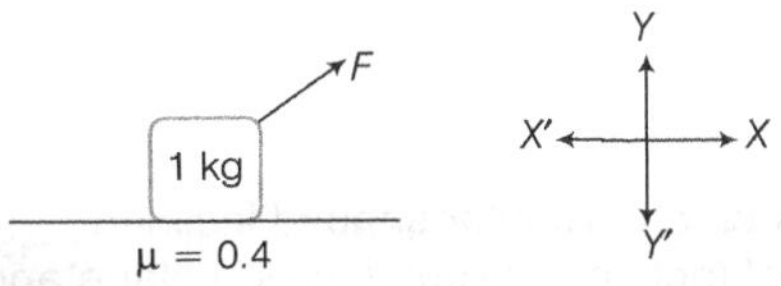

a. the acceleration of the block is zero
b. the frictional force on the block is $\mathbf{f} = -(1\text{ N})\hat{\mathbf{i}}$
c. the frictional force on the block is 2.4 N in magnitude
d. the acceleration of the block is non-zero

25 A labourer lifts a stack of several identical bricks each of mass 1 kg by pressing hard with his hands. The coefficient of static friction for all surfaces is $\mu_s = 0.5$. The minimum normal force applied by each hand to just hold the bricks is 50 N.

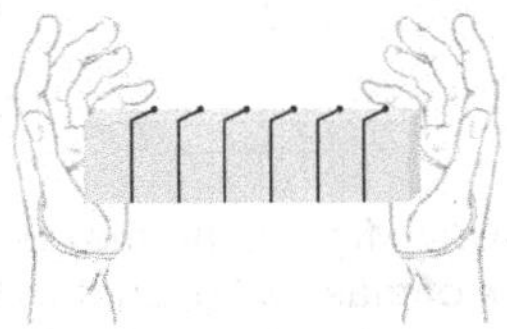

Which of the following statement(s) is/are correct?

a. The number of bricks hold by labourer is 5, if normal contact force by hand is 50 N
b. The number of bricks, hold by labourer is 10, if normal contact force by hand is 50 N
c. If normal contact force applied by labourer is slightly less than 50 N number of bricks hold by labourer is 10
d. If labourer just hold the stack, the contact force applied by each hand on the brick is 55.9 N

Comprehension Type

Working Space

Direction (Q. Nos. 26 and 27) *This section contains a paragraph, describing theory, experiments, data, etc. Two questions related to the paragraph have been given. Each question has only one correct answer among the four given options (a), (b), (c) and (d).*

Passage

Figure shows three identical cylinders each of mass $10\sqrt{3}$ kg. Two lower cylinders are fixed. The coefficient of static friction between upper cylinder and lower cylinder is 0.5. A horizontal force *F* is applied on upper cylinder along its axis. All cylinders are stationary.

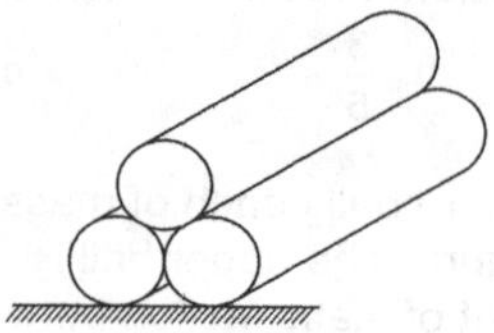

26 The frictional force exerted by each lower cylinder on the upper cylinder is

a. *F* b. *F*/2
c. *F*/3 d. zero

27 Calculate maximum value of *F*, so that upper cylinder remains stationary.

a. 50 N b. 75 N
c. 100 N d. 150 N

One Integer Value Correct Type

Direction (Q. Nos. 28-30) *This section contains 3 questions. Each question, when worked out will result in an integer from 0 to 9 (both inclusive).*

28 A cylinder of mass $m = 10$ g is held at rest by prongs of a forceps. If $\mu_s = 0.01$ be the coefficient of friction, find the minimum normal force (in newton) exerted by the cylinder on one of the prongs.

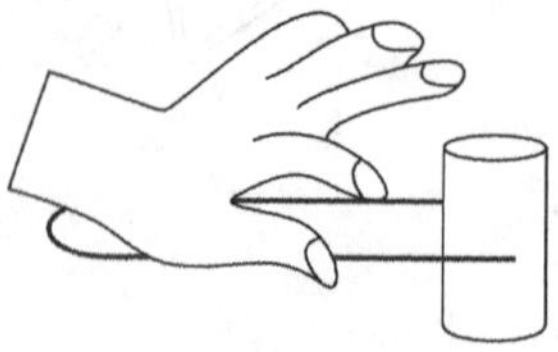

29 A uniform rope of mass 5 kg and length 5 m is on an irregular shaped inclined plane as shown in the figure. The coefficient of friction between inclined plane and the rope is μ. If frictional force is 10^x N, then find the value of *x*.

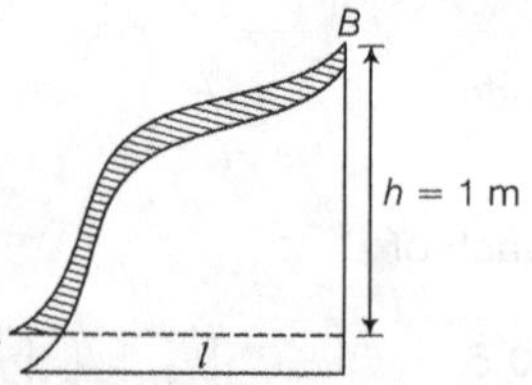

30 In the situation shown in figure, the maximum value of horizontal force *F* to be applied on the block of mass 5 kg, so that block *A* of mass 100 kg does not move upward. The coefficient of static friction for contact surfaces of the block *A* is 0.1 but the block *B* is placed on smooth horizontal surface. If $F = (774.6)^x$ N, then find the value of *x*.

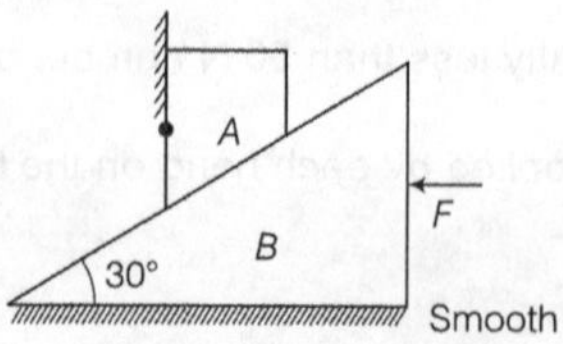

DPP-2 Basic Concepts of Kinetic Friction

Subjective Questions

Working Space

Direction (Q. Nos. 1-2) *These questions are subjective in nature, need to be solved completely on notebook.*

1 State True or False.

i. Kinetic frictional forces oppose the relative motion between contact surfaces.

ii. If the kinetic frictional force on a book is in backward direction, then the equal magnitude of the kinetic frictional force acting on the table is in forward direction.

iii. Kinetic frictional force has a certain formula, $f = \mu_k N$.

iv. In the case of kinetic frictional force, relative velocity between contact surfaces must be non-zero.

v. In the presence of kinetic frictional force, relative acceleration may or may not be zero.

vi. In the presence of kinetic frictional force, relative acceleration between contact surfaces must be non-zero.

vii. If a tyre rolls without slipping over a road, static frictional force comes into play between the tyre and road.

viii. A man is moving without slipping on road, the nature of friction between shoes of man and the road is static friction.

2 Find direction and magnitude of kinetic friction in following cases.

i.

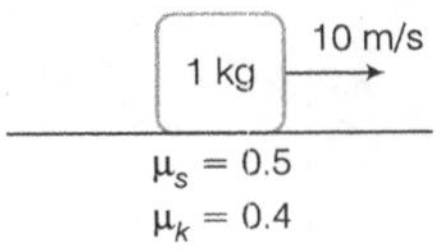

ii.

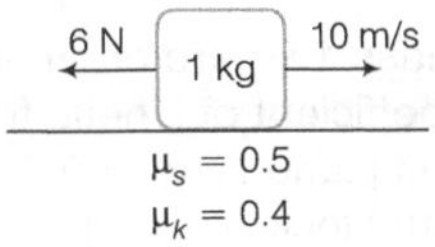

iii.

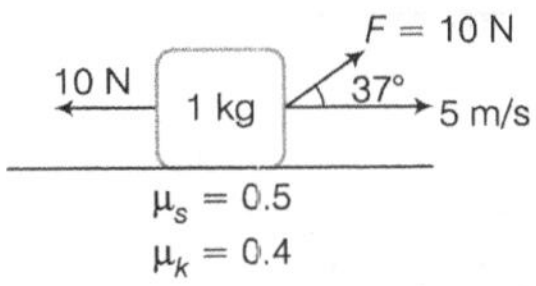

iv.

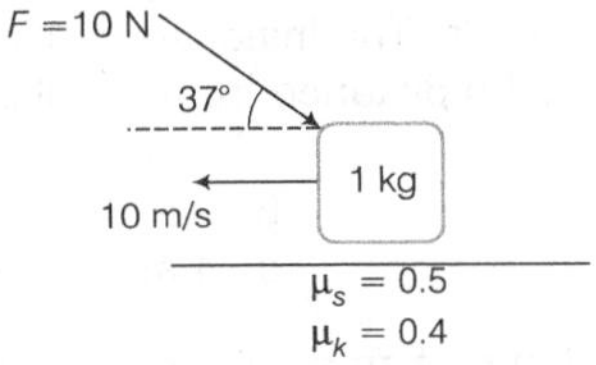

v. A rod of mass 3 kg is sliding on horizontal ground. The one-third part of rod is smooth and other part is rough. The coefficient of kinetic friction between rough part and ground is $\mu_k = 0.5$.

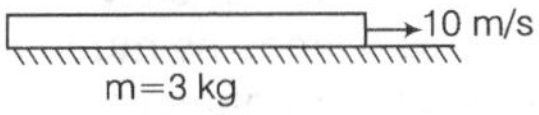

vi. A block of mass 1 kg starts to move up on the inclined plane of inclination of 37°. The coefficient of static friction and kinetic friction between inclined plane and block are $\mu_s = 0.5$ and $\mu_k = 0.4$, respectively.

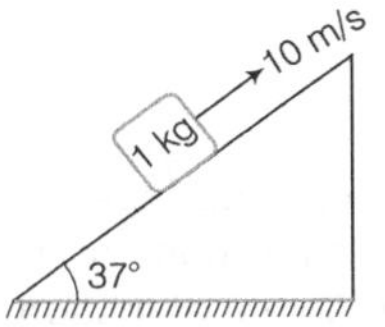

$\mu_s = 0.5$ and $\mu_k = 0.4$

Working Space

Only One Option Correct Type

Direction (Q. Nos. 3-17) *This section contains 15 multiple choice questions. Each question has four choices (a), (b), (c) and (d), out of which ONLY ONE option is correct.*

3 A piece of rough wire bent in the shape of parabola $y = kx^2$ (Y-axis vertical) with a bead of mass m at lowest point. The wire starts to accelerate along negative X-axis. Find the unit vector along the kinetic friction acting on the bead when bead is at distance $\frac{1}{2k}$ from the Y-axis.

a. $\frac{1}{\sqrt{2}}\hat{\mathbf{i}} + \frac{1}{\sqrt{2}}\hat{\mathbf{j}}$
b. $-\frac{1}{2}\hat{\mathbf{i}} - \frac{1}{2}\hat{\mathbf{j}}$
c. $\frac{-1}{\sqrt{2}}\hat{\mathbf{i}} - \frac{1}{\sqrt{2}}\hat{\mathbf{j}}$
d. Data insufficient

4 A block of mass 1 kg kept in a plate is moving horizontally with a velocity $\mathbf{v}_1$. The plate is also moving horizontally with a constant velocity $\mathbf{v}_2 = 3\hat{\mathbf{i}} + 4\hat{\mathbf{j}}$. The coefficient of kinetic friction between the block and the plate is $\mu_k = \frac{1}{\sqrt{2}}$. Choose the correct statement.

a. If $\mathbf{v}_1 = 3\hat{\mathbf{i}} + 4\hat{\mathbf{j}}$, the kinetic friction on the block is zero
b. If $\mathbf{v}_1 = \hat{\mathbf{i}} + 2\hat{\mathbf{j}}$, the kinetic friction on the block is $(-5\hat{\mathbf{i}} - 5\hat{\mathbf{j}})$ N
c. If $\mathbf{v}_1 = \hat{\mathbf{i}} + 2\hat{\mathbf{j}}$, the kinetic friction on the block is $(-2\hat{\mathbf{i}} - 2\hat{\mathbf{j}})$ N
d. If $\mathbf{v}_1 = \hat{\mathbf{i}} + 2\hat{\mathbf{j}}$, the kinetic friction on the plate due to the block is $(5\hat{\mathbf{i}} + 5\hat{\mathbf{j}})$ N

5 In the situation shown in the figure, the coefficient of kinetic friction between block of mass 1 kg and vertical walls is $\mu_k = 0.5$. The block starts from rest and travels a distance of 1 m in 1 s. Find the contact force exerted by each wall on the block.

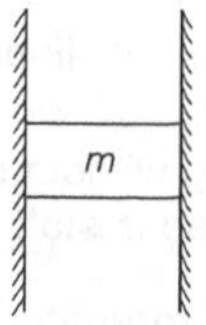

a. 8 N
b. 20 N
c. $4\sqrt{5}$ N
d. $10\sqrt{5}$ N

6 Two blocks A and B each of mass 1 kg are on an inclined plane of inclination 37°. The coefficient of kinetic friction between blocks and the inclined plane is $\mu_k = 0.5$. At $t = 0$, block A is released from rest but block B is projected up along inclined plane with speed 10 m/s. The initial separation between blocks is 6 m. Find the distance from initial position of A, where they will collide.

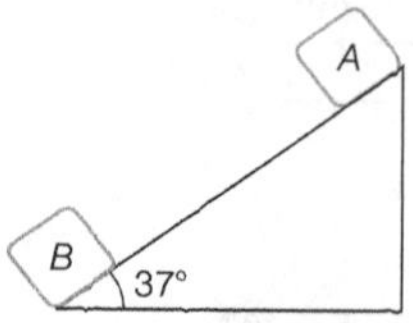

a. 1 m
b. 2 m
c. 3 m
d. 4 m

7 A smooth block is released at rest on a 45° incline and then slides a distance d. The time taken to slide is n times as much to slide on rough incline than on a smooth incline. The coefficient of friction is **[AIEEE 2005]**

a. $\mu_s = 1 - \frac{1}{n^2}$
b. $\mu_s = \sqrt{1 - \frac{1}{n^2}}$
c. $\mu_k = 1 - \frac{1}{n^2}$
d. $\mu_k = \sqrt{1 - \frac{1}{n^2}}$

8 The upper half of an inclined plane with inclination of ϕ is perfectly smooth while the lower half is rough. A body at the top starts from rest will again come to rest at the bottom, the coefficient of friction for the lower half is given by **[AIEEE 2005]**

a. $2 \tan \phi$
b. $\tan \phi$
c. $2 \sin \phi$
d. $2 \cos \phi$

9 Two blocks A and B of equal mass are sliding down along straight parallel lines on an inclined plane of 45°. Their coefficients of kinetic friction are $\mu_A = 0.2$ and $\mu_B = 0.3$, respectively. At $t = 0$, both blocks are at rest and block A is $\sqrt{2}$ m behind block B. Find the time and distance from the initial position, where the front faces of the blocks come in line on the inclined plane as shown in the figure. (use $g = 10\ \text{ms}^{-2}$) **[IIT JEE 2004]**

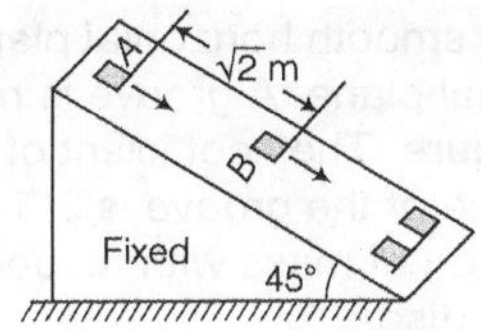

a. 2s, $8\sqrt{2}$ m b. $\sqrt{2}$ s, 7 m c. $\sqrt{2}$ s, $7\sqrt{2}$ m d. 2 s, $7/\sqrt{2}$ m

10 A block of mass m is on an inclined plane of angle θ. The coefficient of friction between the block and the plane is μ and $\tan\theta > \mu$. The block is held stationary by applying a force P parallel to the plane. The direction of force pointing up the plane is taken to be positive. As P is varied from $P_1 = mg(\sin\theta - \mu\cos\theta)$ to $P_2 = mg(\sin\theta + \mu\cos\theta)$, the frictional force f *versus* P graph will be. **[IIT JEE 2010]**

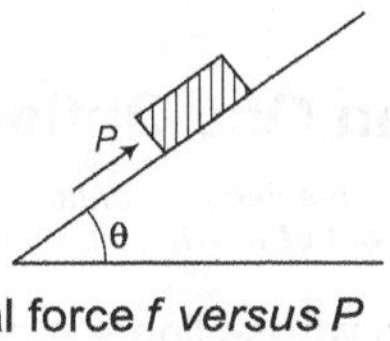

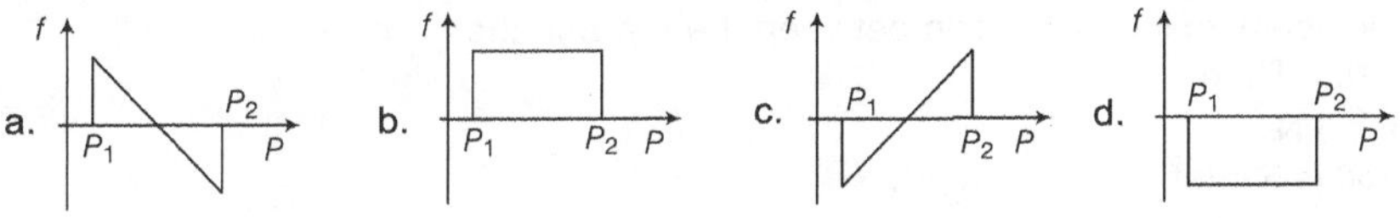

11 A man with his middle finger of length 4 cm vertical is sitting in a car, which is moving horizontally with an acceleration $a_0 = 5\ \text{m/s}^2$. A ring can slide on the finger with coefficient of kinetic friction 0.4. If the ring is released from rest with respect to man from top of the finger. Find the time after which it will reach at the lowest point of the finger.

a. 0.1 s b. 0.2 s c. 0.3 s d. 0.4 s

12 A small block A can slide on upper rough surface of a block B, which is accelerating horizontally with $a_0 = 3\ \text{m/s}^2$. The length of the block B is 2 m. Find the time after which block A will separate from block B. The coefficient of kinetic friction between A and B is $\mu_k = 0.2$.

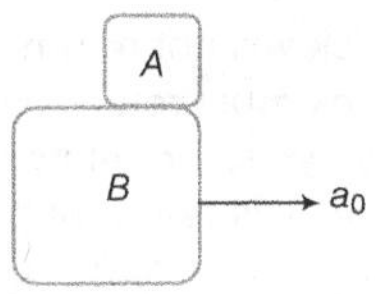

a. 1 s b. 2 s c. 3 s d. 4 s

13 A block A of mass 1 kg is projected along rough horizontal surface from $x = 0$ with speed $v_0 = \sqrt{10}$ m/s. Find the distance travelled by block before coming to rest. The coefficient of kinetic friction between block and the surface is $\mu = bx$, where $b = 1\ \text{m}^{-1}$.

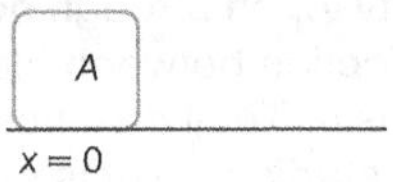

a. 1 m b. 2 m c. 3 m d. 4 m

14 A box rests on a truck moving with a velocity of 2 m/s. The coefficient of friction between box and truck is 0.1. Driver apply the brake and truck starts to decelerate uniformly and stops in 0.5 s. The total distance travelled by the box with respect to truck is

a. 0.75 m b. 0.375 m c. 1.5 m d. 2 m

15 A marble block of mass 2 kg lying on ice, when given a velocity of 6 m/s, is stopped by friction in 10 s. Then, the coefficient of friction is **[AIEEE 2003]**

a. 0.02 b. 0.03 c. 0.06 d. 0.01

16 A raw brick of mass m made of soil placed on an inclined plane of inclination of θ is released from rest. The brick losses mass at constant rate μ kg/m. Find distance travelled in time t. The coefficient of friction is μ_k.

a. $\frac{g}{2}(\sin\theta - \mu_k\cos\theta)t^2$ b. $\frac{g}{2}(\cos\theta - \mu_k\sin\theta)t^2$

c. $g(\sin\theta - \mu_k\cos\theta)t^2$ d. $g(\cos\theta - \mu_k\sin\theta)t^2$

Working Space

17 A disc is kept on a smooth horizontal plane with its plane parallel to horizontal plane. A groove is made in the disc as shown in the figure. The coefficient of friction between mass m and surface of the groove is 2/5 and $\sin\theta = 3/5$. Find the acceleration of mass with respect to the frame of reference of the disc.

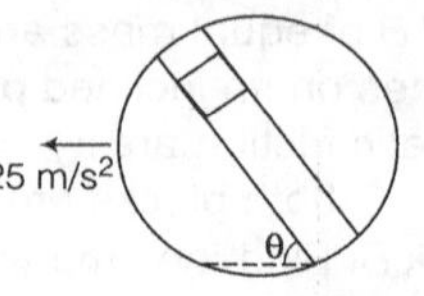

[IIT JEE 2006]

a. $10\ m/s^2$
b. $5\ m/s^2$
c. $25\ cm/s^2$
d. None of these

One or More than One Options Correct Type

Direction (Q. Nos. 18-20) *This section contains 3 multiple choice questions. Each question has four choices (a), (b), (c) and (d), out of which ONE or MORE THAN ONE are correct.*

18 A rod of length 1 m whose linear mass density increases uniformly from 1 kg/m to 2 kg/m, is being pulled along its length by a constant force $F = 15$ N on a rough horizontal surface. The coefficient of kinetic friction between the rod and the horizontal surface is $\mu_k = 0.5$. Then,

a. the friction on the rod is 7.5 N
b. the acceleration of the rod is $5\ m/s^2$
c. the velocity of rod at $t = 2$ s is 20 m/s
d. the distance travelled by the rod before coming in rest is 20 m

19 A block is projected upward along rough inclined plane of inclination of θ from point A. The coefficient of static friction is $\mu_s = \frac{1}{\sqrt{2}}$. Which of the following statement(s) is/are correct

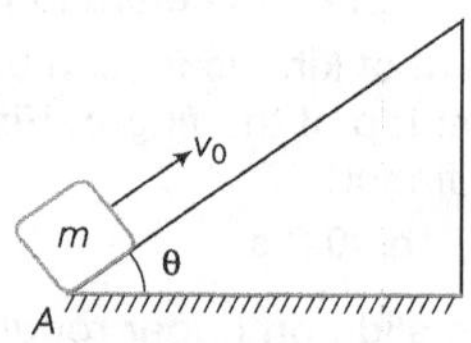

a. If $\theta \le 45°$, the block will not return at point A
b. If $\theta > 45°$, the block must return at point A
c. The block will always return at point A
d. If $\theta \le 45°$, the block will return at point A

20 A constant force F is applied on a block of mass 1 kg moving with velocity v_0 on a rough horizontal surface. The coefficient of friction between block and the horizontal surface is μ. Total distance travelled by block before reaching at starting point is s_0. Choose the correct statements.

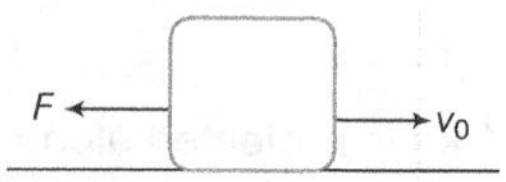

a. If $F \le \mu mg$, $s_0 = \frac{mv_0^2}{2(F+\mu mg)}$
b. If $F \le \mu mg$, $s_0 = \frac{mv_0^2}{F+\mu mg}$
c. If $F > \mu mg$, $s_0 = \frac{mv_0^2}{2(F+\mu mg)}$
d. If $F > \mu mg$, $s_0 = \frac{mv_0^2}{F+\mu mg}$

Comprehension Type

Direction (Q. Nos. 21-23) *This section contains a paragraph, describing theory, experiments, data, etc. Three questions related to the paragraph have been given. Each question has only one correct answer among the four given options (a), (b), (c) and (d).*

Passage

A bus is moving in the East direction with constant velocity of 3 m/s. A small packet enters on the roof of bus with a horizontal velocity 4 m/s in North direction. The coefficient of friction between the packet and roof of the bus is 0.5.

21 The direction of friction on the block is

a. 37° East of South
b. 37° South of East
c. South
d. West

Working Space

22 An observer on roof of a building observes the path of packet as

a. straight line b. parabola c. circle d. hyperbola

23 Find the distance travelled by the packet with respect to bus.

a. 2.5 m b. 5 m c. 2 m d. 10 m

One Integer Value Correct Type

Direction (Q. No. 24-26) This section contains 3 questions. When worked out will result in an integer from 0 to 9 (both inclusive).

24 A block is moving on an inclined plane making an angle of 45° with the horizontal surface and the coefficient of friction is μ. The force required to push up the inclined plane, is 3 times the force required to prevent it from sliding down. If we define $N = 10\mu$, then find the value of N.

25 A groove $AB = 1.16$ m cutted in plane of an inclined plane is of inclination of 37°. A block is touching every surface of the groove, starts slide from point B. The coefficient of kinetic friction between the block and all surfaces of groove is 0.1. Find the time (in second) taken by the block to come at point A.

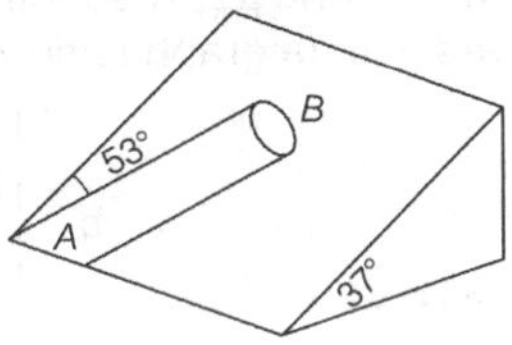

26 A horizontal force is applied on a plank. The acceleration of the plank is $a = (4t)\ \text{m/s}^2$ for $0 \le t \le 2$ s. After the end of acceleration, the plank continues to move at constant speed. The block initially at rest on the plank begins to slip at $t = 1$s and stop slipping at $t = 3$ s. The ratio of coefficient of static friction and kinetic friction between the block and the plank is $\frac{8}{x}$. Find the value of x.

Statement Type

Direction (Q. Nos. 27-30) This section is based on Statement I and Statement II. Select the correct answer from the codes given below.

a. Both Statement I and Statement II are correct and Statement II is the correct explanation of Statement I
b. Both Statement I and Statement II are correct and Statement II is not the correct explanation of Statement I
c. Statement I is correct but Statement II is incorrect
d. Statement II is correct but Statement I is incorrect

27 **Statement I** A bicycle with rubber tyres moves on a wet concrete road has more chance of slip over the road than that of on a dry concrete road.

Statement II The coefficient of friction at dry concrete road is greater than that of wet concrete road.

28 **Statement I** When brake is applied, then the tyres of bicycle slips over the road. In this case, the friction between road and tyre is kinetic friction.

Statement II Frictional force between tyre and ground is always static in nature.

29 **Statement I** If an insect will crawl up on your body, the frictional force between your body surface and insect is static frictional force.

Statement II Static friction does not produce heat.

30 **Statement I** It is easier to pull a heavy object than to push it on a level ground.

Statement II The magnitude of frictional force depends on the nature of the two surfaces in contact.

DPP-3 Analysis of Friction

Only One Option Correct Type

Direction (Q. Nos. 1-4) *This section contains 4 multiple choice questions. Each question has four choices (a), (b), (c) and (d), out of which ONLY ONE option is correct.*

Working Space

1 A block of mass m is projected with speed v_0 along the inclined plane of inclination θ. The coefficients of static friction and kinetic friction are μ_s and μ_k, respectively. Which of the following velocity *versus* time graph is correct?

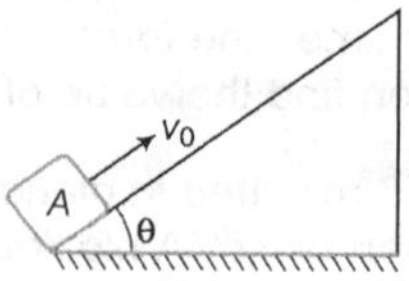

a.
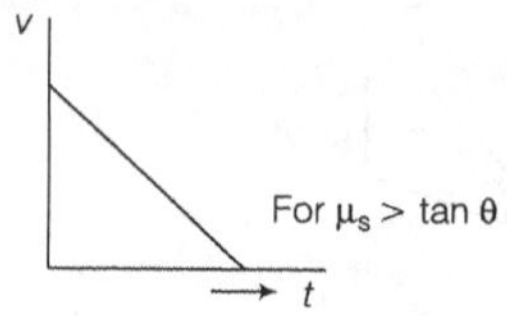

b.
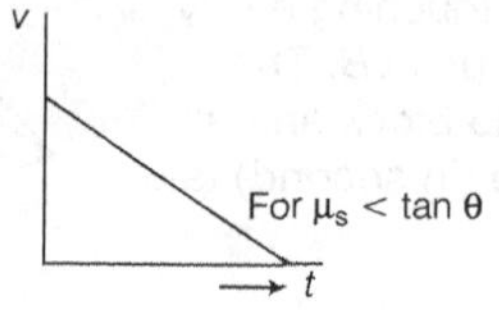

c.
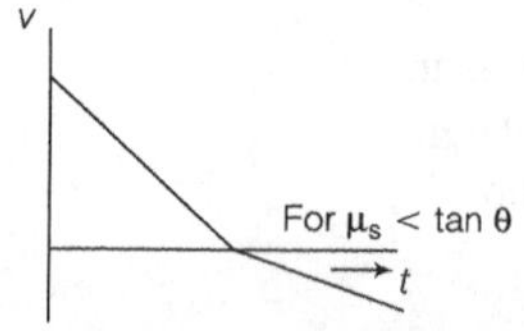

d.
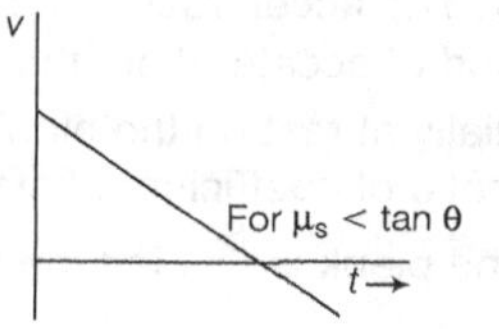

2 In the system of connecting blocks, $m_A = 1$ kg, $m_B = 2$ kg and $m_C = 2$ kg. The inclined plane is smooth while the friction coefficient between block B and horizontal ground is μ. Mark the correct option.

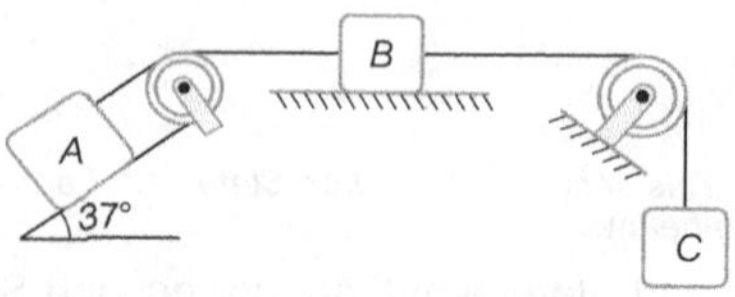

a. If $\mu = 0.8$, the friction on block B is 14 N and the tension in the string connecting blocks A and B is 6 N
b. If $\mu = 0.4$, the acceleration of blocks is 0.8 m/s^2
c. If $\mu = 0.6$, the friction on the block B is zero and tension in the string connecting the blocks B and C is 20 N
d. All of the above

3 In the situation shown in the figure, the coefficient of static and kinetic friction between block of mass 10 kg and horizontal surfaces are 0.5 and 0.4, respectively. Which of the following statements is correct?

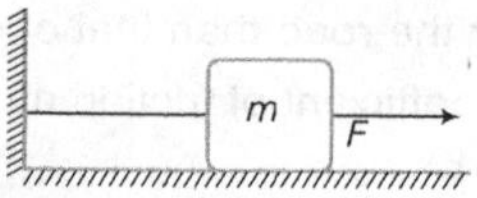

a. If $F = 30$ N, then tension in string is zero and the friction on the block is 30 N in leftward direction along the surface
b. If $F = 50$ N, then tension in string is zero and friction on the block is 50 N
c. If $F = 60$ N, then tension in string is 10 N and the friction on the block is 50 N
d. All of the above

4 A block of mass 5 kg is pressed against a wall (in YZ-plane) with a force $\mathbf{F} = 40\hat{i} + 30\hat{j} + 10\hat{k}$. The vertical upward direction is taken as positive Z-axis. The coefficient of friction is 0.5. The acceleration of the block is

a. zero b. $(3.6\hat{j} - 4.8\hat{k})$ m/s^2 c. $(3.6\hat{i} + 4.8\hat{k})$ m/s^2 d. $(-4\hat{k})$ m/s^2

One or More than One Options Correct Type

Direction *(Q. Nos. 5-12) This section contains 8 multiple choice questions. Each question has four choices (a), (b), (c) and (d), out of which ONE or MORE THAN ONE are correct.*

5 A block of mass 10 kg is placed on rough horizontal surface as shown in the figure. Choose the correct statements.

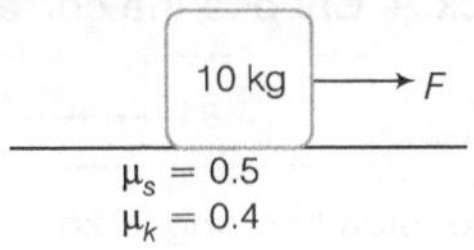

a. If $F = 45$ N, the friction on the block is 45 N and acceleration of the block is zero

b. If $F = 50$ N, the force of friction is 40 N and the acceleration of the block is 1 m/s^2

c. If $F = 60$ N, the force of friction is 40 N and the acceleration of the block is 2 m/s^2

d. If $F = 60$ N, the distance travelled by the block is 1 m in 1 s

6 A block is resting on a board and subjected to no other force than its weight and the contact force of the board. The board can be fixed at any desired value of angle of inclination θ with horizontal. The coefficient of static friction and the coefficient of kinetic friction between the block and the board is μ_s and μ_k, respectively. Which of the following statement is/are correct?

a. If $\theta = 0$, friction on the block is zero and the angle of friction is also zero

b. If $\theta \le \tan^{-1}(\mu_s)$, the block acceleration is zero and the angle of friction is θ

c. If $\theta > \tan^{-1}(\mu_s)$, the block accelerates downward along the board and the angle of friction is $\tan^{-1}(\mu_k)$

d. If $\theta > \tan^{-1}(\mu_s)$, the contact force on the block is $\sqrt{1+\mu_k^2}\, mg \cos\theta$

7 In previous problem, if N is normal reaction exerted by the board on the block, f is friction exerted by the board on the block and F is the contact force exerted by the board on the block. Then, which of the following graphs is/are correct?

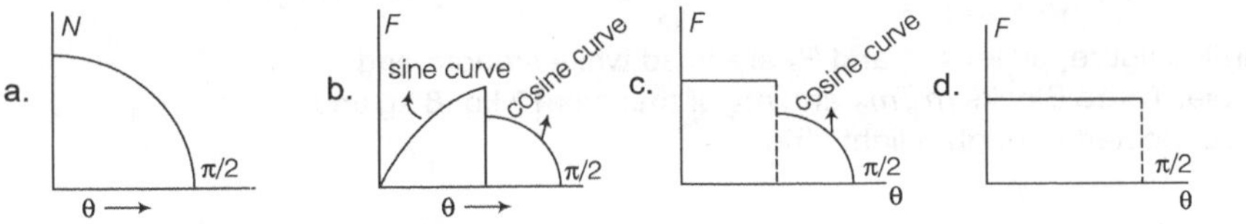

8 The situation shown in the figure, the coefficients of static friction and kinetic friction between mass m_1 and the horizontal surface are μ_s and μ_k, respectively. The blocks are released from rest. Which of the following statements is/are correct.

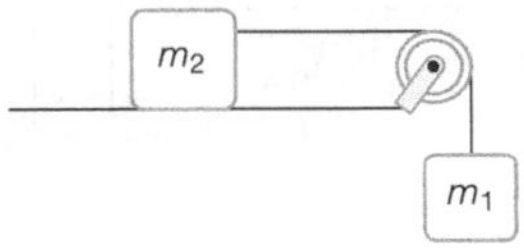

a. If $m_1 \le \mu_s m_2$, the acceleration of the blocks is zero and tension in string is $m_1 g$

b. If $m_1 \le \mu_s m_2$, the friction on the block m_2 is $m_1 g$

c. If $m_1 > \mu_s m_2$, the friction on the block m_2 is $\mu_k m_2 g$

d. If $m_1 > \mu_s m_2$, the acceleration of blocks is $\dfrac{(m_1 - \mu_k m_2)g}{m_1 + m_2}$

9 In the situation shown in the figure, the coefficients of static friction and kinetic friction are μ_s and μ_k, respectively. The blocks are released from rest. Which of the following statements is/are correct:

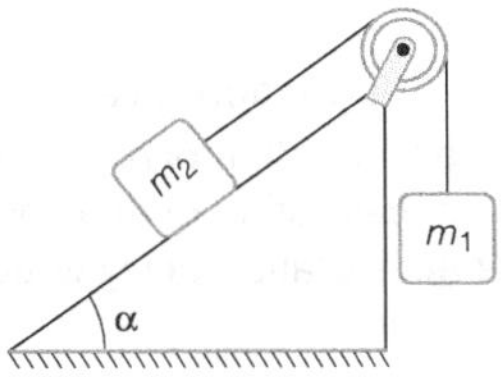

a. If $m_1 g - m_2 g \sin\alpha \le \mu_s m_2 g \cos\alpha$, the tension in string is $m_1 g$ and acceleration of blocks is zero

b. If $m_1 g - m_2 g \sin\alpha \le \mu_s m_2 g \cos\alpha$, the friction on m_2 is $(m_1 g - m_2 g \sin\alpha)$ in downward direction along the inclined plane

c. If $m_2 g \sin\alpha - m_1 g \le \mu_s m_2 g \cos\alpha$, the friction on m_2 is $(m_2 g \sin\alpha - m_1 g)$ in upward direction along inclined plane

d. If $m_1 g - m_2 g \sin\alpha > \mu_s m_2 g \cos\alpha$, the acceleration of mass m_2 is $\left(\dfrac{m_1 g - \mu_k m_2 g \cos\alpha - m_2 g \sin\alpha}{m_1 + m_2}\right)$ is upward direction along inclined plane

10 In the situation shown in the figure, the masses of blocks A and B are 1 kg and 2 kg, respectively. The coefficients of static friction and kinetic friction for all surfaces of contact are 0.5 and 0.4 respectively. A horizontal constant force F is applied on the block A. Choose the correct statements.

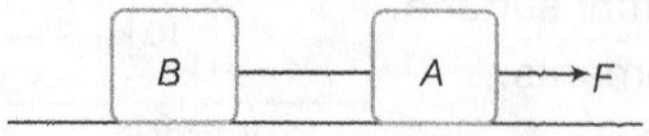

a. If $F = 4$ N, then tension in string is zero and friction on block A is 4 N
b. If $F = 4$ N, then acceleration of blocks is zero and friction on block B is zero
c. If $F = 12$ N, then friction on block A is 5 N and the friction on block B is 7 N
d. If $F = 30$ N, then acceleration of the blocks is 6 m/s^2

11 A block of mass 5 kg is pressed against a vertical wall. The coefficients of static friction and kinetic friction between blocks are 0.5 and 0.4, respectively. Choose the correct statements

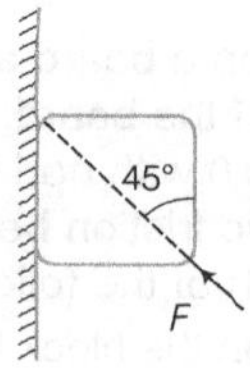

a. If $F = 50\sqrt{2}$ N, then friction on the block is zero
b. If $F = 40\sqrt{2}$ N, then friction on the block is 10 N in upward direction
c. If $F = 20\sqrt{2}$ N, then acceleration of the block is 4.4 m/s^2 in downward direction
d. If $F = 120\sqrt{2}$ N, then acceleration of the block is 4.4 m/s^2 in upward direction

Comprehension Type

Direction *(Q. Nos. 12-23) This section contains 4 paragraphs, each describing theory, experiments, data etc. Twelve questions related to the paragraphs have been given. Each question has only one correct answer among the four given options (a), (b), (c) and (d).*

Passage I

The situation shown in the figure, pulleys P_1 and P_2 are fixed while smooth and light pulley P_3 is movable. Three Blocks m_1, m_2 and m_3 of masses 10 kg, 8 kg and 2 kg respectively, are connected through a light string.

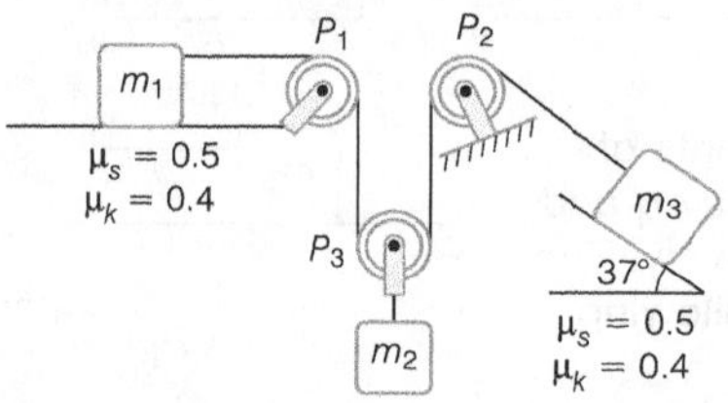

12 With reference to the information given in the passage mark the correct option.
a. m_1 remains stationary but m_3 is moving up along the inclined plane
b. m_2 is moving upward
c. All blocks remain stationary
d. m_1 moves rightward

13 With reference to the information given in the passage mark the correct option.
a. The magnitude of acceleration of m_3 is twice of magnitude of acceleration of m_1
b. The magnitude of acceleration of m_1 is twice of magnitude of acceleration of m_2
c. The magnitude of acceleration of m_3 is equal to magnitude of acceleration of m_1
d. No sufficient information

14 The acceleration of block m_2 is

a. 2 m/s^2 b. 3 m/s^2
c. 2.7 m/s^2 d. 4.8 m/s^2

Working Space

Passage II

The situation shown in the figure $m_1 = 50$ kg, $m_2 = 40$ kg and $m_3 = 20$ kg. The coefficient of friction is μ for both blocks based on inclined plane. The values of $\mu m_1 g\cos\theta_1 = 80$ N, $\mu m_3 g\cos\theta_2 = 30$ N, $m_1 g\sin\theta_1 = 300$ N and $m_3 g\sin\theta_2 = 120$ N.

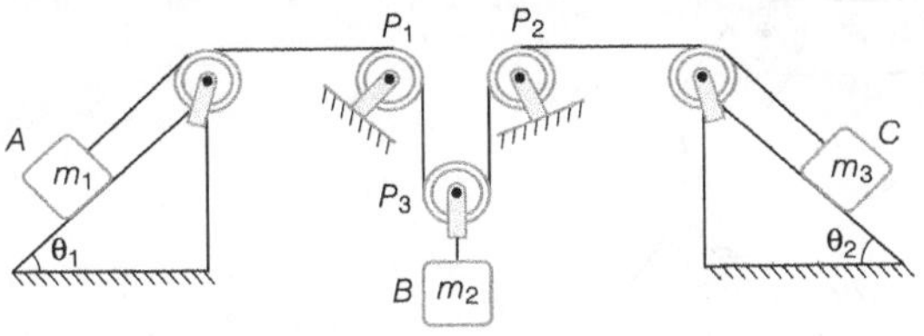

15 with reference to the information given in the passage mark the correct option.

a. The block A moves down along inclined plane and the block C moves up along inclined plane
b. The block B moves upward
c. The block C is stationary but block A moves up along inclined plane
d. All blocks are stationary

16 The tension in the string is

a. zero b. 200 N c. 187.65 N d. 300 N

17 The acceleration of block B is

a. 2 m/s^2 b. 5 m/s^2 c. 0.62 m/s^2 d. 3.72 m/s^2

Passage III

The figure shows two blocks A and B of masses m_1 and m_2 connected by a massless string placed on the two inclined parts of a triangular structure. The coefficient of friction between block A and inclined plane is μ_1 and coefficient of friction between of block m_2 and inclined plane is μ_2. Assume that $m_1 g\sin\theta_1 > m_2 g\sin\theta_2$.

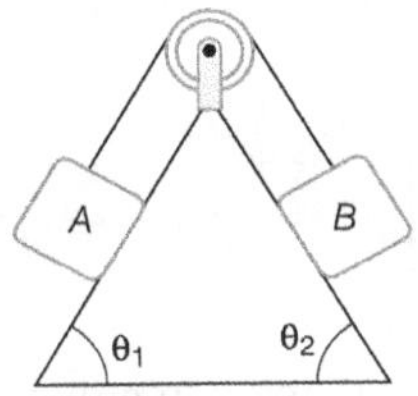

18 If $m_1 g\sin\theta_1 - m_2 g\sin\theta_2 = 50$ N and $\mu_1 m_1 g \cos\theta_1 + \mu_2 m_2 g \cos\theta_2 = 30$ N and $m_1 + m_2 = 12$ kg. The acceleration of block is

a. zero b. 1.67 m/s^2 c. 4.17 m/s^2 d. 6.67 m/s^2

19 If $m_1 g\sin\theta_1 = 90$ N, $\mu_1 m_1 g \cos\theta_1 = 50$ N, $m_2 g\sin\theta_2 = 30$ N, and $\mu_2 m_2 g \cos\theta_2 = 50$ N, then,

a. the friction on block A is 50 N in upward direction along the inclined plane
b. the tension in the string is 40 N
c. the friction on the block B is 10 N in downward direction along the inclined plane
d. All of the above

20 If $m_1 g \sin\theta_1 = 90$ N, $\mu_1 m_1 g \cos\theta_1 = 100$ N, $m_2 g\sin\theta_2 = 30$ N and $\mu_2 m_2 g \cos\theta_2 = 50$ N, then

a. the friction on the block A is 90 N and the friction on the block B is 30 N
b. the tension in the string is non-zero
c. the acceleration of the blocks is non-zero
d. the friction on the block B is downward along inclined plane

Passage IV

In the system of connecting blocks, a block B of mass 3 kg is placed on smooth surface while the coefficient of friction between block A of mass m_1 and inclined plane is μ.

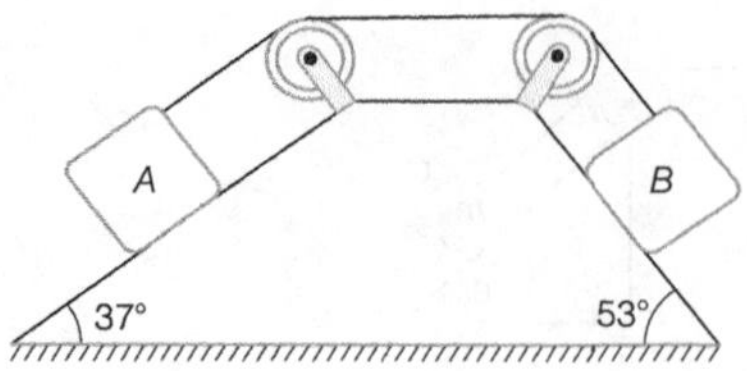

21 If $m_1 = 4$ kg and $\mu = 0.9$, then friction on block A is

a. zero
b. 24 N
c. 28.8 N
d. no sufficient information

22 If $m_1 = 3$ kg and $\mu = 0.5$, then friction on block A is

a. 4 N
b. 6 N
c. 8 N
d. 12 N

23 If $m_1 = 2$ kg and $\mu = 0.5$, then acceleration of blocks is

a. zero
b. 0.8 m/s^2
c. 3.33 m/s^2
d. no sufficient information

Matching List Type

Direction (Q. Nos. 25 and 26) Choices for the correct combination of elements from Column I and Column II are given as options (a), (b), (c) and (d), out of which one is correct.

24 The situation shown in the figure, blocks are connected by light string. The coefficients of static friction and kinetic friction for all surfaces of contact are 0.5 and 0.4, respectively. ($m_A = 1$ kg, $m_B = 2$ kg, $m_C = 3$ kg and $m_D = m$).

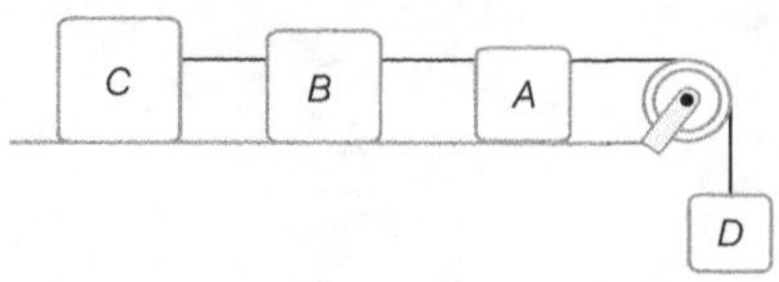

The tension in string when connecting A and B is T_1 and tension in string when connecting B and C is T_2. The friction on block A, B and C are f_A, f_B and f_C, respectively. Then, match the column I with column II and choose the correct options from codes given below:

	Column I		Column II
i.	$m = 0.4$ kg	p.	$T_2 = 0$ and $f_C = 0$
ii.	$m = 0.8$ kg	q.	$T_1 = 0, f_B = 0, f_C = 0$
iii.	$m = 2$ kg	r.	The acceleration of blocks is zero
iv.	$m = 4$ kg	s.	The acceleration of block is 1.6 m / s^2
		t.	$f_A \neq 0, f_B \neq 0$ and $f_C \neq 0$

Codes

	i	ii	iii	iv
a.	p,q,r	p,r	r,t	s,t
b.	p,q	q,t	s,t	p,r
c.	p,q,r	s,t	q,r	p
d.	p,r	q,t	p,s,t	q,r

Working Space

25 A hollow sphere is accelerating horizontally, leftward with constant acceleration a_0. Block A always remain stationary with respect to the sphere. The coefficient of static friction between block A and sphere is μ_s. Then, mach the column I with column II.

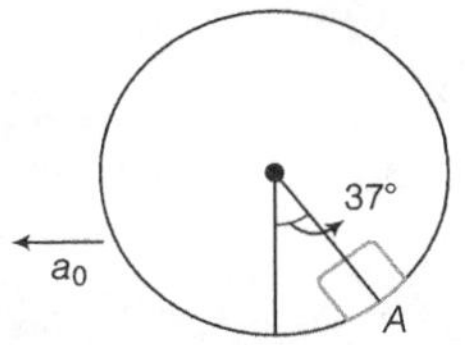

	Column I		Column II
i.	$a_0 = 0$	p.	Friction on block is zero
ii.	$a_0 = g \tan\theta$	q.	Friction on the block is directed downward along surface of sphere
iii.	$a_0 = \dfrac{g(\sin\theta - \mu_s \cos\theta)}{\mu_s \sin\theta + \cos\theta}$	r.	The magnitude of friction on the block is $mg\sin\theta$
iv.	$a_0 = \dfrac{g(\sin\theta + \mu_s \cos\theta)}{\cos\theta - \mu \sin\theta}$	s.	Friction on the block is directed upward along the surface of sphere
		t.	Limiting friction acts on the block

Codes

	i	ii	iii	iv
a.	p	q	r,s	t
b.	r,s	p	s,t	q,t
c.	p,q	q,t	s,t	r,s
d.	p,r,s	r,t	s	q

One Integer Value Correct Type

Direction (Q. No. 27) This section contains 1 question. When worked out will result in an integer from 0 to 9 (both inclusive).

26 A block A of mass 1 kg is pulled up with constant speed by pulling a light rope taken over a fixed pulley as shown in the figure. The coefficient of friction between pulley and the rope is $\frac{2}{\pi}$. The value of F is $9x$ N, then find the value of x. (take, $e = 2.7$)

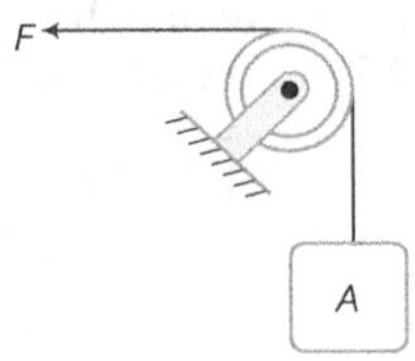

Subjective Questions

Direction (Q. Nos. 28-31) These questions are subjective in nature, need to be solved completely on notebook.

27 In the system of connected blocks shown in the figure, $m_A = 20$ kg, $m_B = 40$ kg and $m_C = 50$ kg. The block A is placed on smooth horizontal surface. The coefficient of friction between block B and ground is μ_2 and that of block C and ground is μ_3.

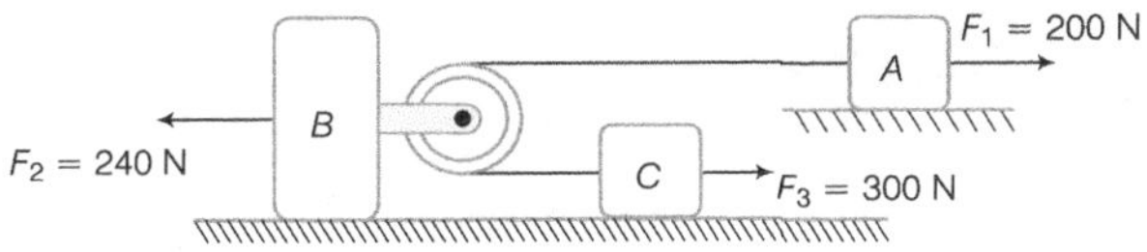

If $\mu_2 m_2 g = 64$ N and $\mu_3 m_3 g = 80$ N, find

i. tension in string connecting A and C and

ii. acceleration of the block A.

Working Space

28 A block of mass $\sqrt{2}$ kg is released from rest inside an inclined right angled trough. The coefficient of friction is μ. Find acceleration of the block if

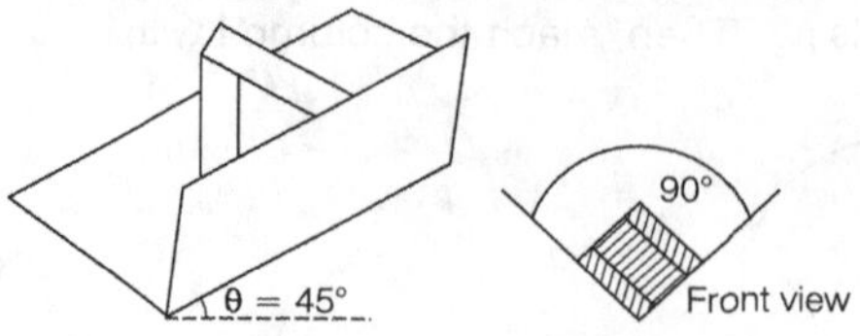

i. $\mu = \frac{1}{\sqrt{2}}$

ii. $\mu = 0.5$

29 Block A of mass m and block B of mass $2\ m$ are placed on a fixed triangular wedge by means of a massless, inextensible string and a frictionless pulley as shown in the figure.

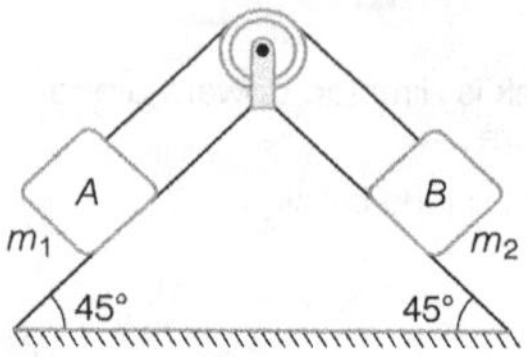

The wedge is inclined at 45° to the horizontal on both sides. The coefficient of friction between block A and the wedge is 2/3 and coefficient of priction between block B and wedges is 1/3. If the blocks A and B are released from rest, find

i. the acceleration of A

ii. tension in the string and

iii. the magnitude and direction of the force of friction acting on A.

30 Two blocks of mass M_1 and M_2 are connected by a massless string, slides down an inclined plane having an angle of inclination of 37°. The masses of the blocks are kg and 2 kg respectively and the coefficients of friction of both blocks with the inclined plane are 0.75 and 0.25, respectively. Assuming the string to be taut, find

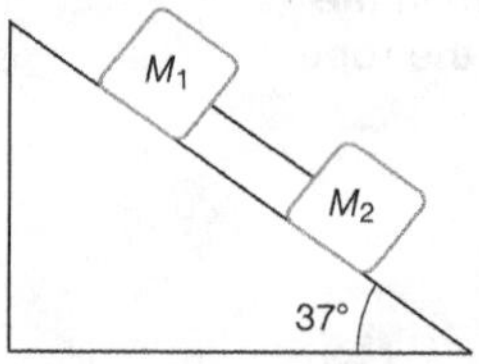

i. the common acceleration of two masses and

ii. the tension in the string. ($\sin 37° = 0.6$, $\cos 37° = 0.8$ and take $g = 9.8\ m/s^2$)

DPP-4 Blocks in Contact

Subjective Questions

Working Space

Direction (Q. Nos. 1-8) *These questions are subjective in nature, need to be solved completely on notebook.*

1 Find maximum value of F, so that block A does not slide on B in following cases.

i. 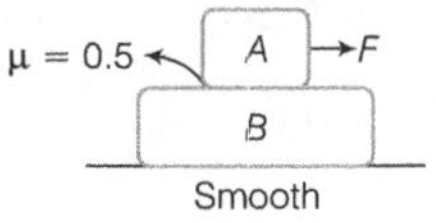

Here, $m_A = 1$ kg, $m_B = 2$ kg, $\mu = 0.5$

ii. 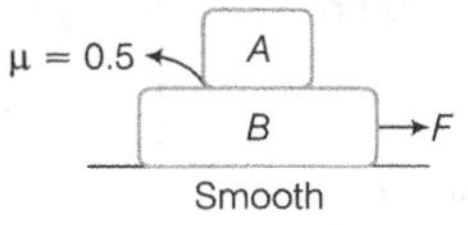

$m_A = 1$ kg, $m_B = 2$ kg, $\mu = 0.5$

iii. 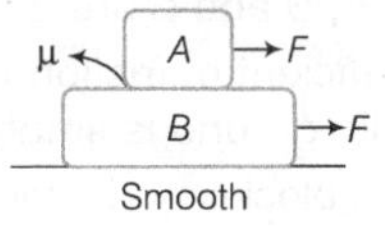

$m_A = 1$ kg, $m_B = 2$ kg, $\mu = 0.5$

iv. 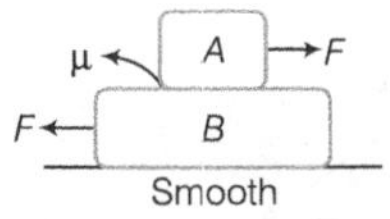

$m_A = 1$ kg, $m_B = 2$ kg, $\mu = 0.5$

v. 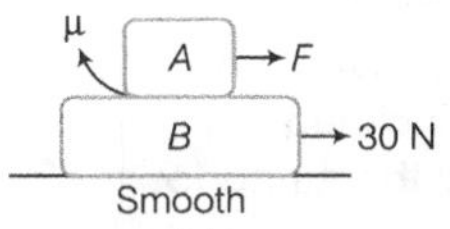

$m_A = 1$ kg, $m_B = 2$ kg, $\mu = 0.5$

vi. 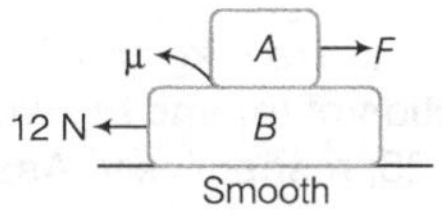

$m_A = 1$ kg, $m_B = 2$ kg, $\mu = 0.5$

vii. 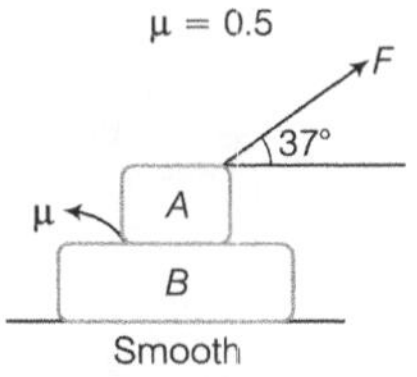

$m_A = 1$ kg, $m_B = 2$ kg, $\mu = 0.5$

viii. 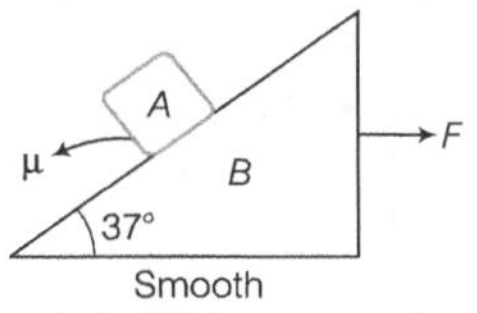

$m_A = 1$ kg, $m_B = 2$ kg, $\mu = 0.5$

2 Find the minimum value of coefficient of friction between blocks A and B, so that A does not slide on B.

i. A block A of mass 1 kg is placed at the bottom of a hollow cube B of mass 2 kg. The coefficient of friction between the cube B and ground is 0.5.

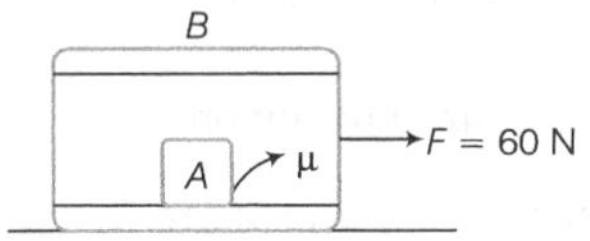

ii. The masses of blocks A and B are 1 kg and 2 kg respectively. The coefficient of friction between ground and block B is 0.3. Blocks are initially at rest. A force of 39 N is applied on block A.

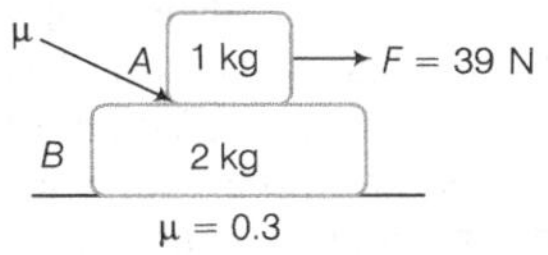

3 Two blocks A and B are initially in contact are placed on an inclined plane as shown in the figure. The coefficient of friction between the block A of mass m_1 and inclined plane is μ_1 and coefficent of friction between the block B of mass m_2 and inclined plane is μ_2. Calculate the acceleration of block B, if

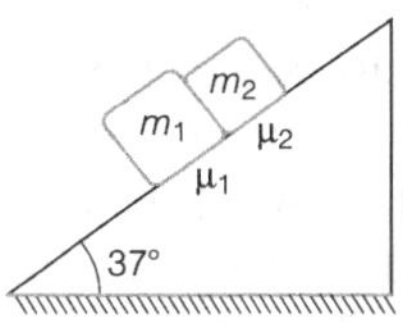

i. $m_1 = 1$ kg, $m_2 = 2$ kg, $\mu_1 = 0.5$ and $\mu_2 = 0$

ii. $m_1 = 2$ kg, $m_2 = 1$ kg, $\mu_1 = \mu_2 = 0.5$

4 A small block is placed on a plate. The plate is placed on a glass vessel. The block lies just above the centre of the vessel as shown in the figure. A boy hits the plate horizontally giving it a large velocity. Suppose the block starts from rest from centre of the vessel. The plate passes swiftly and the block falls in the vessel. If μ_k is coefficient of kinetic friction, then

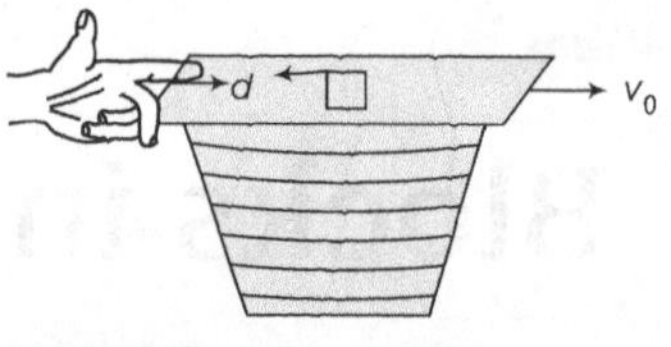

i. prove that the distance moved by the block before falling is $\frac{1}{2}\mu_k g t^2$, where t is the time in which rear edge of the plate crosses the block.

ii. If r is the radius of mouth of vessel and d be the distance of rear edge of plate from the block, find v_0, the minimum velocity given to the plate for successful falling of the block in the vessel. Neglect friction between the plate and the vessel.

5 The masses of block A, B and C are 2 kg, 3 kg and 5 kg respectively. The coefficient of friction for all surfaces of contact is $\mu = 0.5$, while ground is smooth. Find,

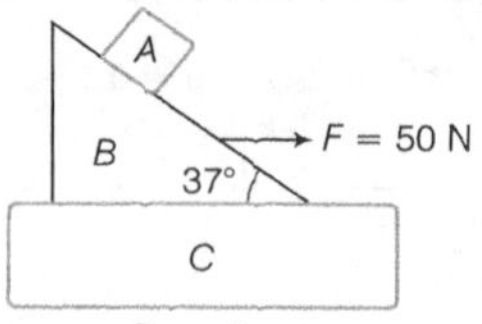

i. acceleration of each blocks   ii. friction on block A

iii. friction on block B

6 Two blocks connected by a massless string slides down an inclined plane having an angle of inclination of 37°. The masses of the two blocks are 4 kg and 2 kg respectively and the coefficients of friction of m_1 and m_2 with the inclined plane are 0.75 and 0.25, respectively. Assuming the string to be taut, find

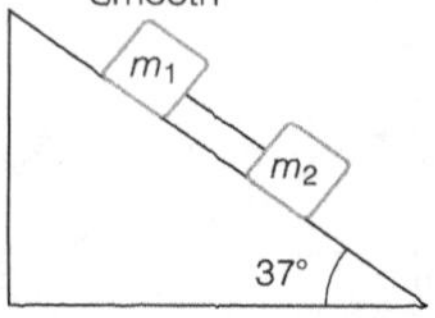

i. the common acceleration of two masses and

ii. the tension in the string. ($\sin 37° = 0.6$, $\cos 37° = 0.8$, take, $g = 9.8\,\text{m/s}^2$)

7 Masses m_1, m_2 and m_3 are connected by strings of negligible mass which passes over massless and frictionless pulleys P_1 and P_2 as shown in the figure. The masses move such that the portion of the string between P_1 and P_2 is parallel to the inclined plane and the portion of the string between P_2 and m_3 is horizontal. The masses m_2 and m_3 are of 4 kg each and the coefficient of kinetic friction between the masses and the surfaces is 0.25. The inclined plane makes an angle of 37° with the horizontal. If the mass m_1 moves downwards with a uniform velocity, find

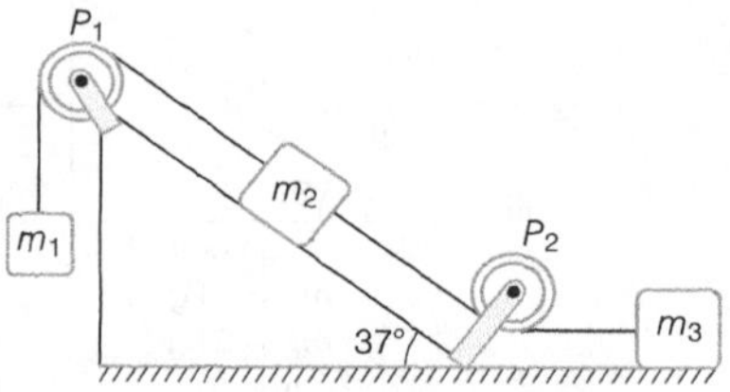

i. the mass of m_1,

ii. the tension in the horizontal portion of the string. (take, $g = 9.8\,\text{m/s}^2$, $\sin 37° \cong 3/5$)

8 In the figure, masses m_1, m_2 and M are 20 kg, 5 kg and 50 kg, respectively. The coefficient of friction between M and ground is zero. The coefficient of friction between m_1 and M and coefficient of the friction between m_2 and ground is 0.3. The pulleys and the strings are massless. The string is perfectly horizontal between P_1 and m_1 and also between P_2 and m_2. **[IIT JEE 2000]**

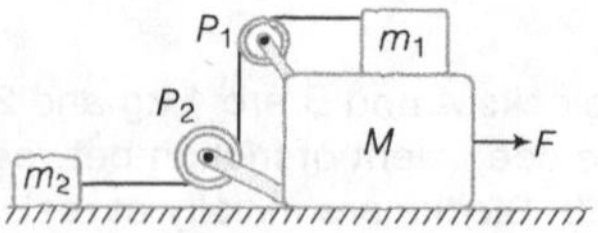

The string is perfectly vertical between P_1 and P_2. An external horizontal force F is applied to the mass M. (take, $g = 10\,\text{m/s}^2$)

i. Draw a free body diagram of mass M, clearly showing all the forces.

ii. Let the magnitude of the force of friction between m_1 and M be f_1 and magnitude of frictional force. between m_2 and ground is f_2. For a particular force F, it is found that $f_1 = 2f_2$, then find f_1 and f_2. Write equations of motion for all the masses. Find tension F in the string and acceleration of the masses.

Only One Option Correct Type

Working Space

Direction (Q. Nos. 9-14) This section contains 6 multiple choice questions. Each question has four choices (a), (b), (c) and (d), out of which ONLY ONE option is correct.

9 If a stationary brick leans against a stationary triangular block on a horizontal surface as shown in the figure. Then,

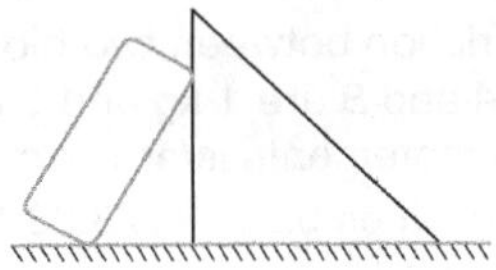

a. the net force on brick is zero
b. the direction of friction acting on brick is same as the friction acting on the triangular block
c. the force on the brick by triangular block is less than that by brick on the triangular block
d. None of the above

10 The mass of triangular block B is 10 kg. The coefficient of friction between A and B is 0.1 and that between triangular block B and ground is 0.4. The maximum value of mass of block A, so that system remains in equilibrium is

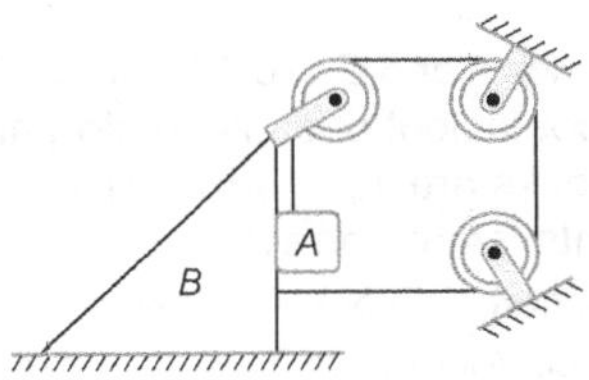

a. 10 kg
b. 3 kg
c. 4 kg
d. 2.5 kg

11 The coefficient of friction between any two surfaces in contact is 0.5. Pulleys and strings are frictionless. Find friction between box A and block B and tension in string connecting box A and block B are respectively (in newton).

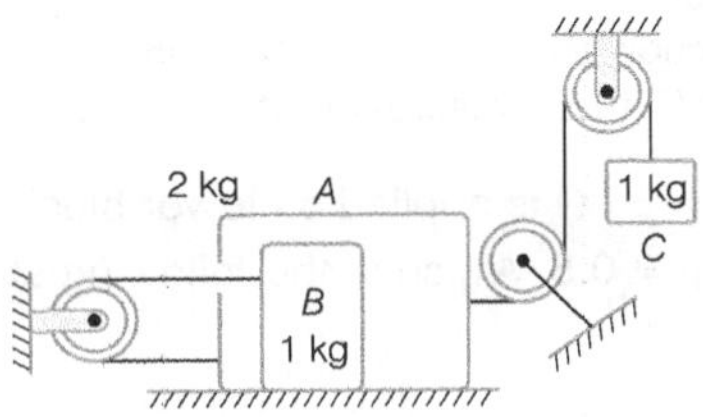

a. 0, 0
b. 0, 5
c. 5, 10
d. 5, 15

12 A block of 5 kg is lying on another block of mass 10 kg. Now, a horizontal force $F = 7.5 \cos^2 \alpha t$ N is applied on 5 kg block, the coefficient of friction between the blocks is $\mu = 0.1$. The floor is smooth and acceleration of 5 kg block w.r.t. 10 kg block at any time is (take $g = 10 \text{ m/s}^2$)

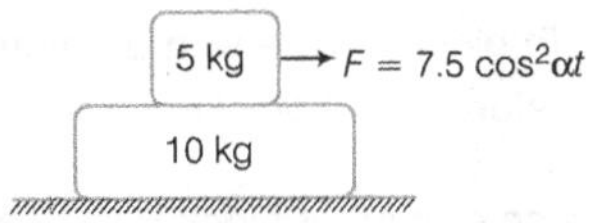

a. zero
b. variable
c. uniform and positive
d. depend upon value of α

13 In the arrangement shown in the figure, there is a friction between the blocks of masses m and $2\,m$ which are in contact. The ground is smooth. The mass of the suspended block is m. The block of mass m which is kept on mass $2m$ is stationary with respect to block of mass $2m$. The force of friction between m and $2\,m$ is (pulleys and strings are light and frictionless)

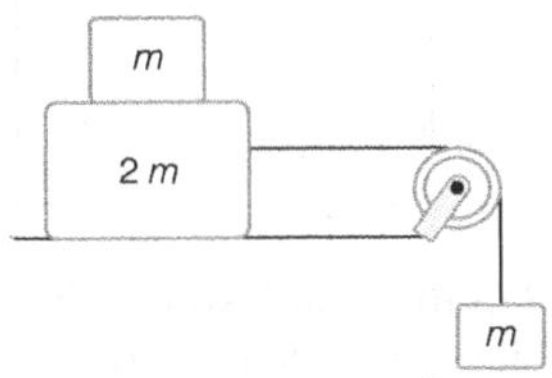

a. $\frac{mg}{2}$
b. $\frac{mg}{\sqrt{2}}$
c. $\frac{mg}{4}$
d. $\frac{mg}{3}$

14 In the situation shown in the figure, the coefficient of static friction for all surfaces of contact is 0.5. Find the maximum value of mass of block A, so that the blocks B, C and D move together.

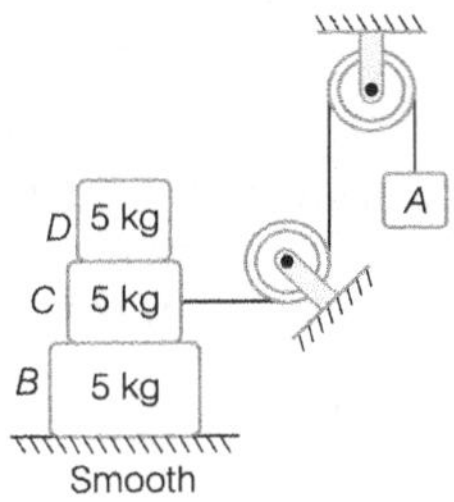

a. 10 kg
b. 15 kg
c. 20 kg
d. 25 kg

One or More than One Options Correct Type

Working Space

Direction (Q. Nos. 15-19) *This section contains 5 multiple choice questions. Each question has four choices (a), (b), (c) and (d), out of which ONE or MORE THAN ONE are/is correct.*

15 The coefficient of friction between two blocks is 0.5. The masses of blocks A and B are 1 kg and 2 kg, respectively. Which of the given statements is/are correct?

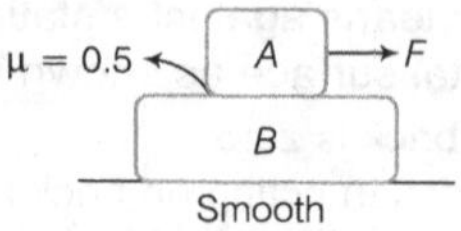

a. If $F = 5$ N, the friction on block B is 3.33 N in rightward direction
b. If $F = 15$ N, the acceleration of block A is 10 m/s^2
c. If $F = 6$ N, the friction on block A is 4 N
d. If $F = 6$ N, the friction on block A is 5 N

16 The masses of blocks A and B are 1 kg and 2 kg, respectively. The coefficient of static friction and kinetic friction between blocks are $\mu_s = 0.5$ and $\mu_k = 0.4$. Which of the given statements is/are correct?

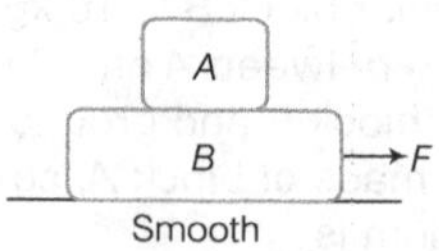

a. If $F = 9$ N, the friction on block A is 3 N in rightward direction
b. If $F = 16$ N, the acceleration of block A is 4 m/s^2 and the block B is 6 m/s^2 in rightward direction
c. If $F = 9$ N, the friction on block A is 5 N in rightward direction
d. For any value of F, the acceleration of B will never be less than that of the block A

17 A variable force $F = 3\,t$ N is applied on lower block B. The coefficient of friction between blocks is $\mu = 0.5$. Which of the following statements is/are correct:

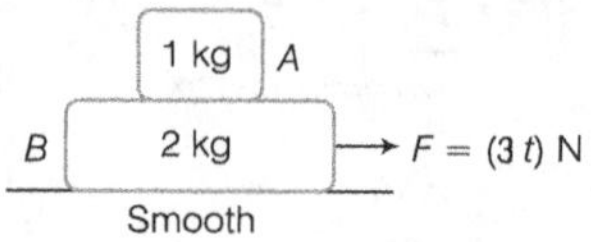

a. For $0 \le t \le 5$ s, both blocks move with a common acceleration
b. For $t = 1$ s, the acceleration of blocks is 1 m/s^2 and friction on block A is 1 N in rightward direction
c. For $t = 10$ s, the acceleration of block B is 12.5 m/s^2
d. For $t = 10$ s, the acceleration of block A is 5 m/s^2

18 In previous problem, velocity-time graph for block A is

a. 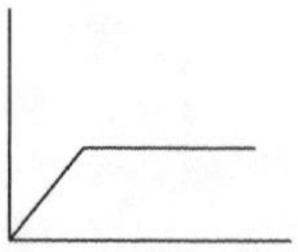 b. 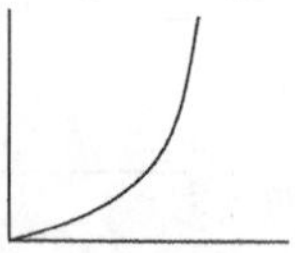 c. 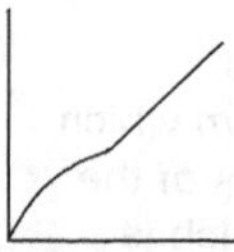 d. 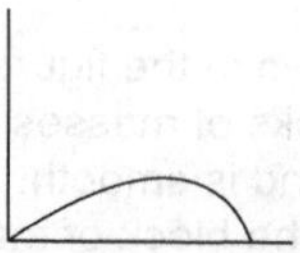

19 Two blocks A and B of masses 2 kg and 3 kg are placed in contact on horizontal surface. The coefficient of static friction and kinetic friction between blocks and ground are 0.5 and 0.4, respectively. Which of the following statements is/are correct?

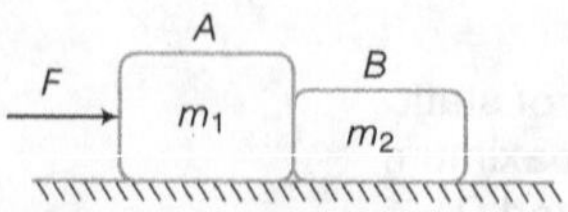

a. If $F = 5$ N, the friction on block A is 5 N and friction on block B is zero
b. If $F = 15$ N, the friction on block A is 10 N and the friction on the block B is 5 N
c. If $F = 40$ N, the force exerted by block A on block B is 24 N and acceleration of the blocks is 4 m/s^2
d. None of the above

Comprehension Type

Working Space

Direction (Q. Nos. 20-26) *This section contains three paragraphs, each describing theory, experiments, data, etc. Seven questions related to the paragraph have been given. Each question has only one correct answer among the four given options (a), (b), (c) and (d).*

Passage I

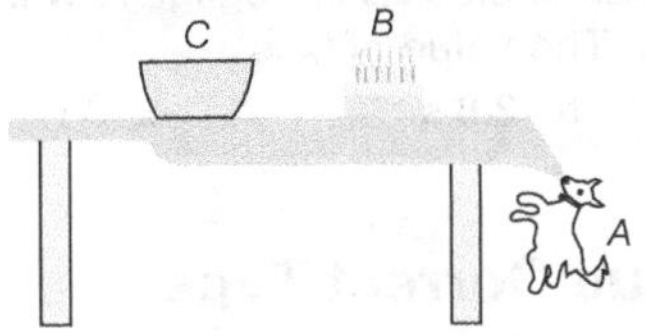

In the birthday party of Rishu, a birthday cake (M_B = 3 kg) and a punch bowl (M_C = 7 kg) rest on a table cloth with a table height of 2 m as shown in the figure. His puppy named Teddy (M_A = 5 kg) jumps up and grabs the edge of a table cloth and hangs off the edge at the table. The friction between table cloth and the bowl and between the table cloth and the cake plate is such that the bowl and plate do not move relative to the table cloth. The table cloth sides across the polished table with a coefficient of sliding friction 0.20. (g = 10 m/s^2)

20 Acceleration of the system is

a. 10/3 m/s^2 b. 3 m/s^2 c. 2 m/s^2 d. 2.4 m/s^2

21 The tension in the table cloth between the puppy and the cake is

a. 20 N b. 40 N c. 28 N d. 14 N

22 The tension in the table cloth between the cake and bowl is

a. 40 N b. 14 N c. 12 N d. 28 N

Passage II

The masses of blocks *A*, *B* and *C* are 1 kg, 2 kg and 3 kg. The coefficient of friction between contact surfaces of blocks is 0.5.

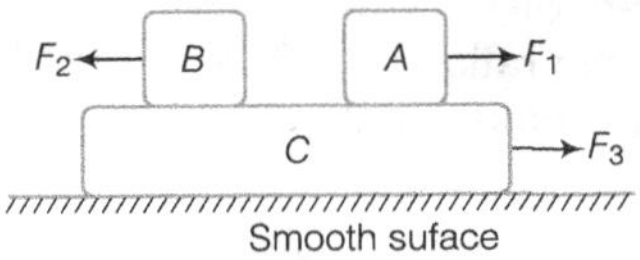

23 If $F_1 = 1$ N, $F_2 = 5$ N and F_3 = 6 N, then

a. the blocks move with common acceleration of magnitude $\frac{1}{3}$ m/s^2

b. the friction on block *A* is $\frac{2}{3}$ N, in leftward direction

c. the friction on block *B* is 5.67 N in rightward direction

d. All of the above

24 If F_1 = 3 N, F_2 = 20 N and F_3 = 10 N, then

a. the acceleration of block *B* is 5 m/s^2 in leftward direction

b. the acceleration of block *C* is 5.75 m / s^2 in rightward direction

c. the friction on block *A* is 2.75 N in rightward direction

d. None of the above

Passage III

Two blocks *A* and *B* are shown in the figure. The coefficient of friciton between *A* and *B* is 0.1 and cofficient of friction between ground and block *A* is 0.2. The system set in motion with a velocity of 5 m/s at t = 0. (g = 10 m/s^2)

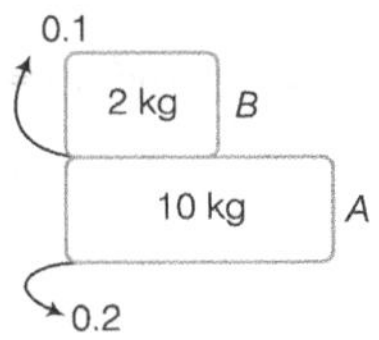

Working Space

25 Suppose that $t = t_0$, the block B drops from block A. At $t = t_1$ $(0 < t_1 \le t_0)$ the force of friction on B is

a. 2 N in leftward
b. 2 N in rightward
c. Zero
d. 24 N in rightward

26 Assume that the size of block B is negligible with respect to that of A. The length of block A is 1.6 m. The value of t_0 is

a. 1.63 s
b. 2.8 s
c. 3 s
d. ∞

One Integer Value Correct Type

Direction (Q. No. 27-29) This section contains 3 questions. When worked out will result in an integer from 0 to 9 (both inclusive).

27 In the given figure, the magnitude of force of friction between m_1 and M is f_1 and magnitude of force of friction between m_2 and ground is f_2. For the value of $F = 10n$ N, $f_1 = 2f_2$ (take, $g = 10\ \text{m/s}^2$). Find the value of n. (Here, $m_1 = 20$ kg, $m_2 = 5$ kg and $M = 50$ kg)

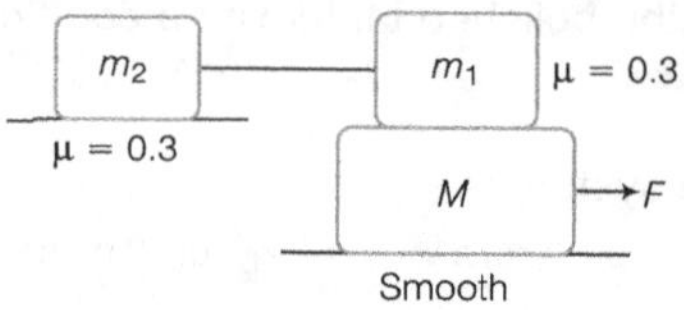

28 The situation shown in the figure, a block A of mass 2 kg is placed inside a box B of mass 2 kg. The coefficient of static and kinetic friction between the blocks and also between box and ground are $\mu_s = 0.6$ and $\mu_k = 0.4$. If $F = 10x$ N applied on the box B, the block A begins to slide relative to the box. Find the value of x.

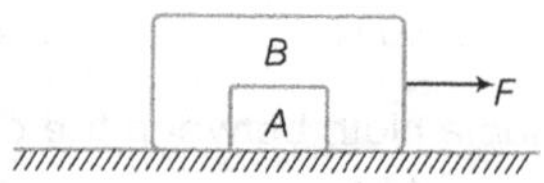

29 Two blocks A and B have coefficient of friction $\mu_2 = 0.5$ between them the block A and wall have coefficient of friction $\mu_1 = 0.25$. The applied force F is 100 N. If the ratio of magnitude of acceleration of blocks A and B is n. Find the value of n.

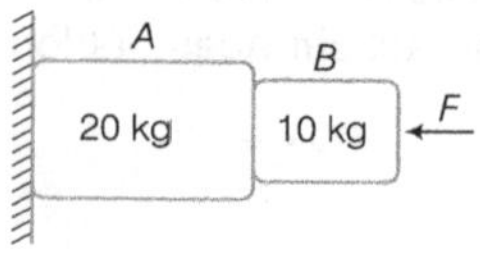

Matching List Type

Direction (Q. Nos. 30-34) Choices for the correct combination of elements from Column I and Column II are given as options (a), (b), (c) and (d), out of which one is correct.

30 The situation shown in the figure, blocks are released from rest.

Here, $m_A = 1$ kg and $m_B = 2$ kg,

Match the column I with column II and choose the correct option from codes given below.

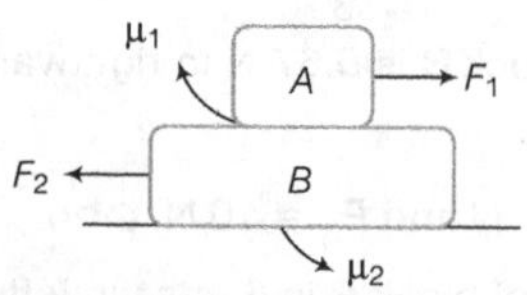

	Column I		Column II
i.	$F_1 = F_2 = 4$ N, $\mu_1 = 0.5$ and $\mu_2 = 0$	p.	Static friction between blocks.
ii.	$F_1 = 2$ N, $F_2 = 12$ N, $\mu_1 = 0.5$ and $\mu_2 = 0$	q.	Kinetic friction between blocks.
iii.	$F_1 = 5$ N, $F_2 = 0$, $\mu_1 = 0.1$ and $\mu_2 = 0.3$	r.	The acceleration of block A is greater than that of block B.
iv.	$F_1 = 0$, $F_2 = 12$ N, $\mu_1 = 0.4$ and $\mu_2 = 0.1$	s.	Acceleration of block B is zero.

Codes

	i	ii	iii	iv
a.	p,s	q	q,r,s	p
b.	q,s	p	q,r,s	p
c.	p,s	q,r	p	q
d.	p,r	q,s	q	p

Working Space

31 Two masses m_1 and m_2 are connected by a light rod. The coefficient of friction between m_1 and inclined plane is μ_1 and that for m_2 and the inclined plane is μ_2.

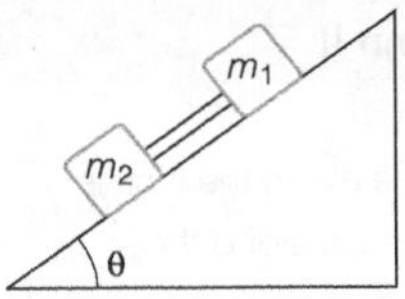

Match the Column I with Column II and choose the correct options from codes given below.

	Column I		Column II
i.	$\mu_1 = 0, \mu_2 = 0, m_1 > m_2$	p.	Blocks remain stationary.
ii.	$m_1 = m_2 = m, \mu_1 > \mu_2, \sin\theta \le \frac{(\mu_1 + \mu_2)\cos\theta}{2}$	q.	Acceleration of blocks is $g\sin\theta$.
iii.	$m_1 = m_2 = m, \mu_1 < \mu_2$, and $\sin\theta > \left(\frac{\mu_1 + \mu_2}{2}\right)\cos\theta$	r.	Tension on m_1 is directed towards m_2.
iv.	$(m_1 + m_2)\sin\theta \le (\mu_1 m_1 + \mu_2 m_2)\cos\theta$ and $\mu_1 \ne 0, \mu_2 = 0$	s.	Tension in rod on m_1 is directed down the inclined plane.

Codes

	i	ii	iii	iv
a.	q,r	p,q	s,r	p,r
b.	p,r	q,r	p,r	q,r
c.	q,s	p,s	q,s	p,s
d.	p,s	q,s	p,s	q,s

32 The situation shown in the figure, the coefficient of friction between box B of mass 20 kg and block A of mass 10 kg is 0.3 and between box B and surface is 0.1. Match the column I with column II and choose the correct options from codes given below:

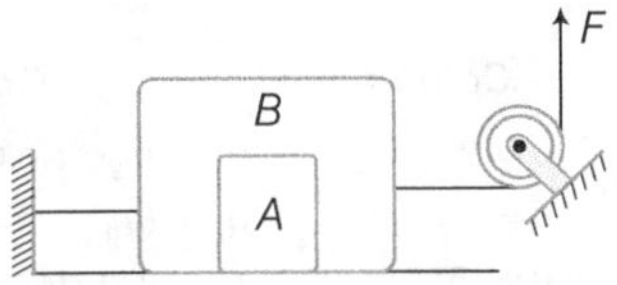

	Column I		Column II
i.	Tension in left string is $T = 0$	p.	$F = 20$ N
ii.	Friction between A and $B = 20$ N	q.	$F = 30$ N
iii.	Friction between block and surface = 30 N	r.	$F = 40$ N
iv.	Acceleration of A = zero	s.	$F = 0$

Codes

	i	ii	iii	iv
a.	p,q,r,s	p	q,r	p,q,s
b.	p,q,r,s	q	r	p,q,s
c.	p,s	p	q	r
d.	q,r	q	q,r	p,q,s

33 For the arrangement as shown in the figure, the coefficient of kinetic friction between A and B and between B and incline is same. Block A of mass 3 kg is sliding down with constant velocity. Block B is having a mass of 1 kg. (take $g = 10$ m/s^2).

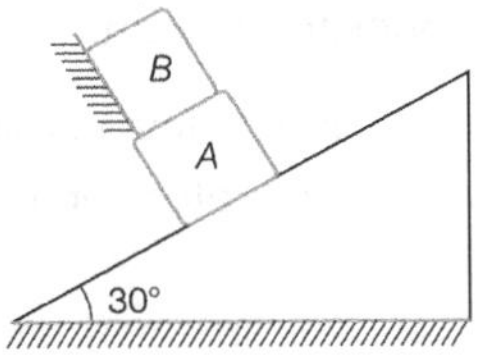

For this situation, match the entries for Column I with the entries of Column II and choose the correct option from codes given below.

Working Space

	Column I		Column II
i.	Friction force on *B* has	p.	magnitude 12 N
ii.	Total friction force on *A* has	q.	direction upwards along the incline
iii.	Friction force on the incline has	r.	direction downwards along the incline
iv.	Friction force on *A* due to the incline has	s.	magnitude 15 N

Codes

	i	ii	iii	iv
a.	r	p,q	q,s	p,r
b.	r	q,s	r	q
c.	q	q,s	p,r	p,q
d.	r	r,s	p,r	p,q

34 A block of mass $m_1 = 1$ kg another mass $m_2 = 2$ kg are placed together (see in the figure) on an inclined plane with angle of inclination θ. Various values of θ are given in Column I. The coefficient of friction between the block m_1 and the plane is always zero. The coefficient of static and dynamic friction between the block m_2 and the plane are equal to $\mu = 0.3$.

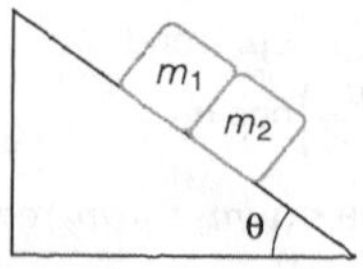

[JEE Advanced 2014]

In Column II, expressions for the friction on the block m_2 are given. Match the correct expression for the friction in Column II with the angles given in Column I, and choose the correct option. The acceleration due to gravity is denoted by *g*. [take $\tan(5.5°) \approx 0.1$, $\tan(11.5°) \approx 0.2$, $\tan(16.5°) \approx 0.3$]

	Column I		Column II
i.	$\theta = 5°$	p.	$m_2 g \sin\theta$
ii.	$\theta = 10°$	q.	$(m_1 + m_2) g \sin\theta$
iii.	$\theta = 15°$	r.	$\mu m_2 g \cos\theta$
iv.	$\theta = 20°$	s.	$\mu(m_1 + m_2) g \cos\theta$

Codes

	i	ii	iii	iv
a.	p	p	p	r
b.	q	q	q	r
c.	q	q	q	s
d.	q	q	r	r

Statement Type

Direction (Q. Nos. 35 and 36) *This section is based on Statement I and Statement II. Select the correct answer from the codes given below.*

a. Both Statement I and Statement II are correct and Statement II is the correct explanation of Statement I
b. Both Statement I and Statement II are correct and Statement II is not the correct explanation of Statement I
c. Statement I is correct but Statement II is incorrect
d. Statement I is incorrect but Statement II is correct

35 **Statement I** A cloth covers a table. Some dishes are kept on it. The cloth can be pulled out without dislodging the dishes from the table.

Statement II For every action, there is an equal and opposite reaction.

36 **Statement I** It is easier to pull a heavy object than to push it on a level ground.

Statement II The magnitude of frictional force depends on the nature of the two surfaces in contact.

DPP-5 Friction in Daily Life

Conceptual Question

Direction (Q. No. 1) This question is conceptual in nature, need to be solved completely on notebook.

Working Space

1 **Situation :** The coefficient of static friction between feet of PT Usha of mass m and the ground is μ_s. Mark the correct option(s) for the following questions.

i. The frictional force between feet of PT Usha and ground is
a. static b. dynamic c. both d. None of these

ii. If she walks with constant velocity, the frictional force on her is
a. $\mu_s mg$ b. zero c. $> \mu_s mg$ d. None of these

iii. Assume that she starts from rest and speed up continuously. The direction of friction is
a. in the direction of her velocity b. in opposite direction of her velocity
c. zero d. None of these

iv. In previous problem, if she accelerates with constant acceleration a_0. The value of frictional force acting on her legs is
a. $f = ma_0$ b. $f = 0$ c. $f < ma_0$ d. None of these

v. The maximum value of frictional force f is
a. $\mu_s mg$ b. $< \mu_s mg$ c. $> \mu_s mg$ d. None of these

vi. The maximum value of her acceleration a_0 is
a. Zero b. $\mu_s g$ c. $> \mu_s g$ d. None of these

vii. If she starts from rest, the minimum time in which she gains a velocity v_0 is
a. zero b. $v_0/\mu_s g$ c. $< v_0/\mu g$ d. None of these

viii. In previous problem, the minimum time taken by her to complete a race of 100 m is (take, $g = 10\ \text{m/s}^2$, $\mu_s = 0.8$)
a. 10 s b. 5 s c. 20 s d. None of these

ix. In previous problem, suppose she takes exactly minimum time to complete the 100 m race. Find next minimum time in which she will come in rest.
a. 5 s b. 10 s c. 20 s d. zero

x. In previous problem, during the period of deceleration, the frictional force is
a. static friction acting in the direction of her velocity
b. dynamic friction acting in the direction of her velocity
c. static friction acting in the opposite direction of her velocity
d. dynamic friction acting in opposite direction of her velocity

Only One Option Correct Type

Direction (Q. Nos. 2-13) This section contains 12 multiple choice questions. Each question has four choices (a), (b), (c) and (d), out of which ONLY ONE option is correct.

2 A car is moving with speed v_0 as shown in the figure. The coefficient of friction between road and tyre of car is $\mu_s < \tan\theta$. When car enters on the inclined plane of length l, brake is applied by driver. Find minimum speed of car at the bottom of inclined plane.

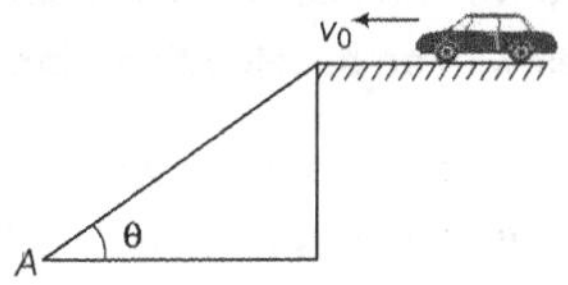

a. $\sqrt{v_0^2 + 2gl(\sin\theta + \mu_s \cos\theta)}$ b. $\sqrt{v_0^2 - 2gl(\sin\theta - \mu_s \cos\theta)}$
c. $\sqrt{v_0^2 + 2gl(\sin\theta - \mu_s \cos\theta)}$ d. None of these

Working Space

3 A car starts with speed v_0 and moves on a horizontal road in which the coefficient of static friction between road and the tyres increases linearly with distance x as $\mu = \mu_0 x$. During journey, car moves with maximum possible acceleration. The speed of car as a function of x is

a. $\sqrt{v_0^2 + \mu_0 g x^2}$ b. $\sqrt{v_0^2 - \mu_0 g x^2}$

c. $\sqrt{2v_0^2 + \mu_0 g x^2}$ d. None of these

4 A car is travelling with a speed of 10 m/s. Suddenly the brakes are applied by driver causing the tyres to skid. Find stopping distance. The coefficient of friction between tyres and road is $\mu = 0.5$.

a. 10 m b. 20 m c. 30 m d. 40 m

5 An insect crawls from top of a fixed hemispherical umbrella of radius $\frac{30}{\pi}$ cm very slowly. The coefficient of static friction between the insect and surface of umbrella is $\mu_s = \frac{1}{\sqrt{3}}$. Find distance travelled by the insect before sliping.

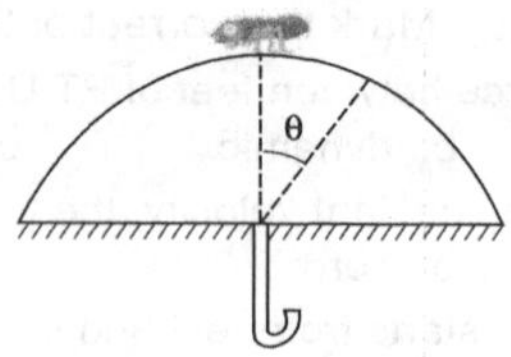

a. 1 cm b. 2 cm c. 3 cm d. 5 cm

6 An insect crawls up a hemispherical surface very slowly as shown in the figure. The coefficient of static friction between the surface and the insect is 1/3. If the line joining the centre of the hemispherical surface to the insect makes an angle α with the vertical, the maximum possible value of α is given by

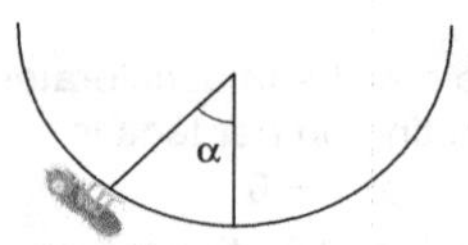

[IIT JEE 2001]

a. $\cot\alpha = 3$ b. $\tan\alpha = 3$

c. $\sec\alpha = 3$ d. $\operatorname{cosec}\alpha = 3$

7 A light plank is placed on a smooth horizontal surface. The plank is in contact of a sugar cube of mass 4 g. The coefficient of static friction between sugar cube and the horizontal ground is 0.001. A bug of mass 1 g is moving on the plank away from the sugar cube. Find the maximum acceleration of the bug, so that cube does not move.

a. $10\,\text{cm/s}^2$ b. $2\,\text{cm/s}^2$ c. $3\,\text{cm/s}^2$ d. $4\,\text{cm/s}^2$

8 Three bugs A, B and C of masses 3 g, 5 g and 2 g respectively are moving on a horizontal cardboard placed on smooth table. Bug A is moving with an acceleration of $1\,\text{cm/s}^2$ in East direction, while bug C is moving with an acceleration of $2\,\text{cm/s}^2$ in North direction. Bug B is moving such that the cardboard remains stationary. Mark the correct option.

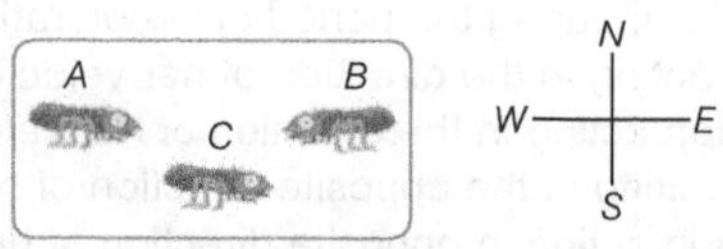

a. The bug B remains stationary

b. The bug B moves with an acceleration $2.5\,\text{cm/s}^2$ in a direction 53° South of West

c. The bug B moves with an acceleration $1\,\text{cm/s}^2$ in a direction 53° South of West

d. The bug B moves with an acceleration $1\,\text{cm/s}^2$ in South-West direction

9 A monkey of mass 10 kg slides down with constant velocity on a light rope. The coefficient of kinetic friction is $\mu_k = 0.5$. The contact force exerted by monkey on the rope is

a. 100 N b. 50 N

c. 250 N d. 1000 N

Working Space

10 In a tug war team *A* of total mass m_1 is found to be winning over the team *B* of total mass m_2. The coefficient of friction for team *A* and ground is μ_1 and that for team *B* and ground is μ_2, then

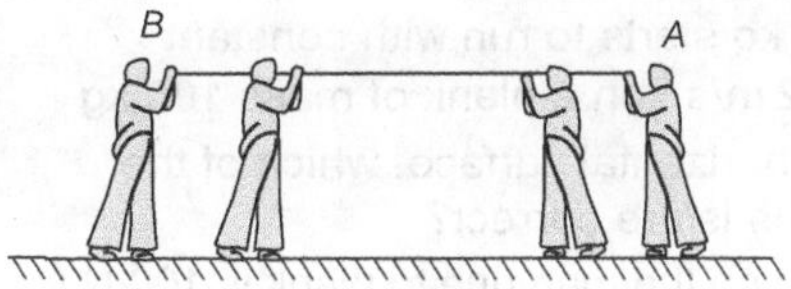

a. $\mu_1 m_1 > \mu_2 m_2$
b. $\mu_1 m_1 = \mu_2 m_2$
c. $\mu_1 m_1 < \mu_2 m_2$
d. No sufficient information

11 Suppose you are standing in a compartment of moving metro train. The moving train decelerates to stop. Friction on your feet is

a. in the direction of motion of train
b. in the opposite direction of motion of train
c. in perpendicular direction of motion of train
d. No sufficient information

12 A horse try to pull a wagon as shown in the figure. The mass of horse is m_1 and that of wagon is m_2. The coefficient of static friction between horse and road is μ_1 and coefficient of static friction between wagon and the road is μ_2. In which of the following cases, wagon is certainly not pulled by the horse?

a. $\mu_1 < \mu_2, m_1 < m_2$
b. $\mu_1 > \mu_2, m_1 < m_2$
c. $\mu_1 < \mu_2, m_1 > m_2$
d. $\mu_1 > \mu_2, m_1 > m_2$

13 During padding of a bicycle, the force of friction exerted by the ground on the two wheels is such that it acts **[IIT JEE 1990]**

a. in the backward direction on the front wheel and in the forward direction on the rear wheel
b. in the forward direction on the front wheel and in the backward direction on the rear wheel
c. in the backward direction on both the front and the rear wheels
d. in the forward direction on both the front and the rear wheels

One or More than One Options Correct Type

Direction (Q. Nos. 14-17) This section contains 4 multiple choice questions. Each question has four choices (a), (b), (c) and (d), out of which ONE or MORE THAN ONE are correct.

14 The maximum speed of a horse is 16 m/s. The coefficient of static friction between feet of the horse and the horizontal ground is 0.8. If horse starts from rest, then

a. it takes at least 2 s to achieve its full speed
b. it takes at least 4 s to complete a race of 112 m
c. it takes at least $\sqrt{28}$ s to complete a race of 112 m
d. it takes at least 8 s to complete a race of 112 m

15 A boy of mass 20 kg is running up without slipping on an inclined plane of inclination 37°. The coefficient of static friction between feet of the boy and the inclined plane is $\mu_s = 0.9$.

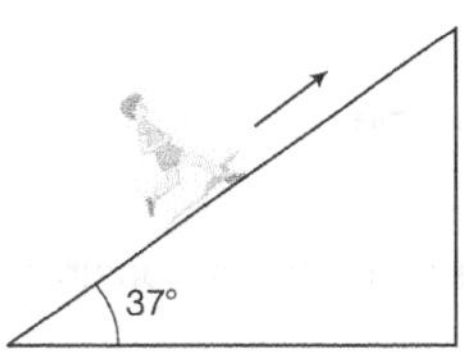

Which of the following statements is/are correct?

a. If boy is running with constant velocity, the friction on him exerted by inclined plane is 120 N up along inclined plane
b. If boy is running with an acceleration 10 m/s^2, the friction on him exerted by the inclined plane is 140 N up along inclined plane
c. The maximum acceleration of the boy is 1.2 m/s^2
d. None of the above

16 A man of mass 50 kg starts to run with constant acceleration $a_0 = 2\ m/s^2$ on a plank of mass 100 kg placed on smooth horizontal surface. which of the following statements is/are correct?

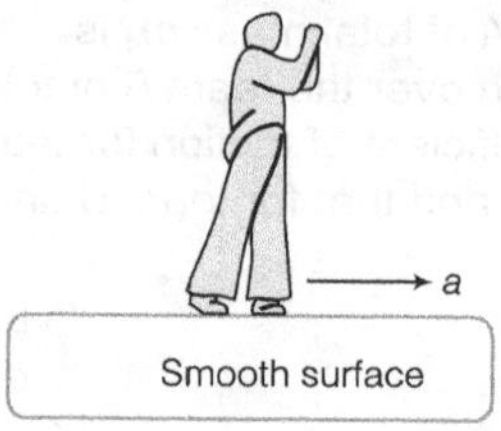

a. The static friction on the man due to plank is 100 N in the direction of his acceleration
b. The acceleration of the plank is 1 m/s^2 in opposite direction of acceleration of man
c. The coefficient of static friction between man and plank is at least 0.2
d. Plank remains stationary

17 A monkey of mass 10 kg is climbing up on a light rope with an acceleration of 2 m/s^2. Which of the following statements is/are correct?

a. The static friction exerted by rope on monkey is in downward direction
b. The static friction exerted by rope on monkey is 120 N in upward direction
c. If the coefficient of friction between monkey and rope is 0.8, the normal contact force exerted by rope on monkey is at least 150 N
d. None of the above

Comprehension Type

Direction (Q. Nos. 18-20) This section contains a paragraph, describing theory, experiments, data, etc. Three questions related to the paragraph have been given. Each question has only one correct answer among the four given options (a), (b), (c) and (d).

Passage

A rat is at the top of a block, having rough inclined planes. The block is placed on smooth horizontal surface. Two cats *A* of mass 20 kg and *B* of mass 40 kg are near the bottom of the block. They see the rat sitting at top of the block, whose inclined planes are rough simultaneously and start moving towards the rat along the inclined planes without sliping. So that the triangular block remains stationary. The coefficient of friction between legs of cats and the inclined planes are $\mu_s = 2$. (assume that rat is always stationary with respect to the block)

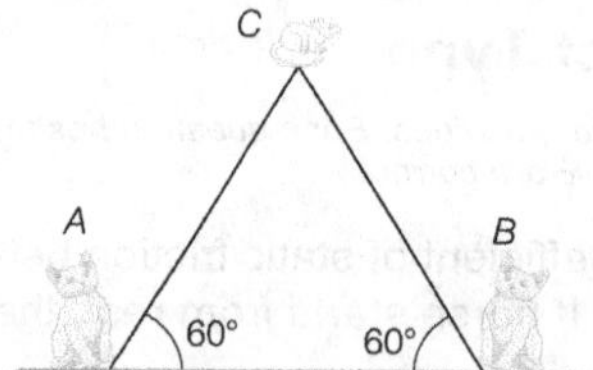

18 Which cat will catch the rat, if both want to catch the rat as soon as possible?
a. Cat *A*
b. Cat *B*
c. both at the same time
d. Neither cat *A* nor cat *B*

19 Both want to reach as soon as possible. The ratio of acceleration of cat *A* and cat *B* is

a. $\frac{1}{4}$
b. $\frac{1}{3}$
c. $\frac{1}{2}$
d. 2

Working Space

20 What is the difference of the horizontal components of the forces of friction on cats *A* and *B* when both cats are at rest near the top of block?

a. 10 N b. 20 N
c. 50 N d. $50\sqrt{3}$ N

One Integer Value Correct Type

Direction (Q. Nos. 21 and 22) This section contains 2 questions. When worked out will result in an integer from 0 to 9 (both inclusive).

21 A straw of mass 1 g and length 4 cm lies on a smooth horizontal surface. A bug of mass 3 g starts to move from one end of the straw. The coefficient of friction between straw and the bug is $\mu = 0.05$. The minimum time in which it can cross the straw length is $\sqrt{100n}$ s. Find the value of n.

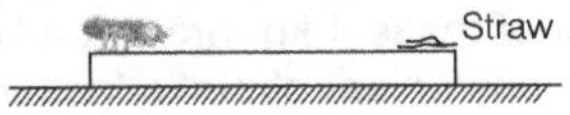

22 A monkey of mass $10\sqrt{3}$ kg is climbing up the corner formed by two intersection of two vertical perpendicular walls. The coefficient of friction is $\sqrt{2}$. The minimum force (in newton) with which the monkey had to push on the walls whilst climbing is $3x$ N, find the value of x.

Statement Type

Direction (Q. Nos. 23 and 24) This section is based on Statement I and Statement II. Select the correct answer from the codes given below.

a. Both Statement I and Statement II are correct and Statement II is the correct explanation of Statement I
b. Both Statement I and Statement II are correct and Statement II is not the correct explanation of Statement I
c. Statement I is correct but Statement II is incorrect
d. Statement II is correct but Statement I is incorrect

23 **Statement I** The application of brakes on the driving wheels of a bus means the reversal of the direction of friction between the wheels and the ground.

Statement II The application of brakes means the application of force on the axle in the opposite direction. Consequently the frictional force is also reversed.

24 **Statement I** If a monkey climbs up on a fixed vertical rope, the normal contact force between monkey and rope is zero.

Statement II Static frictional force is exerted by monkey on the rope.

Subjective Questions

Direction (Q. Nos. 25-28) These questions are subjective in nature, need to be solved completely on notebook.

25 A horse rider of mass 50 kg on his horse of mass 250 kg whips it. The horse is running on the rough horizontal surface without slipping. The horse is speed up rightward with constant acceleration a_0. The coefficient of static friction between horse rider and the horse is 0.5.

i. Find maximum acceleration of horse, so that horse rider does not slip over the horse.
ii. If horse is running with the acceleration calculated in part (i). Find the friction exerted by ground on the horse.

Working Space

26 In the case of paddling, the rear wheel of a bicycle feels forward static friction of $\frac{1}{5}$ of the weight of cyclist system by the horizontal road. The static friction on the forewheel is in backward direction. The value of static friction on forewheel is $\frac{1}{20}$ of the weight of cyclist system. If cyclist starts from rest,

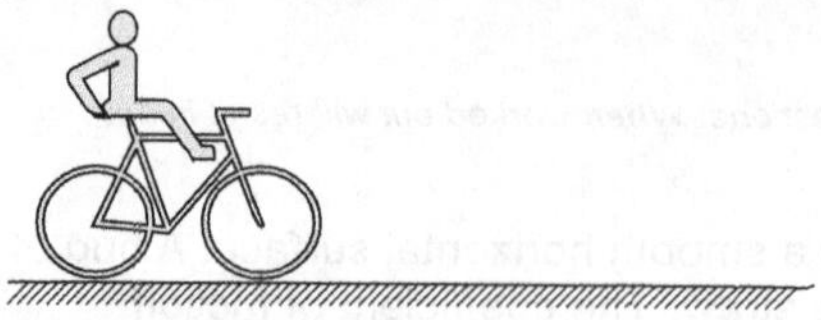

i. find speed after 10 s.
ii. find the distance covered in 2 min.

27 A plank of mass $m_1 = 10$ kg is connected to a block of mass 4 kg through a light string passing over a smooth and fixed pulley. The coefficient of static friction between plank and horizontal surface is 0.05. A boy of mass 10 kg runs rightward on the plank without sliping. The plank remains stationary during motion of boy.

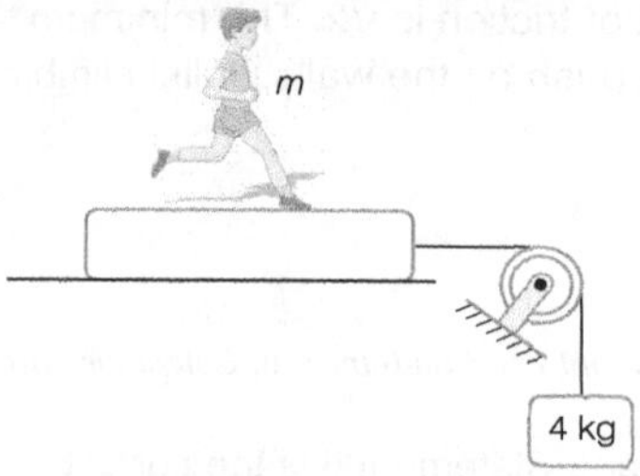

i. Find minimum acceleration of the boy.
ii. Find maximum acceleration of the boy.

28 A rat of mass 10 g runs down on a plate of 50 g placed on a rough inclined plane, so that the plate does not slip over the inclined plane. The coefficient of static friction between the plate and inclined plane is $\mu_S = 0.5$.

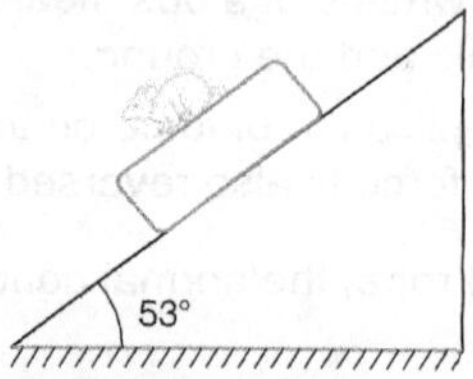

i. Find minimum acceleration of rat.
ii. Find the maximum acceleration of rat.

Revisal Problems
for JEE Main

Only One Option Correct Type

Working Space

Direction (Q. Nos. 1-20) *This section contains 20 multiple choice questions. Each question has four choices (a), (b), (c) and (d), out of which ONLY ONE option is correct.*

1 If you push your pillow by a horizontal force **F** on your bed, pillow does not move, but gets deformed. If at the time of maximum deformation, the frictional force by the bed on pillow is $\mathbf{f}_1$. Then,

a. $\mathbf{F} = \mathbf{f}_1$
b. $|\mathbf{F}| > |\mathbf{f}_1|$
c. $\mathbf{F} + \mathbf{f}_1 = 0$
d. $|\mathbf{f}_1| < |\mathbf{F}|$

2 A block A of mass 10 kg is placed on a rough horizontal surface. The coefficient of friction between the block A and horizontal surface is $\mu = 0.04$. At $t = 0$, the block is given velocity of 10 m/s in rightward direction along the surface. Choose the correct graph for block A.

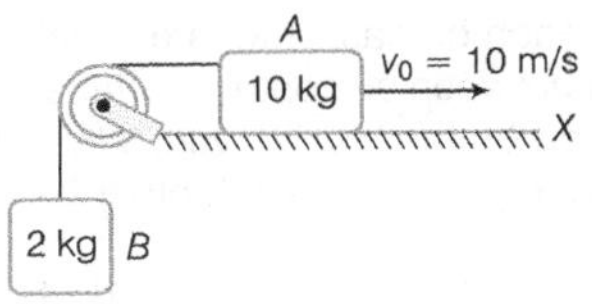

a.
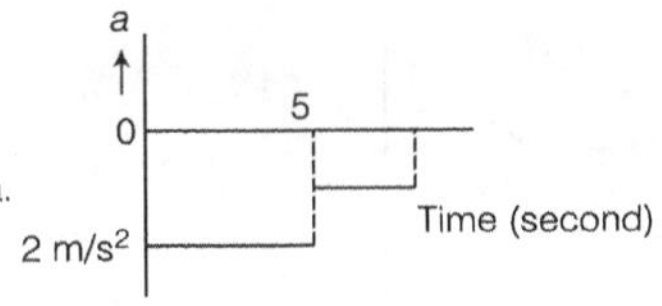

b.
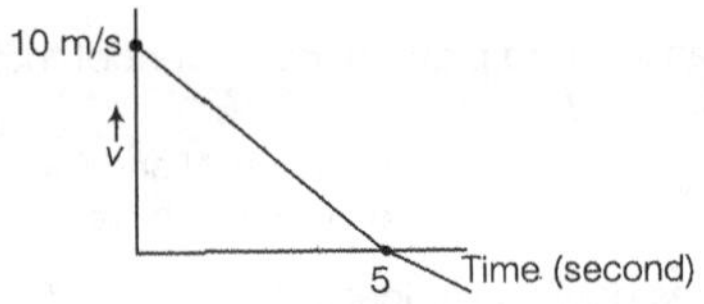

c.
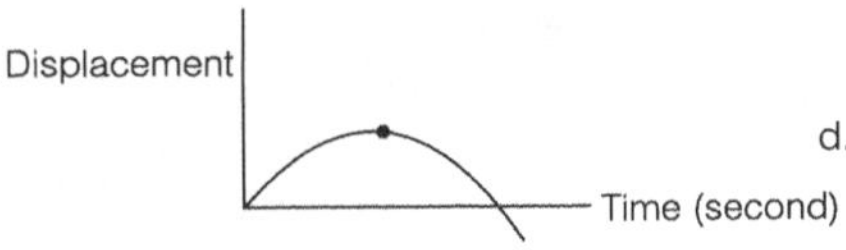

d. All of these

3 A connecting block system is shown in figure. If f_1 and f_3 represent forces of friction on block A and block C, respectively. The tension in light string connecting blocks A and B is T_1 while the tension in the light string connecting blocks C and D is T_2. Mark the correct option.

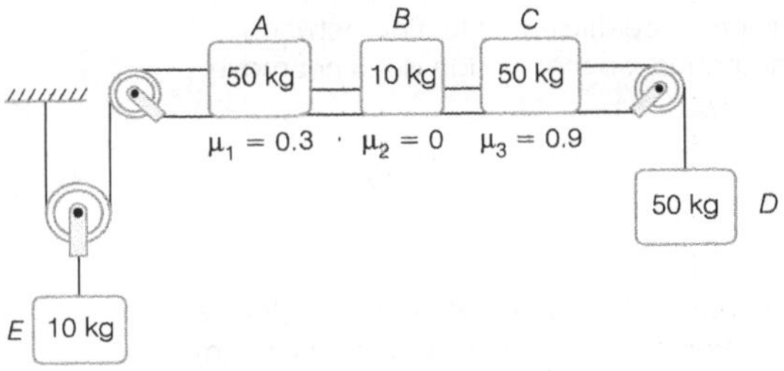

a. $T_1 = 50$ N and $T_2 = 500$ N
b. $f_1 = 0$ and $f_3 = 450$ N
c. $f_1 = 50$ N and $T_1 = 50$ N
d. Both (a) and (b)

4 A small block of mass 1 kg is kept on a long flat part of the truck. The coefficient of static friction and kinetic friction between the block and the surface of truck are $\mu_s = 0.3$ and $\mu_k = 0.2$. Truck starts from rest with an acceleration of 4 m / s^2. Find distance travelled by the block in one second.

a. 1 m
b. 2 m
c. 3 m
d. 4 m

5 A block of mass 1 kg initially at $x = 0$ is acted upon by a force $F = (10 - x)$ N. The coefficient of friction between the block and the horizontal surface is $\mu = 0.5$. The maximum speed of the block during forward motion is

a. 1 m/s
b. 5 m/s
c. 10 m/s
d. 15 m/s

Working Space

6 A rope of length $L = 5$ m is placed on smooth horizontal surface. For $x > L$, the surface is rough having coefficient of friction $\mu = 0.5$. Find minimum velocity v_0, imparted to the rope horizontally, so that the rope completely enter the rough horizontal surface.

a. 1 m/s　　b. 2 m/s
c. 4 m/s　　d. 5 m/s

7 In the situation shown in figure, the spring constant of the spring is 1 N/m and the horizontal surface beyond the spring region is smooth. The coefficient of friction between the block of mass 1 kg and horizontal surface is 0.05for the spring region. Initially, the block is moving with a speed of $\sqrt{2}$ m/s towards the spring. The maximum compression in the spring is

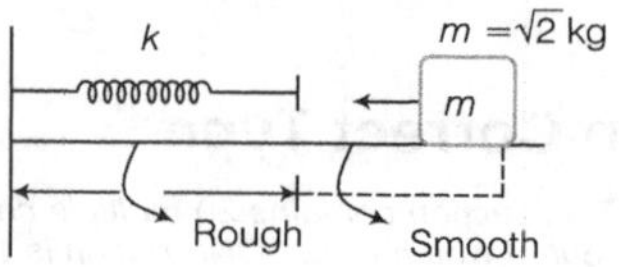

a. 1 m　　b. 2 m
c. 3 m　　d. 4 m

8 In the situation, a cubical block is moving horizontally with an acceleration $a_0 = 5\ m/s^2$.

The blocks A and B, each of mass 1 kg are initially 5 m apart. The coefficient of friction between all contact surfaces of blocks is 1. A force of 20 N is applied on the block B vertically upwards at $t = 0$. The time 't' at which they collide to each other is

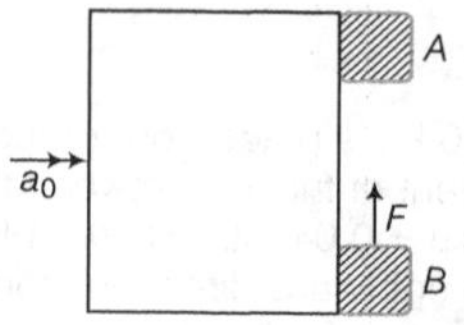

a. 1 s　　b. 2 s
c. 3 s　　d. 4 s

9 A block of 1 kg is at rest on an inclined plane of coefficient of friction μ. The friction force on the block is f. Then,

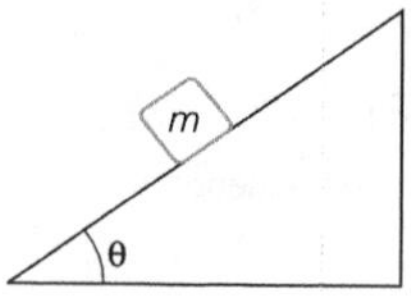

a. $N < 10$ N and $f < 10$ N　　b. $N = 10$ N and $f = 10$ N
c. $N > 10$ N and $f > 10$ N　　d. None of these

10 A monkey of mass 50 kg slides down a vertical pole of length 50 m in 10 s. Find the force exerted by monkey on the pole.

a. 900 N　　b. $900\sqrt{2}$ N
c. 450 N　　d. None of these

11 A boy of mass 20 kg slides on a wedge of mass 30 kg as shown in the figure. The coefficient of friction between the boy and the wedge is 0.6. Find minimum coefficient of friction between the wedge and horizontal ground, so that wedge does not move during sliding of the body.

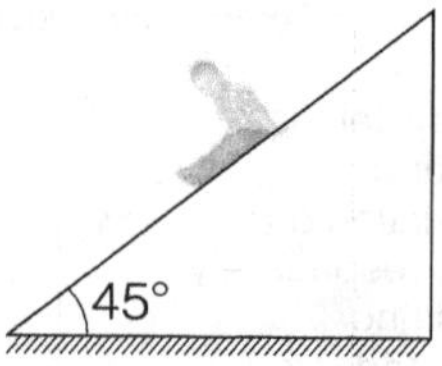

a. 1　　b. $\frac{10}{17}$
c. $\frac{2}{23}$　　d. $\sqrt{2}$

12 A baby is sitting on the horizontal surface of triangular block B, which is sliding on a frictionless inclined plane. Find minimum coefficient of friction between baby and the triangular block B, so that baby remains stationary with respect to the triangular block B.

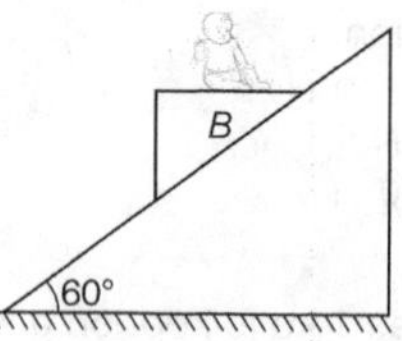

a. Zero　　b. $\frac{1}{\sqrt{3}}$
c. 1　　d. $\sqrt{3}$

13 The situation shown in figure, $m_A = 5$ kg, $m_B = 10$ kg and $m_C = 2$ kg. The coefficient of friction between A and B is 0.9 and that of between B and C is 0.1. The inclined plane is frictionless. Choose the correct option(s).

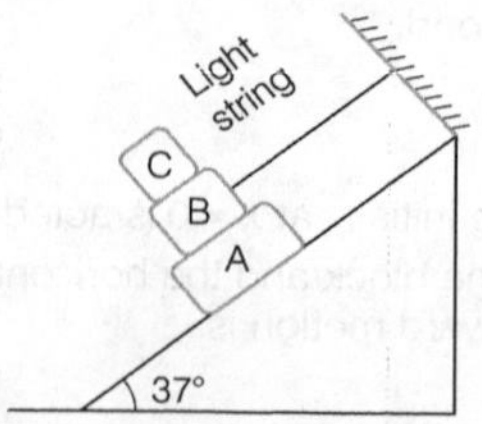

a. The acceleration of block C is $5.2\ m/s^2$
b. The frictional force on block A is 30 N
c. The tension in the light string is 91.6 N
d. All of the above

Working Space

14 Sixty identical books are placed on a horizontal rough tables as shown in the figure. The mass of each book is 1 kg. The coefficient of friction for all contact surfaces of books is 0.5. An IIT aspirant of mass 50 kg standing on ground try to pull the lower book horizontally. The coefficient friction between his hand and the lower books is so large that his hands do not slip out the book. The coefficient of friction between his legs and ground is 0.5. Mark the correct option.

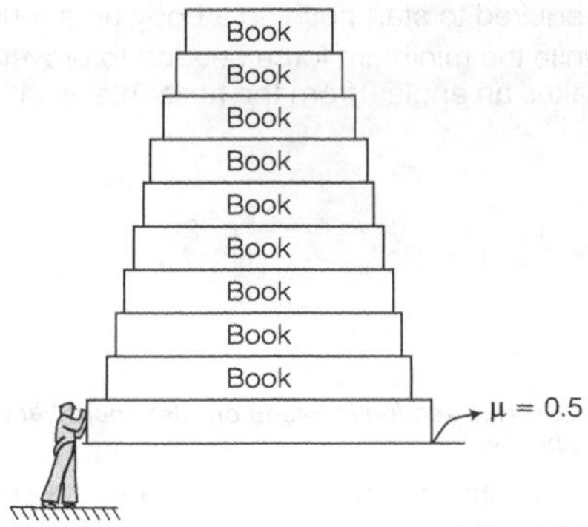

a. All books are sliding together
b. All books always remain stationary
c. He will not pull out the lower book
d. Both (b) and (c)

15 The n identical blocks each of mass 1 kg are connected through string as shown in the figure. The mass of hanging block is also 1 kg . The horizontal surface is rough and provides coefficient of friction $\mu = 0.8$ for all blocks. Mark the correct option.

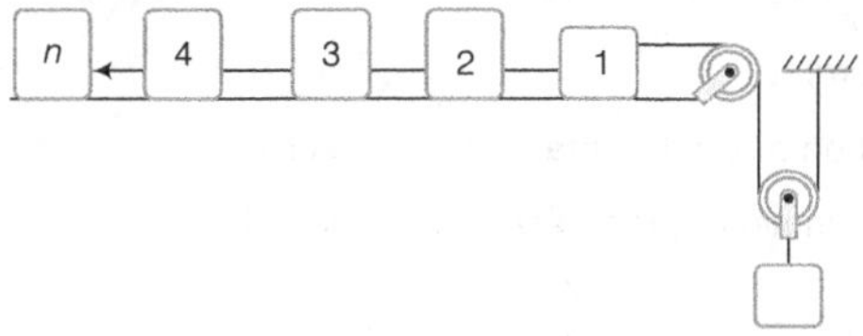

a. The friction on first block is 5 N
b. The frictional forces on the blocks except first block are zero
c. The tensions in the strings connecting the numbered blocks are zero
d. All of the above

16 In the figure shows,

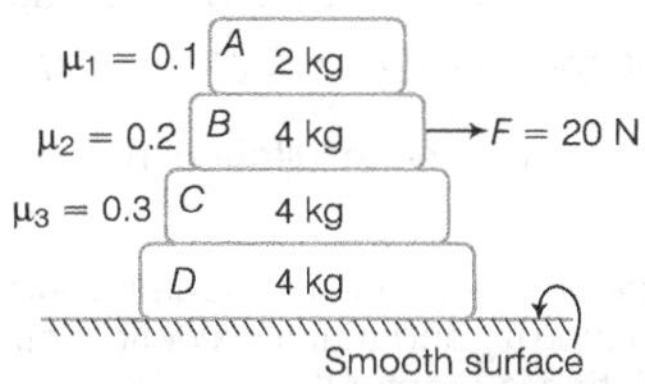

a. all blocks move together with an acceleration 1 m/s^2
b. friction on block D exerted by block C is 10 N is rightward direction
c. friction on block B exerted by block C is 14 N in leftward direction
d. All of the above

17 The situation shown in the figure, find the minimum value of coefficient of friction, so that all blocks move together.

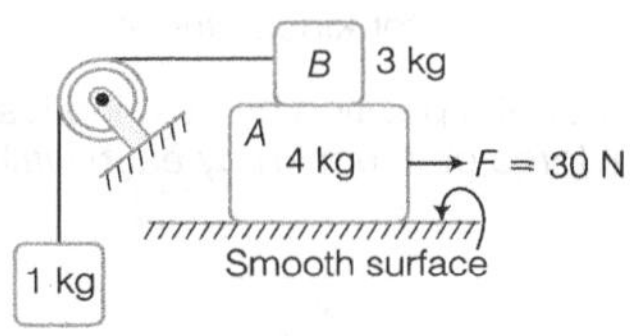

a. 1 b. $\frac{1}{2}$ c. $\frac{1}{3}$ d. $\frac{2}{3}$

18 A marble block of mass 2 kg lying on ice, when given a velocity of 6 ms^{-1} is stopped by friction in 10 s. Then, the coefficient of friction is **[AIEEE 2003]**

a. 0.02 b. 0.03 c. 0.06 d. 0.01

19 A block of mass m is placed on surface with a vertical cross-section given by $y = \frac{x^3}{6}$. If the coefficient of friction is 0.5, the maximum height above the ground at which the block can be placed without slipping is **[JEE Main 2014]**

a. $\frac{1}{6}$ m b. $\frac{2}{3}$ m
c. $\frac{1}{3}$ m d. $\frac{1}{2}$ m

20 The minimum force required to start pushing a body up a rough (frictional coefficient μ) inclined plane is F_1 while the minimum force needed to prevent it from sliding down is F_2. If the inclined plane makes an angle θ from the horizontal such that $\tan\theta = 2\mu$, then the ratio $\frac{F_1}{F_2}$ is **[AIEEE 2011]**

a. 4 b. 1 c. 2 d. 3

Working Space

Statement Type

Direction (Q. Nos. 21 and 26) ***This section is based on Statement I and Statement II. Select the correct answer from the codes given below.***

a. Both Statements I and Statements II are correct and Statement II is the correct explanation for Statement I
b. Both Statements I and Statements II are correct and Statement II is not a correct explanation for
Statement I
c. Statement I is correct but Statement II is incorrect
d. Statement I is incorrect but Statement II is correct

21 **Statement I** Frictional force on a block placed on a rough surface may be zero.

Statement II If tendency of relative motion or relative motion between surfaces is absent, frictional force is zero.

22 **Statement I** In the situation shown in the figure acceleration of block *A* is always greater or equal to the block *B*.

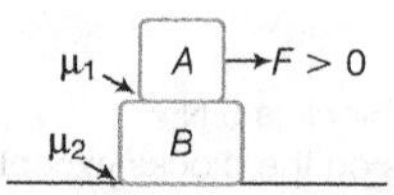

Statement II The friction by ground on block *B* may be zero in the given situation.

23 **Statement I** If a block of mass *m* is placed on a rough horizontal surface, the contact force *F* on the block is lying as $mg \le F \le \sqrt{1+\mu^2}\, mg$.

Statement II Contact force is the resultant of normal reaction and the force of friction on the block.

24 **Statement I** If a block on a horizontal surface having coefficient of static friction μ_s and that for kinetic friction is μ_k. The angle of static friction is greater than angle of kinetic friction.

Statement II The coefficient of static friction is greater than that for kinetic friction.

25 **Statement I** If a man of 50 kg balances himself in a horizontal position by pushing his hands and back against two vertical walls, the frictional force on the man by each wall is 250 N.

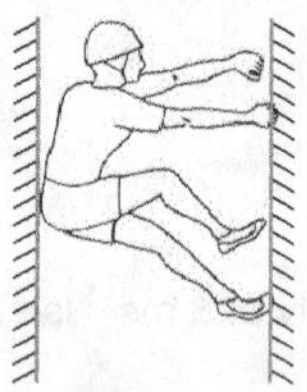

Statement II The weight of man is balanced by frictional forces by walls on the man.

26 **Statement I** If you move up a block on rough wall by applying a force *F*, frictional force on the block is $\mu_k mg_s$, where μ_k is the coefficient of kinetic friction.

Statement II The kinetic frictional force on a block is $f = \mu_k N$, where *N* is normal contact force between contact surfaces.

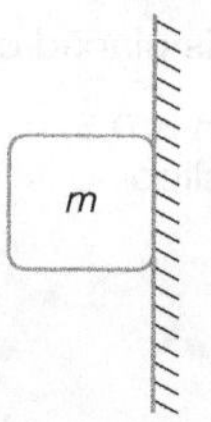

Revisal Problems
for JEE Advanced

Only One Option Correct Type

Working Space

Direction (Q. Nos. 1-6) *This section contains 6 multiple choice questions. Each question has four choices (a), (b), (c) and (d), out of which ONLY ONE option is correct.*

1 A light string is stretched round one half of the circumference of a fixed cylindrical pulley as shown in the figure. As a result of friction, the string does not slip on the cylinder when the masses of the blocks acting on its ends fulfil the inequality $\frac{m_A}{2} \le m_B \le 2m_A$.

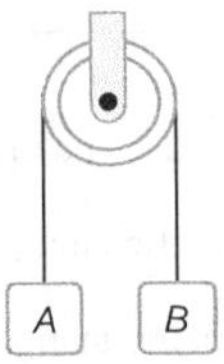

The coefficient of friction between the string and the pulley is

a. m_2 b. $\frac{\ln 2}{\pi}$ c. 0.5 d. $\frac{\pi}{\ln 2}$

2 Four blocks placed in contact on an inclined plane as shown in the figure. The coefficients of friction between inclined plane and these blocks are μ_1, μ_2, μ_3 and μ_4 respectively with $\mu_1 > \mu_2 > \mu_3 > \mu_4$. The minimum value of $\tan\alpha$ at which the blocks start sliding down is

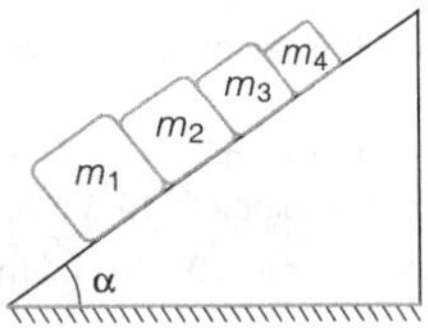

a. $\frac{m_1m_2 + m_2m_3 + m_3m_4}{(\mu_1m_1 + \mu_2m_2 + \mu_3m_3 + \mu_4m_4)(m_1 + m_2 + m_3 + m_4)}$

b. $\frac{\mu_1m_1 + \mu_2m_2 + \mu_3m_3 + \mu_4m_4}{m_1 + m_2 + m_3 + m_4}$

c. $\mu_1 + \mu_2 + \mu_3 + \mu_4$

d. $\frac{m_1 + m_2 + m_3 + m_4}{\mu_1m_1 + \mu_2m_2 + \mu_3m_3 + \mu_4m_4}$

3 Blocks *A* and *B* of masses m_1 and m_2 are rest on the fixed wedge. The coefficient of friction between the wedge and the blocks is $\mu = 0.5$. If there is no motion, then

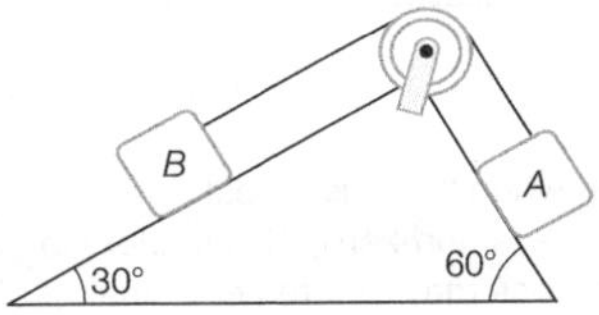

a. $\left(\frac{2-\sqrt{3}}{1+2\sqrt{3}}\right) \le \frac{m_1}{m_2} \le \left(\frac{2+\sqrt{3}}{2\sqrt{3}-1}\right)$

b. $\frac{\sqrt{3}-1}{1+2\sqrt{3}} \le \frac{m_1}{m_2} \le \frac{2+\sqrt{3}}{2\sqrt{3}-1}$

c. $\frac{m_1}{m_2} \le \sqrt{3}$

d. $\frac{m_1}{m_2} \ge \sqrt{3}$

4 In the figure shown, find velocity of the block at $t = 2.5$ s.

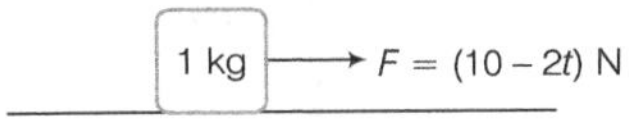

a. Zero b. 6.25 m/s c. 10 m/s d. 12.5 m/s

5 If engine of a bus moving along a rough horizontal surface is shut off, then

a. decelerating force on bus is equal to friction between tyre and road
b. decelerating force on bus is equal to resistance in rotating part
c. net decelerating force is equal to friction plus resistance
d. None of the above

6 Find maximum acceleration of monkey of mass 5 kg, so that block remains stationary.

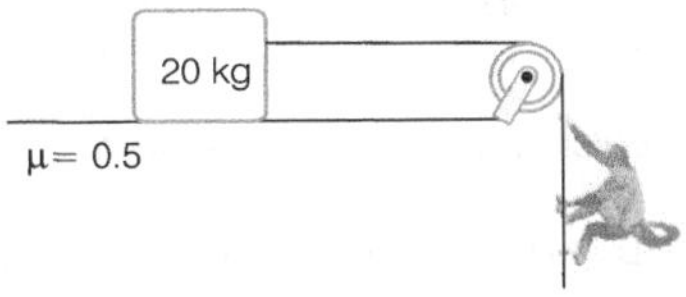

a. $5\,m/s^2$ b. $10\,m/s^2$
c. $20\,m/s^2$ d. $6.67\,m/s^2$

Working Space

One or More than One Options Correct Type

Direction **(Q. Nos. 7-10)** ***This section contains 4 multiple choice questions. Each question has four choices (a), (b), (c) and (d) out of which ONE or MORE THAN ONE are correct.***

7 A block is projected with a velocity 10 m/s on a rough horizontal surface. It covers a distance x in time 2 s. Which of following cases are possible?

a. $0 < x \le 10$ m
b. 10 m $< x <$ 20 m
c. $x > 20$ m
d. None of these

8 The coefficient of friction between block A and ground is μ. Other blocks are smooth. If $0 < F \le F_0$, blocks do not move. If $F > F_0$, block start to move

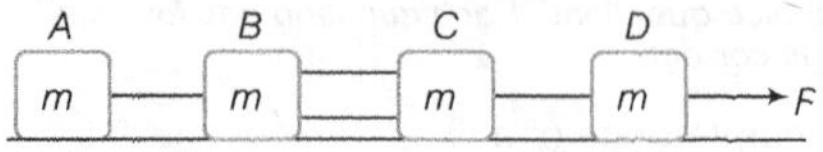

a. $F_0 = \mu mg$
b. at $F = \frac{F_0}{2}$, the frictional force on block A is $\frac{F_0}{2}$
c. the tension in one of the string between B and C is $\frac{F}{2}$
d. the tension in one of the string between B and C is $\frac{F_0}{2}$ for $F = F_0$

9 A block of mass 2 kg rests on a horizontal plane having coefficient of friction $\mu = 0.5$. At $t = 0$, a horizontal force **F** is applied that varies as $F = kt$, where $k = 2$ N/s. Mark the correct option(s).

a. At $t = 5$ s, block starts to move
b. At $t = 10$ s, velocity of the block is 12.5 m/s
c. Distance travelled by block in 10 s is $\frac{125}{6}$ m
d. None of the above

10 A man of mass 50 kg pulls a block mass 100 kg with a light string (as shown in the figure).

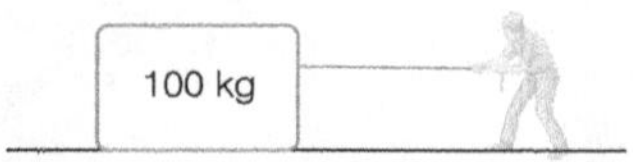

The coefficient of friction is 0.5 for all contact surfaces. Which of the following statements is/are corrects?

a. Man can move even when block is in rest
b. If acceleration of block is non-zero, the acceleration of man is greater than that of the block
c. The acceleration of block may be greater than that of man
d. None of the above

Comprehension Type

Direction **(Q. Nos. 11-16)** ***This section contains 6 paragraphs, each describing theory, experiments, data, etc. Six questions related to the paragraphs have been given. Each question has only one correct answer among the four given options (a), (b), (c) and (d).***

Passage I

A cup of tea of total mass 50 g is kept on smooth table. A light plastic plate is placed on the mouth of cup. A coin of mass 10 g is placed on the plate. The coefficient of friction between each surface of contact of the plate is 0.5. A boy pulls the plate horizontally by applying a force F. (assume that cup does not topple during motion).

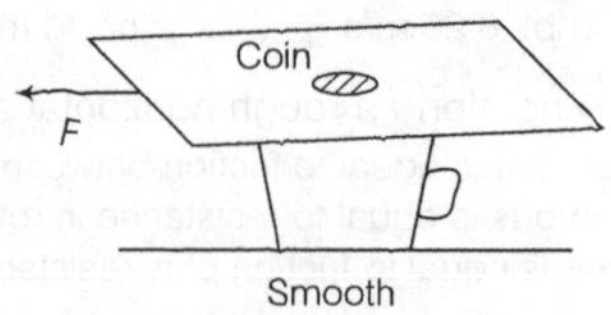

11 Find the maximum value of F for which the cup does not slide with respect to the plate.

a. 0.1 N
b. 0.2 N
c. 0.06 N
d. 0.02 N

12 Find the value of F for which the plate slides relative to the cup but does not relative to the coin.

a. 0.06 N $< F <$ 0.1 N
b. 1 N $< F <$ 2 N
c. 0.06 N $< F <$ 0.5 N
d. 0.06 N $< F <$ 0.5 N

Passage II

A long plank is connected with a motor through a long string. The motor wraps 20 m length of string per second.

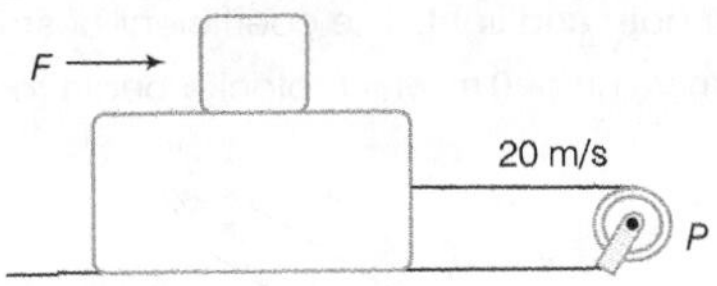

A block of mass 10 kg is gently put on the plank at $t = 0$ as shown in the figure. A constant force of 15 N is applied on the block continuously during its motion. The coefficient of kinetic and static friction between the plank and the block are 0.25 and 0.4, respectively.

13 The nature of friction acting on the block during its motion is

a. static only
b. kinetic only
c. sometime kinetic and sometime static
d. cannot be predicted

14 The magnitude of acceleration of the block at $t = 6$ s if frictional force is just 25 N .

a. 1 m/s^2
b. 2 m/s^2
c. 3 m/s^2
d. 4 m/s^2

Passage III

The coefficient of friction between inclined plane and the block C is 0.5. The masses of blocks A, B and C are 1 kg, 2 kg and 3 kg, respectively. All other surfaces are smooth.

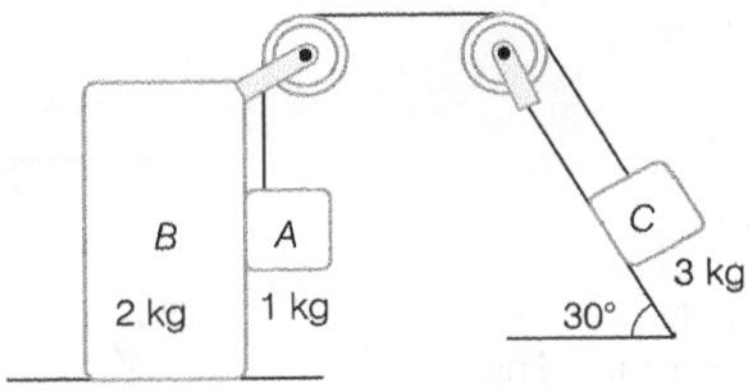

15 The acceleration of block C is

a. Zero
b. $\frac{g}{4}$
c. $\frac{\sqrt{2}\, g}{4}$
d. $\frac{g}{2}$

16 The acceleration of block A is

a. zero
b. $\frac{g}{4}$
c. $\frac{\sqrt{2}\, g}{4}$
d. 2/16

One Integer Value Correct Type

Direction **(Q. No. 17-20)** ***This section contains 4 questions. When worked out will result in an integer from 0 to 9 (both inclusive).***

17 A man of mass m is stationary by pulling a light string. The string makes an angle of 37° with the vertical wall as shown in the figure. The coefficient of static friction between legs of man and the vertical wall is $\mu = 0.5$. The maximum value of $\left(\frac{m}{M}\right)$ is $\frac{11x}{30}$. Find the value of x.

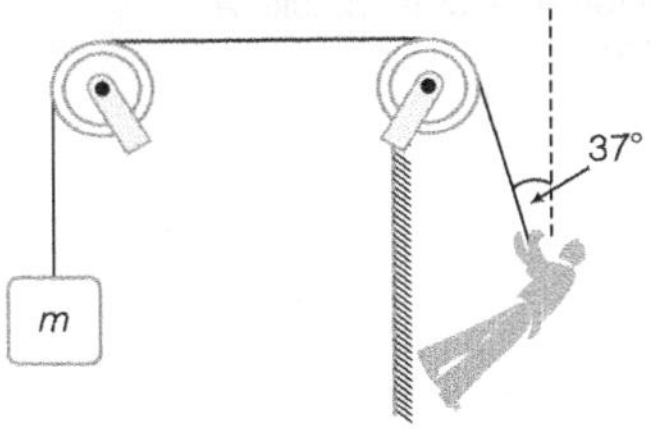

Working Space

18 Two blocks A and B each of mass 1 kg are connected by light string as shown in the figure. Pulleys are smooth and light. The coefficient of static friction for all surfaces is $\mu = 0.6$. The minimum value of θ at which blocks begin to slide is $\frac{\pi}{n}$ rad. Find the value of n.

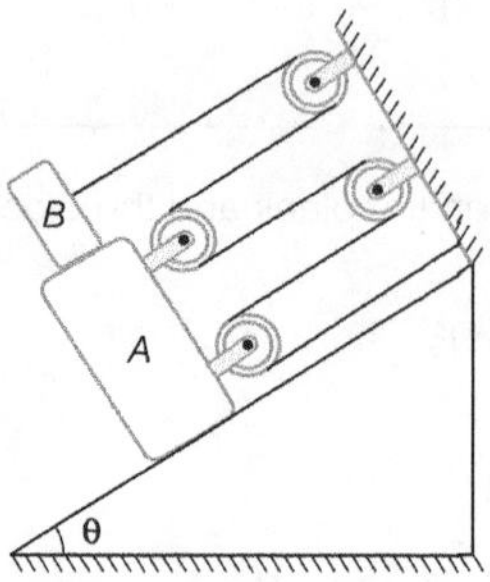

19 The situation shown in the figure, all strings and pulleys are ideal. If tension in the string connecting blocks A and C is $\frac{1350}{11n}$ N. Find the value of n.

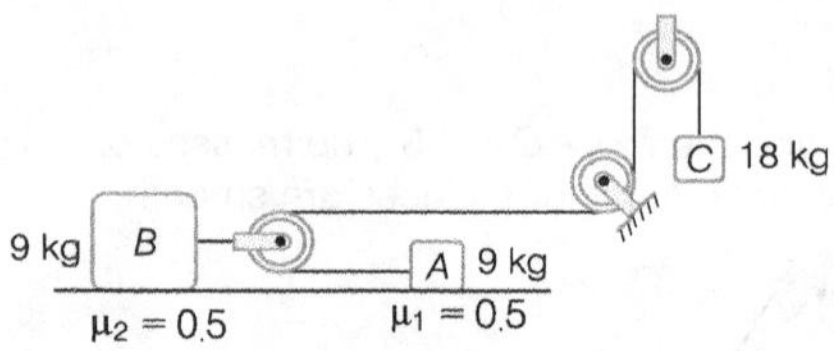

20 A small block is kept on a rough inclined plane of inclination θ. The inclined plane is just sufficiently rough with friction coefficient that the block would fall and slide down with a uniform speed as the result of even the slightest jerk. Now, the block is pushed horizontally with an initial speed $v_0 = 2$ m/s. The block slides down the slope with a velocity that changes in both magnitude and direction. Find the final speed of the block.

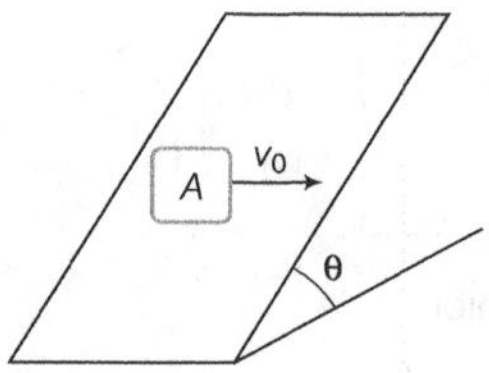

Matching List Type

Direction ***(Q. No. 21) Choices for the correct combination of elements from Column I and Column II are given as options (a), (b), (c) and (d), out of which one is correct.***

21 Two blocks A and B of masses 2 kg and 3 kg are placed on horizontal surface. The coefficient of friction between block A and horizontal surface is 0.5 while the block B is smooth. A force $F = (\sqrt{2}\,t)$ N is applied on block A as shown in the figure.

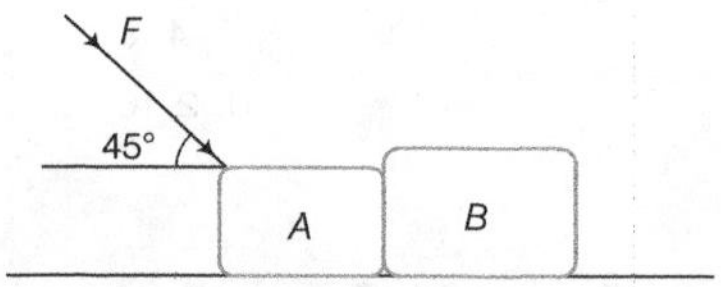

Match the Column I with Column II and choose the correct option from the codes given below.

	Column I		Column II
i.	$t = 4$ s	p.	The acceleration of block B is zero.
ii.	$t = 20$ s	q.	Friction on block A is more than 20 N.
iii.	$t = 40$ s	r.	The acceleration of block A is 2 m/s^2.
iv.	$t = 60$ s	s.	Normal contact force between blocks A and B is non-zero.

Codes

	i	ii	iii	iv
a.	p	p	p,q	q,r,s
b.	p	p	p,s	q,r,s
c.	p,q	p,q	p,s	q,r,s
d.	p	p	q,r	q,r,s

JEE Main & AIEEE Archive

(Compilation of Last 13 Years Questions)

1 A block of mass m is placed on a surface with a vertical cross-section given by $y = x^3/6$. If the coefficient of friction is 0.5, the maximum height above the ground at which the block can be placed without slipping is **(JEE Main 2014)**

a. $\frac{1}{6}$ m b. $\frac{2}{3}$ m

c. $\frac{1}{3}$ m d. $\frac{1}{2}$ m

2 The minimum force required to start pushing a body up a rough (frictional coefficient μ) inclined plane is F_1 while the minimum force needed to prevent it from sliding down is F_2. If the inclined plane makes an angle θ from the horizontal such that $\tan\theta = 2\mu$, then the ratio $\frac{F_1}{F_2}$ is **(AIEEE 2011)**

a. 4 b. 1 c. 2 d. 3

3 Consider a car is moving on a straight road with a speed of 100 ms^{-1}. The distance at which car can be stopped, is $[\mu_k = 0.5]$ **(AIEEE 2005)**

a. 800 m b. 1000 m

c. 100 m d. 400 m

4 A smooth block is released at rest on a 45° incline and then slides a distance d. The time taken to slide is n times as much to slide on rough incline than on a smooth incline. The coefficient of friction is **(AIEEE 2005)**

a. $\mu_k = 1 - \frac{1}{n^2}$ b. $\mu_k = \sqrt{1 - \frac{1}{n^2}}$

c. $\mu_s = 1 - \frac{1}{n^2}$ d. $\mu_s = \sqrt{1 - \frac{1}{n^2}}$

5 The upper half of an inclined plane with inclination ϕ is perfectly smooth, while the lower half is rough. A body starting from rest at the top will again come to rest at the bottom, if the coefficient of friction for the lower half is given by **(AIEEE 2005)**

a. $2\sin\phi$ b. $2\cos\phi$

c. $2\tan\phi$ d. $\tan\phi$

6 A block rests on a rough inclined plane making an angle of 30° with the horizontal. The coefficient of static friction between the block and the plane is 0.8. If the frictional force on the block is 10 N, the mass of the block (in kg) is ($g = 10$ m/s^2) **(AIEEE 2004)**

a. 2.0 b. 4.0 c. 1.6 d. 2.5

7 A horizontal force of 10 N is necessary to just hold a block stationary against a wall. The coefficient of friction between the block and the wall is 0.2. The weight of the block is **(AIEEE 2003)**

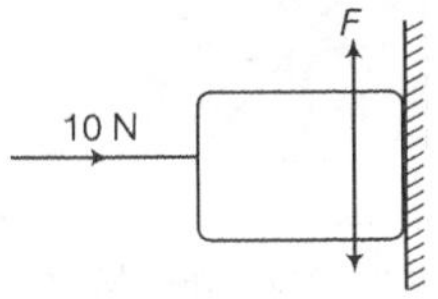

a. 20 N b. 50 N c. 100 N d. 2 N

8 A marble block of mass 2 kg lying on ice when given a velocity of 6 ms^{-1} is stopped by friction in 10 s. Then, the coefficient of friction is **(AIEEE 2003)**

a. 0.02 b. 0.03

c. 0.06 d. 0.01

9 The minimum velocity (in ms^{-1}) with which a car driver must traverse a flat curve of radius 150 m and coefficient of friction 0.6 to avoid skidding is **(AIEEE 2002)**

a. 60 b. 30

c. 15 d. 25

10 Three identical blocks of mass $m = 2$ kg are drawn by a force $F = 10.2$ N with an acceleration of 0.6 ms^{-2} on a frictionless surface, then what is the tension (in N) in the string between the blocks B and C? **(AIEEE 2002)**

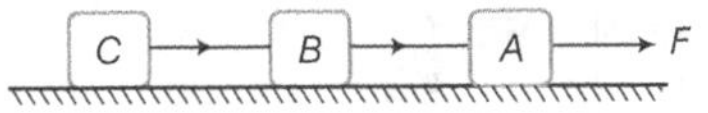

a. 9.2 b. 7.8

c. 4 d. 9.8

JEE Advanced & IIT JEE Archive

(Compilation of Last 10 Years Questions)

1 A block of mass $m_1 = 1$ kg and another mass $m_2 = 2$ kg are placed together (see figure) on an inclined plane with angle of inclination θ. Various values of θ are given in Column I. The coefficient of friction between the block m_1 and the plane is always zero. The coefficient of static and dynamic friction between the block m_2 and the plane are equal to $\mu = 0.3$.

(2014 Adv., Matching Type)

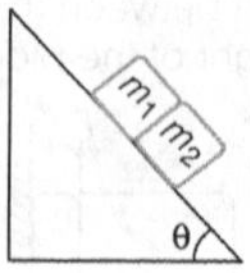

In Column II, expressions for the friction on the block m_2 are given. Match the correct expression of the friction in Column II with the angles given in Column I and choose the correct option. The acceleration due to gravity is denoted by g.

[Useful information $\tan(5.5^\circ) \approx 0.1$, $\tan(11.5^\circ) \approx 0.2$, $\tan(16.5^\circ) \approx 0.3$]

	Column I		Column II
i.	$\theta = 5^\circ$	p.	$m_2 g \sin\theta$
ii.	$\theta = 10^\circ$	q.	$(m_1 + m_2) g \sin\theta$
iii.	$\theta = 15^\circ$	r.	$\mu m_2 g \cos\theta$
iv.	$\theta = 20^\circ$	s.	$\mu (m_1 + m_2) g \cos\theta$

Codes

	i	ii	iii	iv
a.	p	p	p	r
b.	q	q	q	r
c.	q	q	q	s
d.	q	q	r	r

2 A block is moving on an inclined plane making an angle 45° with the horizontal and the coefficient of friction is μ. The force required to just push it up the inclined plane is 3 times the force required to just prevent it from sliding down. If we define $N = 10\mu$, then the value of N is **(2011, Integer Type)**

3 A block of mass m is on an inclined plane of angle θ. The coefficient of friction between the block and the plane is μ and $\tan\theta > \mu$. The block is held stationary by applying a force P parallel to the plane. The direction of force pointing up the plane is taken to be positive. As P is varied from $P_1 = mg(\sin\theta - \mu\cos\theta)$ to $P_2 = mg(\sin\theta + \mu\cos\theta)$, the frictional force f versus P graph will look like **(2010, Only One Option Correct Type)**

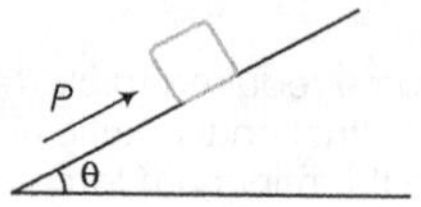

a.
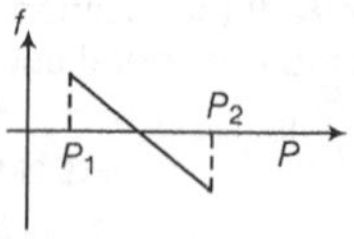

b.
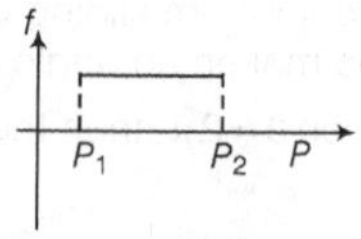

c.
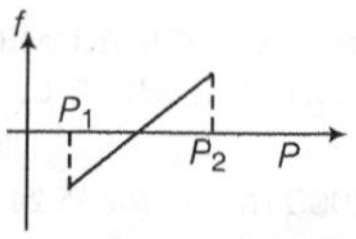

d.
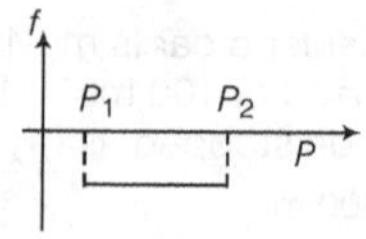

4 **Statement I** It is easier to pull a heavy object than to push it on a level ground.

Statement II The magnitude of frictional force depends on the nature of the two surfaces in contact.

(2008, Statement Type)

a. Both Statement I and Statement II are correct and Statement II is the correct explanation of Statement I
b. Both Statement I and Statement II are correct and Statement II is not the correct explanation of Statement I
c. Statement I is correct but Statement II is incorrect
d. Statement II is correct but Statement I is incorrect

5 A circular disc with a groove along its diameter is placed horizontally. A block of mass 1 kg is placed as shown. The coefficient of friction between the block and all surfaces of groove in contact is $\mu = 2/5$. The disc has an acceleration of 25 m/s^2. Find the acceleration of the block with respect to disc. **(2006, Descriptive Type)**

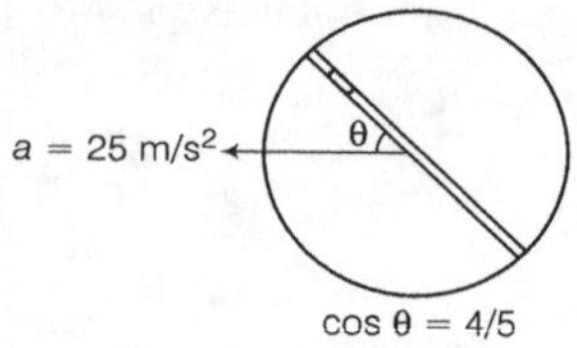

$\cos\theta = 4/5$
$\sin\theta = 3/5$

JEE Archive

3. Circular Motion

DPP
DAILY PRACTICE PROBLEMS

DPP-1 Kinematics of Circular Motion

Conceptual Questions

Working Space

Direction (*Q. Nos.* 1 *and* 2) *These questions are conceptual in nature, need to be solved completely on notebook.*

1 Fill in the blanks.

i. If a particle moves on a circular path, a line passing through the centre of the circle and perpendicular to the plane of circular path is ....

ii. If angular displacement is infinitesimal, it is a ...... quantity.

iii. A large angular displacement is a ...... quantity.

iv. A large angular displacement does not obey ...... law of vector addition.

v. If a particle moves on a circular path, it is an example of an ...... motion.

vi. If a particle moves on a circular path in XY-plane in anti-clockwise direction, then unit vector along the direction of angular velocity is ....

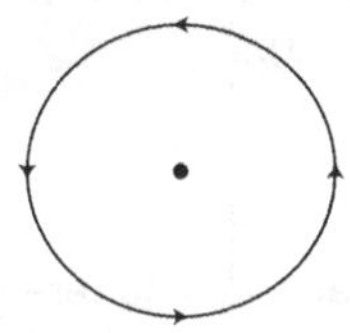

vii. If a particle moves on a circular path as shown in the figure below. The correct relation between angular velocity ω, linear velocity v and position vector **r** of the particle is ....

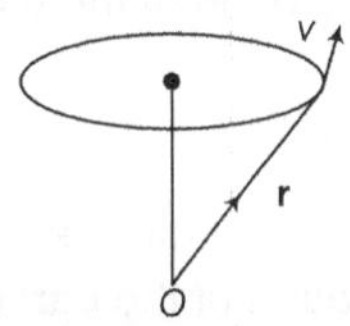

viii. If a particle is moving on a circular path of radius R with constant speed, then acceleration of the particle is ....

ix. If a particle is moving with variable speed, then acceleration of particle is ...... directed towards the centre of the circular path.

x. The rate of change of speed is ....

2 If a particle moves on a circular path of radius r such that angular position $\theta = at^2 - bt + c$, where θ is in radian and t is in second. Then, mark the correct option(s) for the following questions.

i. Initial angular position at $t = 0$ is

a. $\theta_i = c$ b. $\theta_i = a$ c. $\theta_i = -b$ d. None of these

ii. The angular displacement from $t = 0$ to $t = t$ is

a. $\Delta\theta = at^2 - bt + c$ b. $\Delta\theta = at^2 - bt$

c. c d. None of these

iii. The angular velocity of the particle at instant t is

a. $\omega = 2at - b$ b. $2at$

c. $-b$ d. None of these

iv. The speed at instant t is

a. $v = (2at - bt^2)r$ b. $v = (2at - b)r$

c. $(2at - b)/r$ d. None of these

v. The angular acceleration of the particle is

a. $\alpha = 2b$ b. $\alpha = 2a$
c. 0 d. None of these

vi. The SI unit of angular acceleration is

a. rad/s b. rad/s^2
c. rad/s^3 d. None of these

vii. The tangential acceleration of the particle is

a. $a_t = br$ b. $a_t = 2ar$
c. zero d. None of these

Working Space

Only One Option Correct Type

Direction (Q. Nos. 3-13) *This section contains 11 multiple choice questions. Each question has four choices (a), (b), (c) and (d), out of which ONLY ONE option is correct.*

3 Find angular velocity of the earth rotating about its own axis.

a. $\frac{\pi}{12}$ rad/s b. $\frac{\pi}{12}$ rad/h
c. $\frac{\pi}{3}$ rad/h d. None of these

4 Find acceleration of the earth revolving around the sun in a circular orbit of radius 1.5×10^{11} m $(\pi^2 \approx 9.8)$.

a. 58.8 m/yr^2 b. 29.4 m/yr^2
c. 40 m/yr^2 d. None of these

5 Assume that the earth is a sphere of radius 6400 km/h. A particle is located at the equator. The acceleration of the particle is

a. 8×10^5 m/h^2 b. 4.36×10^5 m/h^2
c. 9×10^5 m/h^2 d. 3×10^5 m/h^2

6 In previous problem, a particle is located at latitude of 60° N on the surface of the earth. The acceleration of the particle is

a. 4.36×10^5 m/h^2 b. 1.88×10^5 m/h^2 c. 2.18×10^5 m/h^2 d. None of these

7 A particle is moving on a circular path. The angle turned by radius vector is $\theta = \theta_0(1 - e^{-t/3})$, where θ is in radian and t is in second. The angular acceleration at $t = 0$ is

a. $\frac{1}{3}$ rad/s^2 b. $-\frac{1}{9}$ rad/s^2 c. 10 rad/s^2 d. $-\frac{1}{6}$ rad/s^2

8 A point P moves in a counter-clockwise direction on a circular path as shown in the figure. The movement of P is such that it sweeps out a length $s = t^3 + 5$, where s is in metre and t is in second. The radius of the path is 20 m. The acceleration of P when, $t = 2$ s is nearly, is **[AIEEE 2010]**

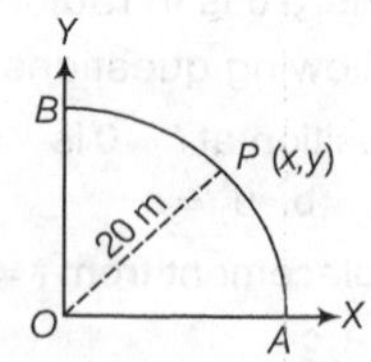

a. 13 ms^{-2} b. 12 ms^{-2}
c. 7.2 ms^{-2} d. 14 ms^{-2}

9 A mass m is fixed at $x = a$ on a parabolic wire with its axis vertical and vertex at the origin as shown in the figure. The equation of parabola is $x^2 = 4ay$. The wire frame is rotating with constant angular velocity ω about Y-axis. The acceleration of the bead is

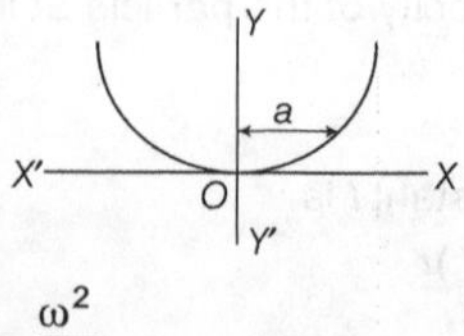

a. $\frac{a}{4}\omega^2$ b. $a\omega^2$ c. zero d. $\frac{\omega^2}{a}$

Working Space

10 A particle P starts from rest with constant angular acceleration α in anti-clockwise direction from point P on a circular path of radius R. The acceleration of the particle at $t = \sqrt{\pi/\alpha}$ is

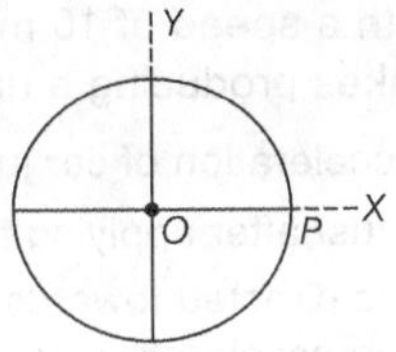

a. $R\alpha\,\hat{\mathbf{i}} + \pi R\alpha\,\hat{\mathbf{j}}$ b. $R\alpha\,\hat{\mathbf{k}} + \pi R\alpha\,\hat{\mathbf{j}}$

c. $-R\alpha\,\hat{\mathbf{i}} - \pi R\alpha\,\hat{\mathbf{j}}$ d. None of these

11 Which of the following statements is false for a particle moving in a circle with a constant angular speed? **[AIEEE 2004]**

a. The velocity vector is tangent to the circle
b. The acceleration vector is tangent to the circle
c. The acceleration vector points to the centre of the circle
d. The velocity and acceleration vectors are perpendicular to each other

12 A particle moves on a circular path of radius 0.2 m with an angular acceleration 0.25 rad/s^2. The particle starts from rest, find the time at which the magnitude of radial acceleration and the magnitude of tangential acceleration are same.

a. 1 s b. 2 s c. 3 s d. 4 s

13 For a particle in a uniform circular motion, the acceleration **a** at a point $P(R, \theta)$ on the circle of radius R is (here, θ is measured from the X-axis) **[AIEEE 2010]**

a. $-\frac{v^2}{R}\cos\theta\,\hat{\mathbf{i}} + \frac{v^2}{R}\sin\theta\,\hat{\mathbf{j}}$ b. $-\frac{v^2}{R}\sin\theta\,\hat{\mathbf{i}} + \frac{v^2}{R}\cos\theta\,\hat{\mathbf{j}}$

c. $-\frac{v^2}{R}\cos\theta\,\hat{\mathbf{i}} - \frac{v^2}{R}\sin\theta\,\hat{\mathbf{j}}$ d. $\frac{v^2}{R}\hat{\mathbf{i}} + \frac{v^2}{R}\hat{\mathbf{j}}$

One or More than One Options Correct Type

Direction (Q. Nos. 14-19) This section contains 6 multiple choice questions. Each question has four choices (a), (b), (c) and (d) out of which ONE or MORE THAN ONE are correct.

14 A particle is moving on a circular path of radius 2 m with constant angular velocity ω. In time $t = \frac{\pi}{3\omega}$,

a. the magnitude of displacement is 2 m b. the magnitude of displacement is $\frac{4\pi}{3}$ m

c. distance travelled by particle is $\frac{2\pi}{3}$ m d. distance travelled by particle is 2 m

15 A cyclist is travelling on a straight road with constant speed v_0. The radius of wheels is R. The wheels of bicycle are rotating with constant angular velocity ω. Mark the correct options.

a. The speed of top point of front wheel measured by cyclist is $R\omega$
b. The speed of top point is $v_0 + R\omega$
c. The speed of top point is $v_0 - R\omega$
d. The speed of top point is $\sqrt{v_0^2 + R^2\omega^2}$

16 A ceiling fan is turning at $\frac{1}{\pi}$ rpm. A bug is moving on the blade with velocity $v_0 = 1$ cm/s towards the axis of the fan with respect to blade. Then,

a. the speed of the bug at distance 30 cm from the axis of fan is $\sqrt{2}$ cm/s
b. the acceleration of the bug at distance 30 cm from axis of fan is $\frac{1}{30}$ cm/s^2
c. the acceleration of the bug at distance 30 cm from axis of fan is 1 cm/s^2
d. None of the above

Working Space

17 A car is moving with a speed of 10 m/s on a circular path of radius 25 m. Driver of car applies the brakes producing a uniform deceleration of 3 m/s^2. Then,

a. the centripetal acceleration of car just after applying the brake is 4 m/s^2
b. the acceleration just after applying the brake is 5 m/s^2
c. the acceleration is directed towards the centre just after applying the brake
d. the angle between acceleration and velocity just after applying the brake is 127°

18 A particle is moving on a circular path of radius 1 m. The angle turned by the radius vector of the particle is $\theta = (3t + 4t^2)$, where t is in second and θ is in radian. Then,

a. the centripetal acceleration at $t = 1$ s is 121 m/s^2
b. the tangential acceleration at $t = 15$ is 8 m/s^2
c. the acceleration of the particle at $t = 1$ s is 121.26 m/s^2
d. None of the above

19 If a particle is moving on a circular path with increasing speed, then which of the following options are correct?

a. Acceleration and velocity never be perpendicular to each other
b. Acceleration is always directed towards centre
c. Angular acceleration and angular velocity are along the axis of rotation in the same direction
d. Tangential acceleration and angular acceleration are always perpendicular to each other

Comprehension Type

Direction *(Q. Nos. 20 and 21) This section contains a paragraph describing theory, experiments, data, etc. Two questions related to the paragraph have been given. Each question has only one correct answer among the four given options (a), (b), (c) and (d).*

Passage

A small bead is fixed on a circular loop of radius R as shown in the figure below. The loop is rotating about YY'-axis with constant angular acceleration α. The loop starts from the rest.

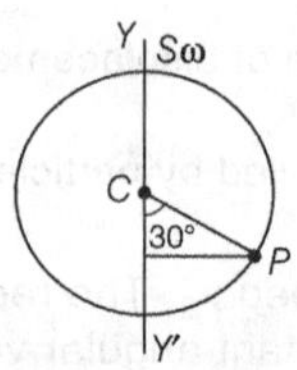

20 The Bead is in circular motion, then

a. the speed of the bead at instant t is $v = R\alpha t$
b. the speed of the bead at instant t is $v = \frac{R\alpha t}{2}$
c. the acceleration of the bead at instant t is $\frac{R}{2}\alpha\sqrt{1+\alpha^2 t^4}$
d. Both (b) and (c) are correct

21 A bug is moving along the axis of the circular loop with a constant velocity $\mathbf{v} = v_0\hat{\mathbf{j}}$. The mass of bead is m. At $t = 0$, the bead is at a point on positive X-axis. Find the momentum of the bead seen by the bug at instant, $t = \sqrt{\frac{\pi}{2\alpha}}$.

a. $-\frac{mR}{4}\sqrt{\pi\alpha}\,\hat{\mathbf{i}} - mv_0\hat{\mathbf{j}} - \frac{mR}{4}\sqrt{\pi\alpha}\,\hat{\mathbf{k}}$
b. $\frac{mR}{4}\sqrt{\pi\alpha}\,\hat{\mathbf{i}} + mv_0\hat{\mathbf{j}} + \frac{mR}{4}\sqrt{\pi\alpha}\,\hat{\mathbf{k}}$
c. $-mR\sqrt{\pi\alpha}\,\hat{\mathbf{i}} - mv_0\hat{\mathbf{j}} - mR\sqrt{\pi\alpha}\,\hat{\mathbf{k}}$
d. None of these

Working Space

One Integer Value Correct Type

Direction (Q. Nos. 22-24) This section contains 3 questions. When worked out will result in an integer from 0 to 9 (both inclusive).

22 A particle is moving on a circular path of radius 1 m. It starts from rest, its tangential acceleration is $\sqrt{3}\,t$. The time after which the acceleration of the particle makes an angle of 30° with centripetal acceleration is $(n)^{2/3}$ s. Find the value of n.

23 A particle moves on a circular path with constant angular velocity $\omega = \sqrt{2}$ rad/s about the line $x = \frac{y}{2} = \frac{z}{2}$, find the speed (in m/s) of the particle at the instant, it passes through the point (2 m, 3 m, 5 m).

24 A particle is moving on a circular path of radius 0.5 m with a tangential deceleration that is always half of the centripetal acceleration. The initial velocity is 20 m/s. If speed of the particle at $t = 5$ s is $\frac{4x}{101}$ m/s then, find the value of x.

Matching List Type

Direction (Q. No. 25) Choices for the correct combination of elements from Column I and Column II are given as options (a), (b), (c) and (d), out of which one is correct.

25 A particle moves on a circular path of radius 7 m with constant speed 7 m/s. Match the Column I with column II and mark the correct option given below.

	Column I		Column II
i.	The magnitude of displacement in metre of the particle during travelling the quarter of circle.	p.	6.3
ii.	The magnitude of average velocity in m/s during travelling quarter of circle.	q.	7
iii.	The magnitude of acceleration in m/s^2 of the particle.	r.	$\frac{\pi}{2}$
iv.	The magnitude of average acceleration in m/s^2 during travelling a quarter of circle.	s.	$7\sqrt{2}$

Codes

	i	ii	iii	iv		i	ii	iii	iv
a.	s	p	q	p	b.	p	q	r	s
c.	q	p	s	r	d.	s	r	p	q

Statement Type

Direction (Q. Nos. 26 and 27) This section is based on Statement I and Statement II. Select the correct answer from the codes given below.

a. Both Statement I and Statement II are correct and Statement II is the correct explanation of Statement I
b. Both Statement I and Statement II are correct and Statement II is not the correct explanation of Statement I
c. Statement I is correct but Statement II is incorrect
d. Statement II is correct but Statement I is incorrect

26 **Statement I** In decelerated circular motion, magnitude of acceleration is lesser than the magnitude of radial acceleration.

Statement II The acceleration of the particle in the case of circular motion is found by resultant of radial acceleration and tangential acceleration.

Working Space

27 **Statement I** In an atom, an electron can move around the nucleus in two circular orbits of radii r and $4r$. The ratio of the times taken by it to complete one revolution in two orbits is 1: 8.

Statement II Time period T and radius r of circular path of electron case of atom is related as $T^2 \propto r^3$.

Subjective Questions

Direction (Q. Nos. 28-30) *These questions are subjective in nature, need to be solved completely on notebook.*

28 A particle is moving on a circular path whose angular velocity is $\omega_0 e^{-\frac{t}{t_0}}$. (given, $\omega_0 = 3.6 \times 10^3$ rpm and $t_0 = 10^4$ s)

i. How long does it take for the particle to reach half its original angular velocity?
ii. Find the angular acceleration at $t = 0$.
iii. How many revolutions does the particle turn during the time $t = 0$ to $t = t$?

29 A particle starts from rest on a circular path with an angular acceleration $\alpha = (t^2 - t + 1)$, where t is in second and α is in rad/s^2.

i. Find angular velocity at $t = 1$ s.
ii. Find the angular position as a function of time.
iii. Find mean angular acceleration from $t = 0$ to $t = 1$ s.

30 A particle of mass m, moving in a circular path of radius R with a constant speed v_2 is located at point $(2R, 0)$ at time $t = 0$ and a man starts moving with a velocity v_1 along the positive Y-axis from origin at time $t = 0$. Calculate the linear momentum of the particle with respect to man as a function of time. **[IIT JEE 2003]**

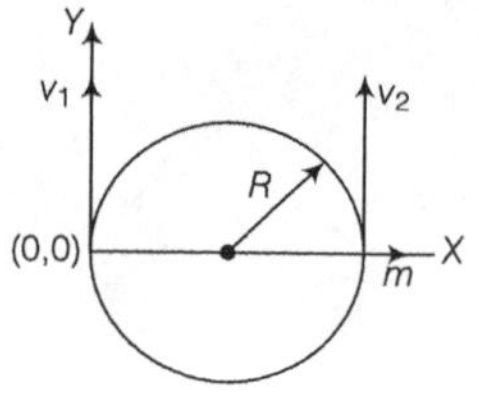

DPP-2 Uniform Circular Motion

Conceptual Questions

Working Space

Direction (Q. Nos. 1-3) *These questions are conceptual in nature, need to be solved completely on notebook.*

1 Mark the correct option/options.

i. If a particle moves on a circular path with constant speed, then its motion is known as uniform circular motion. In the case of a uniform circular motion,

a. tangential acceleration is zero

b. radial acceleration is zero

c. resultant force is directed towards centre

d. dynamic equations are $\Sigma F_x = \frac{mv^2}{r} = mr\omega^2$, $\Sigma F_y = 0, \Sigma F_z = 0$

Here, F_x = sum of components of forces towards centre,

F_y = sum of components of forces along velocity and

F_z = sum of components of forces perpendicular to the plane of circular path and kinematic equations are $\theta = \omega t$ and $s = vt$

ii. A planet is moving in a circle around the sun, then

a. the resultant force acting on the planet is gravitational attraction of the sun

b. the direction of gravitational force on the planet is towards the centre of the circular path

c. this is an example of a uniform circular motion

d. if F is gravitational force on the planet due to the sun, then $F = mv^2/r$, where r is the radius of circle and v is the speed of the planet

iii. In the Bohr model of hydrogen atom, the electron is moving on a circular path of radius 5.3×10^{-11} m around the fixed proton. The electrostatic attractive force between electron and proton is $\frac{9 \times 10^9 (1.6 \times 10^{-19})^2}{r^2}$, where r is the distance in SI between them.

a. Proton attracts the electron towards it by electrostatic force

b. The electrostatic force on electron provides centripetal force

c. Tangential acceleration is zero

d. This is an example of a uniform circular motion

iv. In previous problem, the dynamic equation for the motion of the electron is

a. $\Sigma F_x = \frac{9 \times 10^9 (1.6 \times 10^{-19})^2}{r^2} = \frac{mv^2}{r}$

where, m = mass of electron and r = radius of circular path

b. $\Sigma F_x = 0$

c. $\Sigma F_x = \frac{9 \times 10^9 (1.6 \times 10^{-19})^2}{r^2} = \frac{mv^2}{r^2}$

d. None of the above

v. In previous problem, the mass of the electron is 9.1×10^{-31} kg, the speed of the electron is

a. zero b. 2.2×10^6 m/s c. 4.4×10^6 m/s d. None of these

Working Space

2 Find normal reaction at point *A* in the following cases,

i. A block of mass *m* is moving with velocity v_0 on top point *A* on a vertical circle of radius *R*.

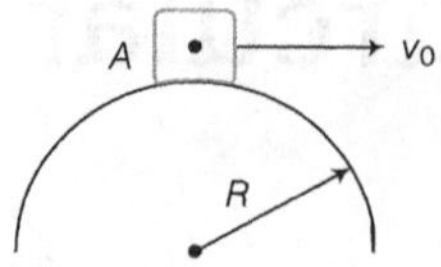

ii. A block of mass *m* is moving with velocity v_0 at lowest point *A* of a vertical circle of radius *R*.

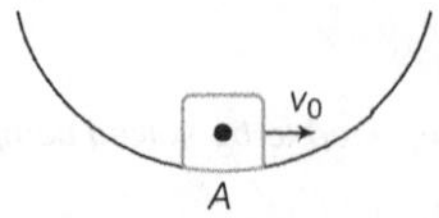

iii. A block of mass *m* is moving with velocity v_0 at point *A* as shown in the figure.

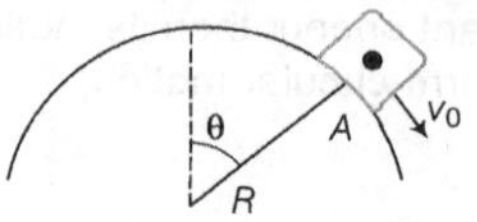

iv. A block of mass *m* is moving with velocity v_0 at point *A* as shown in the figure.

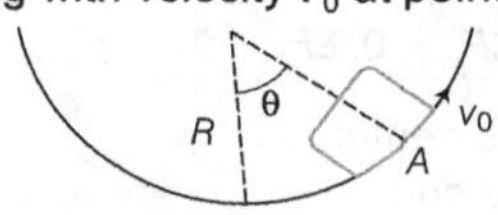

v. A bead of mass *m* is attached to one end of a spring of natural length *R* and spring constant $\sqrt{2}\,mg$. The other end is fixed at point *B* of a vertical circular wire of radius $(\sqrt{2}+1)$ m. Find normal reaction on bead after released.

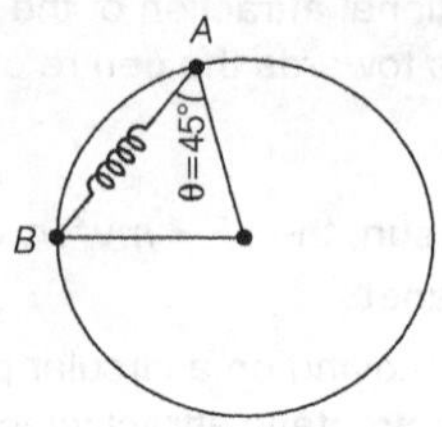

3 Find tension in light string or light rod of length *l* at the instant as shown in the figure in the following cases.

i.

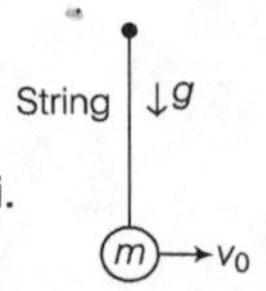

ii.

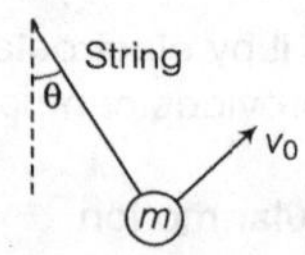

iii.

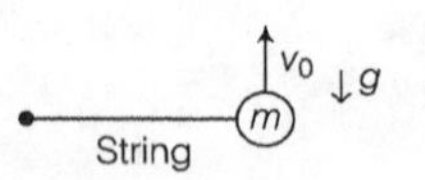

iv.

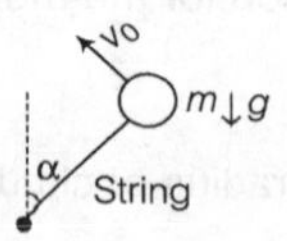

v.

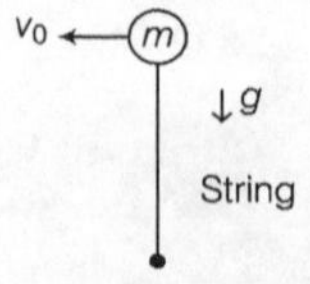

vi.

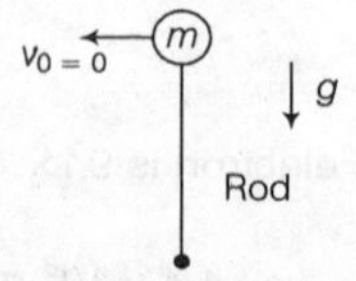

Only One Option Correct Type

Working Space

Direction (Q. Nos. 4-18) *This section contains 15 multiple choice questions. Each question has four choices (a), (b), (c) and (d), out of which ONLY ONE option is correct.*

4 A block of mass $\frac{1}{\pi^2}$ kg is constrained to move in a circular path and makes one revolution in 1 s. It is attached to the centre of circle by a light spring of spring constant 5 N/m and natural length 1 m. The radius of the circular path is

a. 1 m b. 1.1 m c. 3 m d. 5 m

5 A bead of mass m rests on a vertical smooth parabola $x^2 = cy$ at $(-a, b)$ as shown in the figure. The required value of angular velocity ω is

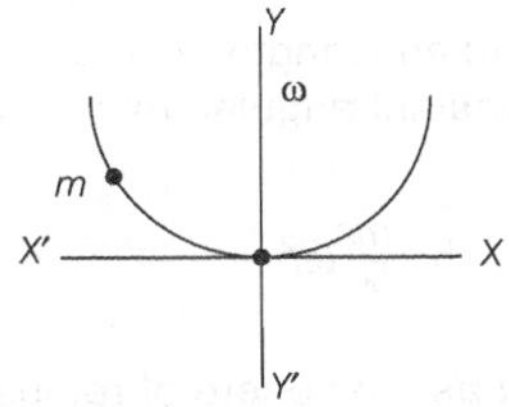

a. $\sqrt{\frac{2g}{a}}$ b. $\sqrt{\frac{2g}{c}}$

c. $\sqrt{\frac{2g}{b}}$ d. $\sqrt{\frac{2gc}{ab}}$

6 A small block mass 1 kg is connected by a light string passing through a hole in a smooth table with a block of mass 5 kg which hangs vertically. Find the angular velocity of the block on the table moving in a circle of radius 0.1 m in order to keep the other block at rest.

a. 10 rad/s b. $10\sqrt{5}$ rad/s c. $10\sqrt{2}$ rad/s d. $5\sqrt{2}$ rad/s

7 A block of mass 1 kg attached to a light string and whirled around in a horizontal circle of radius 1 m with angular velocity 10 rad/s. Find minimum tension in string so that string does not break.

a. 50 N b. 100 N c. 200 N d. 500 N

8 A small block of mass 10 g is fastened to one end of a light string and a second block m_2 of mass 20 g to the middle point, the other end of the string fastened to a fixed point on a smooth horizontal table. The blocks are then projected such that the two portions of the string are always in the same straight line and the blocks describe horizontal circles on the table as shown in the figure below.

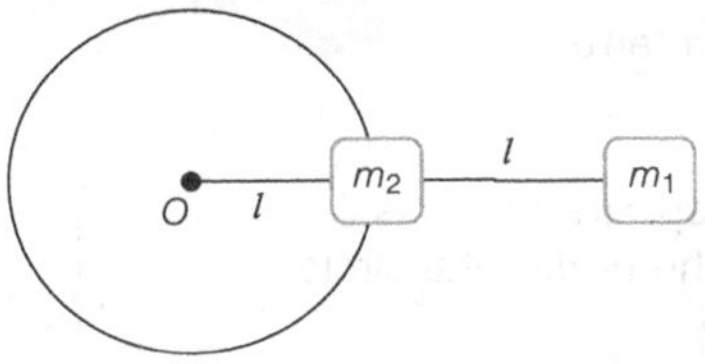

The tension in string portion between blocks is T_1 and tension between point O and the block of mass m_2 is T_2. The value of $\frac{T_1}{T_2}$ is

a. 1 b. 2 c. 0.5 d. 0.25

9 A phonograph turntable rotates at constant angular velocity $\sqrt{5}$ rad/s. A coin is placed at distance 0.5 m from the centre of the phonograph. Find the minimum value of coefficient of static friction so that the coin does not slip over the phonograph.

a. 0.25 b. 0.5 c. 0.75 d. 0.8

10 A circular ring of mass 3.14 kg and radius 0.05 m is rotating with constant angular velocity 10 rad/s about its centre. The tension in the wire of ring is

a. 1 N b. 2 N c. 2.5 N d. 5 N

Working Space

11 A bug rests on vertical part *BC* of the frame *ABC*. The frame is rotating with constant angular velocity ω about ZZ'-axis. The length of portion *AB* of the wire is l. Find possible value of coefficient of static friction between the wire and bug so that bug does not slip over the wire.

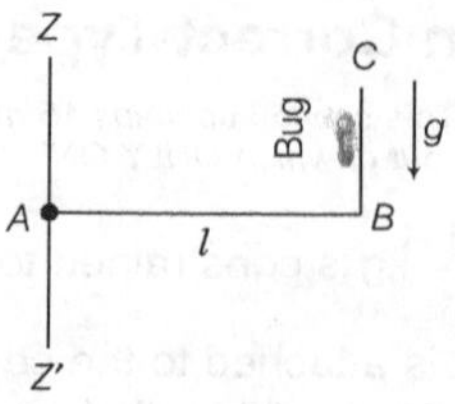

a. $\frac{2g}{l\omega^2}$ b. $\frac{g}{\omega^2}$

c. $\frac{g}{2l\omega^2}$ d. $\frac{g}{l\omega^2}$

12 A thin rod of mass *m* and length *l* is rotating in the horizontal plane about its one of the ends with constant angular velocity ω. The tension at the middle point of the rod is

a. $ml\omega^2$ b. $\frac{m}{4}l\omega^2$ c. $\frac{3m}{8}l\omega^2$ d. $\frac{ml\omega^2}{8}$

13 If earth is assumed as the sphere of radius 6400 km. The ratio of centripetal force on a body at equator and the true weight of the body is

a. 1 b. $\frac{1}{303}$ c. $\frac{1}{5}$ d. $\frac{1}{10}$

14 The maximum velocity (in ms^{-1}) with which a car driver must traverse a flat curve of radius 150 m and coefficient of friction 0.6 to avoid skidding is

a. 60 b. 30 c. 15 d. 25

15 A car is travelling on a circular track of radius *r*. The centre of gravity is at height *h* above the track. The distance between the wheels is *d*. The coefficient of friction between tyres and the track is μ. The value of *h* at which the car may be on the point of both slipping as well as overturning simultaneously

a. $\frac{d}{\mu}$ b. $\frac{d}{2\mu}$ c. $\frac{d}{8\mu}$ d. *d*

16 In previous problem, if track is banked at an angle α, the value of *h* is

a. $\frac{d}{\mu \tan\alpha}$ b. $\frac{d\tan\alpha}{\mu}$

c. $\frac{d}{2\mu}$ d. $\frac{d\tan\alpha}{2\mu}$

17 Two wires *AC* and *BC* are tied at *C* of small sphere of mass 5 kg which revolves at a constant speed *v* in the horizontal circle of radius 1.6 m. The minimum value of *v* is

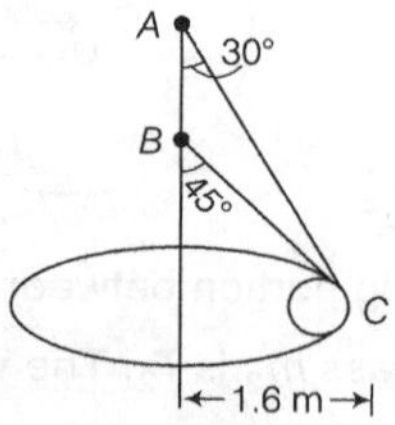

a. 3.01 m/s b. 4.01 m/s

c. 8.2 m/s d. 3.96 m/s

18 The small spherical balls are free to move on the inner surface of the rotating spherical chamber of radius 0.2 m. If the balls reach a steady state at angular position $\theta = 45°$, then angular velocity ω of device is

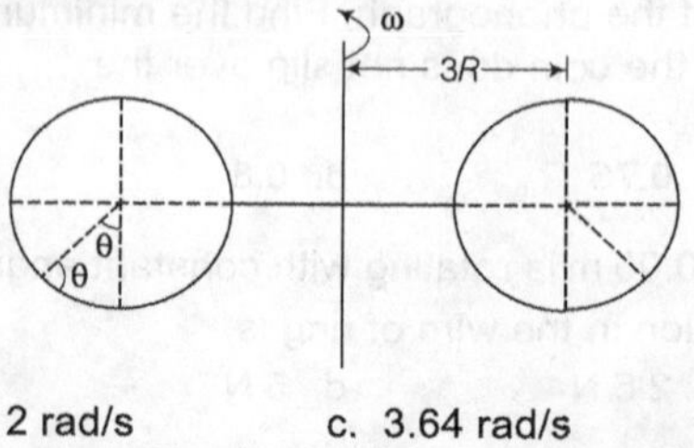

a. 8 rad/s b. 2 rad/s c. 3.64 rad/s d. 9.34 rad/s

One or More than One Options Correct Type

Working Space

Direction (Q. Nos. 19 and 20) *This section contains 2 multiple choice questions. Each question has four choices (a), (b), (c) and (d) out of which ONE or MORE THAN ONE are correct.*

19 An aeroplane is flying at constant speed of 100 m/s in a vertical circle of radius 1000 m. The force exerted by the seat on the pilot of mass 100 kg at lowest point of the trajectory is N_A and that at the highest point is N_B. Then,

a. $N_A = 0, N_B = 2000\,N$
b. $N_A = N_B = 2000\,N$
c. $N_A = 2000\,N$
d. $N_B = 0$

20 A cyclist is moving on a circular path of radius $\sqrt{3}$ m. The coefficient of friction between tyre and ground is $\mu_s = \frac{1}{\sqrt{3}}$.

a. The maximum speed of cyclist to safe riding is $\sqrt{10}$ m/s
b. The maximum inclination with vertical is 30°
c. Friction provides required centripetal acceleration
d. The line of action of net contact force is passing through the centre of gravity of cyclist system

Comprehension Type

Direction (Q. Nos. 21 and 22) *This section contains a paragraph, describing theory, experiments, data, etc. Two questions related to the paragraph have been given. Each question has only one correct answer among the four given options (a), (b), (c) and (d).*

Passage

The length of each light rod is l. The mass of each small ball B and D is m. Balls are connected with a light smooth sleeve at point A as shown in the figure below. The frame is rotating with constant angular velocity ω. A force F_0 is applied on the sleeve.

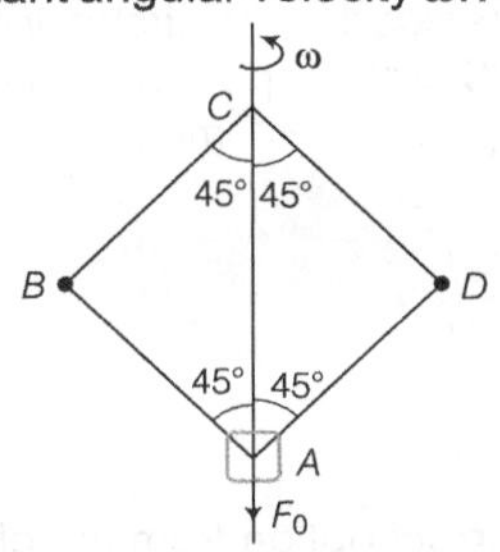

21 If tension in AB wire is T_1 and that in wire BC is T_2. Then the value of $(T_2 - T_1)$ is

a. mg
b. $\sqrt{2}\,mg$
c. $2\,mg$
d. $\frac{mg}{\sqrt{2}}$

22 The required force F_0 on the sleeve is

a. $\sqrt{2}\,ml\omega^2 - 2\,mg$
b. $\sqrt{2}\,ml\omega^2$
c. $2\,mg$
d. $\sqrt{2}\,ml\omega^2 + 2\,mg$

One Integer Value Correct Type

Working Space

Direction *(Q. Nos. 23 and 24) This section contains 2 questions. When worked out will result in an integer from 0 to 9 (both inclusive).*

23 A small block of mass 1 kg is held by a light string rests on a smooth inclined plane which can turn about ZZ'-axis with an angular velocity 2 rad/s as shown in the figure. The block is at distance $\frac{1}{\sqrt{2}}$ m from the point O. The tension in the string is $(32)^{\frac{1}{x}}$ N. Find the value of x.

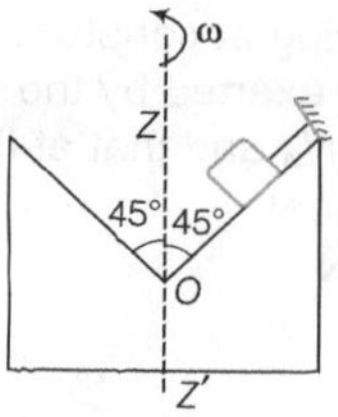

24 In the situation shown in the figure, a block A is placed on rough horizontal fixed surface and connected with a ball B. The block A is in limiting equilibrium. The coefficient of friction between block A and rough surface is $\mu = 0.5$. The ball B is $l = 0.5$ m distance below the fixed pulley. The ball is now imparted horizontally with speed $v_0 = 1$ m/s. The initial acceleration of the block A is $\frac{2}{n}$ m/s^2. Find the value of n.

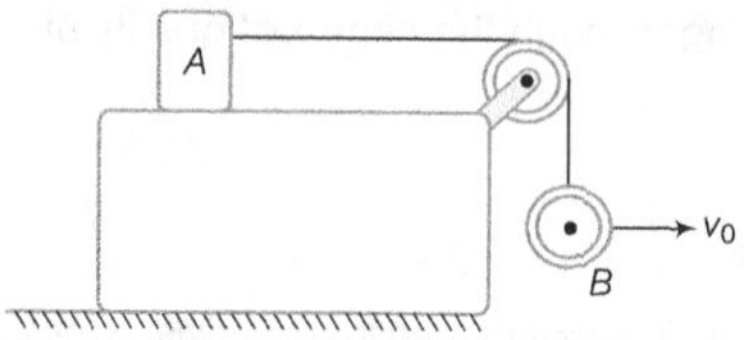

Matching List Type

Direction *(Q. No. 25) Choices for the correct combination of elements from Column I and Column II are given as options (a), (b), (c) and (d) out of which one is correct.*

25 A rough wire is bent into two circular parts of equal radius 1 m as shown in figure below. Each part is a quarter of circle. Points O and O' are respective centres. A bug of mass 1 kg travels on the wire with constant speed $v_0 = 1$ m/s.

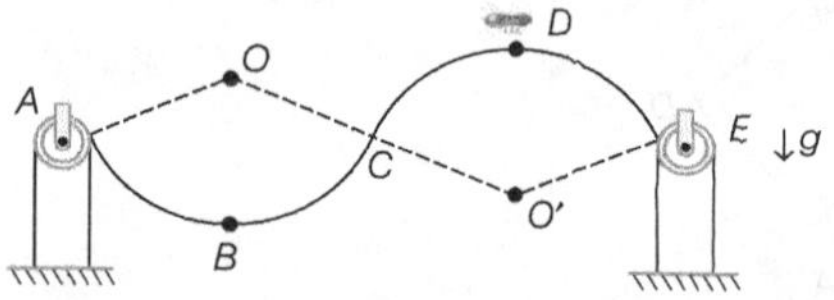

Match the Column I with Column II and select the correct option from the given codes.

	Column I		Column II
i.	Normal reaction in newton on bug at B.	p.	zero
ii.	Friction in newton on bug at D.	q.	$(5\sqrt{2}-1)$
iii.	Normal reaction in newton on bug just before crossing the point C.	r.	$\frac{10}{10-\sqrt{2}}$
iv.	Minimum coefficient of friction between bug and the wire so that bug does not slip over the wire.	s.	11

Codes

	i	ii	iii	iv
a.	s	p	q	r
b.	p	q	r	s
c.	q	p	s	r
d.	s	r	p	q

Working Space

Statement Type

Direction (Q. Nos. 26 -28) *This section is based on Statement I and Statement II. Select the correct answer from the codes given below.*

a. Both Statement I and Statement II are correct and Statement II is the correct explanation of Statement I
b. Both Statement I and Statement II are correct and Statement II is not the correct explanation of Statement I
c. Statement I is correct but Statement II is incorrect
d. Statement II is correct but Statement I is incorrect

26 **Statement I** A particle is moving on a circular path with centripetal acceleration $a_r = 2t^2$. The motion of the particle is a uniform circular motion.

Statement II In the case of a uniform circular motion, net force is always directed towards centre of the circular path.

27 **Statement I** If the body comes to rest on a circular path, i.e. $\mathbf{v} \to 0$, the body will move along radius towards the centre and if centripetal acceleration vanishes, the body will fly off tangentially if speed is non-zero.

Statement II For a uniform circular motion, tangential velocity and centripetal acceleration are necessary.

28 **Statement I** A person applies a constant force **F** on a particle of mass m and find that the particle moves on a circular path of radius r with constant speed as seen from inertial reference frame. The other force on the particle is $\mathbf{F}_2$.

Statement II The resultant of **F** and $\mathbf{F}_2$ is always directed towards the centre of circular path.

Analytical Questions

Direction (Q. Nos. 29-31) *These questions are subjective in nature, need to be solved completely on notebook.*

29 A small sphere of mass 0.2 kg is attached to a light string of length 1.3 m whose upper end is fixed to ceiling. The sphere is made to describe a horizontal circle of radius 0.5 m.

i. Calculate the time taken by the ball to complete one revolution.
ii. Find the tension in the string.

30 A hemispherical bowl of radius 0.1 m is rotating about its own axis (which is vertical) with an angular velocity ω. A particle of mass 10^{-2} kg on the frictionless inner surface of the bowl is also rotating with the same ω. The particle is at a height h from the bottom of bowl.

i. Obtain the relation between h and ω. What is the minimum value of ω needed in order to have a non-zero value of h?
ii. It is desired to measure g (acceleration due to gravity) using the set up by measuring h accurately. Assuming that R and ω are known precisely and that the least count in the measurement of h is 10^{-4} m. What is the minimum possible error Δg in the measured value of g?

31 A circular track of radius R is banked so that there is no friction at speed v_0 of the vehicle of mass m. Find the friction when the vehicle is moving with speed v.

i. $v < v_0$
ii. $v > v_0$

DPP-3 Non-Uniform Circular Motion

Only One Option Correct Type

Working Space

Direction (Q. Nos. 1-7) *This section contains 7 multiple choice questions. Each question has four choices (a), (b), (c) and (d) out of which ONLY ONE option is correct.*

1 Which of the following best represents the plot of the centripetal acceleration a_r against speed v for a particle moving with an increasing speed on a circular path.

a. 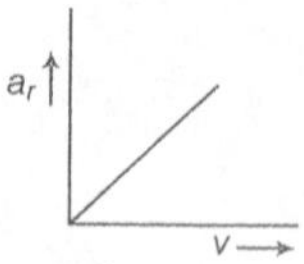

b. 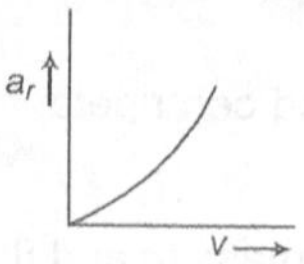

c. 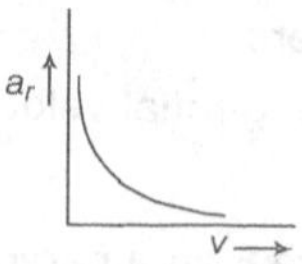

d. 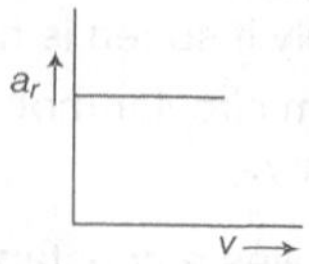

2 An angular ring with inner and outer radii R_1 and R_2 is rolling without slipping with a uniform angular speed. The ratio of the forces experienced by the two particles situated on the inner and outer parts of the ring F_1/F_2 is **[AIEEE 2005]**

a. $\frac{R_2}{R_1}$ b. $\left(\frac{R_1}{R_2}\right)^2$ c. 1 d. $\frac{R_1}{R_2}$

3 A block of mass m is gently placed in a circular groove in a turntable. Vertical walls of groove are smooth while coefficient of kinetic friction between bottom surface of groove and the block is $\mu = 0.2$. The turntable is rotating rapidly about the vertical axis passing through the centre. Find the speed gained by the block in 2 s.

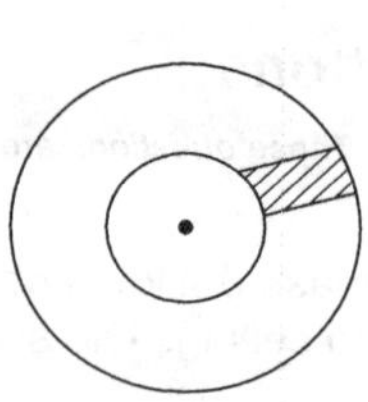

a. 1 m/s b. 2 m/s
c. 3 m/s d. 4 m/s

4 In the previous problem, if bottom surface is smooth while the coefficient of kinetic friction between outer vertical wall of groove and the block is μ. The width of block is lesser than the width of the groove. The block starts with speed v_0 along the groove. Find the speed of the block after one revolution is.

a. $2\pi\mu v_0$ b. $v = v_0 e^{2\pi\mu}$
c. $v = v_0 e^{-2\pi\mu}$ d. $v = v_0 e^{-2\pi}$

5 A small block is shot into each of the four tracks as shown in figure below. Each of the tracks rises to the same height. The speed with which the block enters the track is the same in all cases. At the highest point of the track, the normal reaction is maximum in **[IIT JEE 2001]**

a. 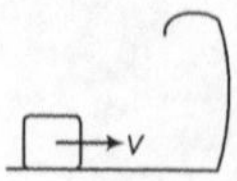

b. 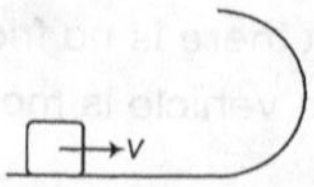

c. 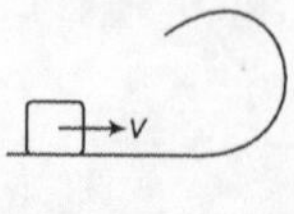

d. 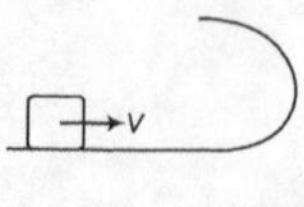

Working Space

6 A simple pendulum of length L and mass (bob) M oscillating in a plane about a vertical line between angular limits $-\phi$ and $+\phi$. For an angular displacement θ ($|\theta| < \phi$) tension in the string and the velocity of the bob are T and v, respectively. The relation(s) hold good under above conditions is/are.

a. $T \cos\theta = Mg$

b. $T - Mg\cos\theta = \dfrac{Mv^2}{L}$

c. The magnitude of the tangential acceleration of the $|a_T| = g\sin\theta$

d. $T = Mg\cos\theta$

7 A long horizontal rod has a bead which can slide along its length and is initially placed at a distance L from one end A with a constant angular acceleration α. If the coefficient of friction between the rod end bead is μ and gravity is neglected, then the time after which the bead starts slipping is

a. $\sqrt{\dfrac{\mu}{\alpha}}$ b. $\dfrac{\mu}{\sqrt{\alpha}}$ c. $\dfrac{1}{\sqrt{\mu\alpha}}$ d. infinitesimal

One or More than One Options Correct Type

Direction (Q. Nos. 8-11) *This section contains 4 multiple choice questions. Each question has four choices (a), (b), (c) and (d), out of which ONE or MORE THAN ONE are correct.*

8 If a particle moves on a circular path with variable speed, its motion is known as non-uniform circular motion. In the case of non-uniform circular motion, which of the following statements are correct?

a. Radial acceleration is non-zero

b. Tangential acceleration is non-zero

c. Net force is not directed towards the centre of circular path

d. Dynamic equations are

$\Sigma F_x = ma_r = mv^2/r = mr\omega^2,\ \Sigma F_y = ma_t = mr\alpha,\ \Sigma F_z = 0$

Here, ΣF_x = the sum of components of forces towards the centre of circular path.

ΣF_y = the sum of components of forces in the direction of velocity and

ΣF_z = the sum of components of forces perpendicular to the plane of circular path.

Kinematic equations are

Case I If α is non-zero constant, then $\theta = \omega_0 t + (1/2)\alpha t^2,\ \omega = \omega_0 + \alpha t,\ \omega^2 = \omega_0^2 + 2\alpha\theta$

Case II If α is variable, then

$$\alpha = \frac{d\omega}{dt},\ \int d\omega = \int \alpha dt,\ \alpha = \frac{\omega d\omega}{d\theta},\ \int \omega d\omega = \int \alpha d\theta,\ \omega = \frac{d\theta}{dt},\ \int d\theta = \int \omega dt$$

9 The situation shown in the figure below, which of the following are true?

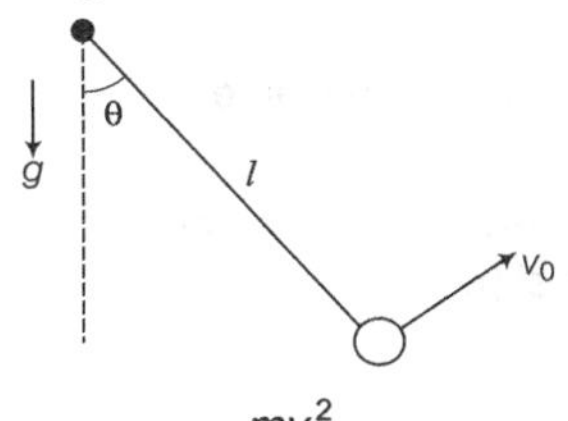

a. The tension in string is $mg\cos\theta + \dfrac{mv_0^2}{l}$

b. The magnitude of tangential acceleration of bob is $g\sin\theta$

c. The acceleration **a** of the bob is as indicated in the figure

d. The angular acceleration of the bob is $\dfrac{g}{l}\sin\theta$

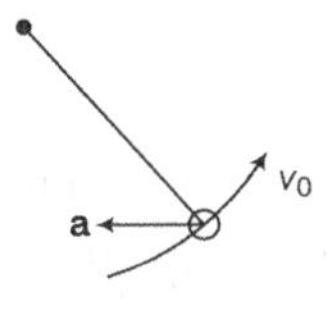

Working Space

10 A block of mass 1 kg is hanging as shown in the figure below. Just after cutting the spring,

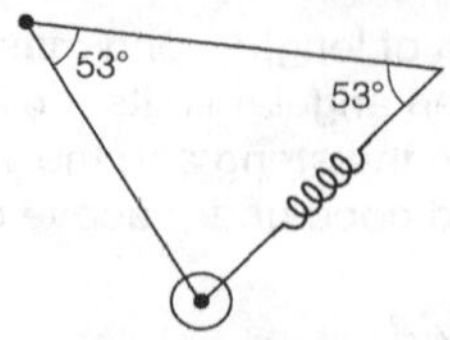

a. the tension in the light string is 8 N
b. the acceleration of the ball is 6 m/s^2 directed downward
c. the acceleration of the ball is 6 m/s^2 directed perpendicular to the string
d. the acceleration of the ball is 10 m/s^2

11 A cyclist is moving on a circular path of radius R with increasing speed. The rate of increase of speed is a_0. The coefficient of friction between tyre and horizontal ground is μ.
a. Friction is directed towards centre
b. Maximum angle of inclination with vertical is $\tan^{-1}\mu$
c. Friction is not directed towards centre
d. Friction provides tangential as well as centripetal acceleration

Comprehension Type

Direction (Q. Nos. 12-16) This section contains 2 paragraphs, each describing theory, experiments, data, etc. Five questions related to the paragraphs have been given. Each question has only one correct answer among the four given options (a), (b), (c) and (d).

Passage I

A horizontal circular platform is free to rotate about its centre. A block of mass m is placed at distance l from centre on the platform. The coefficient of friction between platform and the block is μ.

12 If space is gravity free, then find the time after which the block starts to slip.
a. After one rotation of platform
b. After half rotation of platform
c. After two revolutions of platform
d. In infinitesimally small time

13 If gravity is present and platform is rotating with constant angular velocity ω, then find the maximum value of ω for which block does not slip.
a. $\sqrt{\frac{\mu g}{l}}$
b. $\sqrt{\frac{2\mu g}{l}}$
c. $\sqrt{\frac{3\mu g}{l}}$
d. None of these

Passage II

A bead of mass m is threaded on a light rod that is pivoted to a fixed point O at one end. The bead is at distance l from the fixed point O. The coefficient of friction between bead and rod is μ. The space is gravity free.

14 If rod is moving with constant angular velocity in a gravity free space, then find the time after which bead starts to slip.
a. $\sqrt{\frac{\mu l}{g}}$
b. $\sqrt{\frac{g}{l}}$
c. $\sqrt{\frac{l}{g}}$
d. Infinitesimally small time

15 Now rod starts to rotate with constant angular acceleration α in same gravity free region. Find the time after which the bead starts to slip.
a. $\sqrt{\mu\alpha}$
b. $\sqrt{\frac{\mu}{\alpha}}$
c. $\sqrt{\frac{2\mu}{\alpha}}$
d. Infinitesimally small time

16 Repeat the above problem, if gravity is present.
a. $\sqrt{\frac{\mu}{l\alpha^2}(g^2 + l^2\alpha^2)^{1/4}}$
b. $\sqrt{\frac{\mu}{\alpha}}$
c. $\sqrt{\frac{\mu l}{g}}$
d. None of these

Statement Type

Working Space

Direction (Q. Nos. 17 and 18) *This section is based on Statement I and Statement II. Select the correct answer from the codes given below.*

a. Both Statement I and Statement II are correct and Statement II is the correct explanation of Statement I
b. Both Statement I and Statement II are correct and Statement II is not the correct explanation of Statement I
c. Statement I is correct but Statement II is incorrect
d. Statement II is correct but Statement I is incorrect

17 **Statement I** A particle of mass m begins to slide down a fixed smooth hemisphere from the top. Its tangential acceleration is $\frac{\sqrt{5}}{3}g$ when it breaks off the sphere.

Statement II At the time of breaking the surface of hemisphere, acceleration of the particle is g.

18 **Statement I** A ball tied with a string is whirled along a vertical circle. The acceleration is always directed towards the centre of circular path.

Statement II In the case of non-uniform circular motion, acceleration is not directed towards the centre of circular path.

One Integer Value Correct Type

Direction (Q. Nos. 19-23) *This section contains 5 questions. When worked out will result in an integer from 0 to 9 (both inclusive).*

19 A bug is moving on horizontal circular path of radius R. It starts from rest with a tangential acceleration of constant magnitude a_0. The coefficient of static friction between the bug and horizontal path is μ. The time after which the bug will skid is $\frac{\{R^2(\mu^2 g^2 - a_0^2)\}^{1/n}}{a_0}$. Find the value of n.

20 A toy car starts from rest along the circular track of radius R on a horizontal board with a tangential acceleration a_0. At the same instant, board starts from rest and acceleration upward with an acceleration a_1. The coefficient of friction between board and tyres of the toy car is μ. The speed of the car relative to board at which it will skid is $[R^2\{\mu^n(g + a_1)^2 - a_0^2\}]^{1/4}$. Find the value of n.

21 A car of mass m starts from rest and is moving on a circular path of radius R. Such that its angular velocity ω varies with angular displacement θ of car as $\omega = b\sqrt{\frac{\theta}{R}}$, where b is constant. The friction acting on the car as function of θ is $\sqrt{b^4\theta^n + \frac{b^4}{4}}$. Find the value of n.

22 A vehicle of mass m starts moving along a horizontal circle without slipping, so that centripetal acceleration is $a_r = bt^2$, where b is constant. At $t = t_0$, the vehicle skids. The friction on the car as function of time t $(< t_0)$ is $nm\sqrt{b^2t^4 + bR}$. Find the value of n.

23 The situation shown in the figure, a ball of mass (m_1) 1 kg is in touch with a triangular block of mass (m_2) 2 kg placed on smooth horizontal surface. The ball is suspended through a light string from point O. The system is released from rest. The initial acceleration of triangular block is a_2 in m/s^2. Find the value of $2a_2$.

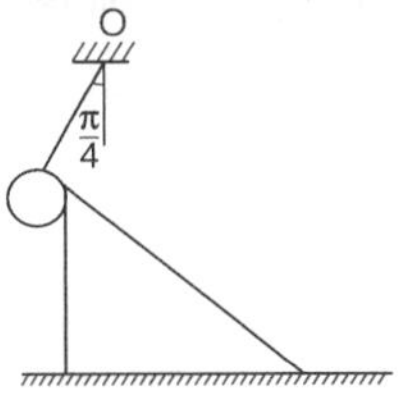

Matching List Type

Working Space

Direction (Q. No. 24) Choices for the correct combination of elements from Column I and Column II are given as options (a), (b), (c) and (d), out of which one is correct.

24 Match the Column I with Column II and select the correct option from the given codes.

	Column I		Column II
i.	Motion of a particle on circular path with increasing speed ($v \neq 0$)	p.	Acceleration is directed towards centre
ii.	Motion of a car on circular path with constant speed	q.	Acceleration is not directed towards centre
iii.	Motion of cyclist on circular path with decreasing speed ($v \neq 0$)	r.	Centripetal acceleration is non-zero
iv.	Motion of a particle on a circular path with constant angular acceleration	s.	Friction is not directed towards centre

Codes

	i	ii	iii	iv
a.	q,r	p,r	q,r,s	q,r
b.	p,q	r,s	p	s
c.	p	q	r	s
d.	q,r	p,s	q,p	q

Analytical Questions

Direction (Q. Nos. 25 and 26) These questions are analytical in nature, need to be solved completely on notebook.

25 A toy car is moving on a circular path of radius 10 m on an inclined plane of inclination 30°. The coefficient of friction between the tyres and the inclined plane is $\mu = \frac{2}{\sqrt{3}}$. Find maximum constant speed of car such that car does not slip. Assume that the width of a car is negligible with respect to radius of circular path.

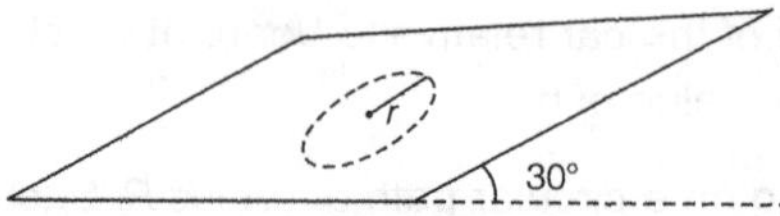

26 A smooth semicircular wire track of radius R is fixed in a vertical plane as shown in the figure below. One end of a massless spring of natural length $3R/4$ is attached to the lowest point O of the wire track. A small ring of mass m which can slide on the track is attached to the other end of the spring. The ring is held stationary at point P such that the spring makes an angle 60° with the vertical. The spring constant $k = mg/R$. Consider the instant when the ring is making an angle 60° with the vertical. The spring is released

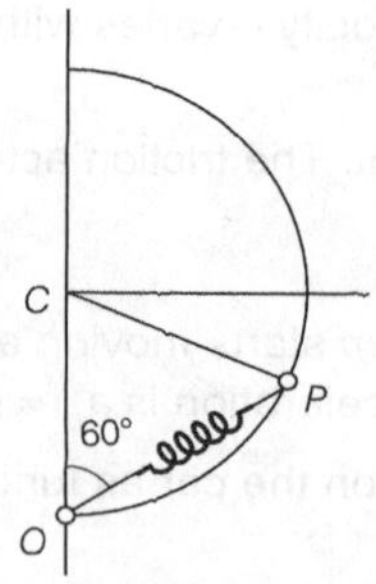

i. Draw the free body diagram of the ring.
ii. Determine the tangential acceleration of the ring and the normal reaction.

DPP-4 Centrifugal Force

Only One Option Correct Type

Working Space

Direction (*Q. Nos. 1-10*) *This section contains 10 multiple choice questions. Each question has four choices (a), (b), (c) and (d), out of which ONLY ONE option is correct.*

1 A block of mass m is placed stationary on a rough rotating platform as shown in the figure below. The free body diagram of the block in the frame of platform is

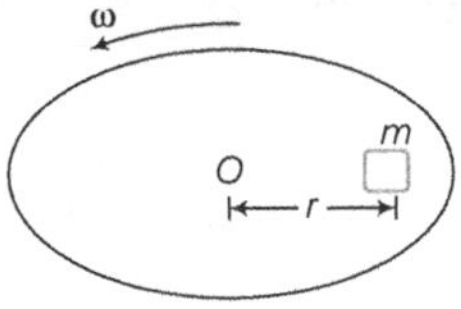

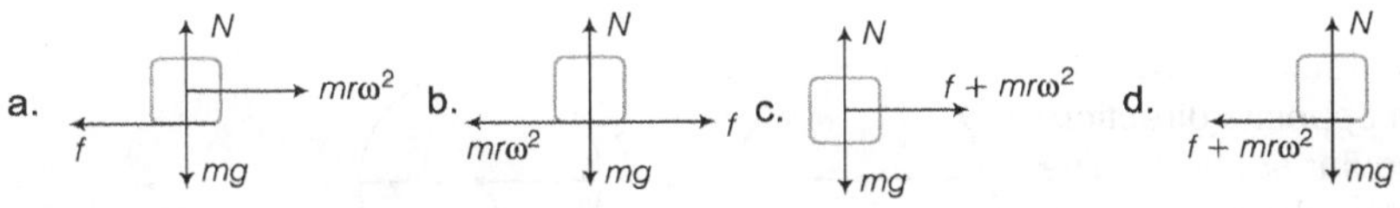

2 A particle of mass m rotates with a uniform angular velocity ω_1. It is viewed from a frame rotating with constant angular velocity ω. The centrifugal force on a particle is proportional to

a. $m\omega_1^2$ b. $m\omega^2$

c. $m(\omega_1 + \omega)^2$ d. $m\omega\omega_1$

3 A horizontal rectangular pipe is rotating with constant angular velocity of 5 rad/s. The vertical faces of the pipe are smooth but bottom face is rough. The coefficient of friction between bottom face and a block of mass 1 kg is $\mu = 0.5$. Find the acceleration of the block relative to pipe, if it is at distance 1 m from the point O.

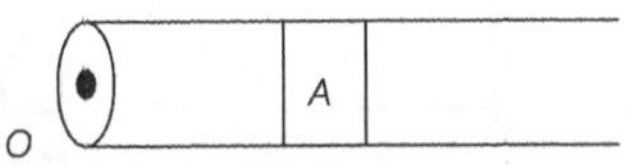

a. 25 m/s^2 b. 5 m/s^2 c. 20 m/s^2 d. 10 m/s^2

4 A platform rotates with constant angular velocity of 1 rad/s. Two masses $m_1 = 2$ kg and $m_2 = 1$ kg $(m_1 > m_2)$ slide without friction in a groove passing through the centre of platform. The initial acceleration of mass m_1 is

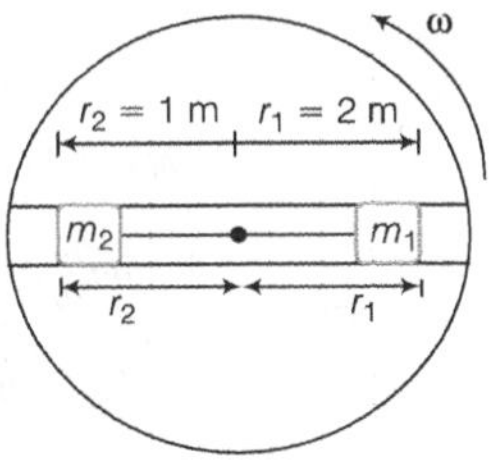

a. 5 m/s^2 b. 1 m/s^2

c. $\sqrt{5}$ m/s^2 d. 4 m/s^2

5 A toy car is driven on a large revolving platform which rotates with constant angular speed 1 rad/s. At $t = 0$, the car leaves the origin and follows a groove radially outward on the platform with constant speed $v_0 = 1$ m/s. The mass of the car is m and the coefficient of friction between the wheels of car and bottom face of groove is 0.5. Find the time after which car starts to roll along the groove.

a. 1 s b. 3 s

c. 5 s d. 10 s

Working Space

6 The situation shown in the figure, a ball of mass $\sqrt{2}$ kg connected to a vertical revolving rod by two light strings of length $\sqrt{2}$ m, each making an angle of 45° with the rod. Both the rod and ball are revolving with angular velocity of $\sqrt{10}$ rad/s.
The tension in upper string is

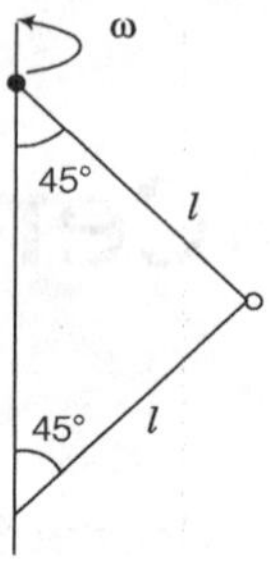

a. 20 N
b. 40 N
c. 50 N
d. zero

7 A ball of mass m hangs by a light rod of length l which rotates with constant angular velocity ω as shown in the figure below. The ball moves with constant speed in a horizontal circular path of constant radius R. The rod makes an angle α with the vertical surface. Which of the following $\cos\alpha$ *versus* ω graph is possible.

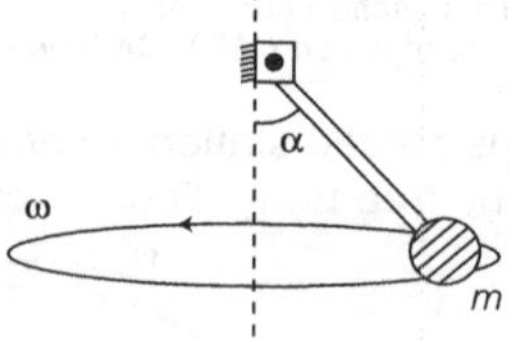

a. b. c. d.

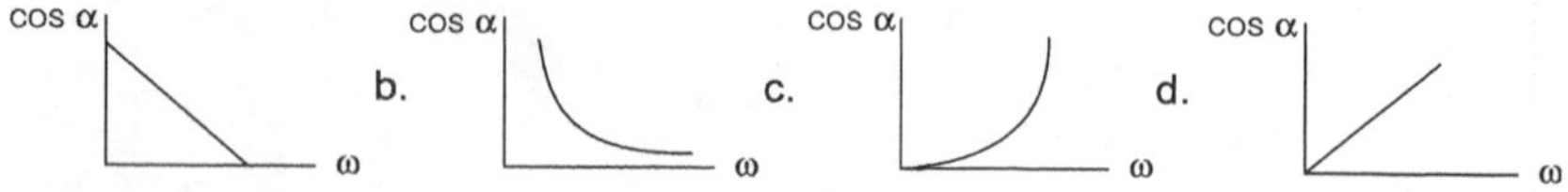

8 Two identical discs of same radius R are rotating about their axes in opposite directions with the same constant angular speed ω. The discs are in the same horizontal plane.
At time $t = 0$, the points P and Q are facing each other as shown in the figure. The relative speed between the two points P and Q is v_r. In one time period T of rotation of the discs, v_r as a function of time is best represented by **[IIT JEE 2012]**

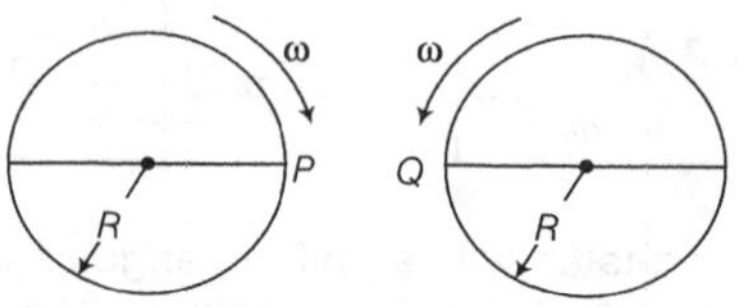

a. b. c. d.

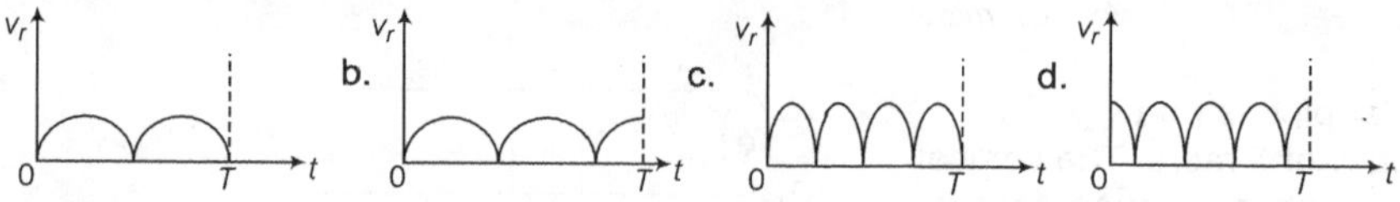

9 Consider a disc rotating in the horizontal plane with a constant angular speed ω about its centre O. The disc has a shaded region on one side of the diameter and an unshaded region on the other side as shown in the figure. When the disc is in the orientation as shown, two pebbles P and Q are simultaneously projected at an angle towards R.

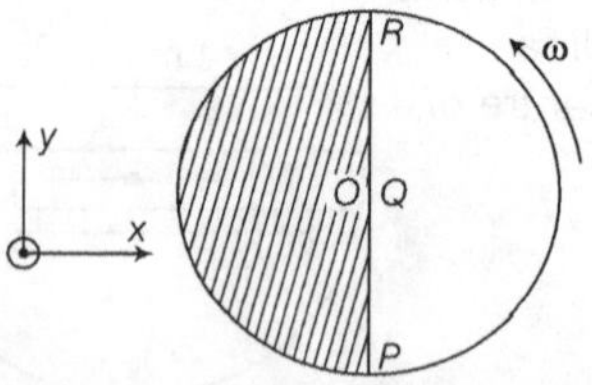

The velocity of projection is in the yz-plane and is same for both pebbles with respect to the disc. Assume that (i) they land back on the disc before the disc has completed 1/8 rotation, (ii) their range is less than half the disc radius and (iii) ω remains constant throughout. Then, **[IIT JEE 2012]**

a. P lands in the shaded region and Q in the unshaded region
b. P lands in the unshaded region and Q in the shaded region
c. both P and Q land in the unshaded region
d. both P and Q land in the shaded region

Working Space

10 A table with smooth horizontal surface is turning with constant angular velocity 2 rad/ s about point *O*. Two grooves are made on the surface along the radial lines. Smooth blocks *A* and *B* of masses 1 kg and 2 kg respectively, connected with light strings are gently placed in the grooves as shown in the figure below. Find initial tension in string, if both blocks are initial at the distance 3 m from the point *O*.

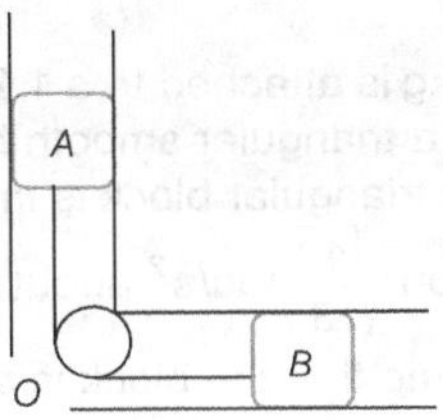

a. 10 N b. 15 N c. 16 N d. 20 N

One or More than One Options Correct Type

Direction *(Q. Nos. 11-13) This section contains 3 multiple choice questions. Each question has four choices (a), (b), (c) and (d), out of which ONE or MORE THAN ONE are correct.*

11 The length of light string *AB* is $\sqrt{2}$ m. The platform is rotating with angular speed $\sqrt{10}$ rad/s about the point *O*. The ball of mass 1 kg is stationary with respect to platform.

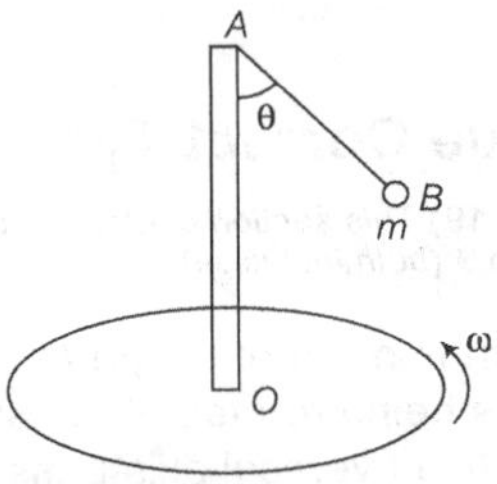

a. The value of $\theta = 45°$
b. The tension in string is $10\sqrt{2}$ N
c. The centrifugal force on the ball in the frame of platform is along the string
d. None of the above

12 A car is moving in a circular horizontal track of radius 10 m with a constant speed of 10 m/s. A plumb bob is suspended from the roof of the car by a light rigid rod. The angle made by the rod with the vertical is (take, $g = 10 \text{ m/s}^2$)

a. zero b. 30°
c. 45° d. non-zero

13 A car enters a turn whose radius is *R*. The road is banked at an angle θ and the coefficient of friction between wheels and road is μ. If car does not skid sideways at any speed. For this condition which of the following combinations is not possible

a. $\mu = 1, \theta = 45°$ b. $\mu = 0.5, \theta = 45°$
c. $\mu = 0.6, \theta = 45°$ d. $\tan\theta = \mu$

Comprehension Type

Direction *(Q. Nos. 14-17) This section contains 2 paragraphs, each describing theory, experiments, data etc. Four questions related to the paragraph have been given. Each question has only one correct answer among the four given options (a), (b), (c) and (d).*

Passage I

A block of mass 1 kg is placed on a turntable rotating with constant angular velocity ω about point O. The coefficient of friction between turntable and the block is 0.4. The length of the string is 1 m.

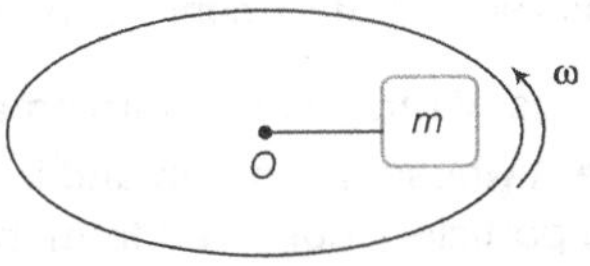

14 Find the maximum value of ω for which tension in string is zero.

a. 2 rad/s b. 4 rad/s
c. 6 rad/s d. 8 rad/s

15 The tension in string for $\omega = 4$ rad/s is

a. 4 N b. 8 N
c. zero d. 12 N

Working Space

Passage II

A block of mass 1 kg is attached to a 1.25 m long light string tied to the apex of a triangular smooth block as shown in the figure. At $t = 0$, the triangular block is imparted a constant angular acceleration $\sqrt{\frac{40}{3}}$ rad/s^2 about an axis along its vertical side. Assume that the block is stationary with respect to triangular block.

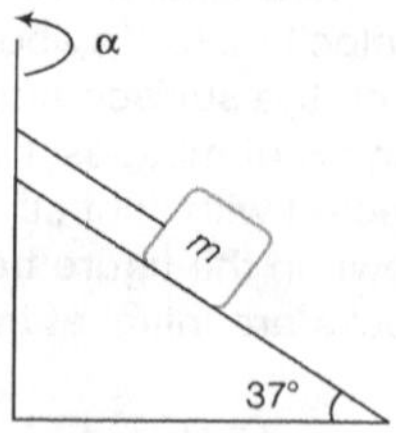

16 At what time will the block leaves contact with the triangular block?

a. 0.5 s   b. 1 s   c. 2 s   d. 3 s

17 What would be tension in the string at the instant when the block leaves the contact of triangular block?

a. 10 N   b. 16.67 N   c. 30 N   d. 40 N

One Integer Value Correct Type

Direction (*Q. Nos. 18 and 19*) *This section contains 2 questions. Each question, when worked out will result in an integer from 0 to 9 (both inclusive).*

18 A horizontal smooth thin rod of length l wearing a sleeve of mass m is being rotated at constant angular velocity ω about a fixed vertical axis passing through one of its ends. Initially, the sleeve is at distance $r = r_0$ from the axis. The speed with which sleeve leaves the rod is $\sqrt{nl^2\omega^2 - \omega^2 r_0^2}$. Find the value of n.

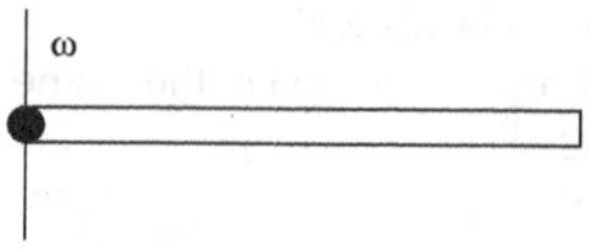

19 A small block slides on a frictionless table. It is constrained to move inside a ring of radius 1m fixed to the table. At $t = 0$, the block is moving along circumference inside the ring with speed $v_0 = 10$ m/s. The coefficient of friction between block and ring is $\mu = 0.1$. Find the velocity of the block (in m/s) at $t = 1$ s.

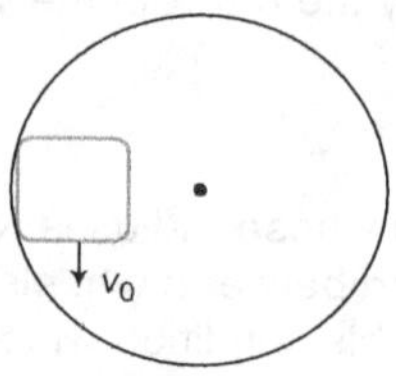

Statement Type

Direction (Q. Nos. 20 and 21) *This section is based on Statement I and Statement II. Select the correct answer from the codes given below.*

a. Both Statement I and Statement II are correct and Statement II is the correct explanation of Statement I
b. Both Statement I and Statement II are correct and Statement II is not the correct explanation of Statement I
c. Statement I is correct but Statement II is incorrect
d. Statement II is correct but Statement I is incorrect

20 **Statement I** Two rings of radii R and r turning at angular velocities ω_1 and ω_2. Consider a particle A whose mass is m and is fixed at upper ring. The centrifugal force acting on the particle A observed from the frame of lower ring is $mr\omega_1^2$.

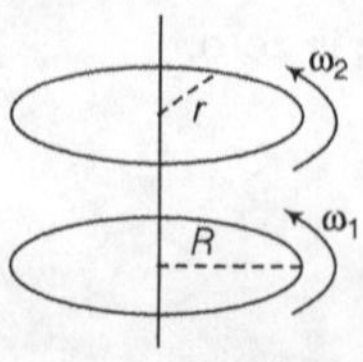

Statement II Centrifugal force is a pseudo force.

Working Space

21 **Statement I** A rough disc is rotating with constant angular velocity ω. In the frame of disc, centrifugal force on the block is balanced by frictional force if block is stationary in the frame of disc.

Statement II In the frame of ground, frictional force on block provides required centripetal force.

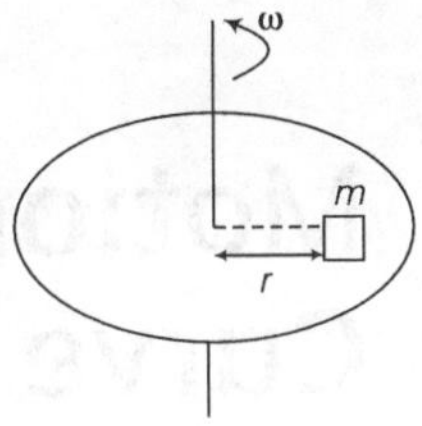

Matching List Type

Direction (Q. No. 22) Choices for the correct combination of elements from Column I and Column II are given as options (a), (b), (c) and (d), out of which one is correct.

22 In the given figure the path of a car that travels at constant speed through five circular path arcs and corresponding centrifugal force on driver in the frame of car is F_1, F_2, F_3, F_4 and F_5.

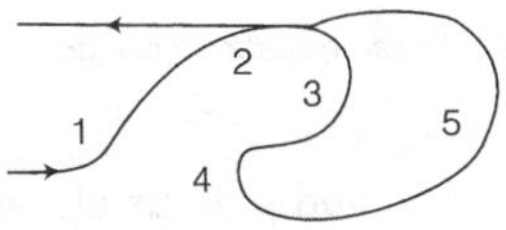

Match the Column I with Column II and select the correct option from the codes given below.

	Column I		Column II
i.	$\frac{\lvert F4\rvert}{\lvert F_3\rvert}$	p.	1
ii.	$\frac{\lvert F_2\rvert}{\lvert F_5\rvert}$	q.	< 1
iii.	$\frac{\lvert F_1\rvert}{\lvert F_4\rvert}$	r.	> 1
iv.	$\frac{\lvert F_2\rvert}{\lvert F_4\rvert}$	s.	zero

Codes

	i	ii	iii	iv		i	ii	iii	iv
a.	r	p	q	q	b.	p	q	r	s
c.	p,q	r,s	p	q	d.	s	p	r	q

Analytical Question

Direction (Q. No. 23) These questions are analytical in nature, need to be solved completely on notebook.

23 Two blocks of masses $m_1 = 10$ kg and $m_2 = 5$ kg are connected to each other by a massless inextensible string of length 0.3 m and are placed along a diameter of turntable. The coefficient of friction between the table and m_1 is 0.5 while there is no friction between m_2 and the table. The table is rotating with an angular velocity of 10 rad/s about a vertical axis passing through its centre *O*. The masses are placed along the diameter of the table on either side of the centre *O* such that the mass m_1 is at a distance of 0.124 m from centre *O*. The masses are observed to be at rest with respect to an observer on the turntable.

i. Calculate the frictional force on m_1.

ii. What should be the minimum angular speed of the turntable so that the masses will slip from this position?

iii. How should the masses be placed with the string remaining taut so that there is no frictional force acting on the mass m_1?

DPP-5 Motion on General Curve

Conceptual Questions

Direction (Q. Nos. 1 and 2) *These questions are conceptual in nature, need to be solved completely on notebook.*

Working Space

1 A particle moves in a curved path as shown in the figure. (Here, v = speed, a = acceleration)

Answer the following questions on the basis of information given above.

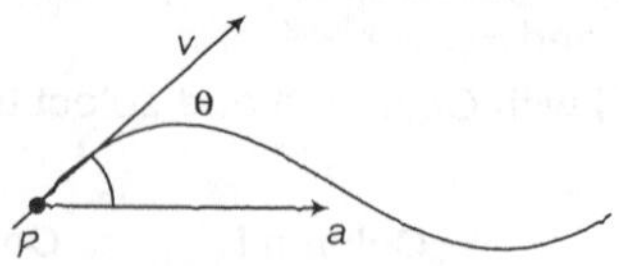

i. At point P, find tangential acceleration

a. $a_t = a\cos\theta$ b. $a_t = a\sin\theta$ c. $a_t = a\tan\theta$ d. None of these

ii. The value of $d|\mathbf{v}|/dt$ at the point P is

a. $a\sin\theta$ b. $a\cos\theta$ c. zero d. None of these

iii. The radial acceleration at point P is

a. zero b. $a\cos\theta$ c. $a\sin\theta$ d. None of these

iv. If radius of curvature at point P is r. Show the radius of curvature.

a. 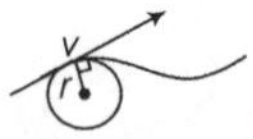 b.  c. 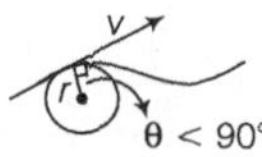

d. None of these

v. The correct relation at point P is

a. $a_r = a\cos\theta = v^2/r$ b. $a_r = a\sin\theta = v^2/r$

c. $a_r = a\cos\theta = v^2/r\sin\theta$ d. None of the above

vi. The radius of curvature at the point P is

a. $r = \dfrac{v^2}{a\sin\theta}$ b. $r = \dfrac{v^2}{av\sin\theta}$ c. $r = \dfrac{v^3}{|\mathbf{a}\times\mathbf{v}|}$ d. None of these

2 A particle is projected with speed u at an angle α with horizontal.

Answer the following questions on the basis of information given above.

i. The tangential acceleration at the projection point is

a. $a_t = g\cos\alpha$ b. $a_t = -g\sin\alpha$ c. $a_t = 0$ d. None of these

ii. The radial acceleration at the projection point is

a. $a_r = 0$ b. $a_r = -g\sin\alpha$ c. $a_r = g\cos\alpha$ d. None of these

iii. The radius of curvature at the projection point is

a. $r = \dfrac{u^2}{g\cos\alpha}$ b. $r = \dfrac{u^2}{g\sin\alpha}$ c. zero d. None of these

iv. Draw radius of curvature at maximum height.

a.  b. 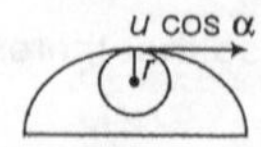

c. 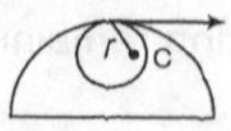 d. None of these

v. Tangential acceleration at the maximum height is

a. $a_t = g$ b. $a_t = g\cos\alpha$

c. zero d. None of these

Working Space

vi. Radial acceleration at the maximum height is

a. $a_r = g$ b. $a_r = 0$ b. $a_r = g\sin\alpha$ d. None of these

vii. If radius of curvature at the maximum height is r, then correct relation is

a. $g = \dfrac{u^2\cos^2\alpha}{r}$ b. $g\sin\alpha = \dfrac{u^2\cos^2\alpha}{r}$

c. $g = \dfrac{u^2\sin^2\alpha}{r}$ d. None of these

viii. The radius of curvature at the maximum height is

a. $r = \dfrac{u^2\sin^2\alpha}{g}$ b. $r = \dfrac{u^2\cos^2\alpha}{g}$ c. $r = 0$ d. None of these

Only One Option Correct Type

Direction (Q. Nos. 3-11) *This section contains 9 multiple choice questions. Each question has four choices (a), (b), (c) and (d), out of which ONLY ONE option is correct.*

3 A particle is moving along Y-axis with a speed of 10 m/s. Find angular velocity of the particle about at (3 m, 0) when the particle is at (0, 4 m).

a. 1.2 rad/s b. 4 rad/s c. 8 rad/s d. 10 rad/s

4 The velocity of a particle at A (1 m, 2 m, 2 m) is $(3\hat{i} + 4\hat{j})$ m/s. Find the angular velocity of the particle about at point P(2 m, 4 m, 4 m).

a. 1 rad/s b. 2 rad/s c. 3.4 rad/s d. 8 rad/s

5 A spotlight S rotates in a horizontal plane with constant angular velocity of 0.1 rad/s. The spot of light P moves along the wall at a distance of 3 m. The velocity of the spot P when $\theta = 45°$ as shown in the figure is

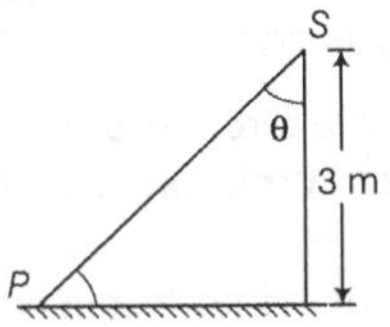

a. 0.6 m/s b. 6 m/s

c. 0.5 m/s d. 5 m/s

6 Two moving particles P and Q are 10 m apart at a certain instant. The velocity of P is 8 m/s making an angle of 30° with the line joining P and Q and that of Q is 6 m/s making an angle of 30° with PQ as shown in the figure.

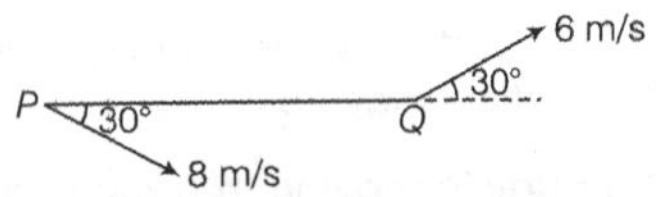

Then, angular velocity of P with respect to Q is

a. zero b. 0.1 rad/s

c. 0.4 rad/s d. 0.7 rad/s

7 A smooth wire bent in the form of parabola $x^2 = 4by$ as shown in the figure. A particle is moving with speed of $v_0 = \sqrt{2gb}$ at the vertex. Find radius of curvature of vertex, if the normal reaction on the particle at the vertex is $2mg$.

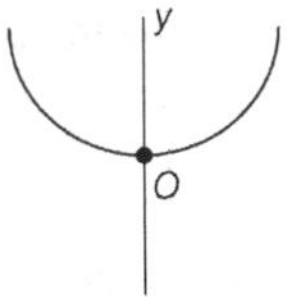

a. b b. $2b$

c. $3b$ d. $4b$

8 A particle is moving on a elliptical path whose position at instant t is $\mathbf{r} = a\cos\omega t\,\hat{i} + b\sin\omega t\,\hat{j}$. The radius of curvature at the point A as shown in the figure is

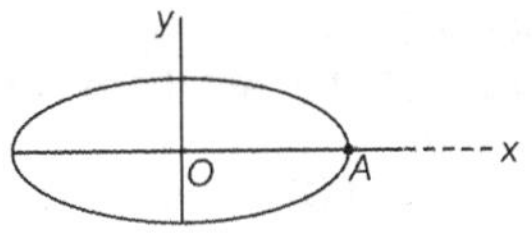

a. $\dfrac{a^2}{b}$ b. $\dfrac{b^2}{a}$

c. $\dfrac{2a^2}{b}$ d. $\dfrac{b^2}{2a}$

Working Space

9 A bug crawls at a constant rate v_0 along the spoke of a bicycle wheel of radius R, the bicycle is moving with a speed u. Find the magnitude of component of acceleration along the spoke in the form of r, the distance from centre of wheel.

a. $\frac{ru^2}{R^2}$ b. $\frac{rv_0^2}{R^2}$ c. $\frac{r}{R^2}(u^2 + v_0^2)$ d. None of these

10 A particle is acted upon by a force of constant magnitude which is always perpendicular to the velocity of the particle. The motion of the particle takes place in a plane, if follows that **[AIEEE 2004]**

a. its velocity is constant
b. its acceleration is constant
c. its kinetic energy is constant
d. it moves in a straight line

11 A particle is moving with a velocity $\mathbf{v} = k(y\hat{\mathbf{i}} + x\hat{\mathbf{j}})$, where k is a constant. The general equation for its path is **[AIEEE 2010]**

a. $y = x^2 + c$ b. $y^2 = x + c$ c. $xy = c$ d. $y^2 = x^2 + c$

One or More than One Options Correct Type

Direction (Q. Nos. 12-14) *This section contains 3 multiple choice questions. Each question has four choices (a), (b), (c) and (d), out of which ONE or MORE THAN ONE are correct.*

12 A particle is acted upon by a force of constant magnitude which is always perpendicular to the velocity of the particle. The motion of the particle takes place in a plane. It follows that

a. its velocity is constant
b. its acceleration is constant
c. its kinetic energy is constant
d. it moves in a circular path

13 A particle moves on a circular path of radius R, so that its angular velocity about a fixed point on the circumference of the circular path is constant and is equal to ω. Then,

a. the acceleration of the particle is variable
b. the acceleration of the particle is constant in magnitude and is equal to $4R\omega^2$
c. the acceleration of the particle varies in magnitude
d. acceleration of the particle may be zero

14 The coordinates of a particle moving in a plane are given by $x(t) = a\cos(pt)$ and $y(t) = b\sin(pt)$ where a, $b\,(< a)$ and p are positive constants of appropriate dimensions. Then,

a. the path of the particle is an ellipse
b. the velocity and acceleration of the particle are normal to each other at $t = \pi/2p$
c. the acceleration of the particle is always directed towards a focus
d. the distance travelled by the particle in time interval at $t = 0$ to $t = \pi/2p$ is a

Statement Type

Direction (Q. Nos. 15 and 16) *This section is based on Statement I and Statement II. Select the correct answer from the codes given below.*

a. Both Statement I and Statement II are correct and Statement II is the correct explanation of Statement I
b. Both Statement I and Statement II are correct and Statement II is not the correct explanation of Statement I
c. Statement I is correct but Statement II is incorrect
d. Statement II is correct but Statement I is incorrect

15 **Statement I** A particle moves on a parabolic path with constant speed. The acceleration of the particle is zero.

Statement II Tangential acceleration changes the speed of the particle.

16 **Statement I** In the case of projectile motion, the resultant of normal and tangential acceleration is constant in magnitude and direction.

Statement II In the case of projectile motion, the resultant acceleration is due to gravity.

One Integer Value Correct Type

Direction *(Q. Nos. 17-21) This section contains 5 questions. When worked out will result in an integer from 0 to 9 (both inclusive).*

17 The instantaneous velocity of point B of the given rod of length 0.5 m is 3 m/s in the represented direction. The angular velocity of the rod for minimum velocity of end A is ω rad/s. Find the value of ω.

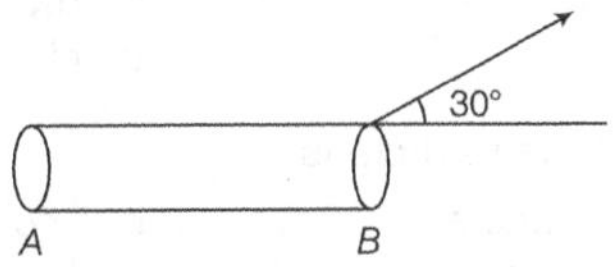

18 A particle starts from rest at the origin and moves along the positive branch of curve $y^2 = 4x^3$, so that the distance s measured from origin along the curve varies with time t according to $s = 2t^3$, where x, y and s are in metre and t is in second. The tangential acceleration in m/s^2 at $t = 1$ s is $3x$. Find the value of x.

19 A particle of mass 1 kg moves with a constant speed $v = 1$ m/s along the curve ACB in vertical plane whose equation is $y = 0.5\cos\frac{2\pi x}{l}$, where x and y are in metre and $l = 1$ m.

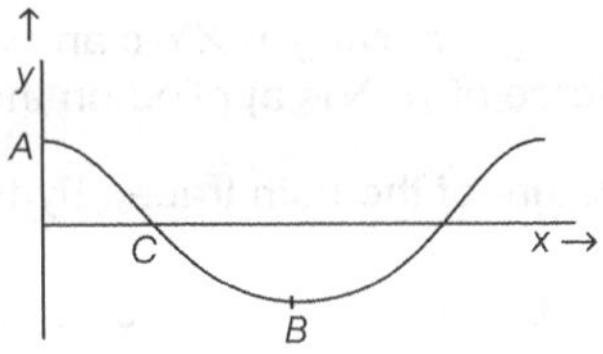

The normal reaction on particle at lowest point B is $5n$ N (take $\pi^2 = g$). Find the value of n.

Hint *Radius of curvature,* $r = \dfrac{\left[1+\left(\frac{dy}{dx}\right)^2\right]^{3/2}}{\frac{d^2y}{dx^2}}$

20 A car travels along a depression in a road. The equation of depression being $x^2 = 200y$, where x and y are in metre. The car travels with a constant speed of 20 m/s. Find the acceleration in m/s^2 at deepest point.

Hint *Radius of curvature,* $r = \dfrac{\left[1+\left(\frac{dy}{dx}\right)^2\right]^{3/2}}{\frac{d^2y}{dx^2}}$

21 There is a vertical parabolic shaped wire. The highest point of the wire is 2.5 m above the ground as shown in the figure. An insect of mass 5 g is moving on the wire with constant speed of $\sqrt{5}$ m/s. The normal reaction on insect at highest point is $5n \times 10^{-3}$ N. Find the value of n.

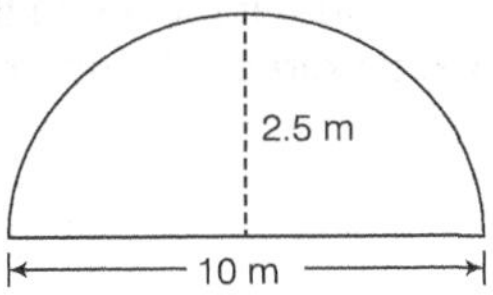

Comprehension Type

Working Space

Direction (Q. Nos. 22-33) This section contains 5 paragraphs, each describing theory, experiments, data etc. Twelve questions related to the paragraphs which has been given. Each question has only one correct answer among the four given options (a), (b), (c) and (d).

Passage I

If a particle is moving along a helix around Z-axis, then position of the particle is $\mathbf{r} = b \sin \omega t\, \hat{\mathbf{i}} + b \cos \omega t\, \hat{\mathbf{j}} + ct\hat{\mathbf{k}}$.

22 The trajectory of the particle seen by an observer O moving with a velocity $c\hat{\mathbf{k}}$ is

a. elliptical
b. parabolic
c. helix
d. circle in XY-plane having radius b

23 The acceleration of the particle is

a. $\omega^2 b$ directed in XY-plane
b. $\omega^2 b$ directed in XZ-plane
c. ωc directed in a YZ-plane
d. None of the above

24 The angle between acceleration and velocity at an instant t is

a. 0
b. $\frac{\pi}{2}$
c. $\frac{\omega t}{2}$
d. ωt

Passage II

A particle of mass 1 kg is moving in XY-plane with a velocity 6 m/s at an angle 53° with X-axis. A force of 15 N is applied on the particle along X-axis at $t = 0$.

25 Find radius of curvature of the path traced by the particle just after applying the force.

a. 1 m
b. 2 m
c. 3 m
d. None of these

26 The rate of change of speed of the particle just after applying the force is

a. 9 m/s^2
b. 12 m/s^2
c. 8 m/s^2
d. 20 m/s^2

Passage III

A particle of mass m is moving with constant velocity v_0 along positive X-axis. At $t = 0$, a force $\mathbf{F} = \mathbf{v} \times \mathbf{c}$ starts to act, where $\mathbf{c}$ is vector of constant magnitude directed towards negative Z-axis.

27 The path of the particle is

a. straight line
b. circular
c. elliptical
d. parabolic

28 At $t = \frac{m\pi}{6c}$, the velocity of the particle is

a. $\frac{\sqrt{3}}{2} v_0\, \hat{\mathbf{i}} + \frac{v_0}{2} \hat{\mathbf{j}}$
b. $\frac{v_0}{2} \hat{\mathbf{i}} + \frac{\sqrt{3}}{2} v_0 \hat{\mathbf{j}}$
c. $v_0\, \hat{\mathbf{i}} + \frac{v_0}{2} \hat{\mathbf{j}}$
d. $\frac{v_0}{2} \hat{\mathbf{i}} + \frac{v_0}{2} \hat{\mathbf{j}}$

Passage IV

A particle of mass 1 kg is lying at the origin. At $t = 0$, it is given a velocity $v_0 = 1$ m/s at an angle 30° with Y-axis in the XY-plane, so that it moves in the influence of a force $\mathbf{F} = \mathbf{v} \times \mathbf{c}$, where $\mathbf{c}$ is a constant vector directed along positive X-axis. The magnitude of $\mathbf{c}$ is π.

29 The path of the particle is

a. helical with constant pitch
b. circular
c. elliptical
d. parabolic

30 The x-coordinate of the particle at $t = 2$ s is

a. 1 m
b. 2 m
c. 3 m
d. 4 m

Working Space

Passage V

A force $\mathbf{F} = b\hat{\mathbf{k}} + \mathbf{v} \times \mathbf{c}$ is acting on a particle of mass m. $\mathbf{c}$ is a constant vector directed along positive Z-axis. At $t = 0$, the particle starts with velocity v_0 along positive X-axis.

31 Find z-coordinate of the particle, when it crosses Z-axis first time.

a. $\frac{v_0 c}{2\pi b}$
b. $\frac{2\pi^2 mb}{c}$
c. $\frac{2\pi mb}{c}$
d. $\frac{2mb}{c}$

32 Find the angle between velocity vector $\mathbf{v}$ and Z-axis at the moment, it crosses the Z-axis first time.

a. $\tan^{-1}\left(\frac{v_0 c}{2\pi b}\right)$
b. $\tan^{-1}\left(\frac{v_0 c}{b}\right)$
c. $\tan^{-1}\left(\frac{v_0 c}{\pi b}\right)$
d. None of these

33 Suppose that a plane screen oriented at right angles to the X-axis is located at a distance l from the origin. Find the equation of trace that the particle leaves on the screen for $|y| << l$.

a. Circle
b. Helix
c. Parabola
d. Ellipse

Matching List Type

Direction (Q. No. 32) Choices for the correct combination of elements from Column I and Column II are given as options (a), (b), (c) and (d), out of which one is correct.

34 The position of a particle at an instant t is $\mathbf{r} = t^3\hat{\mathbf{i}} + t^2\hat{\mathbf{j}} + t\hat{\mathbf{k}}$, (SI units).

Match the Column I with Column II and select the correct option from the codes given below.

	Column I		Column II
i.	Acceleration in m/s^2 at $t = 1$ s	p.	5.88
ii.	Radius of curvature in metre at $t = 1$ s	q.	6
iii.	The centripetal or normal acceleration in m/s^2 at $t = 1$ s	r.	6.32
iv.	The tangential acceleration in m/s^2 at $t = 1$ s	s.	2.33

Codes

	i	ii	iii	iv		i	ii	iii	iv
a.	p	q	r	s	b.	r	q	s	p
c.	p	r	q	s	d.	r	s	p	q

SET - 1

Revisal Problems for JEE Main

Only One Option Correct Type

Working Space

Direction (Q. Nos. 1-18) *This section contains 18 multiple choice questions. Each question has four choices (a), (b), (c) and (d), out of which ONLY ONE option is correct.*

1 A particle is moving as such $\mathbf{r} = b \sin \omega t\hat{\mathbf{i}} + b \cos \omega t\hat{\mathbf{j}}$.
The trajectory of the particle is

a. straight line b. parabola c. circle d. ellipse

2 A particle of mass m moves on a circular path of radius r with constant speed in the influence of the force $\mathbf{v} \times \mathbf{b}$, where $\mathbf{b}$ is a constant vector. The speed v of the particle is

a. $\frac{rb}{m}$ b. $\frac{2rb}{m}$ c. $\frac{rb}{2\,m}$ d. $\frac{3rb}{4m}$

3 A bug is moving on inner surface of a fixed vertical circular wire of radius R with a constant speed v_0. Mark the correct option.

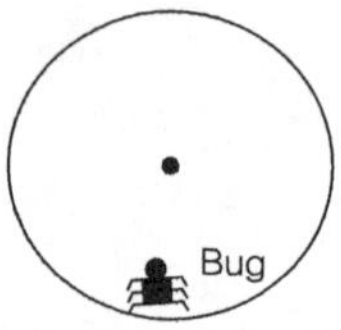

a. Friction on bug at top and bottom is zero
b. The frictional force on bug is always zero
c. The frictional force on bug never be zero
d. Normal reaction on bug provides required centripetal acceleration

4 A particle is moving with constant speed inside a fixed hollow sphere of radius $\frac{1}{15}$ m, describes a horizontal circle at a distance $\frac{1}{30}$ m below its centre. The speed of the particle is

a. 1 m/s b. 2 m/s c. 3 m/s d. 4 m/s

5 A car starts from rest on a circular path with a constant magnitude tangential acceleration. Slipping does not take place in one revolution. The angle between acceleration and velocity at the end of one revolution is α. Find the value of $\frac{\tan \alpha}{2\pi}$.

a. 1 b. 2
c. 3 d. 4

6 A particle is rotating with a constant speed at inner surface of a fixed hollow sphere. The normal reaction on the blocks at points A, B, C and D are N_1, N_2, N_3 and N_4, respectively. Then,

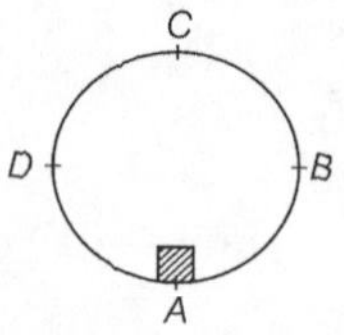

a. $N_1 = N_2 = N_3 = N_4$ b. $N_1 > N_2 > N_3, N_2 = N_4$
c. $N_1 > N_3 > N_2, N_2 = N_4$ d. $N_1 < N_2 < N_3, N_2 = N_4$

7 An insect is moving with constant speed on the path shown in figure.

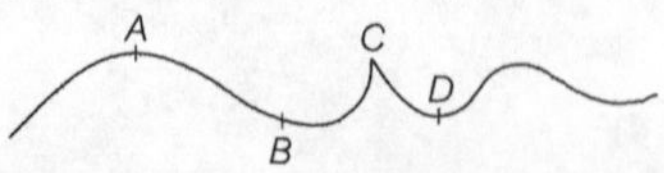

At which point normal reaction on the insect is minimum?

a. A b. B c. C d. D

Working Space

8 A smooth block loosely fits in a horizontal fixed circular pipe. The block is moving with constant speed inside the pipe. The block is in contact with

a. inner wall
b. outer wall
c. neither inner nor outer wall
d. None of these

9 A ball of mass 1 kg is hanging from a fixed point through a light string. The maximum tension in the string is 10 N. If block starts to move with a velocity of 1 m/s, then string

a. may break
b. must break
c. never break
d. None of these

10 After a good meal at a party you wash your hands and shakes hand in air to detach water from hand. The radius of path is 0.04 m and speed of finger tip is 4 m/s. The centripetal acceleration of a drop before leaving the finger tip is

a. 80 m/s^2
b. 400 m/s^2
c. 100 m/s^2
d. 200 m/s^2

11 A particle is moving on a circular path of radius 18 m with an angular velocity $\omega = \omega_0 t$ where, $\omega_0 = \frac{1}{9}$ rad/s^2. The acceleration of the particle at $t = 3$ s is

a. 2 m/s^2
b. 3 m/s^2
c. $2\sqrt{2}$ m/s^2
d. 10 m/s^2

12 A circular horizontal ring of radius R is in horizontal plane. A light string of length $L = 2\pi R$ is tied at its one end on the ring. Keeping the string horizontal, the particle of mass m is projected along the smooth surface with a speed v_0 as shown in the figure. Find the tension in string when string covers half portion of the ring.

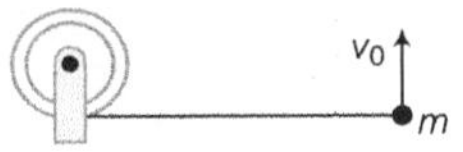

a. $\frac{mv_0^2}{R}$
b. $\frac{mv_0^2}{2\pi R}$
c. $\frac{mv_0^2}{\pi R}$
d. $\frac{2mv_0^2}{R}$

13 The situation shown in figure, the block B of mass 2 kg in hanging vertically downward and the block A is revolving around it. The acceleration of the block A at steady state is

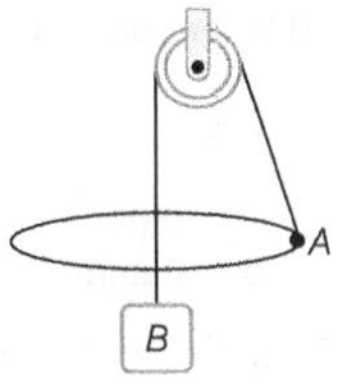

a. 10 N
b. $10\sqrt{3}$ N
c. $20\sqrt{3}$ N
d. $30\sqrt{3}$ N

14 A small ball of mass 1 kg is suspended by two light strings A and B as shown in the figure below. The ball is stationary. Just after cutting the string A, then

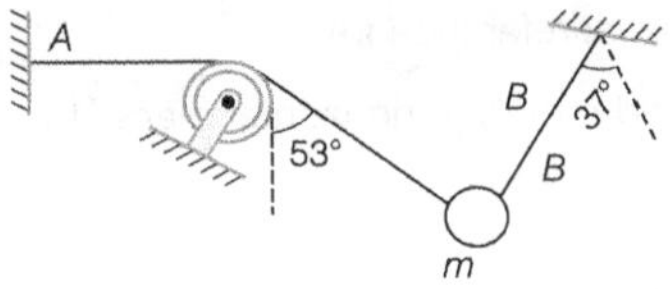

a. the tension in string B is 8 N
b. the acceleration of the ball is 6 m/s^2
c. the component of acceleration of the ball along the string is non-zero
d. Both (a) and (b) are correct

15 A ball is thrown at an angle 37° with horizontal with a speed of 10 m/s. Find angular velocity of the ball about the point of projection when it is at highest point.

a. 0.55 rad/s
b. 1 rad/s
c. 2 rad/s
d. 4 rad/s

16 A bug is moving along the inner surface of a fixed transparent semicircular track of radius $R = 2$ cm with a constant speed of $v_0 = \frac{\pi}{2}$ cm/s. Sun rays are vertical.

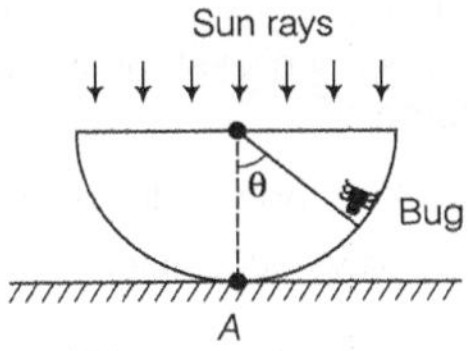

The bug starts from rest from point A at $t = 0$, the speed of its shadow on horizontal clean surface at $t = 1$ s is

a. $\frac{\pi}{2}$ cm/s
b. $\frac{\pi}{2\sqrt{2}}$ cm/s
c. $\frac{\pi}{\sqrt{2}}$ cm/s
d. $\frac{\pi}{4}$ cm/s

Working Space

17 A wheel has eight equally spaced very thin spokes. It is rotating about vertical axis passing through the centre (as shown in figure) with 5π rad/s. A vertical arrow pass the gap between the spokes without hitting the spoke. The maximum possible length of arrow is

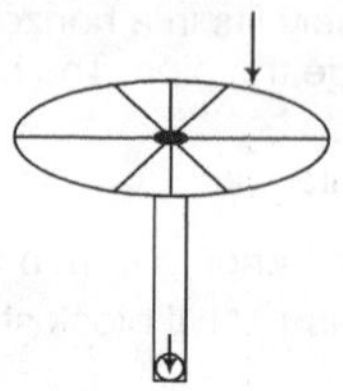

a. 0.3 m
b. $\frac{1}{80}$ m
c. $\frac{1}{20}$ m
d. $\frac{1}{10}$ m

18 A ball of mass 1 kg is connected with a *L*-shaped rod rotating with constant angular velocity $\omega = \sqrt{10}$ rad/s through a light string *CD*. The length of part *BC* of the rod is 0.5 m. The length of string *CD* is

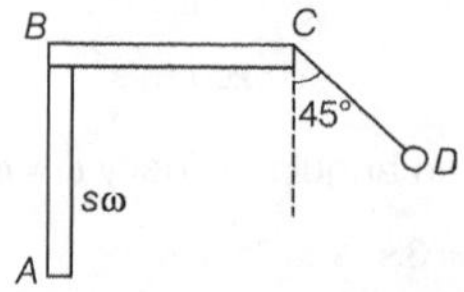

a. 0.5 m
b. $\frac{1}{2\sqrt{2}}$ m
c. $\frac{1}{\sqrt{2}}$ m
d. $\sqrt{2}$ m

Statement Type

Direction *(Q. Nos. 19-25) This section is based on Statement I and Statement II. Select the correct answer from the codes given below.*

a. Both Statement I and Statement II are correct and Statement II is the correct explanation of Statement I
b. Both Statement I and Statement II are correct and Statement II is not the correct explanation of Statement I
c. Statement I is correct but Statement II is incorrect
d. Statement II is correct but Statement I is incorrect

19 **Statement I** Centrifugal force obeys Newton's third law of motion.
Statement II To every action, there is equal and opposite reaction.

20 **Statement I** Pseudo forces are proportional to mass of the body.
Statement II Pseudo forces on a body is zero in inertial reference frame.

21 **Statement I** A sprit level is tied to a string and whirled rapidly in horizontal plane, the bubble will lie at the inner edge.
Statement II The acceleration of bubble is zero.

22 **Statement I** An electron moving around nucleus in hydrogen atom is in equilibrium.
Statement II Electrostatic force exerted by nucleus on electron provides centripetal acceleration.

23 **Statement I** When a car passes at high speed over a curved bridge across a river, it leaves the highest point tangentially.
Statement II Due to inertia of car, it will leave the ground tangentially.

24 **Statement I** If a cyclist is moving on horizontal circular path, resultant of normal reaction and friction is passing through centre of gravity.
Statement II In the case of cyclist motion on horizontal circular path, friction provides required centripetal force.

25 **Statement I** In the case of non-uniform circular motion with increasing speed, the angle between acceleration and velocity is lesser than $\frac{\pi}{2}$.
Statement II In the case of non-uniform circular motion, acceleration is not directed towards the centre of circular path.

SET - 2

Revisal Problems
for JEE Advanced

Only One Option Correct Type

Working Space

Direction (Q. Nos. 1-7) *This section contains 7 multiple choice questions. Each question has four choices (a), (b), (c) and (d), out of which ONLY ONE option is correct.*

1 A ball of mass 1 kg is revolving in horizontal circle with the help of an elastic massless string of constant 100 N m^{-1}. The string makes an angle of 60° with the vertical. The elongation in the elastic string is

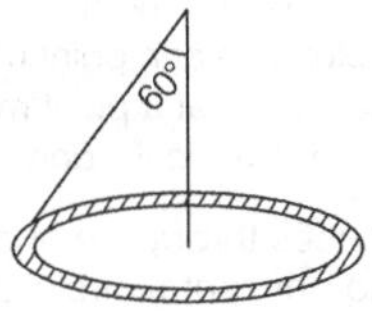

a. 0.1 m
b. 0.2 m
c. 0.3 m
d. 0.4 m

2 A closed rope of mass 1 kg is attached to a vertical shaft by means of a light string and rotates with a constant angular velocity. The string makes an angle of 60° with the vertical. Find the tension in the string.

a. 10 N b. 20 N c. 30 N d. 40 N

3 The coefficient of friction between a horizontal *L*-shaped frame and a sleeve of mass 1 kg at the end of the rod is 0.5. The whole system is rotating with constant angular velocity ω about a vertical axis passing through the point *O*. When $\omega \le \omega_0$, then sleeve does not slip over the rod. Find normal contact force on the sleeve for $\omega = \omega_0$.

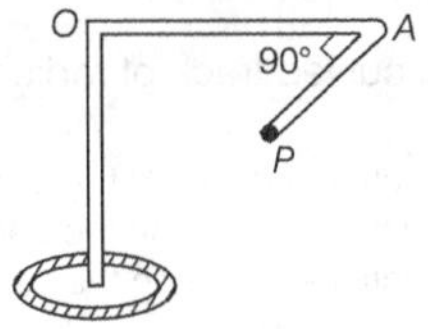

a. $20\sqrt{3}$ N b. $\frac{20}{\sqrt{3}}$ N
c. 20 N d. $10\sqrt{3}$ N

4 A particle of mass *m* starts to slide from top of a smooth cylindrical arc of radius *R* as shown in the figure. Find maximum value of θ_0, so that the particle does not leave the contact of circular arc during sliding on it.

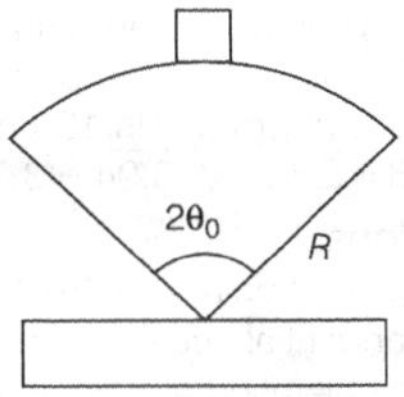

a. $\sin^{-1}\left(\frac{2}{3}\right)$ b. $\cos^{-1}\left(\frac{2}{3}\right)$
c. $\cos^{-1}\left(\frac{1}{3}\right)$ d. $\cos^{-1}\left(\frac{1}{6}\right)$

5 A ball is shot into the end *A* of hollow, circular pipe fixed to a smooth horizontal table to leave the other and *B* at high speed. The ball's diameter is slightly less than the internal diameter of the tube. After the ball leaves the pipe, what force(s) act on it?

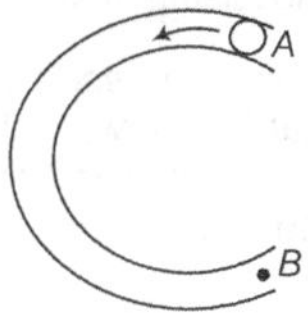

a. The weight of the ball
b. The weight and normal reaction by the table
c. Centripetal force
d. Centrifugal force

6 In the case of projectile motion under gravity,

a. radius of curvature is minimum at the point of projection
b. radius of curvature is minimum at maximum height
c. radius of curvature remains constant
d. radius of curvature is minimum at landing point

7 A simple pendulum is suspended from a fixed point O through a light string of length l. It is projected horizontally from lowest point with speed $u = \sqrt{2gl}$. Find the angle with lowest vertical at which acceleration is directed horizontal.

a. 45° b. $\tan^{-1}\sqrt{2}$ c. $\sin^{-1}\frac{1}{\sqrt{3}}$ d. $\cot^{-1}\sqrt{2}$

Working Space

One or More than One Options Correct Type

Direction (Q. Nos. 8-14) *This section contains 3 multiple choice questions. Each question has four choices (a), (b), (c) and (d), out of which ONE or MORE THAN ONE are correct.*

8 A small ball of mass m is tied to a vertical light rod of length l. The ball is projected horizontally with a velocity v_0.

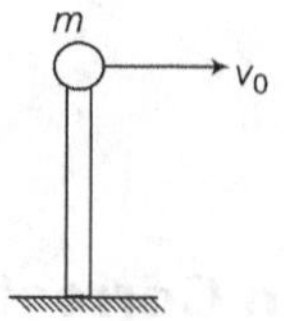

a. If $v_0 = \sqrt{gl}$, then initial tension in rod is zero
b. If $v_0 < \sqrt{gl}$, then rod is initially under compression
c. If $v_0 > \sqrt{gl}$, then rod is initially under tension
d. At any value of v_0, the rod is initially in the condition of tension

9 A small ball of mass m is suspended by a light string is first taken aside so that the string forms the right angle with vertical and then released.

a. At lowest point tension is maximum
b. During motion, acceleration at a point may be horizontal
c. During motion, acceleration at a point may be vertical
d. The initial acceleration is vertically downward

10 A single wire ACB passes through a ring C, which revolves at a constant speed in the horizontal circle of radius r ?

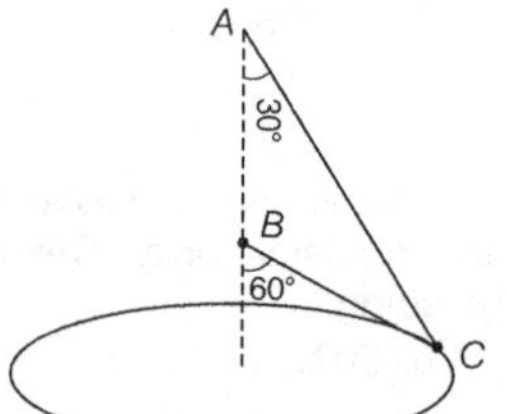

a. The angular speed of revolution is $\sqrt{\frac{g}{r}}$
b. The tension in the wire is $\left(\frac{2mg}{\sqrt{3}+1}\right)$
c. The tension in the wire may be zero
d. In vertical direction, net force is zero

11 A car is moving on a curved track of radius 10 m banked at an angle of 45° with speed v without slipping.

a. If $v = 10$ m/s, then frictional force on the car due to track is zero
b. If $v > 10$ m/s, then frictional force on the car due to tracks is non-zero
c. If $v < 10$ m/s, then frictional force on the car due to track is non-zero
d. None of the above

12 The driver of a car moving on road ABC applies uniform break throughout its motion. The speed of car is 27 m/s at the bottom A and 13 m/s at the top point C. The distance of C from point A along ABC is 80 m. The acceleration of car is 4.5 m/s^2 at point A and radius of curvature at point C is 130 m path portion DBE is straight line. (take $\sqrt{8} = 2.83$, $\sqrt{13.94} = 3.7$)

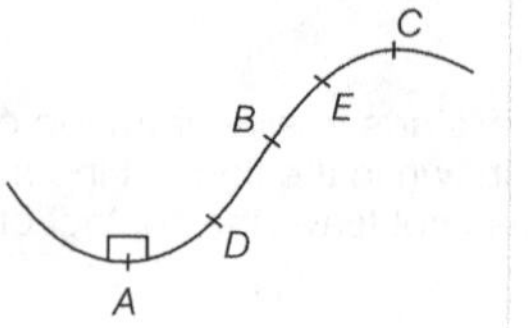

Mark the correct options.

a. The radius of curvature at point A is 257.6 m
b. The magnitude of tangential acceleration is 3.5 m/s^2
c. The magnitude of acceleration at point B is 3.5 m/s^2
d. The magnitude of acceleration at point C is 3.7 m/s^2

13 Two small smooth balls of masses 10 kg and 15 kg are connected by a light spring and supported on a turntable (shown in figure). When the turntable is in rest, the tension in spring is 500 N and the balls exert the same force on each of the stops A and B. When turntable starts to rotate with constant angular velocity 2π rad/s. (take $\pi^2 = 10$) Then,

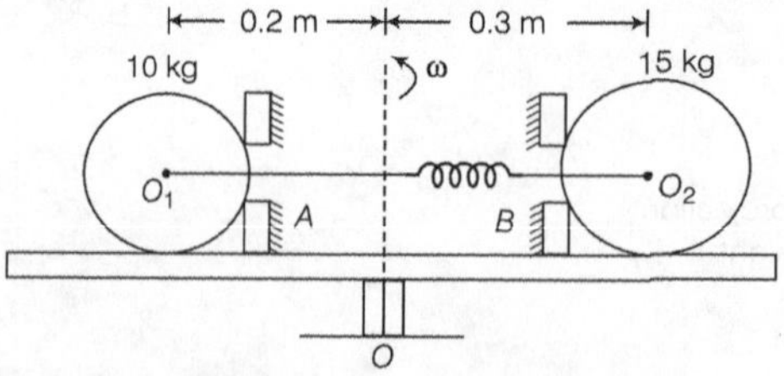

a. the force exerted by the ball of mass 10 kg on stop A is 420 N
b. the force exerted by the ball of mass 15 kg on stop B is 320 N
c. the force exerted by the ball of mass 10 kg on stop A is 500 N
d. None of the above

14 A small ball A of mass 10 kg is suspended in a vertical plane by two light strings as shown in figure below.

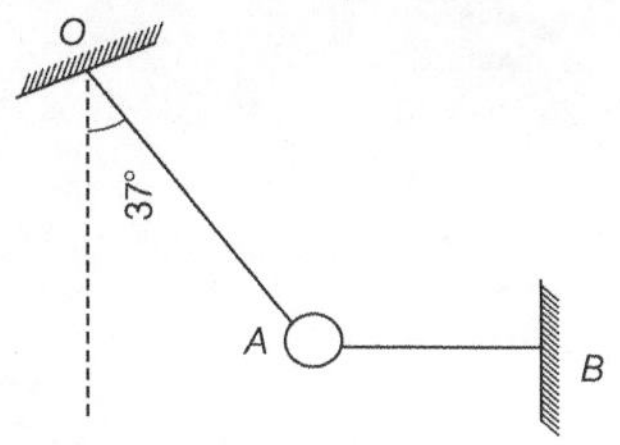

a. The tension in string OA is 12.5 N
b. The tension in string OA just after cutting horizontal string AB is 12.5 N
c. The tension in string OA just after cutting horizontal string AB is 8 N
d. None of the above

Working Space

Comprehension Type

Direction (Q. Nos. 15-20) ***This section contains 2 paragraphs, each describing theory, experiments, data, etc. Six questions related to the paragraph have been given. Each question has only one correct answer among the four given options (a), (b), (c) and (d).***

Passage I

A small smooth block of mass m is projected with speed $v_0 = \sqrt{2gR}$ from bottom of a fixed sphere of radius R, so that the block moves in a vertical circular path.

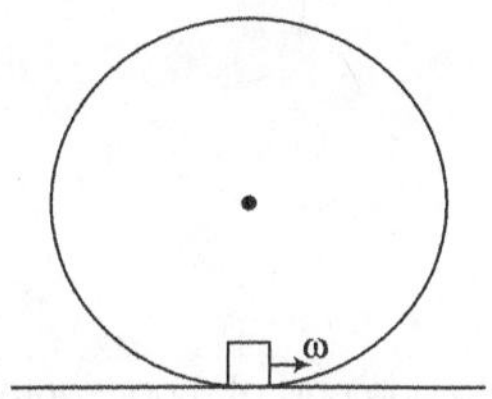

15 The speed of the block as function of θ, the angle of deflection from the lowest vertical.

a. $\sqrt{gR\cos\theta}$ b. $\sqrt{2gR\cos\theta}$ c. $\sqrt{3gR\cos\theta}$ d. $\sqrt{4gR\cos\theta}$

16 Normal reaction at the moment when vertical component of the block's velocity is maximum.

a. mg b. $\sqrt{2}\,mg$ c. $\sqrt{3}\,mg$ d. $2\,mg$

17 The angle θ between the thread and the lowest vertical at the moment when the total acceleration vector of the block is directed horizontally.

a. $\frac{\pi}{4}$
b. $\frac{\pi}{3}$
c. $\tan^{-1}(\sqrt{2})$
d. $\tan^{-1}\left(\frac{1}{\sqrt{2}}\right)$

Passage II

Two nails A and B remain in the slot of yoke that is moving with constant velocity of 10 cm/s. Further more the nails cannot leave circular slot also over which the yoke is moving. The radius of the circular slot is $R = 1$ m.

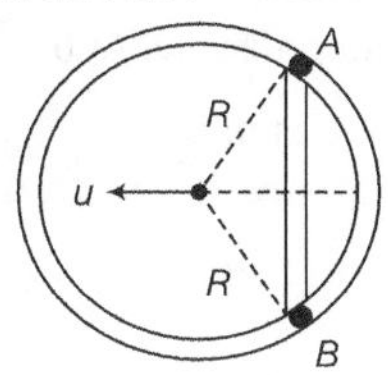

18 The speed of nails at $\theta = 60°$ is

a. 10 cm/s
b. 20 cm/s
c. 40 cm/s
d. 50 cm/s

Working Space

19 Centripetal acceleration of nails A at $\theta = 60^\circ$ is

a. $1\,m/s^2$
b. $2\,m/s^2$
c. $3\,m/s^2$
d. $4\,m/s^2$

20 The tangential acceleration of the nails is

a. $32\,m/s^2$
b. $4\sqrt{3}\,m/s^2$
c. $20\,m/s^2$
d. $10\,m/s^2$

One Integer Value Correct Type

Direction *(Q. Nos. 21-27) This section contains 7 questions. When worked out will result in an integer from 0 to 9 (both inclusive).*

21 A ball suspended by a light rod swings in a vertical plane, so that its acceleration magnitude at the extreme and the lowest position are equal. If the thread deflection angle is θ, then find the value of $\tan\frac{\theta}{2}$.

22 A particle of mass m is placed at top of circular arc of radius R. The circular arc is fixed on a plank moving on a horizontal surface with constant acceleration a_0 as shown in the figure below. For the value of $a_0 = \left(\frac{30\sqrt{3} - 40}{n}\right)\,m/s^2$, the particle does not loose contact during motion on circular arc (take, $\theta_0 = 30^\circ$). Find the value of n.

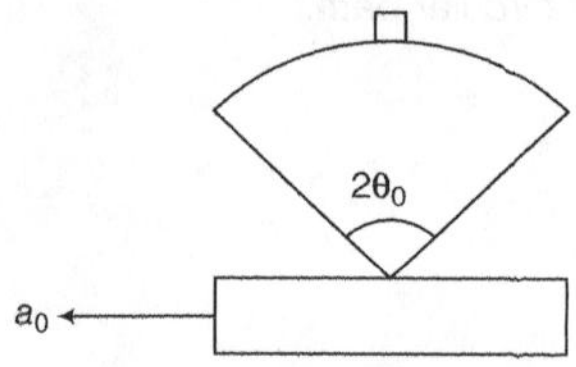

23 A ball of mass m is placed on a smooth table is attached to one end of a horizontal light rod of length L. The other end is attached to vertical shaft. The shaft is given an initial angular velocity ω_0. The air resistance on the ball is $-b\,\mathbf{v}$. The tension in rod as a function of distance s moved by the ball is $mL\left(n\omega_0 - \frac{bs}{mL}\right)^2$. Find the value of n.

24 A particle is moving in XY-plane whose velocity at instant t is $\mathbf{v} = by\hat{\mathbf{i}} + c\hat{\mathbf{j}}$. The normal acceleration is $\frac{nbc^2}{\sqrt{b^2y^2 + c^2}}$. Find the value of n.

25 A semicircular wire of radius 5 m is hinged at point O and a smooth peg A. If the peg A starts from point O and moves with constant speed 8 m/s along the horizontal. Find angular speed in radian per second of the semicircular wire at the instant when $\theta = 37^\circ$.

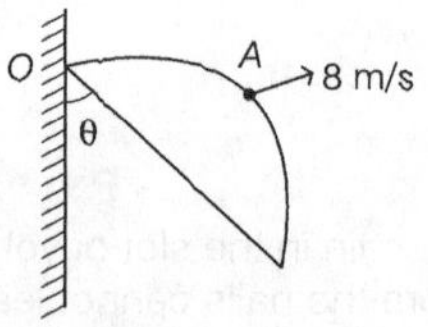

26 Two particles A and B move on coplanar concentric circles of radii 1 m and 2 m with angular velocities 1 rad/s and 0.5 rad/s (in the same sense) respectively. At $\theta = \frac{\pi}{n}$ radian, relative angular velocity between B and A is zero. Find the value of n.

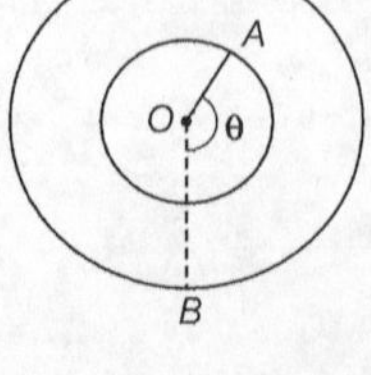

Working Space

27 A particle is moving in the XY-plane with an acceleration, $\mathbf{a} = -(3\ \text{m/s}^2)\hat{\mathbf{i}} - (15\ \text{m/s}^2)\hat{\mathbf{j}}$. The initial velocity of the particle is $\mathbf{v}_0 = (30\hat{\mathbf{i}} + 40\hat{\mathbf{j}})$ m/s. The radius of curvature at $t = 2$ s is $\frac{2197x}{330}$ m. Find the value of x.

Matching List Type

Direction (Q. Nos. 28-31) *Choices for the correct combination of elements from Column I and Column II are given as options (a), (b), (c) and (d), out of which one is correct.*

28 A particle is moving on a circular path with increasing speed as shown in the figure

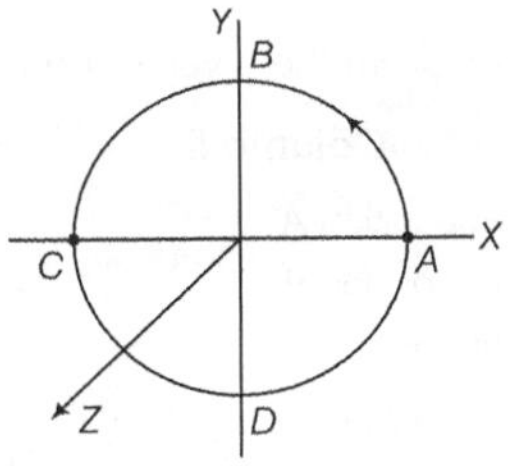

Match the Column I with Column II and select the correct option from the codes given below.

	Column I		Column II
i.	Unit vector along angular velocity	p.	$\hat{\mathbf{k}}$
ii.	Unit vector along angular acceleration	q.	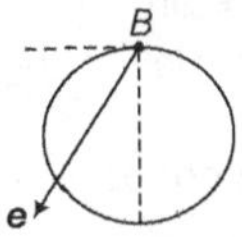
iii.	Unit vector along velocity at point B	r.	$-\hat{\mathbf{i}}$
iv.	Unit vector along acceleration at point B	s.	may be $\frac{-(\hat{\mathbf{i}}+\hat{\mathbf{j}})}{\sqrt{2}}$

Codes

	i	ii	iii	iv
a.	p	q	r	s
b.	s	r	p	q
c.	p	p	r	q,s
d.	p,s	q,s	p,q	r

29 In all cases, $\mathbf{v}$ is velocity of the particle and $\mathbf{F}_0$ and $\mathbf{c}$ as constant vectors.

Match the Column I with Column II and select correct option from the given codes.

	Column I		Column II
i.	A particle is moving in the presence of a force $\mathbf{F} = \mathbf{F}_0 + \mathbf{v} \times \mathbf{c}$ in a gravity free space. Here, $\mathbf{v}$ is the velocity of the particle.	p.	Velocity of particle may be constant.
ii.	A particle is moving in the presence of $\mathbf{F} = \mathbf{F}_0 + \mathbf{v} \times \mathbf{c}$ and the force of gravity.	q.	Path of particle may be straight line.
iii.	A particle is moving in the presence of $\mathbf{F} = \mathbf{v} \times \mathbf{c}$ and the force of gravity.	r.	The path of particle may be circular.
iv.	A particle is moving in the presence of only force of gravity.	s.	Path of particle may be helical.

Codes

	i	ii	iii	iv
a.	p,q	r,s	p	p,q
b.	p,q,s	p,q,r	p	q
c.	p,q,r,s	r,q	q,s	p,s
d.	p,q,s	p,q,r,s	p,q,r,s	q

Working Space

30 A boy of mass m is standing on a weighing machine fixed inside a cabin. Cabin is now rotated in a circular path in a vertical plane about a fixed axis with constant angular velocity ω. The boy rotates in a circular path of radius R and remains at rest with respect to the cabin. F_C represents the contact force on the boy.

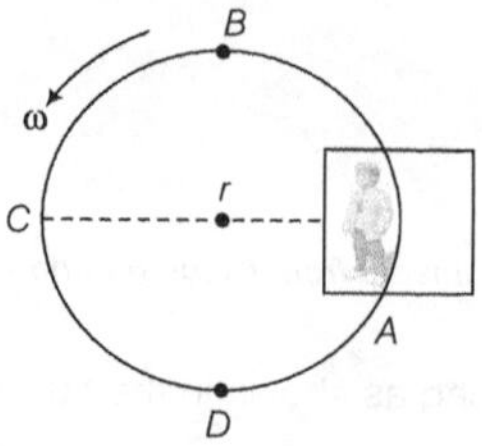

Match the Column I with Column II and select correct option from the given codes.

	Column I		Column II
i.	Maximum frictional force on the boy is	p.	at position A
ii.	Reading of weighing machine is maximum	q.	at position B
iii.	$mg - mr\omega^2 \le F_C \le mg + mr\omega^2$	r.	at position C
iv.	$F_C = m\sqrt{g^2 + r^2\omega^4}$	s.	at position D

Codes

	i	ii	iii	iv
a.	p,r	s	p,q,r,s	p,r
b.	p,q	p,r	p,q,r	p,s
c.	p,q,s	p,s	p,r	p,q
d.	p,q,r	p,q	q,r	p,q,r,s

31 A turn of radius 20 m has angle of banking $\theta = \tan^{-1}(0.5)$. The coefficient of friction between tyre of car and the banked road is $\mu = 0.4$. If car is moving on the banked road without slipping with a speed of v_0.

Match the Column I with Column II and select correct option from the given codes.

	Column I		Column II
i.	If $v_0 = 36$ km/h	p.	Frictional force on a car is zero.
ii.	If $v_0 = 14.7$ km/h	q.	Frictional force on a car is directed towards inner wheels.
iii.	If $v_0 = 20$ km/h	r.	Frictional force is directed towards outer wheel.
iv.	If $v_0 = 54$ km/h	s.	Frictional force is maximum.

Codes

	i	ii	iii	iv
a.	r,s	p	q,s	p,q
b.	p	r,s	r	q,s
c.	q,s	r,s	p,q	s
d.	p	q,s	r	q,s

JEE Main & AIEEE Archive

(Compilation of Last 13 Years Questions)

1 A particle of mass m is at rest at the origin at time $t = 0$. It is subjected to a force $F(t) = F_0 e^{-bt}$ in the X-direction. Its speed $v(t)$ is depicted by which of the following curves? **(AIEEE 2012)**

a. 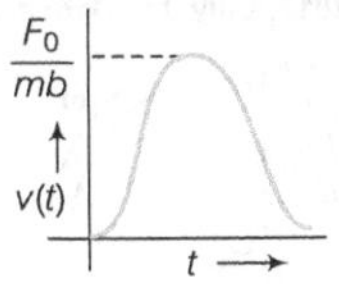

b. 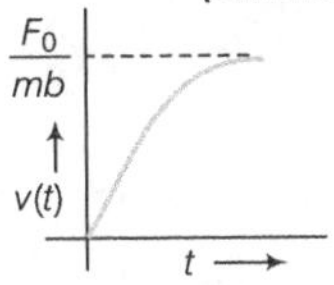

c. 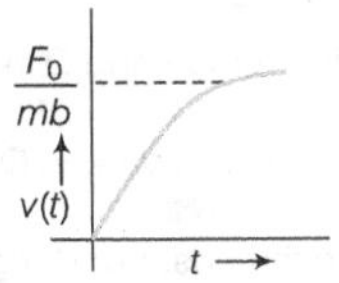

d. 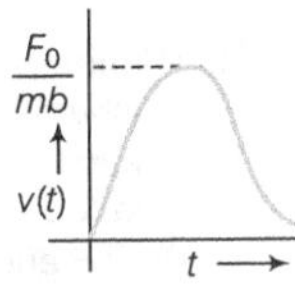

2 A mass m hangs with the help of a string wrapped around a pulley on a frictionless bearing. The pulley has mass m and radius R. Assuming pulley to be a perfect uniform circular disc, the acceleration of the mass m if the string does not slip on the pulley is **(AIEEE 2011)**

a. g b. $\frac{2}{3}g$ c. $\frac{g}{3}$ d. $\frac{3}{2}g$

3 For a particle in a uniform circular motion, the acceleration **a** at a point $P(R, \theta)$ on the circle of radius R is (here, θ is measured from the X-axis) **(AIEEE 2010)**

a. $-\frac{v^2}{R}\cos\theta\,\hat{i} + \frac{v^2}{R}\sin\theta\,\hat{j}$

b. $-\frac{v^2}{R}\sin\theta\,\hat{i} + \frac{v^2}{R}\cos\theta\,\hat{j}$

c. $-\frac{v^2}{R}\cos\theta\,\hat{i} - \frac{v^2}{R}\sin\theta\,\hat{j}$

d. $\frac{v^2}{R}\hat{i} + \frac{v^2}{R}\hat{j}$

4 A particle is moving with velocity $\mathbf{v} = k(y\hat{\mathbf{i}} + x\hat{\mathbf{j}})$, where k is constant. The general equation for its path is **(AIEEE 2010)**

a. $y = x^2 + c$ b. $y^2 = x + c$

c. $xy = c$ d. $y^2 = x^2 + c$

5 A point P moves in a counter-clockwise direction on a circular path as shown in the figure below. The movement of P is such that it sweeps out a length $s = t^3 + 5$, where s is in metre and t is in second. The radius of the path is 20 m. The acceleration of P when, $t = 2$ s is nearly. **(AIEEE 2010)**

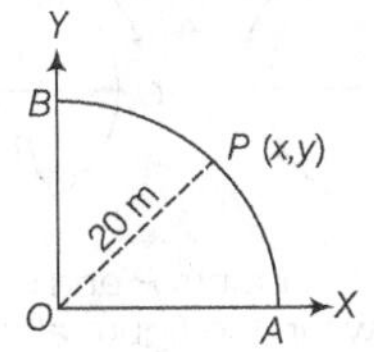

a. 13 ms^{-2} b. 12 ms^{-2}

c. 7.2 ms^{-2} d. 14 ms^{-2}

6 An annular ring with inner and outer radii R_1 and R_2 is rolling without slipping with a uniform angular speed. The ratio of the force experienced by the two particles situated on the inner and outer parts of the ring F_1/F_2 is **(AIEEE 2005)**

a. $\frac{R_2}{R_1}$ b. $\left(\frac{R_1}{R_2}\right)^2$

c. 1 d. $\frac{R_1}{R_2}$

7 Which of the following statements is false for a particle moving in a circle with a constant angular speed? **(AIEEE 2004)**

a. The velocity vector is tangent to the circle
b. The acceleration vector is tangent to the circle
c. The acceleration vector points to the centre of the circle
d. The velocity and acceleration vectors are perpendicular to each other

8 A particle is acted upon by a force of constant magnitude which is always perpendicular to the velocity of the particle. The motion of the particle takes place in a plane, if follows that **(AIEEE 2004)**

a. its velocity is constant
b. its acceleration is constant
c. its kinetic energy is constant
d. it moves in a straight line

JEE Advanced & IIT JEE Archive

(Compilation of Last 10 Years Questions)

1 Two identical discs of same radius R are rotating about their axes in opposite directions with the same constant angular speed ω. The discs are in the same horizontal plane.

(2012, Only One Option Correct Type)

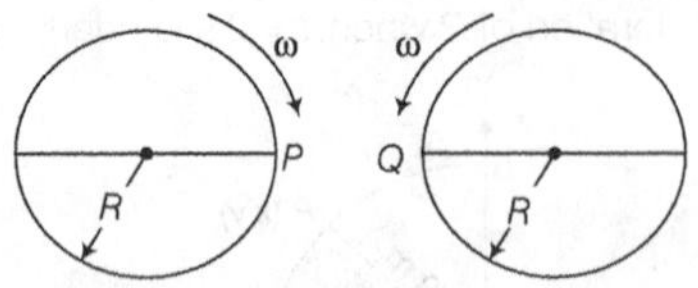

At time $t = 0$, the points P and Q are facing each other as shown in the figure above. The relative speed between the two points P and Q is v_r. In one time period (T) of rotation of the discs, v_r as a function of time is best represented by

a. 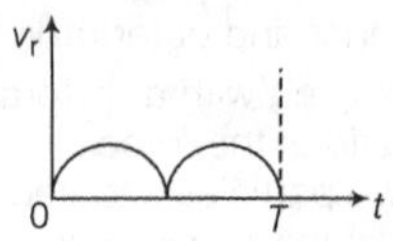

b. 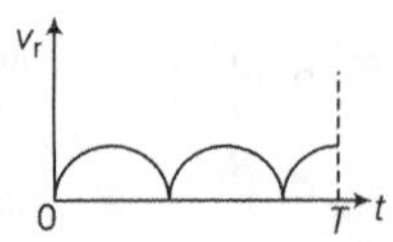

c. 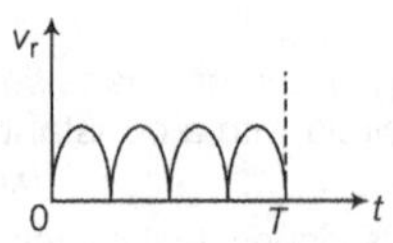

d. 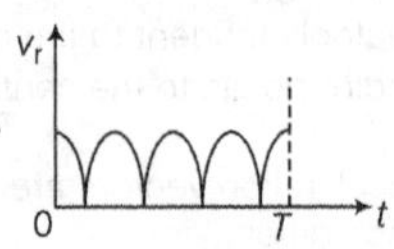

2 Consider a disc rotating in the horizontal plane with a constant angular speed ω about its centre O. The disc has a shaded region on one side of the diameter and an unshaded region on the other side as shown in the figure. When the disc is in the orientation as shown, two pebbles P and Q are simultaneously projected at an angle towards R. The velocity of projection is in the yz-plane and is same for both pebbles with respect to the disc. Assume that (i) they land back on the disc before the disc has completed 1/8 rotation, (ii) their range is less than half the disc radius and (iii) ω remains constant throughout. Then,

(2012, Only One Option Correct Type)

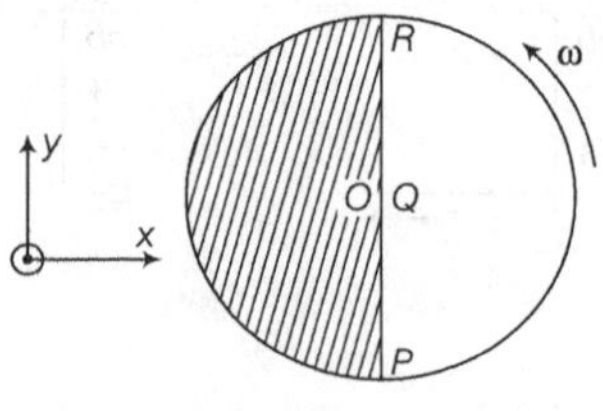

a. P lands in the shaded region and Q in the unshaded region

b. P lands in the unshaded region and Q in the shaded region

c. Both P and Q land in the unshaded region

d. Both P and Q land in the shaded region

3 A ball of mass (m) 0.5 kg is attached to the end of a string having length (L) 0.5 m. The ball is rotated on a horizontal circular path about vertical axis. The maximum tension that the string can bear is 324 N. The maximum possible value of angular velocity of ball (in rad/s) is **(2011, Only One Option Correct Type)**

a. 9 b. 18

c. 27 d. 36

4 A small block is shot into each of the four tracks as shown in the figure below. Each of the tracks raises to the same height. The speed with which the block enters the track is the same in all cases. At the highest point of the track, the normal reaction is maximum in

(2007, Only One Option Correct Type)

a. 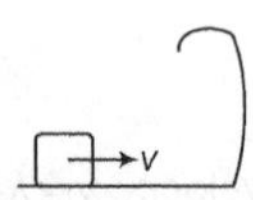

b. 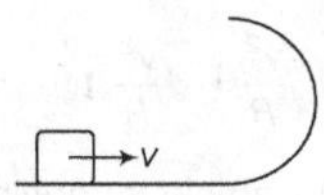

c. 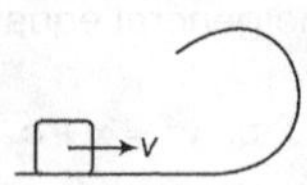

d. 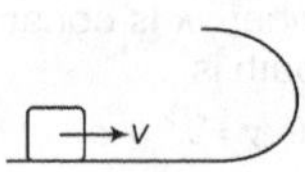

4. Work, Power and Energy

DPP
DAILY PRACTICE PROBLEMS

DPP-1 Kinetic Energy

Conceptual Questions

Working Space

Direction (Q. No. 1) *This question is conceptual in nature, need to be solved completely on notebook.*

1 **Situation** Suppose you are sitting in a bus moving with constant velocity 5 m/s, your mass is 50 kg.

i. Estimate your kinetic energy by yourself.
   a. 650 J b. Zero c. 100 J d. None of these

ii. If your friend is standing on the bus stop. Your kinetic energy measured by your friend is
   a. zero b. 625 J c. 100 J d. None of these

iii. Suppose a car is moving in the opposite direction of the bus with a speed of 2 m/s. Your kinetic energy measured by driver of car is
   a. 1225 J b. 625 J c. zero d. None of these

iv. Mark the correct option.
   a. Kinetic energy depends upon the frame of reference
   b. kinetic energy does not depend upon the direction of motion of body
   c. Kinetic energy never be negative
   d. All of the above

Only One Option Correct Type

Direction (Q. Nos. 2-15) *This section contains 14 multiple choice questions. Each question has four choices (a), (b), (c) and (d) out of which ONLY ONE option is correct.*

2 Your mass together with your bike is 100 kg. You are riding on a straight road with speed of 18 km/h. A car of mass 1000 kg is approaching towards you with a speed of 36 km/h. If you with your bike and the car are taken as a system. Find kinetic energy of the system.
a. 51250 J b. 50000 J c. 1250 J d. 500000 J

3 An athlete in the olympic games covers a distance of 100 m in 10 s. His kinetic energy can be estimated to be in the range. **[AIEEE 2008]**
a. 200 J – 500 J b. 2×10^5 J – 3×10^5 J c. 20000 J – 50000 J d. 2000 J – 5000 J

4 A block of mass 2 kg is free to move along the x-axis. It is at rest and from $t=0$ onwards it is subjected to a time-dependent force $F(t)$ in the x-direction. The force $F(t)$ varies with t as shown in the figure. The kinetic energy of the block after 4.5 s is **[IIT JEE 2010]**

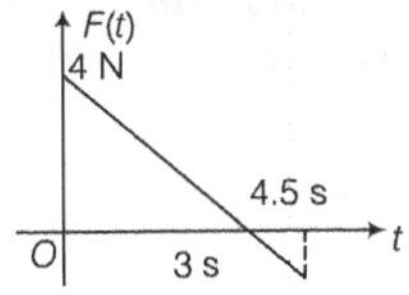

a. 4.50 J b. 7.50 J c. 5.06 J d. 14.06 J

5 A balloon is rising up with constant velocity of 10 m/s at height of 15 m from the ground. At that instant, a stone of mass 1 kg is dropped from it. Find kinetic energy of stone just before reaching at the ground.
a. 100 J b. 200 J c. 300 J d. 400 J

6 A juggler keeps four balls in air. He throws each ball vertically upwards with the same speed at equal interval of time. The maximum height attained by each ball is 20 m. Find kinetic energy of first ball when fourth ball is in hand. Assume that the mass of each ball is 10 g.
a. 0.5 J b. 1 J c. 1.5 J d. 2.5 J

7 A stone of mass 10 g is thrown vertically upwards. Its minimum and maximum distance from a fixed point A is 6 m and $6\sqrt{5}$ m, respectively. When the stone is at the maximum height, it is at a distance 10 m from point A. Find initial kinetic energy of the ball.

a. 1 J b. 2 J c. 3 J d. 4 J

Working Space

8 A particle is projected upward from ground, its kinetic energy *versus* height graph is

a. 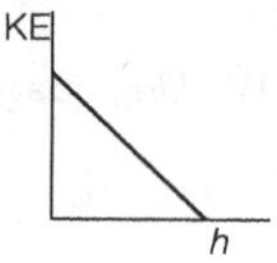

b. 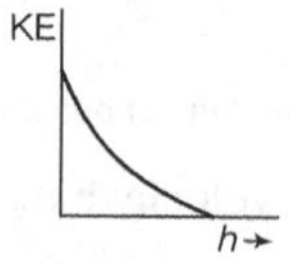

c. 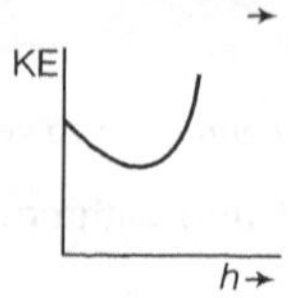

d. 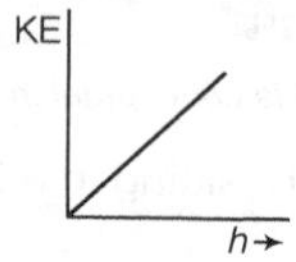

9 A particle is projected at an angle of 60° to the horizontal with a kinetic energy K. The kinetic energy at the highest point is **[AIEEE 2007]**

a. K b. zero c. $K/4$ d. $K/2$

10 A swimmer going downstream overcame a float at a point A, one hour latter it turned back and after some time passed the float at a distance of 6 km from the point A. The speed of flow of river is same everywhere. Find kinetic energy per unit volume of water. (Density of water = 1000 kg/m^3)

a. 347.2 J/m^3 b. 200 J/m^3 c. 380.6 J/m^3 d. 960 J/m^3

11 A particle of mass 1 kg starts moving so that its speed varies as $b\sqrt{s}$ where, $b = 1$ unit (SI unit) and s is distance covered by the particle in time t. Find kinetic energy of particle at $t = 1$s.

a. 1 J b. $\frac{1}{2}$ J c. $\frac{1}{4}$ J d. $\frac{1}{8}$ J

12 A particle moves in a straight line with retardation proportional to its displacement. Its loss of kinetic energy for any displacement x is proportional to **[AIEEE 2004]**

a. x^2 b. e^x c. x d. $\log_e x$

13 A particle of mass 1 kg is moving with a variable speed, kt, where $k = 1\text{m/s}^2$. Find mean kinetic energy from $t = 1$s to $t = 2$ s.

a. 1 J b. $\frac{7}{6}$ J c. $\frac{6}{7}$ J d. $\frac{9}{11}$ J

14 At time $t = 0$, particle starts moving along the x-axis. If its kinetic energy increases uniformly with time t, the net force acting on it must be proportional to **[AIEEE 2011]**

a. $\sqrt{t}$ b. constant c. t d. $1/\sqrt{t}$

15 The mass of heavy spring is m_0. Its ends are projected in opposite direction with speed v_0. Find the initial kinetic energy of the spring.

a. $\frac{1}{2}m_0v_0^2$ b. $\frac{m_0}{6}v_0^2$ c. $2\,m_0v_0^2$ d. $\frac{1}{3}m_0v_0^2$

One or More than One Options Correct Type

Direction (Q. Nos. 16-19) *This section contains 4 multiple choice questions. Each question has four choices (a), (b), (c) and (d), out of which ONE or MORE THAN ONE are correct.*

16 A particle of mass 1 kg is moving on a curved path with constant speed 10 m/s. Which of the following statements is/are correct.

a. The kinetic energy at instant t is 50 J
b. The change in kinetic energy in time 2 s is zero
c. Kinetic energy of the particle may be negative
d. None of the above

Working Space

17 Suppose you are waiting your friends on the platform of Kota station. The train approaches the station with constant speed of 108 km/h. Your mass is 50 kg. Mark the correct options(s).

a. Your kinetic energy is zero
b. Your kinetic energy measured by your friend is 22500 J
c. Your kinetic energy measured by your friend is zero
d. None of the above

18 A block of mass 1 kg starts from rest on a straight road whose acceleration *versus* displacement graph is shown in the figure. Find the increase in kinetic energy if speed increases as it travels a distance 10 m.

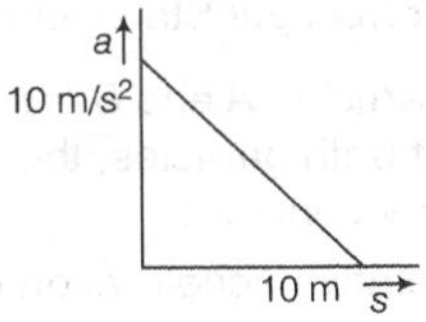

a. Non-zero
b. 20 J
c. 40 J
d. 50 J

19 A particle of mass 1 kg is moving in the *xy*-plane whose position at an instant *t* is $\mathbf{r} = 3\sin 2t\,\hat{\mathbf{i}} + 3(1-\cos 2t)\,\hat{\mathbf{j}}$, where **r** is in metre and *t* is in second. The kinetic energy of the particle at instant *t* is

a. 4.5 J
b. 10 J
c. 18 J
d. does not depend on the time

Only Integer Value Correct Type

Direction (Q. Nos. 20-23) This section contains 4 questions. Each question, when worked out will result in an integer from 0 to 9 (both inclusive).

20 A car of mass 1000 kg starts from rest with an acceleration $5\ \text{m/s}^2$.

i. The change in kinetic energy in $t = 2$ s is $n \times 10^4$ J. Find the value of n.
ii. The change in kinetic energy from $t = 1$s to $t = 2$ s is 375×10^n J. Find the value of n.
iii. The change in kinetic energy when it travels a distance 10 m is 5×10^x J. Find the value of x.

21 A block of mass 1 kg starts with velocity 5 m/s. Its acceleration-time graph is shown in the figure.

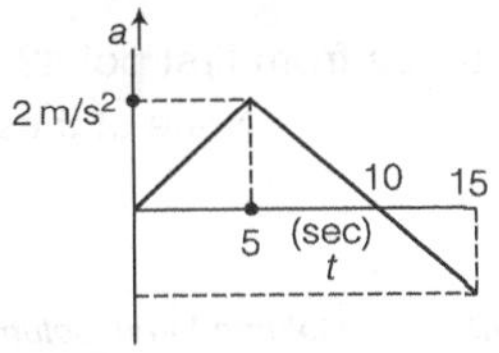

i. The kinetic energy of the particle at $t = 2$ s is 1682×10^n. Find the value of n.
ii. The change in kinetic energy between $t = 0$ to $t = 5$ s is $\frac{75}{x}$ J. Find the value of x.
iii. Find change in kinetic energy from $t = 5$ s to $t = 15$ s.

22 A particle of mass 1 kg is moving on a circular path of radius 1 m. Its kinetic energy is $K = bt^4$, where $b = 1$ J/s. The force acting on the particle at $t = 1$ s is $(2x)^{1/2}$ m/s^2. Find the value of x.

23 Two particles A and B each of mass 1 kg are moving under gravity. Initially, they moved with the same speeds 3 m/s horizontally in opposite directions. The kinetic energies of A as seen by B at the moment when their velocities become mutually perpendicular is $9N$ Joule. Find the value of N.

Working Space

Statement Type

Direction (Q. Nos. 24-26) *This section is based on Statement I and Statement II. Select the correct answer from the codes given below.*

a. Both Statement I and Statement II are correct and Statement II is the correct explanation of Statement I
b. Both Statement I and Statement II are correct and Statement II is not the correct explanation of Statement I
c. Statement I is correct but Statement II is incorrect
d. Statement II is correct but Statement I is incorrect

24 **Statement I** Two particles A and B are in projectile motion under gravity. During projectile motion of both particles, the kinetic energy of the particle A in the frame of particle B remains constant.

Statement II The relative acceleration of particle A with respect to particle B is zero during the projectile motion of both particles.

25 **Statement I** A ceiling fan is turning with angular velocity $10\sqrt{2}$ rad/s. An insect is sticking to the blade at a distance of 1 m from axis of the fan. If mass of insect is 1 g, its kinetic energy is 0.1 J.

Statement II In the frame of ground, insect is moving on a circular path.

26 **Statement I** The kinetic energy of a block of mass 1 kg is 4 Joule in reference frame S_1. Another reference frame S_2 is moving relative to frame S_1 at velocity 2 m/s. The maximum kinetic energy can be observed in frame S_2 measured is 8 J.

Statement II Kinetic energy does not depend on the frame of reference.

Comprehension Type

Direction (Q. Nos. 27 and 28) *This section contains a paragraph, describing theory, experiments, data, etc. Two questions related to the paragraph have been given. Each question has only one correct answer among the four given options (a), (b), (c) and (d).*

A car of mass 1000 kg moving with constant acceleration covers the distance between two points 60 m apart in 6 s. Its speed as it passes the second point is 15 m/s.

27 What was the kinetic energy of car at the first point?

a. 12500 J b. 6250 J c. Zero d. None of these

28 What was the kinetic energy of car at 7.5 m prior distance from first point?

a. Zero b. 12500 J c. 6250 J d. None of these

Matching List Type

Direction (Q. No. 29) *Choices for the correct combination of elements from Column I and Column II are given as options (a), (b), (c) and (d), out of which one is correct.*

29 Match the Column I with the information given in Column II and select the correct option from the codes given below.

	Column I		Column II
i.	A particle is moving along x-axis as such $v_x < 0$ and $a_x > 0$ at $t = 0$	p.	Kinetic energy first decreases, then increases.
ii.	During projectile motion of a particle under gravity	q.	Kinetic energy remains constant.
iii.	A particle is moving on a circular path in the condition of a uniform circular motion.	r.	Kinetic energy continuously decreases during motion.
iv.	A car is moving with a uniform deceleration	s.	Kinetic energy increases continuously during motion.

Codes

	i	ii	iii	iv		i	ii	iii	iv
a.	p	p	q	r	b.	s	r	q	p
c.	p,q	p,r	s	q	d.	p	q	r	s

DPP-2 Work Done by Constant Force

Conceptual Questions

Working Space

Direction *(Q. Nos. 1 and 2) These questions are conceptual in nature, need to be solved completely on notebook.*

1 **Situation** A block of mass m is placed on a rough horizontal surface.

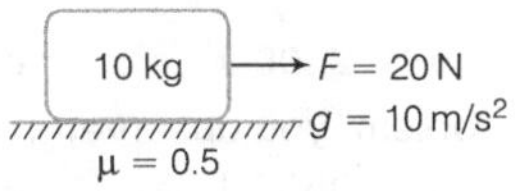

Mark the correct option(s) for the following questions.

i. The work done by friction on the block in time 5 s is

a. zero b. > 0 c. < 0 d. None of these

ii. Suppose an observer is situated in a reference frame S moving with constant velocity 10 m/s in the direction of applied force F on the block. The displacement of the block seen by reference frame S in 5 s is

a. 50 m in leftward direction b. 50 m in rightward direction
c. zero d. None of these

iii. In the previous problem, the work done by friction on the block as seen from reference frame S is

a. zero b. -1000 J c. 1000 J d. None of these

iv. In the previous problem, work done by friction on the ground seen from the reference frame S is

a. zero b. 1000 J c. -1000 J d. None of these

v. In the previous problem, the sum of work done by static friction on the block and the ground as seen from reference frame S is

a. zero b. -1000 J c. 1000 J d. None of these

vi. From above discussion,

a. work done by a force on a block depends upon the frame of reference
b. work done by static friction on a block depends upon the frame of reference
c. total work done by static friction on a system of blocks between which static friction acts is zero in any frame of reference
d. All of the above

2 **Situation** A block of mass 10 kg is moving with a constant velocity 10 m/s in rightward direction on a rough horizontal surface. Mark the correct option(s) for the following questions.

i. The work done by dynamic friction on the block in 5 s is

a. zero b. -2500 J c. $+2500$ J d. None of these

ii. In the previous problem, net work done by frictional forces on contact surfaces on the system in 5 s is

a. zero b. -2500 J c. 2500 J d. None of these

iii. The total work done by dynamic friction on system appears as heat loss. Mathematically, work done by dynamic friction on system $W_f = -\Delta H$. In previous problem, the total heat loss is

a. zero b. 2500 J c. -2500 J d. None of these

Working Space

iv. Suppose you are sitting in a reference frame moving with constant velocity of 10 m/s in the direction of velocity of the block. You observed that
   a. the block is in rest but the ground is moving leftward with a velocity 10 m/s
   b. the block is moving with a velocity of 20 m/s but ground is in rest
   c. both block and ground are moving with velocity 10 m/s
   d. None of the above

v. In the previous problem, you observed that the work done by dynamic friction on the block in 5 s is
   a. – 2500 J b. 2500 J c. zero d. None of these

vi. In the previous problem, you observed that the work done by dynamic friction exerted by the block on ground in 5 s is
   a. zero b. – 2500 J c. 2500 J d. None of these

vii. In the previous problem, if the block and ground are taken as a system, then you observed that the work done by dynamic friction between contact surfaces on the system in 5 s is
   a. zero b. – 2500 J c. 2500 J d. None of these

viii. In the previous problem, you observed that total heat loss is
   a. zero b. – 2500 J c. 2500 J d. None of these

ix. From above discussion,
   a. the work done by kinetic friction on a block depends on the reference frame
   b. total work done by kinetic friction on a system of blocks among which kinetic friction act is negative and the value of the work done is same in all references frame
   c. heat loss by a system does not depend on the reference frame
   d. the work done by internal forces may or may not be zero

Only One Option Correct Type

Direction (Q. Nos. 3-14) *This section contains 12 multiple choice questions. Each question has four choices (a), (b), (c) and (d) out of which ONLY ONE option is correct.*

3 A block of mass 20 kg is moving on a smooth surface by pulling through a light string of tension $10\sqrt{2}$ N. Find the work done by string on the block for a displacement of 2 m along the surface.

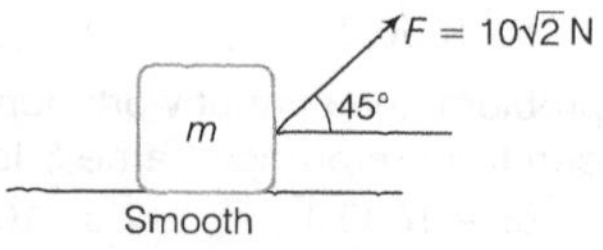

a. 10 J b. $20\sqrt{2}$ J c. 20 J d. $10\sqrt{2}$ J

4 If force $\mathbf{F} = (3\hat{i} + 4\hat{j})$ N is applied on a block, the block displaces from A (1, 1) to B (2, 3). All quantities are in the SI unit. Find work done by the force on the block.

a. 11 J b. – 11 J c. 10 J d. 20 J

5 A force $\mathbf{F} = (5\hat{i} + 3\hat{j} + 2\hat{k})$ N is applied on a particle which displaces from its origin to the point $\mathbf{r} = (2\hat{i} - \hat{j})$ m. The work done on the particle in joule is **[AIEEE 2004]**

a. – 7 b. + 7 c. + 10 d. + 13

6 A force $\mathbf{F} = (3\hat{i} + 4\hat{j})$ N is acting on a block. The block displaces from A (1, 1) to B (2, 3) and finally reaches at C (4, 5). All quantities are in SI unit. Find the work done by the force on the block.

a. 10 J b. 20 J c. 25 J d. 30 J

7 An object is moving in such a way that the point of application of force does not move. Then, which of the following statements is/are correct?

a. The work done by the force may be zero
b. The work done by the force must be zero
c. The work done by the force never be zero
d. The force must be non-conservative

Working Space

8 A triangular block is moving horizontally with constant velocity v_0. A small block A of mass m is fixed on the triangular block. The work done by friction on the block A in time t will be

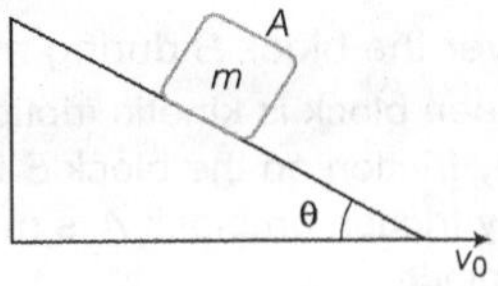

a. zero

b. $-mg\, v_0 t \cos^2 \theta$

c. $-mgv_0 t \sin^2 \theta$

d. $-\dfrac{mg\, v_0 t \sin 2\theta}{2}$

9 A particle moves from (0, 0) to (a, a) to $(2a, 3a)$ and finally reaches at the origin. During motion a constant force $\mathbf{F} = (3\hat{i} + 4\hat{j})$ N is acting on the block. The work done by the force on the block is

a. zero b. > zero c. < zero d. None of these

10 A block of mass 5 kg slides down on an inclined plane of inclination 30° from a height of 10 m from ground. The coefficient of friction between the block and the inclined plane is 0.5. Find the work done by friction on the block.

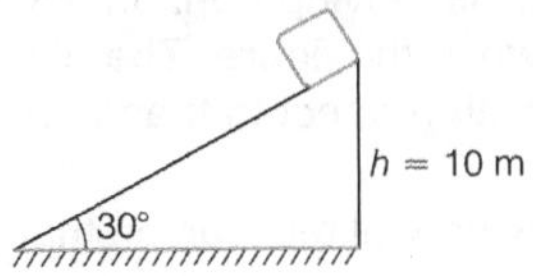

a. $250\sqrt{3}$ J

b. $-250\sqrt{3}$ J

c. 500 J

d. -500 J

11 A block of mass 10 kg is being brought down slowly by a light string through a distance of 2 m. Mark the correct option.

a. Work done by tension is zero

b. Work done by tension is 200 J

c. Work done by tension is -200 J

d. Work done by tension is 100 J

12 A block of mass 1 kg moves from point A (2 m, 3 m, 4 m) to point B (3 m, -4 m, 8 m). During motion of the block, a force $\mathbf{F} = (2)\hat{\mathbf{i}} + (4)\hat{\mathbf{j}} - (10)\hat{\mathbf{k}}$ N acts on it. Find work done by the force $\mathbf{F}$ on the block.

a. 66 J b. -66 J c. -90 J d. 90 J

13 A coolie moves on platform with a box of 20 kg on his head. If he covers a distance of 10 m with an acceleration of 0.5 m/s^2. Find the work done by the coolie on the box during the motion.

a. Zero b. 50 J c. 100 J d. -100 J

14 In previous problem, if coolie is moving with constant velocity of 10 m/s. Find work done by the coolie on the box.

a. Zero b. 50 J c. 100 J d. -100 J

One or More than One Options Correct Type

Direction (Q. Nos. 15-22) This section contains 8 multiple choice questions. Each question has four choices (a), (b), (c) and (d) out of which ONE or MORE THAN ONE are correct.

15 Block A does not slide over block B. During motion,

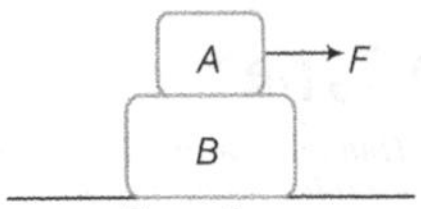

a. the force of friction between blocks is static friction

b. the work done by friction on the block A is negative

c. the work done by friction on the block B is positive

d. if blocks A and B are taken as system, the work done by friction on the system is zero

16 In previous problem,

a. the work done by the friction on the block A depends on the reference frame

b. the work done by the friction on the system (block A + block B) does not depend on the frame of reference

c. no heat loss takes place

d. the work done by the friction on the block B depends on the reference frame

Working Space

17 The block A slip over the block B during motion, then

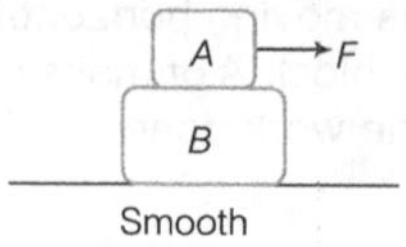

a. the friction between block is kinetic friction
b. the work done by friction on the block B is positive
c. the work done by friction on block A is negative
d. heat loss takes place

18 In the previous problem,
a. the work done by friction on block A depends on the frame of reference
b. the work done by friction on block B does not depend on the frame of reference
c. the work done by friction on the system (block A + block B) are same in all frames of reference
d. heat loss is different in different frames of reference

19 The triangular block is moving horizontally with constant velocity v_0 as shown in the figure. The block A always remain stationary with respect to triangular block. In time t,

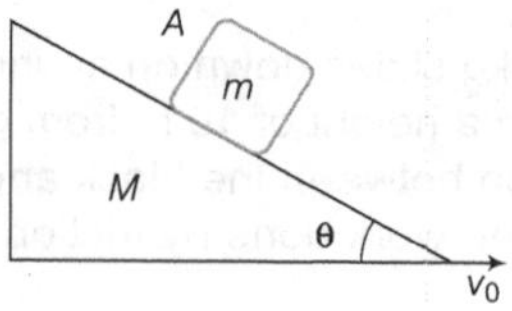

a. the work done by normal reaction on triangular block is $W_1 = -\frac{mg}{2} v_0 t \sin 2\theta$
b. the work done by normal reaction on triangular block is zero
c. the work done by normal reaction on block A is $W_2 = \frac{mg}{2} v_0 t \sin 2\theta$
d. the work done by normal reaction (between block A and the triangular block) is zero on the system (triangular block + block A) in any frame of reference

20 A block of mass 10 kg is moving on a horizontal rough surface by applying a force of 10 N along the surface. The block is moving with a constant velocity. Then
a. the work done by frictional force on the block is – 100 J for displacement of 10 m
b. the net work done on the block for a given displacement is zero
c. the net work done on the block for a given displacement is non-zero
d. None of the above

21 A block of mass 10 kg is placed on a rough horizontal ground. A force of 40 N is applied on the block as shown in the figure. The work done by friction on the block in 2 s is

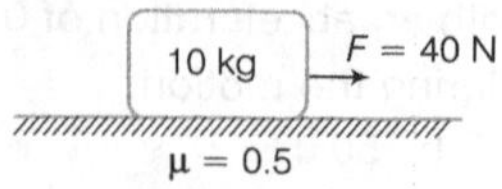

a. zero b. 10 J c. – 10 J d. None of these

22 In previous problem, you are approaching the block with constant velocity 10 m/s in opposite direction of the applied force F. Then, (take, $g = 10 \text{ m/s}^2$)
a. the work done by friction on the block in 2 s is – 800 J
b. the work done by friction on ground is 800 J in 2 s
c. the work done by applied force is 800 J in 2 s
d. the work done by friction on ground is 0 J in 2 s

Comprehension Type

Direction *(Q. Nos. 23-32) This section contains 3 paragraphs, each describing theory, experiments, data, etc. Nine questions related to the paragraphs have been given. Each question has only one correct answer among the four given options (a), (b), (c) and (d).*

Passage I

You and three friends stand at the corners of a square whose sides are 8 m long in the middle of the gym floor as shown in the figure.

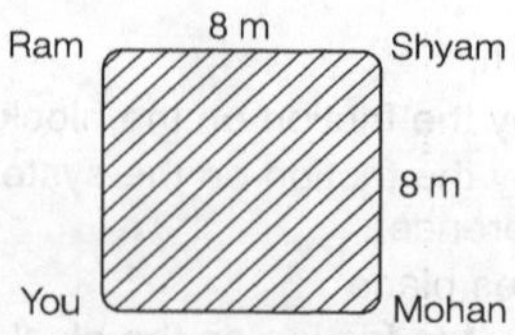

You take your physics book and push it from one person to the other. The mass of book is 2.5 kg and the coefficient of dynamic friction between the book and the floor is 0.3.

23 The book slides from you to Ram and then from Ram to Shyam along the lines connecting these people. The work done by the friction is

a. 10 J b. – 118 J c. – 83 J d. – 50 J

24 You slide the book to Shyam along the diagonal of the square. What is the work done by the friction?

a. – 118 J b. – 83 J c. 30 J d. – 59 J

25 You slide the book to Mohan, then who slides book to you. What is the total work done by the friction during this motion of the book?

a. – 118 J b. – 96 J c. – 83 J d. – 59 J

Passage II

Consider the situation shown in the figure. The system is released from rest and the block *A* travels a distance 5 m in downward direction (take, $g = 10\ \text{m/s}^2$).

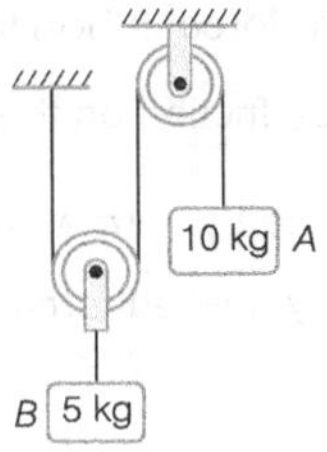

26 Find the work done by tension on the block *B*.

a. Zero b. 116.67 J
c. 375 J d. – 116.67 J

27 Find the total work done by the tension on the system (block *A* + block *B*).

a. Zero b. 116.67 J
c. 375 J d. – 375 J

28 Find net work done on the system.

a. Zero b. 116.67 J
c. 375 J d. – 116.67 J

Passage III

A block of mass 10 kg is moving up on an inclined plane with constant acceleration $2\ \text{m/s}^2$ by applying a constant force *F* as shown in the figure. (take, $g = 10\ \text{m/s}^2$).

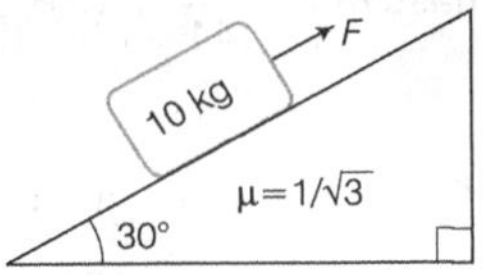

29 The work done on the block by applied force *F* in 2 s is

a. 480 J b. 240 J c. 960 J d. None of these

30 The work done by force of gravity on block in 2 s is

a. 200 J b. zero c. – 200 J d. None of these

31 The value of net work done in time 2 s is

a. zero b. 20 J c. 80 J d. – 80 J

32 From above discussion,

a. net work done on a block or system is equal the algebraic sum of work done on the block or system by each of individual forces (external forces + internal forces) acting on the block or system
b. net work done on a system is equal to the change in kinetic energy of the system
c. the work done by a force on a block is equal to the algebraic sum of work done by the components of the force
d. All of the above

One Integer Value Correct Type

Working Space

Direction *(Q. Nos. 33-38) This section contains 6 questions. Each question, when worked out will result in an integer from 0 to 9 (both inclusive).*

33 A light inextensible string that goes over a smooth fixed pulley as shown in the figure, connects two blocks of masses 0.36 kg and 0.72 kg. Taking $g = 10\ \text{ms}^{-2}$, find the work done (in joule) by string on the block of mass 0.36 kg during the first second after the system is released from rest. **[IIT JEE 2009]**

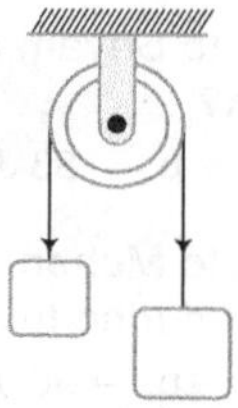

34 A block of mass 10 kg is kept at rest on an inclined plane of inclination 45° is pulled up with an acceleration $2\ \text{m/s}^2$ by applying a constant force in horizontal direction.

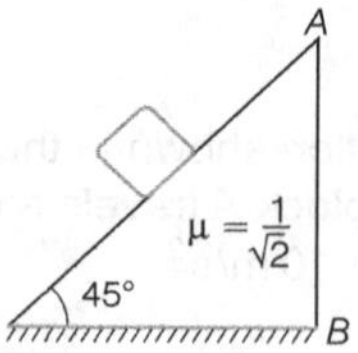

i. The work done by friction on the block in 5 s is -1.25×10^n J. Find the value of n.

ii. Find the work done by normal reaction on the block in 5 s.

iii. The work done by applied force on the block in 5 s is $25\left(70 + \frac{100}{\sqrt{n}}\right)$ J. Find the value of n.

iv. The work done by gravity on the block in 5 s is $\frac{-2500}{\sqrt{n}}$ J. Find the value of n.

v. The net work done on the block in 5 s is 5×10^n J. Find the value of n.

35 All surfaces as shown in the figure are assumed to be frictionless. A block of mass 1 kg slides on the triangular block of mass 10 kg. The system is released from rest. For a horizontal displacement 5 m of the triangular block. (take, $g = 10\ \text{m/s}^2$)

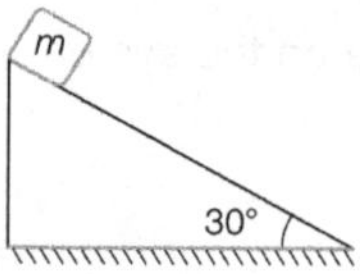

i. The work done by normal reaction by the block of mass m on the triangular block is $\frac{300}{41}\sqrt{n}$ J. Find the value of n.

ii. The work done by normal reaction on the block of mass m is $\frac{-300}{41} \times (3)^{\frac{1}{x}}$ J. Find the value of x.

36 A block A of mass 10 kg is placed over a long plank B of mass 20 kg. The system rests on a smooth horizontal surface. The coefficient of friction between blocks is 0.5. A horizontal constant force of 60 N is applied on the plank B. (take, $g = 10\ \text{m/s}^2$)

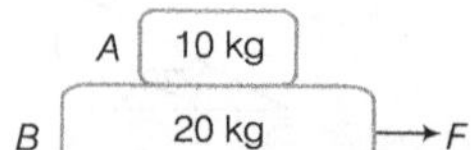

i. If the block A and the plank B are taken as a system, find the work done by the friction on the system in 2 s.

ii. If block A and plank B are taken as system, the net work done on the system in 2 s is $80x$ J. Find the value of x.

37 Water of density $1000\ \text{kg/m}^3$ is to be lifted from a rectangular swimming pool of cross-sectional area $200\ \text{m}^2$ and stored in a cylindrical vessel of radius $\frac{1}{\sqrt{\pi}}$ m and length 200 m whose axis is horizontal and 10 m above the surface of water of the swimming pool. The work done in filling half of the vessel is $10^6\left(10.25 + \frac{n}{3\pi\sqrt{\pi}}\right)$ J. Find the value of n.

Working Space

38 A boy wishes to move the block slowly shown in the figure, a distance of 2 m to the right by either sliding or by tiping over one corner with the least amount of work. The mass of block is 100 kg and coefficient of friction is 0.3. If the least amount of work is 207 nJ. Find the value of n.

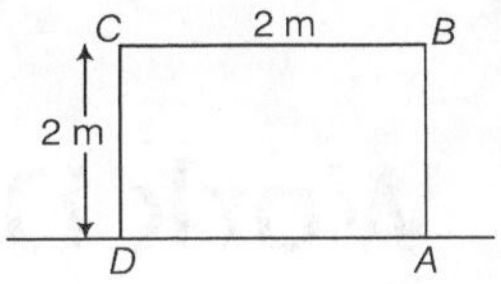

Statement Type

Direction (Q. Nos. 39 and 40) *This section is based on Statement I and Statement II. Select the correct answer from the codes given below.*

a. Both Statement I and Statement II are correct and Statement II is the correct explanation of Statement I
b. Both Statement I and Statement II are correct and Statement II is not the correct explanation of Statement I
c. Statement I is correct but Statement II is incorrect
d. Statement II is correct but Statement I is incorrect

39 **Statement I** During writing, the work done by force exerted by pen on the stationary paper must be zero.

Statement II The paper is stationary.

40 **Statement I** The work done by constant force does not depend on path. It only depends on initial and final position of the particle.

Statement II The work done by constant force is $W = \mathbf{F} \cdot \mathbf{s}$.

Matching List Type

Direction (Q. No. 41) *Choices for the correct combination of elements from Column I and Column II are given as options (a), (b), (c) and (d), out of which one is correct.*

41 Match the Column I and Column II and select the correct option from the codes given below.

	Column I		Column II
i.	In a uniform circular motion, work done by net force is	p.	zero
ii.	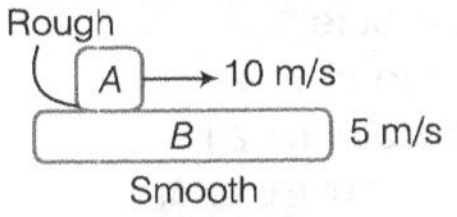 The work done by friction on block B is	q.	non-zero positive
iii.	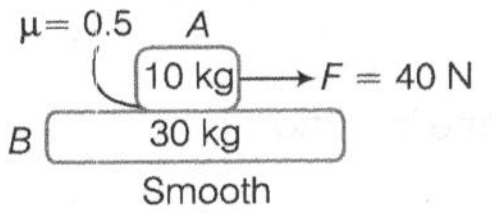 The work done by friction on the system (block A + block B) is	r.	non-zero negative
iv.	In non-uniform circular motion, work done by net force is	s.	either non-zero positive or non-zero negative

Codes

	i	ii	iii	iv		i	ii	iii	iv
a.	p	q	p	s	b.	p	q	r	s
c.	q	p	s	r	d.	p	r	q	s

DPP-3 Work Done by Variable Force

Only One Option Correct Type

Working Space

Direction (Q. Nos. 1-16) *This section contains 16 multiple choice questions. Each question has four choices (a), (b), (c) and (d) out of which ONLY ONE option is correct.*

1 If a force $F = ax^2 + bx + c$ acts on a block along positive x-direction. In the expression of force a, b and c are constant. Find work done by force on the block if block displaces from $x = 0$ to $x = d$.

a. $\frac{ad^3}{3} + \frac{bd^2}{2} + cd$ b. $\frac{ad^3}{2} + bd^2$ c. $\frac{ad^3}{6} + \frac{bd^2}{4} + cd$ d. None of these

2 When a rubber band is stretched by a distance x, it exerts a restoring force of magnitude $F = ax + bx^2$, where a and b are constants. The work done in stretching the unstretched rubber band by L is **[JEE Main 2014]**

a. $aL^2 + bL^3$ b. $\frac{1}{2}(aL^2 + bL^3)$ c. $\frac{aL^2}{2} + \frac{bL^3}{3}$ d. $\frac{1}{2}\left(\frac{aL^2}{2} + \frac{bL^3}{3}\right)$

3 A force $\mathbf{F} = (x^2\hat{\mathbf{i}} + y^2\hat{\mathbf{j}} + z^2\hat{\mathbf{k}})$ N is acting on a particle. The particle displaces from (0, 0, 0) to (1m, 1m, 1m). Find the work done by the force.

a. $\frac{1}{3}$ J b. $\frac{2}{3}$ J c. 1 J d. 2 J

4 A force $\mathbf{F} = (3x^2y\hat{\mathbf{i}} + x^3\hat{\mathbf{j}})$ N is acting on a particle. Find work done by the force during displacement of the particle from (1 m, 1 m) to (2 m, –1 m).

a. 1 J b. –4 J c. 8 J d. –9 J

5 A force $\mathbf{F} = -k(y\hat{\mathbf{i}} + x\hat{\mathbf{j}})$ N (where, k is a positive constant) acts on a particle moving in the XY-plane. Starting from the origin, the particle is taken along the positive X-axis to the point $(a, 0)$ and then parallel to the Y-axis to the point (a, a). The total work done by the force F on the particle is **[IIT JEE 1998]**

a. $-2ka^2$ b. $2ka^2$ c. $-ka^2$ d. ka^2

6 A force $\left(\mathbf{F} = \frac{-ky}{\sqrt{x^2 + y^2}}\hat{\mathbf{i}} + \frac{kx}{\sqrt{x^2 + y^2}}\hat{\mathbf{j}}\right)$ N is applied on a block, whose position is $P(x, y)$. The block moves on a circular path of radius R. Find the work done by the force $\mathbf{F}$ in a complete rotation.

a. $2\pi Rk$ b. πRk c. $4\pi Rk$ d. Zero

7 The work done on a particle of mass m by a force

$$k\left[\frac{x}{(x^2 + y^2)^{3/2}}\hat{\mathbf{i}} + \frac{y}{(x^2 + y^2)^{3/2}}\hat{\mathbf{j}}\right] \text{N}$$

(k is a constant of appropriate dimensions), when the particle is taken from the point $(a, 0)$ to the point (0, a) along a circular path of radius a about the origin in the XY- plane is **[JEE Advanced 2013]**

a. $\frac{2k\pi}{a}$ b. $\frac{k\pi}{a}$ c. $\frac{k\pi}{2a}$ d. zero

Working Space

8 Two blocks each of mass m are placed on a smooth horizontal surface. They attract each other with a force $F = \frac{k}{r}$, where r is the separation between them. Calculate work required to increase the separation between the blocks slowly from a to $2a$.

a. $k \ln 2$ b. $2k \ln 2$ c. $k^2 \ln 2$ d. None of these

9 A force $\left(\mathbf{F} = \frac{b\cos\theta}{r^3}\hat{\mathbf{e}}_r + \frac{b\sin\theta}{r^3}\hat{\mathbf{e}}_\theta\right)$ N is applied on a particle, $\hat{\mathbf{e}}_r$ is a unit vector along r from origin O of coordinate system and $\hat{\mathbf{e}}_\theta$ is a unit vector along increasing angle θ. Find the work done by the force $\mathbf{F}$ when the particle is taken from point $A(a, 0)$ to $B\left(a, \frac{\pi}{2}\right)$ on a circular path of radius a.

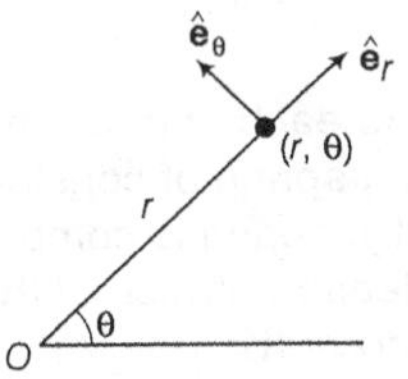

a. $\frac{b}{a}$ b. $\frac{b}{a^2}$ c. $-\frac{b}{a^2}$ d. $-\frac{2b}{a^2}$

10 A circular platform is rotating with constant angular velocity 10 rad/s. A bug starts from point A and reaches at point B through an irregular shape path. During motion of bug, the angular velocity of platform remains constant. Find work done by centrifugal force on the bug. (take, $mr_0^2 = \frac{1}{2} kgm^2$).

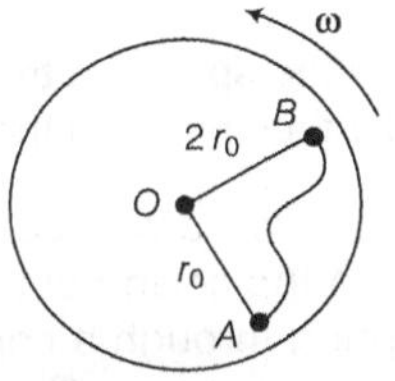

a. 5 J b. 20 J
c. 75 J d. None of these

11 A particle starts from rest from origin and moves along a parabolic path whose equation is $y = x^2$ in the presence of a number of forces. One of the forces acting on the particle is $\mathbf{F} = (x\hat{\mathbf{i}} - 4z\hat{\mathbf{k}})$ N. Find work done by this force $\mathbf{F}$ when the particle moves from origin to $x = 2$ m.

a. Zero b. 2 J c. 1 J d. None of these

12 A block of mass 1 kg is connected with a light spring of constant $k = 100$ N/m. Initially, the spring is compressed upto a compression of 0.2 m and finally comes to an elongation of 0.1 m. Find the work done by the spring on the block.

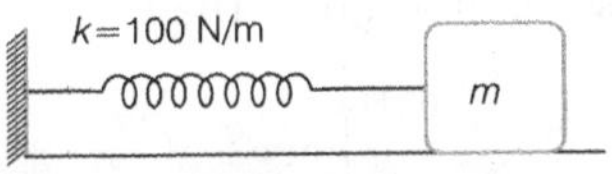

a. 1J b. −1J c. 1.5 J d. −1.5 J

13 A spring of constant $k = 5 \times 10^3$ N/m is stretched initially by 5 cm from the unstretched position. Then, the work required to stretch it further by another 5 cm is **[AIEEE 2003]**

a. 12.50 N-m b. 18.75 N-m c. 25.00 N-m d. 6.25 N-m

14 A spring of force constant 800 N/m has an extension of 5 cm. The work done in extending it from 5 cm to 15 cm is **[AIEEE 2002]**

a. 16 J b. 8 J c. 32 J d. 24 J

15 Work done on a rubber during stretching is

a. positive b. negative c. zero d. None of these

16 A block of mass 1 kg is projected from point A along irregular rough inclined surface and reaches at the point B as shown in the figure. The coefficient of friction between the block and the inclined plane is 0.5. Find work done by the friction on the block if $AC = 1$ m.

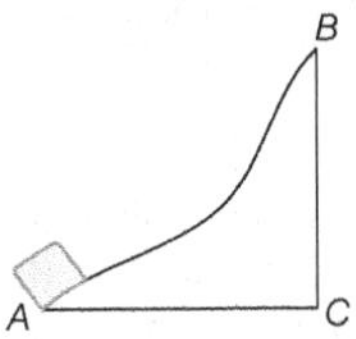

a. −1J b. −2 J
c. −4 J d. −5 J

One or More than One Options Correct Type

Working Space

Direction (Q. Nos. 17-21) *This section contains 5 multiple choice questions. Each question has four choices (a), (b), (c) and (d), out of which ONE or MORE THAN ONE are correct.*

17 A force $F = (y^2 + 2y + 5)$ N acts on a block in positive y-direction. Find the work done by this force during a displacement from $y = -1$ m to $y = +1$ m.

a. 5 J
b. 10.67 J
c. 12.67 J
d. Non-zero positive

18 Two identical blocks each of mass m are connected by a light spring of constant $k = 100$ N/m. Initially, spring is compressed upto 0.2 m and then released. Which of the following statements is/are correct?

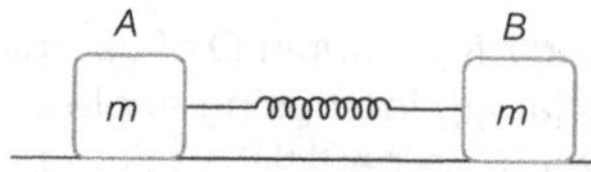

a. The work done by the spring on the block A when the spring comes in natural length is 1 J
b. The work done by the spring on the block B when the spring comes in natural length is 1 J
c. The work done by the spring on the system (block A + block B) is 2 J
d. The work done by the spring on the system (block A + block B) is −2 J

19 An elevator is going upward with constant velocity v_0. One end of light spring of constant k, is fixed to the ceiling of the elevator. A naughty boy inside elevator compresses the spring through a compression x_0 and in the mean time the elevator moves by a distance s. The work done by the naughty boy with respect to ground on the spring is

a. $\frac{1}{2}kx_0^2$
b. $\frac{1}{2}k(x_0 + s)^2$
c. $\frac{1}{2}kx_0 s$
d. same as that in elevator frame

20 A particle moves along X-axis. The force F_x acting on it depends on position x of the particle as shown in the figure. Then,

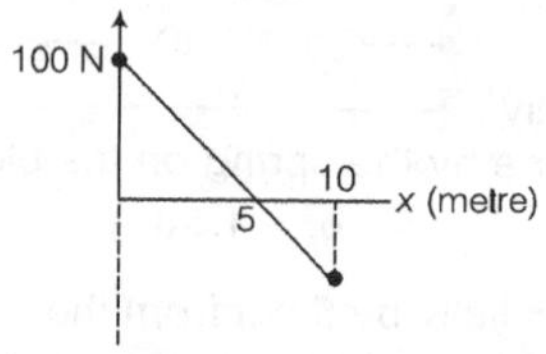

a. the work done by the force F_x for a displacement from $x = 0$ to $x = 5$ m is 250 J
b. the work done by the force F_x for a displacement from $x = 0$ to $x = 10$ m is zero
c. the work done by the force F_x for a displacement from $x = 5$ m to $x = 10$ m is −250 J
d. None of the above

21 Figures show two graphs for a block moving with variable acceleration along X-axis. Mark the correct situation(s).

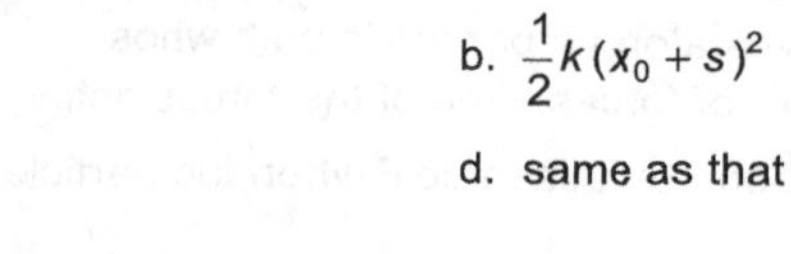

a. The work done in the situation (ii) is zero
b. The work done in the situation (i) is non-zero positive
c. The work done in the situation (i) is non-zero negative, but that of in situation (ii) is zero
d. The work done in both situations are zero

Working Space

One Integer Value Correct Type

Direction *(Q. Nos. 22-26) This section contains 5 questions. Each question, when worked out will result in an integer from 0 to 9 (both inclusive).*

22 A particle of mass 1 kg starting from rest at $x = -10$ m slides on a smooth horizontal surface by applying a single variable force on the particle varies as shown in the figure.

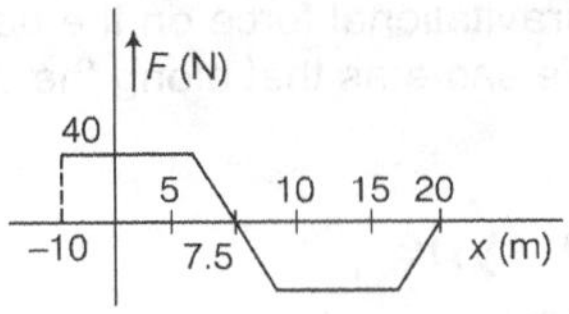

The work done by this force to move the particle

i. from $x = -10$ m to $x = 5$ m is $100n$ J. Find the value of n.
ii. from $x = 0$ to $x = 10$ m is $100n$ J. Find the value of n.
iii. from $x = -10$ m to $x = 20$ m is $100n$ J. Find the value of n.

23 A radial force $\mathbf{F} = br, r < a$; $\mathbf{F} = \frac{b}{r^2}, r > a$ is acting on a particle at distance r from origin. The work done by the force when particle is moving from $r = 0$ to $r = \infty$ is $\frac{a^2 b}{n} + \frac{b}{a}$. Find the value of n.

24 The work done by a force $\mathbf{F} = 2(x+4y)\hat{\mathbf{i}} + 8x\hat{\mathbf{j}}$ N on a particle moving from origin to (4m, 2m, 0) along the path $x^2 = 8y$ is $10n$ J. Find the value of n.

25 Find work done (in joule) by the force $\mathbf{F} = (2xyz\,\hat{\mathbf{i}} + x^2 z\,\hat{\mathbf{j}} + x^2 y\,\hat{\mathbf{k}})$ N on a particle moving from the origin to point C (2m, 1m, 2m) along the path shown in figure.

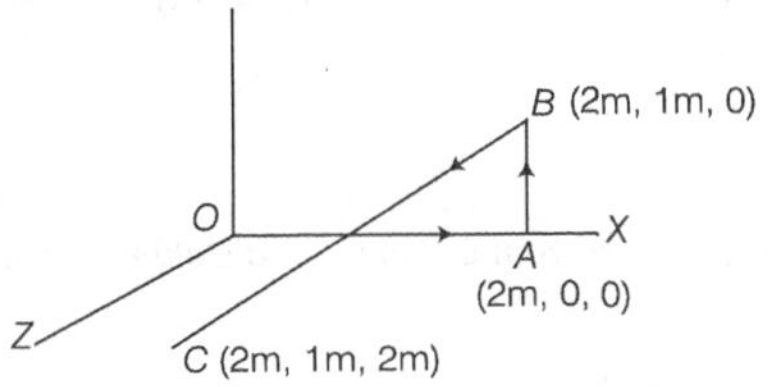

26 A sleeve of mass m is moving along perimeter by applying a constant force of 10 N always directed toward point A. The radius of semicircle is 10 cm. Find work done (in joule) by applied force F during motion of sleeve from B to A.

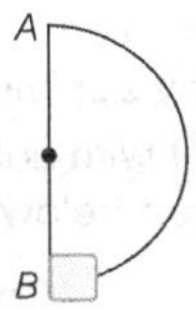

Statement Type

Direction *(Q. Nos. 27-29) This section is based on Statement I and Statement II. Select the correct answer from the codes given below.*

a. Both Statement I and Statement II are correct and Statement II is the correct explanation of Statement I
b. Both Statement I and Statement II are correct and Statement II is not the correct explanation of Statement I
c. Statement I is correct but Statement II is incorrect
d. Statement II is correct but Statement I is incorrect

27 **Statement I** Work is energy transferred to or from an object by means of a force acting on the object.

Statement II The energy transferred to the object is positive work done and energy transferred from object is negative work.

28 **Statement I** The algebraic sum of work of the forces of action and reaction between particles may be non-zero.

Statement II The algebraic sum of work of the forces of action and reaction between particles is independent of reference frame.

Working Space

29 **Statement I** The work done by gravity in bringing a body down from the top to the base along a frictionless inclined plane is the same as the work done in bringing it down along the vertical side.

Statement II The gravitational force on the body during the motion along the inclined plane is the same as that along the vertical side.

Comprehension Type

Direction (Q. Nos. 30 and 31) *This section contains a paragraph, describing theory, experiments, data, etc. Two questions related to the paragraphs have been given. Each question has only one correct answer among the four given options (a), (b), (c) and (d).*

Passage

A smooth wire in the form of a quarter circle of radius 6 m lies in the vertical plane. A sleeve of mass 0.4 mg moves from A to B under the action of forces F_1, F_2 and F_3. F_1 has magnitude 20 N and is always directed towards the point B. F_2 has magnitude 30 N and is always directed horizontally. $F_3 = (15 - 10s)$ N, where s is in metre and is always directed along the tangent to the wire.

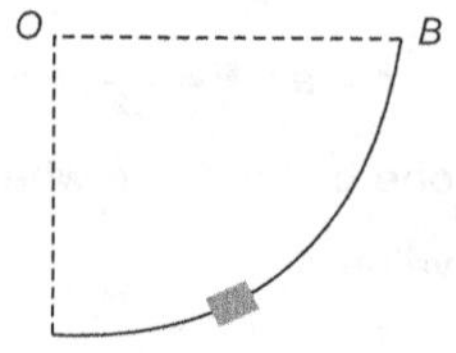

30 The work done by F_1 on the sleeve is

a. −24 J b. $120\sqrt{2}$ J c. −302.8 J d. 23 J

31 The total work done on the sleeve is

a. 11.3 J b. 180 J c. 23 J d. $120\sqrt{2}$ J

Matching List Type

Direction (Q. No. 32) *Choices for the correct combination of elements from Column I and Column II are given as options (a), (b), (c) and (d), out of which one is correct.*

32 Two light springs S_1 and S_2 of spring constants $k_1 = 100$ N/ m and $k_2 = 200$ N/m are connected in series. A man is keeping the springs stretched by applying a force of 50 N.

Match the Column I with column II and choose the correct option from the codes given below.

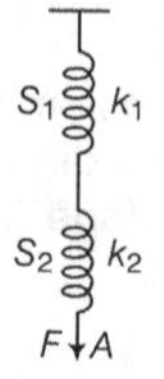

	Column I		Column II
i.	How much work has been done by the spring S_1 on the spring S_2?	p.	18.75 J
ii.	How much work has been done by the spring S_2 on the spring S_1?	q.	−6.25 J
iii.	How much total work has been done by the spring S_1 on both the spring S_2 and the man?	r.	12.5 J
iv.	How much work is done by the man on the spring S_2?	s.	−12.5 J

Codes

	i	ii	iii	iv		i	ii	iii	iv
a.	s	r	q	p	b.	p	q	r	s
c.	q	r	p	s	d.	p	s	q	r

DPP-4 Work-Energy Theorem

Only One Option Correct Type

Working Space

Direction (Q. Nos. 1-9) *This section contains 9 multiple choice questions. Each question has four choices (a), (b), (c), and (d) out of which ONLY ONE option is correct.*

1 When you lower a book of mass m down slowly through a distance h from yourself, the work done by you on the book is

a. mgh b. $-mgh$ c. zero d. $> mgh$

2 A block of mass 1 kg moves along X-axis with variable velocity $v = 3t^2$ (m/s). Find the net work done on the block from $t = 1$s to $t = 2$s.

a. 10 J b. 72 J c. 4.5 J d. 67.5 J

3 A block of mass m moving initially with speed v_0 enters on a rough horizontal surface. Find total distance travelled by the block on the rough surface. Assume the coefficient of friction is μ.

a. $\frac{v_0^2}{\mu g}$ b. $\frac{v_0^2}{2\mu g}$ c. $\frac{v_0^2}{3\mu g}$ d. $\frac{v_0^2}{4\mu g}$

4 A block of mass 1 kg is projected with initial velocity (v_0) of 10 m/s on a rough inclined surface. The block moves up and comes back to its initial position and stops. Find work done by the friction on the block.

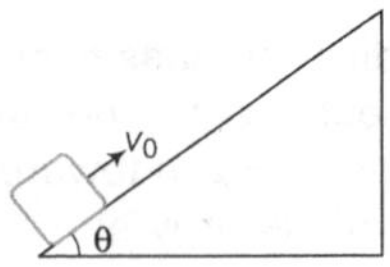

a. zero b. -50 J
c. -100 J d. 50 J

5 A ball of mass 10 g is projected upward with an initial velocity 10 m/s and comes back and finally achieved a velocity of 5 m/s at the point of projection. Find the work done by air resistance. (Neglect buoyancy force due to air)

a. 0.375 J b. 0.275 J c. $-$ 0.375 J d. Zero

6 A block of mass 1 kg is connected with a fixed wall through a light spring of constant $k = 100$ N/m. The surfaces are frictionless. Initially, spring is in relaxed position. A horizontal force F of 50 N is applied on the block. Find the maximum elongation of the spring.

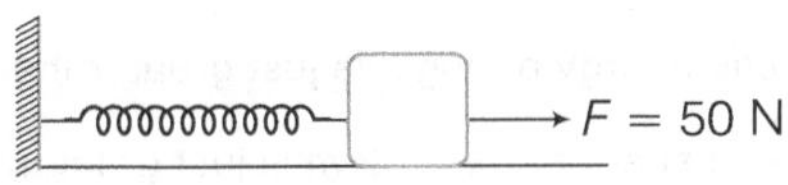

a. 1 m b. 0.5 m c. 0.25 m d. 0.1 m

7 In previous problem, find maximum speed of the block during the motion of the block.

a. 1 m/s b. 2 m/s c. 5 m/s d. 10 m/s

8 A 2 kg block slides on a horizontal floor with a speed of 4 m/s. It strikes a uncompressed spring and compresses it till the block becomes motionless. The kinetic friction force is 15 N and spring constant is 10000 N/m. The spring compresses by **[AIEEE 2007]**

a. 5.5 cm b. 2.5 cm
c. 11.0 cm d. 8.5 cm

Working Space

9 A particle having mass m is released from rest from origin in gravity free space. A force $\mathbf{F} = (F_0)\hat{\mathbf{k}} + (b\,\mathbf{v})\hat{\mathbf{j}}$, where b is constant vector and v is instantaneous speed of the particle, starts to act on the particle at $t = 0$. Find speed of the particle as a function of Z-coordinate.

a. $\sqrt{\frac{F_0 z}{m}}$ b. $\sqrt{\frac{2F_0 z}{m}}$ c. $\sqrt{\frac{2bz}{m}}$ d. $\sqrt{\frac{F_0 z}{mb}}$

One or More than One Options Correct Type

Direction (Q. Nos. 10-14) *This section contains 5 multiple choice questions. Each question has four choices (a), (b), (c) and (d), out of which ONE or MORE THAN ONE are correct.*

10 The acceleration of a particle with speed v is $\mathbf{a} = \mathbf{b} \times \mathbf{v}$, where $\mathbf{b}$ is a constant vector which neither parallel nor anti-parallel to velocity. Then,

a. the velocity of the particle remains constant
b. the work done on the particle is zero
c. the speed of the particle remains constant
d. the work done on the particle is non-zero

11 A woman is pulling a string connected to a block of mass m as shown in the figure along side. During motion of the block,

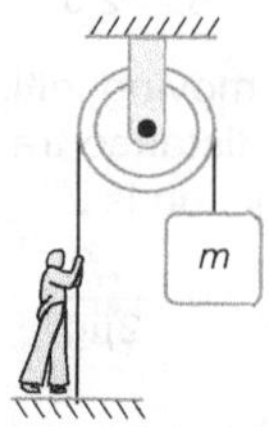

a. the work done by the string on the block is positive
b. the work done by the woman on the string is negative
c. the work done by string on the woman is zero
d. the net work done on the block must be positive

12 The force on a particle of mass m placed on smooth horizontal surface is $F = (2x^2 - 4)$ N directed along positive X-axis. The particle is projected with minimum speed from $x = 2$ m towards mean position along X-axis for crossing the origin, then the kinetic energy at

a. the origin must be zero b. the origin is non-zero
c. $x = \sqrt{2}$ must be zero d. $x = 2$ m may be zero

13 In previous problem, the particle is projected with minimum speed from $x = 2$ m towards mean position along X-axis for crossing the origin, then

a. the minimum initial kinetic energy is $\frac{8}{3}(\sqrt{2} - 1)$
b. the minimum kinetic energy at origin is just greater than $\frac{8\sqrt{2}}{3}$ J
c. the minimum kinetic energy at $x = \sqrt{2}$ m is just greater than zero
d. All of the above

14 The figure shows the x-component of a force $\mathbf{F}$ that acts on a particle moving along the X-axis. If the particle starts from rest from the origin. Then,

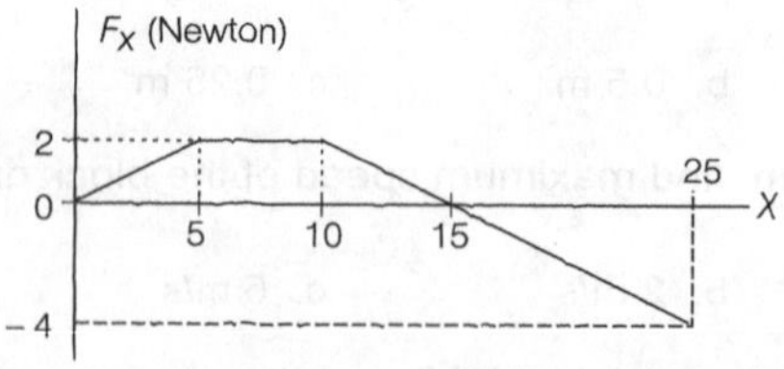

a. the maximum kinetic energy is at $x = 15$ m
b. its speed is zero at $x = 25$ m
c. its kinetic energy is 5 J at $x = 5$ m
d. its kinetic energy is 15 J at $x = 10$ m

Working Space

Statement Type

Direction (Q. Nos. 15 and 16) *This section is based on Statement I and Statement II. Select the correct answer from the codes given below.*

a. Both Statement I and Statement II are correct and Statement II is the correct explanation of Statement I
b. Both Statement I and Statement II are correct and Statement II is not the correct explanation of Statement I
c. Statement I is correct but Statement II is incorrect
d. Statement II is correct but Statement I is incorrect

15 **Statement I** If one of the components of resultant force on a particle is always in the direction of velocity, the kinetic energy continuously increases.

Statement II Net work done on the particle is always positive.

16 **Statement I** A particle is moving on a parabolic curve with constant speed, the net work done on the particle is zero.

Statement II According to work-energy theorem, net work done is equal to the change in kinetic energy.

Comprehension Type

Direction (Q. Nos. 17-20) *This section contains 2 paragraphs, each describing theory, experiments, data, etc. Four questions related to the paragraphs have been given. Each question has only one correct answer among the four given options (a), (b), (c) and (d).*

Passage I

A block of mass m is initially at $x = 0$ is pulled along X-axis by applying the force $F = b - cx$, where b and c are positive constants.

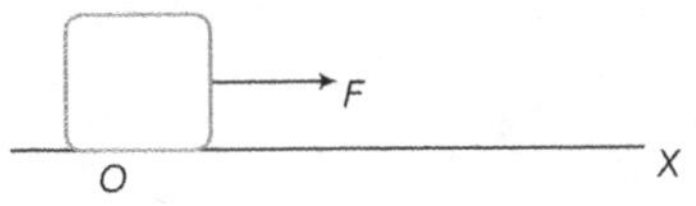

17 The maximum value of x is

a. $\frac{b}{c}$ b. $\frac{b}{2c}$

c. $\frac{2b}{c}$ d. $\frac{b}{3c}$

18 The maximum speed of the block during motion of the block is

a. $\frac{b}{\sqrt{mc}}$ b. $\frac{2b}{\sqrt{mc}}$ c. $\frac{3b}{\sqrt{mc}}$ d. $\frac{4b}{\sqrt{mc}}$

Passage II

In the figures (a) and (b) AC, DG and GF are fixed inclined planes, $BC = EF = x$ and $AB = DE = y$. A small block of mass m is released from the point A. It slides down AC and reaches C with a speed v_C. The same block is released from rest from the point D. It slides down DGF and reaches the point F with speed v_F. The coefficients of kinetic frictions between the block and both the surfaces AC and DGF are μ.

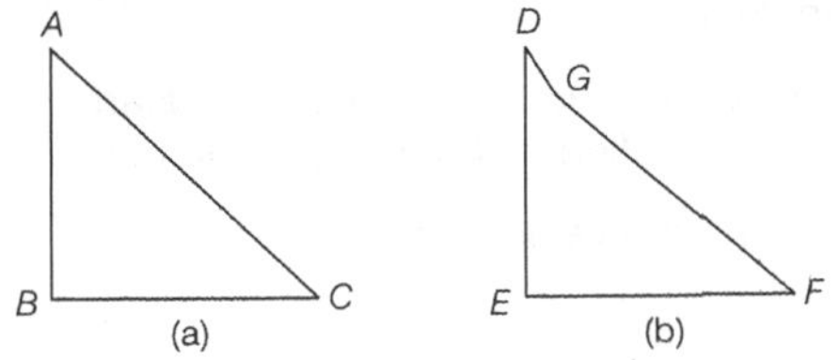

Working Space

19 Find the value of v_C.

a. $\sqrt{2gy - 2\mu gx}$ b. $\sqrt{2gy + 2\mu gx}$

c. $\sqrt{2\mu gx - 2gy}$ d. None of these

20 Mark the correct option.

a. $v_C > v_F$ b. $v_C < v_F$

c. $v_C = v_F$ d. $v_C \geq v_F$

One Integer Value Correct Type

Direction (Q. Nos. 21-27) This section contains 7 questions. Each question, when worked out will result in an integer from 0 to 9 (both inclusive).

21 A particle of mass m is at rest at the origin on a horizontal xy-plane surface. For $0 \leq x \leq l$, the surface is smooth while for $x > l$, the surface is rough with coefficient of friction μ. A force $\mathbf{F} = c(\hat{\mathbf{j}} \times \hat{\mathbf{k}})$, where c is positive constant acts on the particle for the region $0 \leq x \leq l$. The total distance travelled by the particle before coming in rest is $l\left(1+\dfrac{nc}{\mu mg}\right)$. Find the value of n.

22 A block A of mass 2 kg starts slipping down the inclined plane of inclination 37° from a point 4.8 m away from the spring of constant $k = 1000$ N/m. The coefficient of kinetic friction between the block and the inclined plane is 0.5. If the maximum compression in the spring is x_{max}, then find the value of $10x_{max}$ in metre.

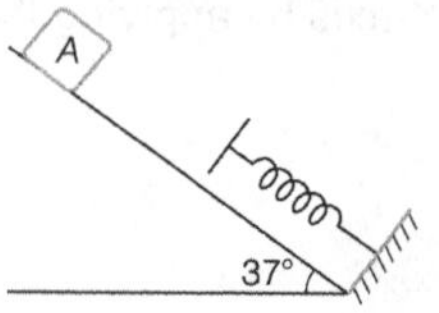

23 In the situation shown in figure, a block of mass 1 kg is attached to a light spring of constant 40 N/m whose other end is fixed to the roof of a building 50 cm above the smooth horizontal surface. Initially, spring is in natural length and vertical. When a force $F = 20\sqrt{3}$ N is applied on the block, the block starts to move. The speed at the instant it breaks off the surface below it is $\sqrt{10n}$ m/s. Find the value of n.

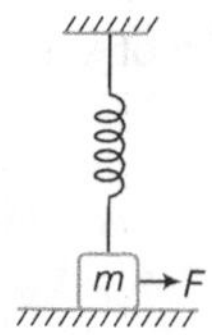

24 A chain of mass m and length $l = 5$ m is placed on a horizontal surface along X-axis as such its end A is at the origin and its end B is at $x = l$. The surface is rough with coefficient of friction is 0.5 from $x = 0$ to $x = l$. For $x > l$, the surface is smooth. Determine the minimum velocity v_0 in m/s with which it must project along positive X-axis so that it lies completely in the frictionless region before coming to rest.

25 A uniform rod of length L and mass M has been placed on a rough horizontal surface. The mass of hanging block is such that the rod is just in the state of rest. If the coefficient of friction varies according to the relation $\mu = kx$, where k is positive constant.

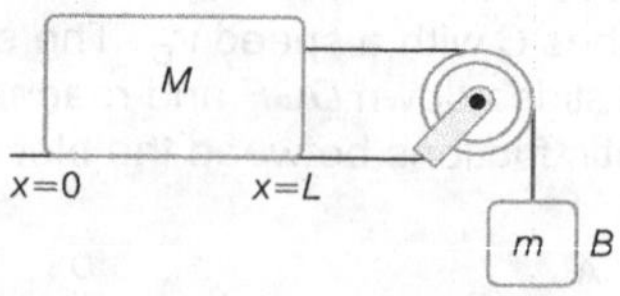

Suppose, friction is present only from $x = 0$ to $x = L$. Due to slight jerk, rod starts to move in rightward direction. The work done by friction in the time in which the rod crosses the region of friction is $-\dfrac{MkgL^2}{n}$. Find the value of n.

Working Space

26 A block of mass 10 kg is suspended through an elastic light rubber thread. An extra force F is applied downward and is slowly increased. When $F = F_0 = 300$ N the thread breaks. If a constant force F_1 is applied at the place of extra force F. The minimum value of F_1 (in newton) such that thread breaks is 1.5×10^n N. Find the value of n.

27 Consider an elliptically shaped rail PQ in the vertical plane with $OP = 3$ m and $OQ = 4$ m. A block of mass 1 kg is pulled along the rail from P to Q with a force of 18 N which is always parallel to line PQ (see figure). Assuming no frictional losses, the kinetic energy of the block when it reaches Q is $(n \times 10)$ J. The value of n is (take, acceleration due to gravity = 10 m/s^2). **[JEE Advanced 2014]**

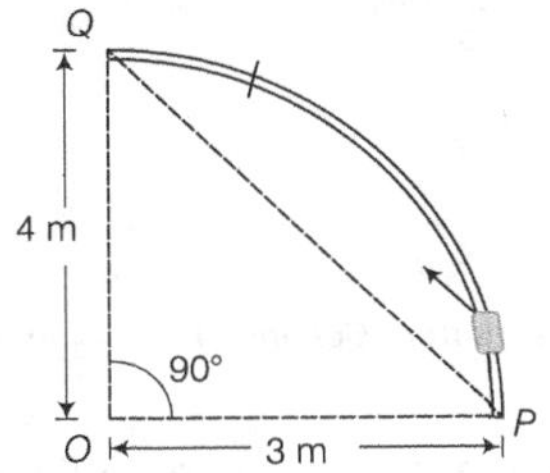

Subjective Questions

Direction (Q. Nos. 28-32) *These questions are subjective in nature, need to be solved completely on notebook.*

28 The displacement x of a particle moving in one dimension under the action of a constant force is related to the time $t = \sqrt{x} + 3$, where x is in metre and t in second. Find

i. the displacement of the particle when its velocity is zero and

ii. the work done by the force in the first 6 s

29 The situation shown in the figure, two blocks A and B, each having mass $m = \sqrt{2}$ kg are connected by a combination of light string and a light spring of constant k. The planes on which blocks A and B are placed are rough having coefficient of friction 0.5. Find the minimum value of constant force F required to move the block B up the inclined plane (take, $g = 10$ m/s^2). Assume that initially spring is relaxed.

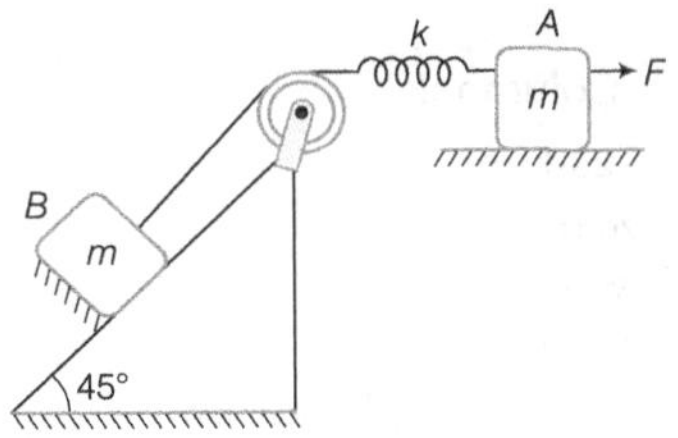

30 A body of mass 2 kg is being dragged with a uniform velocity of 2 m/s on a rough horizontal plane. The coefficient of friction between the body and the surface is 0.20, J = 4.2 J/cal and $g = 9.8$ m/s^2. Calculate the amount of heat generated in 5 s.

31 A lead bullet just melts when stopped by an obstacle. Assuming that 25% of the heat is absorbed by the obstacle, find the velocity of the bullet if its initial temperature is 27°C. (Melting point of lead = 327° C, specific heat of lead = 0.03 cal / g-° C, latent heat of fusion of lead = 6 cal/g , J = 4.2 J/cal)

Working Space

32 A 0.5 kg block slides from the point A (see figure) on a horizontal track with an initial speed of 3 m/s towards a weightless horizontal spring of length 1 m and force constant 2 N/m. The part AB of the track is frictionless and the part BC has the coefficients of static and kinetic friction as 0.22 and 0.2, respectively. If the distance AB and BD are 2 m and 2.14 m respectively, then find the total distance through which the block moves before it comes to rest completely.
(take, $g = 10\ m/s^2$)

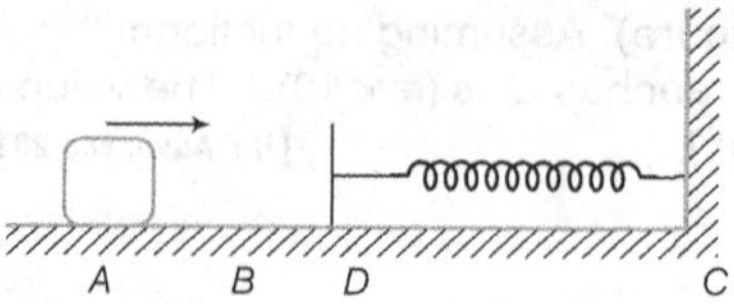

Matching List Type

Direction (Q. No. 33) Choices for the correct combination of elements from Column I and Column II are given as options (a), (b), (c) and (d), out of which one is correct.

33 An observer is situated in a car moving with constant acceleration $a_0 = 5\ m/s^2$. A block A of mass $m = 1$ kg is moving on rough horizontal surface of coefficient of friction $\mu = 0.5$. The car and block both starts from rest. At $t = 1$s, the change in kinetic energy of block A in the frame of car is 50 J.

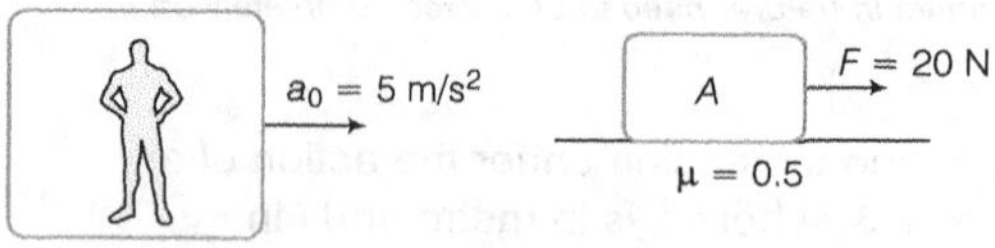

If W_1 = Work done by normal reaction on the block in the frame of car in time $t = 1$s.

W_2 = Work done by gravity on block in the frame of car in time $t = 1$s.

W_3 = Work done by applied force F in the frame of car in time $t = 1$s.

W_4 = Work done by friction on block in the frame of car in time $t = 1$s.

W_5 = Work done by pseudo force on the block in the frame of car in time $t = 1$s.

Match the Column I with Column II and select the correct option from the codes given below.

	Column I		Column II
i.	W_1	p.	−25 J
ii.	W_3	q.	zero
iii.	W_5	r.	50 J
iv.	$W_1 + W_2 + W_3 + W_4 + W_5$	s.	100 J

Codes

	i	ii	iii	iv		i	ii	iii	iv
a.	q	s	p	r	b.	p	q	r	s
c.	s	p	r	q	d.	q	p	r	s

DPP-5 Potential Energy

Conceptual Questions

Working Space

Direction (Q. Nos. 1-4) *These questions are conceptual in nature, need to be solved completely on notebook.*

1 State True or False.

i. The work done by conservative forces are state function.

ii. The work done by conservative forces is not path function.

iii. A particle moves on a closed irregular path as shown in the figure. The particle starts from point A and after several rounds trip the particle finally reaches at the starting point A. The total work done by a force $\mathbf{F}$ on the particle is certainly zero if $\mathbf{F}$ is conservative.

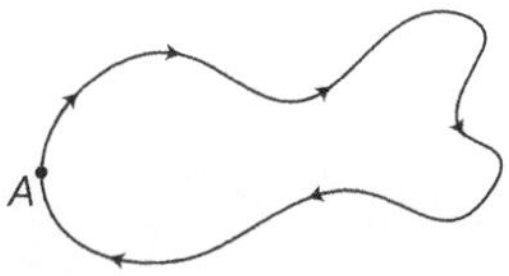

iv. A force is conservative if the total work done by it on a particle is zero when the particle moves around any closed path returning to its initial position, i.e. $\int_c \mathbf{F} \cdot d\mathbf{s} = 0$.

v. A force is conservative if the work done by it on a particle that moves between two fixed points depends only on the position of these points and not on the path followed by the particle.

vi. In one-dimensional motion, forces that depend only on the position are always conservative.

vii. In two-dimensional or three-dimensional motion, it is necessary but not sufficient that a force be a function only of position to be conservative. For example, the dynamic friction on a particle moving on a circular horizontal path is directed tangentially to the circular path. In this case, the direction of dynamic friction depends only on position. Since, the dynamic friction is always in the opposite direction of motion, the total work done by dynamic friction on the particle is negative
(i.e. $-\mu mg \times 2\pi r$) in one complete revolution. Since, the particle is back to its original position and the net work done is non-zero. This force is not conservative.

viii. Velocity independent forces are always conservative, e.g.

a. force of gravity

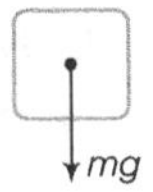

b. spring force

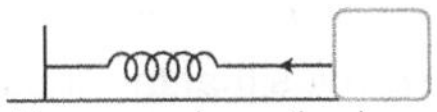

Here x = elongation in spring

c. gravitational force of interaction between particles,
$F = \dfrac{Gm_1m_2}{r^2}$

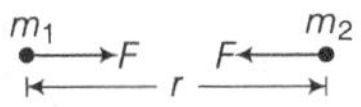

d. electrostatic force of interaction between charges,
$F = \dfrac{1}{4\pi\varepsilon_0}\dfrac{q_1q_2}{r^2}$

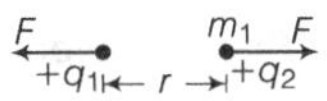

ix. Suppose, you push a book on your table through a distance s. If book moves with constant velocity on the table. For this, you may apply a horizontal force equal to the dynamic friction on the book by the table to just balance the frictional force on the book so that the book does not accelerate or you may apply a greater force so that the book accelerates on the table. The force you apply clearly does not depend only on the position of the book but depends upon your desire and is therefore not a conservative force. Hence, the force applied by human agent such as a push or pull is always non-conservative force.

2 **Situation** In a certain two-dimensional conservative force, the potential energy of a particle is $U = ax^2 + by^2$, where a and b are positive constants of different magnitude.

Mark the correct option(s) for the following questions.

i. The expression for conservative force is

a. $\mathbf{F} = F_x\hat{\mathbf{i}} + F_y\hat{\mathbf{j}} = -2ax\,\hat{\mathbf{i}} - 2by\,\hat{\mathbf{j}}$ b. $\mathbf{F} = 0$

c. $\mathbf{F} = F_{x'}\,\hat{\mathbf{i}} + F_y\hat{\mathbf{j}} = 2ax\hat{\mathbf{i}} + 2by\,\hat{\mathbf{j}}$ d. None of the above

ii. In previous problem, the equation of line of force is

a. $\ln x = \frac{a}{b}\ln y + \ln c$, where c is a constant

b. $x = c\,y^{a/b}$

c. Both a and b

d. None of the above

iii. In previous problem, if $a = b$, draw lines of force.

a. 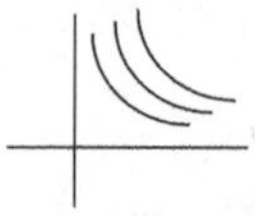 b. 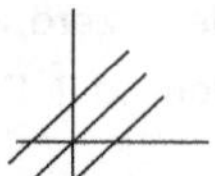 c.  d. 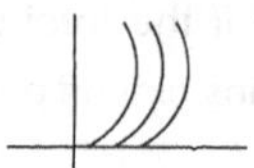

iv. In previous problem, find work done by the conservative force on the particle when the particle moves from point $A(d, d)$ to point $B(2d, 2d)$.

a. $3(a + b)d^2$ b. $4(a + b)d^2$ c. $-3(a + b)d^2$ d. None of these

v. The locus of points of same potential energy of a particle is known as equipotential surface. If the particle is moving on its equipotential surface potential energy does not change. As, we know that the work done by conservative force is $W = -\Delta U$. From this, if the particle moves on a equipotential surface from one point to another point, the work done by conservative force is zero. That is why conservative force is always directed perpendicular to the equipotential surface. The equation of equipotential surface for the given situation is

a. $ax^2 + by^2 =$ constant b. $ax^2 =$ constant

c. $ay^2 =$ constant d. None of these

vi. The shape of equipotential surface is a

a. ellipse b. circle c. parabola d. planar

vii. From above discussion, some students develop some basic concepts.

I. Potential energy exists only for conservative forces.

II. In the equation of lines of force, $x = cy^{a/b}$ for any value of a and b, the lines of force are open curves. From this point of view, it is obvious that the lines of force for conservative forces are open differential curve. If the lines of force are closed differential curve, then the force is non-conservative in nature.

III. The conservative force is always perpendicular to the equipotential surface. So, lines of force cross equipotential surface at right angle.

IV. As we know that, the work done by conservative force is $W = \int \mathbf{F} \cdot d\mathbf{s} = -\Delta U$. From this, the potential energy decreases in the direction of conservative force.

V. In perpendicular direction of conservative force, equipotential surface is directed. So, the potential energy in perpendicular direction of conservative force remains constant.

Working Space

Mark the correct option.

a. Only (V) is incorrect
b. Only (IV) is incorrect
c. (I) and (II) are incorrect
d. All are correct

3 **Situation** If a force $\mathbf{F} = ax\hat{\mathbf{i}} + ay\hat{\mathbf{j}}$, where a is a constant is acting on a particle of mass m.

Mark the correct option(s) for the following questions.

i. The equation of lines of force is

a. $y = kx$, where k is constant
b. $y = kx^2$, where k is constant
c. $y = 0$
d. None of these

ii. The nature of force **F** is

a. conservative b. non-conservative c. Both a and b d. None of these

iii. The expression for potential energy of the particle is

a. $U = -\frac{ax^2}{2} - \frac{ay^2}{2} + k$
b. $U = -\frac{ax^2}{2} + \frac{ay^2}{2} + k$
c. $U = -\frac{ax^2}{2} - \frac{ay^2}{2} + k$
d. $U = -xy + k$

iv. The shape of equipotential surface is a

a. parabolic b. straight line c. circular d. elliptical

v. Find work done by the force **F** during displacement from (0, 0) to (b, b).

a. $W = -\Delta U = -ab^2$
b. $W = ab^2$
c. $W = ab^2/2$
d. None of these

vi. If the potential energy of the particle at origin is taken as zero, then expression for potential energy of the particle is

a. $U = -\frac{ax^2}{2} - \frac{ay^2}{2}$ b. $U = 0$ c. $U = -axy$ d. None of these

4 **Situation** Figure shows a U-x graph for a particle of one-dimensional motion along X-axis.

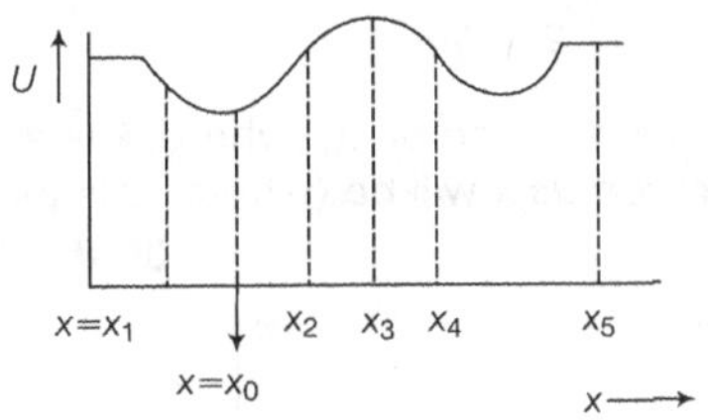

Mark the correct option(s) for the following questions.

i. If a particle is at $x = x_0$, then equilibrium is

a. stable b. unstable c. Both a and b d. None of these

ii. At position $x = x_3$, the particle is

a. at stable equilibrium
b. at unstable equilibrium
c. not in equilibrium
d. None of these

iii. In an interval in which U is constant, the slope of U-x graph is zero so that the force is zero. Such an interval is called one of neutral equilibrium. If a particle is displaced slightly from neutral equilibrium position, then force at displaced position on the particle is zero. Also, potential energy at displaced position is same as that of equilibrium position. The particle is in neutral equilibrium at

a. $x = x_0$ b. $x = x_2$ c. $x = x_3$ d. $x = x_5$

Only One Option Correct Type

Working Space

Direction (Q. Nos. 5-15) *This section contains 11 multiple choice questions. Each question has four choices (a), (b), (c) and (d), out of which ONLY ONE option is correct.*

5 A block is projected upward along a smooth inclined plane of inclination α. Initial gravitational potential energy of the block is zero. Draw gravitational potential energy of the block *versus* time graph during total time of motion of the block on the inclined plane.

a. 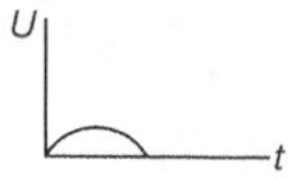
b. 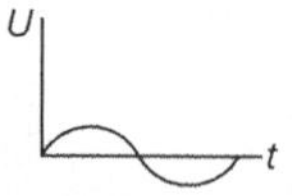
c. 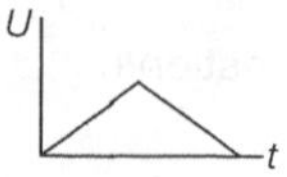
d. 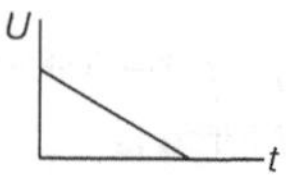

6 A particle which is constrained to move along the X-axis is subjected to a force in the same direction, which varies with the distance x of the particle from the origin as $F = \frac{b}{x^2}$, where b is positive constant. For $x \geq 0$, the functional form of the potential energy $U(x)$ of the particle is

a. 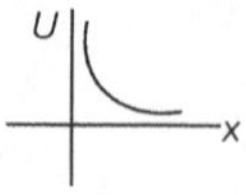
b. 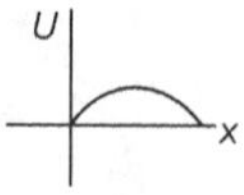
c. 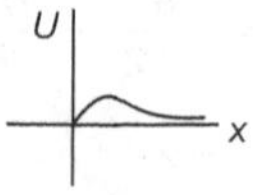
d. 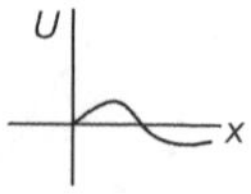

7 A spring of constant 100 N/m is stretched by applying equal forces each of magnitude F at the two ends. The energy stored in the spring is 200 J. Now, spring is cut into two equal parts and one of the part is stretched by applying equal forces each of magnitude F at the two ends. The energy stored is

a. 200 J
b. 100 J
c. 400 J
d. 50 J

8 A point mass m (neglect volume of the point mass) is floating on the surface of water contained by a vertical cylinder. The water is upto height H. Due to leakage at bottom of the cylinder, total water spread out near bottom of cylinder. The total mass of water is M. The work done by gravity is

a. $Mg\frac{H}{2}$
b. $(m+M)gH$
c. $\left(m+\frac{M}{2}\right)gH$
d. $(m+M)g\frac{H}{2}$

9 A particle is placed at the origin and a force $F = kx$ is acting on it, (where, k is a positive constant). If $U(0) = 0$, the graph of $U(x)$ *versus* x will be (where, U is the potential energy of function). **[IIT JEE 2004]**

a. 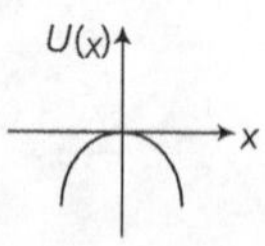
b. 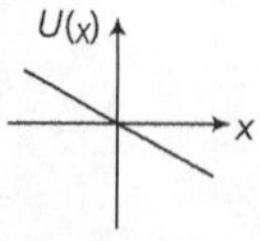
c. 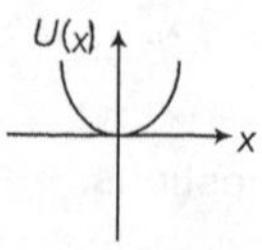
d. 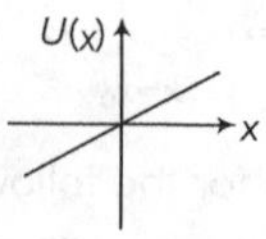

10 A particle which is constrained to move along X-axis is subjected to a force in the same direction which varies with the distance x of the particle from the origin as $F(x) = -kx + ax^3$. Here, k and a are positive constants. For $x \geq 0$, the functional form of the potential energy $U(x)$ of the particle is **[IIT JEE 2002]**

a. 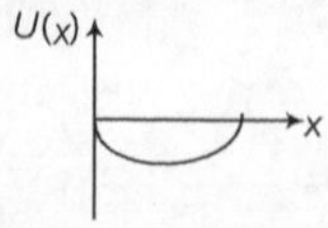
b. 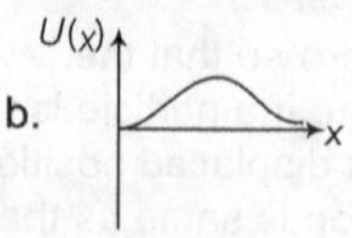
c. 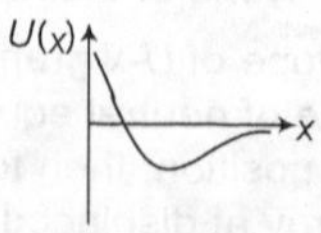
d. 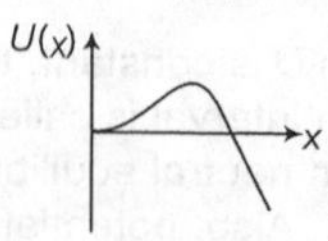

11 n number of identical cubical blocks of edge a and density ρ are lying on the ground. The work done by an external agent in arranging them in a column of height $(n+1)a$ is

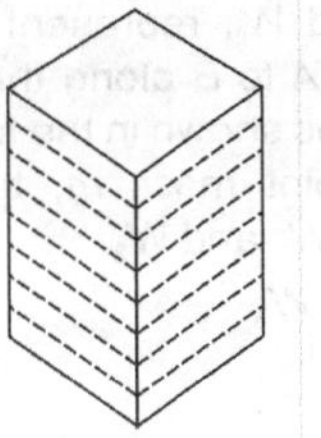

a. $\frac{1}{2}\rho a^4 g n^2$ b. $\frac{1}{2}\rho a^4 (n^2+n)$

c. $\frac{1}{2}(n^2-2)a^4\rho g$ d. $\frac{1}{2}(n^2-1)a^4\rho g$

12 Two small blocks A of mass $\sqrt{3}$ kg and B of mass 1 kg are connected by a light string are placed on a smooth fixed sphere such that the system is in equilibrium for $\angle AOB = \frac{\pi}{2}$.

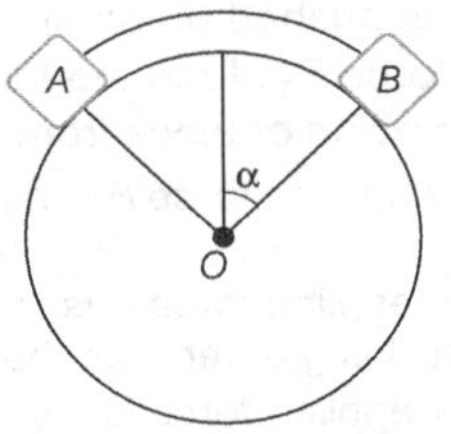

Then,

a. the value of $\alpha = \frac{\pi}{3}$ and equilibrium is stable

b. the value of $\alpha = \frac{\pi}{3}$ and equilibrium is unstable

c. the value of $\alpha = \frac{\pi}{6}$ and equilibrium is stable

d. the value of $\alpha = \frac{\pi}{6}$ and equilibrium is unstable

13 Two flat belts each of mass 20 g are lying on a horizontal surface. First belt is 20 cm long and the second is wider and only 10 cm long. They are kept on a thin wall 10 cm high so that half of them is over the top of wall. What is the ratio of the amount of work done in keeping them on the wall?

a. 1 : 1 b. 2 : 1

c. 2 : 3 d. 1 : 3

14 The elastic energy stored in the spring in the equilibrium position of block is

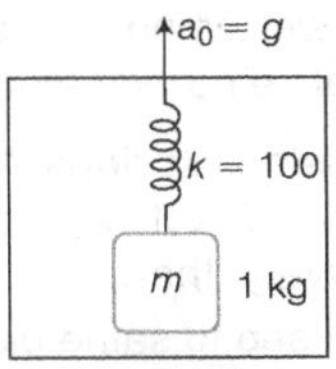

a. zero
b. 0.5 J
c. – 0.5 J
d. 2 J

15 Ideal pulley is smooth and fixed. The inclined planes are also smooth. Balls are in equilibrium. The equilibrium is

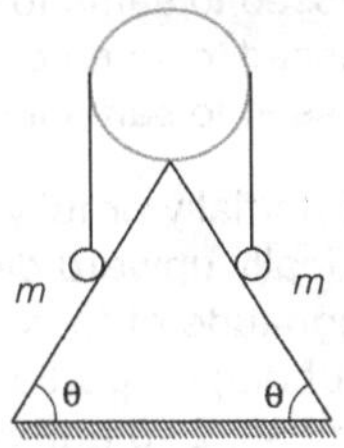

a. stable
b. unstable
c. neutral
d. we cannot predict

One or More than One Options Correct Type

Direction (Q. Nos. 16-23) This section contains 8 multiple choice questions. Each question has four choices (a), (b), (c) and (d), out of which ONE or MORE THAN ONE are correct.

16 i. Figure shows a block of mass m and moves from position A to position B in the presence of force **F** through three different paths. The work done by the force **F** on the block is W_1, W_2 and W_3 for paths 1, 2 and 3, respectively. If

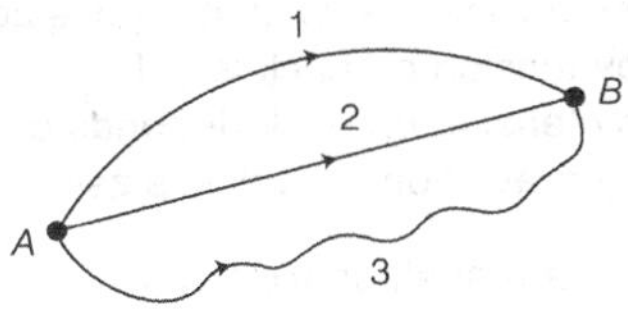

a. $W_1 = W_2 = W_3$, then **F** is certainly conservative
b. $W_1 \neq W_2 \neq W_3$, then **F** is certainly non-conservative
c. $W_1 = W_2 = W_3$, then **F** can never be conservative
d. None of the above

Working Space

ii. If W_1, W_2 and W_3 represent the work done in moving a particle from A to B along three different paths 1, 2 and 3 respectively as shown in the figure below in the gravitational field of a point mass m, then find the correct relation between W_1, W_2 and W_3. **[IIT JEE 2003]**

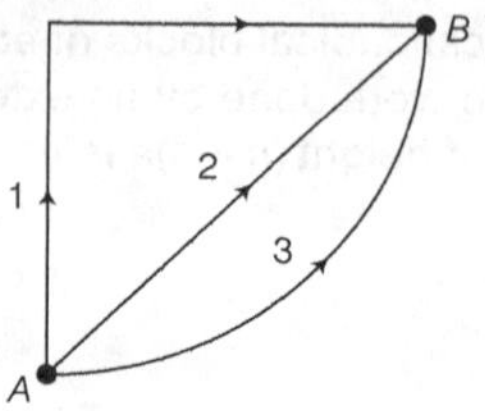

a. $W_1 > W_2 > W_3$ b. $W_1 = W_2 = W_3$
c. $W_1 < W_2 < W_3$ d. $W_2 > W_1 > W_3$

17 A block of mass m is pushed up on smooth irregular inclined plane by applying force F_0 along the tangent on the path of the block. During moving of block from A to B,

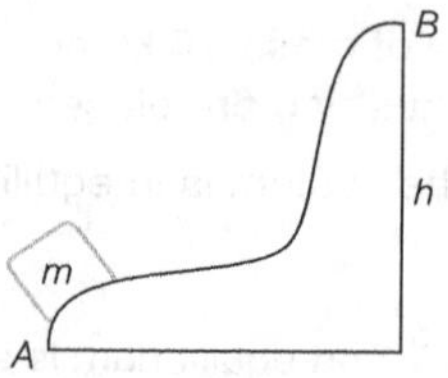

a. the work done by applied force F is mgh if block is moving slowly
b. the work done by applied force F is more than mgh if the speed of block at B is greater than that of point A
c. the work done by applied force is lesser than mgh if the speed of block at B is lesser than that of point A
d. the applied force F is non-conservative in nature

18 In previous problem, the work done by gravity is
a. $W = -mgh$
b. greater than $-mgh$ if speed of block at point A is greater than that of point B
c. less than $-mgh$ if speed of block at point A is lesser than that of point B
d. None of the above

19 A particle moves along a curve of unknown shape but the magnitude of the force is constant and always acts along tangent to the curve. The force is
a. conservative
b. non-conservative
c. conservative as well as non-conservative
d. neither conservative nor non-conservative

20 There are two massless springs A and B of spring constants k_A and k_B respectively and $k_A > k_B$. If W_A and W_B be denoted as work done on A and work done on B respectively, then
a. they are compressed to same distance, $W_A > W_B$
b. they are compressed to same force (upto equilibrium state) $W_A < W_B$
c. they are compressed to same distance, $W_A = W_B$
d. they are compressed to same force (upto equilibrium state) $W_A > W_B$

21 If a body is dipped partially or fully into a fluid at rest, a force is exerted by liquid on the body in vertically upward direction. This force is known as force of buoyancy. The magnitude of force of buoyancy is $F = \rho Vg$. Here, ρ = density of liquid, V = volume of dipped portion of the body.
A cubical brick of mass, 2 kg and side length, 0.5 m is at the bottom of well, completely filled with water. The density of material of brick is equal to that of water. Now, the brick is pulled slowly out of water and is kept at the top surface of well with the help of light string. (Neglect viscous force). (take $g = 10\ \text{m/s}^2$)
a. When brick is inside water, tension in rope is zero
b. The work done by tension on brick is 5 J
c. The work done by tension on brick depends on the depth of well
d. The work done by gravity on the brick is zero

22 A Monkey rises on the massless rope slowly. Then

Monkey

a. the work done by tension on monkey is zero
b. the work done by tension on monkey is positive
c. the work done by internal forces is non-zero
d. we cannot predict any thing

23 A mango falls from a tree on the ground. A person says that the gravitational potential energy of the mango is decreased by 20 J and the other says it is decreased by 40 J. Mark the correct option(s).

a. Both may be incorrect
b. Both may be correct
c. Both must be in different frames of reference
d. Both never be correct

One Integer Value Correct Type

Direction *(Q. Nos. 24-27) This section contains 4 question. When worked out will result in an integer from 0 to 9 (both inclusive).*

24 If X-axis is taken as reference line and Y-axis in vertically upward direction. The potential energy of a parabolic plate $y^2 = kx$ of surface mass density σ as shown in the figure is $\frac{\sigma g a^2 k}{n}$. Find the value of n.

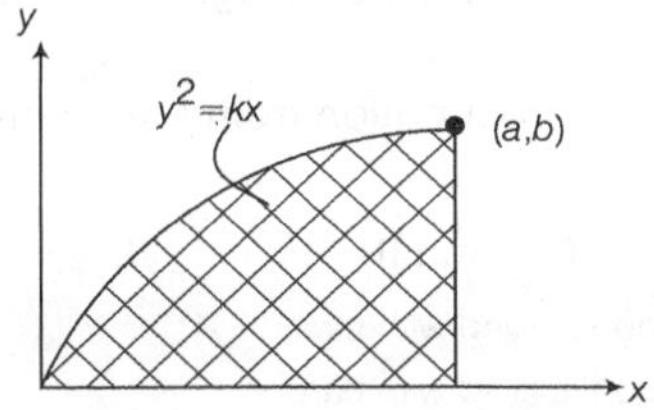

25 A system of blocks A, B and C each of mass 1 kg connected by light springs. The spring constant for each spring is 100 N/m. The lower spring is not fixed at the ground. The whole system is in equilibrium. The increase in gravitational potential energy in lifting block A upward till the lower spring comes in its natural length and the block B is in equilibrium simultaneously is $3n$ J. Find the value of n.

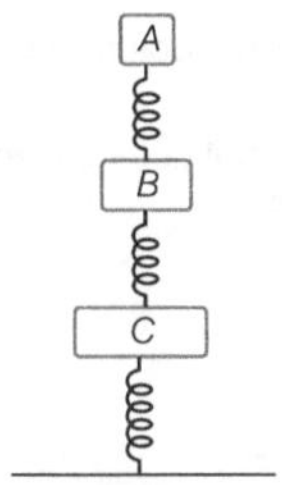

26 Rishu slowly lift his puppy named Teddy by a height of 10 m. The mass of the puppy is 10 kg. The work done on puppy by Rishu is $n \times 10^3$ J. Find the value of n. (take, $g = 10 \text{ m/s}^2$)

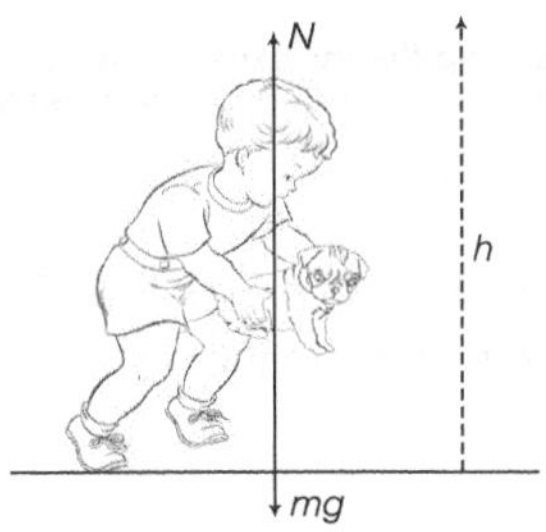

27 If potential energy of a particle of mass 1 kg in a conservative field is $U = \sqrt{2}x^2y$. All are in SI. If at a point defined by position vector $\mathbf{r} = (\hat{\mathbf{i}} + \hat{\mathbf{j}})\,m$, then particle is observed moving perpendicular to $\mathbf{r}$, find the magnitude of tangential acceleration in m/s^2.

Matching List Type

Working Space

Direction (Q. Nos. 28 and 29) Choices for the correct combination of elements from Column I and Column II are given as options (a), (b), (c) and (d), out of which one is correct. Choose the correct option from the codes given below.

28 Match the Column I with Column II and select the correct option from the given Codes.

	Column I		Column II
i.	Work done by conservative force	p.	Path function
ii.	Work done by static friction	q.	State function
iii.	Displacement	r.	May be non-zero
iv.	Work done by non-conservative force	s.	May be zero

Codes

	i	ii	iii	iv
a.	q,s	p,r	q,s	p,r,s
b.	q,r,s	p,r,s	r,s	p,r,s
c.	q,r	p,r	q,s	p,q
d.	s,q	p,s	q	r

29 A balloon is released from rest in air from the roof of a building. (Neglect the mass of balloon surface material).

Match the Column I with Column II and select the correct option from the given Codes.

	Column I		Column II
i.	Potential energy decreases.	p.	If balloon is filled with air.
ii.	Potential energy remains constant.	q.	If balloon is filled with carbon dioxide gas.
iii.	Gravitational potential energy decreases.	r.	If balloon is filled with hydrogen gas.
iv.	Gravitational potential energy increases.	s.	If balloon is filled with steam of water.

Codes

	i	ii	iii	iv
a.	q,r,s	p	q	r,s
b.	p	q	r	s
c.	q	r	p	s
d.	q,r	p,q	q	r,s

Comprehension Type

Direction (Q. Nos. 30-33) This section contains a paragraph, describing theory, experiments, data, etc. Four questions related to the paragraph have been given. Each question has only one correct answer among the four given options (a), (b), (c) and (d).

Passage

The potential energy of force of interaction between atoms in a diatomic molecule is $U = \frac{a}{x^{12}} - \frac{b}{x^6}$, where a and b are positive constants and x is the distance between atoms.

30 The force of interaction between atoms as a function of x is

a. $F =$ zero

b. $F = \frac{6b}{x^7} - \frac{12a}{x^{13}}$

c. $F = \frac{12a}{x^{13}} - \frac{6b}{x^7}$

d. None of these

31 The U-x graph is

a. 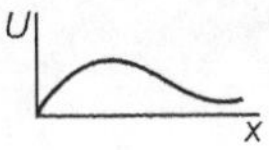

b. 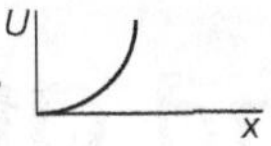

c. 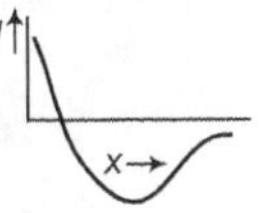

d. None of these

32 The F-x graph is

a. 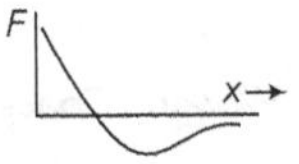

b. 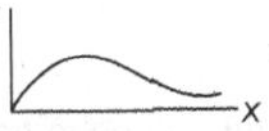

c. 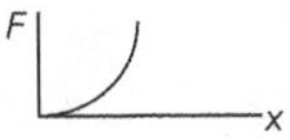

d. 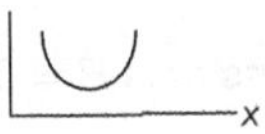

33 At $x = \left(\frac{2a}{b}\right)^{1/6}$, atoms are in

a. stable equilibrium
b. unstable equilibrium
c. neutral equilibrium
d. not in equilibrium

Statement Type

Direction (Q. Nos. 34-36) *This section is based on Statement I and Statement II. Select the correct answer from the codes given below.*

a. Both Statement I and Statement II are correct and Statement II is the correct explanation of Statement I
b. Both Statement I and Statement II are correct and Statement II is not the correct explanation of Statement I
c. Statement I is correct but Statement II is incorrect
d. Statement II is correct but Statement I is incorrect

34 **Statement I** A heavy chain shown in figure (a) is fixed at both ends is pulled down in such a way that it is converted in triangle as shown in figure (b). During pulling, the gravitational potential energy of the chain increases.

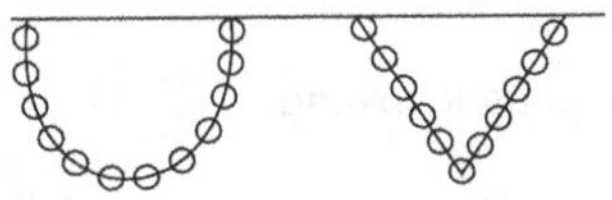

(a) (b)

Statement II During pulling the chain, centre of gravity of chain goes up.

35 **Statement I** The potential energy of a particle in conservative field is $U = (x-3)^2(2-x)$. At $x = 3$ unit, equilibrium is stable.

Statement II In unstable equilibrium, potential energy is maximum at equilibrium position.

36 **Statement I** All the forces which only depend on position are called conservative forces.

Statement II Spring force, $F = -kx$, gravitational force, $F = -\frac{Gm_1 m_2}{r^2}$ and electrostatic force $F = \frac{q_1 q_2}{4\pi \epsilon_0 r^2}$ are conservative in nature.

DPP-6 Power and Other Forms of Energy

Conceptual Questions

Working Space

Direction (*Q. Nos. 1-3*) *This question is conceptual in nature, need to be solved completely on notebook.*

1 **Situation** Power applied to a particle of mass m is $P = 3t^2 - 2t + 1$, where t is in second.

Mark the correct opion(s) for the following questions.

i. The mean power from $t = 0$ to $t = 2$ s is
 a. 11 W b. 3 W c. 10 W d. None of these

ii. The work done on the particle from $t = 0$ to $t = 2$ s is
 a. 22 J b. 6 J c. 11 J d. None of these

iii. The change in kinetic energy from $t = 2$ s to $t = 4$ s is
 a. zero b. 22 J c. 46 J d. None of these

iv. The work done on the particle from $t = 2$ s to $t = 4$ s is
 a. zero b. 22 J c. 46 J d. None of these

2 Choose the correct option(s) for the given questions as following.

i. A body is moved along a straight line by a machine delivering a constant power. The distance moved by the body in time t is proportional to **[AIEEE 2003]**
 a. $t^{3/4}$ b. $t^{3/2}$ c. $t^{1/4}$ d. $t^{1/2}$

ii. A body of mass m is accelerated uniformly from rest to a speed v in a time T. The instantaneous power delivered to the body as a function of time is given by **[AIEEE 2005]**
 a. $\frac{mv^2}{T^2}t$ b. $\frac{mv^2}{T^2}t^2$ c. $\frac{1}{2}\frac{mv^2}{T^2}t$ d. $\frac{1}{2}\frac{mv^2}{T^2}t^2$

3 Choose the correct option(s) for the given questions as following.

i. A truck strikes a tree and comes to rest, then
 a. the internal energy of truck decreases
 b. the internal energy of truck increases
 c. the internal energy of truck remains constant
 d. None of the above

ii. When you turn on your room heater, then
 a. the internal energy of the coil of heater increases
 b. the internal energy of surrounding air increases
 c. coil radiates some energy in the form of electromagnetic radiation
 d. coil radiates some energy in the form of sound energy

iii. Soft steel can be made red hot by hammering continuously on it. This does not work well for hard steel. Why is there such difference?
 a. Due to softness, the deformation in soft steel is more than hard steel under the blow of the hammer. Thus, the positive work done by blow of hammer on the soft steel is more than that of hard steel. Hence, the temperature of soft iron rises rapidly due to more gaining internal energy

b. Soft steel is more conductor than hard steel
c. Both (a) and (b)
d. None of the above

iv. By what transfer mechanism, does energy enter and leave your antenna based set?
a. The energy enters in the form of sound energy and chemical energy
b. The energy leaves in the form of heat energy, sound energy and electromagnetic radiation
c. The energy enters in the form of electric transmission and electromagnetic transmission
d. None of the above

v. You are travelling by your bike freeway at 80 km/h. You now brake to a stop due to congestion in traffic. Mark the correct option(s).
a. The original kinetic energy of your bike has been converted into internal energy of your bike
b. Finally energy will lose as heat into the surrounding air because your brakes are hotter than the surrounding air
c. The temperature of the surrounding air increases
d. None of the above

vi. Generally friction forces tend to decrease the kinetic energy of a body however, friction forces can sometimes increase the kinetic energy of a body. Give example of situations in which friction causes an increase in kinetic energy.
a. If you are in accelerated motion on a level rough surface
b. Your bicycle is in accelerated motion on a street road
c. A book lying on front seat of your car in the condition of accelerated motion on a level road
d. None of the above

Only One Option Correct Type

Direction (Q. Nos. 4-16) *This section contains 13 multiple choice questions. Each question has four choices (a), (b), (c) and (d), out of which ONLY ONE option is correct.*

4 A block of mass 1kg starts to slide down on a fixed inclined plane of inclination 37° with horizontal. The coefficient of friction between the block and inclined plane is 0.5. The rate at which mechanical energy dissipates at $t = 1$ s is

a. 8 J b. 12 J c. 14 J d. 20 J

5 In previous problem, find power of gravitational force at $t = 1$ s.

a. 5 J b. 10 J c. 12 J d. 20 J

6 A uniform rope of mass 1kg and length 1m is lying on the ground. One end of the rope is pulled up by a worker with the constant velocity of 1 m/s. The average power supplied by the worker in pulling the entire rope just off the ground such that the rope becomes vertical. (take $g = 10\ \text{m/s}^2$)

a. 5.5 W b. 6 W c. 10.5 W d. None of these

7 Two blocks are displaced through same distance along rough horizontal surface by applying the forces as shown in the given figure.

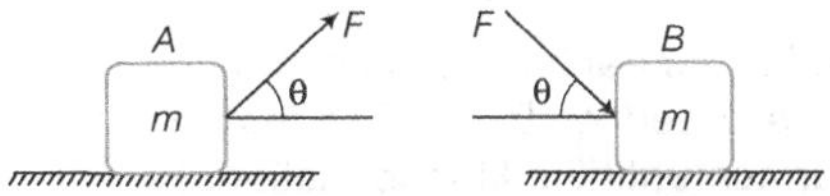

a. Block A takes more time than block B
b. The net work done on both blocks are same
c. The work done by applied force F on block A is more than that of on block B
d. The average power supplied by the force F on the block A is more than that of block B

Working Space

8 A ball is projected upward with an initial speed v_0. Its kinetic energy *versus* time graph during motion of the ball in the air is (neglect air resistance)

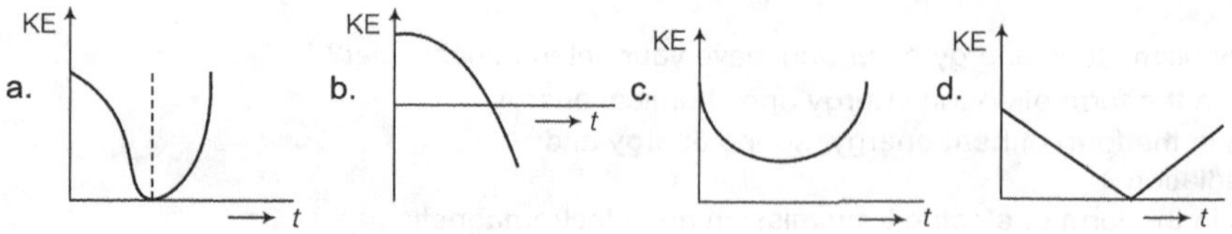

9 In previous problem, the slope of kinetic energy *versus* time graph gives

a. a gravitational potential energy
b. a weight of the ball
c. a power of the ball
d. a work done on the ball by gravitational force

10 The rate of doing work is fast, harassment is more. The phrase is

a. partially correct
b. always correct
c. always incorrect
d. the phrase is correct with some correction

11 A man weighing 60 kg climbs a staircase carrying a 20 kg load on his hand. The staircase has 20 steps and each step has a height of 20 cm. If he takes 20 s to climb, then find his power.

a. 100 W b. 160 W c. 200 W d. 900 W

12 A particle moves with a velocity $(5\hat{i} - 3\hat{j} + 6\hat{k})$ m/s under the influence of a constant force $\mathbf{F} = (10\hat{i} + 10\hat{j} + 20\hat{k})$ N. Find the instantaneous power applied to the particle.

a. 140 W b. 200 W c. 100 W d. 300 W

13 A particle of mass *m* is moving in a circular path of constant radius *r* such that its centripetal acceleration a_c is varying with time *t* as $a_c = k^2rt^2$, where *k* is a constant. The power delivered to the particle by the force acting on it is

a. $2\pi mk^2r^2$ b. mk^2r^2t c. $\frac{(mk^4r^2t^5)}{3}$ d. zero

14 A pump motor is used to deliver water at a certain rate from a given pipe. By what amount the power of the motor should be increased to obtain two times water from the same pipe in the same time.

a. 1 b. 2 c. 4 d. 8

15 A wind-powered generator converts wind energy into electric energy. Assume that the generator converts a fixed fraction of the wind energy intercepted by its blades into electrical energy. For wind speed *v*, the electrical power output will be proportional to **[AIEEE 2000]**

a. v b. v^2 c. v^3 d. v^4

16 Figure shows three blocks each of mass *m* moving initially with the same speed v_0. Block *A* slides on rough horizontal surface until the kinetic frictional force has stopped it. Blocks *B* and *C* slide on rough inclined surfaces until kinetic frictional forces have stopped them.

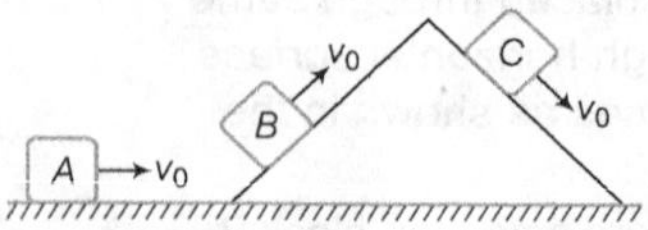

ΔE_1, ΔE_2 and ΔE_3 are loss in mechanical energy of blocks *A*, *B* and *C* respectively. Then,

a. $\Delta E_1 > \Delta E_2 > \Delta E_3$ b. $\Delta E_1 = \Delta E_2 = \Delta E_3$
c. $\Delta E_3 > \Delta E_1 > \Delta E_2$ d. $\Delta E_1 > \Delta E_2 = \Delta E_3$

Working Space

One or More than One Options Correct Type

Direction (Q. Nos. 17-20) This section contains 4 multiple choice questions. Each question has four choices (a), (b), (c) and (d), out of which ONE or MORE THAN ONE are correct.

17 A float is in river. A swimmer is swimming in a river. The velocity of swimmer in water is 4 m/s and the river flow velocity is 3 m/s. Initially, swimmer and float are at the same place. Q_1 energy is expended by swimmer in swimming 30 m downstream ahead of the float. Q_2 energy is expended by swimmer in swimming 30 m towards a bank moving together with the float along the stream. Then,

a. $Q_1 = Q_2$
b. $Q_1 > Q_2$
c. $Q_1 < Q_2$
d. Q_1 and Q_2 are non-zero

18 When a labourer lift a brick and put it gently on upper portion of wall during construction of a building. Then,

a. this is violation of conservation of energy because potential energy of the brick increases, but there is no change in kinetic energy
b. this is not violation of conservative principle of energy because labourer performs internal work done on the cost of lossing chemical energy. Thus, chemical energy of the body of labourer appears as the potential energy of the brick
c. the force of interaction between labourer and the brick is non-conservative in nature
d. the force is a interaction labourer and the brick is in conservative nature

19 For the situation shown in the given figure, you push a block upward along an irregular shape rough inclined plane from point *A* to *B*. Your force does work 200 J on the block + inclined plane + earth system. The kinetic energy of the system increases by 100 J and the gravitational potential energy increases by 30 J. The only other change in energy within the system is for the thermal energy. The increase in the thermal energy is

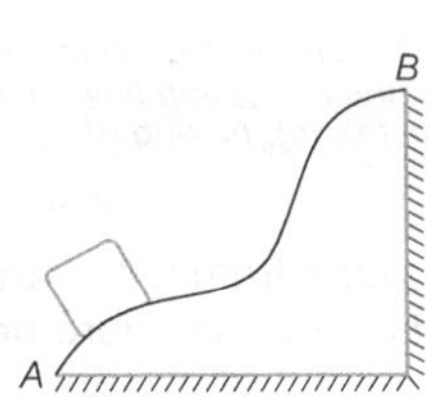

a. zero
b. 70 J
c. 130 J
d. non-zero

20 In the given figure, two blocks *A* and *B* are connected through a light string. Blocks can slide on a fixed rough cubical large block. If you pull downward the block *B* through a rope, then block *A* begins to slide horizontally. The block *A* is smooth, but block *B* is rough. We consider block *B*, cubical block and earth as a system. Your work on the system is 400 J. The system does 80 J work on the block *A*. Within the system, the kinetic energy increases by 250 J and gravitational potential energy decreases by 30 J. The change in the thermal energy within the system is

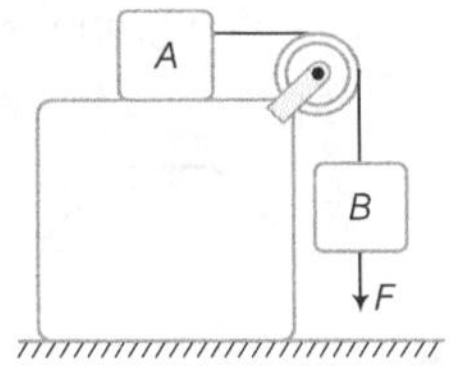

a. 100 J
b. 40 J
c. 260 J
d. non-zero

Statement Type

Direction (Q. Nos. 21-23) This section is based on Statement I and Statement II. Select the correct answer from the codes given below.

a. Both Statement I and Statement II are correct and Statement II is the correct explanation of Statement I
b. Both Statement I and Statement II are correct and Statement II is not the correct explanation of Statement I
c. Statement I is correct but Statement II is incorrect
d. Statement II is correct but Statement I is incorrect

Working Space

21 **Statement I** A block of mass m starts moving on a rough horizontal surface with a velocity v. It stops due to friction between the block and the surface after moving through a certain distance. The surface is now tilted to an angle of 30° with the horizontal and the same block is made to go up on the surface with the same initial velocity v. The decrease in the mechanical energy in the second situation is smaller than that in the first situation.

Statement II The coefficient of friction between the block and the surface decreases with the increase in the angle of inclination. **[IIT JEE 2007]**

22 **Statement I** If a force $\mathbf{F} = \mathbf{v} \times \mathbf{c}$ is applied on a particle, then power supplied by force on the particle is zero.

Statement II $\mathbf{F} = \mathbf{v} \times \mathbf{c}$ is always perpendicular to the velocity of the particle.

23 **Statement I** If you are running on a horizontal road with some acceleration. Your kinetic energy continuously increases due to internal work done in your body on the cost of lossing chemical energy by you.

Statement II If you are running with some acceleration. You are accelerated by static friction exerted by horizontal road on you.

Comprehension Type

Direction (Q. Nos. 24 and 25) This section contains a paragraph, describing theory, experiments, data, etc. Two questions related to the paragraph have been given. Each question has only one correct answer among the four given options (a), (b), (c) and (d).

Passage

A block of mass m starts from rest with an acceleration. The acceleration-time graph of the block is shown in figure below.

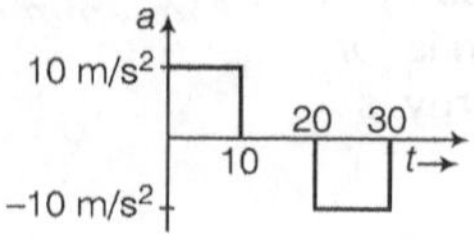

24 The kinetic energy *versus* time graph is

a. 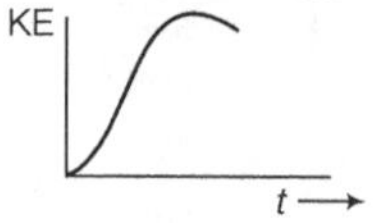

b. 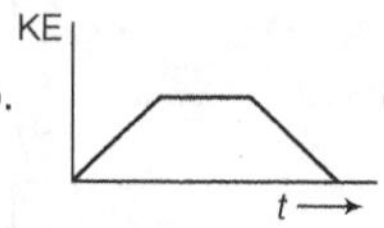

c. 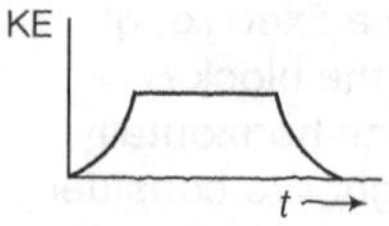

d. 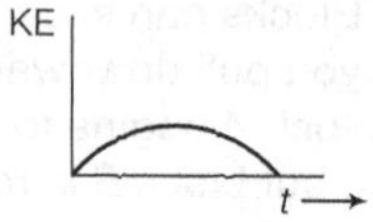

25 The power *versus* time graph is

a. 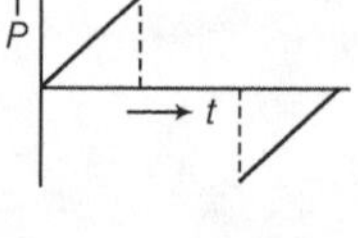

b. 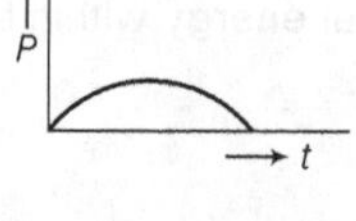

c. 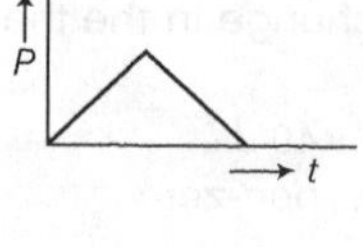

d. 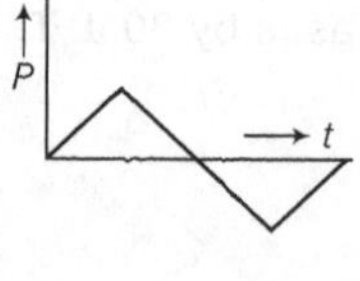

Matching List Type

Direction (Q. No. 26) Choices for the correct combination of elements from Column I and Column II are given as options (a), (b), (c) and (d), out of which one is correct.

26 In the given figure below, the velocity *versus* time graph is shown for a particle of mass 1 kg.

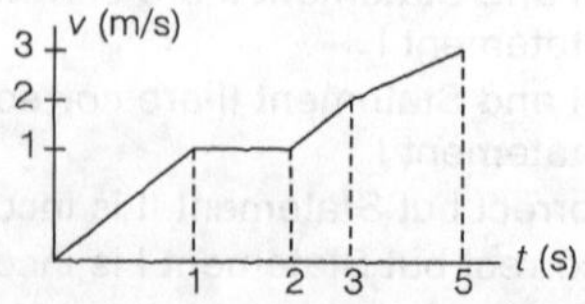

Working Space

Match the Column I and Column II choose the correct option from the given codes.

	Column I		Column II
i.	Work done is maximum	p.	$t = 1$ s to $t = 2$ s
ii.	Power is maximum	q.	from $t = 3$ s to $t = 5$ s
iii.	Work done is zero	r.	from $t = 2$ s to $t = 3$ s
iv.	Force is maximum	s.	$t = 0$ to $t = 1$ s

Codes

	i	ii	iii	iv		i	ii	iii	iv
a.	p	q	r	s	b.	q	p	s	r
c.	q	r	p	r,s	d.	r	q,s	p,s	r,s

One Integer Value Correct Type

Direction (Q. Nos. 27-29) This section contains 3 questions. When worked out will result in an integer from 0 to 9 (both inclusive).

27 A car of mass 2000 kg is driven with constant power $P = 1000$ W. Find the speed of the car (in m/s) at $t = 4$ s.

28 A particle of mass 0.2 kg is moving in one-dimension under a force that delivers a constant power 0.5 W to the particle. If the initial speed (in ms^{-1}) of the particle is zero, then speed (in ms^{-1}) after 5 s is **[JEE Advanced 2003]**

29 Each day human heart takes in and discharges about 7500 L of blood. Assume that the work done by the heart is equal to the work required to lift this amount of blood by 1.63 m. The density of blood is 1.05×10^3 kg/m^3. Heart's power output is $\frac{148}{10^n}$ W. Find the value of n.

Subjective Questions

Direction (Q. Nos. 30 and 31) These questions are subjective in nature, need to be solved completely on notebook.

30 An elevator in the multi-storage building averages speed of 5 m/s in its climb to the top floor 300 m above the ground. Assuming a total load of 1000 kg, what average power must the lifting motor supply. Put your answer in horse power. ($g = 10$ m/s^2)

31 Figure shows a block of mass $m = 2$ kg slides along a horizontal surface with speed $v_0 = 4$ m/s. The block compresses the spring until it momentarily stops. The magnitude of frictional force on the block by the rough surface is 15 N. The spring constant $k = 10^4$ N/m.

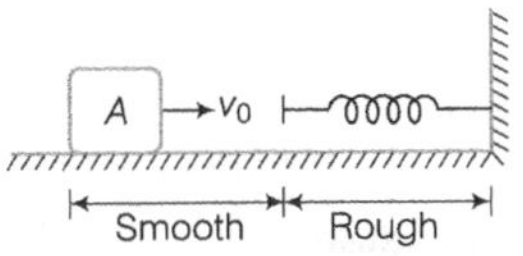

Assuming that the time taken by the spring to go from relaxed state to maximum compression state is $t_0 = 15.17$ s.

i. Find the mean power of spring force during this period.

ii. Find mean power of friction.

iii. Find total mean power.

DPP-7 Mechanical Energy

Only One Option Correct Type

Working Space

Direction *(Q. Nos. 1-21) This section contains 21 multiple choice questions. Each question has four choices (a), (b), (c) and (d), out of which ONLY ONE option is correct.*

1 A ball is projected with speed $v_0 = \sqrt{44}$ m/s at an angle α with horizontal from the roof of a building.

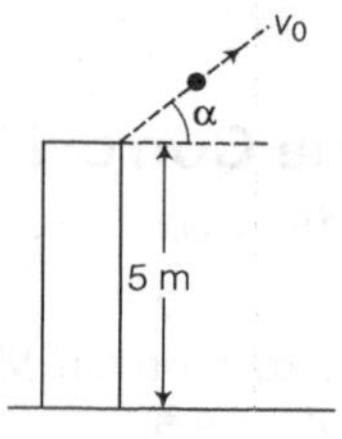

a. The speed at ground is maximum, if $\alpha = \frac{\pi}{4}$

b. The speed at ground is maximum, if $\alpha = -\frac{\pi}{2}$

c. The speed at ground is maximum at $\alpha = 0$

d. The speed at ground is 12 m/s and independent of an angle α

2 A small block of mass 10 g is projected along the smooth path from rest as shown in the given figure below. Find its speed at the point *B*.

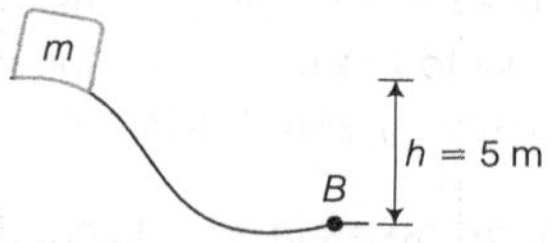

a. 1 m/s   b. 10 m/s
c. 20 m/s   d. None of these

3 You and your dog stand at points *A* and *B* respectively at the ends of a track shown in the given figure below. You take a slice of bread and push it along the friction less track with some initial velocity. With what minimum initial velocity should the slice of bread be pushed, so that it could reach the mouth of dog? ($g = 10$ m/s^2)

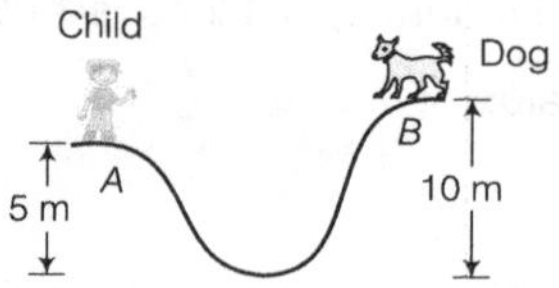

a. 5 m/s   b. 10 m/s   c. zero   d. 20 m/s

4 Two blocks each of mass 1 kg are connected by a light spring of spring constant $k = 100$ N/m. When spring is in natural length, each block is projected in opposite direction with speed of 10 m/s. The work done by spring on each block upto maximum compression is

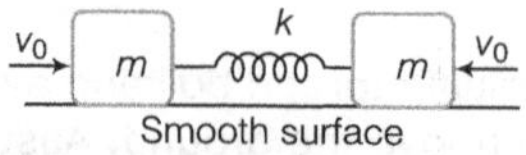

a. 100 J   b. − 100 J   c. 50 J   d. − 50 J

5 In previous problem, the spring force is conservative internal force of the system. The negative workdone by the spring force on the system is equal to

a. change in kinetic energy
b. change in total energy
c. change in potential energy
d. change in kinetic energy of each block

6 A block of mass 1 kg is suspended through spring of constant $k = 100$ N/m. The block is stationary at equilibrium position. Gravity is switch off by some mechanism. The maximum speed of block after switching off gravity is

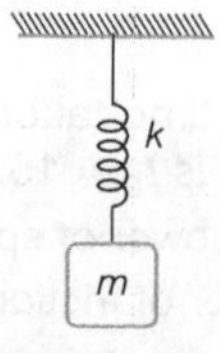

a. 1 m/s   b. zero
c. 2 m/s   d. None of these

Working Space

7 The block of mass M moving on the frictionless horizontal surface collides with the spring of spring constant k and compresses it by length L. Then, maximum momentum of the block after collision is

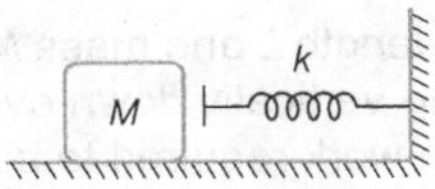

[AIEEE 2005]

a. $L\sqrt{Mk}$ b. $\frac{kL^2}{2M}$ c. zero d. $\frac{ML^2}{k}$

8 A block of mass m is attached to two unstretched springs of constant k_1 and k_2. The other ends are free. There is no friction anywhere. The block is displaced towards left by small distance x_1 and released. The maximum compression in spring of force constant k_2 is x_2. Then, ratio of $\frac{x_1}{x_2}$ is

a. $\frac{k_1}{k_2}$ b. $\frac{k_2}{k_1}$ c. $\sqrt{\frac{k_2}{k_1}}$ d. $\frac{k_2^2}{k_1^2}$

9 A block B is attached to two unstretched springs S_1 and S_2 with spring constants k and $4k$, respectively. The other ends are attached to two supports M_1 and M_2 not attached to the walls. The springs and supports have negligible mass. There is no friction anywhere. The block B is displaced towards wall 1 by a small distance x and released. The block returns and moves a maximum distance y towards wall 2. Displacements x and y are measured with respect to the equilibrium position of block B. Then, ratio $\frac{y}{x}$ is

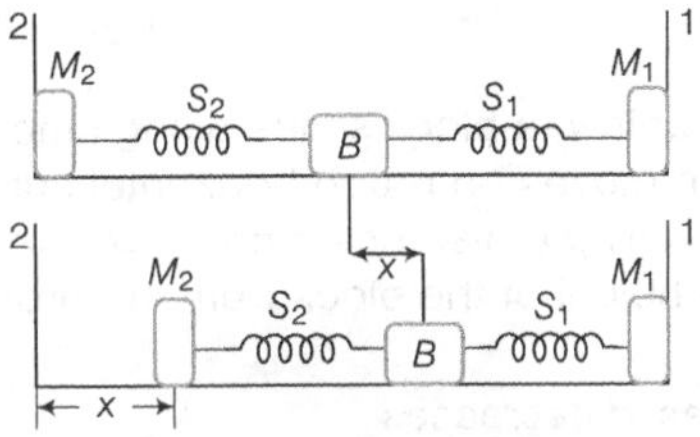

[IIT JEE 2008]

a. 4 b. 2 c. $\frac{1}{2}$ d. $\frac{1}{4}$

10 A block attached with a spring is placed on a smooth horizontal surface. Now, the block and free end A are projected with same speed v_0 in opposite directions along the surface as shown in the given figure below. The maximum elongation in the spring during subsequent motion is

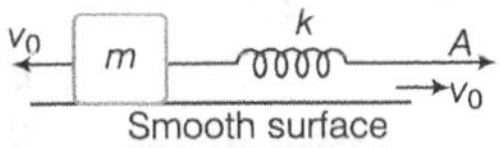

a. $\sqrt{\frac{2m}{k}}v_0$ b. $\sqrt{\frac{4m}{k}}v_0$ c. $\sqrt{\frac{m}{k}}v_0$ d. Insufficient data

11 A block of mass m is connected with the wall of a toy car through a light spring of constant k. The toy car is moving with constant acceleration a_0 (neglect friction). The block is stationary with respect to car. Now, car is abruptly stopped. Then, maximum velocity of the block after stopping the car is

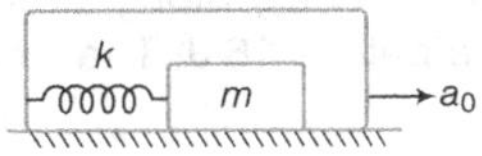

a. zero b. $\sqrt{\frac{k}{m}}a_0$ c. $\sqrt{\frac{m}{k}}a_0$ d. $\sqrt{\frac{m}{2k}}a_0$

12 An ideal spring with spring constant k is hung from the ceiling and a block of mass M is attached to its lower end. The mass is released with the spring initially unstretched. Then, the maximum extension in the spring is [IIT JEE 2002]

a. $\frac{4\,Mg}{k}$ b. $\frac{2\,Mg}{k}$ c. $\frac{Mg}{k}$ d. $\frac{Mg}{2k}$

13 A uniform chain of length 2 m is kept on a table, such that a length of 60 cm hangs freely from the edge of the table. The total mass of chain is 4 kg. What is the work done is pulling the entire chain on the table? [AIEEE 2004]

a. 7.2 J b. 3.6 J c. 120 J d. 1200 J

Working Space

14 A uniform chain of length L and mass M is lying on a smooth table and one-third of its length is hanging vertically down over the edge of the table. If g is acceleration due to gravity, then work required to pull the hanging part on the table is

a. MgL b. $MgL/3$ c. $MgL/9$ d. $MgL/18$

15 A block slides over a rough horizontal surface. The block and ground are taken as a system. For this system, normal reaction and friction are internal forces. The force $F = 20$ N is an external force for the system.

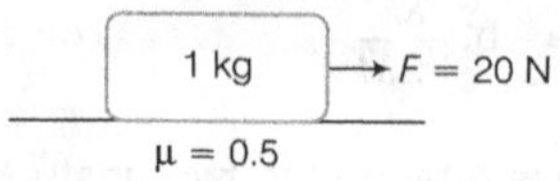

The work done by the net external force $F = 20$ N on the system equals the change in

a. kinetic energy b. total energy
c. potential energy d. mechanical energy

16 In previous problem, the work done by all forces (external + internal) on the system equals the change in

a. kinetic energy b. total energy
c. potential energy d. None of the above

17 In the given figure below, a block slides along smooth inclined plane, then moves on rough horizontal surface and stops after travelling a distance d on horizontal surface. The initial height of the block from the ground is H.

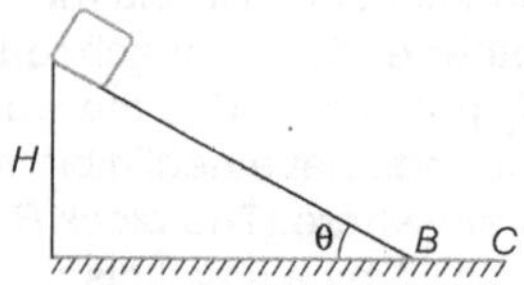

a. When H increases, d decreases
b. When H decreases, d decreases
c. H is always equal to d
d. When mass of the block increases, d increases

18 A bullet fired into a fixed target losses half of its velocity after penetrating 3 cm. How much further it will penetrate before coming to rest, assuming that it faces constant resistance to motion? **[AIEEE 2005]**

a. 3.0 cm b. 2.0 cm
c. 1.5 cm d. 1.0 cm

19 For the situation shown in the given figure one direct path 1 and other curve path 2. Along direct path only a conservative force F_1 acts on the particle. Along curve path, both F_1 and kinetic friction acts on the particle. The change in mechanical energy along direct path is 20 J and that of on curve path is 16 J. The work done by kinetic friction is

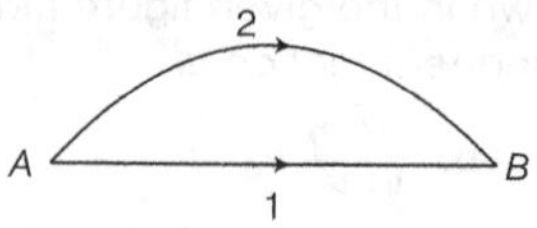

a. – 16 J b. – 4 J
c. – 36 J d. No sufficient information

20 A spherical ball of mass 20 kg is stationary on the top of a hill of height 100 m. If rolls down a smooth surface to the ground, then climb up another hill of height 30 m and final rolls down to a horizontal base at a height of 20 m above the ground. The velocity attained by the ball is **[AIEEE 2005]**

a. 40 m/s b. 20 m/s
c. 10 m/s d. $10\sqrt{30}$ m/s

21 The potential energy of a 1 kg particle free to move along the X-axis is given by $v(x) = \left(\frac{x^4}{4} - \frac{x^2}{2}\right)$ J. The total mechanical energy of the particle is 2 J. Then, the maximum speed (in ms^{-1}) is **[AIEEE 2006]**

a. $\frac{3}{\sqrt{2}}$ b. $\sqrt{2}$ c. $\frac{1}{\sqrt{2}}$ d. 2

One or More than One Options Correct Type

Working Space

Direction (Q. Nos. 22-27) This section contains 6 multiple choice questions. Each question has four choices (a), (b), (c) and (d), out of which ONE or MORE THAN ONE are correct.

22 A particle of potential energy shown in graph is moving towards right. The initial kinetic energy of particle is 21 J at $x = 0$.

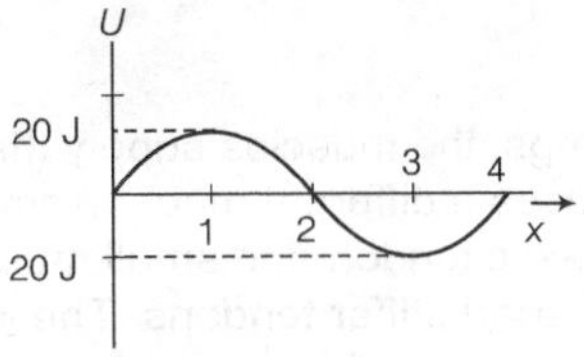

Mark the correct option(s)

a. The particle accelerates for $0 \le x < 1$ m
b. At $x = 1$ m, the particle has minimum KE energy
c. The kinetic energy is 41 J at $x = 3$ m
d. The particle is trapped within $0 < x \le 4$

23 In previous problem, if kinetic energy at $x = 0$ is 19 J. The particle reverses its direction at $x = x_0$. Then

a. $0 < x_0 < 1$ m   b. $0 \le x_0 < 1$ m   c. $0 < x_0 < 4$ m   d. $1 < x_0 < 3$ m

24 The one end of the light spring of constant k is fixed to the wall. During motion of the block on the smooth surface,

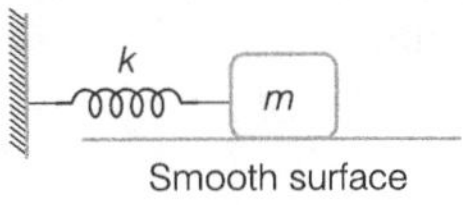

a. if initially spring is compressed and finally comes in relaxed position, then work done by the spring on the block is positive
b. if initially spring is stretched and finally comes in relaxed position, then work done by the spring on the block is positive
c. if initially spring is in relaxed position and finally comes in stretched position, then work done by the spring on block is positive
d. if initially spring is in relaxed position and finally comes in compressed position, then work done by spring on the block is negative

25 A smooth block of mass m is connected to a wall through a light spring of constant k. When the spring is in natural length, the block is projected rightward with speed v_0. Consider the duration in which the spring stretched upto maximum elongation. In the frame of the block,

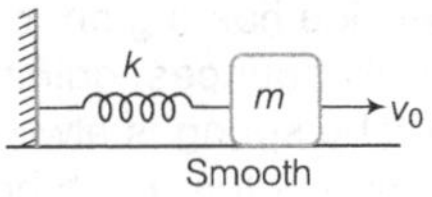

a. the wall is moving toward the block
b. the kinetic energy of block decreases
c. the potential energy of the spring increases due to the positive work by normal reaction between the spring and wall
d. the work done by spring on the block is zero

26 A small block is released from rest from a height of 10 m on a smooth ramp. The block does not fly off in air. ($g = 10$ m/s^2). Friction is absent everywhere.

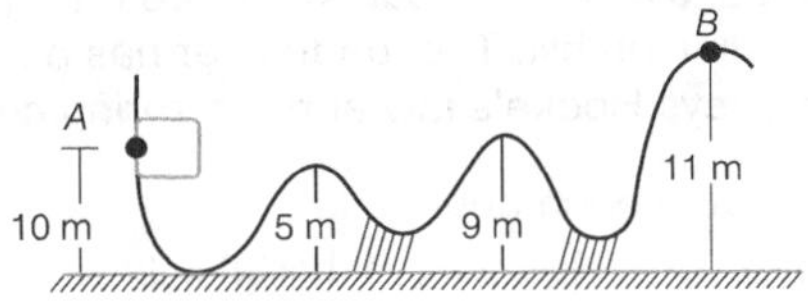

a. The maximum speed of the block during motion is $10\sqrt{2}$ m/s
b. The block repeats its path continuously
c. Total mechanical energy of the block remains constant
d. The block will never reach at the point B

27 A particle moves along X-axis whose potential energy *versus* x-coordinate graph is shown in the given figure. At $x = 0$, the particle is released from rest. Two potential wells A and B are shown in the figure.

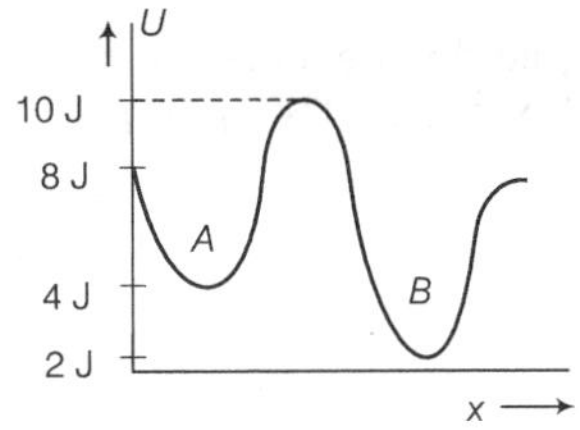

a. The particle will never reach the potential well B
b. The particle will be trapped into the potential well A
c. The maximum kinetic energy during motion of the particle is 4 J
d. The maximum kinetic energy during motion of the particle is 6 J

Comprehension Type

Working Space

Direction *(Q. Nos. 28-32) This section contains 2 paragraphs, each describing theory, experiments, data, etc. Five questions related to the paragraphs have been given. Each question has only one correct answer among the four given options (a), (b), (c) and (d).*

Passage I

When a human jumps, the muscles supply the energy to project the body upward. But the kangaroo uses a different mechanism during jumping. The kangaroo has long, spring like elastic tendon and small muscles in its hind legs. But human has large muscles and short stiffer tendons. The kangaroo folds its legs before jump, using its muscles to stretch the tendons and converting chemical energy into elastic potential energy. The kangaroo then quickly extends its legs, relaxing the tendons like a released spring. The elastic potential energy stored in the tendons supplies the energy for the jump.

28 The hop of kangaroo is 3 m, when it stretches its tendons a distance 10 cm. How high would the hop be after it stretched the tendons 11 cm. Assume that tendons behave as light spring.

a. 2.4 m
b. 3.3 m
c. 3.63 m
d. Data insufficient

29 Suppose the kangaroo has a baby kangaroo in her pouch. The mass of the baby is one-tenth the mass of its mother. The kangaroo without the baby can jump 3.3 m. How high can the kangaroo with its baby jump?

a. 3.5 m
b. 3.3 m
c. 2.8 m
d. 3 m

Passage II

A new roller coaster ride has a giant compressed spring off which the roller coaster car repeatedly bounces, gaining energy and eventually being pushed over a hill, see diagram. The spring is always recompressed a distance of 2 m from its uncompressed position before each bounce.

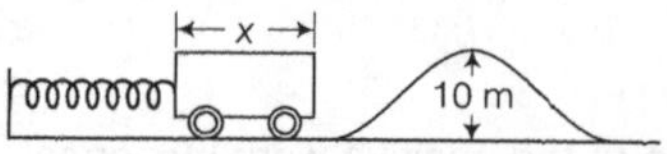

At the start the car is up against the compressed spring, which is then released, pushing the car up the hill. It then rolls back down into the recompressed spring, which is then released, ejecting the car with more energy. This repeats until the car goes over the 10 m high hill. The loaded car has a mass of 1000 kg. Assume the spring obeys Hooke's law and the spring constant $k = 5$ kN/m.

30 The energy gain by car in each bounce is

a. 5 k J
b. 98 k J
c. 10 k J
d. 108 k J

31 How many bounces are needed, including the start off, for the car to go over hill?

a. 1
b. 5
c. 11
d. 10

32 The maximum magnitude of acceleration experienced by the passengers in the car during the last bounce off the spring is

a. 6.32 m/s^2
b. 31.6 m/s^2
c. 10 m/s^2
d. 9.8 m/s^2

Working Space

Statement Type

Direction (Q. Nos. 33-36) *This section is based on Statement I and Statement II. Select the correct answer from the codes given below.*

a. Both Statement I and Statement II are correct and Statement II is the correct explanation of Statement I
b. Both Statement I and Statement II are correct and Statement II is not the correct explanation of Statement I
c. Statement I is correct but Statement II is incorrect
d. Statement II is correct but Statement I is incorrect

33 **Statement I** Two smooth inclined plane of different inclination are shown in the given figure. Blocks *A* is heavier than block *B*. Both are released from rest from same height. Then, both have same speed at the bottom but kinetic energy of the block *A* is greater than that of block *B*.

Statement II The work done by normal reaction on both blocks is zero.

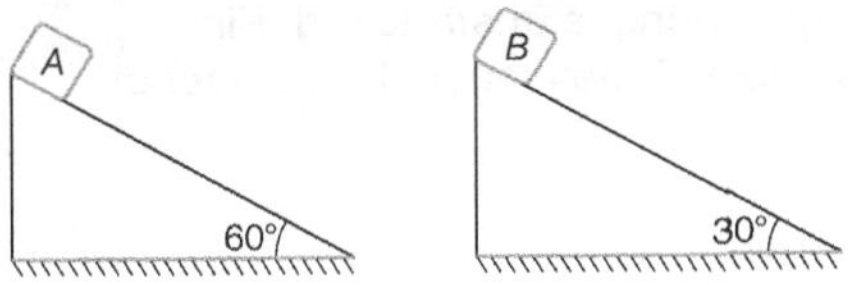

34 **Statement I** A block of mass *m* is released with initial velocity *u* along the smooth irregular shape inclined plane. Its speed at the point *B* is *v*. Then, $v^2 = u^2 + 2gh$

Statement II During motion of block, mechanical energy remains conserved.

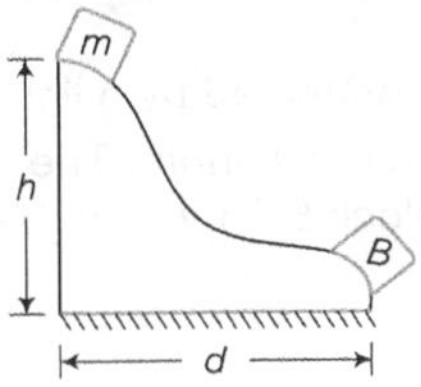

35 **Statement I** A smooth particle of mass *m* is released from rest from top of an irregular shape inclined plane as shown in the given figure. The kinetic energy at bottom is *mgh* and is independent of shape of inclined plane.

Statement II During motion of block, mechanical energy remains conserved.

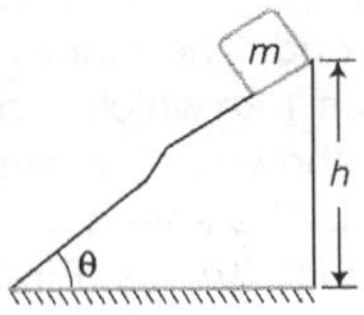

36 **Statement I** A system has same kinetic energy in two different reference frame. Then, mechanical energy of the system is same in both frames.

Statement II Potential energy does not depend on frame of reference.

One Integer Value Correct Type

Direction (Q. Nos. 37-45) *This section contains 9 questions. When worked out will result in an integer from 0 to 9 (both inclusive).*

37 A block of mass, 1 kg is attached to three unstretched springs of spring constants $k_1 = 40$ N/m, $k_2 = 15$ N/m and $k_3 = 45$ N/m as shown in the given figure. The block is displaced towards left on the horizontal smooth surface through a distance 10 cm and is released. Find maximum speed during its motion in m/s.

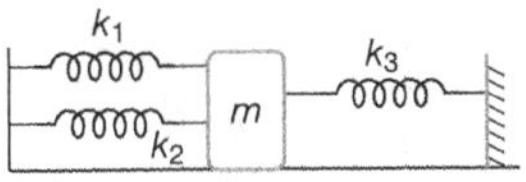

Working Space

38 A block of mass 2 kg is released from rest from a height of $h = 2$ m above the upper end of light spring of spring constant $k = 10$ N/m fixed at lower end. ($g = 10$ m/s^2)

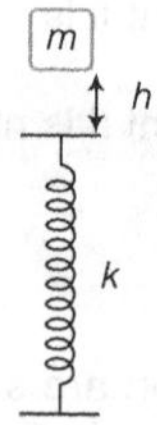

i. Find maximum compression in the spring in metre.
ii. The maximum speed of the block during its motion is $(60)^{\frac{1}{n}}$ m/s. Find the value of n.

39 For the situation shown in the given figure, block A of mass 10 kg is at the ground. The block B is released from rest when the light spring is unstretched. Find minimum mass of the block B in kg such that block B may leave the contact of ground.

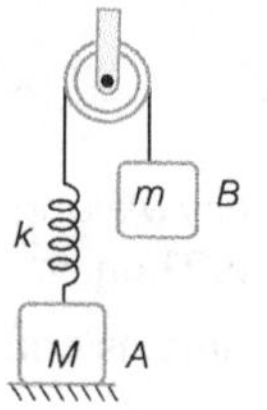

40 Two blocks A and B of equal masses ($m = 10$ kg) are connected by a light spring of spring constant $k = 150$ N/m. The system is in equilibrium. The minimum value of initial downward velocity v_0 of the block B for which the block A bounce up is $\frac{20}{\sqrt{3n}}$ m/s. Find the value of n.

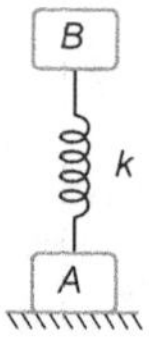

41 The lower part of a light spring of spring constant $k = 240$ N/m and natural length l_0 is fixed at lower end. The other end of the spring is connected with a small smooth ring of mass 1 kg which is allowed to slide on a horizontal rod fixed at a height $l_0 = 1$ m as shown in the given figure. Initially, the spring makes an angle of 60° with the vertical. The speed of ring when spring makes an angle of 37° with vertical at other side is $3n$ m/s. Find the value of n.

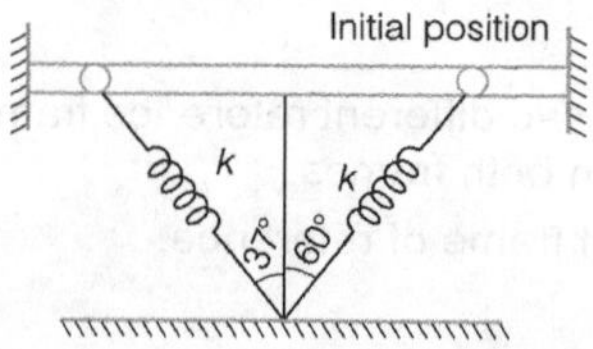

42 Two blocks each of mass 10 kg are connected by a light string passing over a smooth pulley as shown in the given figure. The speed of each block when the block B travels a distance 1 m is v_0 m/s. Find the value of v_0^2.

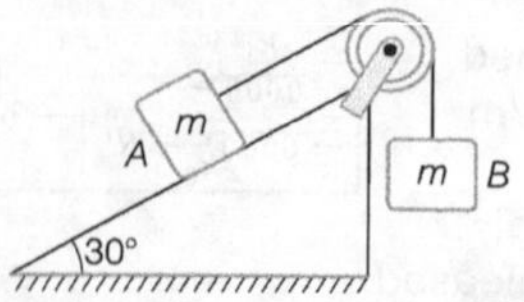

Working Space

43 Consider the situation shown in the given figure. The light smooth pulley C is movable, but the smooth pulley D is fixed. The ratio of speeds of block A and B, when block A moves a distance 2 m is $\frac{v_A}{v_B}$. Find the value of $\frac{v_A}{v_B}$.

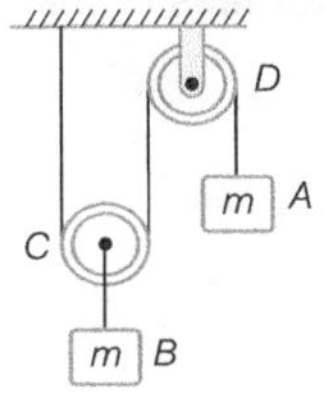

44 A rope of mass m and length $l = 0.9$ m is shown in the given figure. The pulley is smooth and its radius is negligible. The initial distance of end A from the pulley is $\frac{l}{3}$. Find the speed of the rope in m/s when the end A reaches at the pulley.

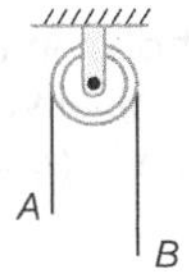

45 For the situation shown in the given figure, the horizontal surface is rough. The coefficient of friction between the block of mass $m_1 = 4$ kg and the horizontal surface is $\mu = 0.5$. The blocks are connected with combination of light spring of spring constant k and a light string. The block of mass m_2 is released from rest, when spring is in its natural length. Find minimum value of m_2 in kg such that the block of mass m_1 will starts move.

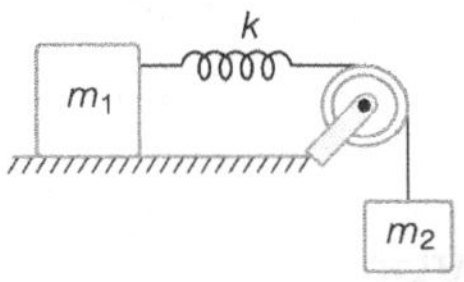

Subjective Questions

Direction (Q. Nos. 46-50) *These questions are subjective in nature, need to be solved completely on notebook.*

46 For the situation shown in the given figure, springs are attached to fixed walls. Their spring constants are $k_1 = 100$ N/m and $k_2 = 50$ N/m. A block of mass 1 kg is projected towards rightward with a speed of 10 m/s.

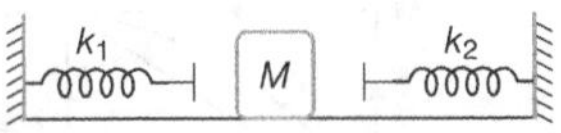

i. Find the ratio of maximum compression in springs.
ii. Find the magnitude of maximum deceleration of the block its motion.

47 A block of mass 1 kg is pressed against a horizontal spring of a spring constant 100 N/m fixed at one end to compress the spring 10 cm. When the block is released, the block moves horizontally on the smooth surface.

i. Find maximum speed of the block.
ii. Find the compression in spring when its speed is half of its maximum speed.

48 For the situation shown in the given figure, initially the light spring of spring constant k is in its natural length. Assuming no friction between the inclined plane and the block.

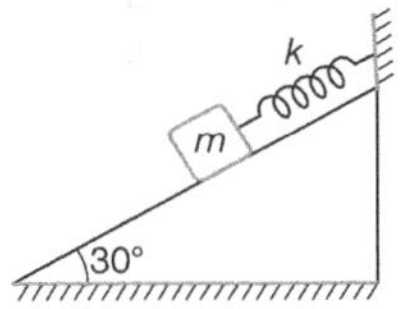

i. Find the maximum elongation in the spring.
ii. Find the maximum speed of the block during its motion.

Working Space

49 A string, with one end fixed on a rigid wall, passing over a fixed frictionless pulley at a distance of 2 m from the wall, has a point mass $M = 2$ kg attached to it at a distance of 1 m, from the wall. A mass $m = 0.5$ kg attached at the free end is held at rest to that the string is horizontal between the wall and the pulley and vertical beyond the pulley. What will be the speed with which the mass M will hit the wall, when the mass m is released? (take $g = 9.8$ m/s^2).

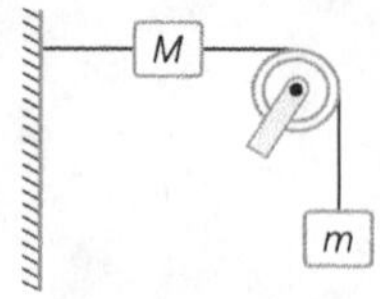

50 For the situation shown in the given figure, a particle is projected along smooth inclined plane with initial velocity 30 m/s. How far away from the point O will the particle hit the ground? ($g = 10$ m/s^2).

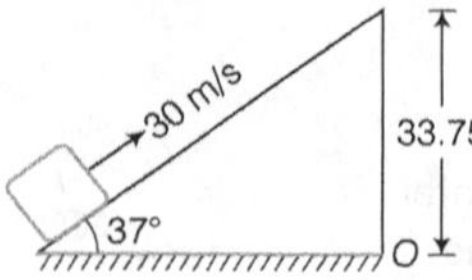

Matching List Type

Direction (Q. No. 51) Choices for the correct combination of elements from Column I and Column II are given as options (a), (b), (c) and (d), out of which one is correct.

51 A particle can be moved in XY-plane from O to P along different paths as shown in figure. Constant force $\mathbf{F} = (4\hat{i} + 6\hat{j})$ N is acting on the particle. Considering the action of only the given force.

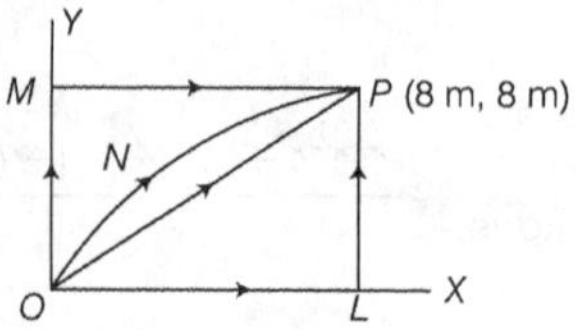

Match the Column I with Column II. Mark the correct option code.

	Column I		Column II
i.	For path OP	p.	Work done by the force is 80 J.
ii.	For path OLP	q.	Mechanical energy is conserved.
iii.	For path OMP	r.	Velocity is not constant.
iv.	For path ONP	s.	Speed at P is greater than that of point O.

Codes

	i	ii	iii	iv
a.	p,q,r,s	p,q,r,s	p,q,r,s	p,q,r,s
b.	p	q	r	s
c.	p,q	r,s	p,q,s	p,q
d.	q	r	p	s

DPP-8 Vertical Circular Motion

Conceptual Questions

Working Space

Direction (Q. Nos. 1-6) *These questions are conceptual in nature, need to be solved completely on notebook.*

1 Fill in the blanks.

i. One end of a string of length l is attached to a fixed point O and a small ball of mass m attached to other end. If ball is projected horizontally from lowest point to complete the vertical circle, then tension at lowest point is......times the weight of the ball.

ii. A small ball of mass m is suspended from a fixed point from a string of length $l = 2$ m. The ball is projected horizontally with speed $u = \sqrt{20}$ m/s from lowest point. The distance travelled by the ball when the ball will return to lowest point after some time is......m.

iii. If a ball of mass $m = 1$ kg is suspended from a fixed point through a light string of length $l = 1$ m. When ball is projected horizontally from lowest point with a velocity $u = 10$ m/s, then ratio of tensions at horizontal position of string and that at highest point is...... .

2 **Situation** A ball of mass m tied with a light string is moving on a vertical circle of radius l and centre d at the point O.

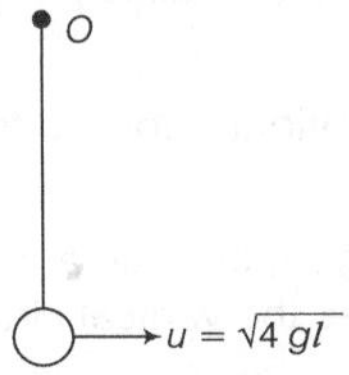

Mark the correct option(s) for the following questions.

i. In the given figure the ball,

a. completes the vertical circle
b. leaves the vertical circle above the half circle
c. oscillates about lowest point
d. keeps the mechanical energy constant during motion

ii. The value of θ at the point P is

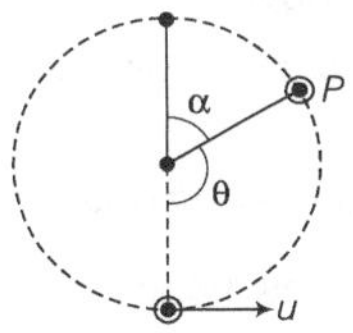

a. α
b. $\theta = \pi - \alpha$
c. $\frac{\theta}{2}$
d. None of these

iii. In previous problem, height of the ball at point P with respect to lowest position is

a. $h = l - l\cos\alpha$
b. $h = l/2 + (l/2)\cos\alpha$
c. $h = l + l\cos\alpha$
d. None of these

iv. In previous problem, the velocity of ball at the point P is

a. $v = \sqrt{u^2 + 2gh}$
b. $v = \sqrt{u^2 - 2gh}$
c. $v = \sqrt{2gl(1 - \cos\alpha)}$
d. None of these

v. In previous problem, the tension in the string when ball is at the point P is

a. $T = mg\cos\theta + \frac{mv^2}{l}$

b. $T = -mg\cos\alpha + \frac{m}{l}(u^2 - 2gh)$

c. $T = -mg\cos\alpha + \frac{m}{l}\{u^2 - 2gl(1 + \cos\alpha)\}$

d. $T = 2mg - 3mg\cos\alpha$

Working Space

vi. In previous problem, tension in string at point P

a. may be zero b. never be zero c. must be zero d. None of these

vii. If $T = 0$ at point P, then value of $\cos\alpha$ is

a. $\cos\alpha = 0$ b. $\cos\alpha = 1/2$ c. $\cos\alpha = 2/3$ d. None of these

viii. In previous problem, the velocity of the ball at the point P is

a. $v = v_0 = \sqrt{(2/3)\,gl}$ b. $v = v_0 = \sqrt{gl}$

c. $v = v_0 = \sqrt{3gl}$ d. None of these

ix. Let at $t = t_0$, the ball leaves the vertical circle at point P. The path of the ball just after leaving the vertical circle is

a. circular b. parabolic c. elliptical d. None of these

x. In previous problem, the maximum height attained by the ball with respect to horizontal line passing through point P is

a. $H = \dfrac{v_0^2 \sin^2\alpha}{2g}$ b. $H = \dfrac{5l}{27}$ c. $H = \dfrac{2l}{3}$ d. None of these

xi. The maximum height attained by the ball with respect to lowest point of the vertical circle is

a. $h_{max} = l + l\cos\alpha + \dfrac{v_0^2 \sin^2\alpha}{2g}$ b. $h_{max} = l - l\cos\alpha + \dfrac{v_0^2 \sin^2\alpha}{2g}$

c. $h_{max} = \dfrac{50\,l}{27}$ d. None of these

3 Situation A small smooth ball of mass m is imparted horizontally with speed u at the lowest point of a fixed hollow sphere of radius $l = \left(\dfrac{1}{2+\sqrt{3}}\right)$ m. During subsequent motion, the ball is passing through the centre O of the hollow sphere. (where, $g = \pi^2$)

Mark the correct option(s) for the following questions.

i. During motion

a. the ball always moves on a vertical circle whose centre is at O

b. the ball leaves the vertical circle at a point and then moves on a parabolic path under gravity

c. the ball oscillates about lowest point

d. None of the above

ii. Suppose the velocity of the ball at point O is v_0. The correct relation is

a. $U_i + T_i = U_f + T_f$ b. $(1/2)mu^2 = mgl + (1/2)mv_0^2$

c. $(1/2)mu^2 + mgl = (1/2)mv_0^2$ d. None of these

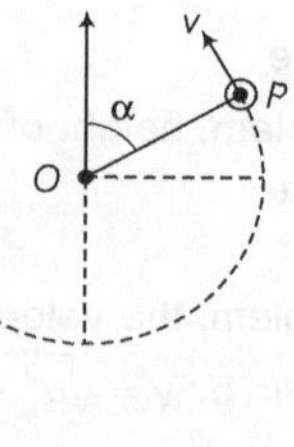

iii. Suppose ball leaves the vertical circle at point P, the height of point P from lowest position is

a. $h = l + l\cos\alpha$

b. $h = l + l/2\cos\alpha$

c. $h = l/2 + l\cos\alpha$

d. None of these

iv. In previous problem, the velocity at point P is

a. $v = \sqrt{u^2 - 2gh}$ b. $v = \sqrt{u^2 - 2gl\,(1 + \cos\alpha)}$

c. $v = \sqrt{u^2 + 2gl\,(1 + \cos\alpha)}$ d. None of these

v. If P is taken as origin, X-axis in horizontally leftward and Y-axis vertically upward. Then, coordinates of the point O is

a. $x = l\sin\alpha$ and $y = -l\cos\alpha$ b. $x = -l\sin\alpha$ and $y = l\cos\alpha$

c. $x = 0$ and $y = l\sin\alpha$ d. None of these

Working Space

vi. In previous problem, the equation of trajectory of the ball after leaving the vertical circle is

a. $y = x \tan\alpha + \dfrac{gx^2}{2v^2\cos^2\alpha}$ b. $y = x \tan\alpha - \dfrac{gx^2}{2v^2\cos^2\alpha}$

c. $y = x \tan\alpha - \dfrac{gx^2}{2u^2\cos^2\alpha}$ d. None of these

vii. The correct relation for position of the ball at point O is

a. $-l\cos\alpha = l\sin\alpha\tan\alpha - \dfrac{gl^2\sin^2\alpha}{2v^2\cos^2\alpha}$

b. $-l\cos\alpha = \dfrac{l\sin^2\alpha}{\cos\alpha} - \dfrac{gl^2\tan^2\alpha}{2v^2}$

c. $-l\cos\alpha = \dfrac{l\sin^2\alpha}{\cos\alpha} - \dfrac{gl^2\tan^2\alpha}{2\{u^2 - 2gl(1+\cos\alpha)}$

d. None of the above

viii. The value of u is

a. 9.8 m/s b. 3.14 m/s c. 6.28 m/s d. None of these

4 **Situation** A small ball of mass m is connected to a light rod and is whirled in a vertical circle whose centre is at the fixed end of the rod as shown in the given figure.

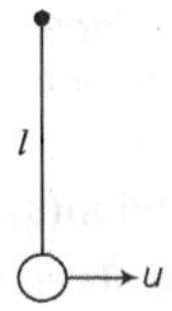

Mark the correct option(s) for the given questions.

i. The force diagram of the ball at the lowest point is

a. 

b. 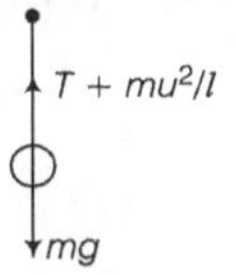

c. 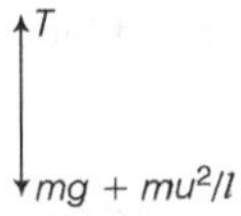

d. None of these

ii. The tension in rod at the lowest point of the ball is

a. $T = mg + mu^2/l$ b. $T = mg - mu^2/l$

c. $T = 0$ d. None of these

iii. The rod makes an angle θ with lowest vertical at instant t, the velocity of ball at instant t is

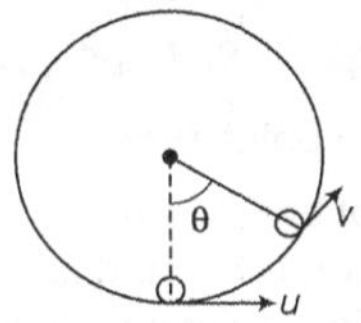

a. $v = \sqrt{u^2 + 2gh}$, where $h = l - l\cos\theta$

b. $v = \sqrt{u^2 - 2gh}$, where $h = l - l\cos\theta$

c. $v = \sqrt{2gh}$, where $h = l - l\cos\theta$

d. None of the above

iv. In previous problem, the expression for tension at instant t is

a. $T = mg\cos\theta + mv^2/l$

b. $T = mg\cos\theta + m(u^2 - 2gh)/l$

c. $T = mg\cos\theta + m\{u^2 - 2g(l - l\cos\theta)\}/l$

d. None of the above

v. At top point, the velocity of the ball is

a. $v = \sqrt{u^2 - 2gh}, h = 2l$ b. $v = \sqrt{u^2 - 2gh}, h = l$

c. $v = \sqrt{u^2 - 4gl}$ d. None of these

vi. Suppose that at top point, velocity is zero. Then, at topmost point,

a. $u = \sqrt{4gl}$ b. $T = mg\cos\pi + mv^2/l$

c. $T = -mg$ d. None of these

Working Space

vii. The minimum value of u to just complete the vertical circle is

a. $u_{min} = \sqrt{4gl}$ b. $u_{min} = \sqrt{5gl}$

c. $u_{min} = \sqrt{gl}$ d. None of these

viii. For u_{min}, tension force exerted by rod on the ball at top is

a. mg and is directed radially inward b. mg and is directed radially outward

c. zero d. None of the above

ix. For $u < \sqrt{4gl}$, the ball

a. does not complete vertical circle b. completes the vertical circle

c. may leave the vertical circle d. None of the above

x. For $u = \sqrt{3gl}$, the maximum height attained by the ball is

a. $h = 2l$ b. $3l/2$

c. l d. None of these

5 **Situation** Similar motion is found in the case of a ball moving inside a vertical pipe as shown in the given figure.

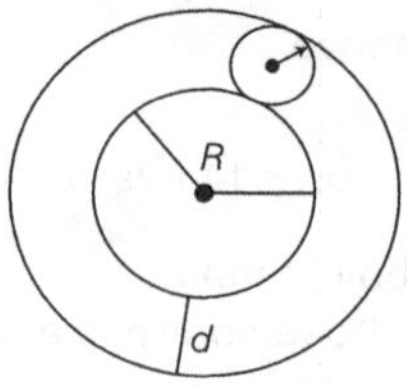

In this case, tension is replaced by normal reaction. If normal reaction is positive, it is directed radially inward and the ball is in touch of outer surface. If normal reaction is negative, then normal reaction is directed radially outward and the ball is in the touch of inner surface.

Mark the correct option(s) for the given questions.

i. The radius of circular path followed by the ball is

a. $r = R$ b. $r = R + d/2$

c. $r = R, R >> d$ d. None of these

ii. At lowest point, the normal reaction is

a. zero

b. $N = mg + mu^2/r$ and ball is in touch of inner surface

c. $N = mg + mu^2/r$ and the ball is in touch of outer surface

d. None of the above

iii. The minimum value of v to complete the vertical circle if $R >> d$ is

a. $u_{min} = \sqrt{5gr}$ b. $u_{min} = \sqrt{4gr}$ c. $u_{min} = \sqrt{2gr}$ d. None of these

iv. If $u = \sqrt{4gr}$ at top point, then

a. the velocity of ball is zero

b. normal reaction is mg and is directed radially outward

c. normal reaction is mg and is directed radially inward

d. the ball is in touch of inner surface

v. At an angle θ with lowest vertical,

a. $N = mg\cos\theta + mv^2/r$ b. $v^2 = u^2 - 2gh$, where $h = r - r\cos\theta$

c. $v^2 = u^2 + 2gh$, where $h = r - r\cos\theta$ d. None of these

6 **Situation** A circular vertical tube of mass M and radius r is placed on a horizontal surface. Two small spheres, each of mass m, just fit in the tube are released from rest from the top. If α gives the angle between radius vector of either ball with the upper vertical.

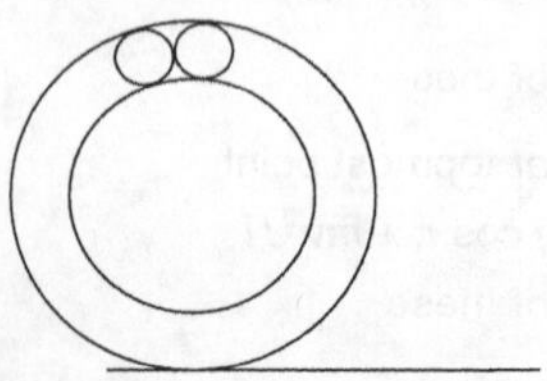

Mark the correct option(s) for the following questions.

i. The position of each ball at an angle $\alpha < \pi/2$ is

a. 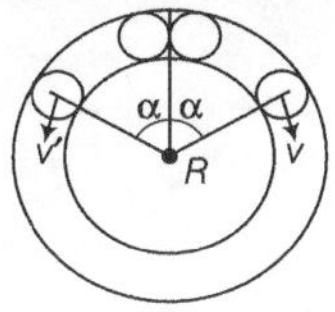

b. 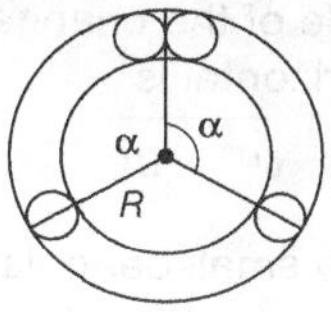

c. 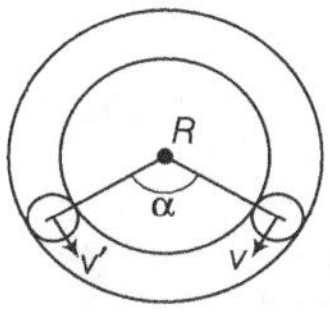

d. None of these

ii. In previous problem, the normal reaction on each ball at the position α is

a. zero
b. $N = mg \cos \theta + mv^2/R$
c. $N = -mg \cos \alpha + mv^2/R$
d. $N = mg \cos \alpha + mv^2/R$

iii. In previous problem, velocity of each ball for $\alpha = 60°$ is

a. $v = \sqrt{gR}$
b. $\sqrt{2gR}$
c. zero
d. None of these

iv. Force diagram of the tube for $\alpha = 60°$ is

a. 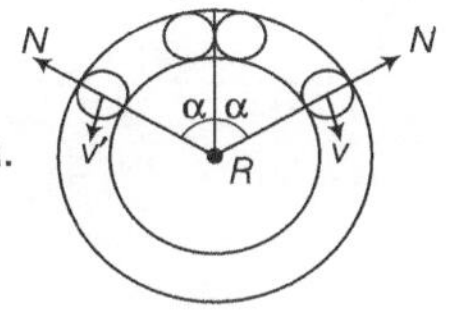

b. 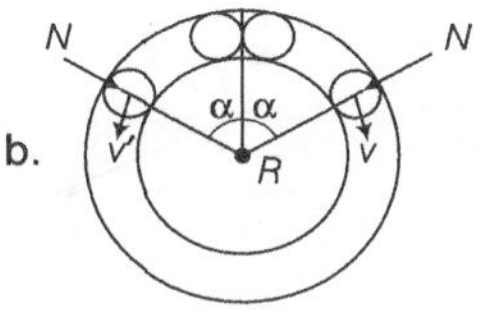

c. 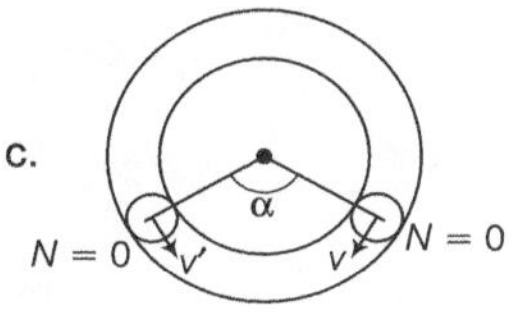

d. None of these

v. In previous problem, if tube bounce up at an angle $\alpha = 60°$. Then,

a. $2N \cos 60° \geq Mg$
b. $2N \cos 60° \leq Mg$
c. $2N \sin 60° \geq Mg$
d. None of these

vi. In previous problem, condition for bounce up to the tube is

a. $m/M \geq 2$
b. $m/M \leq 2$
c. $m/M \geq 1/2$
d. None of these

Only One Option Correct Type

Direction (Q. Nos. 7-14) This section contains 8 multiple choice questions. Each question has four choices (a), (b), (c) and (d), out of which ONLY ONE option is correct.

7 A mass of M kg is suspended by a weightless string. The horizontal force that is required to displace it until the string makes an angle of 45° with the initial vertical direction is **[AIEEE 2006]**

a. $Mg(\sqrt{2}+1)$
b. $Mg\sqrt{2}$
c. $\dfrac{Mg}{\sqrt{2}}$
d. $Mg(\sqrt{2}-1)$

8 A bob of mass M is suspended by a massless string of length L. The horizontal velocity v at position A is just sufficient to make it reach at the point B. The angle θ at which the speed of the bob is half of that at point A, satisfies **[IIT JEE 2008]**

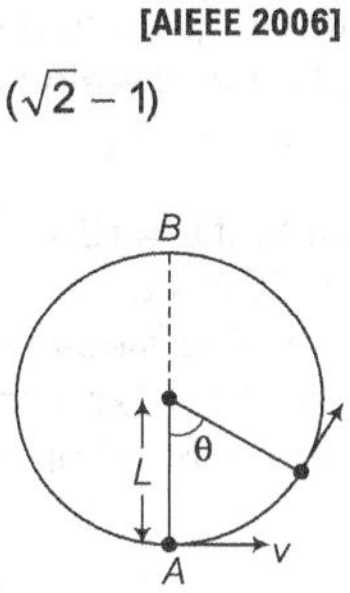

a. $\theta = \dfrac{\pi}{4}$
b. $\dfrac{\pi}{4} < \theta < \dfrac{\pi}{2}$
c. $\dfrac{\pi}{2} < \theta < \dfrac{3\pi}{4}$
d. $\dfrac{3\pi}{4} < \theta < \pi$

Working Space

9 A stone tied to a string of length l is whirled in a vertical circle with the other end of the string at the centre. At a certain instant of time, the stone is at its lowest position and has a speed u. The magnitude of the change in its velocity as it reaches a position, where the string is horizontal is

a. $\sqrt{u^2 - 2gl}$ b. $\sqrt{2gl}$ c. $\sqrt{u^2 - gl}$ d. $\sqrt{2(u^2 - gl)}$

10 A wire, which passes through the hole in a small bead, is bent in the form of quarter of a circle. The wire is fixed vertically on ground as shown in the figure. The bead is released from near to top of the wire and it slides along the wire without friction. As, the bead moves from A to B, the force it applies on the wire is **[IIT JEE Advanced 2014]**

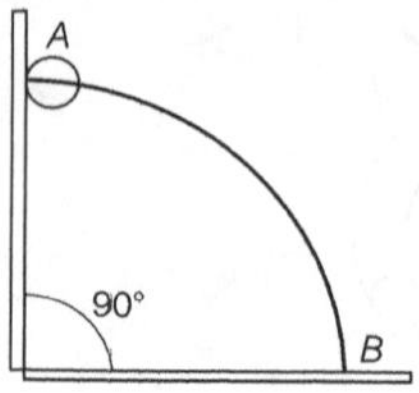

a. always radially outwards
b. always radially inwards
c. radially outwards initially and radially inwards later
d. radially inwards initially and radially outwards later

11 A spherical ball of mass m is connected with a light rod of length l as shown in the figure. The ball is released from highest point. The angle made of rode with upward vertical is θ at an instant. The tension *versus* $\cos\theta$ graph for $0 \le \theta \le \pi$, is

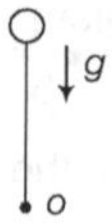

a.
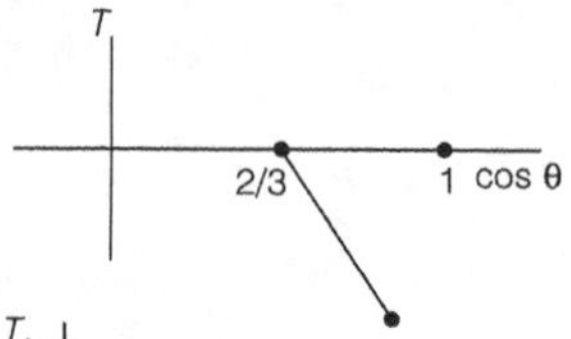

b.
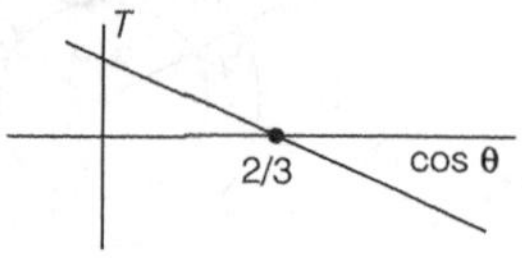

c.
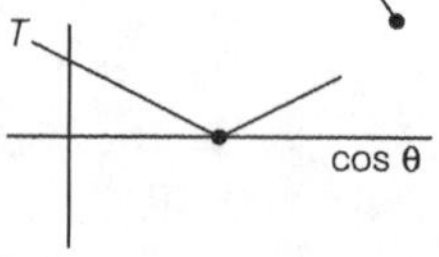

d.
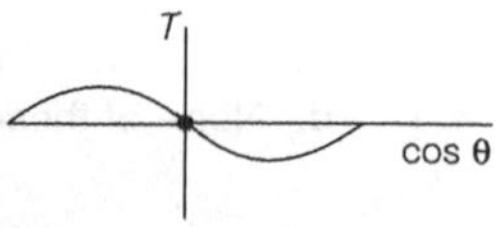

12 Two particles A and B having same mass, A is connected to a light spring of natural length l_A and particle B is connected to a light string of length l_B ($l_A < l_B$). When spring becomes vertical, its length becomes l_B. Both are released from rest from horizontal position (shown in the figure). Mark the correct option(s).

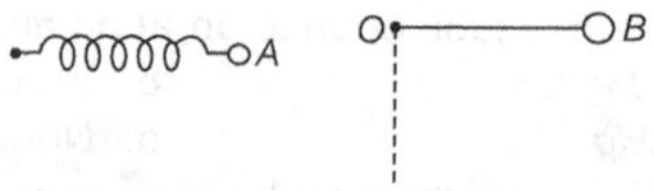

a. At lowest position, particle A has more speed than that of particle B
b. At lowest position, particle A has lesser speed than B
c. At the lowest position, both have same speed
d. None of the above

13 A small block of mass m starts from rest from point A at a height 4 m. The radius of circular path is 1 m. The block slides down on smooth circular path and reaches at point B. The ratio of normal reaction and weight at the point B is

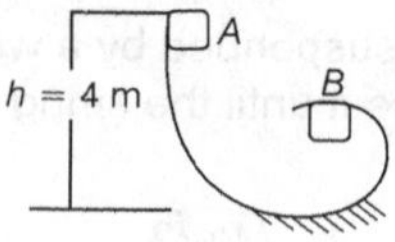

a. 1 b. 2 c. 3 d. 4

14 A light spring of constant $k = (2 + \sqrt{2})$ N/m and natural length $\sqrt{2}R$ is fixed at point O. A sleeve of mass $m = 1$ kg is attached to other end of spring. The sleeve starts from rest from point A and moves along a vertical fixed smooth circular wire of radius $R = 5$ m. Find the normal reaction on sleeve at the point B.

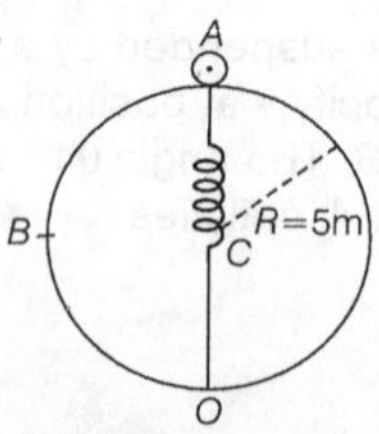

a. 10 N b. 20 N
c. 30 N d. 40 N

Working Space

One or More than One Options Correct Type

Direction (Q. Nos. 15-19) *This section contains 5 multiple choice questions. Each question has four choices (a), (b), (c) and (d), out of which ONE or MORE THAN ONE are correct.*

15 A small ball of mass 1 kg is given vertical velocity of magnitude $v_0 = 5$ m/s and it swings in a vertical plane by the help of a light string of length 2 m fixed at one end. The breaking strength of string is 20 N.

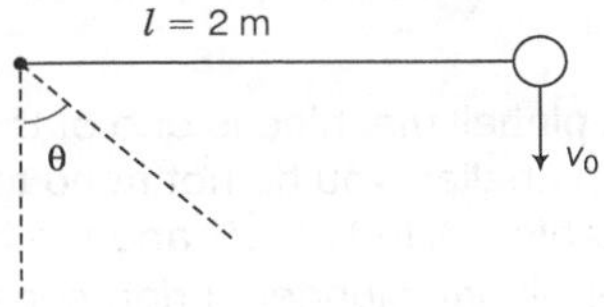

a. String does not break before reaching at lowest point

b. String breakes before reaching at lowest point

c. The acceleration of the ball just after angle $\theta = \cos^{-1}\frac{1}{4}$ is g

d. None of the above

16 A block of mass 10 kg is pulled along a smooth surface in the form of arc of radius 10 m. The applied force F is 200 N as shown in figure. If the block starts from point A, then

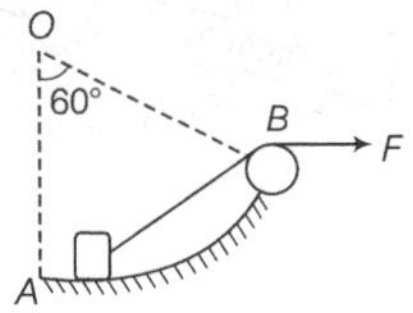

a. the speed at point B is 15.7 m/s

b. the speed at point B is 27 m/s

c. the work done by gravity during moving from A to B is negative

d. the work done by tension on block is zero

17 A small particle of mass m is given an initial velocity v_0 tangent to the horizontal rim of a smooth cone at a radius r_0 from the vertical centre line as shown in figure. As, the particle slides to point B, a vertical distance h below A and a distance r from vertical centre line, its velocity v is inclined at some angle with the horizontal tangent to the cone through the point B.

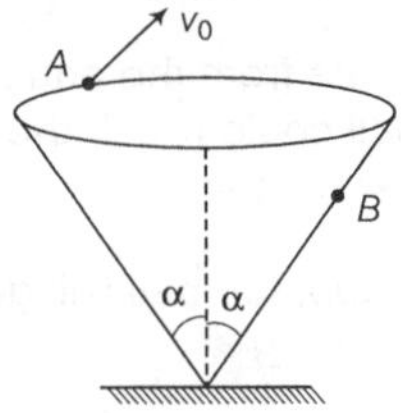

a. The speed of particle at point B is $\sqrt{v_0^2 + 2gh}$

b. The minimum value of v_0 for which particle will be moving in a horizontal circle of radius r_0 is $\sqrt{gr_0 \cot\alpha}$

c. The speed of particle at point B is $\sqrt{v_0^2 - 2gh}$

d. None of the above

18 A particle of mass m is moving on a circular path of radius r in the influence of force $\mathbf{F} = \frac{-k}{r_2}\hat{\mathbf{r}}$, where k is a positive constant.

a. Kinetic energy of particle is $\frac{k}{r}$

b. Kinetic energy of particle is $\frac{k}{2r}$

c. Potential energy of particle is $\frac{-k}{r}$

d. Total mechanical energy of particle is $\frac{-k}{2r}$

19 A ball of mass $m = 1$ kg is tied to a string of length $l = 1$m and a horizontal force is applied on it, which initially, is zero and gradually increases until string becomes inclined at an angle $\theta = 60°$ with the vertical as shown in figure. Find the work done by the force.

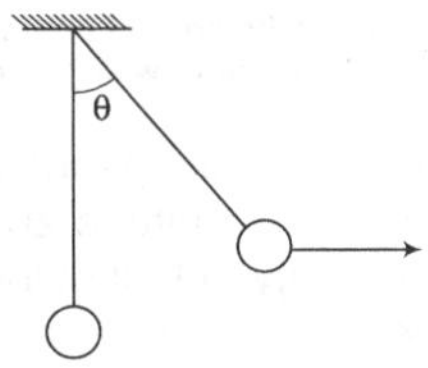

a. Zero

b. Non-zero

c. 5 J

d. 10 J

Comprehension Type

Working Space

Direction (Q. Nos. 20-24) *This section contains 2 paragraphs, each describing theory, experiments, data, etc. Five questions related to the paragraphs have been given. Each question has only one correct answer among the four given options (a), (b), (c) and (d).*

Passage I

Pinball pastimes A pinball machine is one of the ancient 'arcade' games (see figure). As a good pinballer, you do not need to pull the spring knob too hard. In this machine, the table is tilted at 15°, and the ball has a diameter of 3.0 cm and a mass of 100 g. The pinball-plunger spring constant is 200 N m^{-1}. The ball assumed not to roll and the contact between the ball with the table is frictionless. Assume that all thicknesses not specifically indicated are negligibly small.

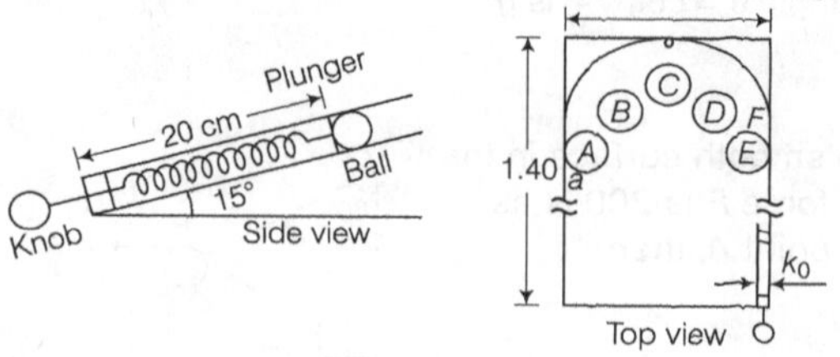

20 Assume that ball is a point of mass. The speed of ball at the top of semi-circle is (take, $g = 9.79$ m/s^2)

a. 0.85 m/s b. 10 m/s c. 200 m/s d. 0.285 m/s

21 The minimum distance from the equilibrium point of the spring that one has to pull the knob, so that ball could reach the top of the semi-circle is

a. 0.333 cm b. 0.85 cm c. 5.9 cm d. 1.170 m

22 After it reaches the top, the ball will go at bumper

a. *A* b. *B* c. *C* or *D* d. *E*

Passage II

A small block of mass 1 kg is released from rest at the top of a rough track. The track is a circular arc of radius 40 m. The block slides along the track without toppling and a frictional force acts on it in the direction opposite to the instantaneous velocity. The work done in overcoming the friction up to the point *Q* as shown in the given figure, is 150 J. (take the acceleration due to gravity, $g = 10$ ms^{-2})

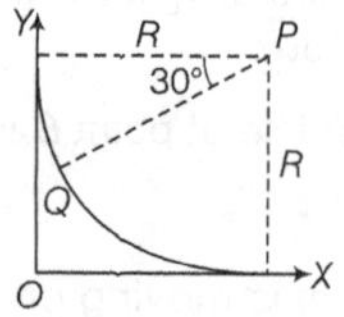

[JEE Advanced 2013]

23 The speed of the block when it reaches the point *Q* is

a. 5 ms^{-1} b. 10 ms^{-1} c. $10\sqrt{3}$ ms^{-1} d. 20 ms^{-1}

24 The magnitude of the normal reaction that acts on the block at the point *Q* is

a. 7.5 N b. 8.6 N c. 11.5 N d. 22.5 N

Matching List Type

Direction (Q. No. 25) *Choices for the correct combination of elements from Column I and Column II are given as options (a), (b), (c) and (d), out of which one is correct.*

25 A small ball of mass 1 kg is suspended through a light string of length 1 m. A constant force of magnitude 20 N is in always acting on the ball. The minimum velocity of ball at lowest point to complete the vertical circle is given in column II.

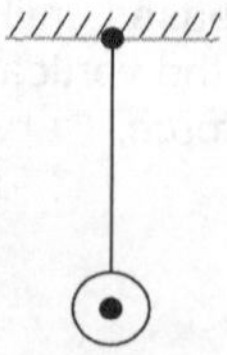

Working Space

Match the Column I with Coumn II and mark the correct option from the given codes.

	Column I		Column II
i.	If force is directed downward.	p.	$\sqrt{50}$ m/s
ii.	If force is directed upward.	q.	$\sqrt{150}$ m/s
iii.	If force is directed horizontally.	r.	$\sqrt{10}$ m/s
iv.	If force is zero.	s.	$\sqrt{30\sqrt{5}+20}$ m/s

Codes

	i	ii	iii	iv		i	ii	iii	iv
a.	q	r	s	p	b.	p	q	r	s
c.	s	p	q	r	d.	p	q	s	r

Statement Type

Direction (Q. Nos. 26-28) This section is based on Statement I and Statement II. Select the correct answer from the codes given below.

a. Both Statement I and Statement II are correct and Statement II is the correct explanation of Statement I
b. Both Statement I and Statement II are correct and Statement II is not the correct explanation of Statement I
c. Statement I is correct but Statement II is incorrect
d. Statement II is correct but Statement I is incorrect

26 **Statement I** When ball is released from rest at an angle θ with vertical, the tension in string at lowest point is maximum.

Statement II Work done by tension on the ball is zero.

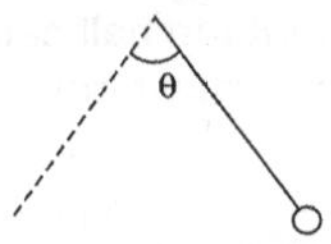

27 **Statement I** The ball is connected to a light rod and is released from rest, the tension in rod never be zero.

Statement II During motion mechanical energy of ball remains conserved.

28 **Statement I** A ball starts from top of a vertical circular pipe and finally comes to rest at lowest point. The loss in gravitational potential energy appears as thermal energy.

Statement II The work done by normal reaction is zero.

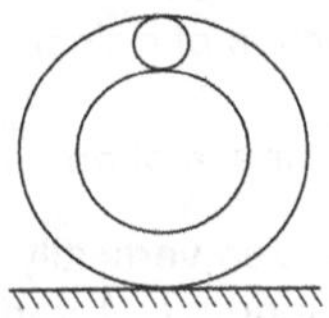

One Integer Value Correct Type

Direction (Q. Nos. 29-34) This section contains 6 questions. When worked out will result in an integer from 0 to 9 (both inclusive).

29 A light string of length 1 m is fixed at point A on a fixed cylinder of radius $R = \frac{l}{2\pi}$ m. The other end of the string is attached to a ball of mass 1 kg. The ball is imparted with a velocity (v_0) of 10 m/s horizontally. The tension in the string when the ball is at the lowest point is $30x$ N. Find the value of x.

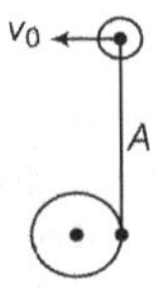

30 A small ball of mass 1 kg is suspended from a fixed point by a light string of length 5 m. It is thrown from the lowest point horizontally with a speed of $10\sqrt{2}$ m. The speed of ball, just before the string slacking is $\frac{10}{\sqrt{n}}$ m/s. Find the value of n.

Working Space

31 A small smooth block is projected horizontally from point C with speed v_0 inside a vertical fixed loop of radius 0.2 m from which a symmetrical section of angle 120° has been removed. The value of v_0 such that the block, after leaving contact at point A and flying through air reaches point B and continuous the loop is $(10)^{\frac{1}{x}}$ m/s. Find the value of x.

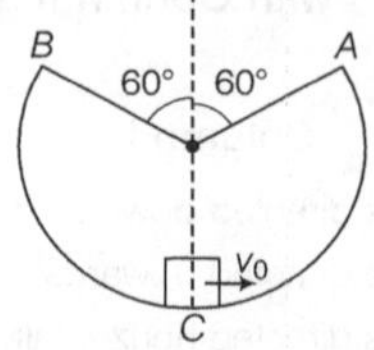

32 A circular track of radius 1 m as shown in the given figure. A small block is moving on horizontal part of the track with speed $v_0 = \sqrt{44}$ m/s. The value of $\sin\theta$ at which it leaves the track is $\frac{x}{5}$. Find the value of x.

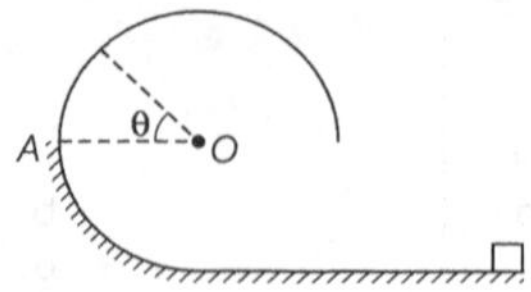

33 A smooth hollow fixed hemisphere has a radius R of 23 m. A vertical smooth circular track of centre O is as such $OC = r$. The block is released from rest from point A. The minimum value of r, so that block complete its turn about O is
$3n$ m. Find the value of n (take, $\theta = 37°$).

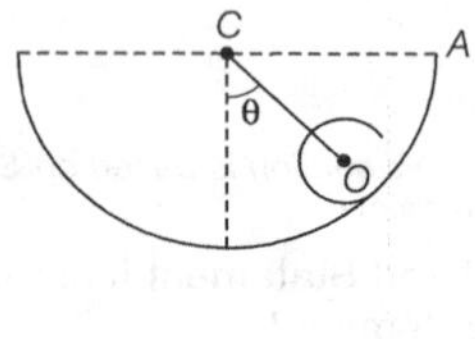

34 A ball of mass 1 kg hangs through a light thread in a car which moves at a speed of $u = \sqrt{50}$ m/s on a horizontal straight road as shown in the figure. The car is brought in rest by a uniform braking. Find maximum stopping distance of car in metre so that the ball can reach of the maximum height of its vertical circular path.

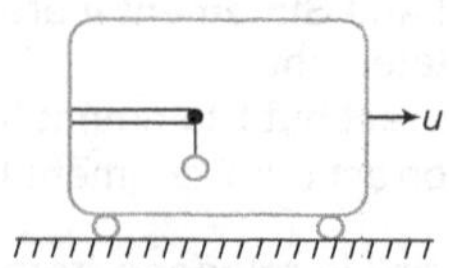

Subjective Questions

Direction (Q. Nos. 35-37) *These questions are subjective in nature, need to be solved completely on notebook.*

35 A small block of mass slides down the circular periphery of a horizontal smooth fixed cylinder of radius 12 cm.

i. Find normal reaction of cylinder on the block when it reaches 37° position.

ii. It will leave the surface at an angle $\theta = \alpha$. Find the value of $\cos\alpha$.

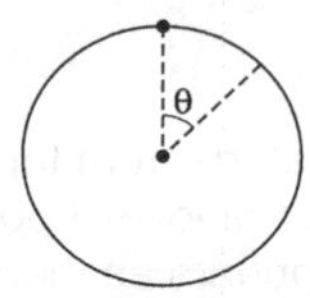

36 A particle is suspended vertically from a point O by an inextensible massless string of length L. A vertical line AB is at a distance $L/8$ from O as shown in the given figure.

The object is given a horizontal velocity u. At some point, its motion ceases to be circular and eventually the object passes through the line AB. At the instant of crossing AB, its velocity is horizontal. Find the value of u.

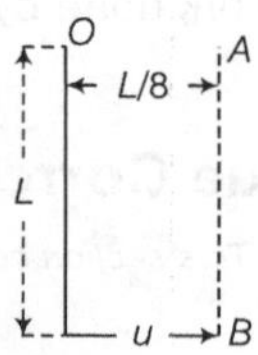

37 A spherical ball of mass m is kept at the highest point in the space between two fixed, concentric spheres A and B (see figure). The smaller sphere A has a radius R and the space between the two spheres has a width d. The ball has a diameter very slightly less than d. All surfaces are frictionless. The ball is given a gentle push (towards the right in the figure). The angle made by the radius vector of the ball with the upward vertical is denoted by θ. Find normal reaction as a function of angle θ.

[IIT JEE 2002]

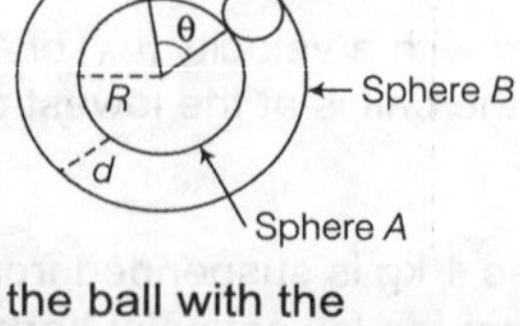

SET - 1

Revisal Problems for JEE Main

Only One Option Correct Type

Working Space

Direction *(Q. Nos. 1-20) This section contains 20 multiple choice questions. Each question has four choices (a), (b), (c) and (d), out of which ONLY ONE option is correct.*

1 The conservation principle of mechanical energy requires that

a. the acceleration should be zero
b. there should be no external forces
c. the motion should be restricted to the gravitational force
d. the force should be conservative

2 A spring gun of a spring constant $k = 4000$ N/m is held in a vertical position. It projects a ball of mass 2 kg at a height of 4 m above its initial compression. The initial compression in the spring is

a. 10 cm b. 20 cm
c. 30 cm d. 40 cm

3 A particle of mass m moves rectilinearly due to force $F = F_0 - bx$, where F_0 and b are positive constants, x is the distance from the point, where the particle was initially at rest. Find the distance covered by the particle till the moment it came stand still.

a. $\frac{F_0}{b}$ b. $\frac{2F_0}{b}$ c. $\frac{3F_0}{b}$ d. $\frac{F_0}{2b}$

4 A conservative force of $10\sqrt{2}$ N acting on a particle of mass m in XY-plane making an angle 45° with X-axis as shown in the given figure. Find work done by conservative force between origin and point A (1 m, 1 m, 0).

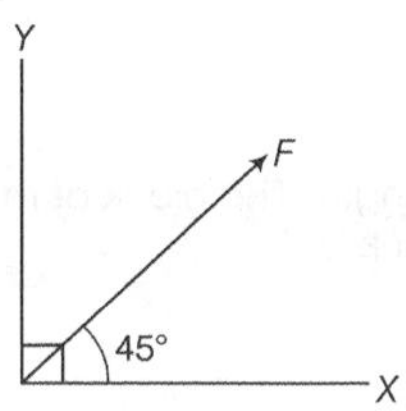

a. Zero b. 10 J c. 20 J d. 30 J

5 A simple pendulum mounted on a car. The car starts to accelerate on horizontal surface with constant acceleration $a_0 = \frac{g}{\sqrt{3}}$. The maximum deflection of pendulum from vertical is

a. 30° b. 45° c. 60° d. 90°

6 A collar of mass 10 kg is released from rest from position A. When spring is in natural length and slides along a rough vertical fixed rod. The block comes in rest at point B, 4 m below the point A. The spring constant of spring is 100 N/m and natural length of spring is 3 m. Find work done by friction.

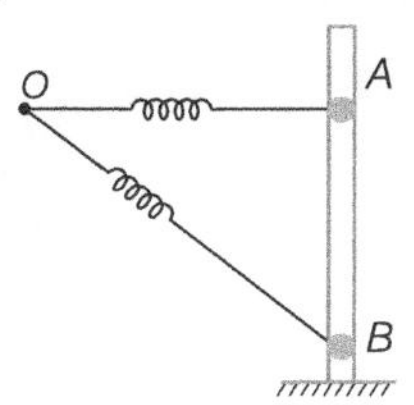

a. -100 J b. -200 J c. -300 J d. -400 J

Working Space

7 A conservative force of magnitude 100 N is directed along the line $y = 3 + x$ is acting on a block of mass $m = 1$ kg. Find work done by the conservative force, when block displaces from $A(3, 1)$ to $B(1, 3)$.

a. Zero
b. 100 J
c. $200\sqrt{2}$ J
d. $400\sqrt{2}$ J

8 The potential energy of a particle on the line $y = 2x$ are same at every point. The conservative force on the particle at $(1, 2)$ may be

a. $4\hat{\mathbf{i}} + 3\hat{\mathbf{j}}$
b. $4\hat{\mathbf{i}} + 8\hat{\mathbf{j}}$
c. $8\hat{\mathbf{i}} + 4\hat{\mathbf{j}}$
d. $-8\hat{\mathbf{i}} + 4\hat{\mathbf{j}}$

9 The conservative force on a particle is $\mathbf{F} = (3x^2 + yz)\hat{\mathbf{i}} + (3y^2 + xz)\hat{\mathbf{j}} + (3z^2 + xy)\hat{\mathbf{k}}$. The expression of potential energy U is

a. $x^4y^2z + c$
b. $-(x^3 + y^3 + z^3 + xyz) + c$
c. $x^2y + k$
d. None of these

10 Drag force on a vehicle is proportional to its speed. If power output from engine is triple, then the maximum speed of the car

a. is also doubled
b. increases by factor of $\sqrt{3}$
c. is unchanged
d. is tripled

11 A small block A of mass 1 kg is placed over a horizontal plate B of mass 2 kg which is placed on a smooth horizontal surface. Block A is being moved continuously with velocity 1 m/s as shown in the figure and the plate is moved continuously with speed 1 m/s towards the direction shown in the figure. If the coefficient of friction between blocks is 0.1, the power loss due to friction

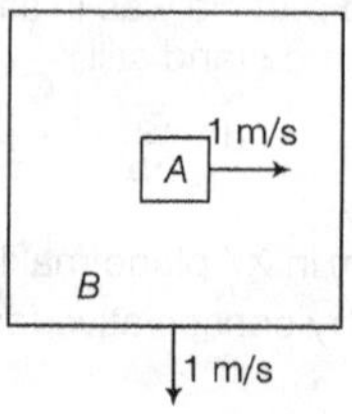

a. 0.5 J/s
b. 1 J/s
c. $\sqrt{2}$ J/s
d. 2 J/s

12 Initially, spring of spring constant 100 N/m is in its natural length. The block of mass 1 kg released from rest, the maximum power of weight of the block is

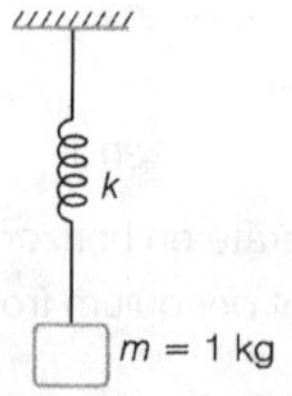

a. 5 W
b. 10 W
c. 20 W
d. 40 W

13 The work done by the force $\mathbf{F} = bx\hat{\mathbf{i}} + by\,\hat{\mathbf{j}}$ around the path shown in the given figure, is

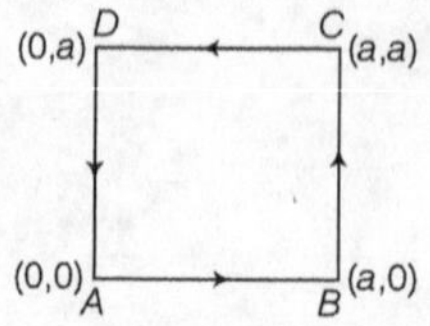

a. zero
b. $\frac{2}{3}a^3b$
c. a^3b
d. $\frac{4}{3}a^3$

14 Variation of potential energy along X-direction is shown in the graph. The correct statement about conservative force is

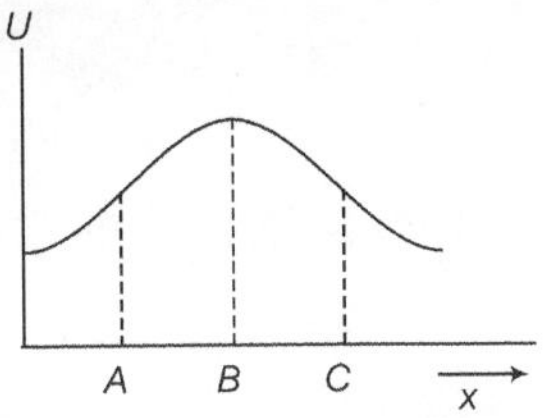

a. x-component of conservative force at point B is maximum
b. x-component of conservative force at point A is towards positive X-axis
c. x-component of conservative force at point C is along negative X-axis
d. x-component of conservative force at point C is along positive X-axis

15 A small block of mass m is lying at rest at point D of a hollow smooth cylinder moving horizontally with constant acceleration a_0. The minimum value of horizontal acceleration a_0 of the cylinder, so that the block can just reach at the point A is

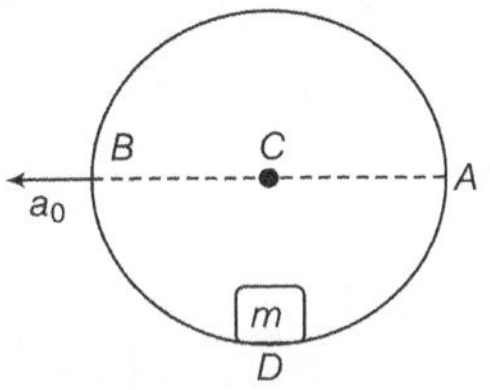

a. $g/3$ b. $g/2$ c. g d. $2g$

16 Inside a smooth spherical cavity particle A can slide freely. The block having this cavity is moving horizontally with constant acceleration $a_0 = 10\ m/s^2$.

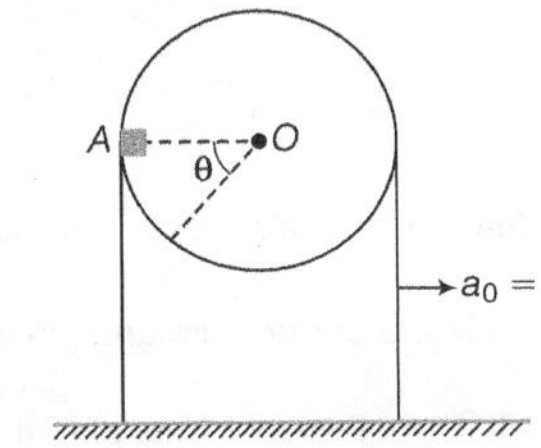

The particle is released from its initial position as shown in the figure. The angle θ with the vertical, when the particle will have maximum speed with respect to the block is

a. 30° b. 45° c. 60° d. 90°

17 A smooth chain of mass 6 kg and length 10 m is placed on a horizontal table. Due to a small jerk, chain starts to slide down from rest. Find the speed of chain when chain is slip off completely.

a. 1 m/s b. 5 m/s c. 10 m/s d. 15 m/s

18 The potential energy of a particle at A, B, C, D and P are $U_A = 2$ J, $U_B = 5$ J, $U_C = 8$ J, $U_D = 5$ J and $U_P = 5$ J. The conservative force on the particle at point P is

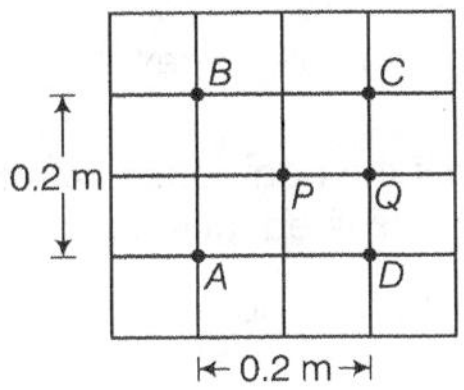

a. 10 N along PQ
b. $15\sqrt{2}$ N along PA
c. 5 N along PC
d. 5 N along PA

19 In the situation shown in the figure, the mass of each block is 10 kg. The system is released from rest, find the speed of block *B* after it has fallen a distance of 10 m.

Working Space

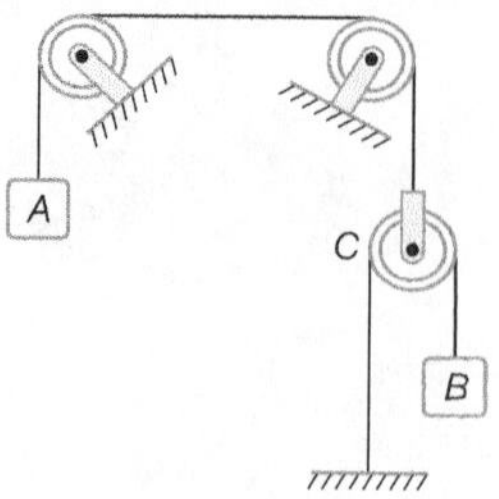

a. $\sqrt{320}$ m/s
b. $\sqrt{80}$ m/s
c. $\sqrt{20}$ m/s
d. $\sqrt{10}$ m/s

20 In the situation shown in the figure, $m_1 = m_2 = 10$ kg, $H = 0.04$ m and spring constant is $k = 200$ N/m. The system is initially in equilibrium, what work perform to bring m_1 on floor slowly.

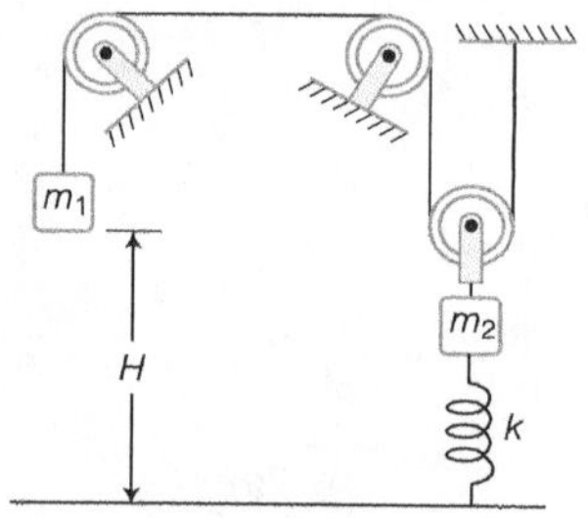

a. 0.1 J
b. 0.2 J
c. 0.3 J
d. 0.4 J

Statement Type

Direction *(Q. Nos. 21-25)* ***This section is based on Statement I and Statement II. Select the correct answer from the codes given below.***

a. Both Statement I and Statement II are correct and Statement II is the correct explanation of Statement I
b. Both Statement I and Statement II are correct and Statement II is not the correct explanation of Statement I
c. Statement I is correct but Statement II is incorrect
d. Statement II is correct but Statement I is incorrect

21 **Statement I** If potential energy of a body while moving in a certain path is constant, then conservative force on the body must be zero.

Statement II $\mathbf{F} = \frac{-\partial U}{\partial x}\hat{\mathbf{i}} - \frac{\partial U}{\partial y}\hat{\mathbf{j}} - \frac{\partial U}{\partial z}\hat{\mathbf{k}}$

22 **Statement I** In the absence of air resistance, during projectile motion, mechanical energy conservation principle is applicable.

Statement II Weight of a block is conservative force.

23 **Statement I** Work-energy theorem is applicable in non-inertial reference frame, if work done by pseudo force is taken into account.

Statement II Mechanical energy may be conserved even if the non-conservative forces are present on system provided the work done by these forces is equal to zero.

24 **Statement I** The rate of change of kinetic energy is equal to total power.

Statement II Net work done = Change in kinetic energy.

25 **Statement I** The force depends on desire is non-conservative in nature.

Statement II Velocity independent forces are conservative in nature.

SET - 2

Revisal Problems for JEE Advanced

Only One Option Correct Type

Working Space

Direction (Q. Nos. 1-5) ***This section contains 5 multiple choice questions. Each question has four choices (a), (b), (c) and (d), out of which ONLY ONE option is correct.***

1 Figure shows three paths connecting points *a* and *b*. A single force **F** does the indicated work on a particle moving along each path in the indicated direction on the basis of this information.

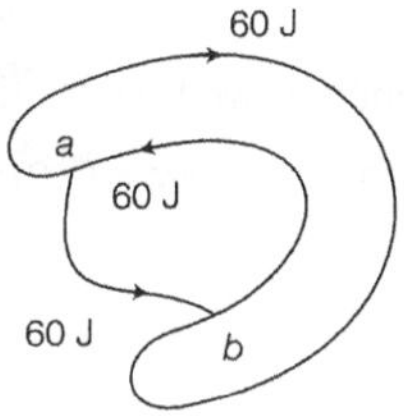

a. **F** must be conservative
b. **F** may be conservation
c. **F** must be non-conservative
d. No sufficient information

2 You lift a book slowly from the ground and keep it on your study table. Choose the correct statement(s).

a. The work done by you on the book is positive but force exerted by you on the book is non-conservative
b. The work done by you depends on weight of book but independent of path of the book
c. The work done by you depends on height of table from ground but independent of time taken by you is doing so
d. The total work done on book is zero but work done by weight of book is negative

3 Efficiency of a car engine measured by an engineer is found to be 12% in an experiment. After the experiment, it was found that the petrol used was duplicate having 4% lesser efficiency than pure petrol. If pure petrol is used, then find actual efficiency of car.

a. 8%
b. 12%
c. 12.5%
d. 25%

4 A ball with volume 1 cm^3 is sinking at velocity of 0.01 m/s in a closed vessel filled with viscous liquid. Find kinetic energy of viscous liquid, if its density is 2 g/cm^3.

a. 1 erg
b. 2 erg
c. 3 erg
d. 4 erg

5 A hollow sphere of radius 1 m is filled with 3 kg water is fixed on the mouth of a cylindrical vessel. Water starts to leak from point *P*. Find work done by gravity when water completely drops in the cylinder *AOB* is horizontal line.

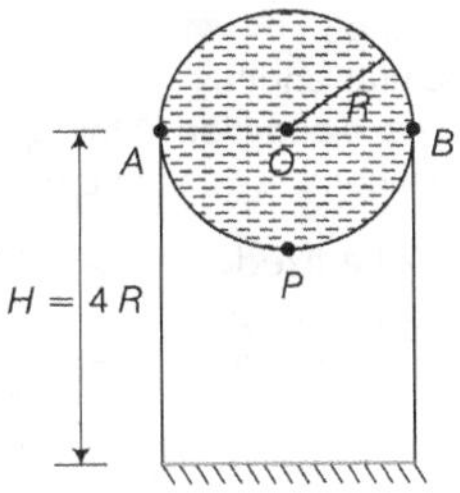

a. 10 J
b. 40 J
c. − 50 J
d. 80 J

One or More than One Options Correct Type

Working Space

Direction *(Q. Nos. 6-11) This section contains 7 multiple choice questions. Each question has four choices (a), (b), (c) and (d), out of which ONE or MORE THAN ONE are correct.*

6 Two smooth balls each of mass m are connected through a light rod as shown in the figure. They are released from rest. At the instant when upper ball leaves the vertical wall,

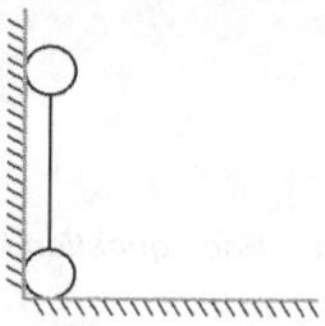

a. acceleration of upper ball is equal to acceleration due to gravity
b. the acceleration of lower ball is zero
c. the normal reaction on lower ball is mg
d. the normal reaction on lower ball is zero

7 A particle of mass 1 kg moving in XY-plane has its potential energy given by $U = (-3x + 4y)$ J. The particle is initially at $\left(-\frac{3}{2}\text{m}, 2\text{m}\right)$ and has velocity $\mathbf{v}_0 = (-3\hat{\mathbf{i}} + 4\hat{\mathbf{j}})$ m/s.

Which of the following statement(s) is/are correct?
a. The speed of the particle at $t = 1$ s is zero
b. The acceleration of the particle is 5 m/s^2 in magnitude
c. The position of the particle at, $t = 1$ s is $(-3, 4)$ m
d. The displacement of the particle in one second is $\left(-\frac{3}{2}\hat{\mathbf{i}} + 2\hat{\mathbf{j}}\right)$ m

8 A block of mass 2 kg is hanging from a light, smooth pulley through a light string. The kinetic energy of block increases 16 J in 2 s by applying a constant force F on one end of upper string. Then,

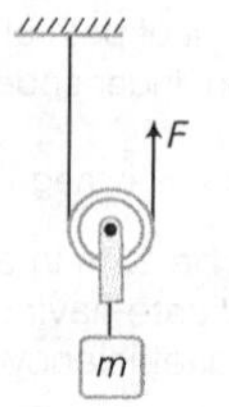

a. the force F may be 12 N
b. the force F may be 8 N
c. the potential energy of block may increases
d. the potential energy of the block may decreases

9 If work done by light spring on the block is non-zero positive. The possible cases are

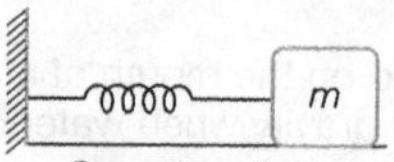

Smooth surface

a. the spring is initially compressed and finally comes in natural length
b. the spring is initially in extension state and finally comes in natural length
c. the initial compression in spring is lesser in magnitude than final elongation in the spring
d. the initial elongation in spring is greater in magnitude than final compression in the spring

10 A particle of mass m initially at the point A at the centre level of a fixed smooth hollow sphere of radius R is given an upward velocity v_0 along the internal surface of the sphere. The particle is just able to complete the vertical circle. Then,

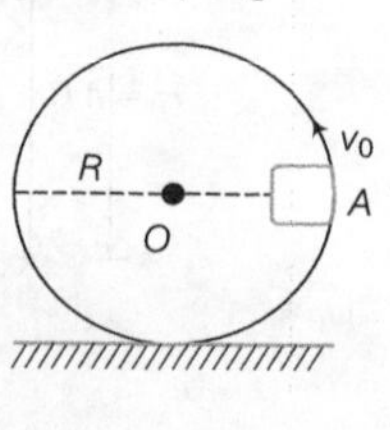

a. the value of v_0 is $\sqrt{5gR}$
b. the value of v_0 is $\sqrt{3gR}$
c. the velocity of the particle at the highest point is zero
d. the normal reaction at the highest point is zero

Working Space

11 In the figure, two smooth blocks *A* and *B* each of mass 5 kg are released from rest. The spring constant is $k = 100\ N/m$. Mark the correct option(s).

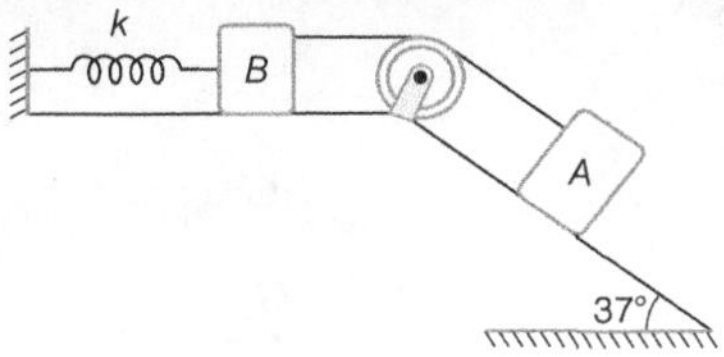

a. The maximum speed of block *B* is $\frac{3}{\sqrt{10}}$ m / s
b. The maximum elongation in the spring is 0.6 m
c. The acceleration of block *A* at the time of maximum speed is zero
d. None of the above

Comprehension Type

Direction *(Q. Nos. 12-17) This section contains 2 paragraphs, each describing theory, experiments, data, etc. Six questions related to the paragraphs. Each question has only one correct answer among the four given options (a), (b), (c) and (d).*

Passage I

A particle *A* of mass 5kg is thrown along a rough horizontal plank with a velocity 10 m/s.

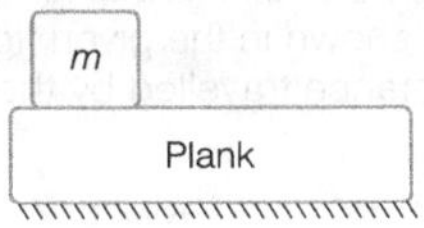

12 Find work done by plank on the particle by the time the particle comes to rest.

a. Zero
b. – 250 J
c. 250 J
d. Cannot be determined

13 If a reference frame *S* is moving with velocity 10 m/s in the direction of motion of particle *A*. Find work done by friction in the frame *S*.

a. Zero
b. – 250 J
c. 250 J
d. Cannot be determined

14 Mark the correct option.

a. Heat loss in ground frame is greater than that of in *S*-frame
b. Heat loss in ground frame is same as that of in *S*-frame
c. Work done by plank on particle is same in ground as well as in *S*-frame
d. The net work done by friction in *S*-frame is zero

Passage II

The two particles of mass *m* and 2 *m* respectively are connected by a rigid rod of negligible mass and slide with negligible friction in a circular path of radius 10 m inside the vertical circular ring. If the system is released from rest at $\theta = 0°$, determine (take, $g = 10\ m/s^2$)

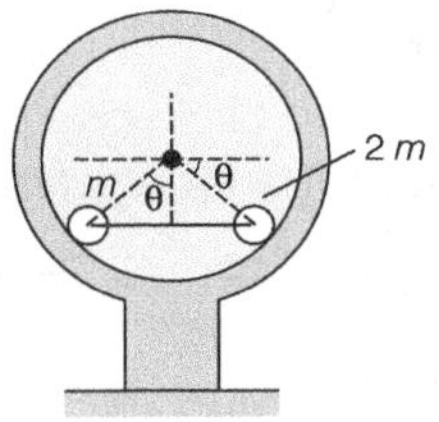

15 The velocity *v* of the particle (in cm/s) when the rod passes the horizontal position, is

a. 864
b. 707
c. 908
d. 800

Working Space

16 The maximum velocity v_{max} of the particle (in cm/s) is

a. 864
b. 707
c. 908
d. 800

17 At $\theta = \theta_{max}$, find the value of $2\sin\theta + \cos\theta - 1$ is

a. 2
b. 1.5
c. 0
d. 0.5

One Integer Value Correct Type

Direction (Q. Nos. 18-22) *This section contains 5 questions. When worked out will result in an integer from 0 to 9 (both inclusive).*

18 The situation shown in the given figure, the block A is initially at rest. The speed of block A (in m/s) when two segments of the thread between the pulleys make right angle is $5x$. Find the value of x. Assume that $l = 10(3 + 2\sqrt{2})$ m.

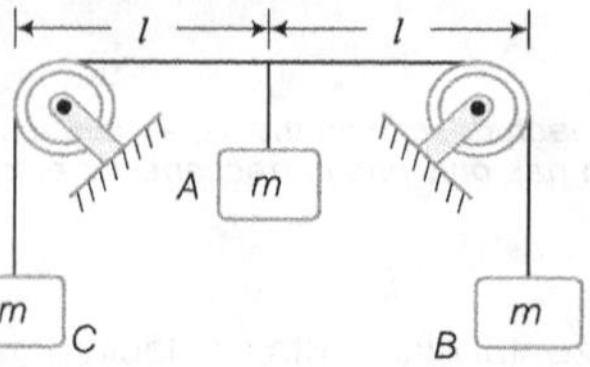

19 A small block is moving on rough horizontal surface of variable coefficient of kinetic friction. The velocities of the block at different positions are shown in the given figure. When the velocity of the block is 4 m/s at $x = 0$. Find the distance travelled by the block (in metre) before stop.

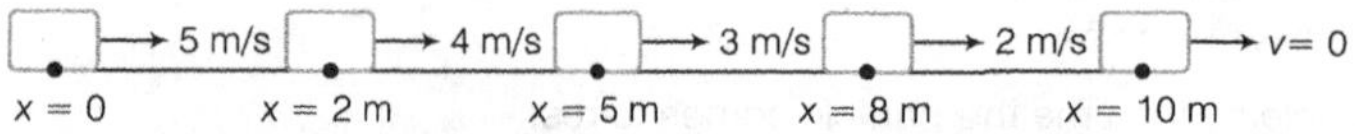

20 In the given figure, the lower light pulley is pulled down with constant velocity $v_0 = 0.75$ m/s by a naughty boy. When a rough block of mass $m = 5$ kg is gently placed on the plank, it slides for a while, and finally moves together with the plank. The work done by naughty boy during the sliding of the block is $\frac{45}{n}$ J. Find the value of n.

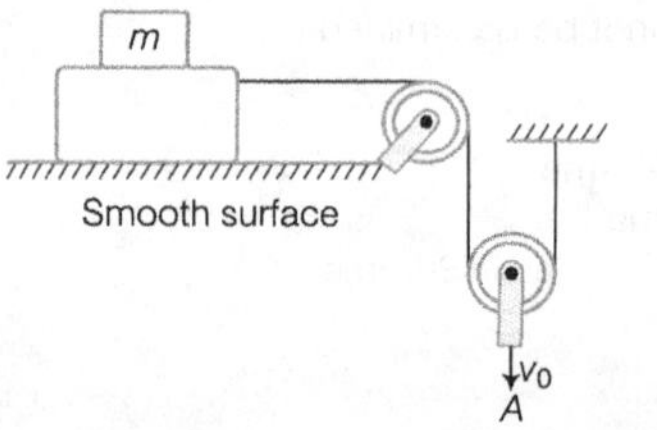

21 A circular smooth turn table is turning at an angular speed ω about a vertical axis passing through centre O. A groove is made on the surface along a radius and a particle is gently placed inside the groove at a distance r_0 from the centre O as shown in the given figure. The work done by external agent maintaining the constant angular velocity of the turn table when the particle slips to a distance $2r_0$ from the centre is $nmr_0^2\omega^2$. Find the value of n.

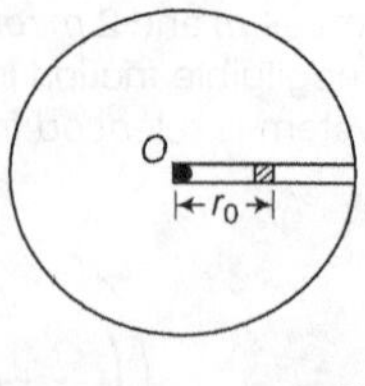

22 A block of mass 0.18 kg is attached to spring of force constant 2 N/m. The coefficient of friction between the block and the floor is 0.1. Initially the block is at rest and the spring is unstretched. An impulse is given to the block as shown in the given figure. The block slides a distance of 0.06 m and comes to rest for the first time. The initial velocity of the block (in m/s) is $v = \frac{N}{10}$. Then N is **[IIT JEE 2011]**

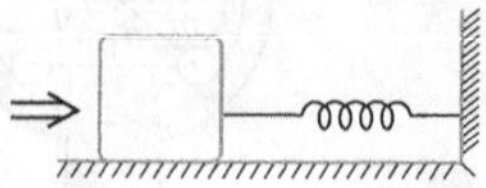

Working Space

Matching List Type

Direction *(Q. Nos. 23 and 24) Choices for the correct combination of elements from Column I and Column II are given as options (a), (b), (c) and (d), out of which one is correct.*

23 Match the Column I with Coumn II and mark the correct option from the given codes.

	Column I		Column II
i.	A block is projected with velocity v_0 on rough horizontal surface having constant coefficient of friction μ 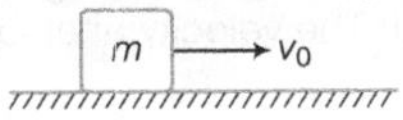	p.	Mechanical energy remains conserved
ii.	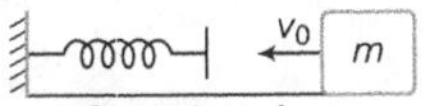 Smooth surface	q.	Work-energy theorem is applicable
iii.	A pendulum is released from rest from horizontal position in vertical plane 	r.	Kinematics formula are applicable
iv.	A block is in projectile motion in absence of air.	s.	Heat loss takes place

Codes

	i	ii	iii	iv
a.	q,r,s	p,q	p,q	p,q,r
b.	p,q	q,r	p,s	q
c.	p	q	r	s
d.	p	r	s	q

24 We move a particle along Z-axis, first outward $z = 1\,\text{m}$ to $z = 4\,\text{m}$ and then back to $z = 1\,\text{m}$, while an external force acts on it. The force is directed along Z-axis and its z-components can have different values for the outward trip and for the return trip. The net work done on the particle by the external force for round trip is given in Column II for each of four situations in Column I.

Match the Column I with Coumn II and mark the correct option from the given codes.

	Column I			Column II
	Outward	Inward		
i.	$+3\,\text{N}$	$+3\,\text{N}$	p.	0
ii.	$+5\,\text{N}$	$+5\,\text{N}$	q.	18 J
iii.	$+2x\,\text{N}$	$-2x\,\text{N}$	r.	30 J
iv.	$3x^2\,\text{N}$	$3x^2\,\text{N}$	s.	Force is conservative in nature

Codes

	i	ii	iii	iv
a.	q	p	r	p
b.	q	p,s	r	p,s
c.	p	q	r	s
d.	p	r	s	q

JEE Main & AIEEE Archive

(Compilation of Last 13 Years Questions)

1 When a rubber band is stretched by a distance x, it exerts a restoring force of magnitude $F = ax + bx^2$, where a and b are constants. The work done in stretching the unstretched rubber band by L is **(JEE Main 2014)**

a. $aL^2 + bL^3$ b. $\frac{1}{2}(aL^2 + bL^3)$

c. $\frac{aL^2}{2} + \frac{bL^3}{3}$ d. $\frac{1}{2}\left(\frac{aL^2}{2} + \frac{bL^3}{3}\right)$

2 At time $t = 0$, particle starts moving along the X-axis. If its kinetic energy increases uniformly with time t, the net force acting on it must be proportional to **(AIEEE 2011)**

a. $\sqrt{t}$ b. constant c. t d. $1/\sqrt{t}$

3 An athlete in the olympic games covers a distance of 100 m in 10 s. His kinetic energy can be estimated to be in the range **(AIEEE 2008)**

a. 200 J – 500 J b. 2×10^5 J – 3×10^5 J

c. 20000 J – 50000 J d. 2000 J – 5000 J

4 A 2 kg block slides on a horizontal floor with a speed of 4 m/s. It strikes uncompressed spring and compresses it till the block is motionless. The kinetic friction force is 15 N and spring constant is 10000 N/m. The spring compresses by **(AIEEE 2007)**

a. 5.5 cm b. 2.5 cm c. 11.0 cm d. 8.5 cm

5 The potential energy of a 1 kg particle free to move along the X-axis is given by

$$V(x) = \left(\frac{x^4}{4} - \frac{x^2}{2}\right) \text{ J}$$

The total mechanical energy of the particle is 2 J. Then, the maximum speed (in ms^{-1}) is **(AIEEE 2006)**

a. $\frac{3}{\sqrt{2}}$ b. $\sqrt{2}$ c. $\frac{1}{\sqrt{2}}$ d. 2

6 A bullet fired into a fixed target losses half of its velocity after penetrating 3 cm. How much further it will penetrate before coming to rest, assuming that it faces constant resistance to motion? **(AIEEE 2005)**

a. 3.0 cm b. 2.0 cm

c. 1.5 cm d. 1.0 cm

7 A body of mass m is accelerated uniformly from rest to a speed v in a time T. The instantaneous power delivered to the body as a function of time, is given by **(AIEEE 2005)**

a. $\frac{mv^2}{T^2}t$ b. $\frac{mv^2}{T^2}t^2$

c. $\frac{1}{2}\frac{mv^2}{T^2}t$ d. $\frac{1}{2}\frac{mv^2}{T^2}t^2$

8 A spherical ball of mass 20 kg is stationary at the top of a hill of height 100 m. It rolls down a smooth surface to the ground, then climbs up another hill of height 30 m and finally rolls down to a horizontal base at a height of 20 m above the ground. The velocity attained by the ball is **(AIEEE 2005)**

a. 40 m/s b. 20 m/s

c. 10 m/s d. $10\sqrt{30}$ m/s

9 The block of mass M moving on the frictionless horizontal surface collides with the spring of spring constant k and compresses it by length L. The maximum momentum of the block after collision is **(AIEEE 2005)**

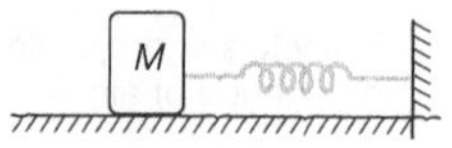

a. $\sqrt{MkL}$ b. $\frac{kL^2}{2M}$ c. zero d. $\frac{ML^2}{k}$

10 A particle moves in a straight line with retardation proportional to its displacement. Its loss of kinetic energy for any displacement x is proportional to **(AIEEE 2004)**

a. x^2 b. e^x c. x d. $\log_e x$

11 A uniform chain of length 2 m is kept on a table such that a length of 60 cm hangs freely from the edge of the table. The total mass of the chain is 4 kg. What is the work done in pulling the entire chain on the table? **(AIEEE 2004)**

a. 7.2 J b. 3.6 J c. 120 J d. 1200 J

12 A force $\mathbf{F} = (5\hat{\mathbf{i}} + 3\hat{\mathbf{j}} + 2\hat{\mathbf{k}})$ N is applied over a particle which displaces it from its origin to the point $\mathbf{r} = (2\hat{\mathbf{i}} - \hat{\mathbf{j}})$ m. The work done on the particle in joule is **(AIEEE 2004)**

a. −7 b. +7 c. +10 d. +13

13 A spring of spring constant 5×10^3 N/m is stretched initially by 5 cm from the unstretched position. Then, the work required to stretch it further by another 5 cm is **(AIEEE 2003)**

a. 12.50 N-m b. 18.75 N-m

c. 25.00 N-m d. 6.25 N-m

14 A body is moved along a straight line by a machine delivering a constant power. The distance moved by the body in time t is proportional to **(AIEEE 2003)**

a. $t^{3/4}$ b. $t^{3/2}$ c. $t^{1/4}$ d. $t^{1/2}$

15 A spring of force constant 800 N/m has an extension of 5 cm. The work done in extending it from 5 cm to 15 cm is **(AIEEE 2002)**

a. 16 J b. 8 J c. 32 J d. 24 J

JEE Advanced & IIT JEE Archive

(Compilation of Last 10 Years Questions)

1 A tennis ball is dropped on a horizontal smooth surface. It bounces back to its original position after hitting the surface. The force on the ball during the collision is proportional to the length of compression of the ball. Which one of the following sketches describes the variation of its kinetic energy K with time t most appropriately? The figures are only illustrative and not to the scale. **(2014 Adv., Only One Option Correct Type)**

a. 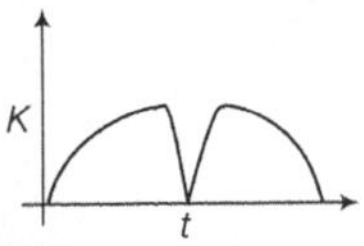

b. 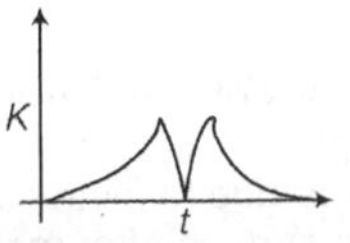

c. 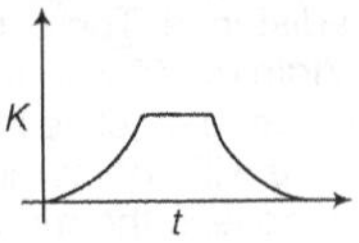

d. 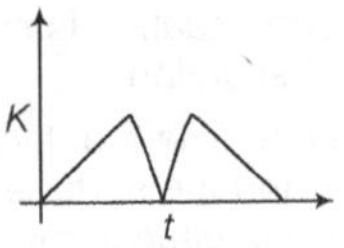

2 Consider an elliptically shaped rail PQ in the vertical plane with $OP = 3\,m$ and $OQ = 4\,m$. A block of mass 1 kg is pulled along the rail from P to Q with a force of 18 N, which is always parallel to line PQ (see figure). Assuming no frictional losses, the kinetic energy of the block when it reaches Q is $(n \times 10)$ J. The value of n is (take, acceleration due to gravity = 10 ms^{-2}) **(2014 Adv., Integer Type)**

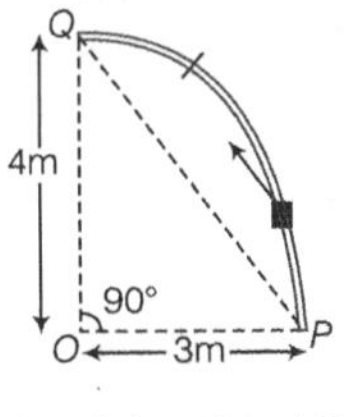

3 A pulse of light of duration 100 ns is absorbed completely by a small object initially at rest. Power of the pulse is 30 mV and the speed of light is 3×10^8 ms^{-1}. The final momentum of the object is **(2013 Adv., Only One Option Correct Type)**

a. 0.3×10^{-17} kg-ms^{-1}
b. 1.0×10^{-17} kg-ms^{-1}
c. 3.0×10^{-17} kg-ms^{-1}
d. 9.0×10^{-17} kg-ms^{-1}

4 The work done on a particle of mass m by a force, $K\left[\frac{x}{(x^2+y^2)^{3/2}}\,\hat{i} + \frac{y}{(x^2+y^2)^{3/2}}\,\hat{j}\right]$

(K being a constant of appropriate dimensions), when the particle is taken from the point $(a, 0)$ to the point $(0, a)$ along a circular path of radius a about the origin in the xy-plane is **(2013 Adv., Only One Option Correct Type)**

a. $\frac{2K\pi}{a}$ b. $\frac{K\pi}{a}$
c. $\frac{K\pi}{2a}$ d. 0

Passage for Q. Nos. (5-6)

A small block of mass 1 kg is released from rest at the top of a rough track. The track is a circular arc of radius 40 m. The block slides along the track without toppling and a frictional force acts on it in the direction opposite to the instantaneous velocity. The work done in overcoming the friction up to the point Q, as shown in the figure, is 150 J. (take the acceleration due to gravity, $g = 10$ ms^{-2}) **(2013 Adv., Comprehension Type)**

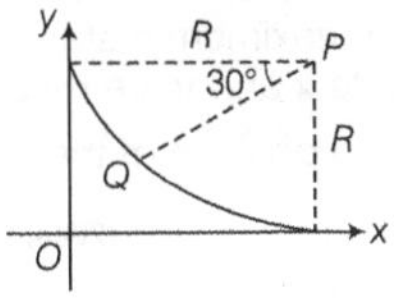

5 The speed of the block when it reaches the point Q is

a. 5 ms^{-1} b. 10 ms^{-1}
c. $10\sqrt{3}$ ms^{-1} d. 20 ms^{-1}

6 The magnitude of the normal reaction that acts on the block at the point Q is

a. 7.5 N
b. 8.6 N
c. 11.5 N
d. 22.5 N

7 A block of mass 0.18 kg is attached to a spring of force constant 2 N/m. The coefficient of friction between the block and the floor is 0.1. Initially the block is at rest and the spring is unstretched.

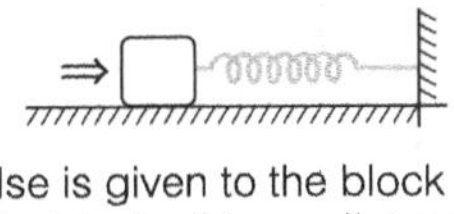

An impulse is given to the block as shown in the figure. The block slides a distance of 0.06 m and comes to rest for the first time. The initial velocity of the block in m/s is $v = \frac{N}{10}$. Then, N is **(2011, Descriptive Type)**

8 A block of mass 2 kg is free to move along the X-axis. It is at rest and from $t = 0$ onwards it is subjected to a time-dependent force $F(t)$ in the x-direction. The force $F(t)$ varies with t as shown in the figure. The kinetic energy of the block after 4.5 s is

(2010, Only One Option Correct Type)

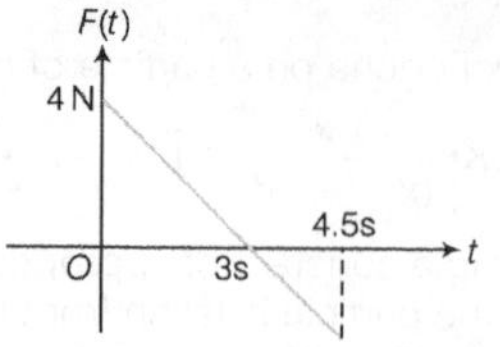

a. 4.50 J b. 7.50 J
c. 5.06 J d. 14.06 J

9 A light inextensible string that goes over a smooth fixed pulley as shown in the figure connects two blocks of masses 0.36 kg and 0.72 kg. Taking $g = 10\ \text{ms}^{-2}$, find the work done (in Joule) by string on the block of mass 0.36 kg during the first second after the system is released from rest.

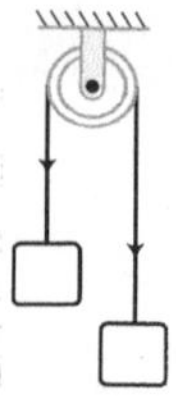

(2009, Descriptive Type)

10 A block B is attached to two unstretched springs S_1 and S_2 with spring constants k and $4k$, respectively. The other ends are attached to two supports M_1 and M_2 not attached to the walls. The springs and supports have negligible mass. There is no friction anywhere. The block B is displaced towards wall 1 by a small distance x and released. The block returns and moves a maximum distance y towards wall 2. Displacements x and y are measured with respect to the equilibrium position of the block B. The ratio $\frac{y}{x}$ is

(2008, Only One Option Correct Type)

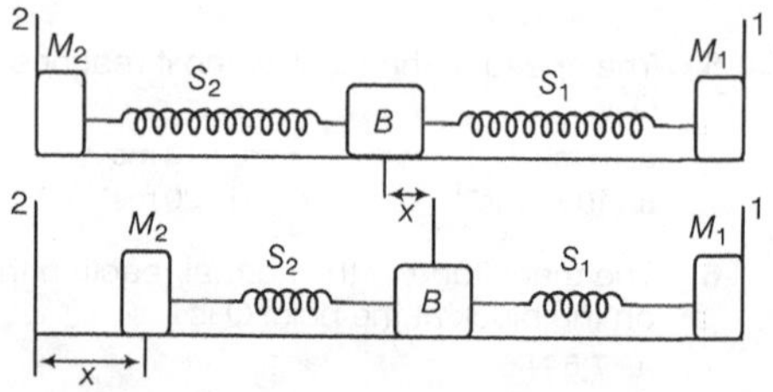

a. 4 b. 2 c. $\frac{1}{2}$ d. $\frac{1}{4}$

11 A bob of mass M is suspended by a massless string of length L. The horizontal velocity v at position A is just sufficient to make it reach the point B. The angle θ at which the speed of the bob is half of that at A, satisfies

(2008, Only One Option Correct Type)

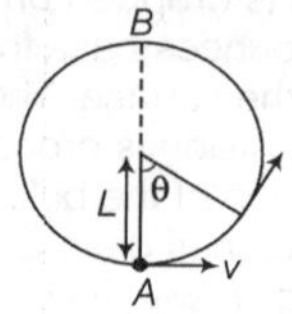

a. $\theta = \frac{\pi}{4}$

b. $\frac{\pi}{4} < \theta < \frac{\pi}{2}$

c. $\frac{\pi}{2} < \theta < \frac{3\pi}{4}$

d. $\frac{3\pi}{4} < \theta < \pi$

12 **Statement I** A block of mass m starts moving on a rough horizontal surface with a velocity v. It stops due to friction between the block and the surface after moving through a certain distance. The surface is now tilted to an angle of 30° with the horizontal and the same block is made to go up on the surface with the same initial velocity v. The decrease in the mechanical energy in the second situation is smaller than that in the first situation.

Statement II The coefficient of friction between the block and the surface decreases with the increase in the angle of inclination. **(2007, Statement Type)**

a. Both Statement I and Statement II are correct and Statement II is the correct explanation of Statement I
b. Both Statement I and Statement II are correct and Statement II is not the correct explanation of Statement I
c. Statement I is correct but Statement II is incorrect
d. Statement II is correct but Statement I is incorrect

Answers *with* Explanations

Answers

1. Laws of Motion

DPP-1 Normal Reaction

1. (i) mg, (ii) 120 N, (iii) 80 N, (iv) 180 N 2. (i) 10 N, (ii) 20 N, (iii) $20/\sqrt{3}$ N, (iv) 10 N 3. *d*
4. *b* 5. *a* 6. *c* 7. *b* 8. *a* 9. *c* 10. *b* 11. *a* 12. *c* 13. *c*
14. *a* 15. *b* 16. *a,b* 17. *a,b* 18. *b,c* 19. *b,d* 20. *b* 21. *c* 22. *a* 23. *d*
24. *d* 25. *a* 26. *a* 27. *b* 28. 2 29. 3

DPP-2 Relation between Force and Acceleration

1. (i) 1 m/s^2, (ii) 2 m/s^2, (iii) 6 m/s^2, (iv) 10 m/s^2 2. *d* 3. *d* 4. *a* 5. *d* 6. *b*
7. *a* 8. *b* 9. *c* 10. *c* 11. *d* 12. *b* 13. *d* 14. *a,b,c* 15. *b,d*
16. *a,c,d* 17. *a* 18. *c* 19. *b* 20. *d* 21. *c* 22. *b* 23. *a* 24. *c* 25. 3
26. 1 27. 6 28. *a*

DPP-3 Tension in String

1. (i) 100 N, (ii) $\frac{m_1 F}{m_1 + m_2}$, (iii) 66.67 N, (iv) 20 N, (v) 5 N, (vi) 8 N
2. (i) 63.3 N, (ii) 50 N, (iii) $\frac{32\sqrt{5}}{3}$ N, (iv) $\sqrt{3}\left(1+\frac{1}{\sqrt{2}}\right)g$
3. (i) 3 : 2, (ii) 3 : 1, (iii) 5 : 2, (iv) 23 : 57 4. (i) 1 : 1 : 1, (ii) 1 : 1 : 2 cos α, (iii) 3 : 4 : 5, (iv) 5 : 3 : 4
5. (i) 10 N, (ii) 30 N, (iii) 30 N, (iv) 50 N, (v) 12 N 6. *b* 7. *b* 8. *a* 9. *c* 10. *c*
11. *c* 12. *d* 13. *c* 14. *a,b,c* 15. *a,b,c,d* 16. 2 17. 5 18. 1 19. *b*
20. *c* 21. *a* 22. *d* 23. *b* 24. *a* 25. *b* 26. *b* 27. *d* 28. *a*

DPP-4 Constraint Motion or Dependent Motion

1. (i) $u = v$, (ii) $u\cos\theta_2 = v\cos\theta_1$, (iii) $u = v\cos\theta$, (iv) $u\cos\theta = v$, (v) $v = \frac{xu}{\sqrt{x^2 + l^2}}$, (vi) $v = \frac{4}{3}u$
2. (i) $v = 2u$, (ii) $u = 2v$, (iii) $u = 2v$, (iv) $u = 2v_B$
3. (i) $v_1 + v_2 = 2v_3$, (ii) $v_2 + v_3 = 2v_1$, (iii) $v_1 = \frac{2v_3 + v_2}{2}$, (iv) $v_2 = v_3 = 2v_1$
4. (i) $a_2 = 3a_1$, (ii) $a_A = \frac{3}{2}a_B$, (iii) $a_A = \frac{4}{5}a_B$, (iv) $a_A = \frac{3}{2}a_B$, (v) $a_A = \frac{a_0 + a_B}{2}$, (vi) $a_B = 8a_A$
5. (i) $v_B = 2v_A\cos\alpha$, (ii) $v_B = \frac{v_A}{\sqrt{2}}$, (iii) $v_A = v_B$, (iv) $v_B = \left(\frac{s + \sqrt{2}x}{x + \sqrt{2}s}\right)v_A$, (v) $v_B = \frac{2x}{\sqrt{x^2 + h^2}}v_A$
6. (i) $a_1 = \frac{F}{4m_2 + m_1}$, $a_2 = \frac{2F}{m_1 + 4m_2}$, (ii) $a_1 = \frac{6g}{17}$, $a_2 = \frac{9g}{17}$, (iii) $a_A = 2.8$ m/s^2, $a_B = 5.6$ m/s^2;
(iv) $a_1 = a_2 = a_3 = \frac{2}{3}$ m/s^2, (v) $a_1 = \frac{240}{29}$ m/s^2,
7. (i) $a_B = \sqrt{5}\,a_A$, (ii) $v_B = v_A\sqrt{\frac{2x^2 + h^2}{x^2 + h^2}}$, (iii) $a_1 = 2$ m/s^2 8. *c* 9. *b* 10. *a* 11. *a*
12. *a* 13. *d* 14. *c* 15. *b* 16. *c* 17. *a,b* 18. *a,b* 19. *a* 20. *b* 21. *c*
22. *b* 23. *a* 24. 2 25. 2

DPP-5 Wedge Constraints

1. *b* 2. *b* 3. *d* 4. *c* 5. *d* 6. *d* 7. *b* 8. *b* 9. *c*
10. *a,b,c* 11. *a,b* 12. *c,d* 13. *c,d* 14. *a,d* 15. (i) $a_1 = \frac{3}{4}a_2$, (ii) 3 m/s^2; (iii) 100 N
16. (i) $a_A = \frac{3a_B}{5}$, (ii) 5.29 m/s^2, (iii) 4.71 N 17. (i) $\frac{9}{4}$ m/s^2, (ii) 5 N (iii) 3 N, (iv) $\frac{16}{31}$ kg
18. (i) $a_1 = \sqrt{3}a_3$, (ii) 2s, (iii) $10\sqrt{3}$ m/s 19. (i) $a_2 = \frac{3}{5}a_1$, (ii) 7 m/s^2, (iii) 1 N 20. 3.75 m/s^2
21. (i) 4 N, (ii) 2 m/s^2

DPP-6 Spring

1. (i) 0.2 m, (ii) 0.2 m, (iii) 0.2 m, (iv) 0.2 m
2. (i) 100 N/m, (ii) 200 N/m, (iii) $k_{eq} = k_1 + k_2$, (iv) 200 N/m, (v) 10 N/m
3. (i) 0.5 m, (ii) 0.4 m, (iii) 0.6 m
4. (i) 400 N/m, (ii) 600 N/m, 300 N/m, 200 N/m, (iii) 3 k, (iv) $\frac{5}{2}$k
5. (i) 15 m/s^2, (ii) 10 m/s^2, (iii) 5 m/s^2, (iv) 10 m/s^2
6. (i) $a_c = 0, a_B = 10\ m/s^2$, (ii) $a_1 = 10\ m/s^2, a_2 = 0$; (iii) g, 40 m/s^2 (iv) $-g/2$, (v) $a_1 = 15\ m/s^2, a_2 = 0$, (vi) $a_1 = 0, a_2 = 5\ m/s^2$; (vii) $a_1 = a_2 = a_3 = 0$
7. (i) 10 kg, (ii) 10 kg both, (iii) 12 kg both, (iv) 2 kg, (v) 2.4 kg, (vi)24 N
8. (i) $\sqrt{\frac{m}{k}}v_0$, (ii) $\frac{2F}{k}$, (iii) $\frac{2Mg}{k}$ 9. *c* 10. *b* 11. *c* 12. *b* 13. *a* 14. *c* 15. *b*

16. *d* 17. *b* 18. *c,d* 19. *a,b,d* 20. *a* 21. *b* 22. *b* 23. 1 24. 2 25. 8
26. *a*

DPP-7 Pseudo Force

1. (i) $-150\,\hat{i} - 400\,\hat{j}$, (ii) 0, (iii) Non-inertial, (iv) False, (v) non-inertial
2. (i) 2 N, (ii) 7.5 m/s^2, (iii) 13.3 m/s^2, (iv) $g\tan\alpha$, (v) 0.5 m
3. (i) 2 s, (ii) 2 s, (iii) 0.2 s (iv) $\frac{u^2 \sin 2\alpha}{g\cos\theta} + \frac{4u^2 \sin^2\alpha \sin\theta}{g\cos^2\theta}$
4. (i) 45°, (ii) $\tan^{-1}(2/3)$, (iii) 2 : 3, (iv) 4 N, (v) (a) 2 g, (b) 50 N

5. *c* 6. *a* 7. *d* 8. *a* 9. *b* 10. *a,b,c* 11. *a,b* 12. *a,b,d* 13. *b,c* 14. *c*
15. *b* 16. *a* 17. *b* 18. *a* 19. 4 20. 4 21. 0 22. 2 23. 2

DPP-8 Problems Related to Climb on a String

1. (i) 100 N, (ii) 200 N, (iii) 60 N, (iv) 56 N
2. (i) 10 m/s^2, (ii) 10 m/s^2, (iii) 2 m/s^2
3. (i) 180 N, (ii) 15 m/s^2, (iii) 30 m/s^2
4. (i) 10 m/s^2; (ii) 12 m/s^2; (iii) 8 m/s^2

5. *c* 6. *b* 7. *b* 8. *c* 9. *d* 10. *c* 11. *c* 12. *a* 13. *a,c* 14. *b,d*
15. *b,d* 16. *b,d* 17. *c* 18. *b* 19. *a* 20. 2

Revisal Problems (JEE Main)

1. *c* 2. *a* 3. *c* 4. *c* 5. *b* 6. *b* 7. *c* 8. *b* 9. *c* 10. *c*
11. *a* 12. *d* 13. *b* 14. *c* 15. *a* 16. *a* 17. *b* 18. *c* 19. *c* 20. *d*

Revisal Problems (JEE Advanced)

1. *c* 2. *a* 3. *b* 4. *b* 5. *b* 6. *c* 7. *d* 8. *a* 9. *a,b,c,d*
10. *a,b,c,d* 11. *a.b.c,d* 12. *a,b,c* 13. *a,b,d* 14. *b* 15. *a* 16. *b* 17. *d*
18. 6 19. 5 20. 3 21. 4 22. 3 23. *b* 24. *b* 25. *d* 26. *d* 27. *a*
28. *a*

JEE Main & AIEEE Archive

1. *b* 2. *d* 3. *c* 4. *d* 5. *d* 6. *c* 7. *d* 8. *d* 9. *d* 10. *d*
11. *d* 12. *a* 13. *d* 14. *c* 15. *d* 16. *a* 17. *a* 18. *c* 19. *b* 20. *a*
21. *b* 22. *c*

JEE Advanced & IIT JEE Archive

1. *d* 2. *c* 3. *a* 4. *a,c* 5. *d* 6. *a,c* 7. *b* 8. *b* 9. *b* 10. *a*
11. *d* 12. 5

2. Friction

DPP-1 Basic Concept of Static Friction

1. (i) True, (ii) True, (iii) True, (iv) True, (v) True, (vi) True, (vii) True, (viii) False, (ix) True, (x) True, (ix) True
2. (i) 10 N, (ii) 102 N, (iii) 100.50 N, (iv) 46.82 N, (v) 31.6 N, (vi) 20 N, (vii) 67.08 N
3. (i) 10 N, (ii) $10\sqrt{2}$ N, (iii) 2 N, (iv) 15.62 N
4. (i) 50 N, (ii) 15 N, (iii) 25 N, (iv) 20 N, (v) 2 N
5. (i) 5 m / s^2 (ii) 1 m / s^2 (iii) 5 N

6. *a* 7. *b* 8. *d* 9. *a* 10. *a* 11. *a* 12. *a* 13. *a* 14. *a* 15. *a* 16. *b* 17. *c* 18. *c* 19. *b* 20. *b* 21. *d* 22. *b* 23. *a,b* 24. *a, b* 25. *a, d* 26. *b* 27. *c* 28. 5 29. 1 30. 1

DPP-2 Basic Concept of Kinetic Friction

1. (i) True, (ii) True, (iii) True, (iv) False, (v) True, (vi) False, (vii) True, (viii) False
2. (i) 4 N, (ii) 4 N, (iii) 1.6 N, (iv) 6.4 N, (v) 10 N, (vi) 3.2 N

3. *c* 4. *a* 5. *c* 6. *a* 7. *c* 8. *a* 9. *a* 10. *a* 11. *a* 12. *b* 13. *a* 14. *c* 15. *c* 16. *a* 17. *a* 18. *a,b* 19. *a,b* 20. *a,d* 21. *a* 22. *b* 23. *a* 24. 5 25. 1 26. 6 27. *a* 28. *c* 29. *b* 30. *c*

DPP-3 Analysis of Friction

1. *c* 2. *a* 3. *d* 4. *b* 5. *a,c,d* 6. *a,b,c,d* 7. *a,b,c* 8. *a,b,c,d* 9. *a,b,c,d* 10. *a,b,c,d* 11. *a,b,c,d* 12. *a* 13. *a* 14. *c* 15. *a* 16. *c* 17. *c* 18. *b* 19. *d* 20. *a* 21. *a* 22. *b* 23. *b* 24. *a* 25. *b* 26. 3

27. (i) 174.1 N (ii) 1.3 m / s^2
28. (i) 0 (ii) $5(\sqrt{2}-1)$ m / s^2
29. (i) zero, (ii) $\frac{2\sqrt{2}mg}{3}$, (iii) $\frac{mg}{3\sqrt{2}}$
30. (i) 1.30 m / s^2, (ii) 5.24 N

DPP-4 Blocks in Contact

1. (i) 7.5 N, (ii) 15 N, (iii) 15 N, (iv) 5 N, (v) 22.5 N, (vi) 1.5 N, (vii) 6 N, (viii) 4.54 N
2. (i) 1.5, (ii) 2.9
3. (i) 4.67 m / s^2, (ii) 1 m / s^2
4. (i) $\frac{1}{2}\mu_k gt^2$, (ii) $v_0 \geq \mu_k g\left(1+\frac{d}{2r}\right)$
5. (i) 5 m / s^2, (ii) 4 N, (iii) 25 N,
6. 5.2 N
7. (i) 4.2 kg, (ii) 9.8 N
8. (ii) $f_1 = 30$ N, $f_2 = 15$ N, $T = 18$ N, $a = \frac{3}{5}$ m / s^2

9. *a* 10. *d* 11. *a* 12. *a* 13. *c* 14. *b* 15. *a,b,c* 16. *a,b,d* 17. *a,b,c,d* 18. *b* 19. *a,b,c* 20. *c* 21. *b* 22. *d* 23. *d* 24. *a* 25. *a* 26. *a* 27. 6 28. 4 29. 1 30. *a* 31. *a* 32. *a* 33. *b* 34. *d* 35. *b* 36. *b*

DPP-5 Friction in Daily Life

1. (i) a, (ii) b, (iii) a, (iv) a, (v) a, (vi) b, (vii) b, (viii).b, (ix) a, (x) c

2. *c* 3. *a* 4. *a* 5. *d* 6. *a* 7. *d* 8. *c* 9. *a* 10. *a* 11. *b* 12. *a* 13. *a* 14. *a,d* 15. *a,b,c* 16. *a,b,c* 17. *a,b* 18. *a* 19. *d* 20. *d* 21. 6 22. 3 23. *a* 24. *b*

25. (i) 5 m / s^2, (ii) 1500 N
26. (i) 15 m/s, (ii) 10.8 km
27. (i) 3 m / s^2, (ii) 5 m / s^2
28. (i) 30 m / s^2, (ii) 66 m / s^2

Revisal Problems (JEE Main)

1. *c* 2. *d* 3. *b* 4. *a* 5. *b* 6. *d* 7. *a* 8. *a* 9. *a* 10. *c* 11. *c* 12. *d* 13. *d* 14. *d* 15. *d* 16. *d* 17. *d* 18. *c* 19. *a* 20. *d* 21. *a* 22. *c* 23. *a* 24. *a* 25. *a* 26. *d*

Revisal Problems (JEE Advanced)

1. *b* 2. *b* 3. *a* 4. *b* 5. *a* 6. *b* 7. *a,b* 8. *a,b,d* 9. *a,b,c* 10. *a,b* 11. *c* 12. *a* 13. *a* 14. *a* 15. *b* 16. b 17. 3 18. 4 19. 1 20. 1 21. *a*

JEE Main & AIEEE Archive

1. *a* 2. *d* 3. *b* 4. *a* 5. *c* 6. *a* 7. *d* 8. *c* 9. *b* 10. *b*

JEE Advanced & IIT JEE Archive

1. *d* 2. 5 3. *a* 4. *b* 5. 10

3. Circular Motion

DPP-1 Kinematics of Circular Motion

1. (i) axis (ii) vector (iii) scalar (iv) commutative (v) accelerated (vi) $\hat{\mathbf{k}}$ (vii) $\mathbf{v} = \omega \times r$ (viii) directed towards the centre of circular path (ix) not (x) tangential acceleration

2. (i) *a* (ii) *b* (iii) *a* (iv) *b* (v) *b* (vi) *b* (vii) *b*

3. *b* 4. *a* 5. *b* 6. *c* 7. *b* 8. *d* 9. *b* 10. *c*

11. *b* 12. *b* 13. *c* 14. *a,c* 15. *a,b* 16. *a,b* 17. *a,b,d* 18. *a,b,c* 19. *a,c,d* 20. *d*

21. *a* 22. 2 23. 2 24. 5 25. *a* 26. *d* 27. *a*

28. (i) $10^4 \ln 2$ s (ii) $-1.2\pi \times 10^{-2}$ rad/s^2 (iii) $\dfrac{0.6\times 10^3 [1-e^{-t/t_0}]}{2\pi}$ revolutions

29. (i) $\dfrac{5}{6}$ rad/s (ii) $\dfrac{5}{12}$ rad (iii) $\dfrac{5}{6}$ rad/s^2 30. $-mv_2 \sin\dfrac{v_2}{R}t\,\hat{\mathbf{i}} + m\left(\dfrac{v_2 \cos v_2 t}{R} - v_1\right)\hat{\mathbf{j}}$

DPP-2 Uniform Circular Motion

1. (i) a,c,d (ii) a,b,c,d (iii) a,b,c,d (iv) a (iv) b

2. (i) $mg - \dfrac{mv^2}{R}$ (ii) $mg + \dfrac{mv^2}{R}$ (iii) $mg\cos\theta - \dfrac{mv^2}{R}$ (iv) $mg\cos\theta + \dfrac{mv^2}{R}$ (v) $2mg$

3. (i) $mg + \dfrac{mv_0^2}{l}$ (ii) $mg\cos\theta + \dfrac{mv_0^2}{l}$ (iii) $\dfrac{mv_0^2}{l}$ (iv) $\dfrac{mv_0^2}{l} - mg\cos\alpha$ (v) $\dfrac{mv_0^2}{l} - mg$ (vi) mg

4. *d* 5. *b* 6. *b* 7. *b* 8. *c* 9. *a* 10. *c* 11. *d* 12. *c* 13. *b*

14. *b* 15. *b* 16. *c* 17. *d* 18. *c* 19. *c,d* 20. *a,b,c,d*

21. *b* 22. *a* 23. 2 24. 3 25. *a* 26. *d* 27. *a* 28. *a*

29. (i) $\dfrac{2\pi}{5}\sqrt{3}$ s (ii) 2.17 N 30. (i) $h = R - \dfrac{g}{\omega^2}$, 9.89 rad/s (ii) 9.8×10^{-3} m/s^2

31. (i) f is positive, friction is towards outer wheels
(ii) f is negative, friction is towards inner wheels

DPP-3 Non-Uniform Circular Motion

1. *b* 2. *d* 3. *d* 4. *b* 5. *a* 6. *b* 7. *a* 8. *a,b,c,d* 9. *a,b,c,d*

10. *a,c* 11. *c,d* 12. *d* 13. *a* 14. *d* 15. *b* 16. *a* 17. *b* 18. *d* 19. 4

20. 2 21. 2 22. 1 23. 5 24. *a* 25. $5\sqrt{2}$ m/s 26. (ii) $\dfrac{5\sqrt{3}}{8}g$, $\dfrac{3mg}{8}$

DPP-4 Centrifugal Force

1. *a* 2. *b* 3. *c* 4. *c* 5. *c* 6. *a* 7. *b* 8. *a* 9. *c* 10. *c*

11. *a,b* 12. *c,d* 13. *a,d* 14. *a* 15. *d* 16. *b* 17. *b* 18. 2 19. 5 20. *a*

21. *b* 22. *a* 23. (i) 36 N (ii) 11.67 rad/s

DPP-5 Motion on General Curve

1. (i) *a* (ii) *b* (iii) *c* (iv) *a* (v) *b* (vi) *a,c*

2. (i) *b* (ii) *c* (iii) *a* (iv) *b* (v) *c* (vi) *a* (vii) *a* (viii) *b*

3. *a* 4. *c* 5. *a* 6. *d* 7. *b* 8. *b* 9. *a* 10. *c* 11. *d* 12. *c,d*

13. *a,b* 14. *a,b,c* 15. *d* 16. *a* 17. 3 18. 4 19. 6 20. 4 21. 9 22. *d*

23. *a* 24. *b* 25. *c* 26. *a* 27. *b* 28. *a* 29. *a* 30. *a* 31. *b* 32. *a*

33. *c* 34. *b*

Revisal Problems (JEE Main)

1. c 2. a 3. a 4. a 5. b 6. b 7. c 8. b 9. b 10. b
11. c 12. c 13. b 14. d 15. a 16. b 17. b 18. c 19. d 20. b
21. c 22. d 23. a 24. b 25. a

Revisal Problems (JEE Advanced)

1. b 2. b 3. b 4. b 5. b 6. b 7. b 8. a,b,c 9. a,b,c,d
10. a,b,d 11. a,b,c 12. a,b,c,d 13. a,b 14. a,c 15. b 16. c 17. c 18. b
19. d 20. b 21. 1/2 22. 3 23. 1 24. 1 25. 1 26. 3 27. 8 28. c
29. d 30. a 31. b

JEE Main & AIEEE Archive

1. c 2. b 3. c 4. d 5. d 6. d 7. b 8. c

JEE Advanced & IIT JEE Archive

1. a 2. c 3. d 4. a 5. $-mv_2 \sin\frac{v_2}{R}t\,\hat{\mathbf{i}} + m\left(v_2\cos\frac{v_2 t}{R} - v_1\right)\hat{\mathbf{j}}$

4. Work, Power and Energy

DPP-1 Kinetic Energy

1. (i) b, (ii) b, (iii) a, (iv) d 2. a 3. d 4. c 5. b 6. a 7. b 8. a
9. c 10. a 11. d 12. a 13. b 14. d 15. b 16. a,b 17. a,b
18. a,d 19. c,d 20. (i) 5, (ii) 2, (iii) 4 21. (i) 2, (ii) 2, (iii) 0 22. 6 23. 2 24. a 25. b
26. c 27. a 28. a 29. a

DPP-2 Work Done by Constant Force

1. (i) a (ii) a (iii) c (iv) c (v) a (vi) d
2. (i) b (ii) b (iii) b (iv) a (v) c (vi) b (vii) b (viii) c (ix) a,b,c,d
3. c 4. a 5. b 6. c 7. b 8. d 9. a 10. b 11. c 12. b
13. c 14. a 15. a,b,c,d 16. a,b,c,d 17. a,b,c,d 18. a,c
19. a,c,d 20. a,b 21. a 22. a,b,c 23. b 24. b 25. a 26. b 27. a 28. c
29. a 30. c 31. c 32. d 33. 8 34. (i) 3, (ii) 0, (iii) 2, (iii) 2, (iv) 2, (v) 2
35. (i) 3, (ii) 2 36. (i) 0, (ii) $x=3$ 37. 4 38. 2 39. a 40. a 41. a

DPP-3 Work Done by Variable Force

1. a 2. c 3. c 4. d 5. c 6. a 7. d 8. a 9. b 10. c
11. b 12. c 13. b 14. b 15. a 16. d 17. b,d 18. a,b,c 19. a,d
20. a,b,c 21. c 22. (i) 6, (ii) 2, (iii) 3 23. 2 24. 8 25. 8 26. 2 27. b
28. b 29. b 30. b 31. c 32. a

DPP-4 Work Energy Theorem

1. b 2. d 3. b 4. b 5. c 6. a 7. c 8. a 9. b 10. b,c
11. a,c 12. b,c 13. a,b,c 14. a,b,c,d 15. a 16. a 17. c 18. a 19. a
20. c 21. 1 22. 2 23. 3 24. 5 25. 3 26. 2 27. 5 28. (a) 9 m, (b) 0
29. 14.57 N 30. 9.33 cal 31. 409.8 m/s 32. 4.24 m 33. a

DPP-5 Potential Energy

1. (i) True (ii) True (iii) True (iv) True (v) True (vi) True (vii) True (viii) True (ix) True
2. (i) a (ii) c (iii) b (iv) c (v) a (vi) a (vii) d
3. (i) a (ii) a (iii) c (iv) c (v) b (vi) a
4. (i) a (ii) a (iii) d 5. a 6. a 7. b 8. c 9. a 10. d 11. a

12. *b* 13. *c* 14. *a* 15. *a* 16. (i) *a,b* (ii) *b* 17. *a,b,c,d* 18. *a* 19. *b*
20. *a,b* 21. *a,b* 22. *a,c* 23. *d* 24. 4 25. 6 26. 1 27. 1 28. *b* 29. *a*
30. *c* 31. *c* 32. *a* 33. *a* 34. *a* 35. *b* 36. *d*

DPP-6 Power and Other form of Energy

1. (i) *b* (ii) *b* (iii) *c* (iv) *c* 2. (i) *b* (ii) *a*
3. (i) *b* (ii) *a,b,c* (iii) *c* (iv) *b,c* (v) *a,b,c* (vi) *a,b,c* 4. *a* 5. *c* 6. *a* 7. *d*
8. *a* 9. *c* 10. *b* 11. *b* 12. *a* 13. *b* 14. *d* 15. *c* 16. *c* 17. *a,d*
18. *b,c* 19. *b,d* 20. *a,d* 21. *c* 22. *a* 23. *b* 24. *c* 25. *a* 26. *c* 27. 2
28. 5 29. 2 30. 67 HP 31. (i) –W (ii) -54.38×10^{-3} W (iii) 1.05 W

DPP-7 Mechanical Energy

1. *d* 2. *b* 3. *b* 4. *d* 5. *c* 6. *a* 7. *a* 8. *c* 9. *c* 10. *b*
11. *c* 12. *b* 13. *b* 14. *d* 15. *b* 16. *a* 17. *b* 18. *d* 19. *b* 20. *a*
21. *a* 22. *b,c* 23. *a* 24. *a,b,d* 25. *c,d* 26. *a,b,c,d* 27. *a,b,c* 28. *c* 29. *d*
30. *c* 31. *d* 32. 33. *b* 34. *a* 35. *a* 36. *a* 37. 1 38. (i) 2 (ii) 2
39. 5 40. 5 41. 5 42. 5 43. 2 44. 2 45. 1 46. (i) $1:\sqrt{2}$ (ii) 100 m/s^2
47. (i) 1 m/s (ii) 8.66 cm 48. (i) $\frac{mg}{k}$, (ii) $\frac{g}{2}\sqrt{m/k}$ 49. 3.29 m/s 50. 43.8 m
51. *a*

DPP-8 Vertical Circular Motion

1. (i) 6 mg, (ii) $\frac{4\pi}{3}$ m, (iii) $\frac{8}{5}$
2. (i) *b,d*, (ii) *b*, (iii) *c*, (iv) *b*, (v) *a,b,c,d*, (vi) *a*, (vii) *c*, (viii) *a*, (ix) *b* (x) *a,b* (xi) *a,c*
3. (i) *b* (ii) *a* (iii) *a* (iv) *a,b* (v) *a* (vi) *b* (vii) *a,b,c* (viii) *b*
4. (i) *a* (ii) *a* (iii) *b* (iv) *a,b,c* (v) *a,c* (vi) *a,c* (vii) *a* (viii) *c* (ix) *a* (x) *b*
5. (i) *b,c* (ii) *c* (iii) *b* (iv) *a,b,d* (v) *a,b*
6. (i) *a* (ii) *c* (iii) *a* (iv) *a* (v) *a* (vi) *a* 7. *d* 8. *c* 9. *d* 10. *d*
11. *b* 12. *b* 13. *c* 14. *c* 15. *b,c* 16. *a,c* 17. *a,b* 18. *b,c,d* 19. *b,c* 20. *a*
21. *c* 22. *a* 23. *b* 24. *a* 25. *a* 26. *b* 27. *d* 28. *b* 29. 9 30. 3
31. 2 32. 4 33. 5 34. 1 35. (i) 6.1 N (ii) $\frac{2}{3}$ 36. $u=\sqrt{gL\left(2+\frac{3\sqrt{3}}{2}\right)}$
37. $N=mg(3\cos\theta-2)$

Revisal Problems (JEE Main)

1. *d* 2. *b* 3. *b* 4. *c* 5. *c* 6. *b* 7. *a* 8. *d* 9. *b* 10. *b*
11. *c* 12. *b* 13. *a* 14. *d* 15. *c* 16. *b* 17. *c* 18. *b* 19. *b* 20. *d*
21. *d* 22. *a* 23. *b* 24. *a* 25. *b*

Revisal Problems (JEE Advanced)

1. *c* 2. *b,c* 3. *c* 4. *a* 5. *d* 6. *a,b,c* 7. *a,b,c,d* 8. *a,b,c,d* 9. *a,b,d* 10. *b,d*
11. *a,b,c* 12. *b* 13. *b* 14. *c* 15. *a* 16. *c* 17. *c* 18. 2 19. 5 20. 4
21. 3 22. 4 23. *a* 24. *b*

JEE Main & AIEEE Archive

1. *c* 2. *d* 3. *d* 4. *a* 5. *a* 6. *d* 7. *a* 8. *a* 9. *a* 10. *a*
11. *b* 12. *b* 13. *b* 14. *b* 15. *b*

JEE Advanced & IIT JEE Archive

1. *b* 2. 5 3. *b* 4. *d* 5. *b* 6. *a* 7. 4 8. *c* 9. 8 10. *c*
11. *d* 12. *c*

1. Laws of Motion

DPP-1 Normal Reaction

1. (i) From free body diagram of the block,

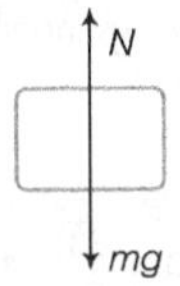

From above diagram, $N = mg$

(ii)

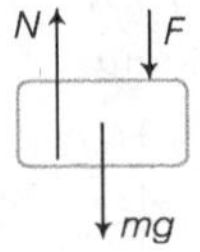

From free body diagram, $N = mg + F = 10 \times 10 + 20 = 120$ N

(iii)

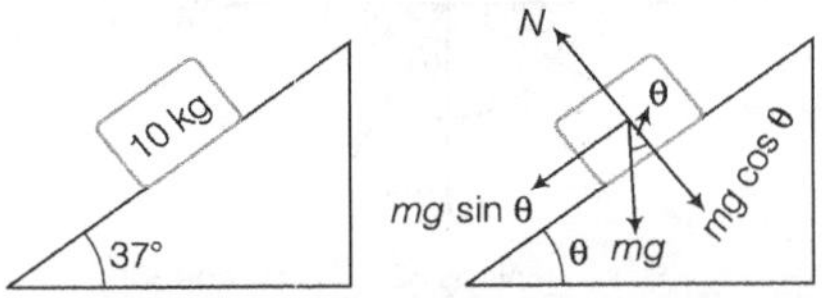

$$N = mg\cos\theta = 10 \times 10 \times \cos 37°$$

$$\approx 10 \times 10 \times \frac{4}{5} = 80 \text{ N} \qquad \left(\cos 37° \approx \frac{4}{5}\right)$$

(iv)

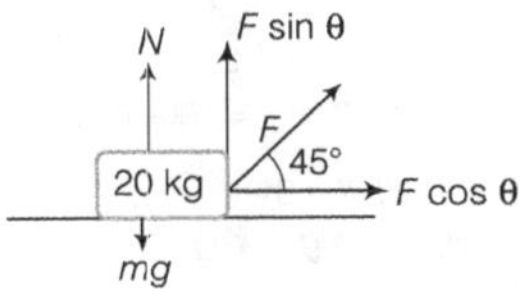

$$N = mg - F\sin\theta = 20 \times 10 - 20\sqrt{2} \times \frac{1}{\sqrt{2}} = 180 \text{ N}$$

2. (i) $N = 10$ N

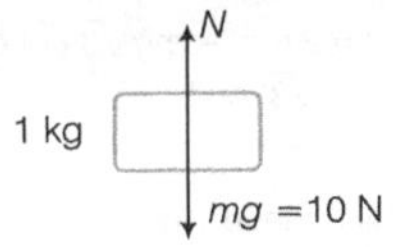

(ii) $N = F = 20$ N

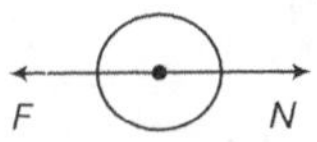

(iii) $N\cos\theta = mg$ and $N\sin\theta = N_0$

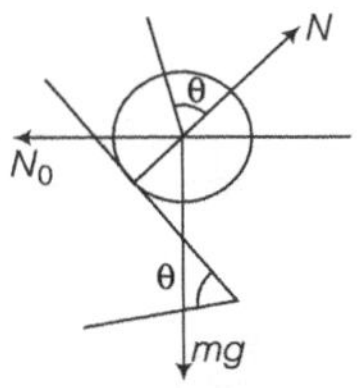

$$\therefore \quad N = \frac{mg}{\cos\theta} = \frac{mg}{\cos 30°} \quad (\because \theta = 30°)$$

$$= \frac{10}{\frac{\sqrt{3}}{2}} = \frac{20}{\sqrt{3}} \text{ N}$$

(iv) $\therefore \; N_0 = 10$ N

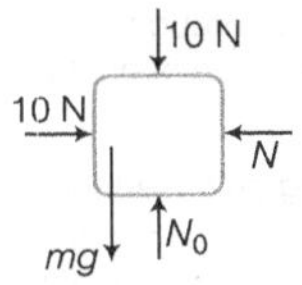

3. (d) Action and reaction forces should be of same nature, i.e. why if action force is gravitational force, the reaction force must be gravitational force.

4. (b) According to Newton's third law of motion, action and reaction forces should be of same magnitude.

5. (a) Action force and reaction force should be of same magnitude. So, the force by wall on you is 10 N.

6. (c) The action and reaction forces are equal in magnitude and opposite in direction. Thus, the force which is exerted by book on you is $\mathbf{F} = -(3\hat{i} + 4\hat{j}) = -3\hat{i} - 4\hat{j}$.

7. (b) When we put two bodies in contact with each other, the atoms at the two surfaces come close to each other and electric and magnetic interaction take place due to charge constituents. This interaction is electromagnetic force. Hence, normal reaction is an example of electromagnetic force.

8. (a) The normal reaction is a component of net contact force perpendicular to the surface of contact.

9. (c) Action and reaction act on two different bodies and also of same nature. So, statements of Agam Goel and Aditi Mehta are correct.

10. (b)

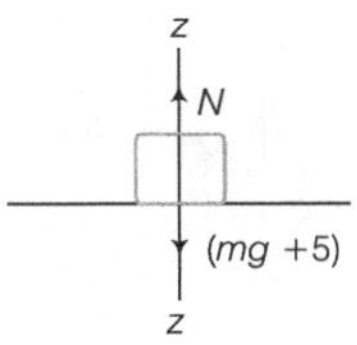

From free body diagram,

$$N = mg + 5$$
$$= 1 \times 10 + 5 = 15 \text{ N}$$

11. (a) From free body diagram of rod,

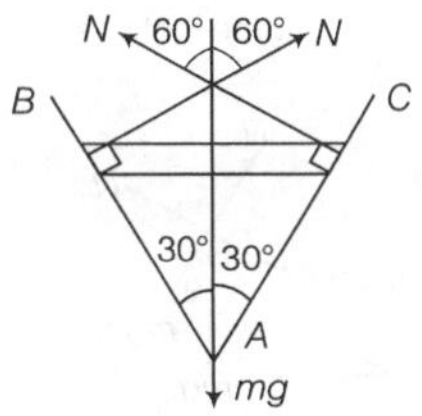

$$2N\cos 60° = mg \qquad (\because g = 10 \text{ m/s}^2)$$

$$\therefore \quad N = \frac{mg}{2\cos 60°} = 100 \text{ N}$$

12. (c) From free body diagram of rod,

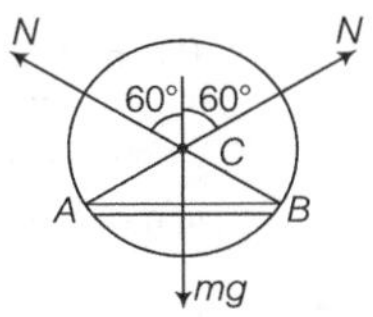

$$2N\cos 60° = mg \quad (\because g = 10 \text{ m/s}^2)$$

$$\therefore \quad N = \frac{mg}{2\cos 60°} = 10 \text{ N}$$

$$\Rightarrow \quad N = \frac{1 \times 10}{2 \times \frac{1}{2}} = 10 \text{ N}$$

13. (c) From free body diagram of light rod,

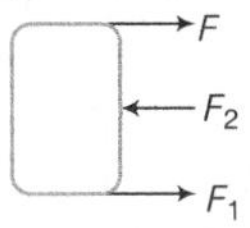

$$\therefore \quad F + F_1 = F_2$$
$$\therefore \quad F_1 < F_2$$

14. (a) mg can be resolved into two forces of equal magnitude F inclined at an angle 60° with each other.

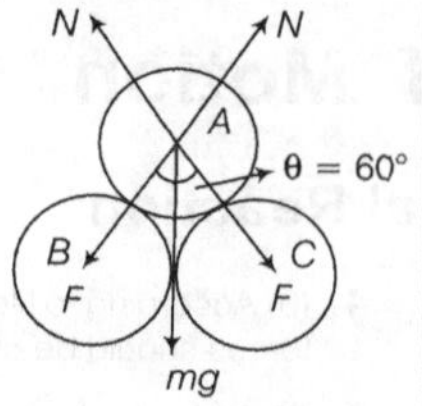

$\therefore \quad \sqrt{F^2 + F^2 + 2F^2 \cos 60^\circ} = \sqrt{3}g$

$\Rightarrow \quad 3F^2 = 3g^2$

$F = g$ or $F = 10$ N

Hence, reaction of B on $A = 10$ N.

15. (b) From free body diagram of the block,

$N_2 = 5t \cos 37^\circ = 4t$

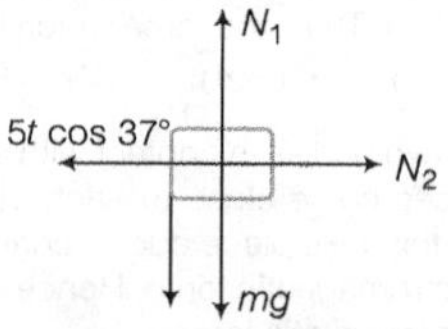

$N_1 = mg + 5t \sin 37^\circ = 10 + 3t$

$\therefore \quad N_1 = 10 + \frac{3}{4} N_2$

16. (a, b) From free body diagram of the block,

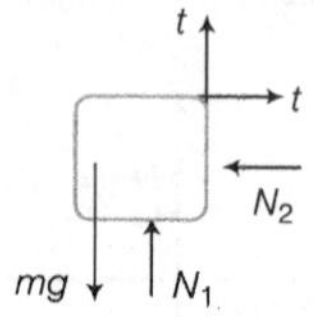

$N_2 = t, N_1 = mg - t$

At $t = 1$ s,

$N_2 = 1$ N, $\quad N_1 = 10 - 1 = 9$ N

At $t = 10$ s,

$N_2 = 10$ N, $\quad N_1 = 10 - 10 = 0$

17. (a,b)

(a) The normal contact force between plank and the block is $N_{wall} = mg \cos \theta$.

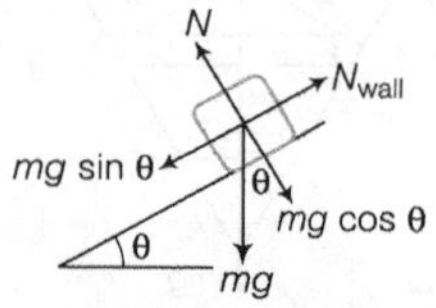

When θ increases, then $\cos \theta$ decreases.

At $\quad \theta = \frac{\pi}{2}$,

$N = 0$

(b) The normal contact force between wall and the block is $N_{wall} = mg \sin \theta$.

When θ increases, then $\sin \theta$ increases.

At $\theta = \frac{\pi}{2}$, $N_{wall} = mg$

18. (b, c) For equilibrium of rod,

$N_1 = f$ and $N_2 = mg$

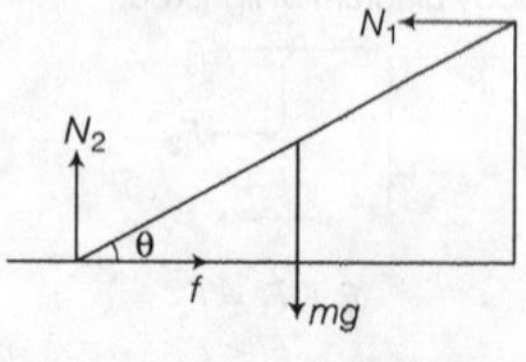

Hence, options (b) and (c) are correct.

19. (b,d) Consider a part of the rope making θ angle with horizontal. The normal reaction on considered part is

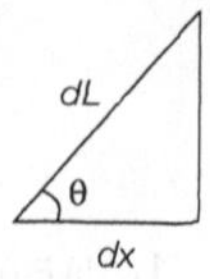

$dN = dmg \cos \theta$

$= \lambda dLg \cos \theta = \lambda g dx$

$\therefore \quad N = \lambda g \int_0^b dx = \lambda g b$

20. (b) For equilibrium of upper cylinder,

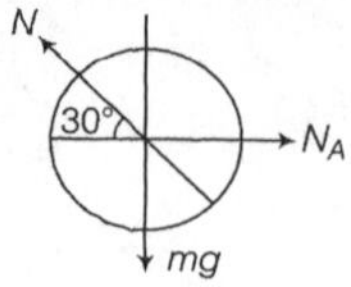

$N \cos 30^\circ = N_A$

and $\quad N \sin 30^\circ = mg$

$N = 2mg$

$\therefore \quad N_A = 200 \times \frac{\sqrt{3}}{2} = 100\sqrt{3}$ N

21. (c) $\quad N_C \cos \alpha + N_B = mg + N \sin 30^\circ \quad$ ...(i)

$N \cos 30^\circ = N_C \sin \alpha \quad$ ...(ii)

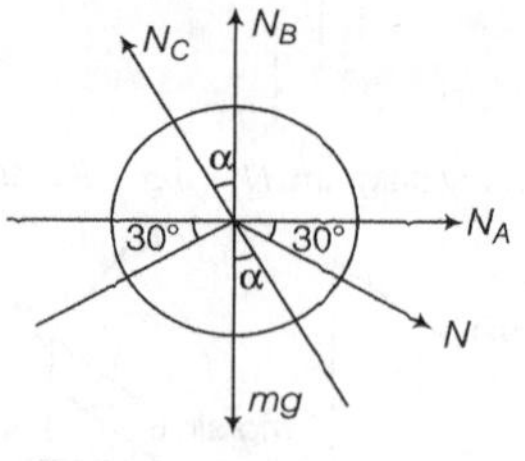

$\therefore \quad N_C = \frac{N \cos 30^\circ}{\sin \alpha} = \frac{2\,mg \cos 30^\circ}{\sin 45^\circ}$

$= \frac{2 \times 10 \times 10 \times \frac{\sqrt{3}}{2}}{\frac{\sqrt{2}}{2}} = 100\sqrt{6}$

$= \sqrt{6}\,mg = 100\sqrt{6}$ N

22. (a) From Eqs. (i) and (ii), we get

$N_B = 2mg - mg\sqrt{3} \cot \alpha = (2 - \sqrt{3})\,100$ N

23. (d) For the least value of α

$N_B = 0$

But, $\quad N_B = 2mg - mg\sqrt{3} \cot \alpha$

$\therefore \quad 0 = 2mg - mg\sqrt{3} \cot \alpha$

$\therefore \quad \tan \alpha = \frac{\sqrt{3}}{2}$

24. (d) Suppose a block is placed inside an elevator moving upward with an acceleration a_0

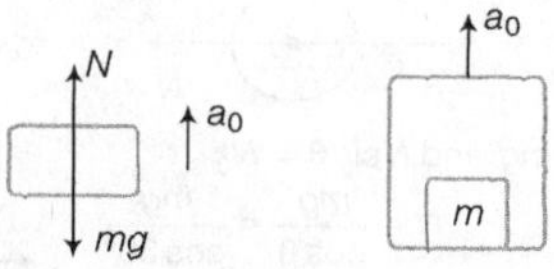

From free body diagram of block,

$N - mg = ma_0$

$\therefore \quad N = mg + ma_0$

Thus, normal reaction may accelerate a block. Hence, Statement I is wrong.

Normal reaction is +ve component of net contact force perpendicular to the surface of contact. Hence, Statement II is the correct.

25. (a) If a man is standing on the surface, $N = mg$. Acceleration due to gravity at the surface of earth is greater than acceleration on the surface of moon. So, the normal reaction on the surface of earth is more normal reaction that on the surface of moon. Hence, Statement I is correct and Statement II is the correct explanation of Statement I.

26. (a) Normal reaction is an electromagnetic force. Electromagnetic interaction is due to the presence of electrons and protons inside atoms of substance. Neutron is neutral. So, it is not responsible for electromagnetic force. Hence, Statement I is correct and Statement II is the correct explanation of Statement I.

27. (b) From free body diagram of C,

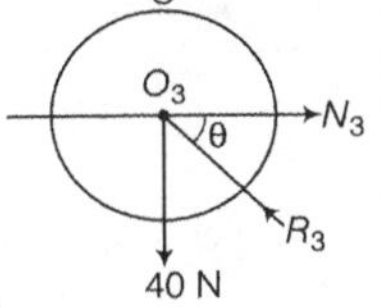

$$R_3 \cos\theta = N_3$$

and $$R_3 \sin\theta = 40$$

or $$R_3 = \frac{40}{\sin\theta} = 51.8 \text{ N.}$$

$\therefore$ $$N_3 = R_3 \cos\theta$$
$$= 51.8 \times \frac{7}{11} = 33 \text{ N}$$

Here, N_3 is normal reaction on C by the wall and R_3 is normal reaction between cylinder B and cylinder C.

From FBD of cylinder B,

$$N_2 = R_2 \cos\phi + R_3 \cos\theta \quad \text{(in horizontal direction)}$$

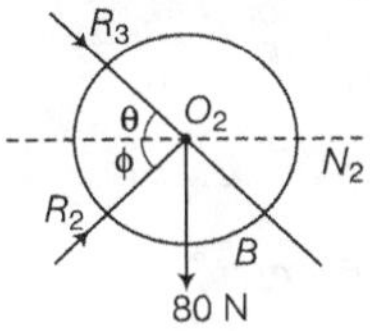

and $$R_2 \sin\phi = 80 + R_3 \sin\theta \quad \text{(in vertical direction)}$$

or $$R_2 \sin\phi = 80 + 40 = 120 \text{ N}$$

$\therefore$ $$R_2 = \frac{120}{3} \times 5 = 200 \text{ N}$$

$\because$ $$N_2 = R_2 \cos\phi + R_3 \cos\theta$$
$$= 200 \times \frac{4}{5} + 51.8 \times \frac{7}{11}$$
$$= 160 + 33 = 193 \text{ N}$$

Here, R_2 is the normal reaction between A and B and N_2 is the force exerted by the wall on the cylinder B.

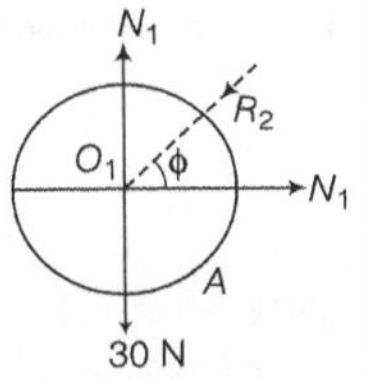

From FBD of cylinder A,

$$R_2 \cos\phi = N_1 \quad \text{(in horizontal direction)}$$

or $$200 \times \frac{4}{5} = N_1$$

$\therefore$ $$N_1 = 160 \text{ N}$$

Here, N_1 is the normal reaction exerted by the vertical wall on cylinder A.

(i) → (r), (ii) → (p), (iii) → (s), (iv) → (q)

28. (2) We take both balls as a system. For equilibrium of system, the component of forces parallel to OD and perpendicular to OD, should be zero. Taking component along OD,

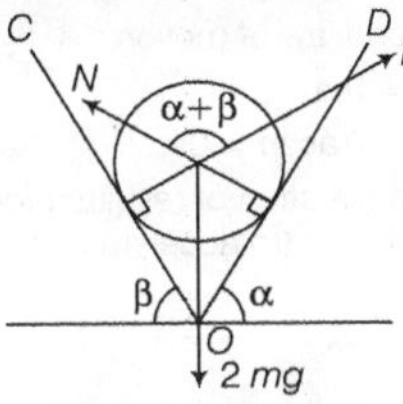

$$N_A \sin(\alpha + \beta) = 2mg \sin\alpha$$

$\Rightarrow$ $$N_A = \frac{2mg \sin\alpha}{\sin(\alpha + \beta)}$$

$\therefore$ $$n = 2$$

29. (3) From free body diagram of cylinder A,

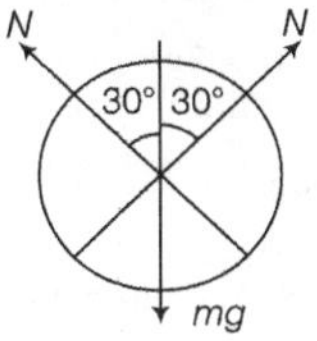

$$2N \cos 30° = mg$$

$\therefore$ $$N = \frac{mg}{\sqrt{3}}$$

From free body diagram of cylinder B,

$$N_0 \cos\theta = mg + N\cos 30°,$$
$$N_0 \sin\theta = N \sin 30°$$
$$\tan\theta = \frac{N \sin 30°}{mg + N \cos 30°}$$

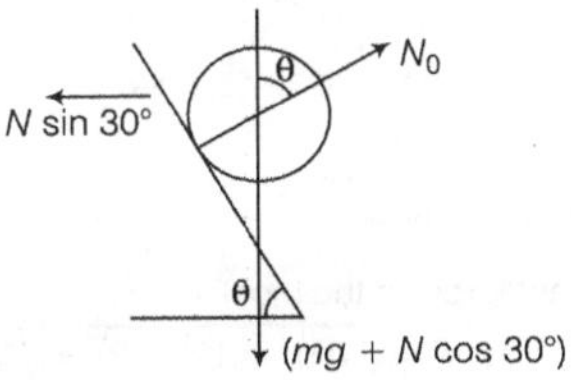

Putting the value of N, $\tan\theta = \dfrac{1}{3\sqrt{3}}$

$\therefore$ $$n = 3$$

DPP-2 **Relation between Force and Acceleration**

1. (i) From free body diagram,

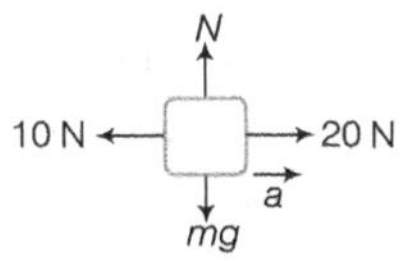

$$20 - 10 = 10a \Rightarrow a = 1 \text{ m/s}^2$$

(ii) From free body diagram,

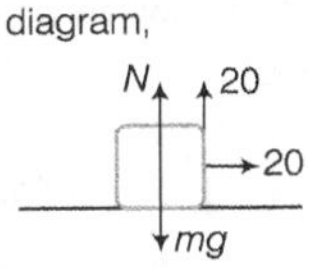

$$20 = 10a$$

$\therefore$ $$a = 2 \text{ m/s}^2$$

(iii) From free body diagram,

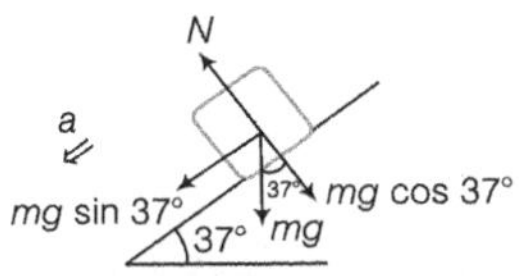

$$mg \sin 37° = ma$$

$\therefore$ $$a = 10 \times \frac{3}{5}$$
$$= 6 \text{ m/s}^2$$

(iv) From free body diagram,

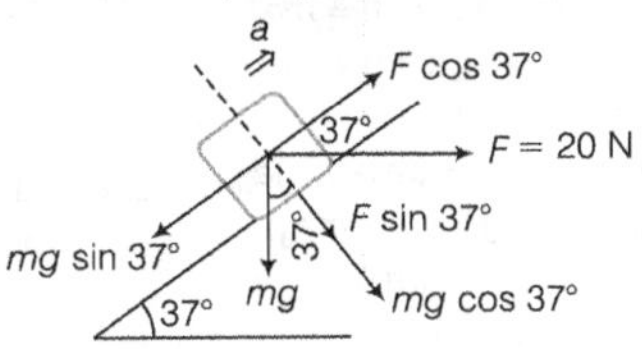

$$20 \cos 37° - mg \sin 37° = ma \quad \text{or} \quad 16 - 6 = 1 \times a$$

$\therefore$ $$a = 10 \text{ m/s}^2$$

2. (d) The resultant force acting on a body is obtained by vector sum of individual forces.

$$\mathbf{F}_R = \mathbf{F}_1 + \mathbf{F}_2 + \cdots\cdots$$

From Newton's second law of motion,

$$F_{ext} = ma$$

If $F_{ext} = 0 \Rightarrow a = 0$ as $m \neq 0$

Acceleration is in the direction of resultant force, hence the perpendicular component (to acceleration) is zero.

3. (d) $a_1 = \dfrac{F}{m_1}$ and $a_2 = \dfrac{F}{m_2}$

and $a = \dfrac{F}{m_1 + m_2} = \dfrac{F}{\dfrac{F}{a_1} + \dfrac{F}{a_2}} = \dfrac{a_1 a_2}{a_1 + a_2}$

4. (a) Acceleration, $a = \dfrac{F}{m} = \dfrac{1}{1} = 1\ \text{m/s}^2$

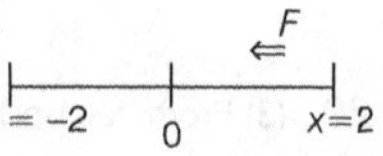

$\therefore \quad 2 = \dfrac{1}{2} \times 1 \times t_0^2$

$\therefore \quad t_0 = 2\text{s}$

$\therefore \quad T = 4t_0 = 8\text{ s}$

5. (d) From free body diagram,

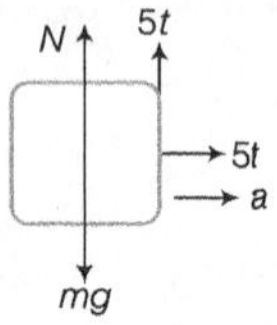

$N + 5t = mg$

$\therefore \quad N = mg - 5t \qquad (g = 10\ \text{m/s}^2)$

For leaving the surface, $N = 0$

$$t = \frac{mg}{5} \text{ or } t = \frac{100}{5} = 20\text{ s}$$

Also, $5t = 10a, \quad a = \dfrac{1}{2}t$

$$\int_0^v dv = \frac{1}{2}\int_0^{20} t\,dt$$

$$v = \frac{1}{2}\left[\frac{t^2}{2}\right]_0^{20} = \frac{1}{2}[200 - 0] = 100\ \text{m/s}$$

6. (b) $\qquad s = \dfrac{1}{2}at^2 \quad \Rightarrow \quad t = \sqrt{\dfrac{2s}{a}}$

$\Rightarrow \quad t \propto \sqrt{s}$

Hence, option (b) is correct.

7. (a) The resultant force on the block is

$$F = \sqrt{(mg\sin\theta)^2 + 8^2} = \sqrt{6^2 + 8^2} = 10\text{ N}$$

Thus, acceleration of the block is

$$a = \frac{F}{m} = \frac{10}{1} = 10\ \text{m/s}^2$$

8. (b) From free body diagram of block,

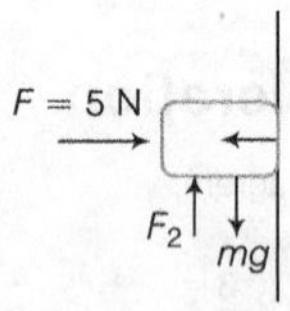

$F_2 = mg = 10\text{ N} \qquad (\because g = 10\ \text{m/s}^2)$

9. (c) Analysing below figure,

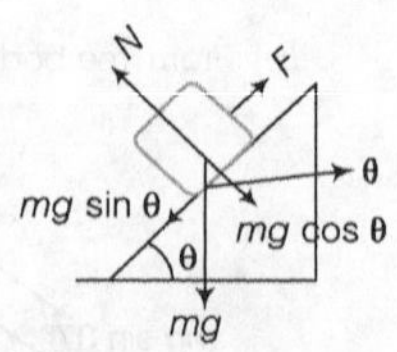

$F - mg\sin\theta = ma_0$

$\therefore \quad F = mg\sin\theta + ma_0$

After putting the block inside box,

$(m + m_0)\,g\sin\theta - F = (m + m_0)a_0$

or $(m + m_0)g\sin\theta - (mg\sin\theta + ma_0) = (m + m_0)a_0$

or $\quad m_0 g\sin\theta = (2m + m_0)\,a_0$

$\therefore \quad m_0 = \dfrac{2ma_0}{(g\sin\theta - a_0)} = \dfrac{2 \times 1 \times 3}{5 - 3} = \dfrac{6}{2} = 3\text{ kg}$

10. (c) Acceleration, $a = \dfrac{F}{m} = 5 - 5t$

or $\qquad \dfrac{dv}{dt} = 5 - 5t$

or $\qquad \int_0^v dv = \int_0^t 5\,dt - \int_0^t 5t\,dt$

$\therefore \qquad v = 5t - \dfrac{5t^2}{2}$

$\therefore \qquad x = \int_0^t v\,dt = \dfrac{5t^2}{2} - \dfrac{5t^3}{6}$

$\therefore$ For x_{max}, $v = 0$

$\therefore \qquad 0 = 5t_0 - \dfrac{5t_0^2}{2}$

$\Rightarrow \qquad t_0 = 2\text{s}$

$\therefore \qquad x_{max} = \dfrac{5}{2}(2)^2 - \dfrac{5}{6}(2)^3 = 10 - \dfrac{20}{3} = \dfrac{10}{3}\text{ m}$

11. (d) $\qquad a_x = \dfrac{dv_x}{dt} = -16\ \text{m/s}^2$

and $\qquad v_y = \dfrac{dy}{dt} = -8t$

$\qquad a_y = \dfrac{dv_y}{dt} = -8$

$\therefore \qquad F = m\sqrt{a_x^2 + a_y^2}$

$\qquad = 10\sqrt{16^2 + 8^2} = 179\text{ N}$

12. (b) The area of force – Time graph = Change in momentum = Impulse

$\therefore \qquad p_f - p_i = \dfrac{1}{2} \times 100 \times 4 = 200$

or $\qquad mv - 0 = 200$

$\therefore \qquad v = 200\ \text{m/s}$

13. (d) Acceleration, $a = \dfrac{30 - 10}{10} = 2\ \text{m/s}^2$

(considering A, B, C as system)

Force diagram of block A,

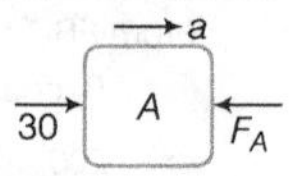

(F_A = contact force)

$30 - F_A = 5 \times 2 = 10$

$\therefore \qquad F_A = 20\text{ N}$

Force diagram of block C,

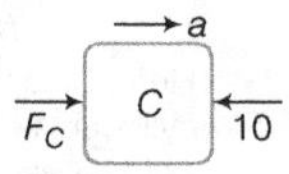

(F_C = contact force)

$F_C - 10 = 3 \times 2 = 6$

$\therefore \qquad F_C = 16$

$\therefore \qquad \dfrac{F_A}{F_C} = \dfrac{20}{16} = \dfrac{5}{4}$

14. (a,b,c) $\qquad F + N = mg$

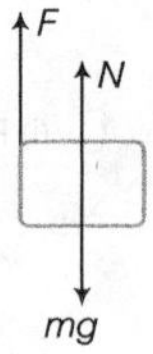

(a) At $t = 2$s, $\quad 4 + N = 10$

$\therefore \qquad N = 6\text{ N}$

(b) $a = \dfrac{F - mg}{m} = \dfrac{2 \times 10 - 10}{1} = 10\ \text{m/s}^2$

$(\because$ at $t = 10$ s, $F = 20$ N)

(c) $a = \dfrac{2t - 10}{m} = 2t - 10$

For $t < 75$ s, $F < mg$

Hence, body does not loose contact and acceleration is zero.

For $t < 5$ s, $a = \left(\dfrac{F - mg}{m}\right)$

or $\qquad \int_0^v dv = \int_5^{10} (2t - 10)dt \qquad \left(\because a = \dfrac{dv}{dt}\right)$

$\qquad = [t^2]_5^{10} - [10t]_5^{10}$

$\qquad = 75 - 50 = 25\ \text{m/s}$

15. (b, d) Free body diagram (FBD) for situation (i),

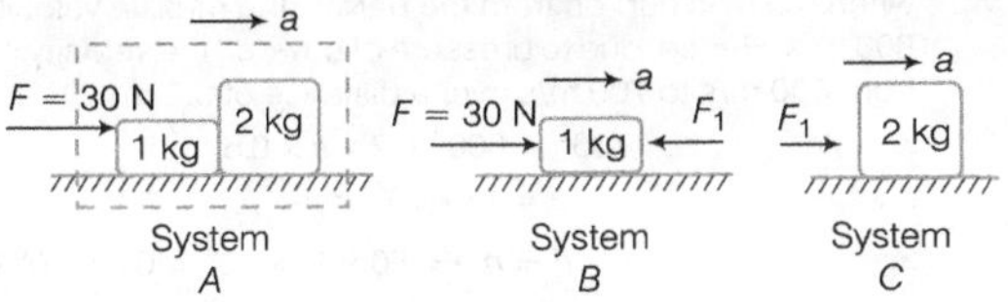

Considering both the masses (1 kg and 2 kg) as a system (*A*), the acceleration of the system is given by

$$F = (m_1 + m_2)a$$

or $$a = \frac{30\text{ N}}{(1+2)\text{ kg}}$$

$$= 10\text{ m/s}^2$$

The acceleration *a* of system is same as that from figure *C*, acceleration of 2 kg mass is

$$F_{ext} = F_1 = ma$$

or $$F_1 = 2 \times 10 = 20\text{ N}$$

Free body diagram for situation (ii),

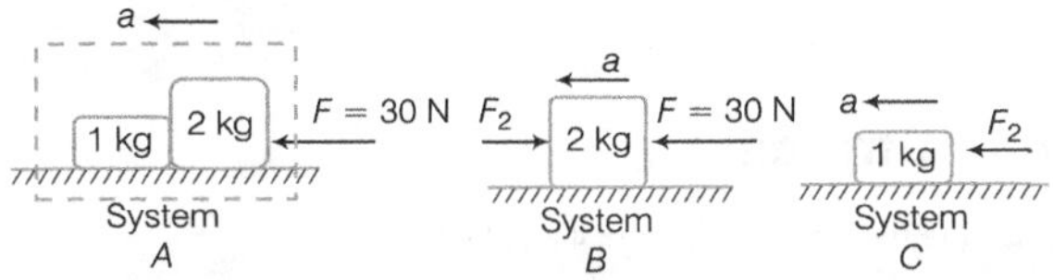

Similarly, from *A* $\quad a = \frac{30}{3} = 10\text{ m/s}^2$

From *C* $\quad F_2 = 1 \times 10 = 10\text{ N}$

Hence, $F_1 > F_2$

16. (a,c,d)

(a) No horizontal force on the block, so the acceleration of the block is zero.

(c) The net horizontal force on the plank is

$$F = 20 - 10 = 10\text{ N}$$

$\therefore$ The acceleration of plank is

$$a = \frac{F}{m} = \frac{10}{2} = 5\text{ m/s}^2$$

(d) $$s = \frac{1}{2}at^2$$

$\Rightarrow$ $$t = \sqrt{\frac{2s}{a}} = \sqrt{\frac{2 \times 2.5}{5}} = 1\text{ s}$$

17. (a) Net force acting on the object

$$\mathbf{F}_{net} = \mathbf{F}_1 + \mathbf{F}_2$$

or $$\mathbf{F}_{net} = (-\hat{\mathbf{i}} - \hat{\mathbf{j}}) + (-8\hat{\mathbf{i}} + 4\hat{\mathbf{j}}) = -9\hat{\mathbf{i}} + 3\hat{\mathbf{j}}$$

$$\mathbf{a}_{net} = \frac{\mathbf{F}_{net}}{m} = \frac{(-9\hat{\mathbf{i}} + 3\hat{\mathbf{j}})}{2}\text{ ms}^{-2} \quad \text{or} \quad \mathbf{v} = \mathbf{u} + \mathbf{a}t$$

At, $t = 10$ s, $\quad \mathbf{v} = 0 + \left(\frac{-9\hat{\mathbf{i}} + 3\hat{\mathbf{j}}}{2}\right)10 \quad$ or $\mathbf{v} = -45\hat{\mathbf{i}} + 15\hat{\mathbf{j}}$

18. (c) At $t = 10$ s, the direction of motion is along the velocity vector.

$$\mathbf{v} = -45\hat{\mathbf{i}} + 15\hat{\mathbf{j}}$$

$\Rightarrow$ $$\tan\theta = \frac{v_y}{v_x} = -\frac{15}{45} = -\frac{1}{3}$$

or $$\theta = \tan^{-1}\left(-\frac{1}{3}\right)$$

19. (b) Using relation

$$\mathbf{s} = \mathbf{u}t + \frac{1}{2}\mathbf{a}t^2$$

$\Rightarrow$ $$\mathbf{s} = \frac{1}{2} \times \left(\frac{-9\hat{\mathbf{i}} + 3\hat{\mathbf{j}}}{2}\right) \times (10)^2 = -225\hat{\mathbf{i}} + 75\hat{\mathbf{j}}$$

$\because$ $$\mathbf{s} = \mathbf{r}_f - \mathbf{r}_i$$

$\mathbf{r}_f$ = Final position vector, $\mathbf{r}_i$ = Initial position vector

or $$\mathbf{r}_f = \mathbf{s} + \mathbf{r}_i$$

$\mathbf{r}_i$ = Initial position vector = $-2\hat{\mathbf{i}} + 4\hat{\mathbf{j}}$

$\Rightarrow$ $$\mathbf{r}_f = (-225\hat{\mathbf{i}} + 75\hat{\mathbf{j}}) + (-2\hat{\mathbf{i}} + 4\hat{\mathbf{j}}) = -227\hat{\mathbf{i}} + 79\hat{\mathbf{j}}$$

20. (d) According to Newton's second law of motion, net external force, $\mathbf{F}_{ext} = m\mathbf{a}$.

The sum of internal forces is always zero. Thus, net internal force on a body is

$$0 = m\mathbf{a} \Rightarrow \mathbf{a} = 0$$

Hence, Statement I is wrong while Statement II is correct.

21. (c) According to Newton's third law, if a block *A* exerts a force **F** on another block *B*, then at the same time, *B* exerts a force (−**F**) on the block *A*. Hence, Statement I is correct. Action and reaction forces act on two different blocks. So, they do not balance each other. Hence, Statement II is incorrect.

22. (b) $\because$ $$\mathbf{F} = m\mathbf{a}$$

$\therefore$ $$a = \frac{F}{m}$$

It means, that more mass less tendency to change the initial state.

So, mass is a measure of inertia. Hence, Statement I is correct.

The change in momentum is known as impulse.

$\because$ $$\mathbf{F} = \frac{\Delta\mathbf{p}}{\Delta t}$$

$$\Delta\mathbf{p} = \mathbf{F}\Delta t$$

Hence, $\Delta\mathbf{p}$ is the change in momentum in time Δt due to applied force **F**.

Hence, Statement II is correct.

23. (a) The seat is decelerated with then deceleration of car. But, due to inertia, you try to maintain your initial state of motion. So, you will fall forward. Hence, Statement II is the correct explanation of Statement I.

24. (c) From free body diagram of box,

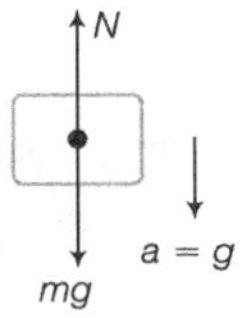

$$mg - N = ma = mg$$

$\therefore N = 0$. Hence, Statement I is correct.

But, true weight of box is *mg*. Hence, Statement II is wrong.

25. (3) $$\mathbf{F}_{net} = \mathbf{F}_{East} + \mathbf{F}_{South} + \mathbf{F}_{down} + \mathbf{F}_0$$

$$= 3\hat{\mathbf{i}} + (-1\hat{\mathbf{j}}) + (-1\hat{\mathbf{k}}) + \mathbf{F}_0$$

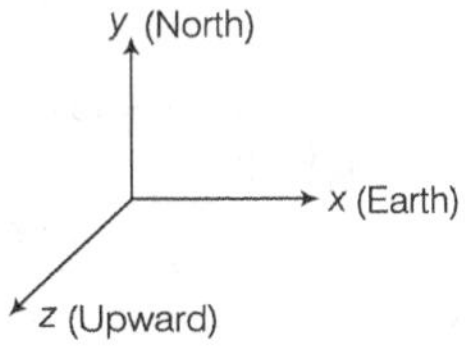

Since, the block moves in South-East direction,

$$\mathbf{F}_{net} = x\hat{\mathbf{i}} - x\hat{\mathbf{j}} + 0\hat{\mathbf{k}} \qquad [\because \tan\theta = -1]$$

or $$x\hat{\mathbf{i}} - x\hat{\mathbf{j}} = 3\hat{\mathbf{i}} - \hat{\mathbf{j}} - \hat{\mathbf{k}} + \mathbf{F}_0$$

or $$\mathbf{F}_0 = (x-3)\hat{\mathbf{i}} + (1-x)\hat{\mathbf{j}} + \hat{\mathbf{k}}$$

$$|F_0| = \sqrt{(x-3)^2 + (1-x)^2 + (1)^2}$$

or $$|F_0|^2 = (x-3)^2 + (1-x)^2 + 1$$

For minimum F_0, $|F_0|^2$ must be minimum.

Differentiating $|F_0|^2$ w.r.t. x, we get

$$\frac{dy}{dx} = 2(x-3) - 2(1-x) \qquad [\text{let } |F_0|^2 = y]$$

For minima, $\frac{dy}{dx} = 0$

or $\quad 2(x-3) - 2(1-x) = 0$

or $\quad 4x + 8 = 0$ or $x = 2$

$$\mathbf{F}_{0(\min)} = (2-3)\hat{\mathbf{i}} + (1-2)\hat{\mathbf{j}} + \hat{\mathbf{k}}$$

$$= (-\hat{\mathbf{i}} - \hat{\mathbf{j}} + \hat{\mathbf{k}})\,\text{N}$$

$\therefore$ $$\mathbf{F}_{0(\min)} = \sqrt{(-1)^2 + (-1)^2 + 1^2}$$

$$= \sqrt{3} = \sqrt{n}$$

$\therefore$ $n = 3$

26. (1) Here, $a = \frac{F}{m} = \frac{bt - cs}{m} = t - s$

or $\frac{da}{dt} = 1 - v$

or $\frac{d^2a}{dt^2} = \frac{-dv}{dt} = -a$

The solution of equation is $a = a_0 \sin \omega t$.

When $\omega = 1$ unit.

27. (6) With full barrel length, the muzzle velocity of block is 700 m/s when 0.5 m is cut -off from the barrel , the muzzle velocity falls to 600 m/s. Hence, due to pressure of powder, the velocity increases from 600 m/s to 700 m/s over a distance of 0.5 m.

$\therefore$ $700^2 = 600^2 + 2 \times a \times 0.5$

$\therefore$ $a = 13 \times 10^4 \text{ m/s}^2$

$\therefore$ $F = ma = 50 \times 13 \times 10^4 = 6.5 \times 10^6$ N

$\therefore$ $n = 6$

28. (a) (i) $N = mg = 800$ N

(i) → (p,q)

(ii) $N - mg = ma$

$\therefore$ $N = mg + ma = 960$ N

(ii) → (r)

(iii) $mg - N = ma$

$\therefore$ $N = mg - ma = 640$ N

(iii) → (s,t)

(iv) $N = mg - ma = 0$

(iv) → (t,u)

DPP-3 Tension in Strings

1. (i) From free body diagram

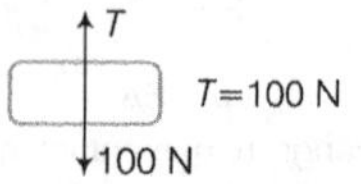

(ii) $N = (m_1 + m_2)g$, $F = (m_1 + m_2)a$

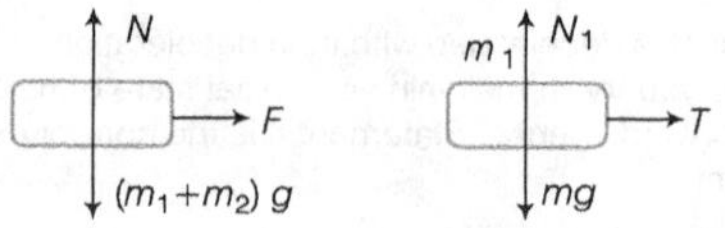

$$T = m_1 a = \frac{m_1 F}{m_1 + m_2}$$

(iii) From free body diagram,

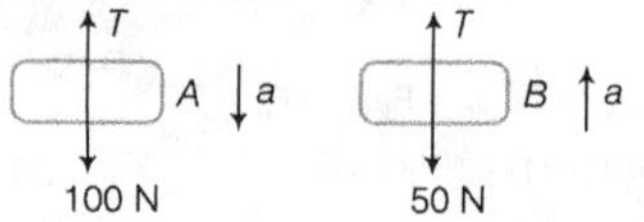

$$T - 50 = 5a \qquad \ldots\text{(i)}$$

$$100 - T = 10a \qquad \ldots\text{(ii)}$$

From Eqs. (i) and (ii), we get

$$T = \frac{200}{3} = 66.67 \text{ N}$$

(iv) $T = F = 20$ N

(v) $T = mg \sin 30° = 5$ N

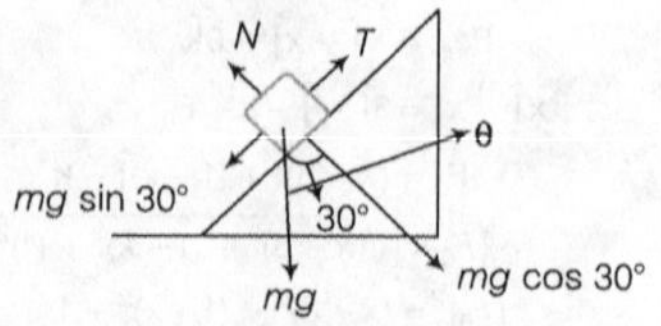

(vi) $$10 - T = 1 \times a \qquad \ldots\text{(i)}$$

and $$T - m_A g \sin 37° = m_A a \qquad \ldots\text{(ii)}$$

From Eqs. (i) and (ii), we get

$$T = 8 \text{ N}$$

2. (i) Free body diagram of 4 kg block,

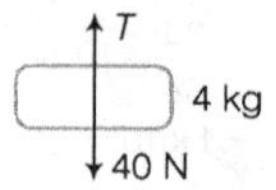

$$40 - T = 4a \qquad \ldots\text{(i)}$$

Free body diagram of 2 kg block,

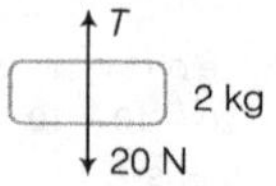

$$T - 20 = 2a \qquad \ldots\text{(ii)}$$

From Eqs. (i) and (ii), we get

$$T = \frac{80}{3} \text{ N}$$

From free body diagram of pulley.

$\therefore$ The clamp force,

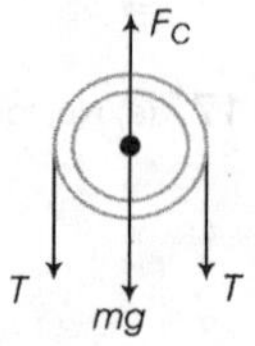

$$F_C = 2T + mg$$

$$= \frac{160}{3} + 10$$

$$= \frac{190}{3} = 63.3 \text{ N}$$

(ii) Since, $T = F = mg$ $\quad (\because a = 0)$

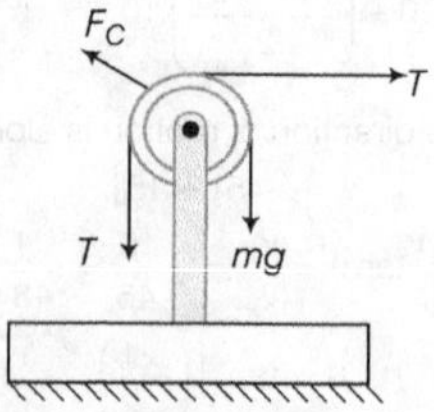

$\therefore$ From free body diagram of the pulley, the clamp force is

$$F_C = \sqrt{(T + mg)^2 + T^2}$$

$$= \sqrt{(40)^2 + (30)^2}$$

$$= 50 \text{ N}$$

(iii) By equilibrium condition, $5g\sin 37° - T = 5a$

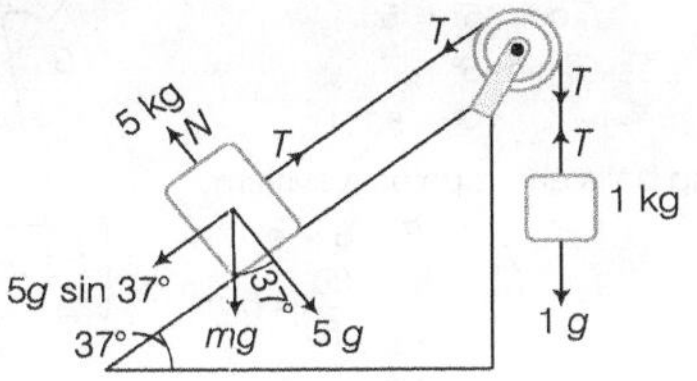

$$5g \times \frac{3}{5} - T = 5a$$

$$3g - T = 5a \quad \text{...(i)}$$

$$T - g = a \quad \text{...(ii)}$$

$$2g = 6a$$

$$\Rightarrow \quad a = \frac{10}{3}\ \text{m/s}^2$$

$$T = 10 + \frac{10}{3} = \frac{40}{3}\ \text{N}$$

Clamp force is

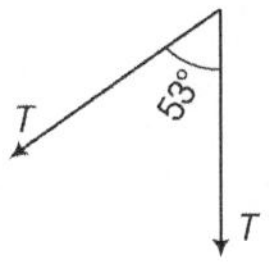

$$F = \sqrt{T^2 + T^2 + 2T^2\cos 53°}$$

$$= \frac{4}{\sqrt{5}}T = \frac{4}{\sqrt{5}} \times \frac{40}{3}$$

$$= \frac{160\sqrt{5}}{3 \times 5} = \frac{32\sqrt{5}}{3}\ \text{N}$$

(iv) From free body diagram of block A,

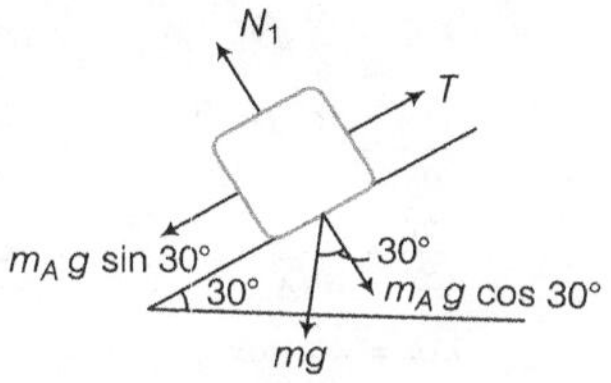

$\therefore \quad m_A g\sin 30° - T = m_A a$

$\therefore \quad \frac{\sqrt{3}}{2}g - T = \sqrt{3}a \quad \text{...(i)}$

From free body diagram of the block B,

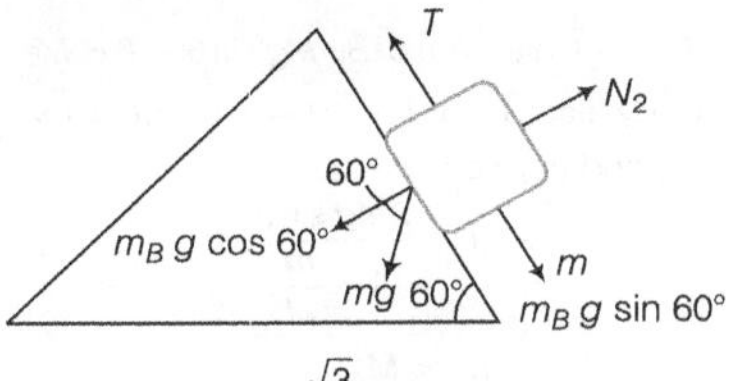

$$T - \frac{\sqrt{3}}{2}g = 1 \times a \quad \text{...(ii)}$$

From Eqs. (i) and (ii), we get

$$a = 0, T = \frac{\sqrt{3}}{2}g$$

From free body diagram of the pulley,

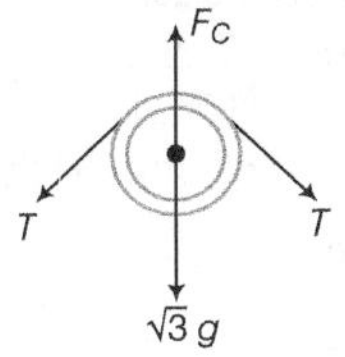

$$F_C = \sqrt{2}T + \sqrt{3}g = \sqrt{\frac{3}{2}}g + \sqrt{3}g$$

$$\therefore \quad F_C = \sqrt{3}\left(1 + \frac{1}{\sqrt{2}}\right)g$$

3. (i) $T_2 - 20 = 0$

$\therefore \quad T_2 = 20\ \text{N}$

and $\quad T_1 = (1 + 2)g = 30\ \text{N}$

$T_1 : T_2 = 3 : 2$

(ii) Taking all blocks in a system,

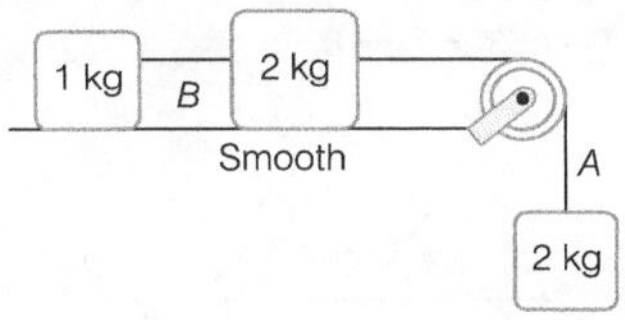

$$20 = 5a$$

$$\therefore \quad a = 4\ \text{m/s}^2$$

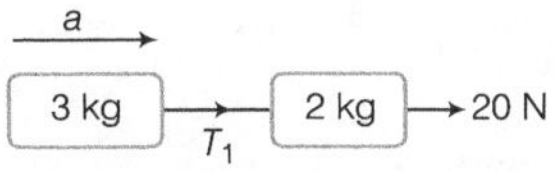

$$T_1 = 3a = 12\ \text{N} \quad \text{and} \quad T_2 = 1 \times a$$

$$\therefore \quad \frac{T_1}{T_2} = \frac{12}{4} = 3$$

(iii)

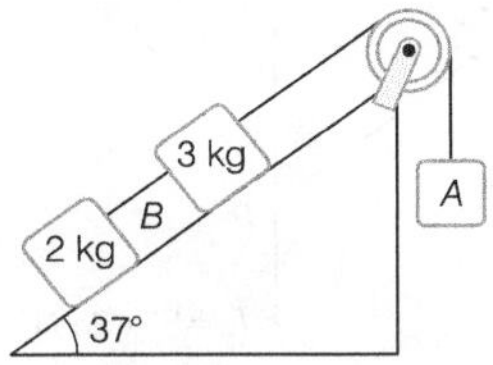

$$100 - T_1 = 10a \quad \text{...(i)}$$

$$T_1 - 50\sin 37° = 5a \quad \text{...(ii)}$$

and $\quad T_2 - 20\sin 37° = 2a \quad \text{...(iii)}$

From Eqs. (i), (ii) and (iii), we get

$$T_1 = \frac{160}{3}\ \text{N}, T_2 = \frac{64}{3}\ \text{N}$$

$$\therefore \quad T_1 : T_2 = 5 : 2$$

(iv) $F - 20\sin 37° - 10 = 3a$

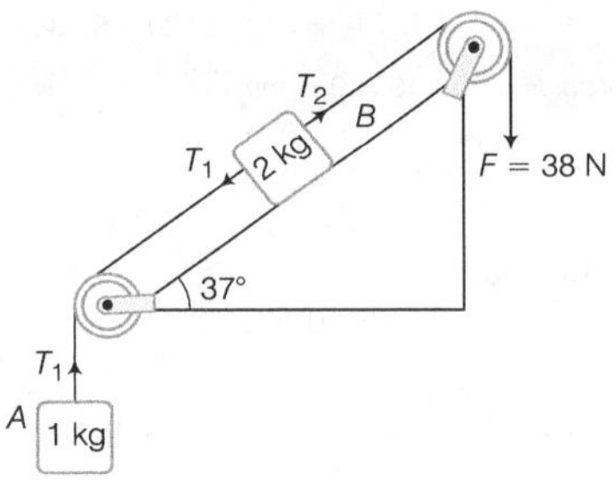

$$\therefore \quad a = \frac{16}{3}\ \text{m/s}^2 \Rightarrow T_2 = F = 38$$

$$T_1 - 10 = 1 \times a \quad \Rightarrow T_1 = 10 + \frac{16}{3} = \frac{46}{3}\ \text{N}$$

$$\therefore \quad T_1 : T_2 = \frac{\frac{46}{3}}{38} = 23 : 57$$

4. (i) $T_C = mg = 20\ \text{N} \Rightarrow T_A = T_B = T$

From free body diagram of point O,

$$2T\cos 60° = T_C = mg$$

$$\therefore \quad T = mg$$

$$\therefore \quad T_A = T_B = T_C$$

$$\therefore \quad T_A : T_B : T_C = 1 : 1 : 1$$

(ii) $F = T_A = T_B$

$$2F\cos\alpha = T_C = mg$$

$\therefore$ $$F = \frac{mg}{2\cos\alpha}$$

$\therefore$ $$T_A : T_B : T_C = 1:1:2\cos\alpha$$

(iii) For equilibrium of point O,

$$T_C \sin 53^\circ = T_B = 8$$

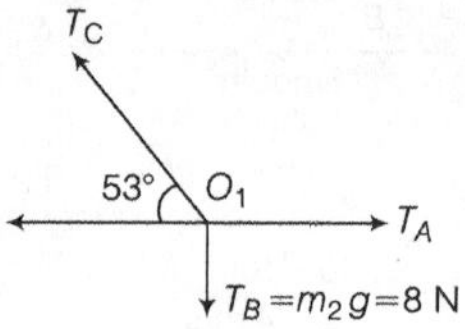

$\therefore$ $$T_C = \frac{8}{\frac{4}{5}} = 10\text{ N}$$

Also, $T_C \cos 53^\circ = T_A$

$\therefore$ $$T_A = 10 \times \frac{3}{5} = 6\text{ N}$$

$\therefore$ $$T_A : T_B : T_C = 6:8:10 = 3:4:5$$

(iv)

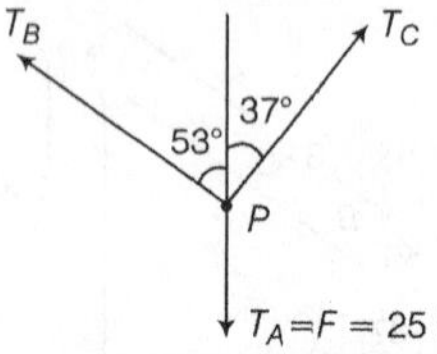

Here, $T_B \sin 53^\circ = T_C \sin 37^\circ$ ...(i)

and $T_B \cos 53^\circ + T_C \cos 37^\circ = T_A = 25$

or $$\frac{T_C \sin 37^\circ}{\sin 53^\circ}\cos 53^\circ + T_C \cos 37^\circ = 25$$

or $T_C = 20$ N

$\therefore$ $$T_B = \frac{T_C \sin 37^\circ}{\sin 53^\circ} = \frac{20 \times \frac{3}{5}}{\frac{4}{5}} = 15\text{ N [from Eq. (i)]}$$

$\therefore$ $$T_A : T_B : T_C = 25:15:20 = 5:3:4$$

5. (i) Net force, $F = 10 - 10 = 0 = ma$

$$a = 0$$
$$10 - T = 0$$
$$T = 10\text{ N}$$

(ii) $40 - 10 = 3a$

$\therefore$ $$a = 10\text{ m/s}^2$$

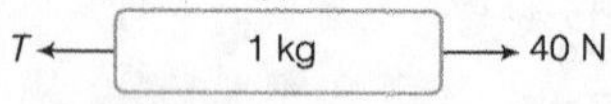

$$40 - T = 1 \times 10$$

$\therefore$ $$T = 30\text{ N}$$

(iii) If a be the net acceleration

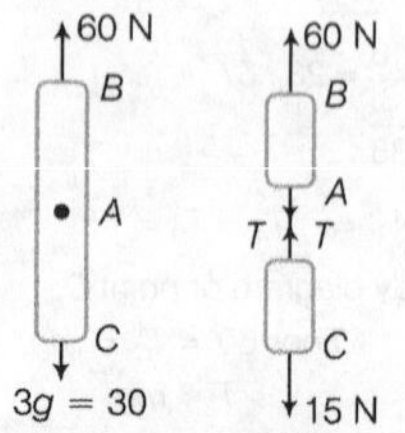

$$60 - 30 = 3a \Rightarrow a = 10\text{ m/s}^2$$

$$60 - T - 15 = 1.5a$$

$$T = 45 - 15 = 30\text{ N}$$

(iv) From free body diagram of half rope,

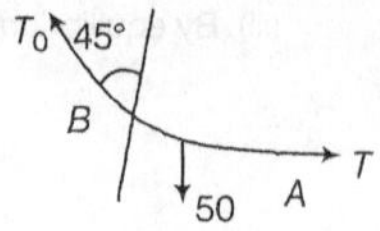

$$T_0 \cos 45^\circ = 50$$

and $$T_0 \sin 45^\circ = T$$

$\therefore$ $$T = 50\text{ N}$$

(v) Taking balls and rope as a system,

$$F = 5 \times a$$

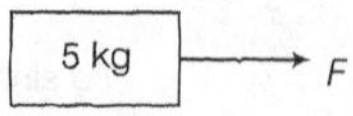

$\therefore$ $$a = \frac{30}{5} = 6\text{ m/s}^2$$

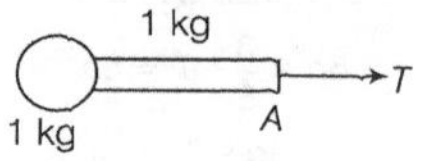

$$T = 2a = 2 \times 6 = 12\text{ N}$$

6. (b) $T_0 = 6g = 60$ N

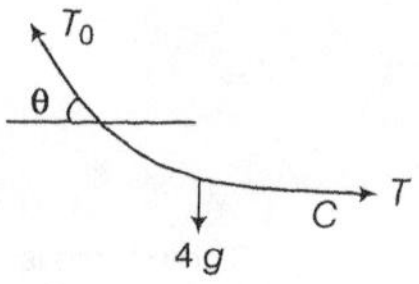

$$T_0 \sin\theta = 4g = 40$$
$$60 \sin\theta = 40$$
$$\sin\theta = \frac{2}{3}$$
$$\cos\theta = \frac{\sqrt{9-4}}{3} = \frac{\sqrt{5}}{3}$$

But, $$T = T_0 \cos\theta = 60\frac{\sqrt{5}}{3} = 20\sqrt{5}\text{ N}$$

7. (b) Here, $$T = ma$$

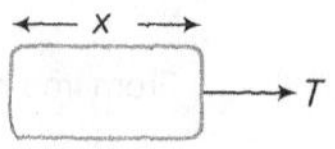

$$10\sqrt{x} = ma$$

$\Rightarrow$ $$m = \frac{10}{a}\sqrt{x}$$

or $$\frac{dm}{dx} = \frac{10}{a}\frac{d(\sqrt{x})}{dx}$$ or $$\frac{dm}{dx} = \frac{10}{a}\frac{1}{2\sqrt{x}}$$

or $$\frac{dm}{dx} = \frac{5}{a\sqrt{x}}$$ or $$dm = \frac{5}{a\sqrt{x}}dx$$

or $$\lambda dx = \frac{5}{a\sqrt{x}}dx \Rightarrow \lambda = \frac{5}{a\sqrt{x}}$$

$\Rightarrow$ $$\lambda \propto \frac{1}{\sqrt{x}}$$

8. (a) $\frac{T}{r} = \frac{100}{0.2} = 500$ N/m

9. (c) From free body diagram of B, $Mg\sin\theta - T = Ma$ ...(i)

From free body diagram of A, $T - mg\sin\theta = ma$ ...(ii)

From Eqs. (i) and (ii), we get

$$T = \frac{2mg\sin\theta}{1 + \frac{m}{M}}$$

But, $m << M$

$\therefore$ $T = 2mg\sin\theta$

10. (c) For equilibrium of ring, $2T = Mg$ ...(i)

Also, $m_2 g - T = m_2 a$ ...(ii)

and $T - m_1 g = m_1 a$ ...(iii)

From Eqs. (i), (ii) and (iii), we get

$$M = \frac{4m_1 m_2}{m_1 + m_2} = 8\text{ kg}$$

11. (c) After sometime,

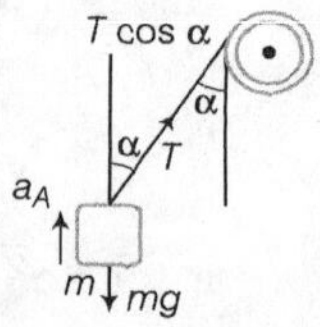

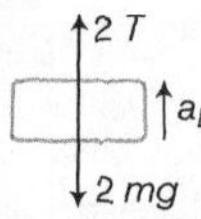

$$\frac{T\cos\alpha - mg}{m} = a_A$$

and $$\frac{2T - 2mg}{2m} = a_B$$

$\therefore$ $a_A < a_B$

$\therefore$ Block A is lower than the block B.

12. (d) Let up to $t = t_1$, the block is accelerated with maximum acceleration, then decelerated with deceleration g.

$\therefore \quad h_1 = \frac{1}{2}\left(\frac{4mg - mg}{m}\right)t_1^2 = \frac{3}{2}gt_1^2$ and $v_0 = 3gt_1$

Also, $\quad 0 = v_0 - g(t - t_1)$

$\therefore \quad 3gt_1 = g(t - t_1)$

$\therefore \quad 4gt_1 = gt$

$\therefore \quad t_1 = \frac{t}{4},\ h_2 = \frac{v_0^2}{2g}$

$\therefore \quad h = h_1 + h_2,$

After solving, $t = 4\,s$

13. (c) The acceleration of blocks is

$$a = \frac{40 - 10}{5} = 6\ \text{m/s}^2$$

After time, $t_0 = \sqrt{\frac{2s}{a}} = \sqrt{\frac{2 \times 3}{6}} = 1\ \text{s}$, small block B strikes the ground. After striking block B with ground, A moves under gravity.

$\therefore \quad v_0 = at_0 = 6 \times 1 = 6\ \text{m/s}$

$\therefore \quad h = \frac{v_0^2}{2g} = \frac{36}{2 \times 10} = 1.8\ \text{m}$

$\therefore$ Total height attained $= 3 + 1.8 = 4.8$ m

14. (a,b,c) From free body diagram of the block,

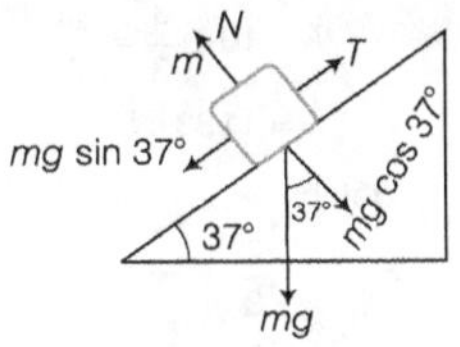

$$T - mg\sin 37° = ma \quad \text{or} \quad T - 360 = 60a \qquad \ldots\text{(i)}$$

From free body diagram of cabin + man,

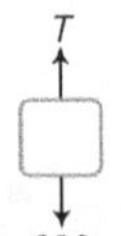

$$600 - T = 60a \qquad \ldots\text{(ii)}$$

From Eqs. (i) and (ii), we get

$$a = 2\ \text{m/s}^2$$

From free body diagram of man,

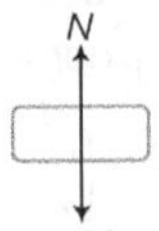

$$500 - N = 50 \times 2$$

$\therefore \quad N = 400\ \text{N} = 40$ kg-wt

15. (a,b,c,d) $T_1 - m_1 g = m_1 a_1$ (for block A)

$$\frac{F}{2} - m_1 g = m_1 a_1$$

$\therefore \quad a_1 = 50\ \text{m/s}^2$ and $T_1 = 2T$

$\therefore \quad T = \frac{F}{4} = 30\ \text{N}$

$\therefore \quad 30 - 20 = 2a_2$ (for block B)

$\therefore \quad a_2 = 5\ \text{m/s}^2$ and $30 - m_3 g = m_3 a_3$ (for block C)

$\therefore \quad a_3 = 0$

16. (2) From the figure,

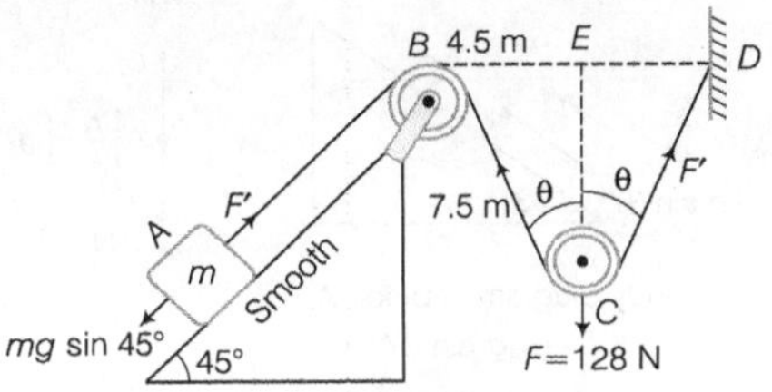

$$EC = \sqrt{(7.5)^2 - (4.5)^2} = 6\ \text{m}$$

$$\sin\theta = \frac{4.5}{7.5} = \frac{3}{5} \Rightarrow \cos\theta = \frac{6.0}{7.5} = \frac{4}{5}$$

$$2F'\cos\theta = 128$$

$\Rightarrow \quad 2F' \times \frac{4}{5} = 128 \Rightarrow F' = \frac{128 \times 5}{8}$

$$= 80\ \text{N}$$

For block at A, $F' = mg\sin 45°$

$$80 = m \times 10 \times \frac{1}{\sqrt{2}} \Rightarrow m = 8\sqrt{2}\ \text{kg}$$

$\therefore \quad \frac{16}{\sqrt{n}} = 8\sqrt{2}$

or $\quad \sqrt{n} = \frac{16}{8\sqrt{2}} = \sqrt{2}$

$\therefore \quad n = 2$

17. (5) Here, $F = 2T$ (from FBD of pulley)

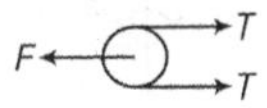

$\therefore \quad T = 5\ \text{N}$

But, $\quad T = ma$

$\therefore \quad a = \frac{T}{m} = \frac{5}{1} = 5\ \text{m/s}^2$

18. (1) ***Case I***

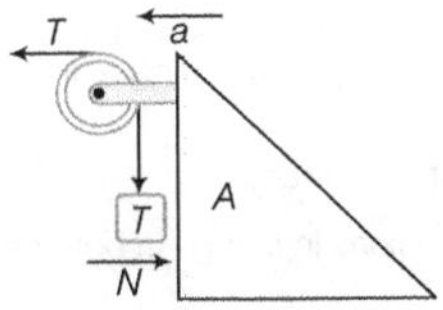

Case I

$$T - N = 40a \qquad \ldots\text{(i)}$$

and $\quad N = 20a \qquad \ldots$(ii)

Also, $\quad m_B g - T = 20a \qquad \ldots$(iii)

From Eqs. (i), (ii) and (iii), we get

$$a = \frac{g}{4}$$

Thus, the acceleration of block B is

$$a_B = \sqrt{a^2 + a^2} = \frac{g}{2\sqrt{2}}$$

Case II $\quad T = 40a$ and $20g - T = 20a$

$\therefore \quad a = \frac{g}{3}$

$\therefore$ The acceleration of block B is

$$a_B = \frac{g}{3}$$

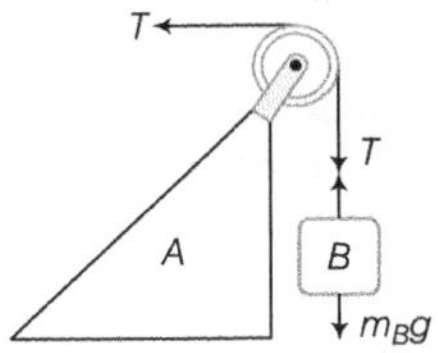

Case II

The required ratio is

$$\frac{\frac{g}{2\sqrt{2}}}{g/3} = \frac{3}{2\sqrt{2}} = \frac{6n}{4\sqrt{2}}$$

$\therefore \quad n = \frac{3 \times 4\sqrt{2}}{2\sqrt{2} \times 6} = 1$

19. (b)

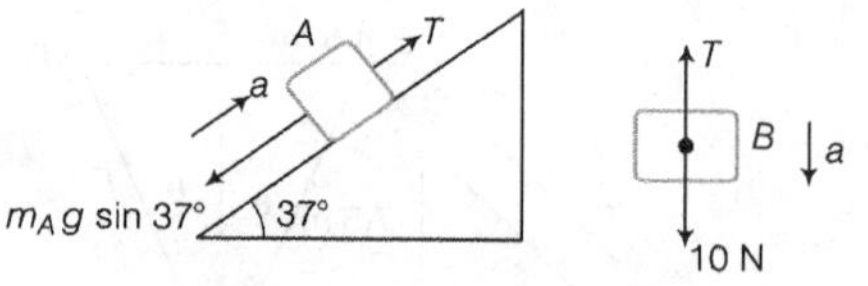

From free body diagram of block A,

$$T - m_A g \sin 37° = ma$$

$$T - 10 \times \frac{3}{5} = 1 \times a$$

or $$T - 6 = a \quad \ldots(i)$$

From free body diagram of block B,

$$10 - T = ma = 1 \times a \quad \ldots(ii)$$

From Eqs. (i) and (ii), we get

$$10 - 6 = 2a$$

$\therefore$ $$a = \frac{4}{2} = 2 \text{ m/s}^2$$

Hence, Statement I is correct.

$\because$ $$a = \frac{F}{m}$$

Force is a vector quantity. So, acceleration is vector quantity.
Hence, Statement II is correct.

20. (c) Here, $2T \cos \theta = mg$

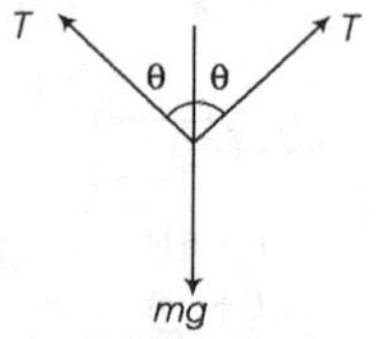

$\therefore$ $$T = \frac{mg}{2 \cos \theta}$$

In Case II, θ is more, so $\cos \theta$ is lesser. It means tension in Case II is more than that in Case I. So, in Case II, string is most likely to break.

Hence, Statement I is correct.

$\because T \propto \frac{1}{\cos \theta}$. So, tensions in strings in both cases are not same.

Hence, Statement II is incorrect.

21. (a) Tension is an electromagnetic force. Electromagnetic force is due to charge in fraction, neutron is a neutral particle. So, neutrons are not responsible for electromagnetic force. Thus, tension in string is zero. Hence, Statement I and Statement II are correct and Statement II is the correct explanation of Statement I.

22. (d) $4T = 900$

$\therefore$ $$T = 225 \text{ N}$$

For man, $$N + 2T = 600$$

$\therefore$ $$N = 600 - 450 = 150 \text{ N} = 15 \text{ kg-wt}$$

23. (b) $4T - 900 = 90a \quad \ldots(i)$

Also, $$N + 2T - 600 = 60a$$

$$600 + 2T - 600 = 60a$$

$\therefore$ $$T = 30a \quad \ldots(ii)$$

From Eqs. (i) and (ii), we get

$$a = 30 \text{ m/s}^2$$

$\therefore$ $$T = 30a = 900 \text{ N}$$

24. (a) The man will leave the weighing machine.

25. (b) When direction of motion will reverse at the instant, the velocity becomes zero.

$\because$ $$v = u + at$$

or $$0 = 2 + a \times 3$$

$\therefore$ $$a = -\frac{2}{3} \text{ m/s}^2$$

Thus, acceleration of A is $\frac{2}{3}$ m/s^2 in rightward direction.

26. (b)

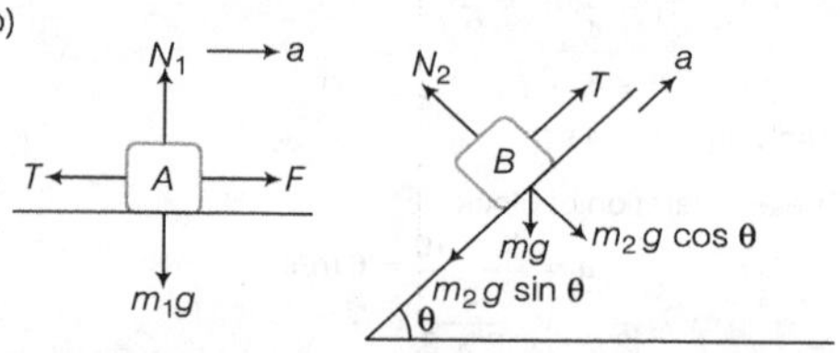

From FBD of A,

$$F - T = m_1 a = 1 \times \frac{2}{3} \quad \ldots(i)$$

From FBD of B, $\ldots(ii)$

$$T - m_2 g \sin \theta = m_2 a$$

$$= 2 \times \frac{2}{3} = \frac{4}{3}$$

or $$T - 10 = \frac{4}{3}$$

$\therefore$ $$T = 10 + \frac{4}{3} = \frac{34}{3}$$

$$= 11.33 \text{ N}$$

27. (d) From Eq. (i), we get

$$F - T = \frac{2}{3}$$

$\therefore$ $$F = \frac{2}{3} + 11.33 = 12 \text{ N}$$

28. (a) $T_A = 10t, T_B = T_C = 5t$

At $t = 2$s,

(i) → (q)

$$20 - 10 = 1 \times a_1$$

$\therefore$ $$a_1 = 10 \text{ m/s}^2$$

$$T_B < m_B g, a_B = 0$$

(ii) → (p)

$$m_C g - T_C = m_C a_C,$$

$$a_C = \frac{20}{3} \text{ m/s}^2$$

For $$\int_0^{v_A} dv \int_1^2 (10t - 10)dt$$

$\therefore$ $$v_A = 10\left(\frac{t^2}{2} - t\right)_1^2$$

$$= 10\left(\frac{1}{2}\right)$$

$$= 5 \text{ m/s}$$

(iii) → (r)

$$C \to 30 - 5t = \frac{dv_C}{dt}$$

$\therefore$ $$v_C = \left[30t - \frac{5t^2}{2}\right]_0^2$$

$$= 60 - 10 = 50 \text{ m/s}$$

(ii) → (s)

DPP-4 Constraint Motion or Dependent Motion

1. (i) $u = v$, since string is taut.

(ii) The component of velocities along string should be same.

$\therefore\ u \cos \theta_2 = v \cos \theta_1$

(iii) $u = v \cos \theta$, since component of velocities along the string must be same.

(iv) $u \cos \theta = v$, since component of velocity along string must be same.

(v) Since, component of velocities along the string must be same. So,

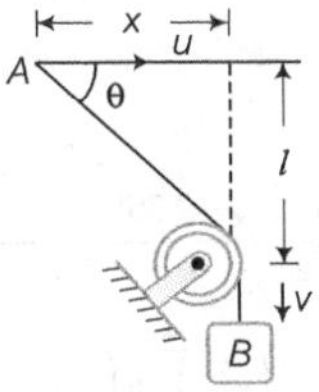

$$v = u \cos \theta = u \frac{x}{\sqrt{x^2 + l^2}} \qquad \left[\because \cos \theta = \frac{x}{\sqrt{x^2 + l^2}}\right]$$

$$v = \frac{xu}{\sqrt{x^2 + l^2}}$$

(vi) The component of velocity is same for all the points on string.

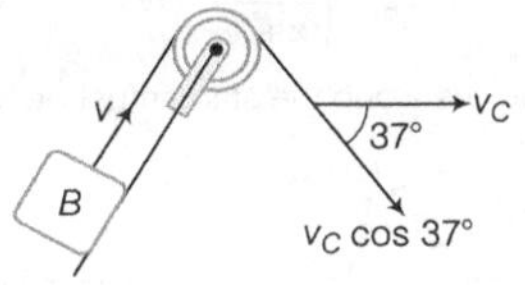

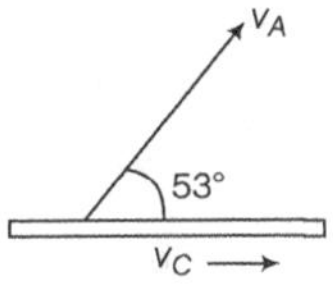

$$v_C \cos 53° = v_A = u$$

$$v_B = v = v_C \cos 37°$$

$$\therefore \quad \frac{u}{v} = \frac{\cos 53°}{\cos 37°} = \frac{3}{4}$$

$$\therefore \quad v = \frac{4}{3} u$$

2 (i) Average velocity on both sides of movable pulley should be same.

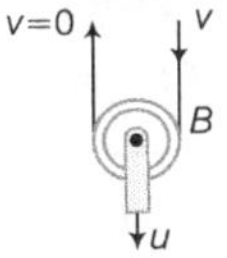

$$\therefore \quad u = \frac{0 + v}{2} \Rightarrow v = 2u$$

(ii) $\therefore$

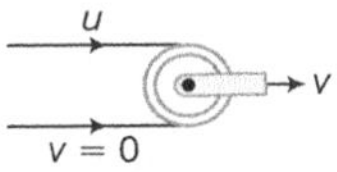

$$\frac{0 + u}{2} = v$$

$$\Rightarrow \quad u = 2v$$

(iii) For pulley B For pulley A

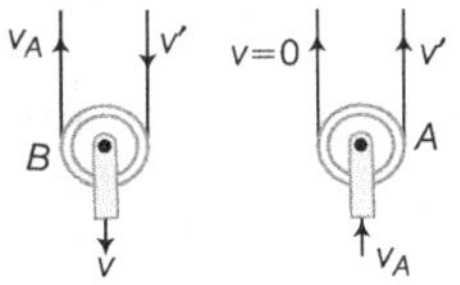

$$\frac{v' - v_A}{2} = v \qquad \ldots \text{(i)}$$

From pulley A, $\quad v_A = \dfrac{0 + v'}{2}$

$$\therefore \quad v' = 2v_A \qquad \ldots \text{(ii)}$$

From Eqs. (i) and (ii), we get

$$v_A = 2v \Rightarrow u = 2v$$

(iv) For pulley A,

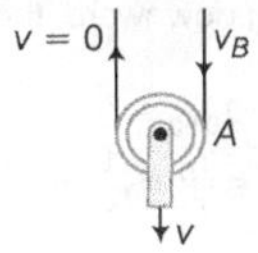

$$v = \frac{0 + v_B}{2} = \frac{u}{4} \qquad \left[\because v_B = \frac{u}{2}\right]$$

$$\therefore \quad u = 4v$$

For pulley B,

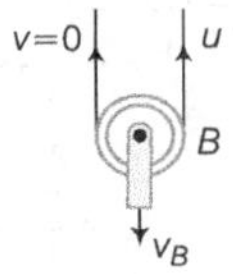

$$v_B = \frac{u + 0}{2}$$

$$\therefore \quad u = 2v_B$$

3. (i) $v_3 = \dfrac{v_1 + v_2}{2}$

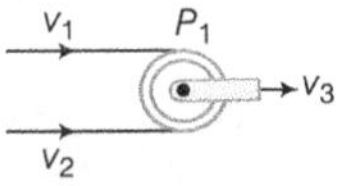

$$\therefore \quad v_1 + v_2 = 2v_3$$

(ii) For pulley P_3,

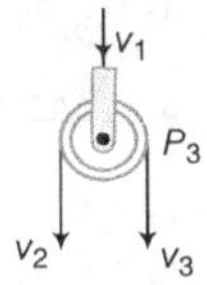

$$v_1 = \frac{v_2 + v_3}{2}$$

$$\Rightarrow \quad v_2 + v_3 = 2v_1$$

(iii) For pulley P_1,

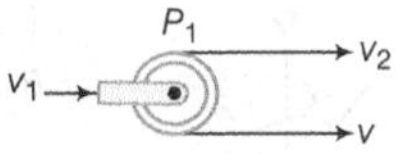

$$v_1 = \frac{v + v_2}{2} \qquad \ldots \text{(i)}$$

For pulley P_3,

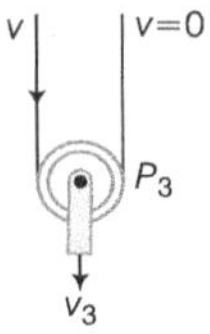

$$\frac{0 + v}{2} = v_3$$

$$\therefore \quad v = 2v_3 \qquad \ldots \text{(ii)}$$

From Eqs. (i) and (ii), we get

$$v_1 = \frac{2v_3 + v_2}{2}$$

(iv) For pulley P_1,

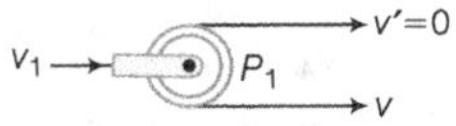

$$\Rightarrow \quad v_1 = \frac{v + v'}{2} \qquad \ldots \text{(i)}$$

$$\text{or} \quad v_1 = \frac{v}{2}$$

For block m_3,

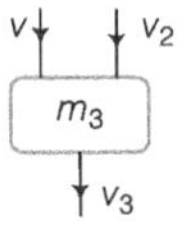

$$v_2 = v_3 = v \qquad \ldots \text{(ii)}$$

From Eqs. (i) and (ii), we get

$$v_2 = v_3 = 2v_1$$

4. (i) If tension forces are internal forces of a system, then $\Sigma \mathbf{T}_i \cdot \mathbf{v}_i = 0$

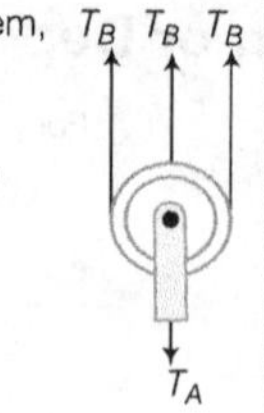

Here $\qquad T_A = 3\,T_B$

If block A is moving downward, then block B is moving leftward.

$\therefore \qquad -T_A v_1 + T_B v_2 = 0$

$\therefore \qquad \dfrac{v_1}{v_2} = \dfrac{T_B}{T_A} = \dfrac{1}{3} \Rightarrow v_2 = 3v_1$

$\therefore \qquad \dfrac{dv_2}{dt} = \dfrac{3dv_1}{dt}$

$\Rightarrow \qquad a_2 = 3a_1$

(ii) $\Sigma \mathbf{T}_i \cdot \mathbf{v}_i = 0$ or $T_B v_B - T_A v_A = 0$

$3T\, v_B - T_A v_A = 0$

$\therefore \qquad 3v_B = 2v_A$

$\therefore \qquad v_A = \dfrac{3}{2} v_B$

$\therefore \qquad a_A = \dfrac{3}{2} a_B$

(iii) $-T_A v_A + T_B v_B = 0$ or $-5T v_A + 4T v_B = 0$

$\therefore \qquad v_A = \dfrac{4}{5} v_B$

$\therefore \qquad a_A = \dfrac{4}{5} a_B$

(iv) $\qquad -3T\, v_B + 2T\, v_A = 0$

$\therefore \qquad 2v_A = 3v_B$

$\therefore \qquad a_A = \dfrac{3}{2} a_B$

(v) $v_A = \dfrac{v_P + v_B}{2} \qquad \Rightarrow a_A = \dfrac{a_0 + a_B}{2}$

(vi) $\qquad T v_B + 8T\, v_A = 0$

$\therefore \qquad v_B = 8\, v_A \quad \Rightarrow \quad a_B = 8\, a_A$

5. (i) Total length of a string remains constant.

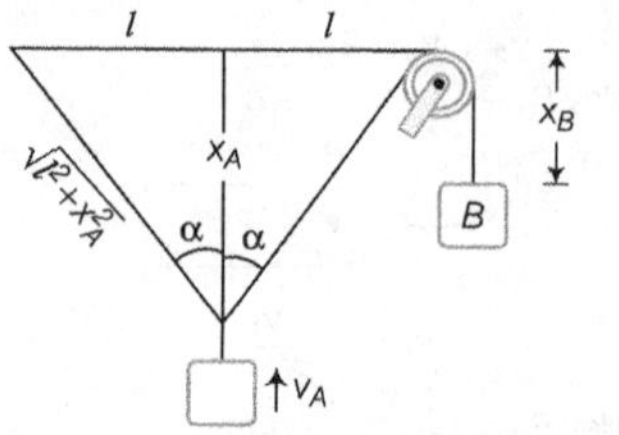

$$2\sqrt{l^2 + x_A^2} + x_B = C$$

$$2\,\frac{1}{\sqrt{l^2 + x_A^2}}\, x_A \frac{dx_A}{dt} + \frac{dx_B}{dt} = 0$$

$$-2\cos\alpha\, v_A + v_B = 0$$

$\therefore \qquad v_B = 2v_A \cos\alpha$

(ii) $v_A : v_{A_2} = 3 : 5$

$\therefore \qquad \dfrac{v_A}{v_{A_2}} = \dfrac{3}{5} \qquad \Rightarrow \qquad v_A = \dfrac{3}{5} v_0$

$v_B = v_A \cos 45°$

$\therefore \qquad v_B = \dfrac{v_A}{\sqrt{2}}$

(iii) Total length of string remains constant.

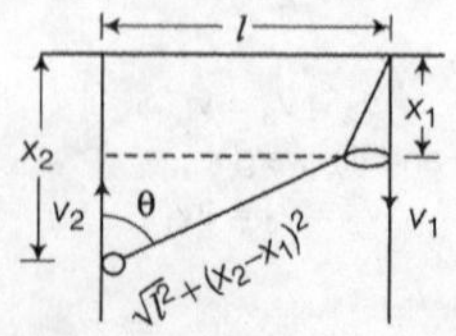

$\therefore \qquad x_1 + \sqrt{l^2 + (x_2 - x_1)^2} = C$

or $\qquad \dfrac{dx_1}{dt} + \dfrac{1}{\sqrt{l^2 + (x_2 - x_1)^2}}(x_2 - x_1)\left(\dfrac{dx_2}{dt} - \dfrac{dx_1}{dt}\right) = 0$

or $\qquad v_1 + \cos\theta\,(-v_2 - v_1) = 0$

$\therefore \qquad v_2 = v_1\left(\dfrac{1 - \cos\theta}{\cos\theta}\right)$

Put $\theta = 60°$ in above equation, $\quad v_2 = v_1$

$\Rightarrow \qquad v_A = v_B$

(iv) $\because$ Length of string = constant

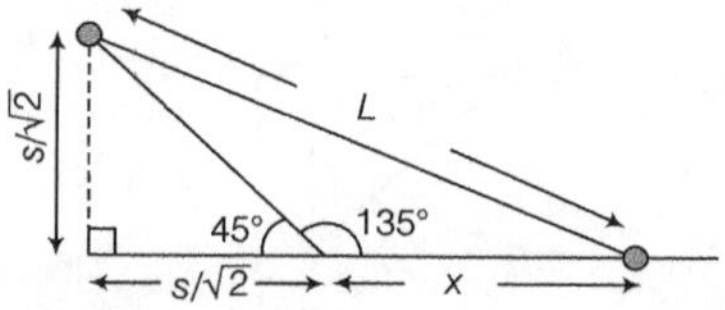

$\Rightarrow \qquad \left(\dfrac{s}{\sqrt{2}}\right)^2 + \left(x + \dfrac{s}{\sqrt{2}}\right)^2 = L^2$

Differentiating w.r.t. time, we get

$$s \cdot (-v_B) + 2\left(x + \frac{s}{\sqrt{2}}\right)\left(v_A - \frac{v_B}{\sqrt{2}}\right) = 0$$

Arranging and solving, we get

$$v_B = \left(\frac{s + \sqrt{2}x}{x + \sqrt{2}s}\right) v_A$$

(v) Since, velocities along the string must be same.

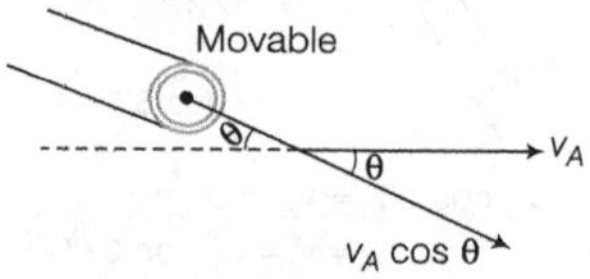

$\Rightarrow \qquad v_A \cos\theta = \dfrac{v_B}{2}$

or $\qquad v_A \dfrac{x}{\sqrt{x^2 + h^2}} = \dfrac{v_B}{2} \qquad \left[\because \cos\theta = \dfrac{x}{\sqrt{x^2 + h^2}}\right]$

or $\qquad v_B = \dfrac{2x}{\sqrt{x^2 + h^2}} v_A$

6. (i)

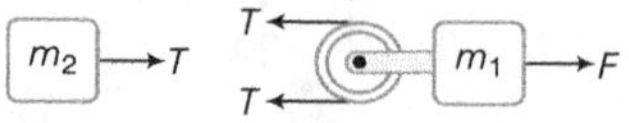

$T = m_2 a_2 \qquad$ ...(i)

and $\qquad F - 2T = m_1 a_1 \qquad$ ...(ii)

Also, $\qquad v_2 = 2v_1$

$\dfrac{dv_2}{dt} = 2\dfrac{dv_1}{dt}$

$\therefore \qquad a_2 = 2a_1 \qquad$ ...(iii)

From Eqs. (i), (ii) and (iii), we get

$$a_1 = \frac{F}{4m_2 + m_1} \quad \text{and} \quad a_2 = \frac{2F}{m_1 + 4m_2}$$

(ii) Here, $\dfrac{v_1}{v_2} = \dfrac{2T}{3T}$

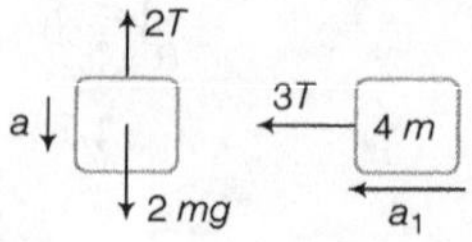

$\therefore \qquad v_2 = \dfrac{3}{2} v_1$

$\therefore \qquad a_2 = \dfrac{3}{2} a_1 \qquad$ ...(i)

$2mg - 2T = 2ma_2 \qquad$ ...(ii)

and $\qquad 3T = 4ma_1 \qquad$ ...(iii)

From Eqs. (i), (ii) and (iii), we get

$$a_1 = \frac{6g}{17}\ \text{m/s}^2 \quad \text{and} \quad a_2 = \frac{9}{17} g$$

(iii) $a_B = 2a_A$

$$m_2g - T = m_2a_B$$

or $\quad 20 - T = 4a_B \quad \ldots(i)$

Also, $\quad 2T - m_1g\sin 37° = m_1a_A$

or $\quad 2T - 12 = 2a_A \quad \ldots(ii)$

From Eqs. (i) and (ii), we get

$$a_A = 2.8\text{ m/s}^2 \quad \text{and} \quad a_B = 5.6\text{ m/s}^2$$

(iv) $m_1g\sin 37° - T = m_1a_1$

or $\quad 12 - T = 2a_1 \quad$ (for block A)

$m_2g\sin 37° - T = 2a_2 \quad$ (for block B)

or $\quad 12 - T = 2a_2 \quad \ldots(i)$

$\therefore \quad a_1 = a_2$

$2T - m_3g = m_3a_3 \quad$ (for block C)

or $\quad 2T - 20 = 2a_3 \quad \ldots(ii)$

Also, $\quad a_3 = \dfrac{a_1 + a_2}{2} = \dfrac{2a_1}{2}$

$\therefore \quad a_1 = a_2 = a_3$

From Eqs. (i) and (ii), we get

$$a_1 = a_2 = a_3 = \frac{2}{3}\text{ m/s}^2$$

(v) $2T = 1 \times a_1 \quad$ (for block A)

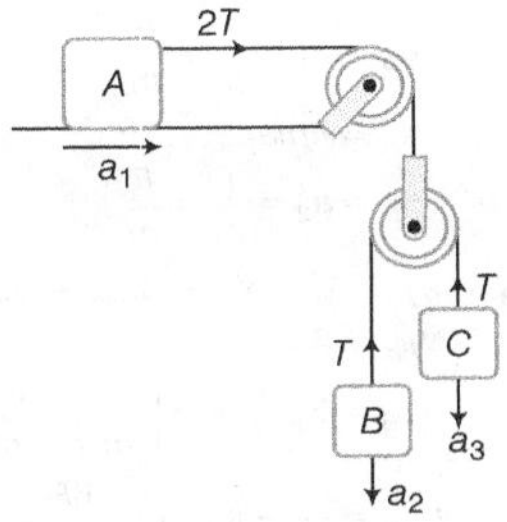

$20 - T = 2a_2 \quad$ (for block B)

$30 - T = 3a_2 \quad$ (for block C)

Also, $\quad a_1 = \dfrac{a_2 + a_3}{2}$

or $\quad 2T = \dfrac{10 - \frac{T}{2} + 10 - \frac{T}{3}}{2}$

$\therefore \quad T = \dfrac{120}{29}\text{ N}$

$\therefore \quad a_1 = 2T = \dfrac{240}{29}\text{ m/s}^2$

7. (i) Total length of string remains constant.

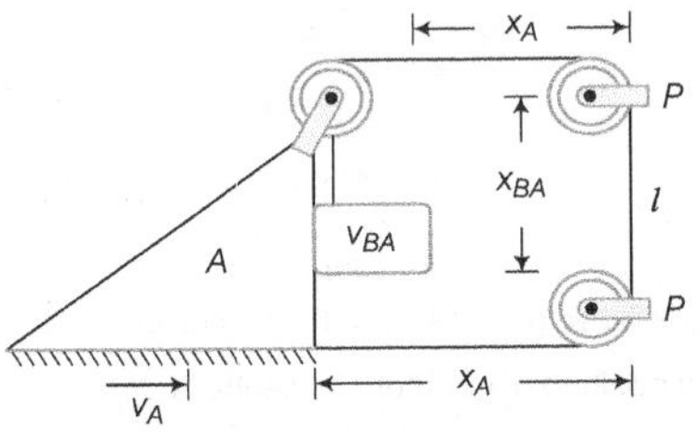

$\therefore \quad x_A + l + x_A + x_{BA} = \text{constant}$

or $\quad -2v_A + v_{BA} = 0$

$\therefore \quad v_{BA} = 2v_A$

or, $\quad a_{BA} = 2a_A$

But $\quad \mathbf{a}_{BA} = \mathbf{a}_B - \mathbf{a}_A$

$\therefore \quad \mathbf{a}_B = \mathbf{a}_{BA} + \mathbf{a}_A$

$\therefore \quad a_B = \sqrt{a_{BA}^2 + a_A^2} = \sqrt{5}\,a_A$

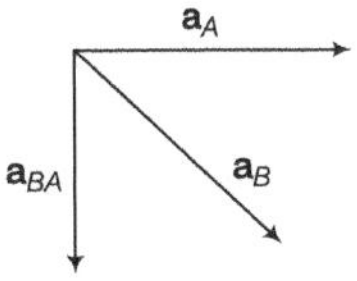

(ii) Total length of string remains constant.

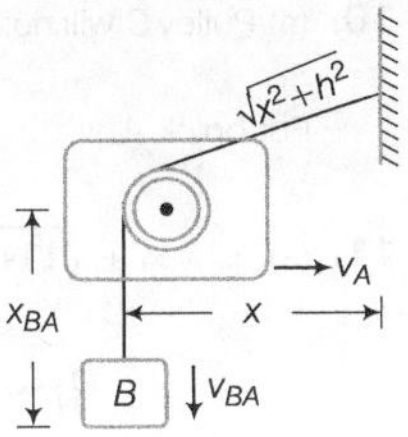

$\therefore \quad x_{BA} + \sqrt{x^2 + h^2} = C$

or $\quad \dfrac{d}{dt}x_{BA} + \dfrac{d}{dt}\sqrt{x^2 + h^2} = 0$

or $\quad v_{BA} + \dfrac{x}{\sqrt{x^2 + h^2}}\dfrac{dx}{dt} = 0$

or $\quad v_{BA} - \dfrac{x}{\sqrt{x^2 + h^2}}v_A = 0$

$\Rightarrow \quad v_{BA} = \dfrac{x}{\sqrt{x^2 + h^2}}v_A$

But, $\quad \mathbf{v}_{BA} = \mathbf{v}_B - \mathbf{v}_A$

$\Rightarrow \quad \mathbf{v}_B = \mathbf{v}_{BA} + \mathbf{v}_A$

$\therefore \quad v_B = \sqrt{v_{BA}^2 + v_A^2} = \sqrt{\dfrac{x^2}{x^2 + h^2}v_A^2 + v_A^2}$

$$= v_A\sqrt{\frac{2x^2 + h^2}{x^2 + h^2}}$$

(iii) $\because \quad x_{rel} + x_1 + x_0 = C$

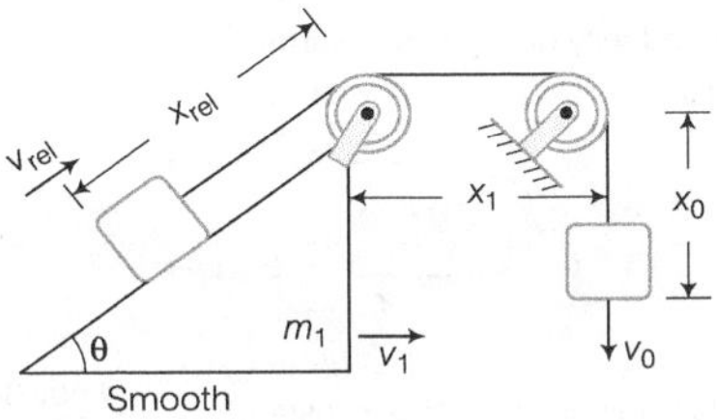

or $\quad \dfrac{dx_{rel}}{dt} + \dfrac{dx_1}{dt} + \dfrac{dx_0}{dt} = 0$

or $\quad -v_{rel} - v_1 + v_0 = 0$

$\therefore \quad v_1 = v_0 - v_{rel}$

$\therefore \quad \dfrac{dv_1}{dt} = \dfrac{dv_0}{dt} - \dfrac{dv_{rel}}{dt}$

$\therefore \quad a_1 = a_0 - a_{rel} = 5 - 3 = 2\text{ m/s}^2$

8. (c) $F = T_2\cos\theta \quad \ldots(i)$

and $\quad T_2\sin\theta = T_1 = 10 \times \dfrac{100}{\sqrt{3}} \quad \ldots(ii)$

$T_2 = 20 \times \dfrac{100}{\sqrt{3}} \quad \ldots(iii)$

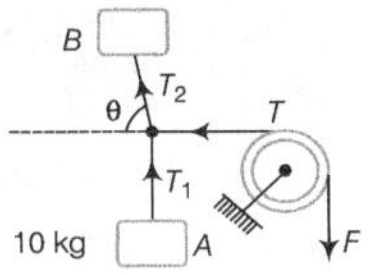

From Eqs. (i), (ii) and (iii), we get

$$F = 1000\text{ N}$$

9. (b) $\quad T\cos\theta = ma \quad \ldots(i)$

$F = 2T\sin\theta \quad \ldots(ii)$

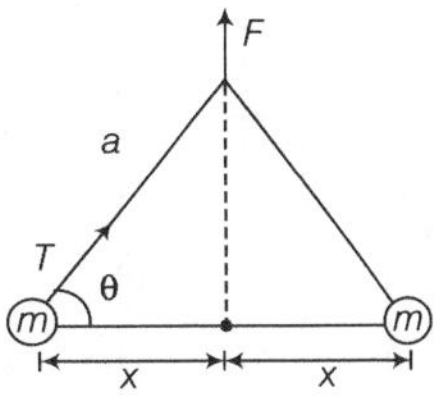

From Eqs. (i) and (ii), we get

$\Rightarrow \quad \dfrac{F}{2}\cot\theta = ma$

or $\quad \dfrac{F}{2}\dfrac{x}{\sqrt{a^2 + x^2}} = ma \quad \left(\because \cot\theta = \dfrac{x}{\sqrt{a^2 - x^2}}\right)$

or $\quad a = \dfrac{Fx}{2m\sqrt{a^2 - x^2}}$

10. (a) Pulley C will not move.

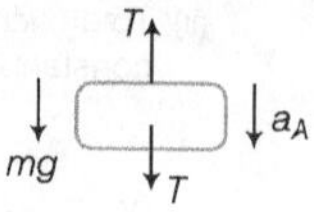

$$\therefore \quad a_B = 0$$

For bolck A, $\quad mg = ma_A$

$$\therefore \quad a_A = g$$

11. (a) $\therefore \quad x_1 + \sqrt{l^2 + (x_2 - x_1)} = C$

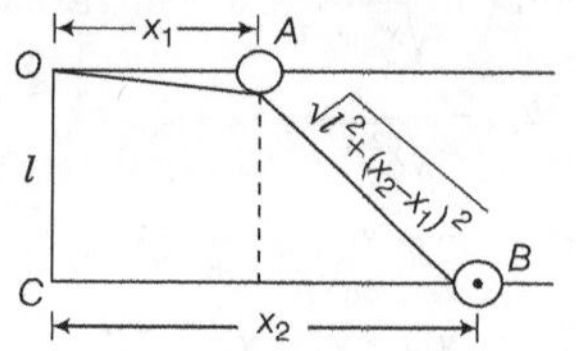

or $\quad \dfrac{dx_1}{dt} + \dfrac{x_2 - x_1}{\sqrt{l^2 + (x_2 - x_1)^2}}\left(\dfrac{dx_2}{dt} - \dfrac{dx_1}{dt}\right) = 0$

or $\quad 10 + \cos 60^\circ\left(\dfrac{dx_2}{dt} - 10\right) = 0$

or $\quad \dfrac{dx_2}{dt} = \dfrac{5 - 10}{\cos 60^\circ} = \dfrac{5-10}{\frac{1}{2}} = -10 \text{ cm/s}$

Thus, the velocity of ring B is towards C.

12. (a) Initially, at, $t = 0$

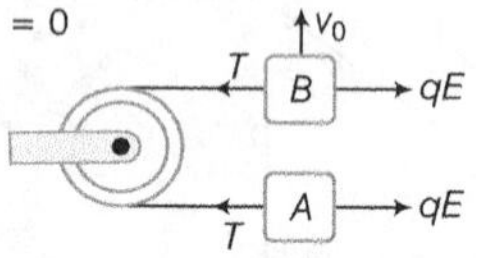

$$\uparrow = qE$$

From free body diagram of B at instant t, $a_B = \dfrac{T\cos\theta - qE}{m}$ in horizontal direction.

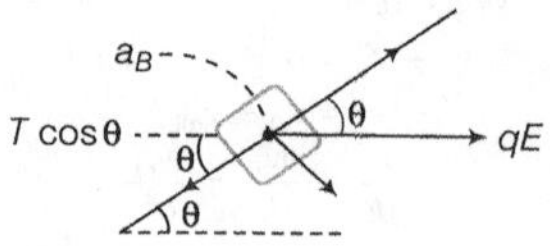

For block A,

$a_A = \dfrac{T - qE}{m}$ in horizontal direction.

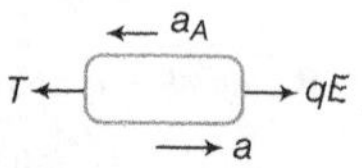

$$\therefore \quad a_A > a_B$$

Thus, block A is closer to pulley after sometime.

13. (d) $v \cos 30^\circ = v_0 \cos 60^\circ$ (along DC)

$$v = \frac{v_0}{\frac{\sqrt{3}}{2}} \times \frac{1}{2}$$

$$= \frac{v_0}{\sqrt{3}}$$

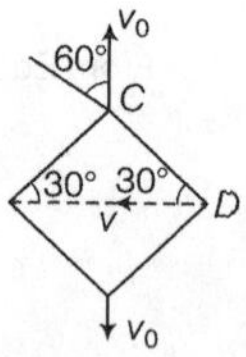

14. (c)

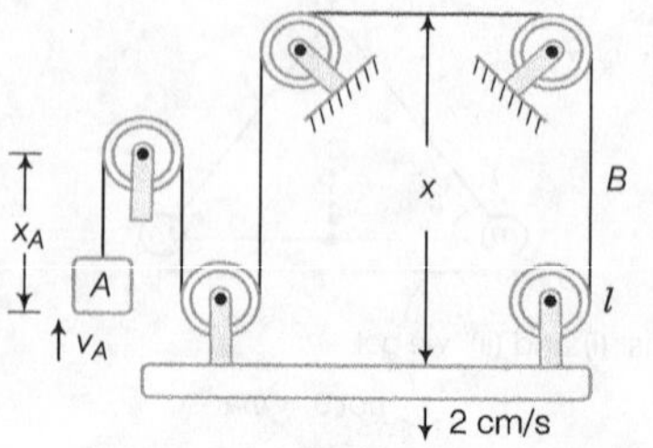

$x_A + 3x + l = C \quad$ or $\quad \dfrac{dx_A}{dt} + \dfrac{3dx}{dt} + \dfrac{dl}{dt} = 0$

$$-v_A + 3 \times 2 + 1 = 0$$

$$\therefore \quad v_A = 7 \text{ cm/s}$$

$\therefore$ Distance moved by block $= 7 \times 2 = 14$ cm

15. (b) From figure,

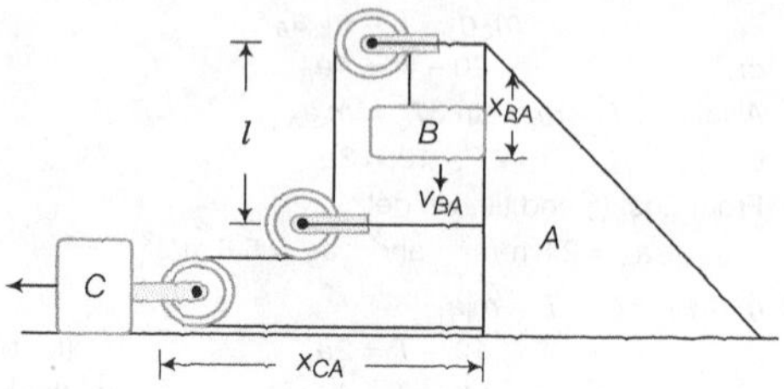

$$2x_{CA} + l + x_{BA} = C \quad \text{or} \quad 2v_{CA} + v_{BA} = 0$$

or $\quad 2\,(2-5) + v_{BA} = 0 \quad$ or $\quad -6 + v_{BA} = 0$

$$\therefore \quad v_{BA} = 6$$

$$\therefore \quad \mathbf{v}_{BA} = \mathbf{v}_B - \mathbf{v}_A \quad \text{or} \quad -6\hat{\mathbf{i}} = \mathbf{v}_B - (-5\hat{\mathbf{i}})$$

$$\therefore \quad \mathbf{v}_B = -5\hat{\mathbf{i}} - 6\hat{\mathbf{j}}$$

16. (c) Here,

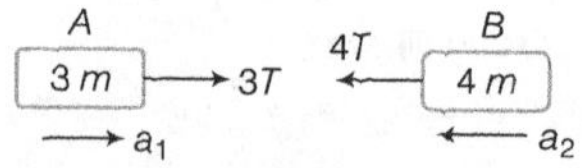

$$3T = 3\,ma_1$$

$\therefore \quad a_A = a_1 = \dfrac{T}{m} = \dfrac{F}{m} \qquad (\because T = F)$

and $\quad 4T = 4ma_2$

$$\therefore \quad a_B = a_2 = \frac{T}{m} = \frac{F}{m}$$

Here, $\quad 3x_{AB} + x_{PB} = C \quad$ or $\quad -3x_{AB} + x_{PB} = 0$

or $\quad v_{PB} = 3\,v_{AB}$

$$a_{PB} = 3\,a_{AB} = 3\left(\frac{2F}{m}\right) = \frac{6F}{m} \quad (\because a_{AB} = a_A - a_B)$$

Here, $\quad a_{PB} = a_P - a_B \quad$ or $\quad \dfrac{6F}{m} = a_P - \dfrac{F}{m}$

$$\therefore \quad a_P = \frac{7F}{m} = 7 \text{ m/s}^2$$

17. (a,b) $mg - 2T = ma$

From free body diagram of P_2 and $P_1 \Rightarrow a_1 = g \downarrow$

Free body diagram (for P_1)

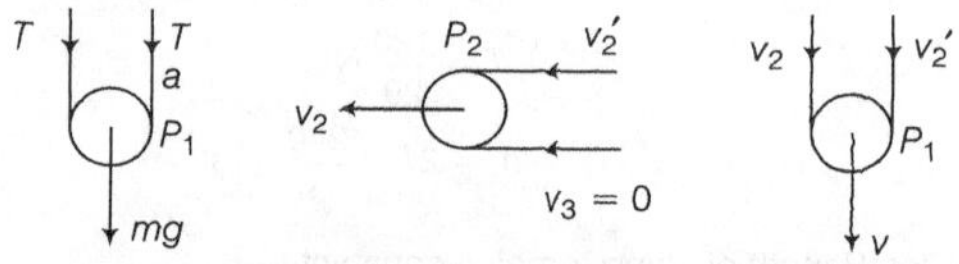

Here, $v_2 = \dfrac{v_2' + 0}{2}$

$$\Rightarrow \quad v_2' = 2v_2 \quad \text{and} \quad \frac{v_2 - v_2'}{2} = v$$

or $\quad \dfrac{v_2 - 2v_2}{2} = v \quad$ or $\quad -v_2 = 2v$

$$\therefore \quad -a_2 = 2a_1 = 2g$$

Pulley P_2 is moving rightward with an acceleration $2g$.

18. (a,b) The position of geometrical centre is

$$X = \frac{x}{2} + \frac{x}{4} = \frac{3x}{4}$$

$$\therefore \quad v_{CM} = \frac{dX}{dt} = \frac{3}{4}\frac{dx}{dt} = \frac{3}{4} \times 8 = 6 \text{ cm/s}$$

and $\quad x_A = \dfrac{x}{2}$

$$\therefore \quad v_A = \frac{1}{2}\frac{dx}{dt}$$

$$= \frac{1}{2} \times 8 = 4 \text{ cm/s}$$

19. (a)
$$T\cos 30^\circ = ma_1 \quad \ldots\text{(i)} \quad \text{(for ring)}$$
$$2mg - T = ma_2 \quad \ldots\text{(ii)} \quad \text{(for block)}$$
$$v_1 \cos\theta = v_2$$
or $$\left(\frac{dv_1}{dt}\right)\cos\theta + v_1\frac{d}{dt}(\cos\theta) = \frac{dv_2}{dt}$$
or $$a_1\cos\theta + 0 \times \frac{d}{dt}(\cos\theta) = a_2$$
$\therefore$ $$a_2 = a_1\cos\theta \quad \ldots\text{(iii)}$$
From Eqs. (i), (ii) and (iii), we get
$$T = 8m = 8 \times 1 = 8\text{ N}$$

20. (b) $N + T\cos 60^\circ = mg$

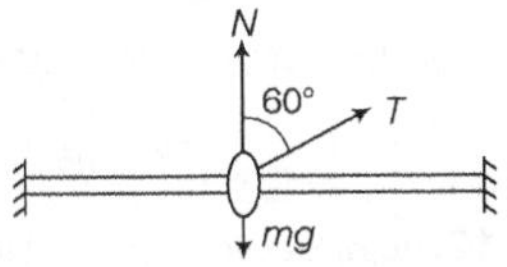

$\therefore$ $$N = mg - T\cos 60^\circ$$
$$= 10 - 8 \times \frac{1}{2}$$
$$= 6\text{ N}$$

21. (c)

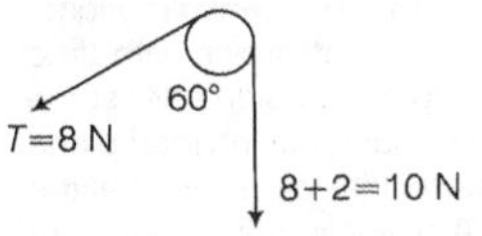

$$F_C = \sqrt{8^2 + 10^2 + 2 \times 8 \times 10\cos 60^\circ}$$
$$= \sqrt{64 + 100 + 80} = \sqrt{244}\text{ N}$$

22. (b) $T = m_2 g = 400$ N

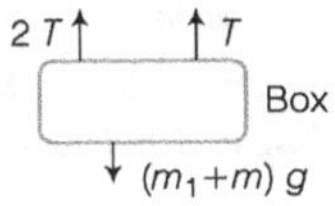

$$3T = m_1 g + mg$$
(here, m = mass of man).
$$1200 - 200 = mg$$
$$1000 = mg$$
For man,

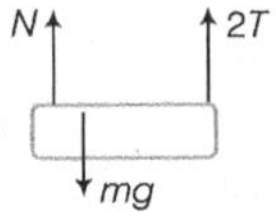

$$N + 2T = mg$$
$\therefore$ $$N = mg - 2T = 1000 - 800$$
$$= 200\text{ N} = 20\text{ kg}$$

23. (a) $m_2 g - T = m_2 a$ or $400 - T = 40 \times 5 = 200$
$\therefore$ $$T = 200\text{ N}$$
$\therefore$ $$3T - 200 - mg = (20 + m)\frac{5}{3} \quad \ldots\text{(i)}$$
and $$N - mg = m\left(\frac{5}{3}\right)$$
$\therefore$ $$N = m\left(g + \frac{5}{3}\right) \quad \ldots\text{(ii)}$$
From Eqs. (i) and (ii), we get
$$400 = 20 \times \frac{5}{3} + m\left(g + \frac{5}{3}\right)$$
$\therefore$ $$N = \frac{1100}{3}\text{ N} = \frac{110}{3}\text{ kg-wt}$$

24. (2) For m_2,
$$T - m_2 g = m_2(2a) \quad \ldots\text{(i)}$$
Free body diagram (FBD) of m_2

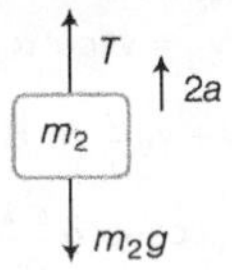

For m_1, $$m_1 g - 2m_2 = m_1(a) \quad \ldots\text{(ii)}$$
Free body diagram (FBD) of m_1

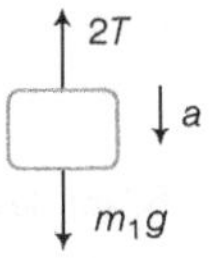

From Eqs. (i) and (ii), we get
$$m_1 g - 2m_2 g = 4m_2 a + m_1 a$$
$\Rightarrow$ $$a = 0\text{ m/s}^2$$
or $$T = m_2 g = 10\text{ N}$$
For m, $2T = mg \Rightarrow m = \frac{20}{10} = 2$ kg

Free body diagram of m

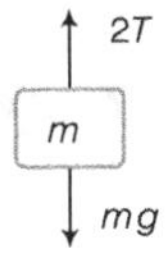

$$mg = 2T$$
$\Rightarrow$ $$m = \frac{2 \times 10}{10}$$
$$= 2\text{ kg}$$

25. (2) In the situation shown in the figure, we will define this infinite system as follows.
Consider it is to be made of n pulleys with a non-zero mass replacing what would have been the $(n + 1)$ pulley. Then, take the limit as $n \to \infty$.

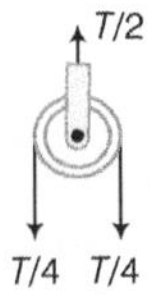

$$\frac{T}{2} - mg = ma_1$$
$$\frac{T}{4} - mg = ma_2$$
$\therefore$ $$\frac{T}{2n} - mg = ma_n$$
Also,
$$2^{n-1}a_1 + 2^{n-2}a_2 + \cdots + a_n = 0$$
or $$2^{n-1}\left(\frac{T}{2} - mg\right) + 2^{n-2}\left(\frac{T}{2^2} - mg\right)$$
$$+ \cdots + \left(\frac{T}{2^n} - mg\right) + \left(\frac{T}{2^n} - mg\right) = 0$$
or $$T(2^{n-2} + 2^{n-4} + 2^{n-6} + \cdots + 2^{-n} + 2^{-n})$$
$$= mg(2^{n-1} + 2^{n-2} + \cdots + 1 + 1)$$
$\Rightarrow$ $$T = 3\,mg$$
$\therefore$ $$a_1 = \frac{g}{2} = 5\text{ m/s}^2$$
or $$x = 2$$

DPP-5 Wedge Constraints

1. (b) The component of velocity along the normal, i.e. along z-direction must be same. So, for motion in XY-plane, z-component of velocity must be zero.

2. (b) The component of velocity along the normal direction, i.e. along z-direction must be same for the table and block so that block maintains contact with table.
$$(v_z)_{\text{block}} = (v_z)_{\text{table}} = 5\text{ m/s}$$

3. (d) The z-component of the velocity of table and block must be same.

4. (c) In the figure, $x = R\cos\alpha$

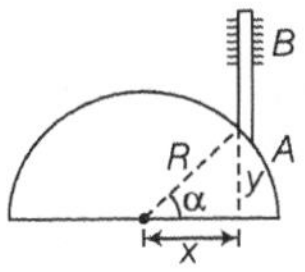

and $$y = R\sin\alpha$$
Here, $$\frac{dx}{dt} = -R\sin\alpha\frac{d\alpha}{dt}$$
or $$-v_0 = -R\sin\alpha\frac{d\alpha}{dt}$$
or $$\frac{d\alpha}{dt} = \frac{v_0}{R\sin\alpha} \quad \ldots\text{(i)}$$
Also, $$y = R\sin\alpha$$
$\therefore$ $$\frac{dy}{dt} = R\cos\alpha\frac{d\alpha}{dt}$$

$$\frac{dy}{dt} = R\cos\alpha \times \frac{v_0}{R\sin\alpha} \qquad \text{[from Eq. (i)]}$$

or $$\frac{dy}{dt} = v_0 \cot\alpha$$

$\therefore$ $$v_B = v_0 \cot\alpha$$

5. (d) $a_B = \frac{dv_B}{dt} = \cot\alpha \frac{dv_0}{dt} + v_0 \frac{d}{dt}\cot\alpha$

or $$a_B = (\cot\alpha)a_0 - v_0 \operatorname{cosec}^2\alpha \frac{d\alpha}{dt}$$

$$= a_0 \cot\alpha - v_0 \operatorname{cosec}^2\alpha \left(\frac{v_0}{R\sin\alpha}\right)$$

$$= a_0 \cot\alpha - \frac{v_0^2}{R}\operatorname{cosec}^3\alpha$$

6. (d) Here, $a_B \cos 53° = a_A \cos 37°$

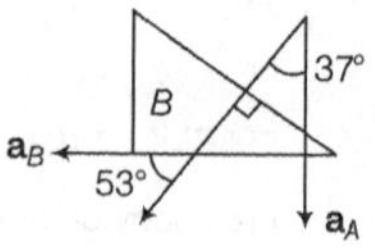

$$a_B \times \frac{3}{5} = 5 \times \frac{4}{5}$$

$\Rightarrow$ $a_B = \frac{20}{3}\text{ m/s}^2 = 6.67\text{ m/s}^2$

7. (b) $v_B = v_A \cos 60°$

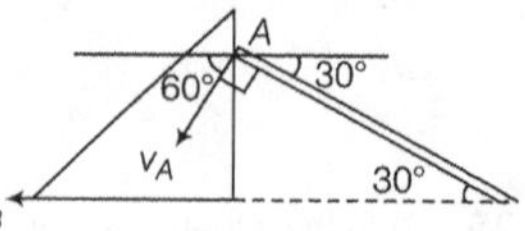

$$v_A = \frac{v_B}{\cos 60°}$$

$\therefore$ $v_A = 2v_B$

$= 2 \times 1 = 2$ cm/s

8. (b) Component of accelerations of cylinder, A and C along $O_1 O_3$ should be same.

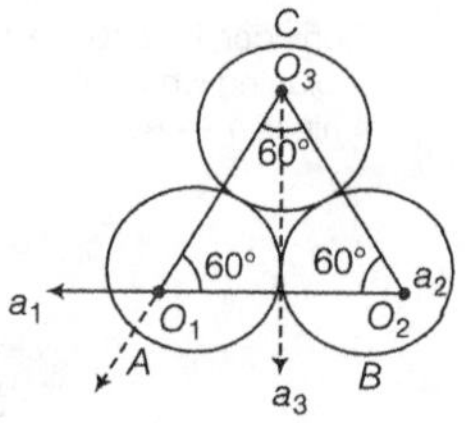

$\therefore$ $a_3 \cos 30° = a_1 \cos 60°$

or $\sqrt{3} \times \frac{\sqrt{3}}{2} = \frac{a_1}{2}$

$\therefore$ $a_1 = 3\text{ m/s}^2$

9. (c) For maintaining contact of cylinder with block A,

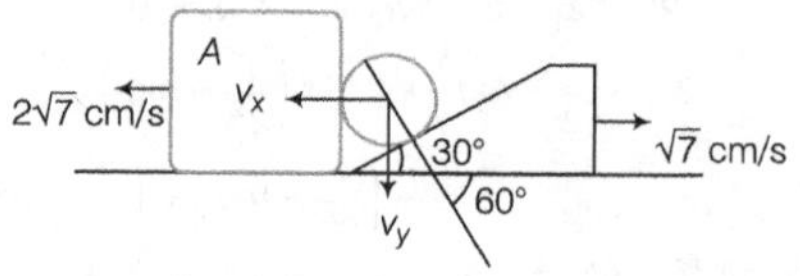

$$v_x = 2\sqrt{7}$$

For maintaining contact of cylinder with block B,

$$\sqrt{7}\cos 60° = v_y \cos 30° - v_x \sin 30°$$

$$\frac{\sqrt{7}}{2} = \frac{\sqrt{3}}{2} v_y - \sqrt{7}$$

$\therefore$ $$v_y = \sqrt{21}$$

$\therefore$ The speed of cylinder is

$$v = \sqrt{v_x^2 + v_y^2} = \sqrt{(2\sqrt{7})^2 + (\sqrt{21})^2} = 7\text{ cm/s}$$

10. (a,b,c) The components of a_{rod} and a_{wedge} along normal to surface of contact should be same.

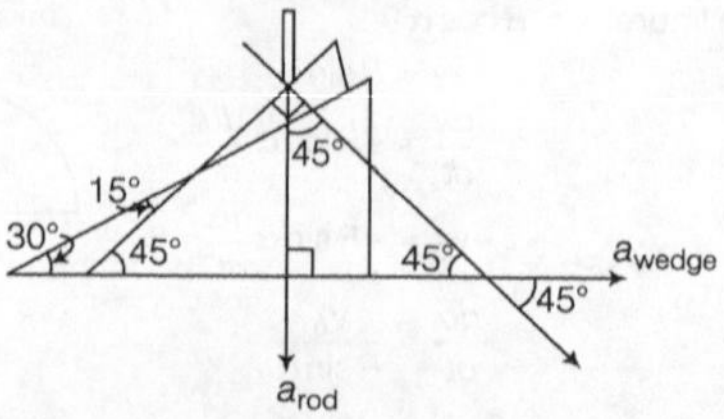

$\therefore$ $a_{wedge} \cos 45° = a_{rod} \cos 45°$

$\therefore$ $a_{wedge} = a_{rod}$

11. (a, b) Along AC, $v_A \cos 60° = v_C \cos 45°$

$\Rightarrow$ $v_A = \sqrt{2}\, v_C = 2$ cm/s (given, $v_C = \sqrt{2}$ cm/s)

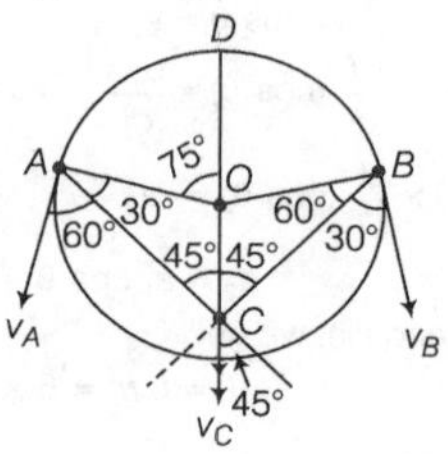

Along BC, $v_B \cos 30° = v_C \cos 45°$

$\therefore$ $$v_B = \frac{v_C}{\sqrt{2}} \times \frac{2}{\sqrt{3}} = \frac{\sqrt{2} \times 2}{\sqrt{2}\sqrt{3}} = \frac{2}{\sqrt{3}}\text{ cm/s}$$

12. (c,d) Sleeves starting at the same time from point A and travelling in different directions, always form a circle that grows with time and has A as its top most point. After some time, the circle will touch the inclined plane with the plane tangential to the circle at the contact point B. A sleeve starting from point A reaches the plane in the shortest time by travelling along the rod AB.

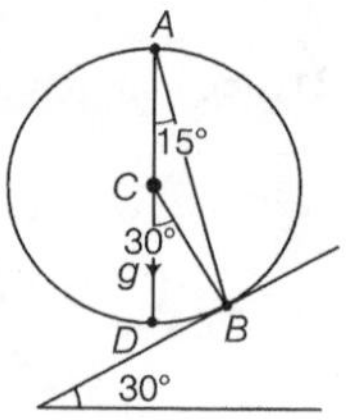

Here, $\angle DCB = 30°$ and C is the centre of circle.

$\therefore$ $\angle CAB = 15°$

13. (c,d) In figure (i), $100 = 2T_{max}$

In figure (ii), $F = 20\,a$ and $2T \cos 60° = 10\,a$

$\therefore$ $$T = 10\,a = \frac{F}{2}$$

$\therefore$ $$T_{max} = \frac{F_{max}}{2} \Rightarrow F_{max} = 100\text{ N}$$

14. (a,d) At any point along the path, the force mg acts downward and normal reaction N acts towards the centre on P.

Let the x and y components of velocity of particle P be v_{Px} and v_{Py}, respectively. From A to D, the horizontal component of the force N is towards right and mg has no horizontal component.

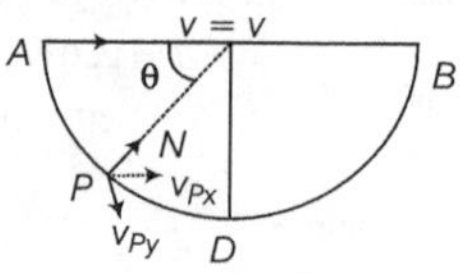

Thus, the net horizontal component is towards right. The horizontal component of velocity, v_{Px} thus increases from v and becomes maximum at D. From D to B, v_{Px} decreases and becomes v at B. It is clear that in the entire path ADB, $v_{Px} > v$.

On the other hand, the velocity of the particle P is always v on the path AB. That is $v_{Qx} = v$.

As the horizontal displacement is the same in the two cases, the particle with larger horizontal velocity will take less time. So, $t_P < t_Q$.

$\therefore$ $$\frac{t_P}{t_Q} < 1$$

15. (i) Along the rod, the component of velocities of A and B should be same.

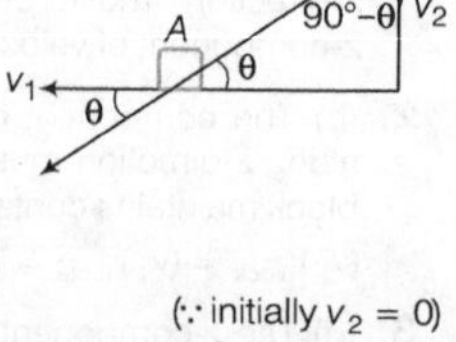

$\therefore$ $v_1 \cos\theta = v_2 \sin\theta$

or $v_1 = v_2 \tan\theta \left(a_2 = \frac{dv_2}{dt}\right)$

or $$\frac{dv_1}{dt} = \tan\theta \frac{dv_2}{dt} + v_2 \frac{d(\tan\theta)}{dt}$$

or $a_1 = a_2 \tan\theta$ ($\because$ initially $v_2 = 0$)

($\because \tan\theta = \frac{3}{4}$ $\therefore \theta = 37°$)

$$a_1 = a_2 \frac{3}{4} = \frac{3}{4} a_2$$

(ii) From free body diagram of block A,

$$F\cos 37° = m_1 a_1$$

or $\frac{4}{5}F = \frac{80}{3}a_1$

or $F = \frac{80}{3}\times\frac{5}{4}\times\frac{3}{4}a_2$

$\therefore$ $F = 25a_2$

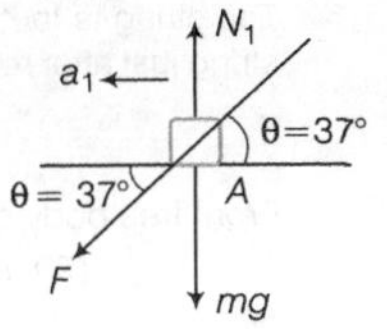

From free body diagram of block B

$$m_2 g - F\cos 53° = m_2 a_2$$

or $100 - 25a_2 \times \frac{3}{5} = 10\,a_2$

or $100 = 25a_2$

$\therefore$ $a_2 = 4\ \text{m/s}^2$

$\therefore$ $a_1 = \frac{3}{4}a_2 = 3\ \text{m/s}^2$

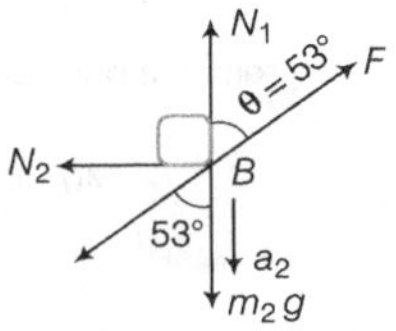

(iii) $\therefore$ $F = 25\,a_2 = 25\times 4 = 100\ \text{N}$

16. (i) Here, $a_A = a_B\cos 53°$

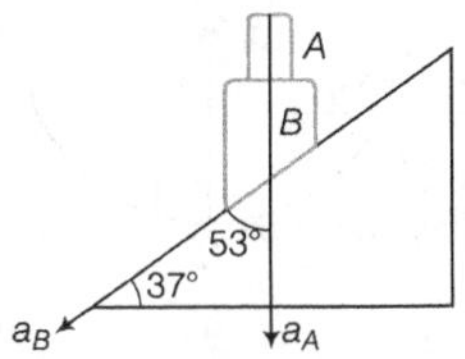

or $a_A = \frac{3}{5}a_B$ ...(i)

(ii) From free body diagram of the block A,

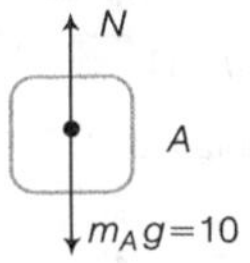

$$10 - N = 1\times a_A \quad \text{...(ii)}$$

Similarly, from free body diagram of the block B,

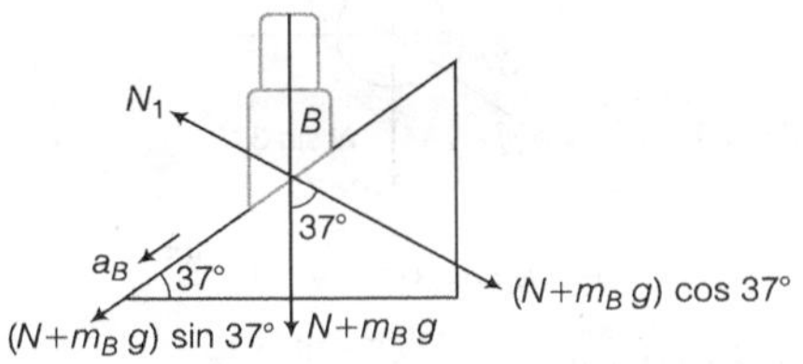

$(N + m_B g)\sin 37° = m_B a_B = 1\times\frac{5}{3}a_A$ (from Eq. (i))

or $(10 - a_A + 10)\frac{3}{5} = \frac{5}{3}a_A$ (from Eq. (ii))

or $20\times\frac{3}{5} = \left(\frac{3}{5}+\frac{5}{3}\right)a_A$

$\therefore$ $a_A = \frac{90}{17}\ \text{m/s}^2 = 5.29\ \text{m/s}^2$

(iii) $10 - N = 5.29 \Rightarrow N = 4.71\ \text{N}$

17. (i) The component of velocities of sphere and block perpendicular to surface of contact should be same.

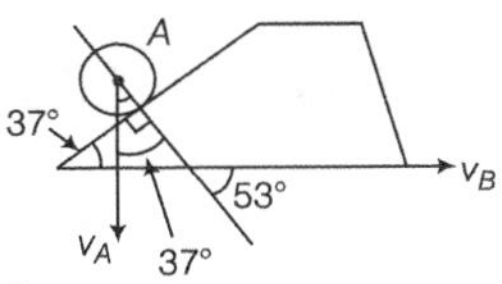

$\therefore$ $v_A\cos 37° = v_B\cos 53°$

or $v_A = \frac{v_B\cos 53°}{\cos 37°} = \frac{v_B\times 3}{5\times\frac{4}{5}} = \frac{3}{4}v_B$

$\because$ $a_A = \frac{dv_A}{dt} = \frac{3}{4}\frac{dv_B}{dt} = \frac{3}{4}a_B$ $\quad\therefore\quad a_A = \frac{3}{4}\times 3 = \frac{9}{4}\ \text{m/s}^2$

(ii) From free body diagram of block B,

$$N\cos 53° = m_B\,a_B = 1\times 3 = 3$$

$$N = \frac{3}{\cos 53°}$$

$\therefore$ $N = \frac{3}{\frac{3}{5}} = 5\ \text{N}$

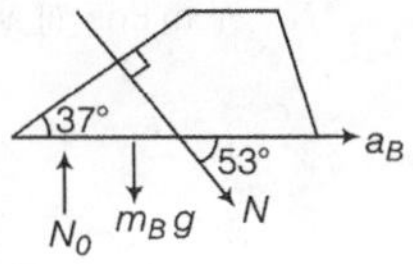

(iii) From free body diagram of the ball A,

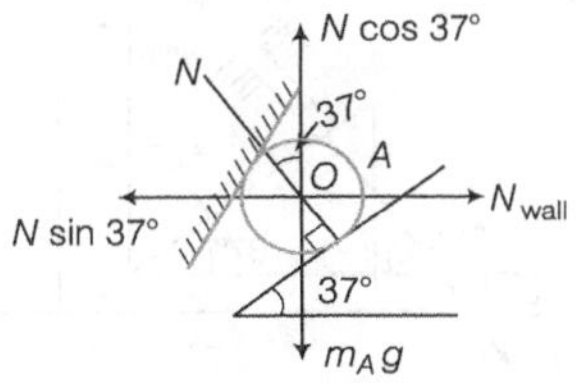

$N_{wall} = N\sin 37°$ (as $\mathbf{a}_x = 0$)

$\therefore$ $N_{wall} = 5\times\frac{3}{5} = 3\ \text{N}$

(iv) Also, $m_A g - N\cos 37° = m_A a_A$

or $m_A g - 4 = m_A\times\frac{9}{4}$

or $m_A\left(10 - \frac{9}{4}\right) = 4$

or $m_A = \frac{16}{31}\ \text{kg}$

18. (i) From free body diagram of m_3,

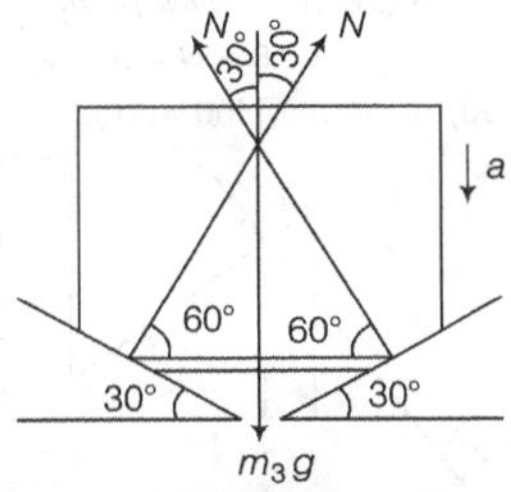

$$m_3 g - 2N\cos 30° = m_3 a_3$$

or $60 - \sqrt{3}\,N = 6a_3$ ...(i)

From constraint relation,

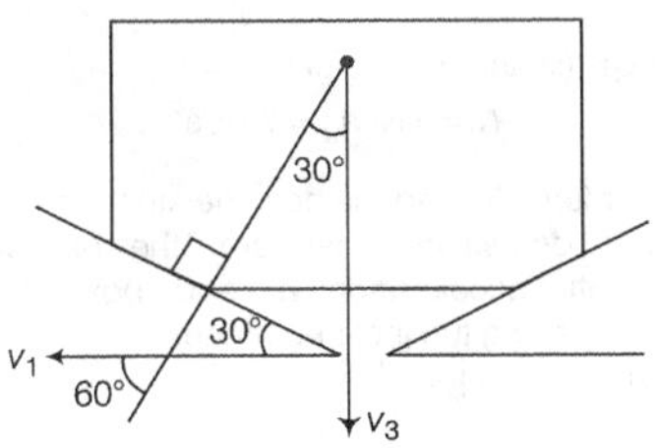

$$v_3\cos 30° = v_1\cos 60°$$

or $\sqrt{3}\,v_3 = v_1$

$\therefore$ $a_1 = \sqrt{3}\,a_3$ ...(ii)

(ii) From free body diagram of m_1,

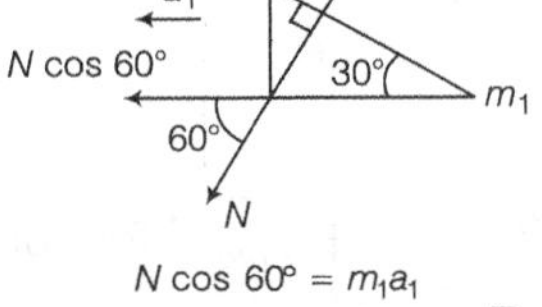

$$N\cos 60° = m_1 a_1$$

or $N = 2a_1 = 2\sqrt{3}\,a_3$ ...(iii)

From Eqs. (i) and (iii), we get

$$a_3 = 5\ \text{m/s}^2$$

$\because$ $$10 = \frac{1}{2} a_3 t^2 \quad \left(\text{using, } s = ut + \frac{1}{2}at^2\right)$$

$\therefore$ $$t = 2\ \text{s}$$

(iii) $v_1 = a_1 t = 5\sqrt{3} \times 2 = 10\sqrt{3}$ m/s

19. (i) The component of accelerations perpendicular to the surface of contact should be same.

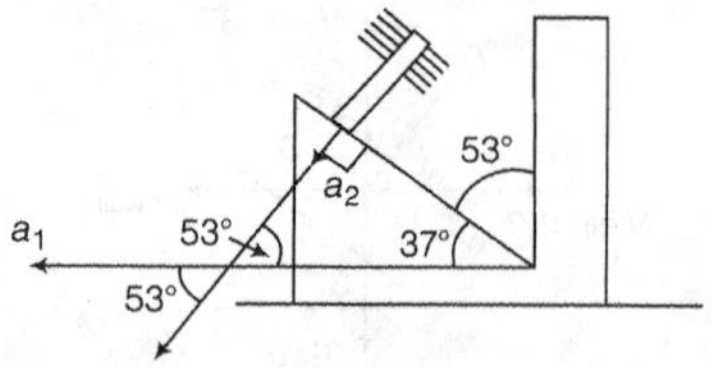

$\therefore$ $$a_2 = a_1 \cos 53°$$

$\therefore$ $$a_2 = \frac{3}{5} a_1 \quad \text{...(i)}$$

(ii) From free body diagram of the rod,

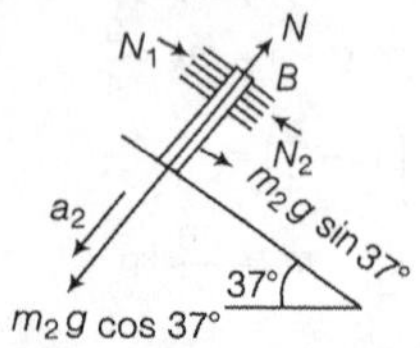

$$m_2 g \cos 37° - N = m_2 a_2 \quad (m_2 = 1\ \text{kg})$$

or $$8 - N = a_2 \quad \text{...(ii)}$$

(iii) From free body diagram of the wedge,

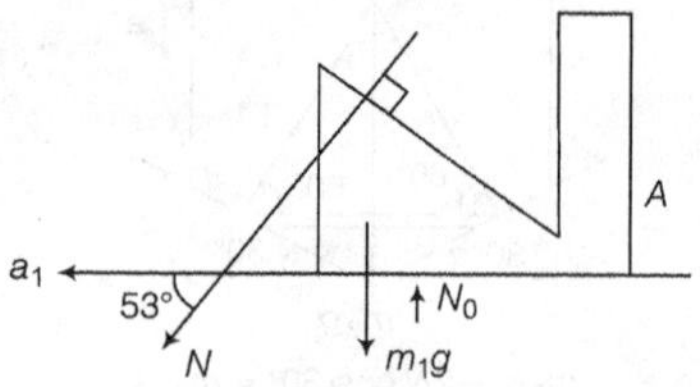

$$N \cos 53° = m_1 a_1$$

or $$N = \frac{a_2}{7} \quad \text{...(iii)}$$

From Eqs. (ii) and (iii), we get

$$N = 1\ \text{N},\ a_2 = 7\ \text{m/s}^2$$

20. Immediately after released the box, the string remains vertical but tension in it decreases, therefore, the ball will accelerate downward with acceleration a_1 and box accelerates with acceleration a_2 along the inclined plane.

Free body diagram of ball,

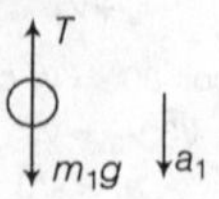

Free body diagram of box,

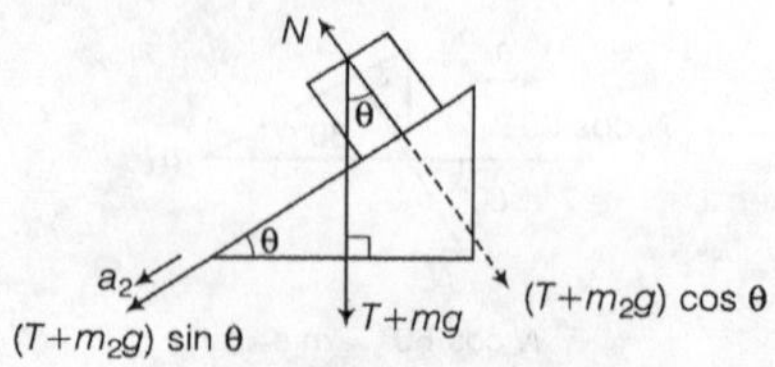

The string is taut, so the component of acceleration of box along string just after released should be equal to acceleration of the ball.

$\therefore$ $$a_2 \sin\theta = a_1 \quad \text{or} \quad a_2 \sin 30° = a_1$$

$\therefore$ $$a_2 = 2a_1$$

From free body diagram of box,

$$(T + m_2 g)\sin\theta = m_2 a_2$$

or $$T + 50 = 2 \times 5a_2 = 20a_1 \quad (\because a_2 = 2a_1)$$

$\therefore$ $$T = 20a_1 - 50$$

From free body diagram of ball,

$$m_1 g - T = m_1 a_1$$

or $$40 - (20a_1 - 50) = 4a_1 \quad \text{or} \quad 90 = 24a_1$$

$\therefore$ $$a_1 = \frac{90}{24} = \frac{15}{4} = 3.75\ \text{m/s}^2$$

21. (i) From figure,

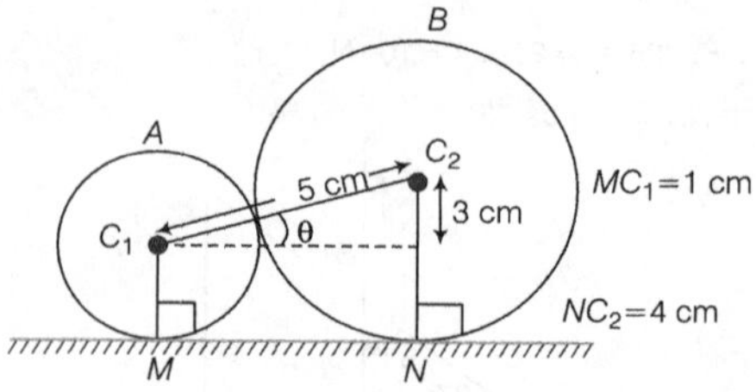

$$\sin\theta = \frac{3}{5}$$

$\Rightarrow$ $$\theta = 37°$$

For leaving the surface,

$$N \cos 53° > m_B g$$

$$N > \frac{5}{3} m_B g \quad \text{...(i)}$$

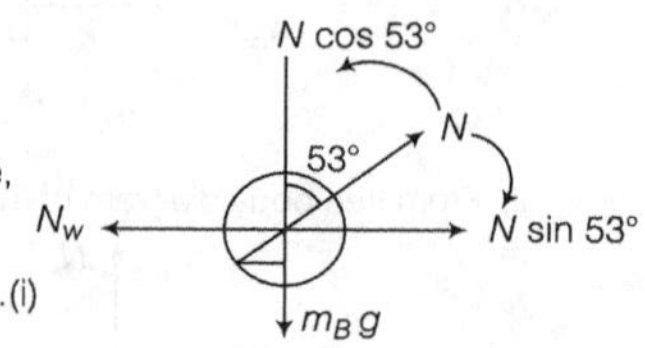

From free body diagram of A,

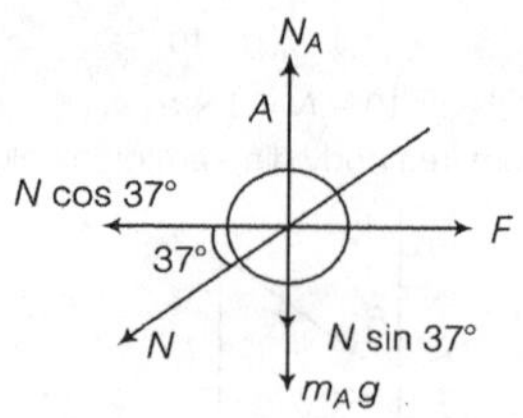

$$F = N\cos 37° \Rightarrow N = \frac{5F}{4} \quad \text{...(ii)}$$

From eq (i) and (ii), $\therefore$ $$\frac{5F}{4} > \frac{5}{3} m_B g$$

$\therefore$ $$F > \frac{4}{3} \times 3 \times 10$$

$\therefore$ Minimum force $= 40\ \text{N} = 10\ F_{min}$ or $F_{min} = 4$ N

(ii) For ball B, $N\cos 53° - m_B g = m_B a_B$

or $$\frac{3}{5}N - 30 = 3a_B \quad \text{...(i)}$$

For ball A, $$F - N\cos 37° = m_A a_A$$

$$44 - \frac{4}{5}N = 4a_A \quad \text{...(ii)}$$

Also, $$a_A \cos 37° = a_B \cos 53° \quad \text{...(iii)}$$

(from constraint relation)

$$a_A \times \frac{4}{5} = a_B \frac{3}{5}$$

$\therefore$ $$a_A = \frac{3}{4} a_B,$$

From Eqs. (i), (ii) and (iii), we get

$$a_B = 2\ \text{m/s}^2$$

DPP-6 Spring

1. (i) $mg = kx_0$, $\quad x_0 = \frac{mg}{k} = \frac{1}{5} = 0.2\text{ m}$

(ii) $F = kx_0$

$\Rightarrow \quad x_0 = \frac{F}{k} = \frac{20}{100} = 0.2\text{ m}$

(iii) $kx = T = mg = 100$

$x = \frac{100}{k}$

$\therefore \quad x = 0.2\text{ m}$

(iv) $kx_0 = m_2 g = T$

$\Rightarrow \quad x_0 = \frac{100}{500} = 0.2\text{ m}$

2. (i) Here, $k_1 = 600$ N/m, $k_2 = 300$ N/m, $k_3 = 200$ N/m

$$\frac{1}{k_{eq}} = \frac{1}{k_1} + \frac{1}{k_2} + \frac{1}{k_3}$$

$\therefore \quad k_{eq} = 100$ N/m

(ii) Here, $k_1 = 50$ N/m, $k_2 = 60$ N/m, $k_3 = 90$ N/m

$k_{eq} = k_1 + k_2 + k_3 = 200$ N/m

(iii) $k_{eq} = k_1 + k_2$

(iv)

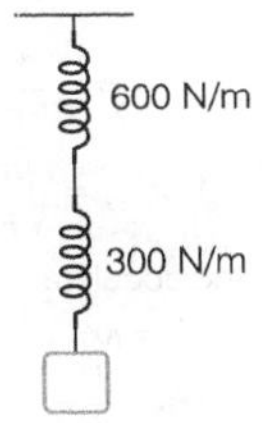

$\therefore \quad k_{eq} = \frac{600 \times 300}{600 + 300} = 200\text{ N/m}$

(v) Adding one more grid does not affect the stiffness of arrangement.

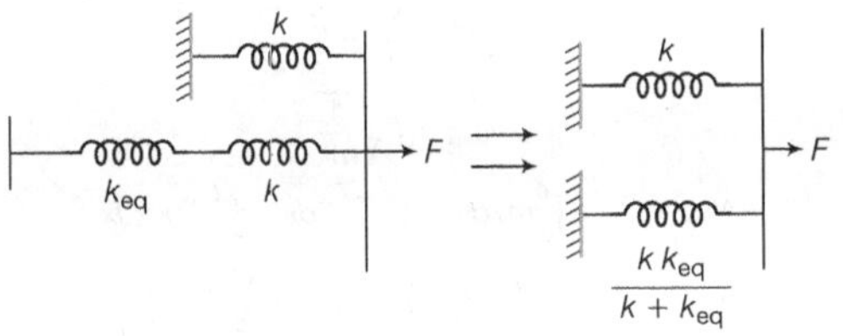

From figure,

$$\therefore \quad k_{eq} = k + \frac{k k_{eq}}{k + k_{eq}}$$

or $\quad k k_{eq} + k_{eq}^2 = k^2 + k k_{eq} + k k_{eq}$

or $\quad k_{eq}^2 - k k_{eq} - k^2 = 0$

$$\therefore \quad k_{eq} = \frac{k \pm \sqrt{k^2 + 4k^2}}{2}$$

$$= \frac{k \pm \sqrt{5}k}{2} = \frac{k + \sqrt{5}k}{2} = 10\text{ N/m}$$

3. (i) $mg = k_{eq} x_0 \Rightarrow 100 = k_{eq} x_0$ $\quad \left(\frac{1}{k_{eq}} = \frac{1}{600} + \frac{1}{300}\right)$

$\Rightarrow \quad x_0 = \frac{100}{200} = 0.5\text{ m}$

(ii) $\quad k x_0 = 2T = 2 \times 100 \Rightarrow x_0 = \frac{200}{500} = 0.4\text{ m}$

But, distance moved by the block to achieve equilibrium is

$s = 2x_0 = 2 \times 0.4$

$= 0.8\text{ m}$

(iii) $k_1 x_1 = T = mg = 100$

$\therefore \quad x_1 = 0.2\text{ m}$ and $2T = k_2 x_2$

$\therefore \quad x_2 = \frac{2mg}{k_2} = \frac{2 \times 10 \times 10}{1000} = 0.2\text{ m}$

The distance moved by the block to achieve equilibrium is

$x = 2x_2 + x_1$

$= 0.4 + 0.2$

$= 0.6\text{ m}$

4. (i) The spring constant is inversely proportional to the natural length of spring.

$\therefore$ The spring constant of each piece is

$k_1 = 200$ N/m

$\therefore \quad k_{eq} = k_1 + k_1 = 400$ N/m

(ii) $l + 2l + 3l = \frac{C}{100}$ or $6l = \frac{C}{100}$

$\therefore \quad \frac{C}{l} = 600\text{ N/m}$

$\therefore \quad k_1 = \frac{C}{l} = 600\text{ N/m}, \; k_2 = \frac{C}{2l} = 300\text{ N/m},$

$k_3 = \frac{C}{3l} = 200\text{ N/m}$

(iii) Spring constant is inversely proportional to its length.

$\because \quad k = \frac{C}{l}$

$\therefore$ Spring constant of each part is $k_1 = \frac{C}{\frac{l}{3}} = \frac{3C}{l} = 3k$

(iv) Given, $l_A = \frac{2}{3} l_B$

$$\therefore \quad k_A = \frac{\left(\frac{2}{3} + 1\right)k}{\frac{2}{3}} = \frac{5}{2}k$$

Note *If a spring is cut into two pieces of lengths l_1 and l_2 such that $l_1 = nl_2$ then, spring constant of first part (l_1), $k_1 = \frac{k(n+1)}{n}$ and spring constant of second part l_2, $k_2 = (n+1)k$ and ratio of spring constants $\frac{k_1}{k_2} = \frac{1}{n}$.*

5. (i) For equilibrium, $\quad F = T + kx_0$

$\therefore \quad kx_0 = 20 - 15 = 5\text{ N}$

When string A is cut,

$F - kx_0 = ma$

$\Rightarrow \quad 20 - 5 = 1 \times a$

$\Rightarrow \quad a = 15\text{ m/s}^2$

(ii) $T = 20 = kx_0$

After cutting the string,

$kx_0 = 2a$

$\Rightarrow \quad 20 = 2a$

$\Rightarrow \quad a = 10\text{ m/s}^2$

(iii) $10 = 2kx_0 \Rightarrow kx_0 = 5\text{ N}$

When the spring A is cut,

$mg - kx_0 = ma$ or $10 - 5 = 1 \times a \Rightarrow a = 5\text{ m/s}^2$

(iv) $T_A \sin 60^\circ = kx \sin 60^\circ$ (horizontal component)

$\therefore \quad T_A = kx$

Also, $\quad T_A \cos 60^\circ + k x \cos 60^\circ = mg$ (vertical component)

$\therefore \quad kx = mg$

Just after cutting the string, resultant of spring force and mg will be equal to the tension in string before cutting.

$\therefore \; T_A = kx = ma$ or $mg = ma, a = g = 10\text{ m/s}^2$

6. (i) $m_C g = k x_0 = T_A \Rightarrow 10 = k x_0 = T_A$

Just after cutting the string, elongation in spring does not change.

$\therefore \quad a_C = 0$ and $kx = m_B a_B$

$\therefore \quad a_B = 10 \text{ m/s}^2$

(ii) $m_2 g = 20 = kx = m_1 g + T_A$

Just after cutting the string A, $T_A = 0$

$\therefore \quad kx - m_1 g = m_1 a_1$

$\therefore \quad a_1 = \frac{20-10}{1} = 10 \text{ m/s}^2, \; a_2 = 0$

(iii) Before cutting the string A,

$$k x_0 = (m_1 + m_2) g$$

$\therefore \quad k x_0 = 50 \text{ N}$

Just after cutting the string,

$$m_2 g = m_2 a_2$$

$\therefore \quad a_2 = g,$ (downward)

$$k x - m_1 g = m_1 a_1 \text{ or } 50 - 10 = a_1$$

$\therefore \quad a_1 = 40 \text{ m/s}^2$ in upward direction

(iv) Initially, $\quad k x_0 = (2m + m) g$ and $T = mg$

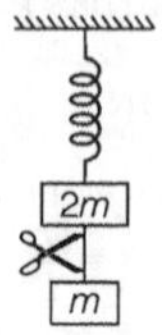

When string is cut, $2mg - kx_0 = 2ma$

$\Rightarrow \quad -mg = 2ma, \; a = \frac{-g}{2}$

(v) Before cutting the string,

$$T = (m_1 + m_2) g = 30 \text{ N}$$

and $\quad k x_0 = m_2 g = 10 \text{ N}$

Just after cutting the string,

$$m_2 g - kx_0 = m_2 a_2$$

or $\quad 10 - 10 = 1 \times a_2$

$\therefore \quad a_2 = 0$

and $\quad m_1 g + kx_0 = m_1 a_1$

or $\quad 20 + 10 = 2 a_1$

$\therefore \quad a_1 = \frac{30}{2} = 15 \text{ m/s}^2$

(vi) $a_1 = 0, \; a_2 = 5 \text{ m/s}^2$

(vii) Before cutting the spring A,

$$k_2 x_2 = (m_1 + m_2) g = 200 \text{ N}$$

Also, $\quad k_1 x_1 = m_1 g = 80 \text{ N}$

$$k_2 x_2 = m_3 g + k_3 x_3$$

$$200 = 30 + k_3 x_3$$

$$170 = k_3 x_3$$

Just after cutting the spring A,

$$k_3 x_3 - m_4 g = m_4 a_4$$

or $\quad 170 - 20 = 2 a_4$

$\therefore \quad a_4 = 75 \text{ m/s}^2$

But, m_3, m_2 and m_1 maintain their equilibrium,

$\therefore \quad a_1 = a_2 = a_3 = 0$

7. (i) $T = kx = mg = 10$ kg

(ii) Both spring balances will read 10 kg.

(iii) $\quad T - mg = ma$

$\therefore \quad T = mg + ma = 100 + 10 \times 2 = 120 \text{ N}$

Here, $\quad T = k_1 x_1 = k_2 x_2$

Both the scales will read 12 kg.

(iv) Here, $\quad T = mg = 20$

or $\quad kx = T = 20 = 2$ kg

Spring balance will read 2 kg.

(v) $30 - T = 3a$ ...(i)

and $\quad T - 20 = 2a$ ...(ii)

$\therefore \quad a = 2 \text{ m/s}^2$

$\therefore \quad T - 20 = 2 \times 2$

$\therefore \quad T = 24 \text{ N}$

But, $T = kx = 24 \text{ N} = 2.4$ kg

Spring balance will read 2.4 kg.

(vi) $kx_0 = mg$

$\therefore \quad 49 = mg$ (when lift is stationary)

When lift is moving,

$$mg - kx = ma$$

$\therefore \quad kx = mg - ma = 49 - \frac{49}{9.8} \times 5$

$$= 49 - 25 = 24 \text{ N}$$

8. (i) $v_{B_{max}} = v_0$

$$-kx = ma$$

or $\quad \int_0^{x_{max}} -\frac{k}{m} x dx = \int_{v_0}^{0} v dv$

$\therefore \quad \frac{v_0^2}{2} = \frac{k}{2m} x_{max}^2$

$\Rightarrow \quad x_{max} = \sqrt{\frac{m}{k}} v_0$

(ii) At maximum speed, the acceleration of the block will be zero.

$\therefore \quad F = kx_0$

$\Rightarrow \quad x_0 = \frac{F}{k}$

$\therefore \quad F - kx = mv \frac{dv}{dx}$

or $\quad \int_0^{v_{max}} mv dv = \int_0^{x_0} F dx - \int_0^{x_0} k x dx$

$$\frac{m v_{max}^2}{2} = \frac{F^2}{k} - \frac{k}{2} \frac{F^2}{k^2} = \frac{F^2}{2k}$$

$\therefore \quad v_{max} = F \sqrt{\frac{1}{mk}}$

Also, $\quad \int_v^0 mv dv = \int_0^{x_{max}} F dx - \int_0^{x_{max}} k x dx$

or $\quad 0 = F x_{max} - \frac{k x_{max}^2}{2}$

$\therefore \quad x_{max} = \frac{2F}{k}$

(iii) Let x_m = maximum extension of spring.

Then, decrease in gravitational potential energy of mass = increase in elastic potential energy of spring.

$$Mgx = \frac{1}{2} kx^2$$

$\Rightarrow \quad x = \frac{2Mg}{k}$

9. (c) $F = kx_1$, $F = 2kx_2$ and $F = 3k x_3$

$\therefore \quad x_1 : x_2 : x_3 = \frac{F}{k} : \frac{F}{2k} : \frac{F}{3k}$

$$= 6 : 3 : 2$$

10. (b) $20 - kx = 2 \times 8$ (for A)

$\therefore \quad kx = 4 \text{ N}$

For block B, $10 - kx = 1 \times a_2$ $\quad (\because T = kx)$

$$10 - 4 = a_2$$

$\therefore \quad a_2 = 6 \text{ m/s}^2$ (downward)

11. (c) The velocity of turns of spring varies linearly from one end to another end.

$\therefore$ $v = bx + C$

At $x = 0, v = -v_0$

$\therefore$ $-v_0 = C$

Also at, $x = L, v = 2v_0$

$\therefore$ $2v_0 = bL + C$ or $2v_0 + v_0 = bL$

$\therefore$ $b = \dfrac{3V_0}{L}$

$\therefore$ $v = \dfrac{3v_0}{L}x - v_0$

12. (b) For equilibrium,

$\therefore$ $kx_1 \sin 60° = kx_3 \sin 60°$

$\Rightarrow$ $x_1 = x_3$

and $kx_1 \cos 60° + k_0x_2 + kx_3 \cos 60° = mg$

or $kx_1 + k_0x_2 = mg = 100$ N

$\therefore$ $kx_1 = 100 - kx_2 = 100 - 1000\dfrac{5}{100} = 50$ N

After breaking,

$$a_x = \frac{kx_1 \sin 60°}{m} = \frac{50\sqrt{3}}{2 \times 10} = \frac{5\sqrt{3}}{2}\ \text{m/s}^2$$

and $$a_y = \frac{mg - kx_1 \cos 60° - k_0x_2}{m}$$

$$a_y = \frac{100 - 25 - 50}{10} = \frac{25}{10} = 2.5\ \text{m/s}^2$$

$\therefore$ $$a = \sqrt{a_x^2 + a_y^2} = \sqrt{\frac{75}{4} + 6.25} = \sqrt{25} = 5\ \text{m/s}^2$$

13. (a) From force diagram, $20g - T = 20a$ ...(i)

$T - 15g = 15a$ ...(ii)

Adding Eqs. (i) and (ii), we get

$5g = 35a$

$\therefore$ $a = \dfrac{5g}{35} = \dfrac{g}{7}$

Let reading of S_2 is T_2 in newton (i.e. tension in S_2 at lower end)

$\therefore$ $18g - T_2 = \dfrac{18g}{7}$

$\therefore$ $T_2 = 18g - \dfrac{18}{7}g = \dfrac{(126 - 18)}{7}g = \dfrac{108}{7}g$

$\therefore$ The reading in kg is given by

$\dfrac{T_2}{g} = \dfrac{108}{7}$ kg

Also, $T_1 = 2T$

$\therefore$ $T_1 = 2(15a + 15g)$ (from figure A)

or $T_1 = 30a + 30g = \dfrac{30g}{7} + 30g = \dfrac{240}{7}g$

But, T_1 is tension in lower end of S_1.

$\therefore$ The reading of S_1 in kg $= \dfrac{T_1}{g} = \dfrac{240}{7}$ kg

14. (c) At equilibrium of the block, the elongation in springs are x_1 and x_2, respectively.

Here, $k_1x_1 = k_2x_2$

Also, $x_1 + x_2 = 2x$

Here, x is distance through which block is descended.

$\therefore$ $x_1 = \dfrac{2k_2x}{k_1 + k_2}$ and $x_2 = \dfrac{2k_1x}{k_1 + k_2}$

For equilibrium, $mg = k_1x_1 + k_2x_2$

or $mg = \dfrac{2k_1k_2x}{k_1 + k_2} + \dfrac{2k_1k_2x}{k_1 + k_2}$

$\therefore$ $x = \dfrac{mg(k_1 + k_2)}{4k_1k_2}$

15. (b) Assume that spring constant of section above the point A is k_1 and spring constant lower section is k_2.

$\therefore$ $k_1 = \dfrac{c}{0.1}$ and $k_2 = \dfrac{c}{0.4}$

But $k = \dfrac{c}{0.5}$

$\therefore$ $k_1 = 5k$ and $k_2 = \dfrac{5}{4}k$

Initially, $k_1x_1 = k_2x_2$

If displacement in the point A is y, then

$k_2(x_2 + y) = F + k_1(x_1 - y)$

or $(k_2 + k_1)y = F$

or $$y = \frac{F}{k_1 + k_2} = \frac{5}{5k + \frac{5}{4}k}$$

$$= \frac{4}{5k} = \frac{4}{5 \times 100}$$

$$= 8 \times 10^{-3}\ \text{m}$$

16. (d) Let N = number of coils

Then, length elapsed per coil $= \dfrac{l}{N}$

For 6 coils, length = $6\ l/N$

For 8 coils, length $= \dfrac{8l}{N}$

$\therefore$ x = displacement of 6th coil's end

$= \dfrac{8l}{N} - \dfrac{6l}{N} = \dfrac{2l}{N}$

If k = stiffness of spring, then stiffness per coil = Nk

As 6 coils are connected in series, stiffness of all 6 coils $= \dfrac{Nk}{6}$

When a mass m is suspended,

$mg = k \cdot x$

$\Rightarrow$ $$m = \frac{kx}{g} = \frac{\frac{Nk}{6} \times \frac{2l}{N}}{g}$$

$$= \frac{k \times 2l}{6 \times g} = \frac{600 \times 2 \times 0.3}{6 \times 10} = 6\ \text{kg}$$

17. (b) The velocity of turns of spring linearly varies from one end to another end.

$\therefore v = bx + c$, where b and c are constants. v (in cm/s) is velocity of a turn at distance x (in metre) from point A.

$\because$ At $x = 0$, $v = v_A = -1$ cm/s

$\therefore$ $-1 = b \times 0 + c$

$\therefore$ $c = -1$

$\therefore$ $v = bx - 1$

At $x = 6$ m, $v_B = 5$ cm/s

$\therefore$ $5 = (b \times 6) - 1$

$\therefore$ $b = 1$

$\therefore$ $v = x - 1$

At $x = 2$ m, $v = 2 - 1 = 1$ cm/s

$\therefore$ $s = vt = 1 \times 5 = 5$ cm towards point B.

18. (c,d) If $mg \geq kx - mg$

When spring is taut, the acceleration of each blocks are same.

If $a_2 = 0, T = mg$

$\therefore$ $a_1 = \dfrac{T + kx - mg}{m} = \dfrac{kx}{m}$

If $mg < kx - mg$, string slacks

and $a_1 = \dfrac{kx - mg}{m}$ and $a_2 = g$

19. (a,b,d) The velocity of thread's free end is 1 cm/s and velocity of thread's end attached with wall is zero.

So, velocity of thread at a distance x metres from wall is proportional to its distance from wall. So, velocity at a distance x from the wall $= xv_0$, where x is in metre and v_0 is in cm/s.

If this speed is greater than caterpillar's speed, then the caterpillar will move away from wall and if speed of caterpillar is more than xv_0, the net velocity of the caterpillar is towards the wall and increases with time.

Limiting condition corresponds to

$$x = \frac{v_{caterpillar}}{v_0} = \frac{1\text{ mm/s}}{1\text{ cm/s}} = \frac{0.1}{1} = 0.1\text{ m} = 10\text{ cm}$$

Starting from this point, caterpillar does not move in either direction.

20. (a) The system is in equilibrium,

$$kx_3 = mg \quad \ldots\text{(i) [from figure (i)]}$$
$$2mg + kx_3 = kx_1 \quad \ldots\text{(ii) [from figure (ii)]}$$
$$\therefore \quad 3mg = kx_1$$

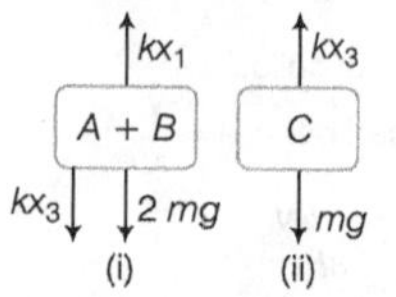

When spring between ceiling and block is cut, then elongation of spring between B and C remains same just after cutting.

$$\therefore \quad a_C = 0 \quad (\because kx_3 = mg)$$

For block $(A + B)$, $\quad kx_3 + 2mg = 3mg$

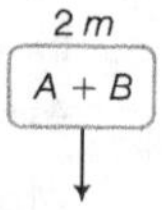

$$\therefore \quad 3mg = 2ma$$
$$\therefore \quad a = \frac{3}{2}g = 15\text{ m/s}^2$$
$$\therefore \quad a_A = a_B = 15\text{ m/s}^2$$

21. (b)

$$mg + kx_3 - T = ma_B$$
$$mg + mg - T = \frac{3mg}{2}$$
$$\therefore \quad T = \frac{mg}{2}$$

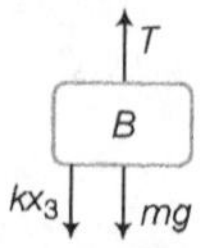

When string between A and B is cut, the elongation in springs do not change just after cutting the string.

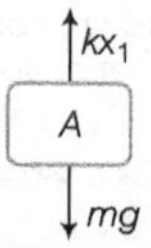

$$mg - kx_1 = ma_A$$
$$mg - 3mg = ma_A \quad (\because kx_1 = 3mg)$$
$$-2mg = ma_A$$
$$\therefore \quad a_A = 2g \quad \text{(upward)}$$

For block B,

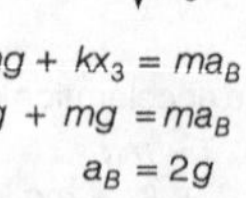

$$\because \quad mg + kx_3 = ma_B$$
$$\text{or} \quad mg + mg = ma_B \quad (\because kx_3 = mg)$$
$$\therefore \quad a_B = 2g \quad \text{(downward)}$$

For bolck C,

$$mg - kx_3 = ma_C$$
$$\text{or} \quad mg - mg = ma_C$$
$$(kx_3 = mg)$$
$$\because \quad a_C = 0$$
$$\Rightarrow \quad T = 0$$

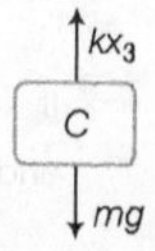

22. (b) When spring between block B and C is cut, then

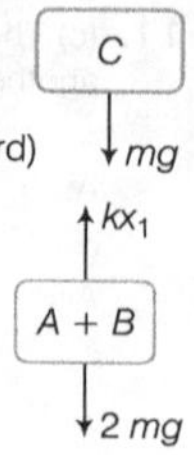

$$mg = ma_C$$
$$\therefore \quad a_C = g \quad \text{(downward)}$$
$$2mg - kx_1 = 2ma_B \quad (\because a_A = a_B)$$
$$2mg - 3mg = 2ma_B \quad (\because kx_1 = 3mg)$$
$$\therefore \quad a_A = a_B = \frac{g}{2} \quad \text{(upward)}$$

23. (1) Let elongation in left spring is x_1 and that of right spring is x_2.

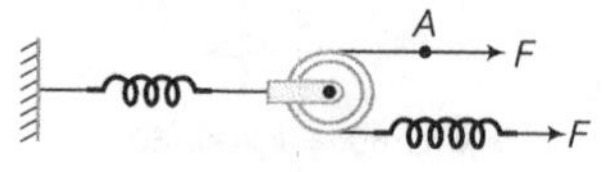

$$\therefore \quad x_0 = 2x_1 + x_2$$

Also,

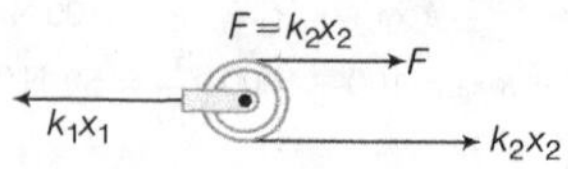

$$\therefore \quad x_2 = \frac{F}{k_2}$$
$$\therefore \quad k_1x_1 = F + k_2x_2 \Rightarrow k_1x_1 = 2F \Rightarrow x_1 = \frac{2F}{k_1}$$
$$\therefore \quad x_0 = \frac{4F}{k_1} + \frac{F}{k_2} = \frac{400}{500} + \frac{100}{500} = 1\text{ m}$$

24. (2) $T_x = kx_1\cos\alpha + kx_2\cos\beta = 8k$

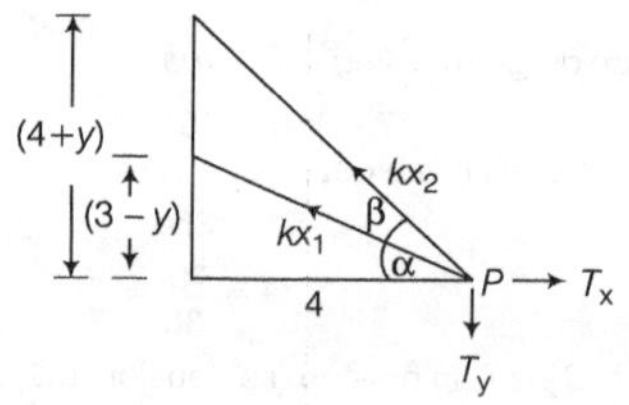

$$T_y = kx_1\sin\alpha + kx_2\sin\beta = (3 - y)k + (3 + y)k = 6k$$
$$\therefore \quad T = \sqrt{T_x^2 + T_y^2} \text{ or } mg = 10k$$
$$\therefore \quad k = 100\text{ N/m} = 1 \times 10^2$$
$$\therefore \quad x = 2$$

25. (8) Let natural length of middle segment is x and that of after segment is y.

Hence, $\quad y^2 = (BC)^2 + x^2$ or $2y\frac{dy}{dx} = 2x$

$$\text{or} \quad ydy = xdx$$
$$\Rightarrow \quad dy = \frac{x}{y}dx = dx\cos\theta$$

Also, $\quad \cos\theta = \frac{x}{y}$

$$\Rightarrow \quad x = y\cos\theta$$

For equilibrium of point P,

$$\because \quad F = k_1\Delta x + 2k_2\Delta y\cos\theta$$
$$\text{or} \quad F = \frac{c}{x}\Delta x + \frac{2c}{y}\Delta y\cos\theta$$
$$\text{or} \quad F = \frac{c}{x}\Delta x + \frac{2c}{x}x\cos^3\theta$$
$$F = T + 2T\cos^3\theta$$
$$\therefore \quad T = \frac{F}{1 + 2\cos^3\theta} = \frac{10}{1 + 2 \times \left(\frac{1}{2}\right)^3} = 8\text{ N}$$

26. (a) (i) → (s), (ii) → (q), (iii) → (s), (iv) → (p)

DPP-7 Pseudo Force

1. (i) $\mathbf{a}_A = \frac{d\mathbf{v}}{dt} = 3\hat{i} + 8t\hat{j} = 3\hat{i} + 8\hat{j}$ at $t = 1$ s

So, the pseudo force is acting on the driver in opposite direction of acceleration of car A.

$\therefore$ Pseudo force $= m(-\mathbf{a}_A) = -50(3\hat{i} + 8\hat{j}) = -150\hat{i} - 400\hat{j}$

(ii) 0

(iii) Non-inertial

(iv) False

(v) Non-inertial, earth is not really an inertial reference frame because of its orbital motion around the sun and its rotation about its own axis, both of which results in centripetal are rotation. However, these are acceleration and are small compared to g and it can be neglected for practical purposes, (exceptions are calculation of trajectories of spacecrafts, satellites, etc).

For this reason, we assume that the earth is an inertial frame, are any other frame attached to it. So, we can say, Newton's laws of motion hold in an inertial system determined for an observer at rest by taking account of only the rotational acceleration due to rotation of earth about its axis and due to motion of the earth in its orbit around sun.

2. (i)

In the frame of block A,

$$T = ma_0 = 1 \times 2 = 2\text{ N}$$

(ii) In the frame of triangular block,

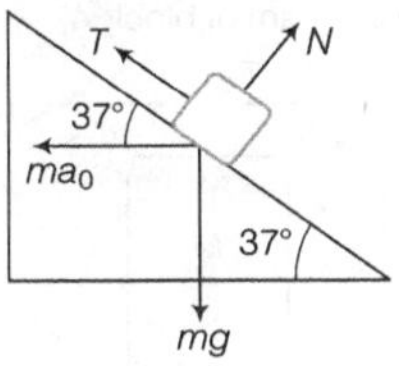

$$T + ma_0 \cos 37° = mg \sin 37°$$

$\therefore$ $a_0 = g \tan 37°$ $(\because T = 0)$

$$= 10 \times \frac{3}{4} = 7.5\text{ m/s}^2$$

(iii) In the frame of triangular block,

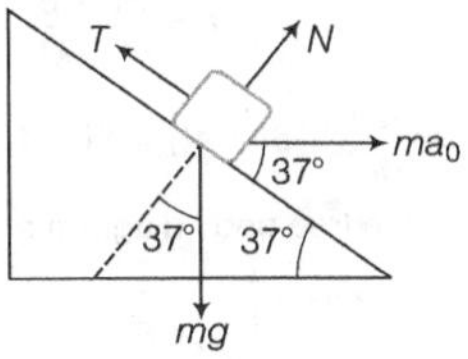

$$N + ma_0 \sin 37° = mg \cos 37°$$

But, $N > 0$

$\therefore$ $a_0 < g \cot 37°$

$\therefore$ $a_{0_{max}} = 10 \times \frac{4}{3} = \frac{40}{3} = 13.3\text{ m/s}^2$

(iv) $ma \cos \alpha = mg \sin \alpha$

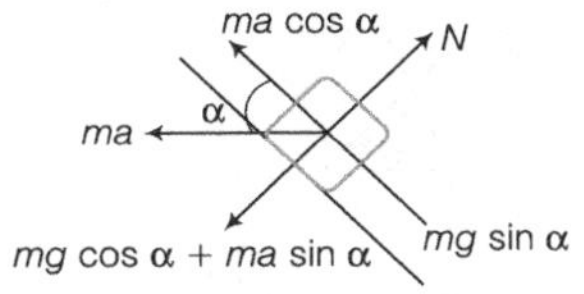

$$a = \frac{g \sin \alpha}{\cos \alpha} = g \tan \alpha$$

(v) $kx_0 = ma_0 \Rightarrow 100x_0 = 10a_0 \Rightarrow x_0 = \frac{50}{100} = 0.5\text{ m}$

3. (i) From free body diagram of block A in the frame of triangular block B,

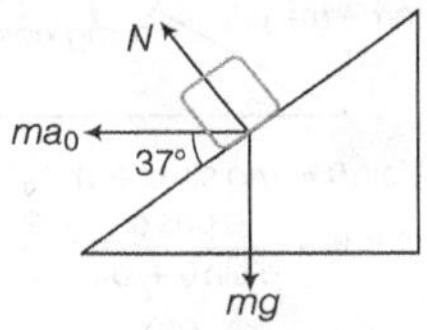

$$ma_0 \cos 37° + mg \sin 37° = ma_{rel}$$

or $5 \times \frac{4}{5} + 10 \times \frac{3}{5} = a_{rel}$ or $a_{rel} = 10\text{ m/s}^2$

$\therefore$ $s_{rel} = u_{rel}t + \frac{1}{2}a_{rel}t^2$

or $20 = 0 + \frac{1}{2} \times 10t^2$

$\therefore$ $t = 2$ s

(ii) Here, $m(g + a_0)\sin 37° = ma_{rel}$

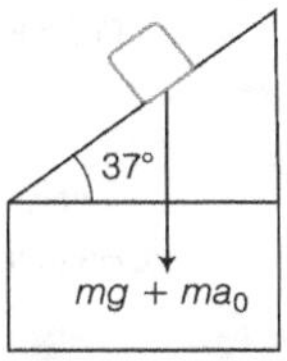

$\therefore$ $a_{rel} = 15 \times \frac{3}{5} = 9\text{ m/s}^2$

$\therefore$ $s_{rel} = u_{rel}t + \frac{1}{2}a_{rel}t^2$

$\therefore$ $18 = 0 + \frac{1}{2} \times 9 \times t^2$

$\therefore$ $t = 2$ s

(iii) Here, $mg \cos 37° = ma_{rel}$

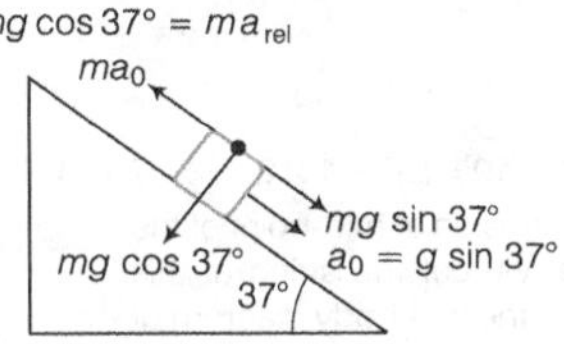

$\therefore$ $a_{rel} = 10 \times \frac{4}{5} = 8\text{ m/s}^2$

$\therefore$ $s_{rel} = u_{rel}t + \frac{1}{2}a_{rel}t^2$

or $0.16 = 0 + \frac{1}{2} \times 8 \times t^2$

$\therefore$ $t = \sqrt{\frac{2 \times 0.16}{8}} = 0.2$ s

(iv) As box a large and particle is not hitting any other surface of box. Time of flight of particle is given by

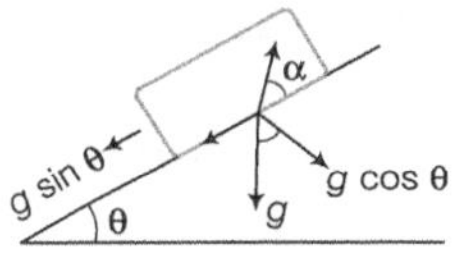

$$T = \frac{2u \sin \alpha}{g \cos \theta}$$

Range of particle is given by

$$s = ut + \frac{1}{2}at^2 \quad (u \text{ and } a \text{ are taken along the plane})$$

where, $a = 2g \sin \theta$ and $t = T$

$$\therefore \quad PQ = u \cos \alpha \frac{2u \sin \alpha}{g \cos \theta} + \frac{1}{2}(2g \sin \theta)\frac{4u^2 \sin^2 \alpha}{g^2 \cos^2 \theta}$$

$$= \frac{u^2 \sin 2\alpha}{g \cos \theta} + \frac{4u^2 \sin^2 \alpha \sin \theta}{g \cos^2 \theta}$$

4. (i) $T \cos \theta = mg$ and $T \sin \theta = ma_0$

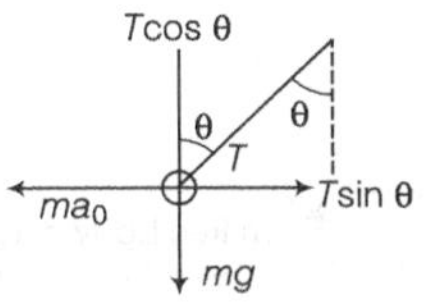

$\therefore$ $\frac{T \sin \theta}{T \cos \theta} = \frac{ma_0}{mg}$

$\therefore$ $\tan \theta = 1$

$\therefore$ $\theta = 45°$

(ii) $T\cos\theta = mg\cos\alpha$

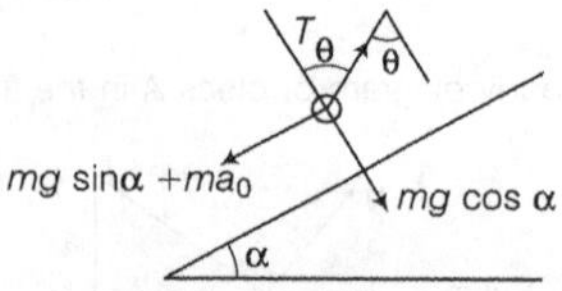

and $\quad T\sin\theta = mg\sin\alpha + ma_0$

$\therefore \quad \tan\theta = \dfrac{g\cos\alpha}{g\sin\theta + a_0} = \dfrac{2}{3}$

$\therefore \quad \theta = \tan^{-1}\left(\dfrac{2}{3}\right)$

(iii) From free body diagram of upper ball,

$$T_1 + m_1a_0 = m_1g + T_2 \quad \text{...(i)}$$

From free body diagram of lower ball,

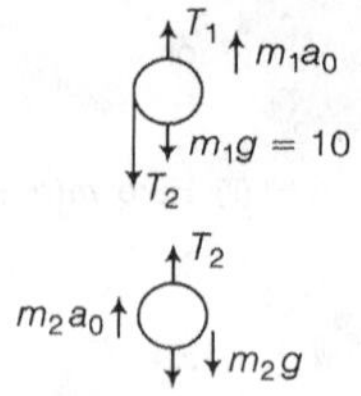

$$T_2 + m_2a_0 = m_2g$$

$\therefore \quad T_2 = m_2g - m_2a_0$

$\therefore \quad T_2 = 20 - 2\times 2 = 16\text{ N}$

From Eq. (i), we get

$$T_1 = 10 + 16 - 1\times 2 = 24\text{ N}$$

$\therefore \quad \dfrac{T_2}{T_1} = \dfrac{16}{24} = 2:3$

(iv) Here, $mg = 2T + ma_0$

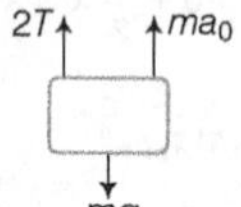

$\Rightarrow \quad 1\times 10 = 2T + 1\times 2 \Rightarrow T = 4\text{ N}$

(v) (a) To find tension at mid-point of the lower wire, we cut the string at this point. Draw the free body diagram of lower portion.

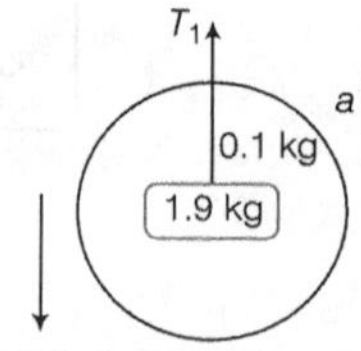

The equation of motion gives

$$T_1 - 2.0g = (2.0)a$$

or $\quad T_1 = (2.0)(g + a)$

$= (2.0)(9.8 + 0.2) = 20\text{ N}$

(b) To find tension at mid-point of upper wire

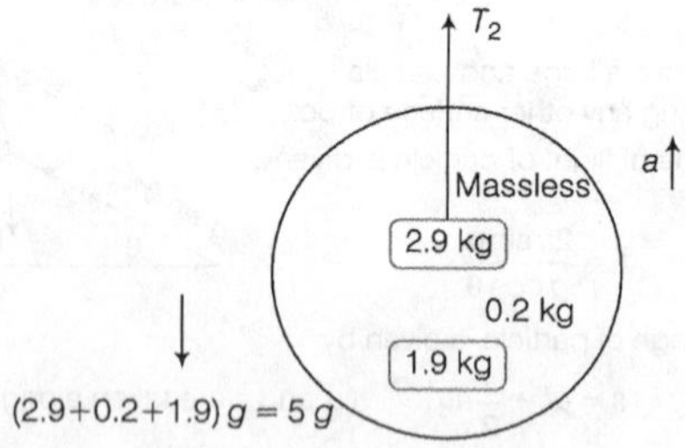

The equation of motion gives

$$T_2 - 5g = 5a$$

$\therefore \quad T_2 = 5(g + a) = 5(9.8 + 0.2) = 50\text{ N}$

5. (c) In the frame of lift

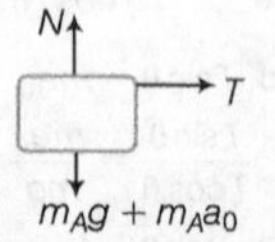

From free body diagram of A,

$$T = m_Aa_{rel} = a_{rel} \quad (\because m_A = 1\text{ kg}) \quad \text{...(i)}$$

From free body diagram of B,

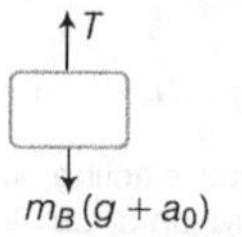

$$m_B(g + a_0) - T = m_Ba_{rel}$$

or $\quad 12 - T = a_{rel} \quad (\because m_B = 1\text{ kg})$

or $\quad 12 - a_{rel} = a_{rel}$ (from Eq. (i))

$\therefore \quad a_{rel} = 6$

$\therefore \quad kx = T = a_{rel} = 6\text{ N}$

6. (a) If wedge is stationary,

$N_1\sin\theta_1 = N_2\sin\theta_2$ (in horizontal direction)

or $\quad m_1g\cos\theta_1\sin\theta_1 = m_2g\cos\theta_2\sin\theta_2$

$\therefore \quad \dfrac{m_1}{m_2} = \dfrac{\sin 2\theta_2}{\sin 2\theta_1}$

7. (d) From free body diagram of ball in the frame of cart

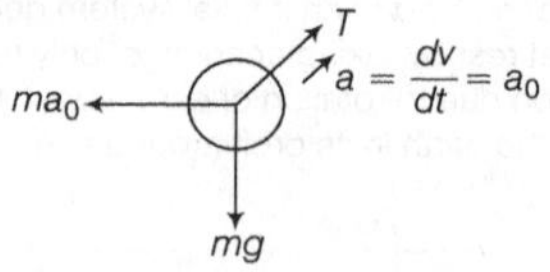

$$T - m\sqrt{g^2 + a_0^2} = ma_0$$

or $\quad T = m\sqrt{g^2 + a_0^2} + ma_0$

8. (a) From free body diagram of block A,

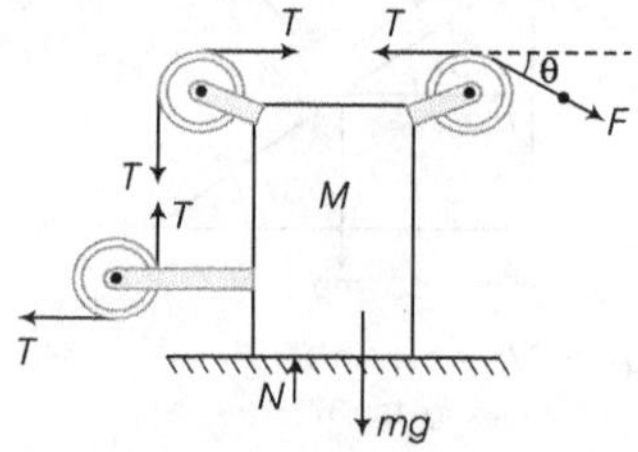

$\because \quad T - F\cos\theta = Ma$

or $\quad F - F\cos\theta = Ma$

$\therefore \quad a = \dfrac{F(1-\cos\theta)}{M}$

9. (b) Acceleration, $\mathbf{a} = \dfrac{d^2\mathbf{r}}{dt^2} = 8\hat{i} - 4t\hat{j} = 8\hat{i} - 4\hat{j}$ (at $t = 1$s)

In the frame of box, the free body diagram of ball

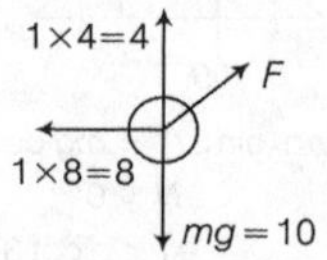

Here, $F = \sqrt{8^2 + (10-4)^2} = 10\text{ N}$

10. (a,b,c) From free body diagram of A,

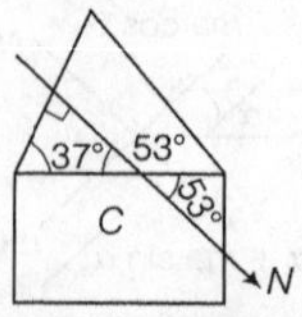

$N\cos 53° = 1\times a_1 \Rightarrow \dfrac{3}{5}N = a_1$

$\therefore \quad N = \dfrac{5}{3}a_1 \quad \text{...(i)}$

From free body diagram of B in the frame of A,

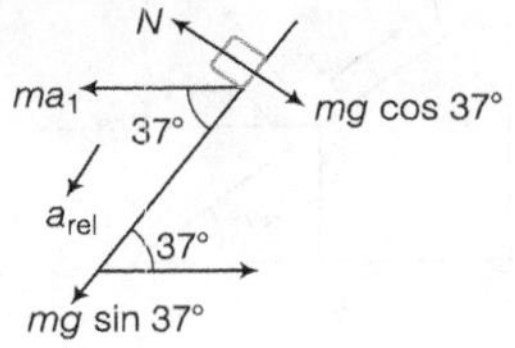

$$N + ma_1 \sin 37° = mg \cos 37°$$

or, $$\frac{5}{3}a_1 + \frac{3}{5}a_1 = 8$$

$\therefore$ $$a_1 = \frac{60}{17} = \frac{6g}{17}$$

Also, $ma_{rel} = ma_1 \cos 37° + mg \sin 37°$

$\therefore$ $a_{rel} = 15g/17$

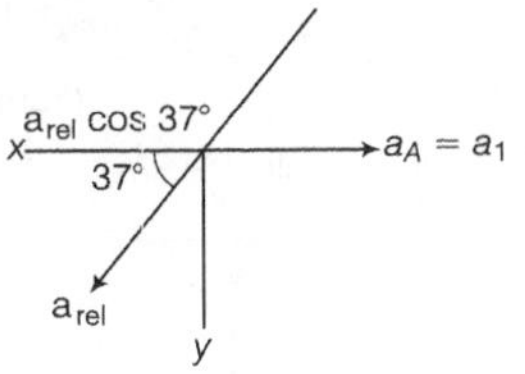

The resultant of a_{rel} and a_1 gives a_B,

$$\tan\theta = \frac{a_y}{a_x} = \frac{a_{rel}\sin 37°}{a_{rel}\cos 37° - a_1}$$

$$= \frac{3}{2} = 1.5$$

11. (a,b)

(a) From free body diagram of A,

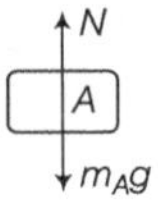

$$m_Ag - N = m_Aa_A$$

Thus, the acceleration of block A is downward.

(b) Free body diagram of block C,

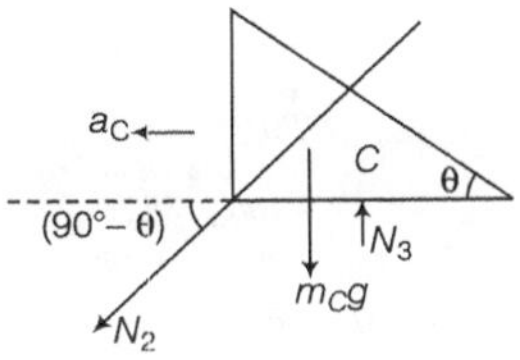

From free body diagram of block C,

$$N_3 = m_Cg + N_2 \cos\theta$$

and $$N_2 \sin\theta = m_Ca_C \quad \text{(in horizontal direction)}$$

Thus, block C is accelerating horizontally leftward.

(c) $\mathbf{a}_{BA} = \mathbf{a}_B - \mathbf{a}_A$

Free body diagram of B in the frame of A,

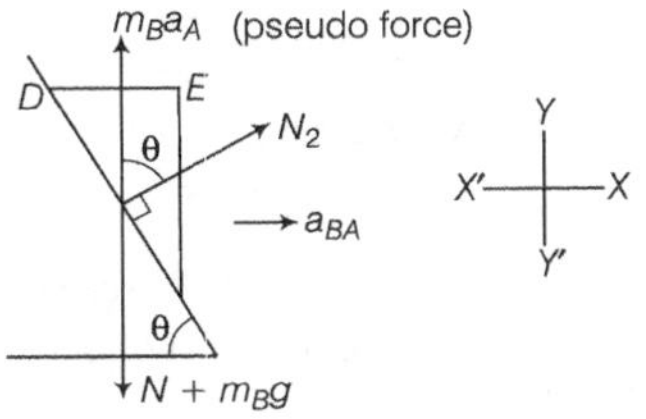

The acceleration of B with respect to A should be along the surface of contact DE.

$\therefore$ $$N_2 \sin\theta = m_Ba_{BA} \quad \text{(in rightward direction)}$$

Hence, option (c) is incorrect.

12. (a,b,d) $N_{AC} \cos 45° = mg$

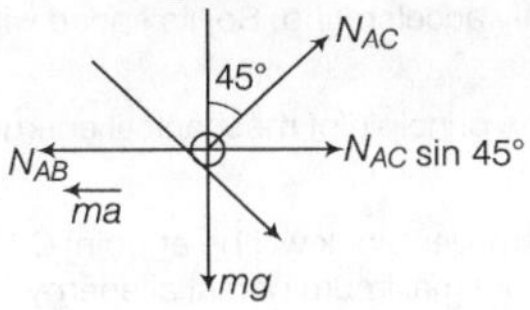

$$N_{AC} = \sqrt{2}\,mg = \sqrt{2} \times \sqrt{2} \times g = 20\text{ N}$$

and $$N_{AB} + ma = N_{AC}\sin 45° = mg$$

$\therefore$ $$N_{AB} = mg - ma$$

For rising, $$N_{AB} = 0,\ a = g$$

$\therefore$ $$2t = 10$$

$\therefore$ $$t = 5\text{ s}$$

13. (b,c) From free body diagram of block E + block F, net force is zero. The accelerations of blocks E and F are zero.

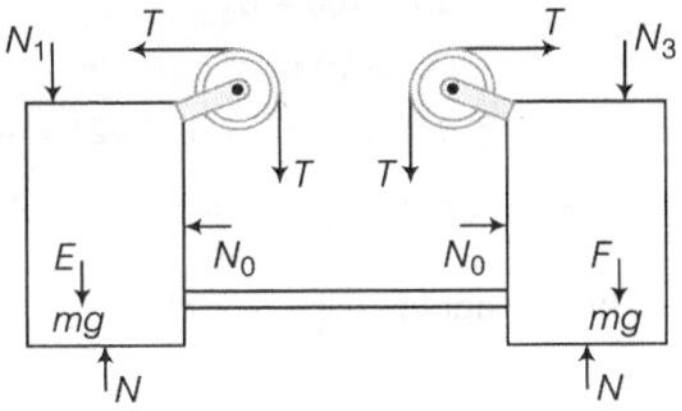

From free body diagram of block A,

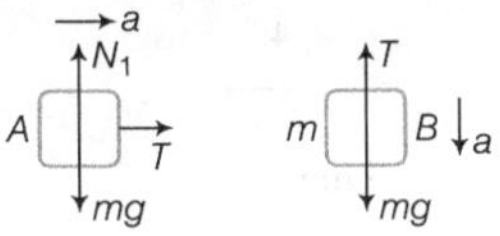

$$T = ma$$

From free body diagram of block B,

$$mg - T = ma$$

$\therefore$ $$mg = 2ma$$

$\therefore$ $$a = g/2$$

Hence, options (b) and (c) are correct.

Explanations for Q. Nos. (14-18)

Let acceleration of m relative to M be a down the face and acceleration of M be b leftwards.

Resultant acceleration of the m is vector sum of these two accelerations as shown in figure. Therefore, horizontal and vertical component of acceleration of m are $(a\cos 37° - b)$ rightwards and $(a\sin 37°)$ downwards, respectively.

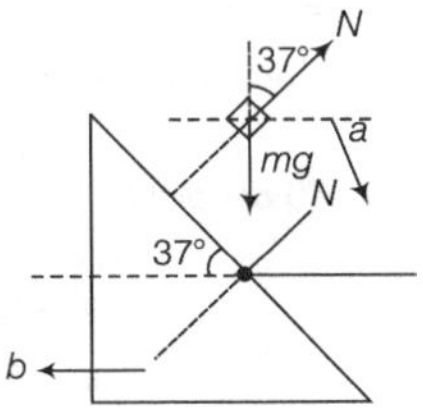

Considering free body diagram,

For horizontal forces on M,

$$N\sin 37° = Mb \quad \text{...(i)}$$

For horizontal forces on m,

$$N\sin 37° = m(a\cos 37° - b) \quad \text{...(ii)}$$

For vertical forces on m,

$$mg - N\cos 37° = ma\sin 37° \quad \text{...(iii)}$$

From above equations, $N = 13$ N

14. (c) The block will move updown periodically. So, centre of mass moves periodically in vertical direction. (In horizontal direction both will have same and opposite acceleration).

15. (b) Solving Eqs. (i), (ii) and (iii), we have

$$b = 2.5\text{ m/s}^2$$

16. (a) Solving Eqs. (i), (ii) and (iii), we have

$$a = 8\text{ m/s}^2$$

17. (b) During the motion of the smaller block on part AB and wedge mass constantly accelerating. So, its speed will be maximum.

or

By conservation principle of mechanical energy, we get the same result.

18. (a) When the smaller block will be at point C then it will has zero kinetic energy but maximum potential energy.

So, by conservation principle of mechanical energy we can say that velocity of bigger block will be zero.

19. (4) As the blocks remain stationary,

$$m_1 g = T = \sqrt{(m_2 g)^2 + (m_2 a_0)^2}$$

or, $\quad m_1 g = m_2 \sqrt{g^2 + a_0^2}$

$\Rightarrow \quad 1.1 \times 10 = 1 \times \sqrt{10^2 + a_0^2}$

$\therefore \quad 121 = 100 + a_0^2$

$\Rightarrow \quad a_0 = \sqrt{21}\ \text{m/s}^2$

$\therefore \quad F = (1.1 + 1 + 1.9)\sqrt{21} = 4\sqrt{21}$ N

$\because$ Given, $\quad F = \dfrac{16\sqrt{21}}{n} \quad \Rightarrow \quad n = 4$

20. (4) From free body diagram of the block B,

$$F - N = ma_0 \qquad \text{...(i)}$$

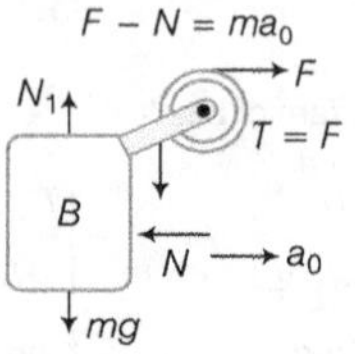

From free body diagram of block A in the frame of block B,

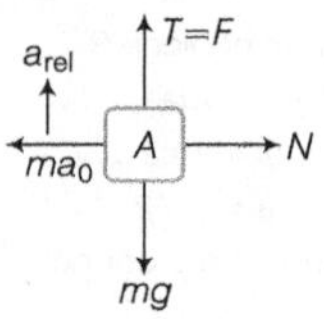

$$N = ma_0 \qquad \text{...(ii)}$$

$$F - mg = ma_{rel}$$

$\Rightarrow \quad 20 - 10 = a_{rel} \Rightarrow a_{rel} = 10\ \text{m/s}^2$

From Eqs. (i) and (ii), we get

$$a_0 = 10\ \text{m/s}^2$$

$\therefore$ For point P

$$a_{rel} = a_P - a_0$$

$\therefore \quad a_P = a_{rel} + a_0 = 10 + 10 = 20\ \text{m/s}^2$

$\because \quad a_P = 5x \quad \Rightarrow \quad x = 4$

21. (0) From free body diagram of block A,

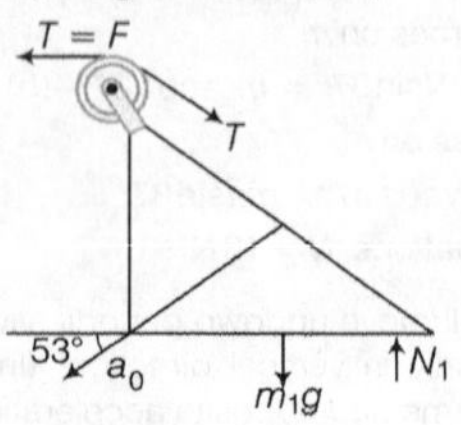

$$T + N\cos 53° - T\cos 37° = m_1 a_0$$

or $\quad 10 + \dfrac{3}{5}N - 8 = a_0$

or $\quad 2 + \dfrac{3}{5}N = a_0 \qquad \text{...(i)}$

From free body diagram of block B in the frame of block A,

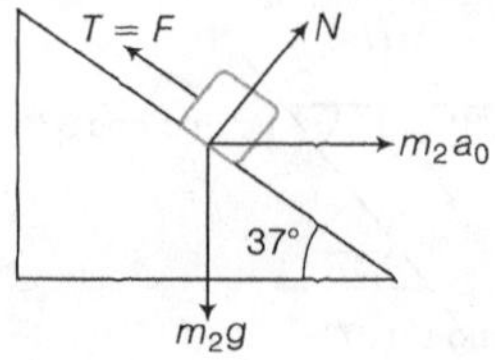

Here, $\quad N + m_2 a_0 \sin 37° = m_2 g \cos 37°$

or $\quad N + \dfrac{3}{5}a_0 = 8 \qquad \text{...(ii)}$

From Eqs. (i) and (ii), we get $a_0 \approx 5\ \text{m/s}^2$

Also, $\quad F - m_2 a_0 \cos 37° - m_2 g \sin 37° = m_2 a_{rel}$

or $\quad 10 - \dfrac{4}{5}a_0 - 6 = a_{rel}$

$\therefore \quad a_{rel} = 0 \quad (\because a_0 = 5\ \text{m/s}^2)$

22. (2) $a_{BA} = a_A$ (from constraint relation)

From free body diagram of block A,

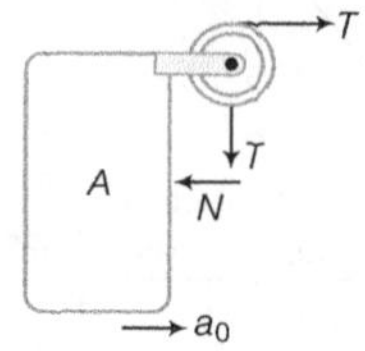

$$T - N = a_0 \qquad \text{...(i)}$$

From free body diagram of block B in the frame of block A,

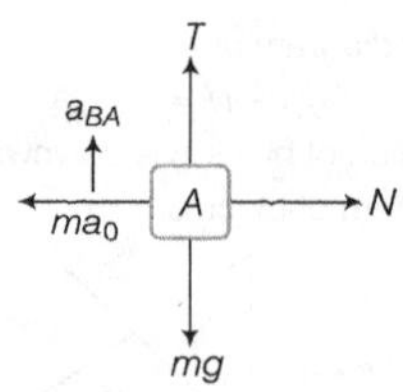

$$N = ma_0 = a_0$$

$$T - mg = ma_{BA} = ma_0$$

or $\quad 2a_0 - g = a_0 \quad$ (from Eq. (i))

$\therefore \quad a_0 = g$

$\therefore \quad \mathbf{a}_{BA} = \mathbf{a}_B - \mathbf{a}_A$

$\therefore \quad a_B = \sqrt{a_{BA}^2 + a_A^2} = \sqrt{2}g$

Hence, the value of $x = 2$.

23. (2) The net acceleration of the collar is along dotted line.

For collar, $\quad mg\sin 30° = ma_0 \cos 30° \quad$ (along the rod)

$$a_0 = g\tan 30° = \frac{g}{\sqrt{3}} \qquad \text{...(i)}$$

Also, $\quad mg\cos 30° - N = ma_0 \sin 30° \qquad \text{...(ii)}$

(perpendicular to rod)

For box, $\quad N\sin 30° = Ma \qquad \text{...(iii)}$

(in horizontal direction)

The component of acceleration perpendicular to the surface of contact should be same.

$$a_0 \sin 30° = a\sin 30°$$

$\therefore \quad a = a_0 = \dfrac{g}{\sqrt{3}}$

From Eqs. (ii) and (iii), we get

$$\frac{m}{M} = 2$$

DPP-8 Problems Related to Climb on a String

1. (i) $T = mg = 100$ N (as the monkey is moving with a constant velocity)

(ii) 200 N

(iii) $T - mg = ma \Rightarrow T = 60$ N

(iv) $mg - T = ma$

$\Rightarrow T = mg - ma = 80 - 24 = 56$ N

2. (i) $T - mg = ma$

$$T_{max} - mg = ma_{max}$$

$$200 - 100 = 10a_{max}$$

$$a_{max} = 10 \text{ m/s}^2$$

(ii) $\frac{T_A}{2} - mg = ma$ or $\frac{200}{2} - 50 = 5a_{max}$

$\therefore a_{max} = 10 \text{ m/s}^2$

(iii) $T_{max} - mg = ma_{max}$

$$720 - 600 = ma_{max}$$

$$a_{max} = \frac{120}{60} = 2 \text{ m/s}^2$$

3. (i) For lower monkey,

$$T_1 - 50 = 5 \times 2$$

$\therefore$ $T_1 = 60$ N

For upper monkey,

$$T_2 - 100 = 10 \times 2$$

$\therefore$ $T_2 = 120$ N

For the block,

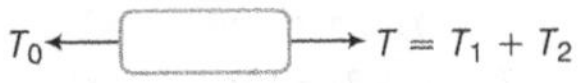

$$T = T_1 + T_2$$

$\therefore$ $T_0 = T = T_1 + T_2 = 60 + 120 = 180$ N

(ii) For lower monkey,

$$T_1 = m_1 g = 50 \text{ N}$$

For upper monkey,

$$T_2 - m_2 g = m_2 a_0$$

$\therefore$ $T_2 = 100 + 10a_0$

For the block,

$$T_1 + T_2 = 150 + 10a_0$$

$\therefore$ $T_{0max} = 150 + 10\, a_{0max}$

or $\frac{300 - 150}{10} = a_{0max}$

$\therefore$ $a_{0max} = 15 \text{ m/s}^2$

(iii) For upper monkey,

$$T_2 = m_2 g = 100 \text{ N}$$

For lower monkey,

$$T_1 = m_1 g + m_1 a_0$$

$$T_1 = 50 + 5a_0$$

For the block,

$$T_0 = T_1 + T_2 = 150 + 5a_0$$

$\therefore$ $a_{0max} = \frac{150}{5} = 30 \text{ m/s}^2$

4. (i) For monkey, $40 - 2T = ma_0 = 4a_0$

For the block, $T = M\,(2a_0) = 2a_0$

$\therefore$ $40 = 8a_0$

$\therefore$ $a_0 = 5 \text{ m/s}^2$

$\therefore$ The acceleration of block is $2a_0 = 10 \text{ m/s}^2$

(ii) If block moves rightward with acceleration a_1. Then, the lower string moves with acceleration $\frac{a_1}{2}$ in downward direction.

If monkey is moving upward in the frame of ground, then

$$2 = a_{monkey} + \frac{a_1}{2}$$

$\therefore$ $a_{monkey} = 2 - \frac{a_1}{2}$

For monkey, $2T - 40 = 4\left(2 - \frac{a_1}{2}\right)$

For the block A, $T = 1 \times a_1$

$\therefore$ $2a_1 - 40 = 8 - 2a_1 \Rightarrow 4a_1 = 48$

$\therefore$ $a_1 = 12 \text{ m/s}^2$

(iii) From solution of (ii), $-2 = a_{monkey} + \frac{a_1}{2}$

$\therefore$ $a_{monkey} = -2 - \frac{a_1}{2} = -\left(2 + \frac{a_1}{2}\right)$

It means that monkey is moving downward with an acceleration $\left(2 + \frac{a_1}{2}\right)$.

$\therefore$ $40 - 2T = 4\left(2 + \frac{a_1}{2}\right)$ and $T = a_1$

$$40 - 2a_1 = 8 + 2a_1 \quad \text{or} \quad 32 = 4a_1$$

$\therefore$ $a_1 = 8 \text{ m/s}^2$

5. (c) For equilibrium, $F_m \cos 37° = F_C \sin\theta$

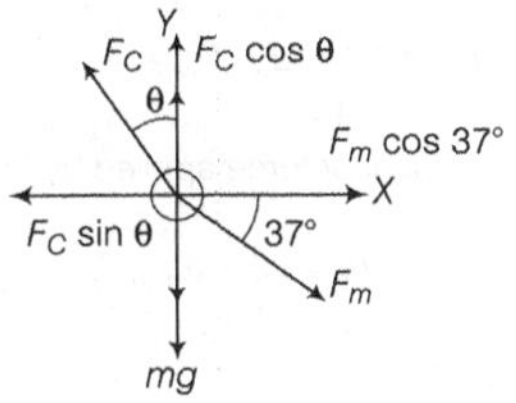

and $mg + F_m \sin 37° = F_C \cos\theta$

$\therefore$ $F_C = \sqrt{(F_m \cos 37°)^2 + (mg + F_m \sin 37°)^2}$

$$= \sqrt{30^2 + 40^2} = 50 \text{ N}$$

6. (b) $2T \cos 37° - mg = ma_{max}$

$\therefore$ $2 \times 200 \times \frac{4}{5} - 100 = 10a_{max}$

or $320 - 100 = 10a_{max}$

$\therefore$ $a_{max} = \frac{220}{10} = 22 \text{ m/s}^2$

$$s = \frac{1}{2}at^2 \Rightarrow \frac{44 \times 2}{22} = t^2$$

$\therefore$ $t = 2$ s

7. (b) $mg - T = ma_0$

or $mg - T = ma_0$ or $T = mg - ma_0$

or $T > 0$ or $mg - ma_0 > 0$

or $g > a_0$ or $a_0 < g$

$\therefore$ $a_{0max} = g$

8. (c) For monkey, $T - m_1 g = m_1 a$ ...(i)

For the block, $2T < m_2 g$

$\therefore$ $T < m_2 g/2$

or $m_1 a + m_1 g < m_2 g/2$ (from Eq. (i))

or $a_{max} = \frac{\left(\frac{m_2}{2} - m_1\right)g}{m_1} = \left(\frac{10 - 5}{5}\right) \times 10 = 10 \text{ m/s}^2$

9. (d) For box system,

$$T - 120 = 12a_0$$

or $156 - 120 = 12a_0$

$\therefore \quad a_0 = \dfrac{36}{12} = 3\ m/s^2$

For boy in the frame of the box,

$$N = m(g + a_0)\cos 37°$$

$$= 10 \times 13 \times \frac{4}{5} = 104\ N$$

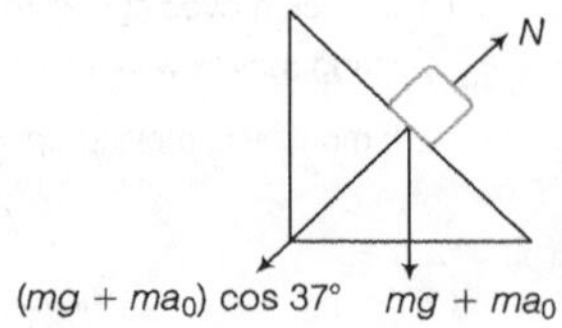

10. (c) $T_C = m_C g = 20\ N$ (for monkey C)

$m_B g - T_B = m_B a_B$ (for monkey B)

$30 - T_B = m_B a_B$

$30 - T_B = 3 \times 1$

$\therefore \quad T_B = 27\ N$

and $T_A - m_A g = m_A a_A$ (for monkey A)

$\therefore \quad T_A = 60\ N$

$\therefore \quad T = T_A + T_B + T_C + Mg = 60 + 27 + 20 + 100$

$= 207\ N \quad [M = m_A + m_B + m_C]$

11. (c) The horizontal component of the force exerted by the bike gives the acceleration to monkey.

$\therefore \quad F = ma = 50 \times 2 = 100\ N$

12. (a) In the horizontal direction,

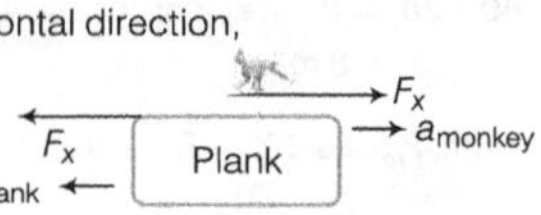

$$5\cos 37° = a_{monkey} + a_{plank}$$

or $\quad 4 = a_{monkey} + a_{plank}$ ...(i)

If horizontal component of force applied by the plank on monkey is F_x.

$\therefore \quad F_x = 20 \times a_{monkey}$ ...(ii)

and $\quad F_x = 60\, a_{plank}$ ...(iii)

From Eqs. (i), (ii) and (iii), we get

$$4 = \frac{F_x}{20} + \frac{F_x}{60}$$

$\therefore \quad F_x = \dfrac{240}{4} = 60\ N$

$\therefore \quad a_{plank} = \dfrac{F_x}{60} = 1\ m/s^2$

13. (a,c) $F = ma = 10 \times 1 = 10\ N$

$$F = ma_{plank}$$

$\therefore \quad a_{plank} = \dfrac{F}{m} = \dfrac{10}{30} = \dfrac{1}{3}\ m/s^2$

14. (b,d) Let force exerted by monkey on bomboo piece along the bomboo is F.

Free body diagram of bomboo.

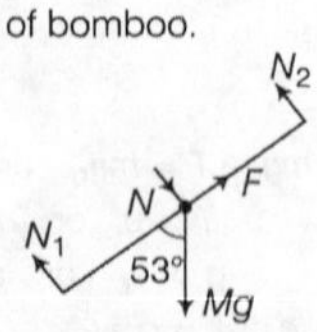

Free body diagram of monkey,

For monkey, $\quad F + mg\cos 53° = ma_0$

For bomboo piece, $\quad Mg\cos 53° = F$

$\therefore \quad ma_0 = (m + M)\, g\cos 53°$

$\therefore \quad a_0 = \left(1 + \dfrac{M}{m}\right) g\cos 53°$

$\therefore \quad a_0 = \dfrac{4}{3} \times 6 = 8\ m/s^2$

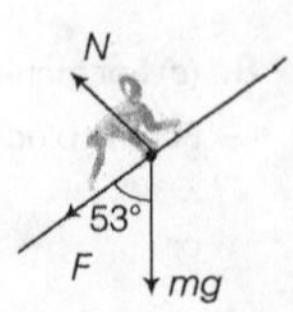

15. (b, d) When the draw is x.

Force $\quad F = \dfrac{100}{0.4} x = 250\, x$

$\therefore \quad a = \dfrac{250\, x}{0.01} = 25000\, x$

or $\quad \dfrac{vdv}{dx} = 25000$ or $\displaystyle\int_0^v vdv = 25000 \int_0^{0.4} xdx$

or $\quad \dfrac{v^2}{2} = 25000 \times 0.08 = 2000$

or $\quad v = \sqrt{4000} = 20\sqrt{10}\ m/s$

16. (b, d)

Free body diagram of monkey A

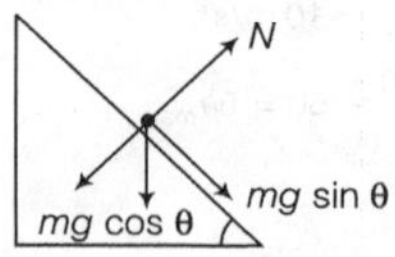

Free body diagram of monkey B

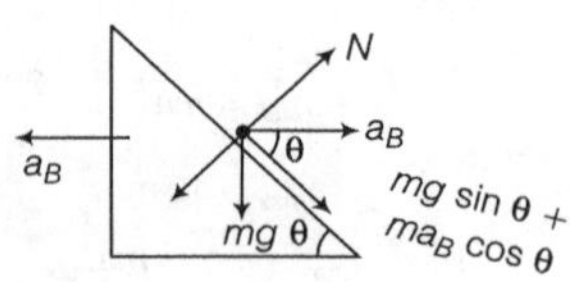

$\Rightarrow a_1 = mg\sin\theta \quad a_B$ = acceleration the wedge

$\Rightarrow \quad a_2 = mg\sin\theta + ma_B\cos\theta$

$a_2 > a_1$

$\therefore$ Acceleration of monkey B > acceleration of monkey A.

17. (c) If a_0 = acceleration of rope

For monkey A, then, $\quad 1 = a_1 - a_0 \Rightarrow a_1 = a_0 + 1$

$T_1 - 200 = 20\, a_1 = 20 + 20\, a_0$

$T_1 = 220 + 20\, a_0$ ...(i)

For monkey B, $\quad 2 = a_2 + a_0 \Rightarrow a_2 = 2 - a_0$

Also, $\quad T_2 - 200 = 20a_2 = 40 - 20\, a_2$

$T_2 = 240 - 20a_2$ ...(ii)

For the block, $\quad T_2 + 10 - T_1 = 10a_0$ ...(iii)

From Eqs. (i), (ii) and (iii), we get $a_0 = 0.6\ m/s^2$

$\therefore$ Acceleration of block C is $0.6\ m/s^2$.

18. (b) For monkey A, $\quad 1 = a_1 - a_0 \Rightarrow a_1 = 1 + a_0$

Also, $\quad T_1 - 200 = 20\, a_1 = 20 + 20a_0$

$\therefore \quad T_1 = 220 + 20a_0$ ...(i)

For monkey B, $\quad 2 = a_2 + a_0 \Rightarrow a_2 = 2 - a_0$

Also, $\quad T_2 - 200 = 20a_2 = 40 - 20\, a_2$

$\therefore \quad T_2 = 240 - 20a_2$ ...(ii)

For the block C, $\quad T_2 + 10 - T_1 = 10a_0$ ...(iii)

From Eqs. (i), (ii) and (iii), we get $a_0 = 0.6\ m/s^2$

$\therefore \quad a_1 = 1.6\ m/s^2$

$\because \quad s = \dfrac{1}{2} a_1 t^2$

$\therefore \quad t = 1\ s$

19. (a) (i) → p, (ii) → p, (iii) → q, (iv) → p

20. (2) Here, $\quad T = T_A + m_D g = T_A + 10$ ...(i)

and $\quad T - (m_B + m_C)g = (m_B + m_C)a$

$T_A + 10 - 150 = 15\, a$

$T_A = 15a + 140$

$T_{Amin} = 15 \times 0 + 140 = 140\ N$

But $\quad T_{min} - m_C g = m_C a \Rightarrow 80 - 50 = 5\, a_{max}$

$\therefore \quad a_{max} = 6\ m/s^2$

$\therefore \quad T_{Amax} = 15a_{max} + 140 = 15 \times 6 + 140 = 90 + 140 = 230\ N$

$\therefore \quad \dfrac{T_{min}}{T_{Amax}} = \dfrac{140}{230} = \dfrac{14}{23} = \dfrac{7x}{23}$

$\therefore \quad x = 2$

Revisal Problems (JEE Main)

1. (c) $a = \dfrac{m_A g}{m_A + m_B} = 5\ \text{m/s}^2$

$v_B = at = 15\ \text{m/s}$

$\therefore \quad v_B t_0 = \dfrac{1}{2} g t_0^2$ or $v_B = \dfrac{g t_0}{2}$ or $15 = \dfrac{10 t_0}{2}$

$\therefore \quad t_0 = 3\ \text{s}$

2. (a) $2T = mg = 20$

$\Rightarrow \quad T = 10\ \text{N}$

Here, $\dfrac{dN}{dL} = \dfrac{T}{r} = \dfrac{10}{10} = 1\ \text{N/cm}$

3. (c) Here, $\sin\theta = \dfrac{a}{\sqrt{3}l} = \dfrac{3}{2\sqrt{3}l} l = \dfrac{\sqrt{3}}{2}$

$\therefore \quad \theta = 60^\circ$

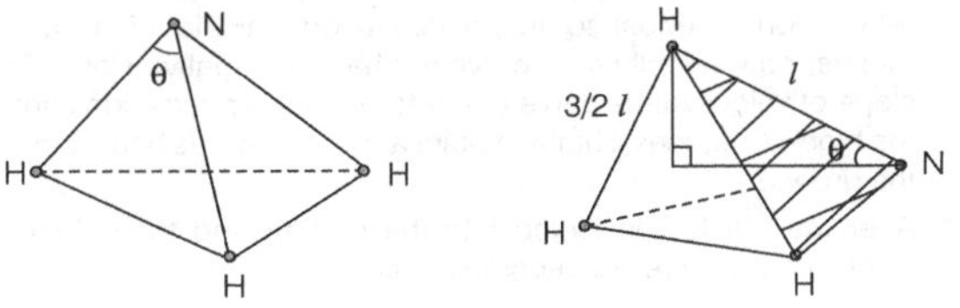

$3F_0 \cos\theta = F$

$\Rightarrow \quad 3F_0 \cos 60^\circ = F$

$\Rightarrow \quad F = \dfrac{3}{2} F_0$

4. (c) Clearly, $2T\cos\theta = Fg$

$\Rightarrow 2(mg)\cos\theta = \sqrt{2}\, mg$

$\cos\theta = \dfrac{1}{\sqrt{2}}, \ \theta = 45^\circ$

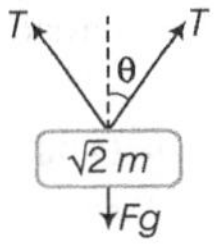

5. (b) $a = \dfrac{m_B g - m_A g}{m_A + m_B + m_C} = \dfrac{20}{5} = 4\ \text{m/s}^2$

$\therefore \quad s = \dfrac{1}{2} a t^2$ or $\sqrt{\dfrac{8 \times 2}{4}} = t$

$\therefore \quad t = 2\ \text{s}$

6. (b) $\mathbf{F} = \dfrac{m}{\Delta t}(\mathbf{v} - \mathbf{u})$

But $|\mathbf{v} - \mathbf{u}| = \sqrt{v^2 + u^2 - 2vu\cos 120^\circ}$

$\therefore \quad |\mathbf{F}| = \dfrac{\sqrt{7} \times \sqrt{7}}{0.01} = 7 \times 10^2 = 700\ \text{N}$

7. (c) $F = m\dfrac{\Delta v}{\Delta t} = \dfrac{150 \times 10^{-3} \times 20}{0.1} = 30\ \text{N}$

8. (b) In the frame of wedge,

$mg \sin 37^\circ = T + m a_0 \cos 37^\circ$

or $\quad T = 6 - \dfrac{4}{5} t$

9. (c) Acceleration, $a = \dfrac{F_0}{m} e^{-bt}$

$\dfrac{dv}{dt} = \dfrac{F_0 e^{-bt}}{m} \Rightarrow dv = \dfrac{F_0 e^{-bt}}{m} dt$

$v = \int_0^t \dfrac{F_0 e^{-bt}}{m} dt = \dfrac{-F_0 e^{-bt}}{mb} + C$

At $\quad t = 0, v = 0$

$0 = \dfrac{-F_0 e^{-b(0)}}{mb} + C \Rightarrow C = \dfrac{F_0}{mb}$

$\therefore \quad v = \dfrac{-F_0 e^{-bt}}{mb} + \dfrac{F_0}{mb}$

10. (c) For block A in the frame of triangular block,

$m_A g \sin 37^\circ + m_A a_0 \cos 37^\circ - T = m_A a_{\text{rel}}$

or $\quad 12 - T = a_{\text{rel}} \quad \ldots\text{(i)}$

For block B in the frame of triangular block,

$T + m_B a_0 \cos 53^\circ - m_B g \sin 53^\circ = m_B a_{\text{rel}}$

$T - 5 = a_{\text{rel}} \quad \ldots\text{(ii)}$

From Eqs. (i) and (ii), $T = 8.5\ \text{N} = 0.85\ \text{kg}$

11. (a) Since, speed is constant. So, the component of resultant force in the direction of velocity should be zero.

12. (d) $F_{AB} + F_{BC} = F_{AC} \Rightarrow F_{AB} + F_{BC} + F_{AC} = F_{AC} + F_{CA} = 0$

As, resultant force is zero, so particle moves with constant velocity.

13. (b) $mg = 2kx \Rightarrow x = \dfrac{mg}{2k} = \dfrac{10}{2 \times 100} \Rightarrow x = \dfrac{1}{20}\ \text{m} = 5\ \text{cm}$

Thus, the distance of the block from pulley is $l = 20 + 5 = 25\ \text{cm}$

14. (c) For the equilibrium of block A, the resultant of upper spring force and $m_A g$ is equal to spring force due to right spring.

Thus, just after cutting the right spring, the resultant force on the block A is equal to previous value of right spring force which is equal to $m_B g$.

$\therefore \quad a_A = \dfrac{m_B g}{m_A} = \dfrac{2g}{1} = 2g$ in leftward direction.

15. (a) $m_1 g - T = m_1 a$ and $T - m_2 g \sin\theta = m_2 a$

$\therefore \quad T = \dfrac{m_1 m_2 g(1 + \sin\theta)}{m_1 + m_2}$

From the concept of number system, if sum of two numbers is constant, the product is maximum, when both numbers are same.

$\therefore$ For T_{max}, $m_1 = m_2 = 5\ \text{kg} \Rightarrow 1 : 1$.

16. (a)

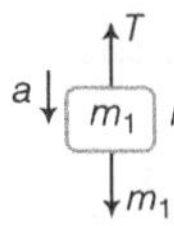

$m_1 g - T = m_1 a$

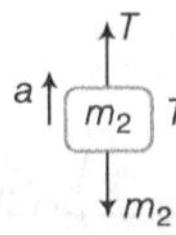

$T - m_2 g = m_2 a$

$(m_1 - m_2)g = (m_1 + m_2)a$

$\Rightarrow \quad a = \dfrac{(m_1 - m_2)g}{(m_1 + m_2)} = \dfrac{0.2}{9.8} \times 9.8 = 0.2\ \text{m/s}^2$

17. (b) $a = \left(\dfrac{m_1 - m_2}{m_1 + m_2}\right) g \Rightarrow \dfrac{g}{8} = \left(\dfrac{\dfrac{m_1}{m_2} - 1}{\dfrac{m_1}{m_2} + 1}\right) g \Rightarrow \dfrac{m_1}{m_2} = \dfrac{9}{7}$

18. (c) $\dfrac{v_A}{v_B} = \dfrac{2T}{3T} \Rightarrow v_A = 8\ \text{m/s}$

But $\dfrac{v_D + v_A}{2} = v_B \Rightarrow v_D + 8 = 24 \Rightarrow v_D = 16\ \text{m/s}$

19. (c) The separation between block B and pulley D, and between pulley C and pulley F decreases. But separation between pulley E and pulley C, and between pulley F and motor increases. Thus,

$-v_B \cos 37^\circ + v_p \cos 37^\circ - v_p \cos 53^\circ + v_A + 3 = 0$

or $\quad -\dfrac{4}{5} v_B + 10 \times \dfrac{4}{5} - 10 \times \dfrac{3}{5} + 10 = 0$

or $\quad \dfrac{4}{5} v_B = 12 \Rightarrow v_B = 12 \times \dfrac{5}{4} = 15\ \text{m/s}$

20. (d) $a_A = g \sin 60^\circ$ and $a_B = g \sin 30^\circ$

The vertical component of a_A is $a_1 = g \sin 60^\circ \cos 30^\circ$

and vertical component of a_B is $a_2 = g \sin 30^\circ \cos 60^\circ$

$\therefore \quad a_{\text{rel}} = a_1 - a_2 = \dfrac{g}{2} = 4.9\ \text{m/s}^2$

Revisal Problems (JEE Advanced)

1. (c) $mg \sin 37° = kx \cos 37°$

$\therefore \quad kx = mg \tan 37° = 2 \times 10 \times \frac{3}{4} = 15 \text{ N}$

$\therefore \quad kx \sin 37° = ma$

or $\quad 15 \times \frac{3}{5} = 2a$

$\therefore \quad a = 4.5 \text{ m/s}^2$

2. (a)

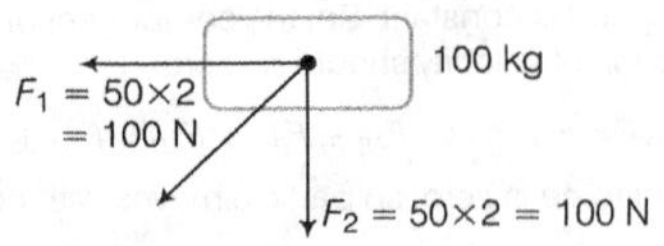

$a = \frac{100\sqrt{2}}{100} = \sqrt{2} \text{ m/s}^2$ in South-West direction

3. (b) In the frame of block C,

$T = mg \sin 37° + ma_0 \cos 37°$ (for block A)

and $2T = 2\,mg$ (for block B)

$\therefore \quad mg = mg \sin 37° + ma_0 \cos 37°$

$\therefore \quad a_0 = 5 \text{ m/s}^2$

4. (b) In the frame of lift, $mg + ma_0 - T = ma_{rel}$

$400 + 40 \times 5 - T = 40 \times 5$

$T = 400 \text{ N}$

$\therefore$ Reading of spring balance

$= 2T = 800 \text{ N} = 80 \text{ kg}$

5. (b)

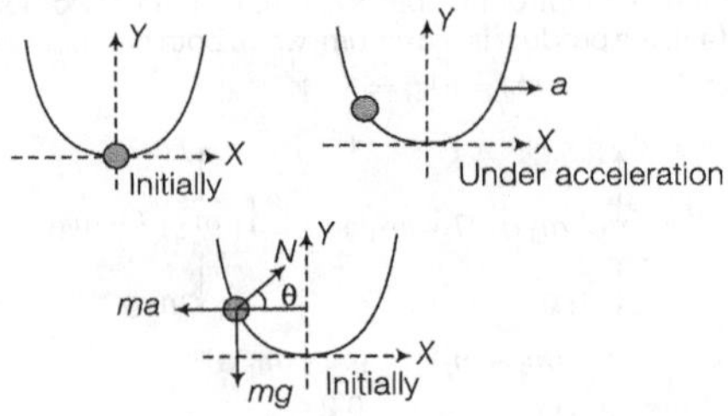

$ma = N\cos\theta, \; mg = N\sin\theta$

$\tan\theta = \frac{g}{a}$

Slope of normal $= \tan\theta = \frac{-1}{dy/dx} = \frac{-1}{-2kx}$

$\frac{g}{a} = \frac{1}{2kx}, \; x = \frac{a}{2gk}$

6. (c) $N\cos\theta = F, \; N\sin\theta = mg$

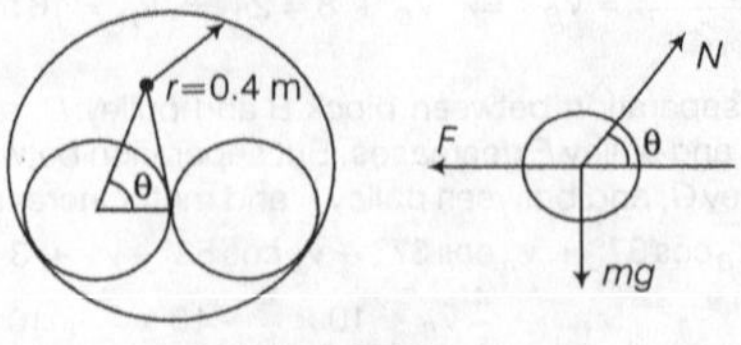

Where, $\quad F =$ contact force

$\tan\theta = \frac{mg}{F}$

$\Rightarrow \quad \frac{mg}{\tan\theta} = F$

$F = \frac{8\sqrt{3} \times 10 \times 0.05}{\sqrt{(0.4)^2 - (0.05)^2}} = 17.45 \text{ N} \approx 17.5 \text{ N}$

7. (d) When the system is released, the block B starts to fall down because of its weight. When B moves downwards, C moves rightwards. During this process, acceleration of these two blocks is equal to $a = \frac{M_B g}{M_B + M_C}$.

This acceleration remains constant till the block B hits the ground. Hence, during this process, velocity of C increases linearly with time t till B hits the ground.

When B hits the ground, it comes to rest while C has rightward velocity. Due to this velocity, block A is jerked into motion and its starts to move up. As soon as, block A starts to move up, the block C starts to retard. The magnitude of retardation is

$$= \frac{M_A g}{M_A + M_C}$$

But M_A and M_B are equal. Therefore, the magnitude of this retardation is equal to the acceleration calculated above. It means, now velocity-time curve will have a negative slope. This slope of velocity-time curve (the retardation) remains constant till the block A reaches it highest point and then comes back and hits the ground.

After the block B is jerked into the motion and then slope of velocity-time curve will become positive.

Hence, option (d) is correct.

Option (c) is wrong because when the string becomes tight, and block A is jerked into the motion, a loss of kinetic energy takes place. Hence, velocity of trolley C decreases slightly.

8. (a) As, acceleration of block A will be more than block B while both having constant speed v_0. Such that block A will touch the ground first.

9. (a,b,c,d) $\quad m_A g - T = m_A a \quad$ or $\quad 10 - T = a$

and $\quad T + m_B g \sin 37° = m_B a \quad$ or $\quad T + 6 = a = 10 - T$

$\therefore \quad T = 8 \text{ N}$

Also, $\quad 10 - T = a$

$\therefore \quad a = 2 \text{ m/s}^2$

$a_{rel} = \sqrt{a_A^2 + a_B^2 - 2a_A a_B \cos 53°}$

$= \sqrt{2^2 + 2^2 - 2 \times 2 \times 2 \times \frac{3}{5}} = \frac{4}{\sqrt{5}} \text{ m/s}^2$

10. (a,b,c,d) For triangular block B,

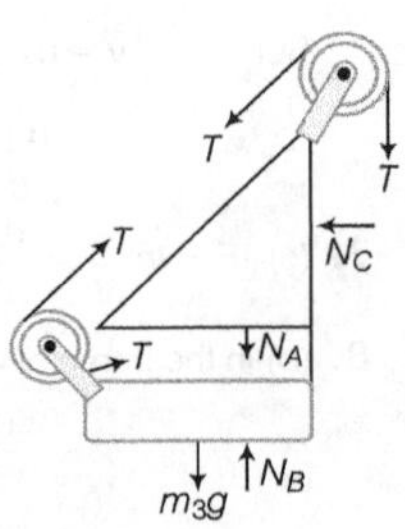

$T - N_C = 5a_0 \quad \text{...(i)}$

$N_B = N_A + T + m_B g$

or $\quad N_B = m_A g + T + m_B g$

$\therefore \quad N_B = 100 + T \quad \text{...(ii)}$

From free body diagram of block A in frame of block B,

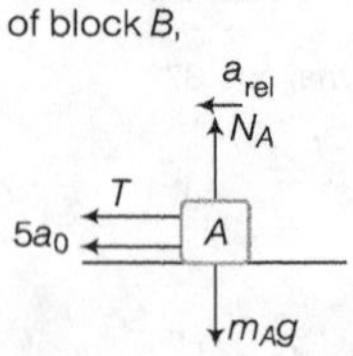

$T + 5a_0 = 5\,a_{rel} \quad \text{... (iii)}$

From free body diagram of block C in the frame of block B

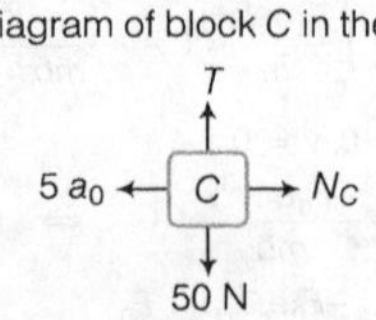

$N_C = 5a_0 \quad \text{...(iv)}$

and $\quad 50 - T = 5\,a_{rel} \quad \text{...(v)}$

From Eqs. (i), (iii), (iv) and (v)

$$T - 5a_0 = 5a_0$$
$$\Rightarrow \quad T = 10\,a_0$$
$$T + 5a_0 = 5a_{rel} \quad \text{(from Eq. (iii))}$$
$$15a_0 = 5a_{rel}$$
$$\Rightarrow \quad a_{rel} = 3a_0$$
$$50 - T = 5a_{rel} \quad \text{(from Eq. (v))}$$
$$50 - 10a_0 = 15a_0$$
$$\Rightarrow \quad 50 = 25a_0$$
$$\therefore \quad a_0 = 2\text{ m/s}^2$$
$$\therefore \quad T = 10 \times 2 = 20\text{ N}$$
$$a_{rel} = 3a_0 = 6\text{ m/s}^2$$
$$N_B = 100 + T = 100 + 20 = 120\text{ N}$$

11. (a,b,c,d) $m_D g = (m_B + m_C + m_D)\,a \Rightarrow a = 5\text{ m/s}^2$

$$T_{BC} = m_B a = 1 \times 5 = 5\text{ N}$$

and $m_D g - T_{CD} = 2 \times 5$

$$\therefore \quad T_{CD} = 20 - 10 = 10\text{ N}$$

12. (a,b,c) $T = (m_B + m_D)\,g = 150$

(b) $T_B - 100 = 10 \times 4$

$$T_B = 140\text{ N} \quad \text{(for child)}$$
$$50 + 140 - T = 5a \quad \text{(for ladder } D\text{)}$$
$$190 - T = 5a$$
$$2T - 300 = 30\frac{a}{2} \quad \text{(for man } A\text{)}$$
$$2T - 300 = 15a$$
$$\Rightarrow \quad a = 3.2\text{ m/s}^2$$
$$\therefore \quad a_{rel} = 0.8\text{ m/s}^2$$
$$s_{rel} = \frac{1}{2}a_{rel}t^2 = 0.4\text{ m}$$

(c)
$$T - 50 - 140 = 5a \quad \text{(for ladder } D\text{)}$$
$$2T = 300 + 30 \times 4 = 420 \quad \text{(for ladder } C\text{)}$$
$$\therefore \quad T = 210$$
$$\therefore \quad 210 - 190 = 5a$$
$$\Rightarrow \quad 20 = 5a$$
$$\Rightarrow \quad a = 4\text{ m/s}^2$$

13. (b,d) According to free body diagram, we get

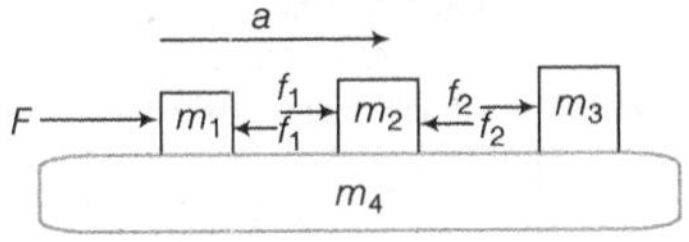

Suppose contact force between m_1 and m_2 be f_1 and between m_2 and m_3 be f_2.

Such that $f_2 = m_3 a$, $\quad f_1 - f_2 = m_2 a$

$F - f_1 = m_1 a$

So, acceleration produced by applying force

i.e. $a = \dfrac{F}{m_1 + m_2 + m_3}$

Then $f_1 = \dfrac{(m_2 + m_3) \times F}{(m_1 + m_2 + m_3)}$ and $f_2 = \dfrac{m_3 \times F}{(m_1 + m_2 + m_3)}$

(b) If we increase masses, i.e. $m_1 < m_2 < m_3$, then $f_1 > f_2$.

(d) As, the normal reaction exerted by ground on m_4 is also constant.

14. (b) Gases do not need to push on anything as gases already expelled from rocket by pushing the rocket.

15. (a) As in vacuum there is nothing for gases to push upon and hence it is impossible to see how a rocket fired in space world work.

16. (b) Here, $T = \left(\dfrac{2m_1m_2}{m_1 + m_2}\right) g \sin 37° = \dfrac{2m_1m_2}{8} \times 10 \times \dfrac{3}{5}$

$$T = \frac{3}{2}m_1m_2$$

or $\quad 22.5 = \dfrac{3}{2}m_1m_2$ or $15 = m_1m_2 = m_1(8 - m_1)$

or $\quad m_1^2 - 8m_1 + 15 = 0$ or $m_1 = \dfrac{8 \pm \sqrt{64 - 60}}{2}$

or $\quad m_1 = \dfrac{8 \pm 2}{2} = 5\text{ kg or }3\text{ kg}$

$\therefore \quad$ Ratio $= \dfrac{3}{5}$

17. (d) $a = \dfrac{(m_1 - m_2)g \sin 37°}{m_1 + m_2} = \dfrac{(5 - 3) \times 10 \times \frac{3}{5}}{8}$

$$= \frac{2 \times 10 \times 3}{5 \times 8} = \frac{3}{2} = 1.5\text{ m/s}^2$$

18. (6) Taking blocks B and C as a system, $N \sin 30° = (m_B + m_C)\,a_0$ (leftward).

Here, N is normal reaction between A and C.

From free body diagram of block B in the frame of block C.

$$m_B g \sin 30° = m_B a_0 \cos 30°$$
$$\therefore \quad a_0 = \frac{10}{\sqrt{3}}\text{ m/s}^2$$

From free body diagram of block A in the frame of block C

$$N + m_A a_0 \sin 30° = m_A g \cos 30°$$

or $\quad \dfrac{(m_B + m_C)a_0}{\sin 30°} + m_A a_0 \sin 30° = m_A g \cos 30°$

or $\quad \dfrac{3 \times 10}{\sqrt{3}} \times 2 + m_A \times \dfrac{10}{2\sqrt{3}} = m_A \times 10 \times \dfrac{\sqrt{3}}{2}$

$$\therefore \quad m_A = 6\text{ kg}$$

19. (5) From free body diagram of triangular block,

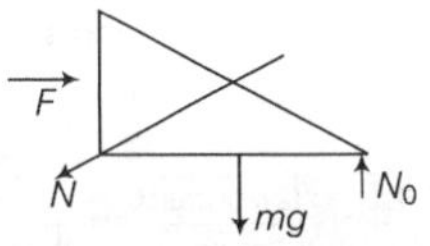

$$F = N \sin 37°$$

From free body diagram of sphere,

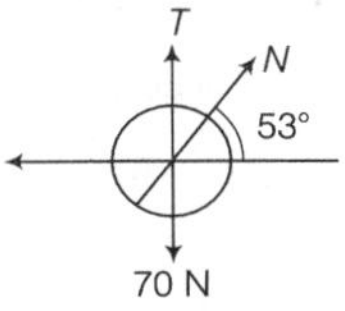

$$T + N \sin 53° = 70$$
$$10 + \frac{F}{\sin 37°}\sin 53° = 70$$
$$\therefore \quad F = 60 \times \frac{3}{5} \times \frac{5}{4} = 45\text{ N}$$
$$\Rightarrow \quad 9n = 45\text{ N} \Rightarrow n = 5$$

20. (3) From free body diagram of block B,

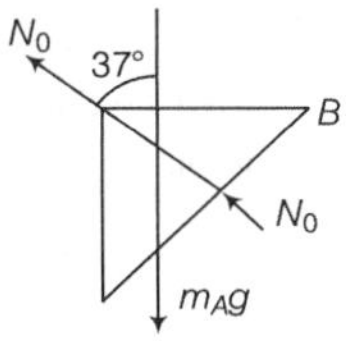

$$N_0 \cos 37° = m_A g = 10$$
$$\therefore \quad N_0 = \frac{10}{4} \times 5 = \frac{50}{4}$$

and $\quad N_0 \sin 37° = 1 \times a_0$

or $\quad \dfrac{50}{4} \times \dfrac{3}{5} = a_0$

$$\Rightarrow \quad a_0 = \frac{30}{4}\text{ m/s}^2$$

For horizontal direction, motion of both blocks can be taken as a system,

$\therefore \quad F \sin 30° = (m_A + m_B)\,a_0$ or $\dfrac{F}{2} = 2 \times \dfrac{30}{4}$

$$\therefore \quad F = 30\text{ N} \Rightarrow 10n = 30$$
$$\Rightarrow \quad n = 3$$

21. (4) For lower laddu, $N_1 = mg + N_3 \sin\theta$

$$N_2 = N_3 \cos\theta \quad \ldots \text{(i)}$$

For upper laddu, $N_4 = N_3 \cos\theta$

and $N_3 \sin\theta = mg \quad \ldots \text{(ii)}$

From Eqs. (i) and (ii), $N_2 = N_3 \cos\theta$

or $N_2 = \dfrac{mg}{\sin\theta} \cos\theta = mg \cot\theta$

$\Rightarrow \quad \cos\theta = \dfrac{d - 2R}{2R}$

$\therefore \quad \cot\theta = \dfrac{d - 2R}{\sqrt{4R^2 - (d^2 - 2R)^2}} = \dfrac{d - 2R}{\sqrt{4R^2 - (d^2 + 4R^2 - 4Rd)}}$

or $\cot\theta = \dfrac{d - 2R}{\sqrt{4Rd - d^2}}$

$\therefore \quad N_2 = mg \dfrac{(d - 2R)}{\sqrt{4Rd - d^2}} \Rightarrow n = 4$

22. (3) $N \sin\theta = 2\, ma_2 \quad \ldots \text{(i)}$

$mg - N \cos\theta = ma_1 \quad \ldots \text{(ii)}$

Also, from constrained relation,

$$a_2 \sin\theta = a_1 \cos\theta$$

$\therefore \quad a_1 = a_2 \tan\theta \quad \ldots \text{(iii)}$

From Eqs. (i), (ii) and (iii), $mg - N \cos\theta = ma_1$

or $mg - \dfrac{2ma_2}{\sin\theta} \cos\theta = ma_2$

or $g - 2a_2 \cot\theta = a_2 \tan\theta$

or $g = (2 \cot\theta + \tan\theta) a_2$

$\therefore \quad a_2 = \dfrac{g}{2 \cot\theta + \tan\theta} = \dfrac{9}{3} = 3 \text{ m/s}^2$

23. (b)

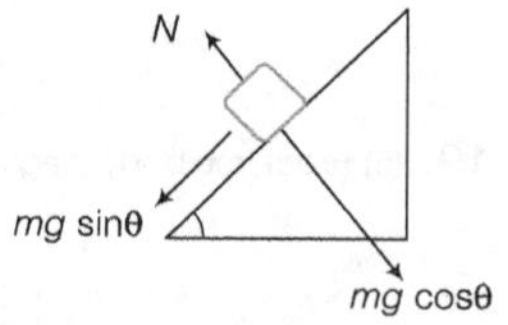

Here, $mg \sin\theta = ma$

$\therefore \quad a = g \sin\theta$

The length of inclined plane is

$l = h \operatorname{cosec}\theta$

$\because \quad s = \dfrac{1}{2} at^2$

$\therefore \quad t = \sqrt{\dfrac{2s}{a}} = \sqrt{\dfrac{2l}{a}} = \sqrt{\dfrac{2h \operatorname{cosec}\theta}{g \sin\theta}} = \sqrt{\dfrac{2h}{g \sin^2\theta}}$

$\Rightarrow a \propto \sin\theta$ and $t \propto \dfrac{1}{\sin\theta}$

Hence, Statements I and II are correct.

24. (c) If net force on a particle is constant, acceleration of particle is constant. If acceleration and velocity are either parallel or anti-parallel, path is straight line. If the acceleration is constant but acceleration and initial velocity are neither parallel nor anti-parallel, path is parabolic. Hence, Statement I is correct but Statement II is incorrect.

25. (d) If a particle is in equilibrium acceleration should be zero. If a particle is in rest at an instant, the acceleration may or may not be zero. Thus, Statement I is incorrect but Statement II is correct.

26. (d) Since, car is accelerating with an acceleration $\mathbf{a}_C = \dfrac{d\mathbf{v}}{dt} = 3\hat{\mathbf{i}}$.

So, the observer is in non-inertial reference frame. So, Newton's law is not valid. Hence, Statement I is incorrect but Statement II is correct.

27. (a) A man runs for a certain distance to maintain his initial inertia. Hence, Statement I is correct and Statement II is correct explanation of Statement I.

28. (a)

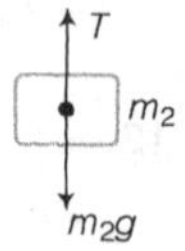

(i) For equilibrium, from force body diagram of m_2

$$m_2 g = T$$

$\therefore \quad T = 200 \text{ N}$

For equilibrium of (man + platform)

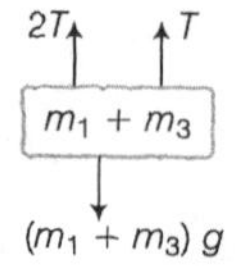

$$3T = (m_1 + m_3) g$$

or $600 = 100 + m_3 g$

$\therefore \quad m_3 g = 500 \text{ N}$

(ii) For upward acceleration,

$$3T > (m_1 + m_3) g$$

$$3T > 600 \text{ N}$$

$\therefore \quad T \geq 200 \text{ N}$

$\therefore$ The force exerted by boy on the string as

$$F = 2T > 400 \text{ N}$$

(iii) For downward acceleration,

$$3T < (m_1 + m_3) g$$

or $T < \dfrac{600}{3}$

$\therefore \quad T < 200 \text{ N}$

$\therefore$ The force exerted by boy on string is

$$F = 2T < 200 \text{ N}$$

(iv) For man, $2T + N = m_3 g$

or $400 + N = 500$

$\therefore \quad N = 100 \text{ N}$

(i) → (q), (ii) → (p,q), (iii) → (r,s), (iv) → (s)

JEE Main & AIEEE Archive

1. (b) From the graph, it is a straight line, so uniform motion. Because of impulse direction of velocity changes as can be seen from the slope of the graph.

Initial velocity, $v_1 = \dfrac{2}{2} = 1 \text{ms}^{-1}$

Final velocity, $v_2 = -\dfrac{2}{2} = -1 \text{ms}^{-1}$

$$\mathbf{p}_i = mv_1 = 0.4 \text{ N-s}$$

$$\mathbf{p}_f = mv_2 = -0.4 \text{ N-s}$$

$$\mathbf{J} = \mathbf{p}_f - \mathbf{p}_i = -0.4 - 0.4 = -0.8 \text{ N-s}$$

$\therefore \quad |\mathbf{J}| = 0.8 \text{ N-s}$ (J = impulse)

2. (d) Force applying on the block

$F = mg \sin\theta$ or $mg \sin\theta = ma$

$\therefore \quad a = g \sin\theta$

where, a is acceleration along the inclined plane.

$\therefore$ Vertical component of acceleration is $g \sin^2\theta$.

$\therefore$ Relative vertical acceleration of A with respect to B is

$$g(\sin^2 60° - \sin^2 30°) = \frac{g}{2} = 4.9 \text{ ms}^{-2}$$

(in vertical direction)

3. (c) Acceleration of system,

$$a = \frac{F}{m + M}$$

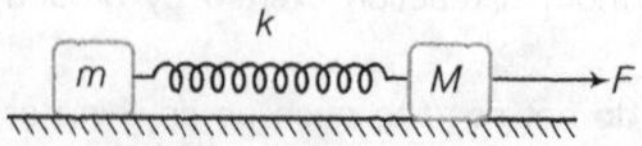

So, force acting on mass,

$$F = ma = \frac{mF}{m + M}$$

4. (d) Here, the constant horizontal force required to take the body from position 1 to position 2 can be calculated by using work-energy theorem. Let us assume that body be taken slowly, so that its speed does not change, then

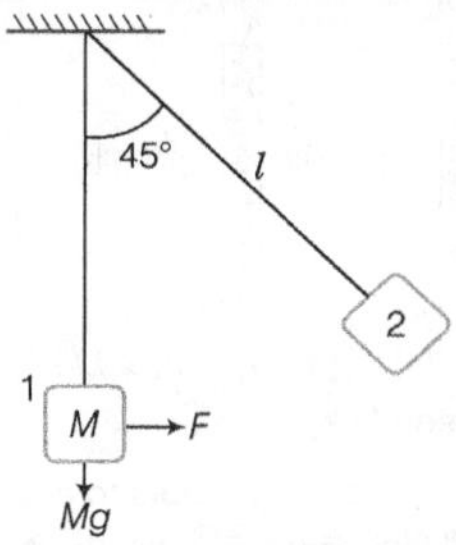

$$\Delta K = 0 = W_F + W_{Mg} + W_{\text{tension}}$$

(symbols have their usual meanings)

$$W_F = F \times l \sin 45^\circ = \frac{Fl}{\sqrt{2}}$$

$$W_{Mg} = Mg(l - l\cos 45^\circ),\ W_{\text{tension}} = 0$$

$$\therefore \quad F = Mg(\sqrt{2} - 1)$$

5. (d) The situation is shown in figure. At initial time, the ball is at P, then under the action of a force (exerted by hand) from P to A and then from A to B, let acceleration of ball during PA be a ms^{-2} (assumed to be constant) in upward direction and velocity of ball at be is v m/s.

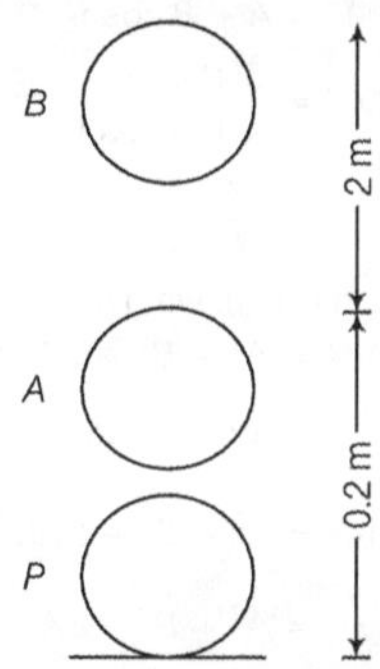

Then, for PA,

$$v^2 = 0^2 + 2a \times 0.2$$

For AB,

$$0 = v^2 - 2 \times g \times 2$$

$$\Rightarrow \quad v^2 = 2g \times 2$$

From above equations,

$$a = 10g = 100 \text{ ms}^{-2}$$

Then, for PA, FBD of ball is

$$F - mg = ma$$

(F is the force exerted by hand on ball)

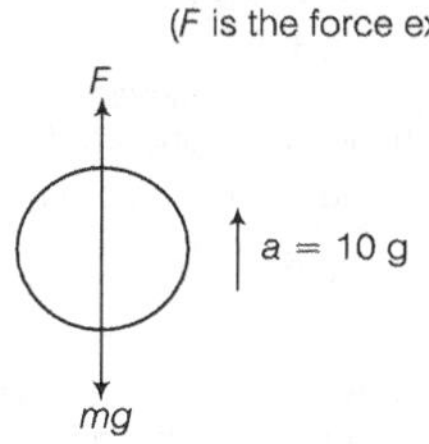

$$\Rightarrow \quad F = m(g + a) = 0.2(11g) = 22 \text{ N}$$

Alternate Method

Using work-energy theorem

$$W_{mg} + W_F = 0$$

$$\Rightarrow \quad -mg \times 2.2 + F \times 0.2 = 0$$

or $\quad F = 22$ N

6. (c) This is the question based on impulse- momentum theorem.

$$|F \cdot \Delta t| = |\text{ Change in momentum}|$$

$$\Rightarrow \quad F \times 0.1 = |p_f - p_i|$$

As the ball will stop after catching

$$p_i = mv_i = 0.15 \times 20 = 3,\ p_f = 0$$

$$\Rightarrow \quad F \times 0.1 = 3$$

$$\Rightarrow \quad F = 30 \text{ N}$$

7. (d) In the frame of wedge, the force diagram of block is shown in figure. From free body diagram of wedge,

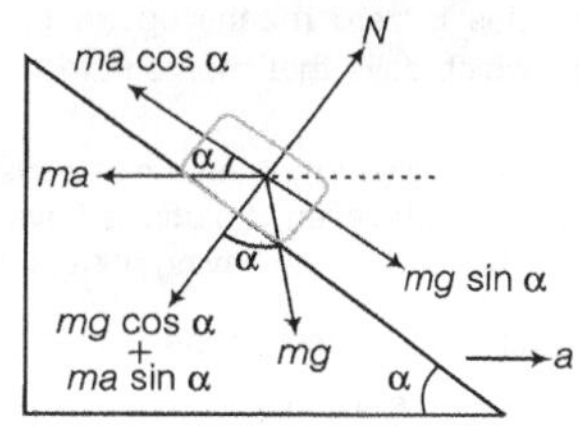

For block to remain stationary,

$$ma\cos\alpha = mg\sin\alpha \quad \text{or} \quad a = g\tan\alpha$$

8. (d) Given, $\quad m = 0.3$ kg, $\quad x = 20$ cm and $k = 15$ N/m

given $\quad F = -kx \quad$ ...(i)

and $\quad F = ma \quad$ ...(ii)

$$\therefore \quad ma = -kx \Rightarrow a = -\frac{15}{0.3} \times 20 \times 10^{-2}$$

$$\Rightarrow \quad a = -\frac{15}{3} \times 2 = -10 \text{ ms}^{-2}$$

Negative sign indicates that acceleration is always towards the mean position.

$\therefore$ Initial acceleration, $a = 10$ ms^{-2}

9. (d) Force exerted by machine gun on man's hand in firing a bullet

= Change in momentum per second on a bullet or rate of change of momentum

$$= \left(\frac{40}{1000}\right) \times 1200 = 48 \text{ N}$$

Force exerted by man on machine gun

= Force exerted on man by machine gun = 144 N

Hence, number of bullets fired $= \frac{144}{48} = 3$

10. (a) On releasing, the motion of the system will be according to figure.

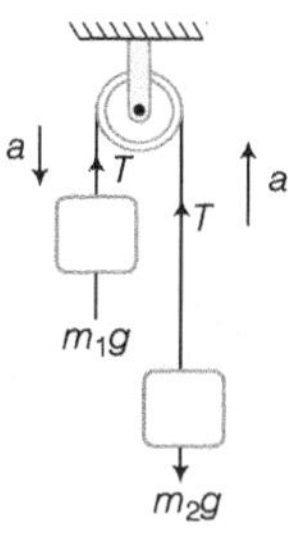

$$m_1g - T = m_1a \quad \text{...(i)}$$

and $\quad T - m_2g = m_2a \quad$ ...(ii)

On solving, $\quad a = \left(\frac{m_1 - m_2}{m_1 + m_2}\right)g \quad$ ...(iii)

Here, $\quad m_1 = 5$ kg, $m_2 = 4.8$ kg, $g = 9.8$ ms^{-2}

$$\therefore \quad a = \left(\frac{5 - 4.8}{5 + 4.8}\right) \times 9.8$$

$$= \frac{0.2}{9.8} \times 9.8 = 0.2 \text{ ms}^{-2}$$

11. (a) In stationary position, spring balance reading $= mg = 49$ N

or $$m = \frac{49}{9.8} = 5 \text{ kg}$$

When lift moves downward,

$$mg - T = ma$$

Reading of balance,

$$T = mg - ma = 5(9.8 - 5)$$
$$= 5 \times 4.8 = 24.0 \text{ N}$$

12. (d) Resultant force is zero, as three forces acting on the particle can be represented in magnitude and direction by three sides of a triangle in same order. Hence, by Newton's second law $\left(\mathbf{F} = m\frac{d\mathbf{v}}{dt}\right)$, the velocity ($\mathbf{v}$) of particle will be same.

13. (c) Here, II is implying I but I is not implying II as kinetic energy of a system of particles is zero means speed of each and every particle is zero which says that momentum of every particle is zero.

But statement I means linear momentum of a system of particles is zero, which may be true even, if particles have equal and opposite momentum and hence having non-zero kinetic energy.

14. (d) Let acceleration of system (rope + block) be a along the direction of applied force. Then,

$$a = \frac{P}{M + m}$$

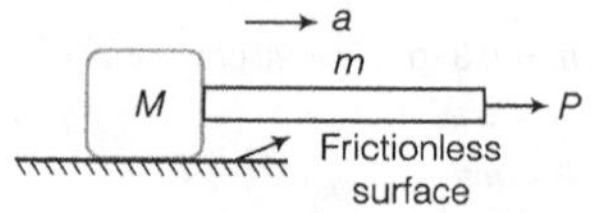

Draw the FBD of block and rope as shown in figure.

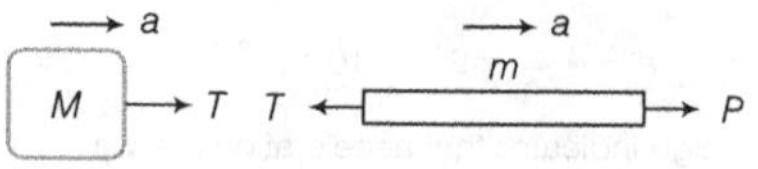

where, T is the required tension in a rope

For block,

$$T = Ma$$

$$\Rightarrow \quad T = \frac{MP}{M + m}$$

15. (a) The arrangement is shown in figure.

Now, draw the free body diagram of the spring balances and block.

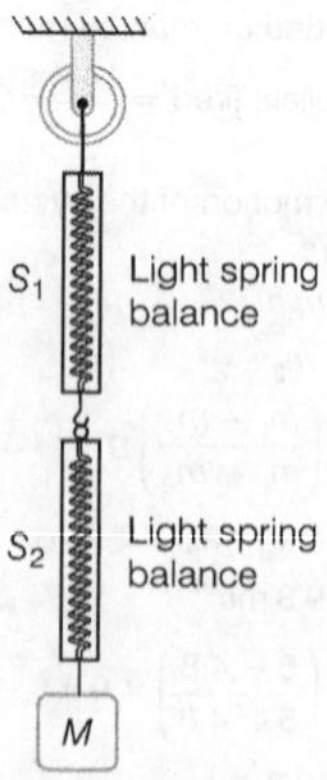

For equilibrium of block,

$$T_1 = Mg$$

where, T_1 = Reading of S_2

For equilibrium of S_2, $\quad T_2 = T_1$

where, T_2 = Reading of S_1

For equilibrium of S_1, $\quad T_2 = T_3$

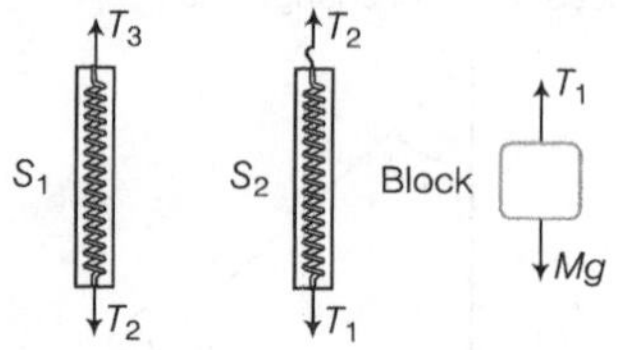

Hence, $\quad T_1 = T_2 = Mg$

So, both scales read M kg.

16. (a) Here, thrust force is responsible to accelerate the rocket, so initial thrust of the blast $= ma = 3.5 \times 10^4 \times 10 = 3.5 \times 10^5$ N

17. (c) Apparent weight of ball,

$$w' = w - R \text{ [where, } R\text{=normal reaction]}$$
$$R = ma \text{ [acting upward]}$$
$$w' = mg - ma = m(g - a)$$

Hence, apparent acceleration in the lift is $g - a$. Now, if the man is standing stationary on the ground, then the apparent acceleration of the falling ball is g.

18. (b) The sum of the two forces

$$A + B = 18 \quad \text{...(i)}$$

$$\tan \alpha = \frac{B \sin \theta}{A + B \cos \theta}$$

$$\Rightarrow \quad \tan 90° = \frac{B \sin \theta}{A + B \cos \theta} \text{ or } A + B\cos\theta = 0$$

$$\Rightarrow \quad \cos \theta = \frac{-A}{B} \quad \text{...(ii)}$$

Squaring both sides in Eq. (i), we get

$$144 = A^2 + B^2 + 2AB\cos\theta$$

On putting $\cos\theta = \frac{-A}{B}$ in above equation, we get

$$144 = A^2 + B^2 + 2AB\left(-\frac{A}{B}\right)$$
$$= A^2 + B^2 - 2A^2$$
$$\Rightarrow \quad 144 = B^2 - A^2$$
$$\Rightarrow \quad 144 = (B - A)\cdot(B + A)$$

On putting $B + A = 18$, we have

$$B - A = \frac{144}{18} = 8$$

Solving $\quad B - A = 8$

$\quad B + A = 18$

We have,

$$B = 13 \text{ N}, A = 5 \text{ N}$$

19. (a) The particle remains stationary under the acting of three forces $\mathbf{F}_1$, $\mathbf{F}_2$ and $\mathbf{F}_3$, it means resultant force is zero.

$$\mathbf{F}_1 + \mathbf{F}_2 + \mathbf{F}_3 = 0$$

Since, in second cases F_1 is removed (in terms of magnitude we are talking now), the forces acting are F_2 and F_3 the resultant of which has the magnitude as F_1, so acceleration of particle is $\frac{F_1}{m}$ in the direction opposite to that of $\mathbf{F}_1$.

20. (b) As the string is inextensible, both masses have the same acceleration a. Also, the pulley is massless and frictionless, hence the tension at both ends of the string is the same. Suppose, the mass m_2 is greater than mass m_1, so the heavier mass m_2 is accelerating downward and the lighter mass m_1 is accelerating upwards.

Therefore, by Newton's second law

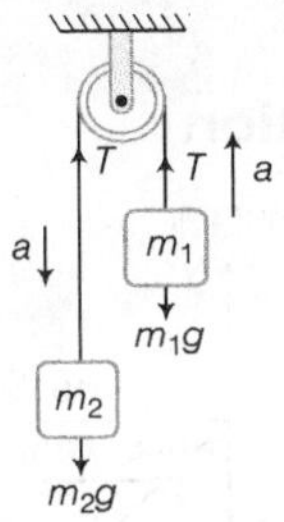

$$T - m_1g = m_1a \quad \ldots(i)$$

and

$$m_2g - T = m_2a \quad \ldots(ii)$$

After solving Eqs. (i) and (ii), we get

$$a = \frac{(m_2 - m_1)}{(m_1 + m_2)} \cdot g = \frac{g}{8} \quad \text{[given]}$$

So,

$$\frac{g}{8} = \frac{m_2(1 - m_1/m_2)}{m_2(1 + m_1/m_2)} \cdot g \quad \ldots(iii)$$

Let

$$\frac{m_1}{m_2} = x$$

Thus, Eq. (iii) becomes

$$\frac{1 - x}{1 + x} = \frac{1}{8} \quad \text{or} \quad x = \frac{7}{9}$$

or

$$\frac{m_2}{m_1} = \frac{9}{7}$$

So, the ratio of the masses is 9 : 7.

21. (c) The free body diagram of the person can be drawn as

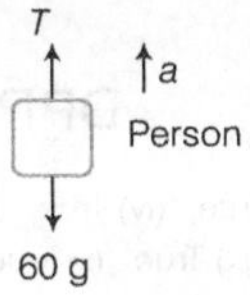

Let the person move up with an acceleration a, then

$$T - 60g = 60a$$

$$\Rightarrow \quad a_{max} = \frac{T_{max} - 60g}{60}$$

or

$$a_{max} = \frac{360 - 60g}{60} \rightarrow \text{negative value}$$

That means, it is not possible to climb up on the rope.

Even in this problem, it is not possible to remain at rest on rope.

Hence, no option is correct.

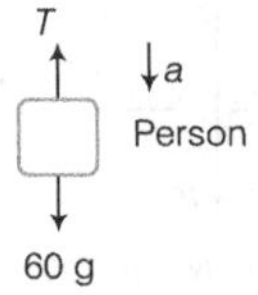

But, if they will ask for the acceleration of climbing down, then

$$60g - T = 60a$$

$$\Rightarrow \quad 60g - T_{max} = 60a_{min}$$

or

$$a_{min} = \frac{60g - 360}{60} = 4 \text{ ms}^{-2}$$

JEE Advanced & IIT JEE Archive

1. (a,c) $w = mg = 0.1 \times 10 = 1$ N

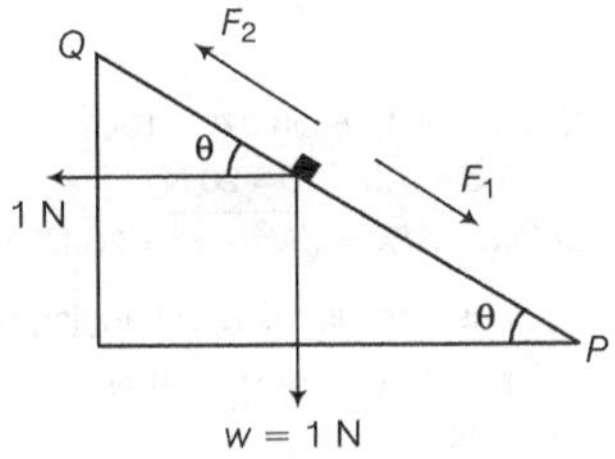

F_1 = component of weight = $1 \cdot \sin\theta = \sin\theta$

F_2 = component of applied force = $1 \cdot \cos\theta = \cos\theta$

Now, at $\theta = 45°$: $F_1 = F_2$ and block remains stationary without the help of friction.

For $\theta > 45°$, $F_1 > F_2$, so friction will act towards Q.

For $\theta < 45°$, $F_2 > F_1$ and friction will act towards P.

2. (b) $N \sin\theta = mg$, $N \cos\theta = ma$

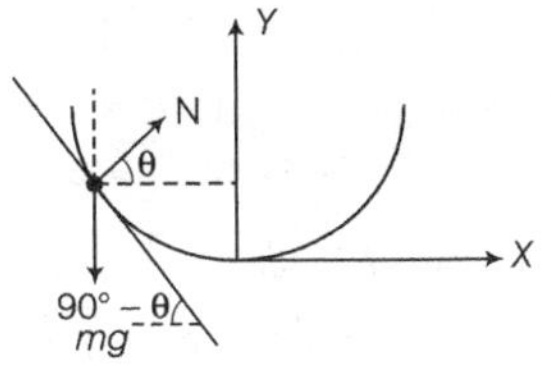

$$\tan\theta = \frac{g}{a}$$

$$\cot\theta = \frac{a}{g} = \tan(90° - \theta)$$

$$= \frac{dy}{dx} = 2kx$$

$$\therefore \quad x = \frac{a}{2kg}$$

3. (b) The cloth can be pulled out without dislodging the dishes from the table due to law of inertia, which is Newton's first law. While, the Statement II is true, but it is Newton's third law.

4. (b) $2T \cos\theta = F$

$$\therefore \quad T = \frac{F}{2} \sec\theta$$

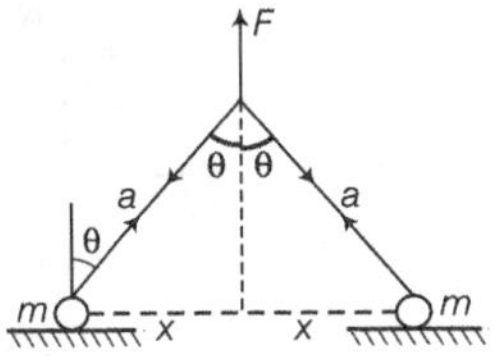

$$\text{Acceleration of particle} = \frac{T \sin\theta}{m}$$

$$= \frac{F \tan\theta}{2m}$$

$$= \frac{F}{2m} \frac{x}{\sqrt{a^2 - x^2}}$$

5. (a) Initially under equilibrium of mass m

$$T = mg$$

Now, the string is cut. Therefore, $T = mg$ force is decreased on mass m upwards and downwards on mass $2m$.

$$\therefore \quad a_m = \frac{mg}{m} = g \quad \text{(downwards)}$$

$$a_{2m} = \frac{mg}{2m} = \frac{g}{2} \quad \text{(upwards)}$$

DPP 2. Friction

DPP-1 Basic Concepts of Static Friction

1. (i) True, (ii) True, (iii) True, (iv) True, (v) True, (vi) True, (vii) True, (viii) False, (ix) True, (x) True, (xi) True

2. (i) Here, horizontal force is zero.

$\because \quad F = f$

$\therefore \quad 0 = f$

$f = 0$

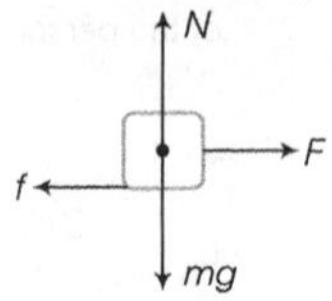

The contact force is $F_c = \sqrt{N^2 + f^2}$

$N = 10\ \text{N}$

(ii) Here, limiting friction is

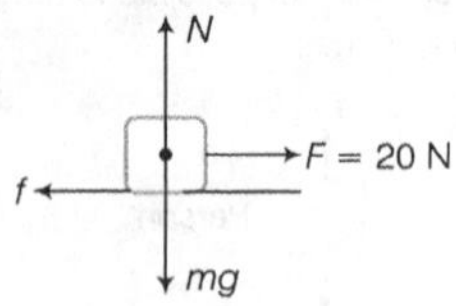

$f_{max} = \mu_s N = \mu_s mg = 50\ \text{N}$

$\because \quad F < f_{max}$

Static friction comes into play.

From free body diagram, $f = F = 20\ \text{N}$

The contact force is $F_c = \sqrt{N^2 + f^2}$

$= \sqrt{100^2 + 20^2} \approx 102\ \text{N}$

(iii) Here, limiting friction is $f_{max} = \mu_s N = 20\ \text{N}$

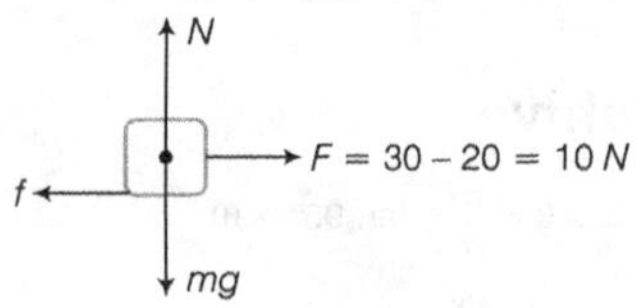

$\because F < f_{max,}$ static friction comes into play.

$\therefore \quad f = F = 10$

The contact force, $F_c = \sqrt{N^2 + f^2}$

$= \sqrt{100^2 + 10^2} = 100.50\ \text{N}$

(iv) Here,

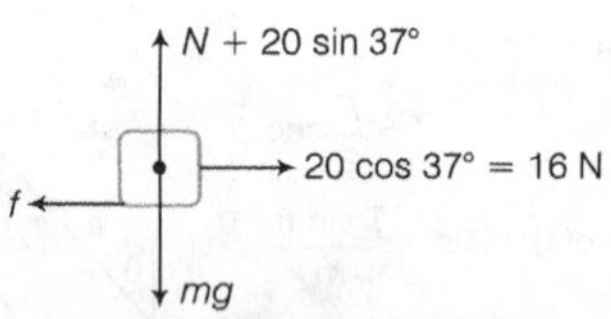

$f_{max} = \mu_s N = 0.5\,(mg - 20 \sin 37°)$

$= 0.5\left(100 - 20 \times \frac{3}{5}\right)$

$= 0.5\,(100 - 12) = 44\ \text{N}$

Here, horizontal applied force is $F = 16\ \text{N} < f_{max}$, static friction comes into play.

$\therefore \quad f = 16\ \text{N}$

$\therefore$ The contact force is $F_c = \sqrt{N^2 + f^2}$

$= \sqrt{44^2 + 16^2}$

$= 46.82\ \text{N}$

(v) From free body diagram,

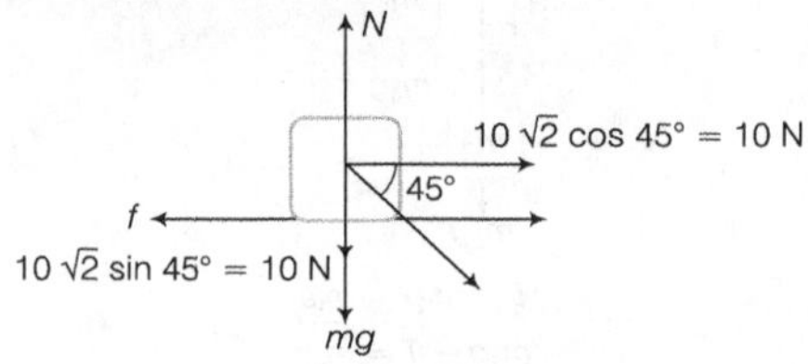

$N = mg + 10\sqrt{2} \sin 45°$

$N = mg + 10 = 20 + 10 = 30\ \text{N}$

$\because \quad f_{max} = \mu_s N = 15\ \text{N}$

Since, horizontal applied force $F = 10\ \text{N}$ is lesser than limiting friction.

$\therefore \quad f = 10\ \text{N}$

The contact force is $F_c = \sqrt{N^2 + f^2}$

$= \sqrt{(30)^2 + (10)^2} = 31.6\ \text{N}$

(vi) From free body diagram of the block,

The horizontal force is

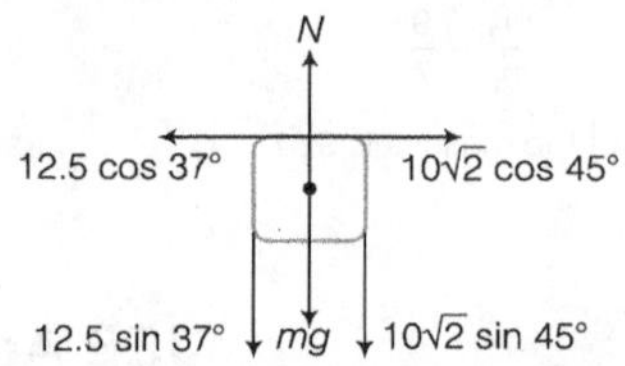

$F = 10\sqrt{2} \cos 45° - 12.5 \cos 37°$

$F = 10 - 12.5 \times \frac{4}{5} = 0$

$\therefore \quad f = 0$

and $\quad N = mg + 12.5 \sin 37° + 10\sqrt{2} \sin 45°$

$= 2.5 + 7.5 + 10 = 20\ \text{N}$

The contact force is $F_C = \sqrt{N^2 + f^2} = 20\ \text{N}$

(vii) Here $f = mg$, where m is mass of hanging portion.

$m = 3\ \text{kg} \Rightarrow f = 30\ \text{N}$

$\therefore$ The contact force is

$F_c = \sqrt{N^2 + f^2} = \sqrt{60^2 + 30^2} = 67.08\ \text{N}$

3. (i) Since, block is stationary with respect to inclined surface. So, static friction comes into play.

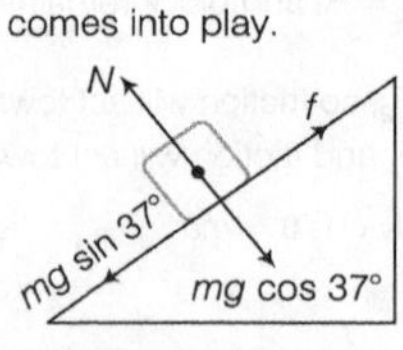

From force diagram, $\quad mg \sin 37° = f$

$\therefore \quad f = 1 \times 10 \times \frac{3}{5} = 6\ \text{N}$

The contact force is $F_c = \sqrt{N^2 + f^2}$

$= \sqrt{(mg \cos 37°)^2 + (mg \sin 37°)^2} = mg$

$= 10\ \text{N}$

(ii) Here, $F \cos 37° = 10 \times \frac{4}{5} = 8$ N

and $mg \sin 37° = 10 \times \frac{3}{5} = 6$ N

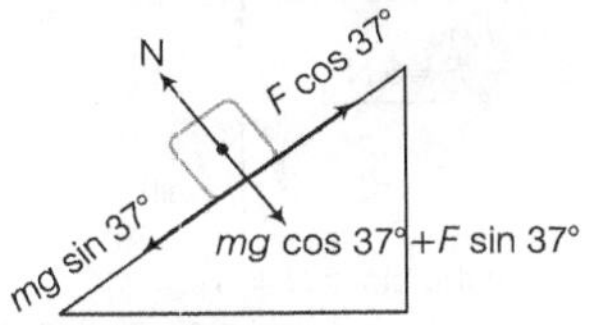

Since, $F \cos 37° > mg \sin 37°$

So, static friction acts on block downward along inclined surface.

For equilibrium of the block,

$$f + mg \sin 37° = F \cos 37°$$

$\therefore$ $$f = 8 - 6 = 2 \text{ N}$$

and $$N = mg \cos 37° + F \sin 37° = 8 + 6 = 14$$

$\therefore$ The contact force is $F_c = \sqrt{N^2 + f^2}$

$$= \sqrt{14^2 + 2^2} = \sqrt{200} = 10\sqrt{2} \text{ N}$$

(iii) Here, $F_3 \cos 37° + F_1 = 10 \times \frac{4}{5} + 18 = 26$ N

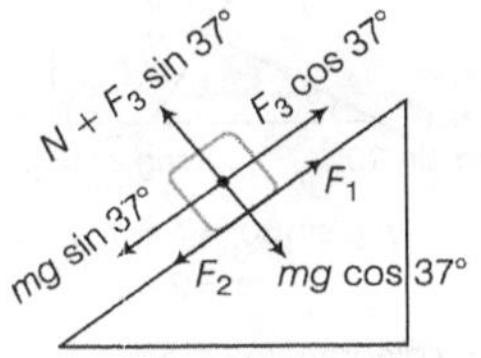

and $mg \sin 37° + F_2 = 6 + 20 = 26$ N

Thus, net force on the block along inclined surface is zero. So, friction is zero.

Also, $N + F_3 \sin 37° = mg \cos 37°$

$\therefore$ $$N = 8 - 6 = 2 \text{ N}$$

Thus, contact force is $F_c = \sqrt{N^2 + f^2} = 2$ N

(iv) The resultant of applied force F and $mg \sin 30°$ is balanced by static friction.

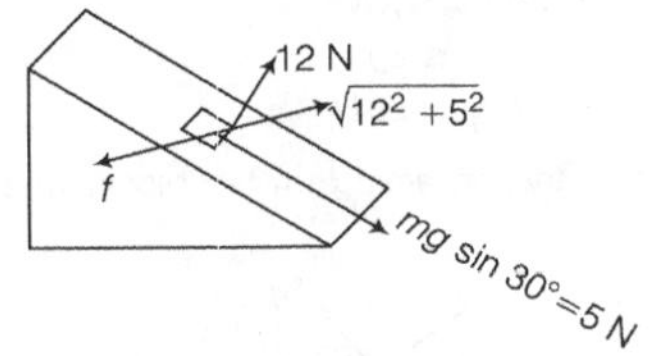

$\therefore$ $$f = \sqrt{(mg \sin 30)^2 + F^2} = \sqrt{5^2 + 12^2} = 13 \text{ N}$$

Also, $N = mg \cos 30° = 5\sqrt{3}$

The contact force is

$$F_c = \sqrt{N^2 + f^2} = \sqrt{75 + 169} = 15.62 \text{ N}$$

4. (i) The limiting friction is the maximum static friction.

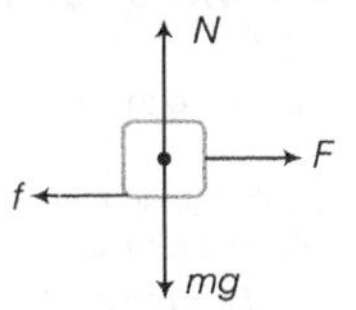

Here, $f_{max} = \mu_s N$ = limiting friction

$$= \mu_s mg = 50 \text{ N}$$

The net force on the block should be zero to hold the block stationary.

From free body diagram of the block,

$$F = f \le f_{max} \quad \text{or} \quad F \le \mu_s N$$

$\therefore$ $$F_{max} = 50 \text{ N}$$

(ii) From free body diagram of the block,

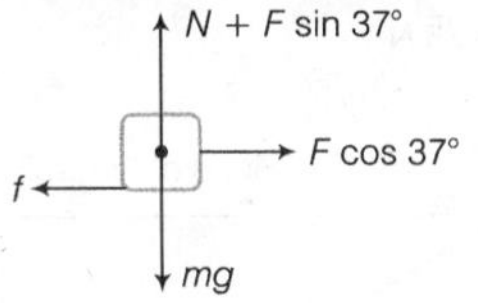

$$N + F \sin 37° = mg \quad \text{...(i)}$$

and $$F \cos 37° = f \le f_{max} \quad \text{...(ii)}$$

or $$\frac{4}{5}F \le \mu_s N$$

or $$\frac{4}{5}F \le \mu_s (mg - F \sin 37°)$$

or $$\frac{4}{5}F + \mu_s \frac{3}{5}F \le \mu_s mg$$

or $$F\left(\frac{4 + 3\mu_s}{5}\right) \le \mu_s mg$$

or $$F \le \frac{5\mu_s mg}{4 + 3\mu_s} = \frac{5 \times 0.3 \times 49}{4.9}$$

$\therefore$ $$F_{max} = 15 \text{ N}$$

(iii) From free body diagram of the block,

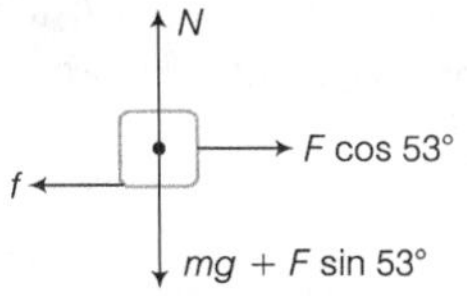

$$N = mg + F \sin 53° = mg + \frac{4}{5}F$$

and $$F \cos 53° = f \le \mu_s N$$

or $$\frac{3}{5}F \le \mu_s mg + \frac{4}{5}\mu_s F$$

or $$F\left(\frac{3}{5} - \frac{4}{5}\mu_s\right) \le \mu_s mg$$

or $$F \le \frac{\mu_s mg}{\frac{3}{5} - \frac{4}{5}\mu_s} = \frac{0.5 \times 1 \times 10}{\frac{3}{5} - \frac{4}{5} \times 0.5}$$

$\therefore$ $$F_{max} = 25 \text{ N}$$

(iv) From free body diagram of the block,

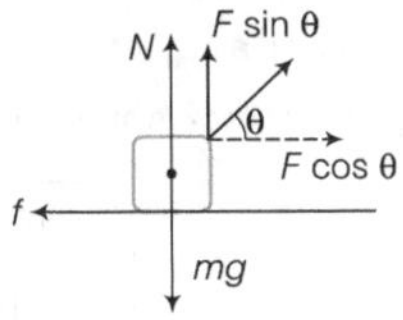

$$N + F \sin \theta = mg \quad \text{and} \quad F \cos \theta = f \le \mu_s N$$

or $$F \cos \theta \le \mu_s (mg - F \sin \theta)$$

or $$F \cos \theta + \mu_s F \sin \theta \le \mu_s mg$$

or $$F \le \frac{\mu_s mg}{\cos \theta + \mu_s \sin \theta}$$

Let $$y = \cos \theta + \mu_s \sin \theta$$

For maxima and minima,

$\therefore$ $$\frac{dy}{d\theta} = -\sin \theta + \mu_s \cos \theta = 0$$

$\therefore$ $$\mu_s = \tan \theta$$

or $\sin\theta = \dfrac{\mu_s}{\sqrt{1+\mu_s^2}}$ and $\cos\theta = \dfrac{1}{\sqrt{1+\mu_s^2}}$

$\therefore$ $$F_{max} = \frac{\mu_s mg}{\sqrt{1+\mu_s^2}} = \frac{\frac{2}{\sqrt{5}}\times 3\times 10}{\sqrt{1+\left(\frac{2}{\sqrt{5}}\right)^2}} = 20 \text{ N}$$

(v) Here, $N = 6\sqrt{5}$ N

$f_{max} = \mu_s N = 3\sqrt{5}$

Plane of the wall

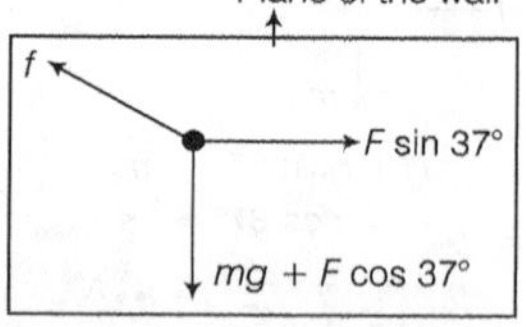

From free body diagram of the block,

$$f = \sqrt{(F\sin 37°)^2 + (mg + F\cos 37°)^2}$$

But $f \le \mu_s N$

or $\sqrt{(F\sin 37°)^2 + (mg + F\cos 37°)^2} \le \mu_s N$

or $$\sqrt{\frac{9}{25}F^2 + \left(5 + \frac{4}{5}F\right)^2} \le 3\sqrt{5}$$

$$F^2 + 8F - 20 = 0$$

or $F \le 2$ N

$\therefore$ $F_{max} = 2$ N

5. (i) From free body diagram of the block B in the frame of block A,

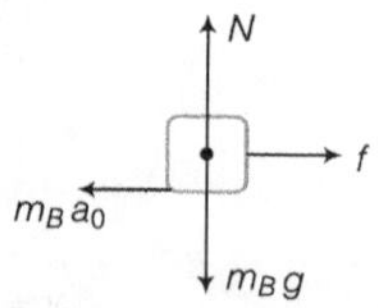

$N = m_B g = 100$ N and $f = m_B a_0$

but $f \le \mu_s N$

or $m_B a_0 \le 0.5 \times 100$

$a_{0\,max} = 5 \text{ m/s}^2$

(ii) From free body diagram of block A in the frame of block B.

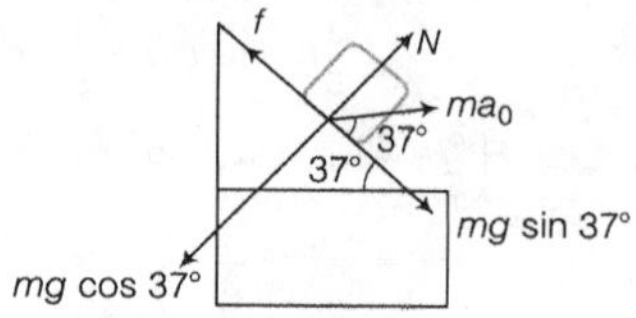

$$N + ma_0 \sin 37° = mg\cos 37°$$

or, $N = mg\cos 37° - ma_0\sin 37°$ ...(i)

and $mg\sin 37° + ma_0\cos 37° = f \le \mu_s N$ ...(ii)

Solving Eqs. (i) and (ii), $a_{0max} = 1 \text{ m/s}^2$

(iii) Free body diagram of block is shown in figure

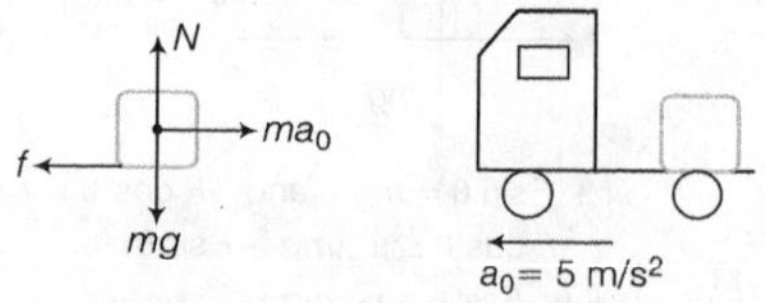

Frictional force acting on the block,

$$f_s = ma_0 = 1\times 5 = 5 \text{ N}$$

as the block is in equilibrium.

6. (a) Since, $\mu mg\cos\theta > mg\sin\theta$

$\therefore$ Force of friction $= mg\sin\theta$

$= 2\times 9.8\times \sin 30° = 9.8$ N

7. (b) Mass of block $m = 0.1$ kg

$\mu = 0.5$

Magnitude of limiting friction force acting on the block $f_{max} = \mu N$

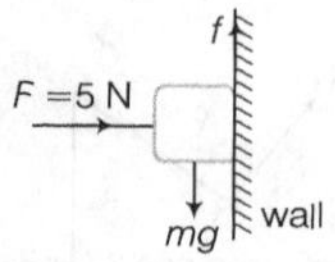

Normal reaction on the block $N = 5$ N

$\Rightarrow$ $f_{max} = 0.5\times 5 = 2.5$ N

As, $f_{max} > mg$ the block will not fall.

$\therefore$ $mg = f$

or $f = 0.1\times 9.8 = 0.98$ N

8. (d) In the given situation, the static friction force will balance the weight of the block

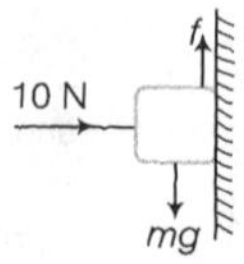

$f_s = mg$

Normal reaction acting on the block $N = 10$ N

$f_s = \mu_s N = 0.2\times 10 = 2$ N

The weight of the block = 2 N

9. (a) From free body diagram of block,

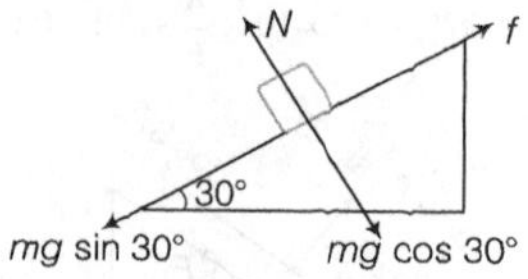

$mg\sin 30° = f = 10$

$\therefore$ $m = 2$ kg

10. (a) From free body diagram the block,

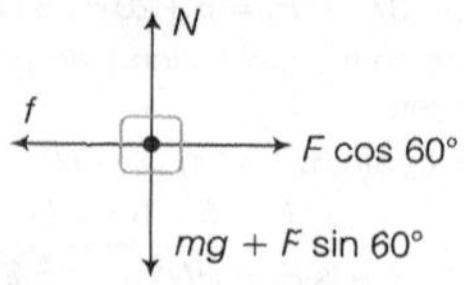

$$N = mg + F\sin 60° = \sqrt{3}g + \frac{\sqrt{3}F}{2}$$ (in vertical direction)

For no motion, $F\cos 60° \le \mu N$

$\therefore$ $F \le 2g$

$\therefore$ $F_{max} = 20$ N

11. (a) The limiting friction acting on the block, when the angle of inclination is 37°

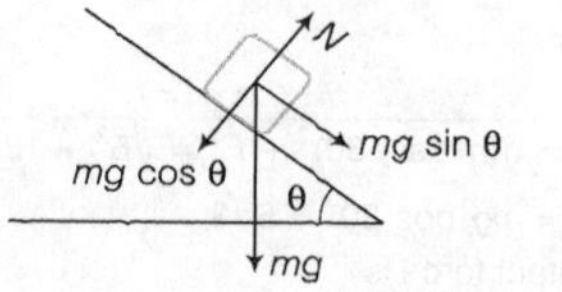

$$f_l = \mu mg\cos 37° = 0.75\times m\times 10\times 0.8 = 6\,m \text{ N}$$

and component of weight acting along the incline $= mg\sin 37°$

$= m\times 10\times 0.6 = 6\,m$ N

This implies that at $\theta = 37°$

The limiting friction will balance the component of weight along the incline.

When $0° < \theta < 37°$, then the static friction will balance the component of weight along the incline.

$f_s = mg\sin\theta$ (for $0° < \theta < 37°$)

$f_s \propto \sin\theta$

Hence, the curve of option (a) represents the above result.

12. (a) $F = f$ = force exerted by the bottle on the cork.

or $$F = f \le \mu_s \left(\frac{dN}{dL}\right) 2\pi r$$

$\Rightarrow$ $$F_{max} = \frac{7}{22} \times 100 \times 2 \times \frac{22}{7} \times 0.05 = 10 \text{ N}$$

13. (a) Limiting friction acting on the block.

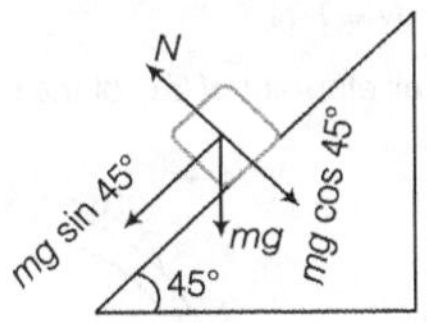

$$f_l = \mu mg \cos 45° = \frac{1}{\sqrt{3}} \times \frac{\sqrt{2}}{10} \times 10 \times \frac{1}{\sqrt{2}} = \frac{1}{\sqrt{3}}$$

Component of weight acting along the incline

$$= mg \sin 45° = \frac{\sqrt{2}}{10} \times 10 \times \frac{1}{\sqrt{2}} = 1$$

As the limiting friction is less than the downward force (component of weight along the inclined plane) that is why the block will slide downward, but if a force F is applied, it will remain stationary.

$\therefore$ $$F + f_l = mg \sin 45°$$

$\therefore$ $$F = 1 - \frac{1}{\sqrt{3}} = \left(\frac{\sqrt{3}-1}{\sqrt{3}}\right) \text{N}$$

14. (a) From free body diagram of the block,

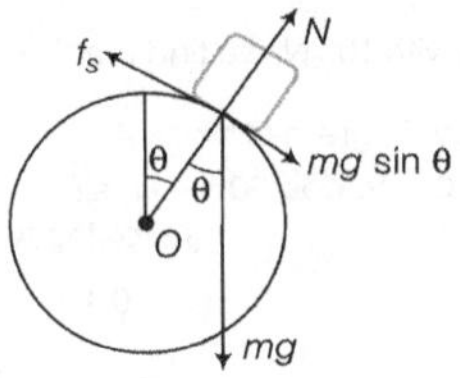

$$mg \sin\theta = \mu_s N \quad \text{(along the tangent on the surface)}$$

and $$N = mg \cos\theta \quad \text{(along the radius)}$$

$\therefore$ $$\tan\theta = \mu_s = \frac{1}{\sqrt{3}}$$

$\therefore$ $$\theta = 30°$$

15. (a) For stationary position,

$$mg \sin\theta = \frac{1}{2} \times \frac{1}{3} mg \cos\theta + \frac{5}{8}\frac{mg}{3} \cos\theta + \frac{3}{8}\frac{m}{3} g \cos\theta$$

$\therefore$ $\tan\theta = \frac{1}{2}$

16. (b) When the block tend to move down the incline plane, force F is applied to stop the block.

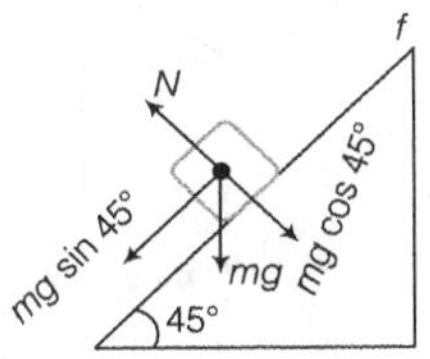

$$F + f = mg \sin 45° \quad \text{...(i)}$$

Let F' be the force applied on the block along the incline to just pull the block up.

Then, $F' = f + mg \sin 45°$

According to problem, $F' = nF$ (given)

$\Rightarrow$ $f + mg \sin 45° = nmg \sin 45° - nf$

or $$(n+1)f = (n-1)\, mg \sin 45°$$

or $$(n+1)\,\mu mg \cos 45° = (n-1)\, mg \sin 45°$$

$\Rightarrow$ $$\mu = \left(\frac{n-1}{n+1}\right)$$

17. (c) From free body diagram of the block in the frame of plate,

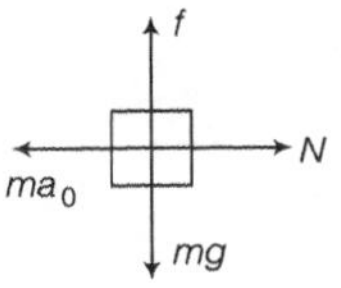

$$N = ma_0 \quad \text{and} \quad f = mg$$

But $$f \le \mu_s N \quad \text{or} \quad mg \le \mu_s a_0$$

$\therefore$ $$\mu_s a_0 \ge mg \quad \text{or} \quad a_0 \ge \frac{mg}{\mu_s}$$

or $$a_{0\,min} = \frac{1 \times 10}{0.5} = 20 \text{ m/s}^2$$

18. (c) The maximum friction on the block is $f_{max} = \mu mg \cos\theta$

The resultant force on the block is $F_0 = \sqrt{(mg \sin\theta)^2 + F^2}$. If block does not slide, then

$$F_0 \le f_{max}$$

or $$\sqrt{(mg \sin\theta)^2 + F^2} \le 8\mu \quad \text{or} \quad \sqrt{36 + 13} \le 8\mu$$

or $$7 \le 8\mu \quad \text{or} \quad \mu \ge \frac{7}{8}$$

19. (b) The mass of considered element is $dm = \frac{m}{l} dx$

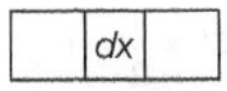

$\therefore$ $$df = \mu dmg = b\frac{m}{l} g\, x\, dx$$

$\therefore$ Total limiting frictional force on the rod is

$$f_{max} = \int_0^l df = \frac{mg\, b}{l} \int_0^l x\, dx = \frac{mgbl}{2}$$

For stationary position of rod,

$$F = f \le f_{max} \Rightarrow F_{max} = f_{max} = \frac{mgbl}{2}$$

20. (b) The scrambled eggs will start sliding, if limiting friction is overcome to weight of the egg. Hence, one have to raise the angle of elevation by angle of repose or slightly more ($\theta \ge \phi$) to start motion.

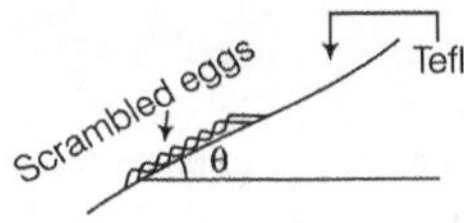

Hence, $\phi = \tan^{-1}\mu_s$ or $\phi = \tan^{-1}(0.04)$

For small θ, $\tan\theta \approx \theta$.

Hence, $$\phi \approx 0.04 \text{ rad} = 0.04 \times \frac{180°}{\pi} \approx 2.23°$$

21. (d) Let linear mass density of rope is λ.

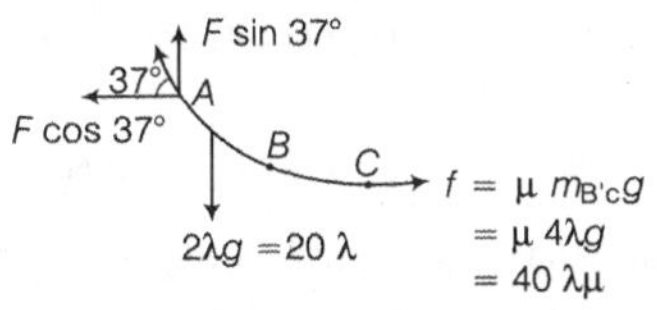

From free body diagram of hanging part,

$$F \cos 37° = 40 \lambda \mu$$

and $$F \sin 37° = 20 \lambda$$

or $$\cot 37° = \frac{40\mu}{20} \quad \text{or} \quad \frac{4}{3} \times \frac{20}{40} = \mu$$

$\therefore$ $$\mu = \frac{2}{3}$$

22. (b) For equilibrium of upper ball,

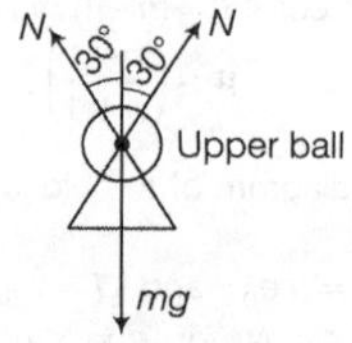

$$2N\cos 30^\circ = mg$$

$$\therefore \quad N = \frac{mg}{2\cos 30^\circ} = \frac{mg}{\sqrt{3}}$$

From free body diagram of leftward ball,

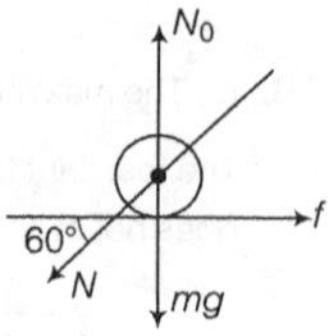

$$N_0 = N\sin 60^\circ + mg$$

(resolving the forces vertically)

$$N\cos 60^\circ = f \le \mu N_0$$

(resolving the forces horizontally)

or $\quad \dfrac{N}{2} \le \mu\left(\dfrac{\sqrt{3}}{2}N + mg\right)$

or $\quad \dfrac{mg}{2\sqrt{3}} \le \mu\dfrac{\sqrt{3}}{2} \times \dfrac{mg}{\sqrt{3}} + \mu mg \quad$ or $\quad \mu \ge \dfrac{1}{3\sqrt{3}}$

$$\mu_{min} = \frac{1}{3\sqrt{3}}$$

23. (a,b) $mg\sin\theta = f$ or $20 \times \dfrac{1}{2} = f$

$\therefore \quad f = 10$ N

Here, contact force is resultant of normal reaction and friction

$\therefore \quad F_c = \sqrt{(mg\sin\theta)^2 + (mg\cos\theta)^2} = mg = 20$ N

24. (a,b) The maximum value of friction is $f_{max} = \mu N$

Here, $\quad N + 4 = mg = 10$

$\therefore \quad N = 6$ N

$\therefore \quad f_{max} = 0.4 \times 6 = 2.4$ N

The horizontal applied force is $F_H = 1$ N.

$\therefore$ $F_H < f_{max}$, so static friction comes into play.

$\therefore$ $F_H = f$ and acceleration of the block is zero.

$\therefore \quad f = 1$ N

$\therefore \quad \mathbf{f} = -(1\text{ N})\hat{\mathbf{i}}$

25. (a, d) All bricks are in identical condition. So, when normal contact force is slightly less than 50 N, all bricks fall simultaneously.

Here, $\quad 2\mu F = n\,mg$

or $\quad n = \dfrac{2\mu F}{mg} = \dfrac{2 \times 0.5 \times 50}{10} = 5$

and contact force, $F_c = \sqrt{F^2 + (\mu F)^2}$

$$= \sqrt{(50)^2 + (0.5 \times 50)^2} = 55.9 \text{ N}$$

Explanations for Q. Nos. (26-27)

From free body diagram of upper cylinder,

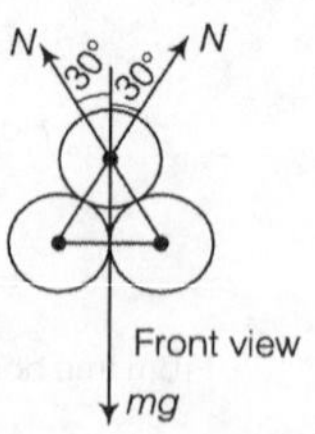

$$2N\cos 30^\circ = mg$$

$$\therefore \quad N = \frac{mg}{2\cos 30^\circ} = \frac{100\sqrt{3}}{\sqrt{3}} = 100 \text{ N}$$

When a horizontal force is applied, along the axis, both lower cylinders exert friction to oppose it.

$\therefore \quad F = 2f$

or $\quad f = \dfrac{F}{2}$ or $f \le \mu N$ or $\dfrac{F}{2} \le \mu N$

$\therefore \quad F_{max} = 2 \times 0.5 \times 100 = 100$ N

26. (b) **27.** (c)

28. (5) The normal reaction of each prong is N.

Weight of the cylinder is balanced by the frictional force between the prongs and the cylinder.

$\therefore \quad mg = \mu_s N + \mu_s N = 2\mu_s N$

$\therefore \quad N = \dfrac{mg}{2\mu_s} = \dfrac{10 \times 10^{-3} \times 10}{2 \times 0.01}$

$\therefore \quad N = 5$ N

29. (1) Consider a small element of 'dl' of the rope,

$$dmg\sin\theta = df$$

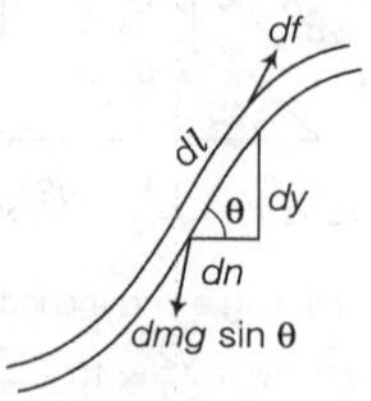

$\Rightarrow \quad \lambda dl g\left(\dfrac{dy}{dl}\right) = df \qquad \left(\because \sin\theta = \dfrac{dy}{dl}\right)$

(where, λ = linear mass density)

$\therefore \quad f = \lambda g \int_o^h dy$

$$= \lambda gh = \frac{mgh}{l}$$

$$= \frac{5 \times 10 \times 1}{5} = 10 \text{ N}$$

$$= 10^1 \text{ N}$$

$\therefore$ On comparing with 10^x N, we find that $x = 1$.

30. (1) From free body diagram of block A,

$$N_2 - 0.1N_1\cos 30^\circ - N_1\sin 30^\circ = 0$$

(resolving two forces horizontally)

$$f_2 = 0.1N_2$$

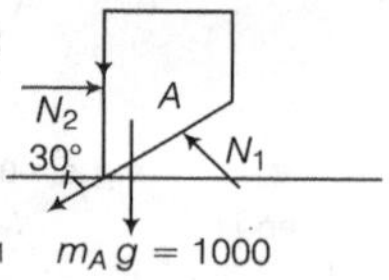

and $\quad f_1 = \mu N_1 = 0.1N_1$

$\therefore \quad N_2 = 0.587\,N_1 \quad$ ...(i)

and $\quad N_1\cos 30^\circ = 1000 + 0.1N_2 + 0.1N_1\sin 30^\circ$

or $\quad N_1\cos 30^\circ = 1000 + 0.1N_2 + \dfrac{0.1}{2}N_1 \quad$ ...(ii)

or $\quad N_1 \times 0.866 = 1000 + 0.1 \times 0.587\,N_1 + 0.05\,N_1$

By solving, $\quad N_1 = 1320.5$

$\therefore \quad f_1 = \mu N_1 = 132.05$

From free body diagram of block B,

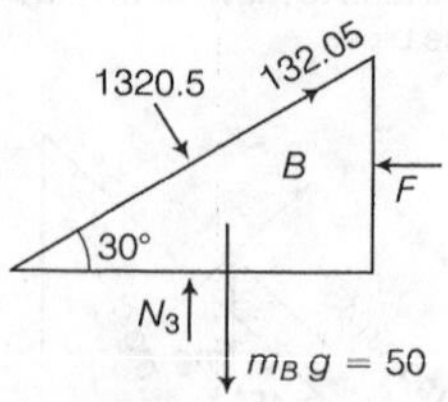

Resolving the forces horizontally,

$$F = 1320.5\sin 30^\circ + 132.05\cos 30^\circ \quad \text{...(iii)}$$

$$= 660.25 + 114.3553$$

$$= 774.6 \text{ N}$$

$\therefore \quad x = 1$

DPP-2 **Basic Concepts of Kinetic Friction**

1. (i) True, kinetic frictional forces always oppose the relative motion between the contact surfaces.

(ii) True, as the table relatively moving backward with respect to the book, hence the frictional force will act in forward direction on the table.

(iii) True, kinetic friction is not adjustable like static friction, hence its magnitude will be certainly $f = \mu_k N$.

(iv) True, kinetic friction comes into existance whenever there is a relative motion between the two surfaces in contact.

(v) True, for kinetic friction there should be net relative velocity, so acceleration may or may not be zero.

(vi) False

(vii) True

(viii) True

2. (i) The formula of kinetic friction is

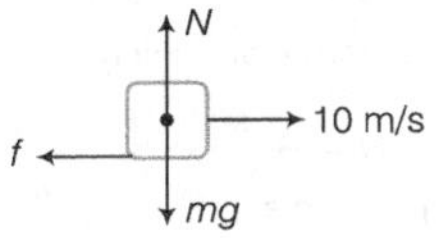

$$f = \mu_k N = 0.4 \times 1 \times 10 = 4 \text{ N}$$

(ii) Kinetic friction is

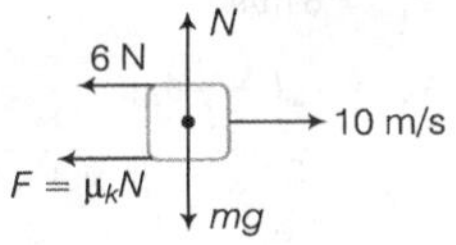

$$f = \mu_k N = 0.4 \times 10 = 4 \text{ N}$$

(iii) From free body diagram of the block,

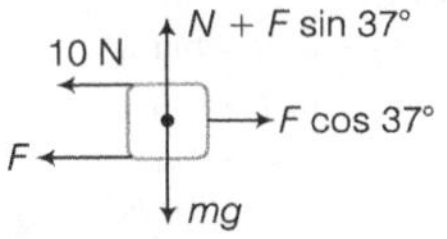

$$N + F \sin 37° = mg$$

$$\therefore \quad N = 10 - 6 = 4 \text{ N}$$

$$\therefore \quad f = \mu_k N = 0.4 \times 4 = 1.6 \text{ N}$$

(iv) From free body diagram of the block,

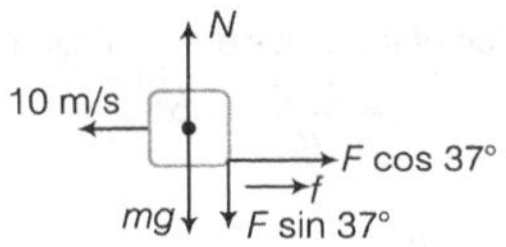

$$N = mg + F \sin 37° = 10 + 10 \times \frac{3}{5} = 16 \text{ N}$$

$$\therefore \quad f = \mu_k N = 0.4 \times 16 = 6.4 \text{ N}$$

(v) Here, $f = \mu_k N = 0.5 \times 2 \times 10 = 10$ N

(vi) From free body diagram of the block,

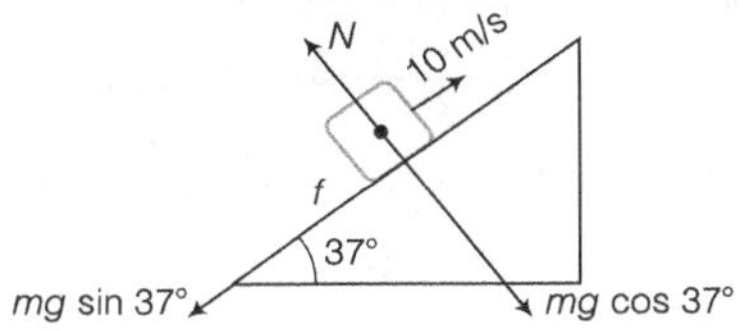

$$N = mg \cos 37°$$

$$\therefore \quad f = \mu_k N = 0.4 \times 1 \times 10 \times \frac{4}{5} = 3.2 \text{ N}$$

3. (c) Given $y = kx^2$

$$\Rightarrow \quad \tan\theta = \frac{dy}{dx} = 2kx = 2k \times \frac{1}{2k} = 1$$

$$\therefore \quad \theta = 45°$$

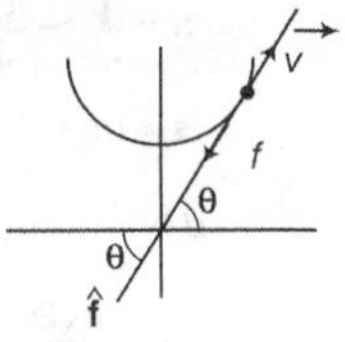

From figure,

$$\hat{\mathbf{f}} = -\cos\theta\,\hat{\mathbf{i}} - \sin\theta\,\hat{\mathbf{j}} = -\frac{1}{\sqrt{2}}\hat{\mathbf{i}} - \frac{1}{\sqrt{2}}\hat{\mathbf{j}}$$

4. (a) If $\mathbf{v}_1 = \mathbf{v}_2 = 3\hat{\mathbf{i}} + 4\hat{\mathbf{j}}$

This means there is no relative motion between block and the plate. Hence there is no kinetic friction between them.

5. (c) As, $\quad s = \frac{1}{2}at^2 = 1\text{m}$

$$\therefore \quad a = 2 \text{ m/s}^2$$

From free body diagram of the block,

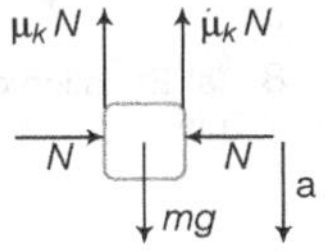

$$mg - 2\mu_k N = ma$$

or $\quad 10 - 2 \times 0.5 \times N = 2$

$$\therefore \quad N = 8 \text{ N}$$

$$\therefore \quad F_c = \sqrt{N^2 + (\mu_k N)^2} = 8\sqrt{1 + 0.25}$$

$$= 8\sqrt{\frac{125}{100}} = 4\sqrt{5} \text{ N}$$

6. (a) From free body diagram of B,

$$mg \sin 37° + \mu_k mg \cos 37° = ma_B$$

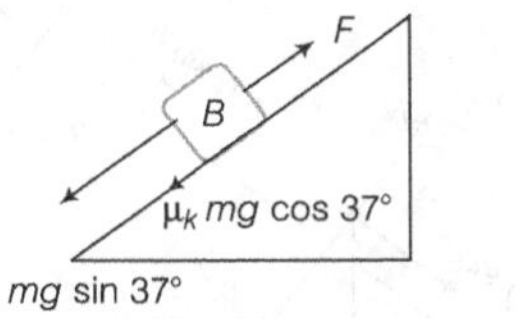

$\therefore \quad a_B = 10 \text{ m/s}^2$ in downward direction along the plane.

From free body diagram of block A,

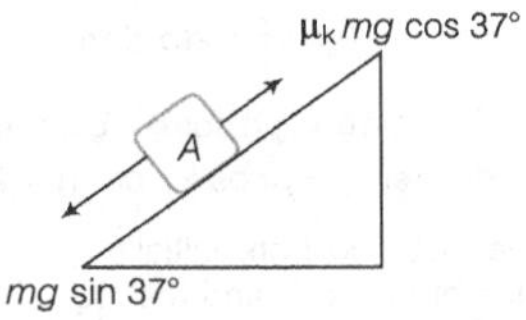

$$mg \sin 37° - \mu_k mg \cos 37° = ma_A$$

$\therefore \quad a_A = 2 \text{ m/s}^2$ in downward direction along the plane.

$$\therefore \quad s_{rel} = u_{rel}t + \frac{1}{2}a_{rel}t^2 \quad \text{or} \quad 6 = 10t - \frac{1}{2} \times 8t^2$$

$$\therefore \quad t = 1\text{s} \Rightarrow s_A = \frac{1}{2}a_A t^2$$

$$\therefore \quad s_A = \frac{1}{2} \times 2 \times 1^2 = 1\text{m}$$

7. (c) For smooth inclined plane,

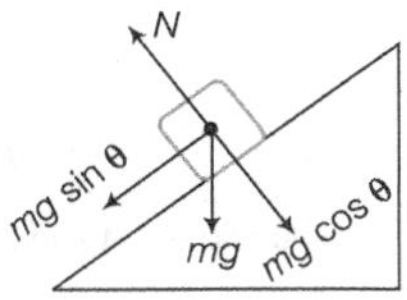

$$a = g\sin\theta$$

$$= g \sin 45°$$

$$= \frac{g}{\sqrt{2}} \text{ m/s}^2$$

Distance = d (given)

Using, $\quad s = ut + \frac{1}{2}at^2$

$$d = \frac{1}{2} \times \frac{g}{\sqrt{2}} t_s^2$$

$$t_s^2 = \frac{2\sqrt{2}d}{g} \quad \text{...(i)}$$

For rough inclined plane,

$$a = g\sin\theta - \mu g\cos\theta$$

$$a = \frac{g}{\sqrt{2}} - \frac{\mu_k g}{\sqrt{2}} = (1-\mu_k)\frac{g}{\sqrt{2}} \quad [\because \theta = 45°]$$

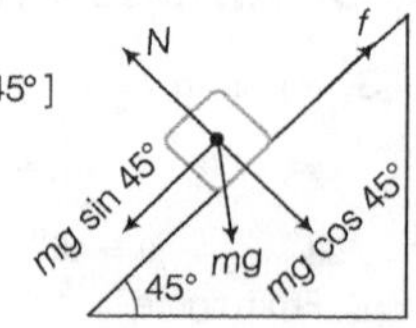

$$s = ut + \frac{1}{2}at^2$$

$$d = \frac{1}{2}(1-\mu_k)\frac{g}{\sqrt{2}}t_R^2$$

$$t_R^2 = \frac{2\sqrt{2}d}{(1-\mu_k)g} \quad ...(ii)$$

According to given problem,

$$t_R = nt_s \Rightarrow t_R^2 = n^2 t_s^2$$

$$\frac{2\sqrt{2}d}{(1-\mu_k)g} = n^2\frac{2\sqrt{2}d}{g} \Rightarrow \frac{1}{1-\mu_k} = n^2$$

$$\mu_k = 1 - \frac{1}{n^2}$$

8. (a) Equation of motion of the block for upper part of the inclined plane.

$$v^2 = u^2 + 2as$$

$$v_1^2 = 0 + 2(g\sin\phi)\cdot\frac{l}{2}$$

$$v_1^2 = gl\sin\phi \quad ...(i)$$

Motion of the block for the lower part of the inclined plane

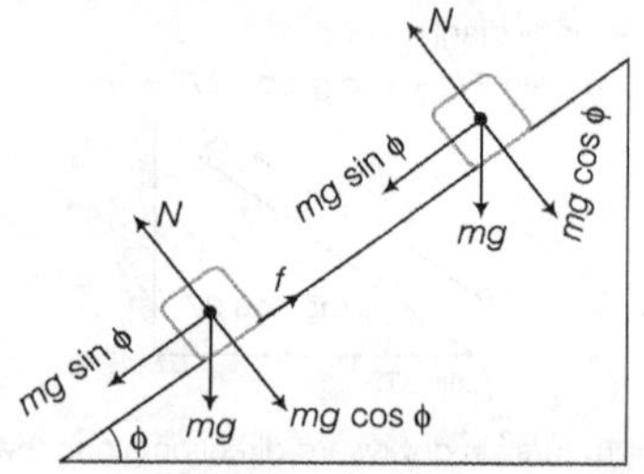

$$v^2 = u^2 + 2as$$

$$0 = v_1^2 + 2(g\sin\phi - \mu g\cos\phi)\frac{l}{2}$$

$\Rightarrow$ $gl\sin\phi = (\mu g\cos\phi - g\sin\phi)l$ [using Eq. (i)]

or $2\sin\phi = \mu\cos\phi$ or $\mu = 2\tan\phi$

9. (a) Let x be the distance from initial position (position of block A) and the position, where the front faces of the blocks come in line.

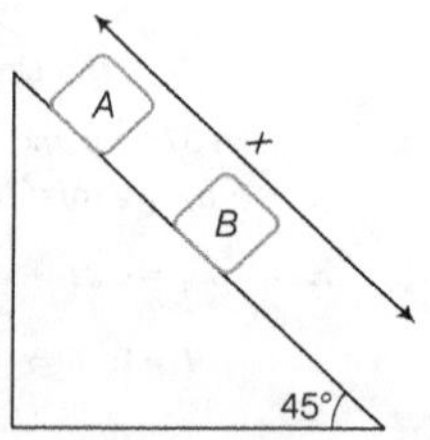

For motion of block A,

$$a_A = g\sin\theta - \mu_A g\cos\theta$$

$$s_A = u_A t + \frac{1}{2}a_A t^2$$

or $x = \frac{1}{2}g(\sin\theta - \mu_A\cos\theta)t^2$ ...(i)

For motion of block B, $a_B = g\sin\theta - \mu_B g\cos\theta$

$$\Rightarrow \quad s_B = u_B t + \frac{1}{2}a_B t^2$$

or $x - \sqrt{2} = \frac{1}{2}g(\sin\theta - u_B\cos\theta)t^2$ ...(ii)

Putting $\theta = 45°$ in Eqs. (i) and (ii), we get

$$x = \frac{1}{2}g\left(\frac{1}{\sqrt{2}} - \frac{\mu_A}{\sqrt{2}}\right)t^2$$

$$x = \frac{g}{2\sqrt{2}}(1-\mu_A)t^2 \quad ...(iii)$$

and $x - \sqrt{2} = \frac{g}{2\sqrt{2}}(1-\mu_B)t^2$ ...(iv)

$$\Rightarrow \quad x = \frac{g}{2\sqrt{2}}(1-\mu_B)t^2 + \sqrt{2}$$

From Eqs. (iii) and (iv), we get

$$\frac{1}{2\sqrt{2}}g(1-\mu_A)t^2 = \frac{1}{2\sqrt{2}}g(1-\mu_B)t^2 + \sqrt{2}$$

$$\left[\frac{1}{2\sqrt{2}}g(1-\mu_A) - \frac{1}{2\sqrt{2}}g(1-\mu_B)\right]t^2 = \sqrt{2}$$

or $\frac{1}{2\sqrt{2}}(3-2)t^2 = \sqrt{2}$

or $t^2 = 2\times 2$

or $t = 2$ s

From Eq. (iii),

$$x = \frac{1}{2\sqrt{2}}g(1-\mu_A)t^2 = \frac{1}{2\sqrt{2}}\times 10\times(1-0.2)\times(2)^2 = 8\sqrt{2} \text{ m}$$

10. (a) If $P_1 = mg(\sin\theta - \mu\cos\theta)$

Then, $f = \mu mg\cos\theta$ in upward direction. If $P = mg\sin\theta$, $f = 0$ and if $P_2 = mg(\sin\theta + \mu\cos\theta)$, $f = \mu mg\cos\theta$ in downward direction.

Hence, graph (a) is correct.

11. (a) From free body diagram of the ring in the frame of car,

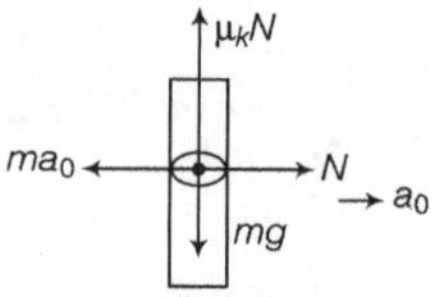

$$N = ma_0$$

and $mg - \mu_k N = ma_{rel}$

or $mg - \mu_k ma_0 = ma_{rel}$

or $10 - 0.4\times 5 = a_{rel}$

$\therefore$ $a_{rel} = 8 \text{ m/s}^2$

$\because$ $s_{rel} = u_{rel}t + \frac{1}{2}a_{rel}t^2$

$\Rightarrow$ $t = \sqrt{\frac{2s_{rel}}{a_{rel}}} = \sqrt{\frac{2\times 4\times 10^{-2}}{8}} = 0.1$ s

12. (b) From free body diagram of block A in the frame of the block B.

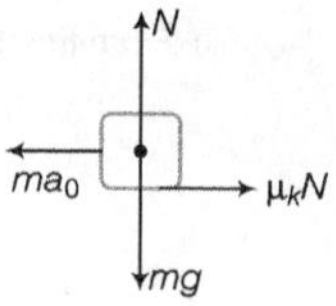

$N = mg$ and $ma_0 - \mu_k N = ma_{rel}$

or $a_0 - \mu_k g = a_{rel}$

or $3 - 2 = a_{rel}$

$\because$ $a_{rel} = 1 \text{ m/s}^2$

and $s_{rel} = u_{rel}t + \frac{1}{2}a_{rel}t^2$

$\therefore$ $t = \sqrt{\frac{2s_{rel}}{a_{rel}}} = \sqrt{\frac{2\times 2}{1}} = 2$ s

13. (a) The retardation of the block due to friction is

$$a = \frac{\mu mg}{m} = gbx$$

$\therefore$ $-v\frac{dv}{dx} = gx$ $\quad(\because b = 1\text{ m}^{-1})$

or $-\int_{v_0}^{0} v\,dv = g\int_0^x x\,dx$ or $\frac{v_0^2}{2} = \frac{gx^2}{2}$

or $v_0 = \sqrt{g}\,x \Rightarrow x = \frac{v_0}{\sqrt{g}} = \frac{\sqrt{10}}{\sqrt{10}} = 1$ m

14. (c) The distance travelled by the truck is

$$s_1 = \left(\frac{2+0}{2}\right)0.5 = 0.5 \text{ m}$$

The acceleration of the box is $a = \frac{\mu mg}{m} = \mu g = 1 \text{ m/s}^2$

$$v^2 = u^2 + 2as_2$$

$\therefore$ $(0)^2 = 2^2 - 2\times 1\cdot s_2$

$\therefore$ $s_2 = \frac{4}{2} = 2$ m

$\therefore$ The distance travelled by the box with respect to the truck is

$$s_{rel} = s_2 - s_1 = 2 - 0.5 = 1.5 \text{ m}$$

15. (c) Here, $t = 10\,s, u = 6\,m/s$

The retardation of the marble block is due to kinetic friction acting on it.

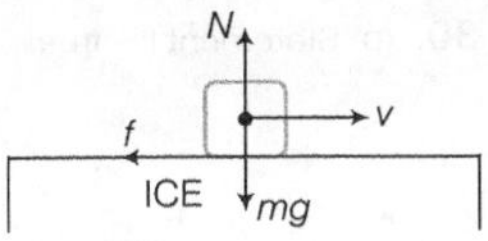

Retardation of the block, $a = \frac{\mu N}{m}$

$\Rightarrow \quad a = \frac{\mu mg}{m} = \mu g$

(where, μ is coefficient of static friction)

Using equation of motion, $v = u + at$

or $\quad 0 = 6 - \mu g \times 10$

or $\quad \mu = \frac{6}{g \times 10} = 0.06$

16. (a) As, $mg \sin\theta - \mu_k mg \cos\theta = ma$

$\therefore \quad a = g\sin\theta - \mu_k g \cos\theta$

$\therefore \quad s = \frac{1}{2}at^2$

$= \frac{g}{2}(\sin\theta - \mu_k \cos\theta)t^2$

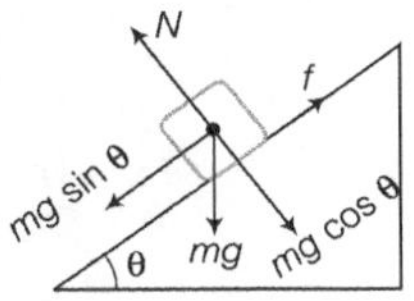

17. (a) Free body diagram of mass with respect to disc is as shown below

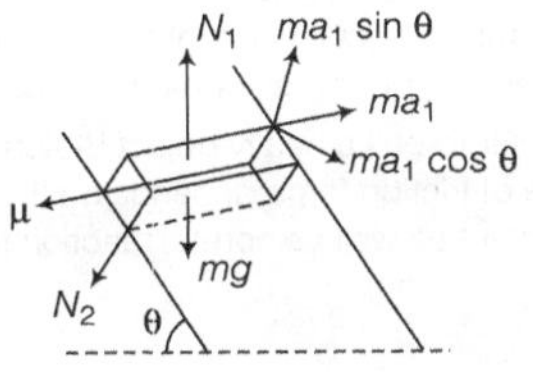

$N_1 = mg$ and $N_2 = ma_0 \sin\theta$ (N_1 is vertically upward and mg is vertically downward)

Acceleration of mass, $a = \frac{ma_0 \cos\theta - \mu N_2 - \mu N_1}{m}$

where, μN_2 friction on the mass due to N_2 and μN_1 frictional force on the mass due to N_1.

$\Rightarrow \quad a = \frac{ma_0 \cos\theta - \mu ma_0 \sin\theta - \mu mg}{m}$

or $\quad a = a_0 \cos\theta - \mu a_0 \sin\theta - \mu g$

or $\quad a = 25 \times \sqrt{1 - \left(\frac{3}{5}\right)^2} - \frac{2}{5} \times 25 \times \frac{3}{5} - \frac{2}{5} \times 10 = 10\ m/s^2$

18. (a,b) The linear mass density of the rod at distance x is $\lambda = 1 + x$

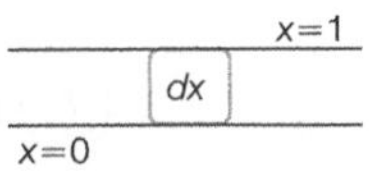

$m = \int dm = \int_0^1 \lambda\, dx = \int_0^1 (1 + x)dx$

$= 1 + \frac{1}{2} = 1.5\ kg$

Thus, friction is $f = \mu_k mg = 1.5 \times 0.5 \times 10 = 7.5\ N$

$\therefore \quad a = \frac{F - 7.5}{m} = \frac{15 - 7.5}{\frac{3}{2}} = \frac{2 \times 7.5}{3} = 5\ m/s^2$

19. (a,b) The angle of repose is $\theta_0 = \tan^{-1}(\mu_s) = 45°$

If $\theta \le \theta_0$, at maximum height, the block will come permanently in rest.

If $\theta > \theta_0$, the block with return at point A.

20. (a,d) If $F \le \mu mg$, the block is only performed in forward motion.

The acceleration of the block in forward motion is, $a = \frac{F + \mu mg}{m}$

$\therefore \quad 0^2 = v_0^2 - 2as_0$

$\therefore \quad s_0 = \frac{v_0^2}{2a} = \frac{mv_0^2}{2(F + \mu mg)}$

If $F > \mu mg$, the block moves in forward direction a distance $s_0 = \frac{mv_0^2}{2(F + \mu mg)}$, then it will come back with an acceleration, $a = \frac{F - \mu mg}{m}$

Thus, total distance travelled by the block to come at initial position is $2s_0 = \frac{mv_0^2}{(F + \mu mg)}$

21. (a) The relative velocity of packet with respect to bus is $\mathbf{u}_{rel} = 4\hat{j} - 3\hat{i}$

The friction on packet acts in opposite direction of $\mathbf{u}_{rel}$.

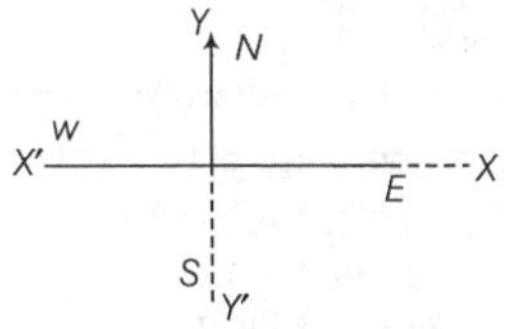

$\therefore$ The unit vector along friction is $\hat{f} = \frac{3\hat{i} - 4\hat{j}}{5}$

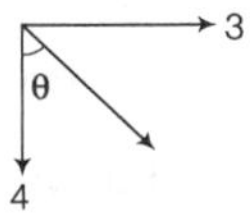

$\therefore$ From figure, $\quad \tan\theta = \frac{3}{4}$

$\Rightarrow \quad \theta = 37°$ East of South

22. (b) Since, acceleration and velocity are neither parallel nor antiparallel, so path is parabolic.

23. (a) $f = \mu mg = 0.5 \times 10\,m = 5\,m$

$\therefore \quad a_{rel} = \frac{f}{m} = 5$

$\therefore \quad v_{rel}^2 = u_{rel}^2 - 2a_{rel}\, s_{rel}$

$0 = 5^2 - 2 \times 5 \times s_{rel}$

$\therefore \quad s_{rel} = \frac{25}{10} = 2.5\,m$

24. (5) Force required to just prevent the block from sliding

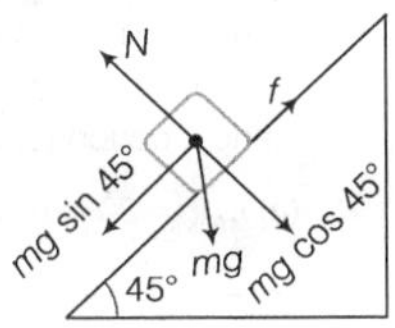

$F_1 = mg \sin 45° - f$

$F_1 = mg \sin 45° - \mu\, mg \cos 45°$

$F_1 = \frac{mg}{\sqrt{2}} - \frac{\mu mg}{\sqrt{2}} = \frac{mg}{\sqrt{2}}(1 - \mu) \quad \ldots(i)$

Force required to just push it up.

$F_2 = mg \sin 45° + f$

$F_2 = mg \sin 45° + \mu mg \cos 45°$

$F_2 = \frac{mg}{\sqrt{2}}(1 + \mu)$

According to given problem $F_2 = 3F_1$

$(1 + \mu) = 3 - 3\mu \Rightarrow \mu = \frac{1}{2}$

$\Rightarrow \quad N = 10\mu = 10 \times \frac{1}{2} = 5$

25. (1) Resolving the forces perpendicular to groove,

$\Rightarrow mg \sin 37° \sin 53° = N$

or $\quad 4.8 = N$

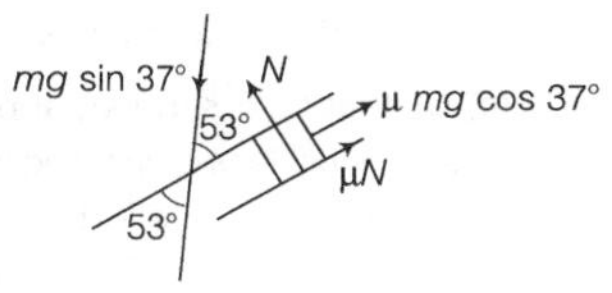

Resolving the forces along the groove,

$mg \sin 37° \cos 53° - \mu mg \cos 37° - \mu N = ma$

or $\quad \frac{18}{5} - 0.8 - 0.48 = a \Rightarrow a = 2.32\ m/s^2$

$\because \quad s = \frac{1}{2}at^2$

$\therefore \quad t = \sqrt{\frac{2s}{a}} = 1\,s$

26. (6) At $t = 1s$, $a = 4\ m/s^2$

Since, plank moves with the plank upto $t = 1$ s.

Hence, static friction comes into between block and plank.

$$f_s = ma$$

But $\quad f_s \le \mu_s mg$

$\therefore \quad a = \mu_s g \Rightarrow 4 = \mu_s g$

or $\quad \mu_s = \frac{4}{g} = \frac{4}{10} = 0.4$

Let the velocity of block at $t = 3$ s is v_0.

The velocity of block at $t = 1$ s is

$$u_0 = 2t^2 = 2 \times (2)^2 = 2\ m/s$$

The velocity of car at $t = 2$ s is

$$v_C = 2t^2 = 2 \times (2)^2 = 8\ m/s$$

The acceleration of work is $\mu_k g$ at $1 < t \le 3$

$\therefore \quad v_0 = u_0 + \mu_k g\,(3-1)$

or $\quad v_0 = u_0 + 2\mu_k g$

or $\quad v_C = u_0 + 2\mu_k g$

$\Rightarrow \quad 8 = 2 + 2\mu_k \times 10$ or $\mu_k = 0.3$

Required ratio, $\quad \frac{\mu_s}{\mu_k} = \frac{0.4}{0.3} = \frac{4}{3} = \frac{8}{6}$

but given, $\quad \frac{\mu_s}{\mu_k} = \frac{8}{x} \Rightarrow \frac{8}{x} = \frac{8}{6}$

$\therefore \quad x = 6$

27. (a) On wet concrete road, the presence of water between the tyre and the road decreases the frictional force.

28. (c) **29.** (b)

30. (c) Statement I is true.

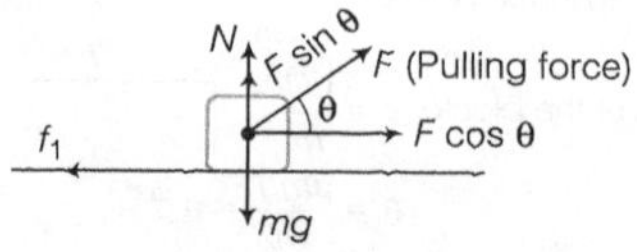

$$N + F\sin\theta = mg$$

$\Rightarrow \quad N = mg - F\sin\theta$

Frictional force acting on the heavy object.

$$f_1 = \mu(mg - F\sin\theta)$$

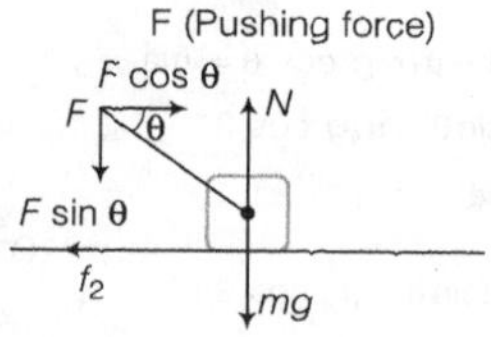

$$N = mg + F\sin\theta$$

$\Rightarrow \quad f_2 = \mu(mg + F\sin\theta)$

It is clear from above calculation that the object will experience less frictional force in pulling as compare to push the object.

Hence, it is easier to pull a heavy object Statement II is false. The magnitude of friction force depends on the nature of the two surfaces in contact as well as normal reaction force between them.

DPP-3 **Analysis of Friction**

1. (c) If $\mu_s \ge \tan\theta$, then $mg\sin\theta \le \mu_s mg\cos\theta$.

Thus, the particle remains stationary at maximum height forever. Hence, option (a) is correct. If $\mu_s < \tan\theta$, then $mg\sin\theta > \mu_s mg\cos\theta$. Thus, particle returns back.

Hence, option (c) is correct.

2. (a) $f_{max} = 20 \times 0.8 = 16$ N

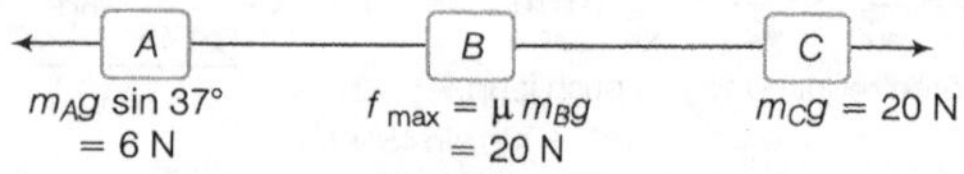

The net applied force F on the block B

= (force due to mass C – force due to mass A)

= 20 – 6 = 14 N

Here, $F < f_{max}$ static friction comes into play.

$\therefore \quad T_{AB} = m_A g \sin 37° = 6$ N

and $\quad T_{BC} = T_{AB} + f_B \quad$ (for block B)

$\Rightarrow \quad 20 = 6 + f_B$

or $\quad f_B = 14$ N

Think the solution for other options as above.

3. (d) **Concept** If $F \le \mu_s mg$, applied force F is balanced by static friction. Thus, tension in string is zero.

If $F > \mu_s mg$, friction is limiting friction. Thus, $F = T + f_{max}$,

$\therefore \quad T = F - \mu_s mg$.

4. (b) The normal contact force on the block is 40 N. The resultant force on the block along the plane of wall is

$$\sqrt{(10-50)^2 + 30^2} = 50\ N$$

Thus, acceleration of the block is

$$a = \frac{50 - \mu N}{m} = \frac{50 - 20}{5} = 6\ m/s^2$$

Here, $\quad \tan\theta = \frac{40}{30} \Rightarrow \theta = 53°$

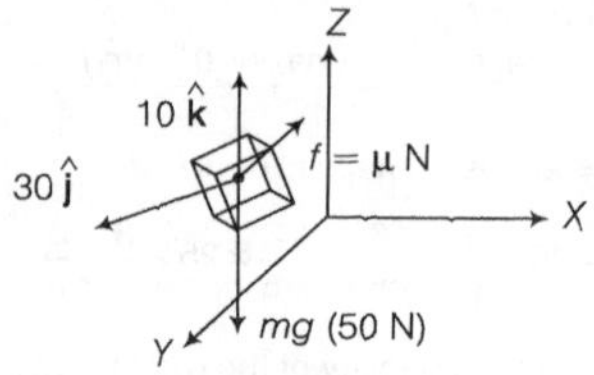

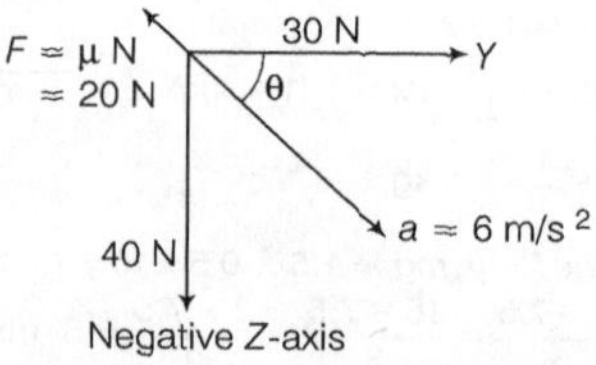

$\therefore \quad a = 6\cos 53°\,\hat{j} - 6\sin 53°\,\hat{k} = 3.6\,\hat{j} - 4.8\,\hat{k}$

5. (a,c,d)

(i) **Concept** If applied force $F \le \mu_s N$, the static friction comes into play.

(ii) If the applied force $F > \mu_s N$, kinetic friction comes into play. From free body diagram of the block, the limiting friction is

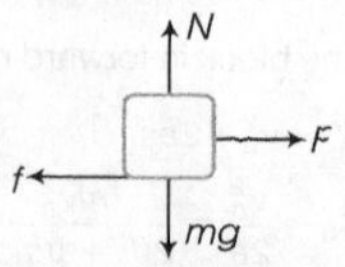

$$f_{max} = \mu_s N = \mu_s mg = 50\ N$$

(a) $F < f_{max}$, the friction is static in nature.

$\therefore \quad F = f = 45$ N

Thus, acceleration is zero.

(b) $\because F = f_{max}$, static friction comes into play.

$\therefore \quad f = 50$ N and the acceleration is zero.

(c) $\because F > f_{max}$, kinetic friction comes into play.

$$\therefore \quad f = f_k = \mu_k N = \mu_k mg = 40 \text{ N}$$

From free body diagram of the block,

$$F - f = ma$$

or $$60 - 40 = 10a$$

$$\therefore \quad a = 2 \text{ m/s}^2$$

(d) $s = \frac{1}{2} a \times t^2 = \frac{1}{2} \times 2 \times 1^2 = 1$ m

6. (a,b,c,d)

(i) **Concept** The angle of repose is $\theta_0 = \tan^{-1}(\mu_s)$.

If $\theta_0 \geq \theta \geq 0$, static friction acts on the block.

If $\frac{\pi}{2} \geq \theta \geq \theta_0$, kinetic friction acts on the block.

(a) The force in horizontal direction is zero. Hence, friction is zero.

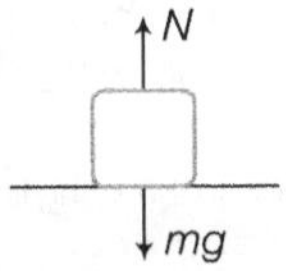

If angle of friction is ϕ, $\tan \phi = \frac{f}{N}$, here ϕ is angle made by normal reaction with contact force.

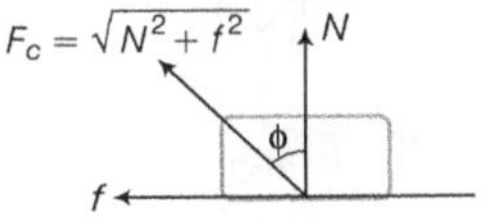

Thus, $\phi = 0$

(b) Here, friction is static in nature.

$$\therefore \quad f = mg \sin\theta$$

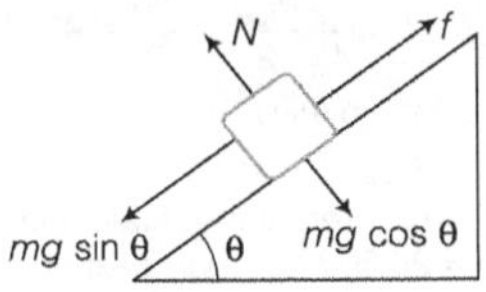

and $$N = mg \cos \theta$$

$$\therefore \quad \tan \phi = \frac{f}{N} = \tan \theta$$

$$\therefore \quad \phi = \theta$$

(c) Here, kinetic friction comes into play.

$$\therefore \quad f = \mu_k N = \mu_k mg \cos \theta$$

$$\therefore \quad \tan \phi = \frac{f}{N} = \mu_k$$

$$\therefore \quad \phi = \tan^{-1}(\mu_k)$$

(d) The contact force is $F_c = \sqrt{N^2 + f^2}$

$$\therefore \quad F_c = \sqrt{(mg \cos \theta)^2 + (\mu_k mg \cos \theta)^2}$$

$$= (mg \cos \theta)\sqrt{1 + \mu_k^2}$$

7. (a,b,c)

8. (a,b,c,d)

(a) As, $$m_1 \leq \mu_s m_2$$

$$\Rightarrow \quad m_1 g \leq \mu_s m_2 g$$

Weight of block m_1 is less than the maximum static friction acting on block m_2.

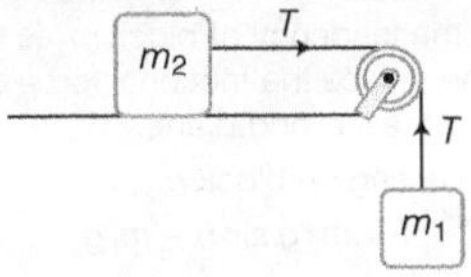

Weight of block m_1 is balanced by the static friction acting on block m_2.

This means the block m_1 will not move downword, hence its acceleration is zero and $T = m_1 g$.

(b) As, $$m_1 \leq \mu_s m_2$$

$$\Rightarrow \quad m_1 g \leq \mu_s m_2 g$$

$$\Rightarrow \quad m_1 g \leq f_{max}$$

Hence, static friction on block m_2 will balance the weight of block m_1.

Friction on block m_2, $f = m_1 g$.

(c) As, $$m_1 > \mu_s m_2$$

$$\Rightarrow \quad m_1 g > \mu_s m_2 g$$

$$\Rightarrow \quad m_1 g > f_{max}$$

As the weight of block m_1 is greater than the limiting friction, the block m_1 will move downward and the block m_2 will slides on the horizontal surface towards right. This means that kinetic friction will act on block $m_2 \Rightarrow f_k = \mu_k m_2 g$

(d) As, $$m_1 > \mu_s m_2$$

$$\Rightarrow \quad m_1 g > \mu_s m_2 g$$

$$\Rightarrow \quad m_1 g > f_{max}$$

For block m_1

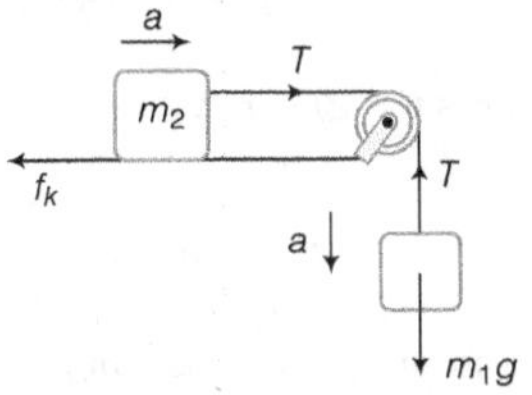

$$m_1 g - T = m_1 a \quad \ldots\text{(i)}$$

and for block m_2

$$T - f_k = m_2 a \quad \ldots\text{(ii)}$$

On solving Eqs. (i) and (ii), $a = \left(\frac{m_1 g - \mu_k m_2 g}{m_1 + m_2}\right)$

9. (a,b,c,d)

(a) As, $m_1 g - m_2 g \sin\alpha \leq \mu_s m_2 g \cos\alpha$

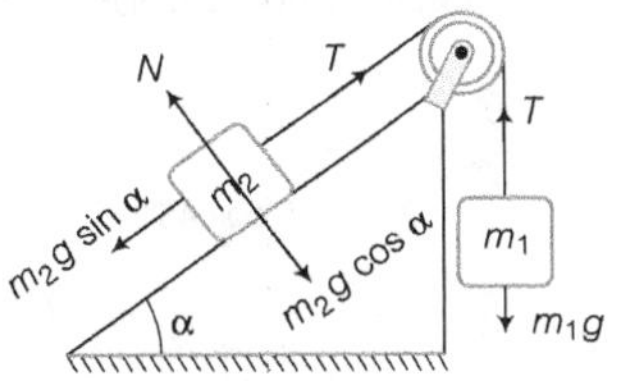

$\Rightarrow m_1 g \leq m_2 g \sin\alpha + \mu_s m_2 g \cos\alpha$

$\Rightarrow$ This means the blocks m_1 and m_2 will not move.

Hence, acceleration of the blocks is zero and $T = m_1 g$.

(b) As, $m_1 g \leq m_2 g \sin\alpha + \mu_s m_2 g \cos\alpha$

It is clear from the figure, that the tendency of block m_1 is to move downward but its weight is balanced by the component of weight of block m_2 ($m_2 g \sin\alpha$) and static friction acting on block m_2 in downward direction along inclined plane.

(c) As, $m_2g\sin\alpha - m_1g \le \mu_s m_2 g\cos\alpha$

$$m_2g\sin\alpha - m_1g \le f_{max}$$

In this case, the tendency of block m_2 is to slide down the inclined plane, hence the frictional force on block m_2 will act upward along the inclined plane.

Static friction acting on block m_2,

$$f = m_2 g\sin\alpha - m_1 g.$$

(d) As, $m_1g - m_2g\sin\alpha > \mu_s m_2 g\cos\alpha$

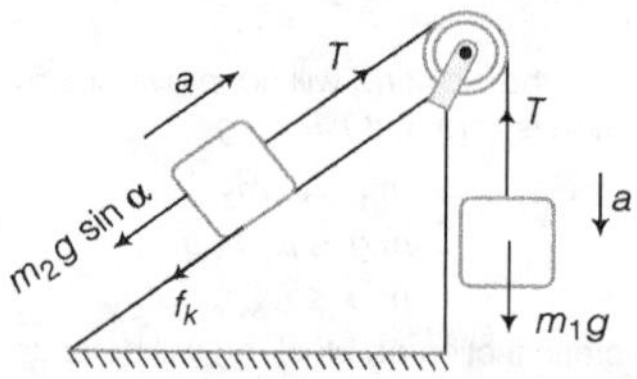

$$\Rightarrow \quad m_1g - m_2g\sin\alpha > f_{max}$$

In this case, the block m_1 will move downward and m_2 will move upward along the inclined plane. As block m_2 will slide upward, the kinetic friction will act on it.

$$m_1g - T = m_1a \quad \ldots(i)$$

and $T - m_2g\sin\alpha - \mu_k m_2g\cos\alpha = m_2a$

or $m_1g - m_1a - m_2g\sin\alpha - \mu_k m_2 g\cos\alpha = m_2a$

or $a = \dfrac{(m_1g - m_2g\sin\alpha - \mu_k m_2g\cos\alpha)}{(m_1 + m_2)}$

10. (a,b,c,d)

Concept

(i) If $F \le \mu_s m_A g$, applied force is balanced by static friction on block A. Thus, tension in string is zero. So, the friction on the block B is zero.

(ii) If $\mu_s m_A g < F < \mu_s m_A g + \mu_s m_B g$, static friction comes into play.

In this case,

$$F = T + \mu_s m_A g \quad \text{(for block A)}$$

$$\therefore \quad T = F - \mu_s m_A g$$

and the friction on the block A is $\mu_s m_A g$.

and $T = f_B$ (for block B)

In this case, the nature of friction is static, the acceleration of blocks is zero.

(iii) If $F > \mu_s m_A g + \mu_s m_B g$, kinetic friction comes into play. In this case, the friction on block A is $\mu_k m_A g$ and that of block B is $\mu_k m_B g$.

For block A, $F - \mu_k m_A g - T = m_A a$

For block B, $T - \mu_k m_B g = m_B a$

11. (a,b,c,d) Here, the upward component of F is $F\cos 45° = 50$ N.

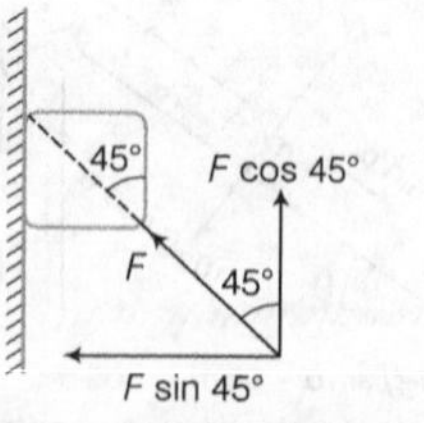

(a) The weight (50 N) of the block is balanced by the upward component of applied force F. Thus, friction is zero.

(b) Here, $N = F\sin 45° = 40$ N

$\therefore \quad f_{max} = \mu_s N = 40 \times 0.5 = 20$ N

The net force in vertical direction is

$$F_v = mg - F\cos 45°$$
$$= 10 < f_{max}$$

So, static friction acts on the block.

$\therefore \quad mg = F\cos 45° + f$

$\therefore \quad f = 10$ N

(c) Here, $f_{max} = \mu_s N = 0.5 \times 20 = 10$ N

$\because \quad mg - F\cos 45° = 30$ N

So, kinetic friction comes into play.

$\therefore \quad f_k = \mu_k N$

or $f_k = \mu_k N = \mu_k F\sin 45°$

$$= 0.4 \times 20\sqrt{2} \times \frac{1}{\sqrt{2}} = 8 \text{ N}$$

and $mg - F\cos 45° - 8 = ma$

$\Rightarrow \quad a = 4.4 \text{ m/s}^2$

(d) Upward force acting on the block is $F\cos 45°$

$$= 120\sqrt{2} \times \frac{1}{\sqrt{2}} = 120 \text{ N}.$$

Kinetic frictional force acting on the block in downward direction,

$$f_k = \mu_k N = \mu_k F\sin 45° = 0.4 \times 120 \times \sqrt{2} \times \frac{1}{\sqrt{2}} = 48 \text{ N}$$

Let a be the acceleration by the block.

$$ma = 120 - (f_{max} + mg)$$

$$\Rightarrow \quad a = \frac{120 - 48 - 50}{5} = 4.4 \text{ m/s}^2$$

12. (a) The maximum tension in the string is T_{max}, if m_2 has tendency to move downward.

$\therefore \quad m_2g - 2T = m_2a$

or $m_2g - 2T_{max} = m_2 a_{min}$

$\therefore \quad T_{max} = \dfrac{m_2g}{2} = \dfrac{80}{2} = 40$ N $\quad (\because a_{min} = 0)$

The maximum friction on m_1 is

$$f_{1\,max} = \mu_s m_1 g = 50 \text{ N}$$

$\because$ $T_{max} < f_{1\,max}$, m_1 remains stationary. The value of $T_{max} - m_3 g\sin\theta = 28$ N and the maximum friction on m_3 is $f_{3max} = \mu_s m_3 g\cos\theta = 0.5 \times 2.0 \times \frac{4}{5} = 8$ N. Hence, m_3 moves up along inclined plane.

13. (a) From constraint relation,

$$a_2 = \frac{a_3 + a_1}{2} = \frac{a_3 + 0}{2}$$

$\because \quad a_3 = 2a_2$

14. (c) From free body diagram of m_2,

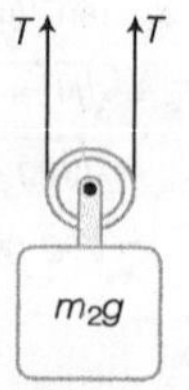

$$m_2g - 2T = m_2a_2$$

or $80 - 2T = 8a_2$

or $40 - T = 4a_2 \quad \ldots(i)$

From free body diagram of m_3,

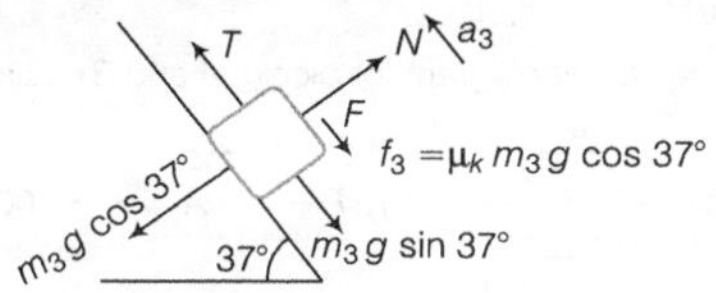

$$T - f_3 - m_3 g \sin 37° = m_3 a_3$$

or $T - \mu_k m_3 g \cos 37° - m_3 g \sin 37° = 2 \times 2a_2 = 4a_2$

or $\quad T - 6.4 - 12 = 4a_2$

or $\quad T - 18.4 = 4a_2 \quad$ ...(ii)

From Eqs. (i) and (ii), $\quad a_2 = 2.7 \text{ m/s}^2$

15. (a) It is not clear that in which direction of the blocks will move. Let us assume that block B has tendency to move downwards.

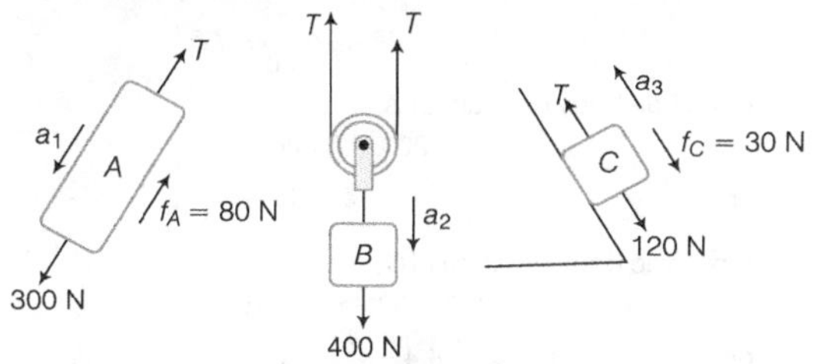

From constraints relation, $a_2 = \dfrac{a_3 - a_1}{2}$

$\therefore \quad 2a_2 = a_3 - a_1 \quad$ ...(i)

From free body diagram of block A,

$$300 - T - 80 = 50a_1 \quad ...(ii)$$

From free body diagram of block B,

$$400 - 2T = 40a_2 \quad ...(iii)$$

From free body diagram of block C,

$$T - 120 - 30 = 20a_3 \quad ...(iv)$$

From Eqs. (i), (ii), (iii) and (iv), we get

$\because \quad 2a_2 = a_3 - a_1$

or $\quad 2\left(10 - \dfrac{T}{20}\right) = \dfrac{T}{20} - 7.5 - \left(\dfrac{220}{50} - \dfrac{T}{50}\right)$

or $\quad 20 + 7.5 + \dfrac{22}{5} = \dfrac{T}{20} + \dfrac{T}{10} + \dfrac{T}{50}$

or $\quad 31.9 = \left(\dfrac{5 + 10 + 2}{100}\right) T$

$\therefore \quad T = \dfrac{3190}{17} = 187.65$ N

From Eqs. (iii), $400 - 2T = 40a_2$

$\therefore \quad a_2 = \dfrac{24.7}{40} = 0.62 \text{ m/s}^2$

Since, acceleration of B is positive. Hence, our assumption is correct.

16. (c) **17.** (c)

18. (b) Given,

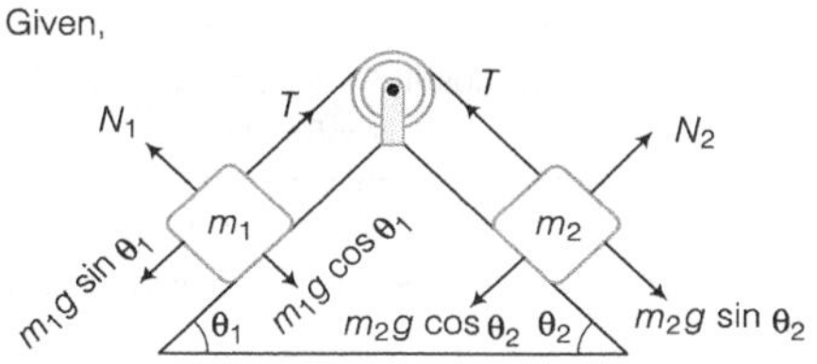

$$m_1 g \sin \theta_1 - m_2 g \sin \theta_2 = 50 \text{ N}$$

$$\mu_1 m_1 g \cos \theta + \mu_2 m_2 g \cos \theta_2 = 30 \text{ N}$$

and $\quad m_1 + m_2 = 12$ kg

From free body diagrams of the blocks, we can write the following equations

$$m_1 g \sin \theta_1 - T - \mu_1 m_1 g \cos \theta_1 = m_1 a \quad ...(i)$$

and $\quad T - m_2 g \sin \theta_2 - \mu_2 m_2 g \cos \theta_2 = m_2 a \quad$ ...(ii)

On adding Eq. (i) and Eq. (ii), we get

$$(m_1 g \sin \theta_1 - m_2 g \sin \theta_2) - (\mu_1 m_1 g \cos \theta_1 + \mu_2 m_2 g \cos \theta_2) = (m_1 + m_2)a$$

$\Rightarrow \quad 50 - 30 = 12 \times a$

or $\quad a = \dfrac{20}{12} = 1.67 \text{ m/s}^2$

19. (d) Here, $F = m_1 g \sin \theta_1 - m_2 g \sin \theta_2 = 90 - 30 = 60$ N

But $f_{max} = \mu_1 m_1 g \cos \theta_1 + \mu_2 m_2 g \cos \theta_2 = 100$ N

Thus, $F < f_{max}$, static friction comes into play.

$\therefore \quad a = 0$

(a) For block A, $\quad m_1 g \sin \theta_1 > \mu_1 m_1 g \cos \theta_1$

$\therefore$ Limiting friction acts on the block A.

$\therefore \quad f_1 = \mu_1 m_1 g \cos \theta_1 = 50$ N

(b) For block A, $\quad m_1 g \sin \theta_1 = f_1 + T$ or $90 = 50 + T$

$\therefore \quad T = 40$ N

(c) For block B, $\quad T = m_2 g \sin \theta_2 + f_2$

or $\quad 40 = 30 + f_2$

$\therefore \quad f_2 = 10$ N

Hence, option (d) is correct.

20. (a) Here, $m_1 g \sin \theta_1 < \mu_1 m_1 g \cos \theta_1$

$\therefore$ Static frictional force balances the component of weight along inclined place, i.e. $m_1 g \sin \theta_1$.

$\therefore$ $T = 0$ and $f_1 = m_1 g \sin \theta_1 = 90$ N

Similarly, $\quad f_2 = m_2 g \sin \theta_2 = 30$ N

Hence, option (a) is correct.

21. (a) Free body diagram of blocks A and B,

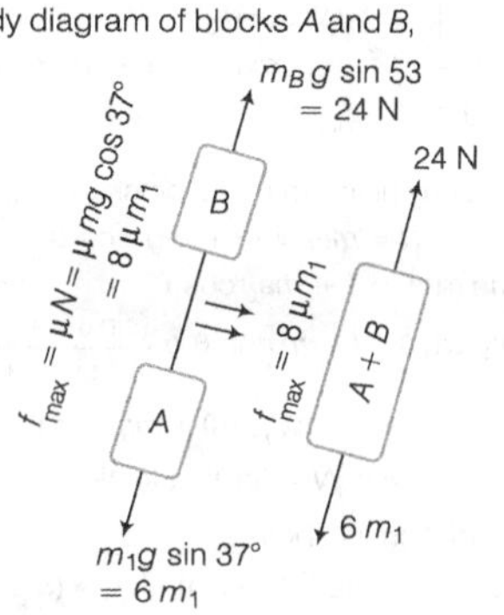

Here, $\quad f_{max} = 8 \mu m_1 = 28.8$ N

The resultant applied force along inclined plane is

$$F = 6m_1 - 24 = 24 - 24 = 0$$

Thus, friction is zero.

22. (b) $f_{max} = 8 \times 0.5 \times 3 = 12$ N

The resultant applied force along inclined plane is $F = 24 - 6m_1 = 6$ N.

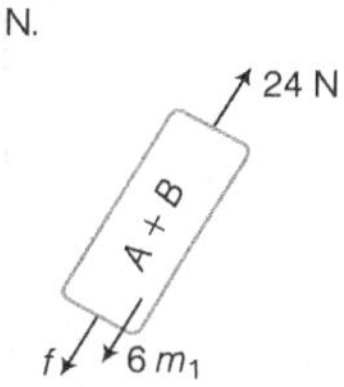

$\because \quad F < f_{max}$, static friction comes into play.

$\therefore \quad 24 = 6m_1 + f$ (from free body diagram)

$\therefore \quad f = 24 - 18 = 6$ N

23. (b) $f_{max} = 8\mu m_1 = 8 \times 0.5 \times 2 = 8$ N

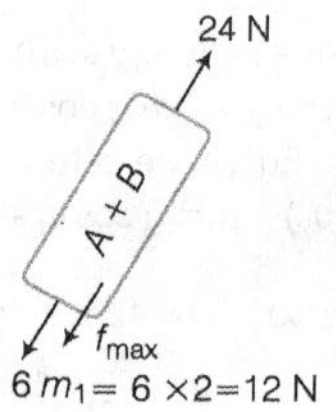

Let a be the acceleration of blocks A and B.

$\Rightarrow \quad 24 - (12 + f_{max}) = (m_1 + m_2)a$

$\Rightarrow \quad a = \frac{4}{5} = 0.8 \text{ m/s}^2$

24. (a) (i) → (p,q,r), (ii) → (p,r), (iii) → (r,t), (iv) → (s,t)

25. (b)

(i) If $a_0 = 0 \Rightarrow f = mg\sin\theta$

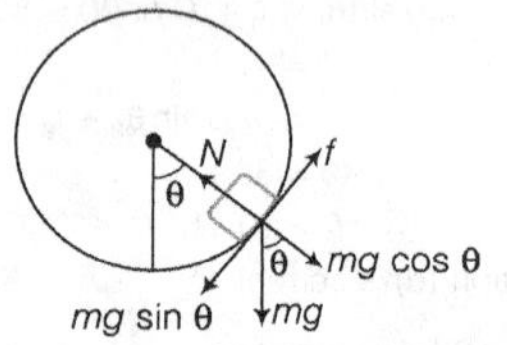

(ii) If $a_0 = g\tan\theta$, $mg\sin\theta$ is balanced by $ma_0\cos\theta$. Thus, friction is zero.

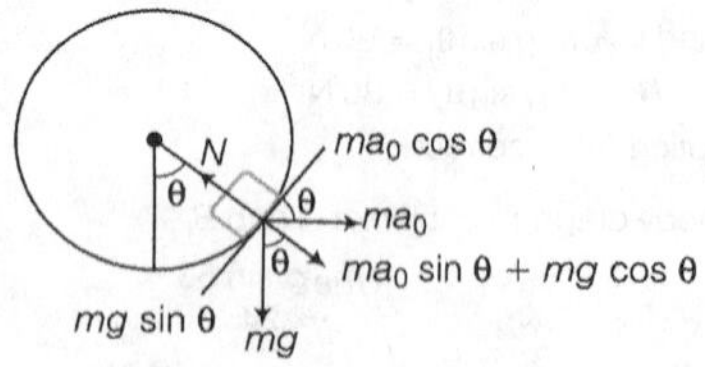

(iii) From free body diagram of the block,

$$N = ma_0\sin\theta + mg\cos\theta$$

and $\quad mg\sin\theta = f + ma_0\cos\theta$

or $\quad mg\sin\theta = f + m\cos\theta\, g\dfrac{(\sin\theta - \mu_s\cos\theta)}{(\mu_s\sin\theta + \cos\theta)}$

$\therefore \quad f = \mu_s(ma_0\sin\theta + mg\cos\theta)$

$\quad = \mu_s N =$ limiting friction

(iv) Think the solution as above.

(i) → (r,s), (ii)→ (p), (iii) → (s,t), (iv) → (q, t)

26. (3) As there is a friction between the rope and pulley, there are different tensions in the rope as shown in figure.

Here, $T_2 = T_1 e^{\mu\alpha}$

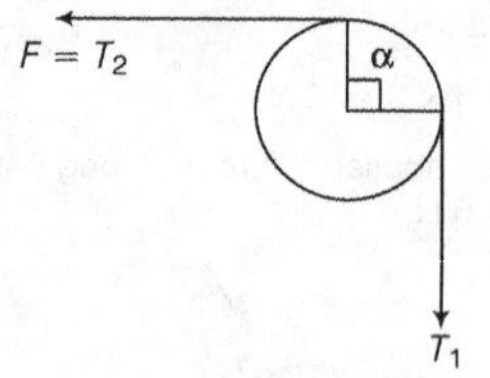

$\Rightarrow \quad F = mge^{\left(\frac{2}{\pi}\times\frac{\pi}{2}\right)} \qquad \left(\because \alpha = \frac{\pi}{2}\right)$

or $\quad F = 10 \times 2.7 = 27$ N

Given, $\quad F = 9x = 27$

$\Rightarrow \quad x = 3$

27. (i) 174.1 N

(ii) 1.3 m/s^2.

Consider free body diagram for blocks A and B as shown below.

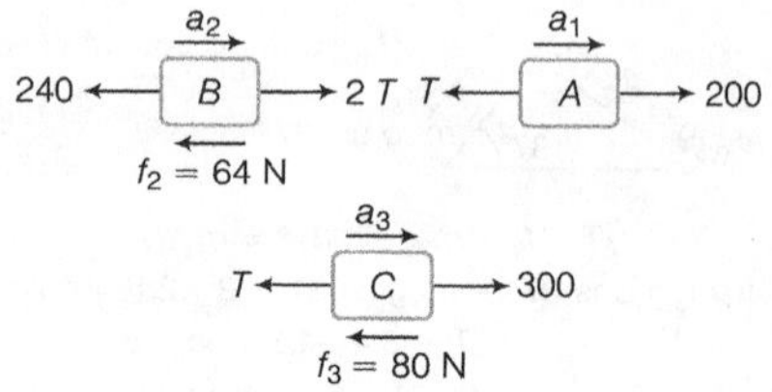

We shall assume that blocks A, B and C move horizontally rightward.

From free body diagram of B,

$$2T - 240 - 64 = 40\,a_2$$

$\therefore \quad a_2 = \frac{T}{20} - 7.6 \qquad \ldots\text{(i)}$

From free body diagram of A,

$$200 - T = 20a_1$$

or $\quad a_1 = 10 - \frac{T}{20} \qquad \ldots\text{(ii)}$

From free body diagram of C,

$$300 - 80 - T = 50a_3$$

or $\quad 4.4 - \frac{T}{50} = a_3 \qquad \ldots\text{(iii)}$

From constraint relation,

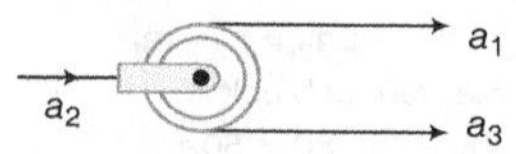

$$a_2 = \frac{a_1 + a_3}{2}$$

$\therefore \quad 2\left(\frac{T}{20} - 7.6\right) = 10 - \frac{T}{20} + 4.4 - \frac{T}{50}$

$\therefore \quad T = \frac{29.6}{0.17} = 174.1$ N

From Eqs. (i) and (ii), we get

$$a_1 = 10 - \frac{174.1}{20} = 1.3 \text{ m/s}^2$$

28. (i) zero

(ii) $5(\sqrt{2} - 1)$ m/s^2

Consider free body diagram of the system as shown below.

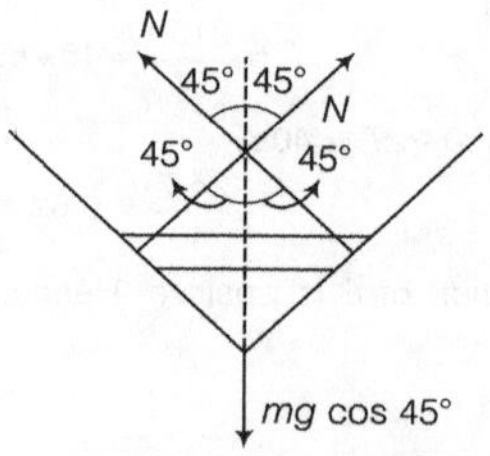

Balancing forces ace have

$$2N\cos 45° = mg\cos 45°$$

$\therefore \quad N = 5\sqrt{2}$ N

(i) The maximum friction by each surface is

$$f_{1\,max} = \mu N = 5 \text{ N}$$

Total friction is $f_{max} = 2\mu N = 10$ N the component of mg along trough is

$$mg\sin 45° = \sqrt{2} \times 10\frac{1}{\sqrt{2}} = 10 \text{ N}$$

$\because \quad mg\sin\theta = f_{max},\ a = 0$

(ii) $f_{max} = 2\mu N = 2 \times \frac{1}{2} \times 5\sqrt{2} = 5\sqrt{2}$ N

Now, $mg \sin\theta > f_{max}$ kinetic friction comes into play.

$\therefore \quad mg \sin\theta - 2\mu N = ma$

or $\quad \sqrt{2} \times g \times \frac{1}{\sqrt{2}} - 5\sqrt{2} = \sqrt{2}a$

$\therefore \quad a = 5(\sqrt{2} - 1)\ \text{m/s}^2$

29. (i) Acceleration of block A

Maximum friction force that can be obtained at A is

$$(f_{max})_A = \mu_A(mg \cos 45°)$$
$$= \frac{2}{3}(mg/\sqrt{2}) = \frac{\sqrt{2}mg}{3}$$

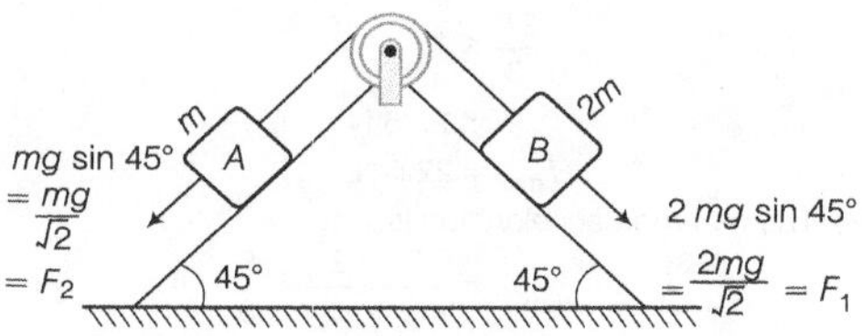

Similarly,

$$(f_{max})_B = \mu_B(2mg \cos 45°)$$
$$= \frac{1}{3}(2mg/\sqrt{2})$$
$$= \frac{\sqrt{2}mg}{3}$$

Therefore, maximum value of friction that can be obtained on the system is

$$(f_{max}) = (f_{max})_A + (f_{max})_B = \frac{2\sqrt{2}\ mg}{3} \quad \ldots\text{(i)}$$

Net pulling force on the system is

$$F = F_1 - F_2 = \frac{2mg}{\sqrt{2}} - \frac{mg}{\sqrt{2}} = \frac{mg}{\sqrt{2}} \quad \ldots\text{(ii)}$$

From Eqs. (i) and (ii), we can see that

Net pulling force $< f_{max}$

Therefore, the system will not move or the acceleration of block A will be zero.

(ii) and (iii) Tension in the string and friction at A

Net pulling force on the system (blocks A and B)

$$F = F_1 - F_2 = mg/\sqrt{2}$$

Therefore, total friction force on the blocks should also be equal to $\frac{mg}{\sqrt{2}}$

or $\quad f_A + f_B = F = mg/\sqrt{2}$

Now, the blocks will start moving from block B first (if they move), therefore, f_B will reach its limiting value first and if still some force is needed, it will be provided by f_A.

Here, $\quad (f_{max})_B < F$

Therefore, f_B will be in its limiting value and rest will be provided by f_A.

Hence, $\quad f_B = (f_{max})_B = \frac{\sqrt{2}mg}{3}$

and $\quad f_A = F - f_B$

$$= \frac{mg}{\sqrt{2}} - \frac{\sqrt{2}mg}{3} = \frac{mg}{3\sqrt{2}} \quad \text{(for iii)}$$

The FBD of the whole system will be as shown in the figure,

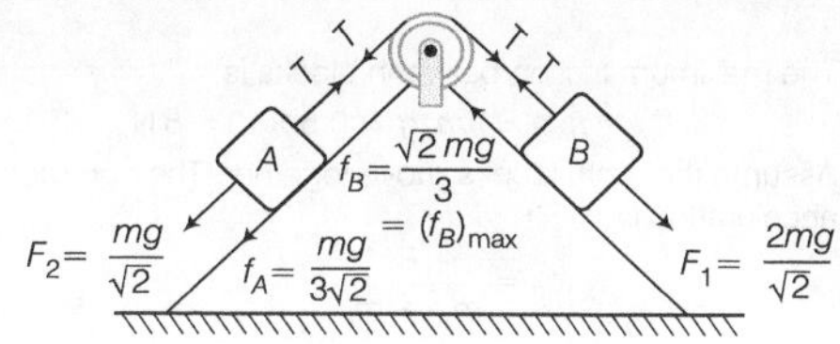

Therefore, friction on A is

$$f_A = mg/3\sqrt{2} \quad \text{(down the plane)}$$

Now, for tension T in the string, we may consider either equilibrium of A or B.

Equilibrium of A gives

$$T = F_2 + f_A = \frac{mg}{\sqrt{2}} + \frac{mg}{3\sqrt{2}}$$
$$= \frac{4mg}{3\sqrt{2}} \text{ or } \frac{2\sqrt{2}mg}{3}$$

Similarly, equilibrium of B gives $T + f_B = F_1$

or $\quad T = F_1 - f_B = \frac{2mg}{\sqrt{2}} - \frac{\sqrt{2}mg}{3}$

$$= \frac{4mg}{3\sqrt{2}} \text{ or } \frac{2\sqrt{2}\ mg}{3}$$

Therefore, tension in the string is $\frac{2\sqrt{2}\ mg}{3}$. (for ii)

30. (i) $1.30\ \text{m/s}^2$

(ii) 5.24 N

From the free body diagrams of blocks M_1 and M_2,

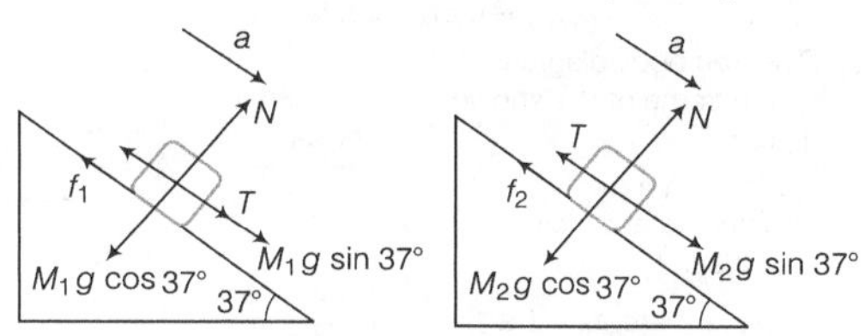

$$M_2g \sin 37° - (T + f_2) = M_2a \quad \ldots\text{(i)}$$
$$M_1g \sin 37° + T - f_1 = M_1a \quad \ldots\text{(ii)}$$

$\Rightarrow \quad (M_1 + M_2)g \sin 37° - (f_1 + f_2) = (M_1 + M_2)a$

or $\quad (M_1 + M_2)g \sin 37° - (\mu_1 M_1 g \cos 37° + \mu_2 M_2 g \cos 37°) = (M_1 + M_2)a$

or $\quad (M_1 + M_2)g \sin 37° - (\mu_1 M_1 + \mu_2 M_2)g \cos 37° = (M_1 + M_2)a$

or $(4 + 2) \times 9.8 \times 0.6 - (0.75 \times 4 + 0.25 \times 2) \times 9.8 \times 0.8 = (4 + 2)\,a$

or $\quad 35.28 - 27.44 = 6a$

$\Rightarrow \quad a = \frac{7.84}{6} = 1.30\ \text{m/s}^2$

Putting $a = 1.30 m/s^2$ in Eq. (i),

$$M_2g \sin 37° - T - f_2 = M_2a$$

or $\quad (2 \times 9.8 \times 0.6) - T - (0.25 \times 2 \times 9.8 \times 0.8) = 2 \times 1.30$

or $\quad 11.76 - T - 3.92 = 2.60$

or $\quad T = 11.76 - 6.52$

$= 5.24$ N

DPP-4 Blocks in Contact

1. (i) The maximum friction between blocks is

$$f_{max} = \mu m_A g = 0.5 \times 10 = 5 \text{ N}$$

Assume that both blocks move together. The common acceleration is

$$a_0 = \frac{F}{m_A + m_B} = \frac{F}{3}$$

The free body diagram of block A in the frame of block B is shown.

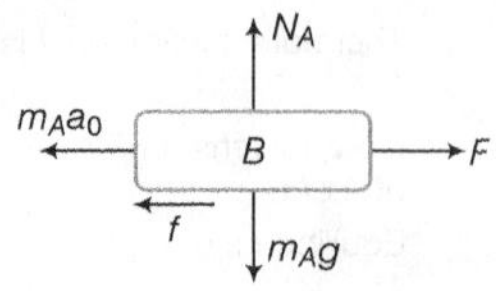

Since, block A does not slide on block B.

The friction between blocks is static in nature. Thus,

$$F - m_A a_0 = f \le f_{max} \quad \text{or} \quad F \le m_A a_0 + f_{max}$$

or $\quad F \le a_0 + 5 \quad$ or $\quad F - \frac{F}{3} \le 5$

$\therefore \quad F \le 7.5$ N

$\therefore \quad F_{max} = 7.5$ N

More discussion (A) If $F \le 7.5$, the blocks move together and the friction between blocks is static in nature.

(B) If $F > 7.5$, blocks A slides on B and friction is kinetic in nature.

(ii) Assume that both blocks move together. The common acceleration is

$$a_0 = \frac{F}{m_A + m_B} = \frac{F}{3}$$

The maximum friction is

$$f_{max} = \mu m_A g = 5 \text{ N}$$

The free body diagram of A in the frame of B is shown in figure.

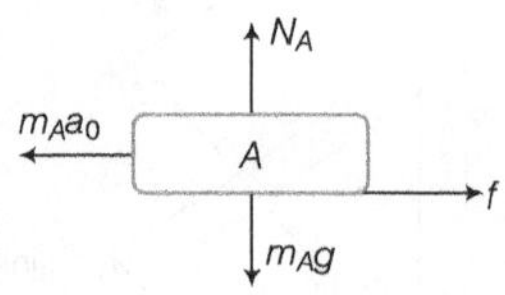

∵ Blocks are moving together, static friction comes into play.

$\therefore \quad m_A a_0 = f \le F_{max} \quad$ or $\quad \frac{F}{3} < 5$

$\therefore \quad F \le 15$

$\therefore \quad F_{max} = 15$ N

(iii) Solution is same as above. The common acceleration is

$$a_0 = \frac{2F}{m_A + m_B} = \frac{2F}{3}$$

The free body diagram of A in the frame of B.

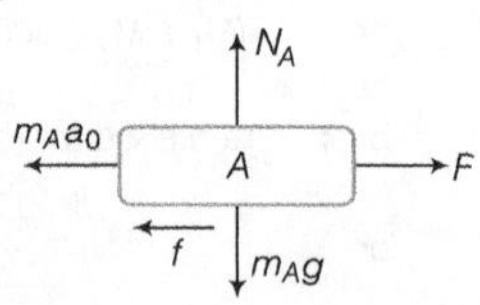

Here, $\quad F - m_A a_0 = f \le \mu m_A g$

or $\quad F - \frac{2F}{3} \le 5$

or $\quad F \le 15$

$\therefore \quad F_{max} = 15$ N

(iv) The common acceleration is

$$a_0 = \frac{F - F}{m_A + m_B} = 0$$

The free body diagram of block A in the frame of block B is shown in figure.

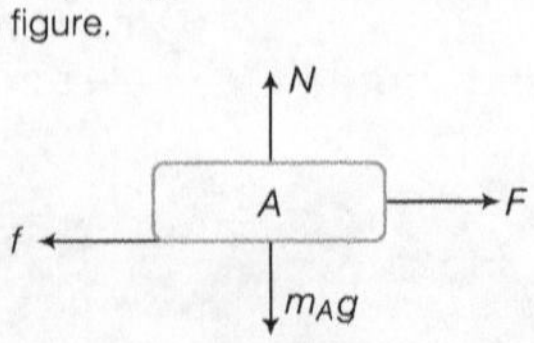

Here, $\quad F = f \le \mu m_A g \quad$ or $\quad F \le 5$

$\therefore \quad F_{max} = 5$ N

(v) The common acceleration is

$$a_0 = \frac{F + 30}{m_A + m_B} = \frac{F + 30}{3} = \frac{F}{3} + 10$$

Free body diagram of block A in frame of block B is as shown.

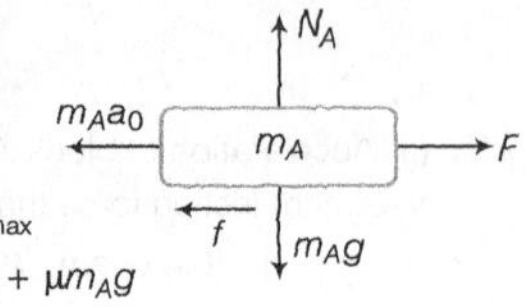

Here,

$$F - m_A a_0 = f \le F_{max}$$

or $\quad F \le m_A a_0 + \mu m_A g$

or $\quad F \le \frac{F}{3} + 10 + 5$

or $\quad \frac{2F}{3} \le 15$

$\therefore \quad F \le 22.5$ N

or $\quad F_{max} = 22.5$ N

(vi) The common acceleration is

$$a_0 = \frac{F - 12}{m_A + m_B} = \frac{F}{3} - 4$$

The free body diagram of block A in the frame of block B is shown in figure.

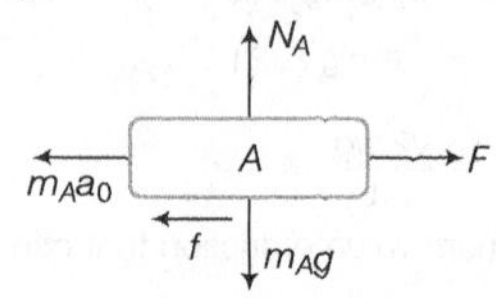

Here, $\quad F - m_A a_0 = f \le f_{max}$

or $\quad F - \left(\frac{F}{3} - 4\right) \le \mu m_A g$

or $\quad \frac{2F}{3} \le 5 - 4 \quad$ or $\quad F \le \frac{3}{2}$

$\therefore \quad F_{max} = 1.5$ N

(vii) The common acceleration is

$$a_0 = \frac{F \cos 37°}{m_A + m_B} = \frac{4}{15} F$$

The normal reaction on block A is

$$N = m_A g - F \sin 37° = 10 - \frac{3}{5} F$$

The maximum friction $= f_{max} \approx \mu N = 5 - \frac{3}{10} F$

The free body diagram of block A in the frame of block B is shown in the figure.

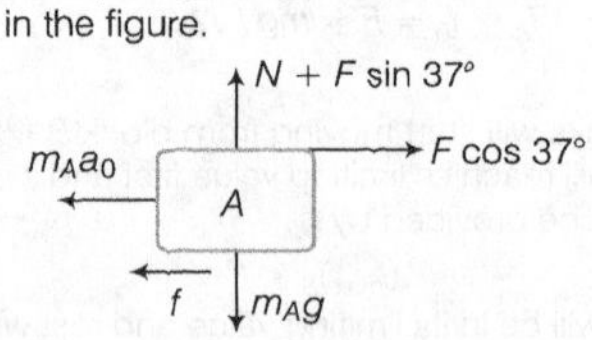

Here, $\quad F \cos 37° - m_A a_0 = f \le f_{max}$

or $\quad \frac{4}{5} F - \frac{4}{15} F \le 5 - \frac{3}{10} F$

or $\quad \left(\frac{4}{5} - \frac{4}{15} + \frac{3}{10}\right) F \le 5$

or $\quad \frac{25}{30} F \le 5$

$\therefore \quad F_{max} = 6$ N

(viii) The common acceleration is

$$a_0 = \frac{F}{m_A + m_B} = \frac{F}{2.5} = \frac{2F}{5}$$

The free body diagram of block A in the frame of B block is shown in figure.

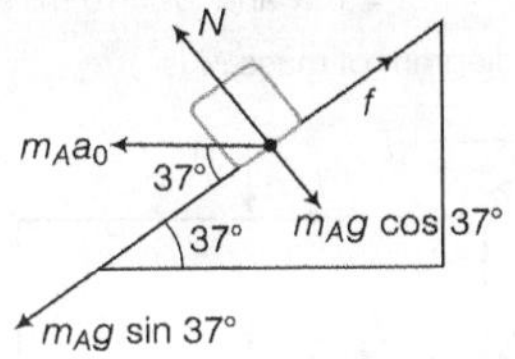

Here, $N + m_A a_0 \sin 37° = m_A g \cos 37°$

$\therefore \quad N = m_A g \cos 37° - m_A a_0 \sin 37° = 4 - \frac{3F}{25}$

and $m_A g \sin 37° + m_A a_0 \cos 37° = f \le \mu N$

or $\quad 3 + \frac{4F}{25} \le 0.5\left(4 - \frac{3F}{25}\right)$

or $\quad \frac{4F}{25} + \frac{3F}{50} \le 1$ or $\frac{4F}{50} \le 1$

$\therefore \quad F \le \frac{50}{11}$

$\Rightarrow \quad F_{max} = 4.54$ N

2. (i) The maximum friction on block A and block B is

$$f_{1\,max} = \mu m_A g = 10\mu$$

The maximum friction between block B and ground is

$$f_{2max} = \mu_B(m_A + m_B)g = 0.5 \times 3 \times 10 = 15 \text{ N}$$

For sliding block B with block A, $F > f_{2max}$

or $\quad F > 15$ N

Thus, common acceleration, $a_0 = \frac{60 - 15}{3} = 15 \text{ m/s}^2$

(considering coefficient of static friction is same as that of kinetic friction)

Free body diagram of block A in frame of block B is as shown below.

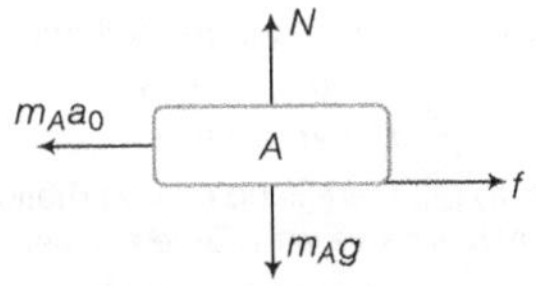

Here, $\quad f = m_A a_0 = 15$ N or $f \le \mu m_A g$

$15 \le 10\mu$ or $\mu \ge 1.5$

$\mu_{min} = 1.5$

(ii) The maximum static friction between ground and block B is

$$f_{2max} = 0.3\,(m_A + m_B)\,g = 9 \text{ N}$$

The maximum friction on block A is

$$f_{1\,max} = \mu m_A g = 10\mu$$

Horizontal applied force is on block A, the acceleration of the block A should be greater or equal to the block B.

The common acceleration of blocks is

$$a_0 = \frac{39 - 9}{3} = 10 \text{ m/s}^2$$

The free body diagram of block A in the frame of block B is shown in figure.

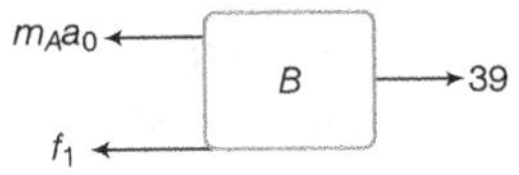

Since, block A does not slide on the block B.

$\therefore \; 39 - m_A a_0 = f_1 \le f_{1\,max}$

or $\quad 39 - 10 \le 10\mu$ or $10\mu \ge 29$

$\therefore \quad \mu \ge 2.9$

$\Rightarrow \quad \mu_{min} = 2.9$

3. (i) $\because \mu_2 < \mu_1$, acceleration of m_2 is greater than m_1 if allowed to move separately. In this case, both blocks are treated as a system.

$$(m_1 + m_2)\,g \sin\theta = 30 \times \frac{3}{5} = 18 \text{ N}$$

But $\quad f_{1\,max} = \mu_1 m_1 g \cos\theta$

$= 0.5 \times 10 \times \frac{4}{5} = 4$ N

$\because \quad (m_1 + m_2)g \sin\theta > f_{1max}$, blocks move together. The acceleration is

$$a = \frac{(m_1 + m_2)\,g \sin\theta - \mu_1 m_1 g \cos\theta}{m_1 + m_2} = \frac{18 - 4}{3} = \frac{14}{3}$$

$= 4.67 \text{ m/s}^2$

(ii) If blocks move separately,

$m_1 g \sin\theta = 12$ N and $m_2\, g \sin\theta = 6$ N

$f_{1\,max} = \mu_1 m_1 g \cos\theta$

and $\quad f_{2max} = \mu_2 m_2 g \cos\theta$

$= 0.5 \times 10 \times \frac{4}{5}$

$= 4 \text{ N} < m_2 g \sin\theta$

$(f_1)_{max} = 8 \text{ N} < m_1 g \sin\theta$

If they move separately,

$$a_1 = \frac{12 - 8}{1} = 4 \text{ m/s}^2$$

and $\quad a_2 = \frac{6 - 4}{2} = 1 \text{ m/s}^2$

$\because \; a_1 > a_2$, thus blocks move separately.

$\therefore \quad a_2 = 1 \text{ m/s}^2$

4. (i) Free body diagram of block over the plate is as shown below,

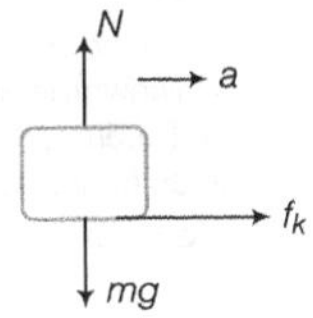

$f_k = \mu_k N = \mu_k mg$

$\Rightarrow \quad a = \frac{f_k}{m} = \mu_k g$

$s = ut + \frac{1}{2}at^2$

or $\quad s = 0 + \frac{1}{2}(\mu_k g)t^2 = \frac{1}{2}\mu_k g t^2$

(ii) Free body diagram of plate w.r.t. block,

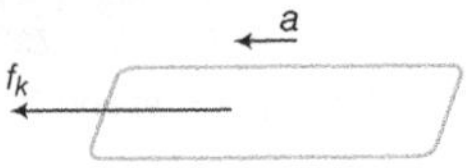

$a = \frac{f_k}{m} = \mu_k g$

Distance travelled w.r.t. block

$= v_0 t + \frac{1}{2}\mu_k g t^2$

So, $\quad v_0 \ge \mu_k g\left(1 + \frac{d}{2r}\right)$

5. The common acceleration is

$$a_0 = \frac{F}{m_A + m_B + m_C} = \frac{50}{10} = 5 \text{ m/s}^2$$

From free body diagram of block A in the frame of block B

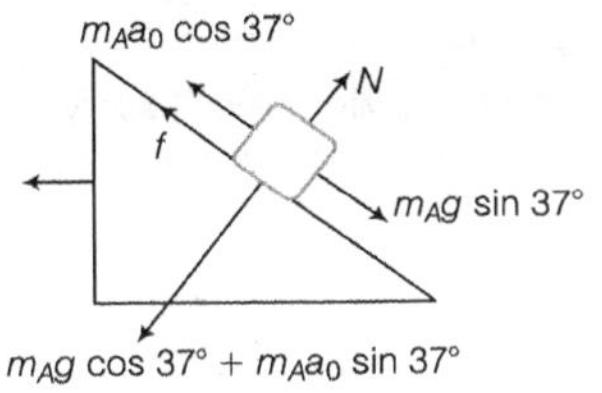

$f_1 = m_A g \sin 37° - m_A a_0 \cos 37°$

or $\quad f_1 = 2 \times 10 \times \frac{3}{5} - 2 \times 5 \times \frac{4}{5} = 12 - 8 = 4$ N

The maximum friction on block A is $f_{1max} = \mu N$

$$= \mu\,(m_A g \cos 37^\circ + m_A a_0 \sin 37^\circ)$$

$$= 0.5\left(2 \times 10 \times \frac{4}{5} + 2 \times 5 \times \frac{3}{5}\right)$$

$$= 0.5\,(16 + 6) = 11 \text{ N}$$

$\because$ $m_A g \sin 37^\circ - m_A a_0 \cos 37^\circ \le f_{1\,max}$ static friction acts on A.

From free body diagram of block $(A + B)$ in the frame of C.

The maximum friction on B due to C is

$$f_{2max} = \mu\,(m_A + m_B)g = 0.5 \times 50 = 25$$

Also, $(m_A + m_B)a_0 = 5 \times 5 = 25$ N

$\therefore$ $(m_A + m_B)a_0 = f_{2max} = 25$ N

$\therefore$ Limiting friction acts between B and C.

(i) $\therefore$ $a_A = a_B = a_C = \dfrac{F}{m_A + m_B + m_C} = \dfrac{50}{10} = 5 \text{ m/s}^2$

(ii) $m_A g \sin 37^\circ = m_A a_0 \cos 37^\circ + f_1$

$\because$ $f_1 = 20 \times \dfrac{3}{5} - 2 \times 5 \times \dfrac{4}{5}$

$$= 12 - 8 = 4 \text{ N}$$

(iii) $f_C = f_{2max} = 25$ N

6. (i) 1.3m/s^2

(ii) 5.2 N

Maximum force of friction between m_1 and inclined plane

$$f_1 = \mu_1 m_1 g \cos\theta = (0.75)\,(4)\,(9.8)\,(0.8) = 23.52 \text{ N}$$

and $m_1 g \sin\theta = (4)\,(9.8)\,(0.6) = 23.52 \text{ N} = F_1$ (say)

Maximum force of friction between m_2 and inclined plane

$$f_2 = \mu_2 m_2 g \cos\theta = (0.25)\,(2)\,(9.8)\,(0.8) = 3.92 \text{ N}$$

and $m_2 g \sin\theta = (2)\,(9.8)\,(0.6) = 11.76 \text{ N} = F_2$ (say)

Both the blocks will be moving downwards with same acceleration a. Different forces acting on two blocks are as shown in the figures.

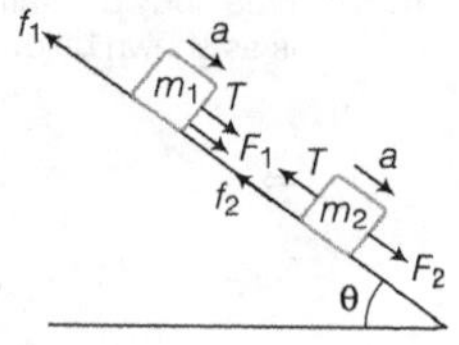

Equation of motion of m_1,

$$T + F_1 - f_1 = m_1 a$$

or $T = 4a$ ...(i)

Equation of motion of m_2,

$$F_2 - T = f_2 = m_2 a$$

or $7.84 - T = 2\,a$ ...(ii)

Solving Eqs. (i) and (ii), we get

$$a = 1.3 \text{ m/s}^2$$

$$T = 5.2 \text{ N}$$

7. Constant velocity means net acceleration of the system is zero. Or net pulling force on the system is zero. While calculating the pulling force, tension forces are not taken into consideration as they are treated internal to the system. Therefore,

(i) $m_1 g = m_2 g \sin 37^\circ + \mu m_2 g \cos 37^\circ + \mu m_3 g$

or $m_1 = m_2 \sin 37^\circ + \mu m_2 \cos 37^\circ + \mu m_3$

Substituting the values

$$m_1 = (4)\left(\frac{3}{5}\right) + (0.25)\,(4)\left(\frac{4}{5}\right) + (0.25)\,(4)$$

$$= 4.2 \text{ kg}$$

(ii) Since, m_3 is moving with uniform velocity

$$T = \mu m_3 g = (0.25)\,(4)\,(9.8) = 9.8 \text{ N}$$

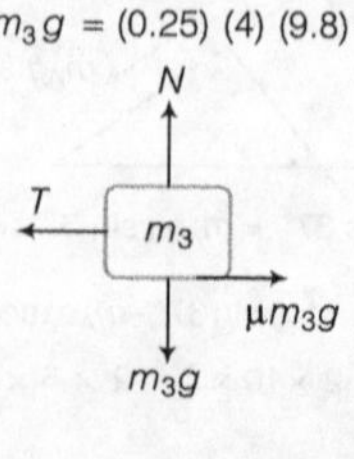

8. Given, $m_1 = 20$ kg, $m_2 = 5$ kg, $M = 50$ kg

$$\mu = 0.3 \text{ and } g = 10 \text{ m/s}^2$$

(i) Free body diagram of mass M is

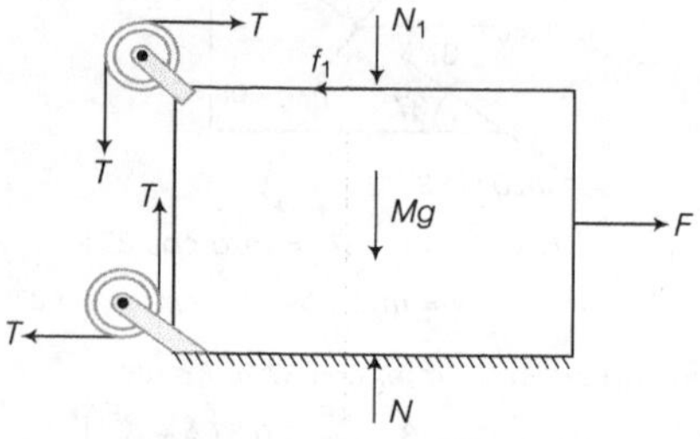

(ii) The maximum value of f_1 is

$$(f_1)_{max} = 0.3 \times 20 \times 10 = 60 \text{ N}$$

The maximum value of f_2 is

$$(f_2)_{max} = (0.3)\,(5)\,(10) = 15 \text{ N}$$

Forces on m_1 and m_2 in horizontal direction are as follows.

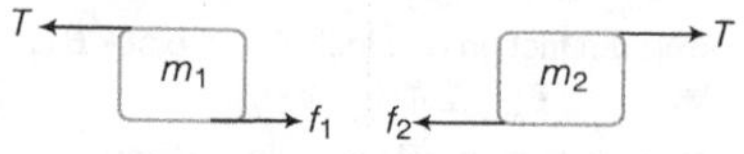

Now, there are only two possibilities :

(a) Either both m_1 and m_2 will remain stationary (w.r.t. ground) or

(b) Both m_1 and m_2 will move (w.r.t. ground)

First case is possible when

$$T \le (f_1)_{max} \text{ or } T \le 60 \text{ N}$$

and $T \le (f_2)_{max}$ or $T \le 15$ N

These conditions will be satisfied when $T \le 15$ N

say $T = 14$ N then $f_1 = f_2 = 14$ N

Therefore, the condition $f_1 = 2f_2$ will not be satisfied. Thus, m_1 and m_2 both can't remain stationary.

In the second case, when m_1 and m_2 both move

$$f_2 = (f_2)_{max} = 15 \text{ N}$$

$\therefore$ $f_1 = 2f_2 = 30$ N

Now, since $f_1 < (f_1)_{max}$, there is no relative motion between m_1 and M, i.e. all the masses move with same acceleration, say a

$\therefore$ $f_2 = 15$ N and $f_1 = 30$ N

Free body diagrams and equations of motion are as follows:

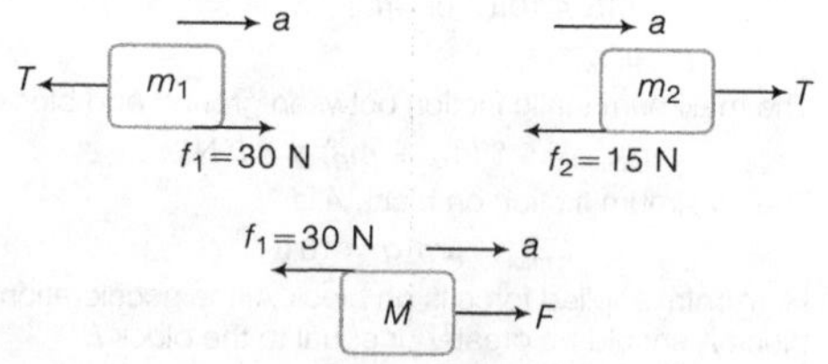

For m_1, $30 - T = 20a$ ...(i)

For m_2, $T - 15 = 5a$ ...(ii)

For M, $F - 30 = 50a$ ...(iii)

Solving these three equations, we get

$$F = 60 \text{ N}$$

$\Rightarrow$ $T = 18$ N

and $a = \dfrac{3}{5} \text{ m/s}^2$

Note *1. Friction always opposes the relative motion between two surfaces in contact.*

2. Whenever there is relative motion between two surfaces in contact, always maximum friction (kinetic) acts, but if there is no relative motion then friction force (f) may be less than its limiting value also. So, don't apply maximum force.

9. (a) Since, brick is stationary. So, net force on brick is zero.

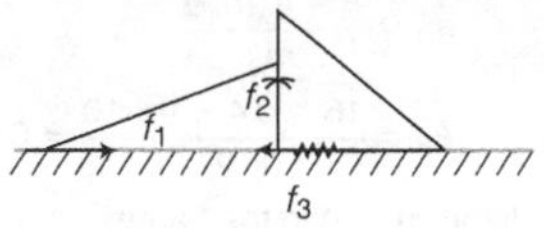

The direction of friction on brick and block is opposite.

10. (d) Free body diagram of blocks A and B is as shown below,

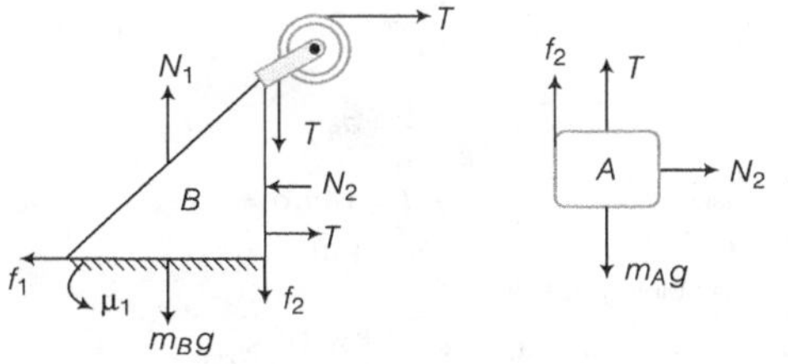

For block A, $\quad T + f_2 = m_Ag \quad$ or $\quad T + \mu_2 N_2 = m_Ag$

Since, block is in equilibrium, $N_2 = 0 \Rightarrow T = m_Ag$ ...(i)

For block B,

Balancing horizontal forces $\quad 2T = f_1$

$\Rightarrow \quad 2T = \mu_1 N_1 = 0.4\, N_1$ ...(ii)

Balancing vertical forces $\quad N_1 = f_2 + T + m_Bg$

$\Rightarrow \quad N_1 = T + m_Bg$ ...(iii)

$[\because f_2 = 0 \text{ as } N_2 = 0]$

From Eqs. (ii) and (iii), we gey

$$2T = 0.4\,(T + m_Bg)$$

$\Rightarrow \quad 1.6\,T = 0.4 \times 10 \times 10 \quad$ or $\quad T = 25$ N

From Eq. (i), $\quad m_A = \frac{T}{g} = \frac{25}{10} = 2.5$ kg

11. (a) The maximum tension in connecting box A and block C is $T_{3\,\max}$. From free body diagram of block C,

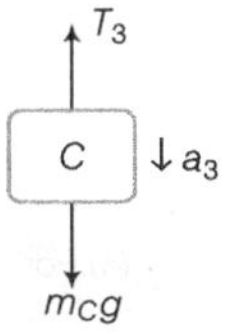

$$m_Cg - T_3 = m_Ca_3$$

or $\quad 10 - T_{3\,\max} = m_Ca_{3\,\min}$

or $\quad 10 - T_{3\max} = 0 \quad (\because a_{3\,\min} = 0)$

$\therefore \quad T_{3\,\max} = 10$ N

The maximum friction by ground on the box is

$$f_{1\max} = 0.5(1+2) \times g = 0.5 \times 30 = 15 \text{ N}$$

$\because T_{3\,\max} < f_{1\,\max}$, hence static friction acts between ground and the box.

The acceleration of block will be zero.

$\therefore \quad T_3 = m_Cg = 10$ N

For box, $T_3 = f_1$ (friction due to ground)

$\therefore \quad f_1 = 10$ N

The tension in connecting box and the block B is zero. Hence, friction between block B and box A is zero.

12. (a) Assuming static friction.

$$F - f = 5a, f = 10a, F = 15a,$$

$\therefore \quad a = 0.5 \cos^2 \alpha t, f = 5 \cos^2 \alpha t$

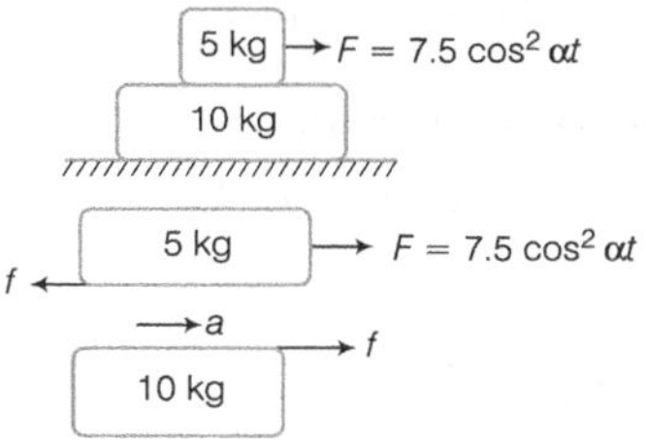

Maximum value of friction is 5 N, when $t = 0$ or $\cos \alpha t = 1$

$$f \le 5$$

$\therefore$ Static friction is applicable, so no relative motion between A and B.

13. (c) $mg = 4\,ma \Rightarrow a = g/4$

$\Rightarrow \quad f_s = ma = mg/4$

14. (b) Free body diagram of block A is as shown below,

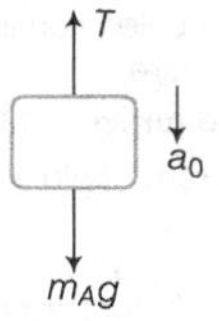

$$m_Ag - T = m_Aa_0 \qquad \text{...(i)}$$

For all the blocks moving together, common acceleration is a_0

$\Rightarrow \quad T = 15a_0$ ...(ii)

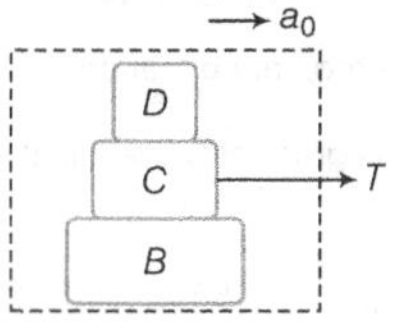

From Eqs. (i) and (ii),

$$m_Ag - 15a_0 = m_Aa_0$$

or $\quad a_0 = \frac{m_Ag}{(15 + m_A)}$

Free body diagram of block D in the frame of reference of block C.

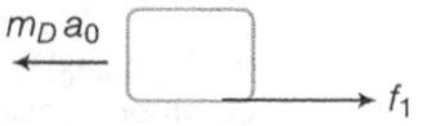

For all blocks moving together,

$$f_1 \le (f_1)_{\max}$$

or $\quad m_Da_0 \le \mu m_Dg \quad$ or $\quad a_0 \le \mu g$

$$\frac{m_Ag}{(15 + m_A)} \le \mu g$$

$$m_A \le (0.5)(15 + m_A)$$

$$m_A - 0.5m_A \le 0.5 \times 15$$

$\therefore \quad m_A \le 15$

$\therefore \quad m_{A\max} = 15$ kg

15. (a,b,c) The common acceleration of the system is $a_0 = \frac{F}{3}$

From free body diagram of block A in the frame of block B,

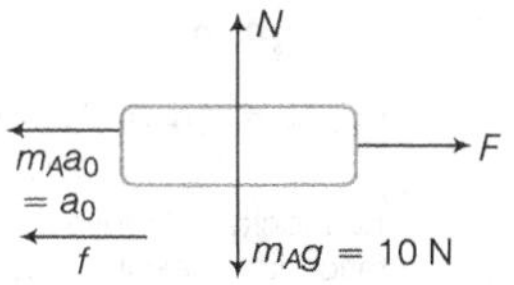

The net applied force is $F - m_Aa_0 \le \mu m_Ag$

$\Rightarrow$ Blocks move with common acceleration and the friction is static.

If $F - m_Aa_0 > \mu m_Ag$

$\Rightarrow$ Blocks have relative motion between them and friction is kinetic.

(a) Here, $\quad F - m_Aa_0 = 5 - \frac{5}{3} = \frac{10}{3}$ N

and $\quad f_{\max} = \mu m_Ag = 5$ N

$\therefore F - m_Aa_0 < f_{\max}$ common acceleration and friction is static.

$\therefore \quad a_0 = \frac{5}{3}$ m/s^2

From free body diagram of block B,

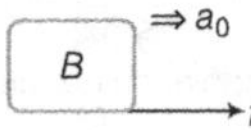

$$f = m_B a_0 = 2 \times \frac{5}{3} = \frac{10}{3} = 3.33 \text{ m/s}^2$$

(b) Here, $a_0 = 15/3 = 5 \text{ m/s}^2$

$$F - m_A a_0 = 15 - 5 = 10 \text{ N} > \mu m_A g$$

Hence, the relative acceleration is found between blocks and friction is kinetic in nature.

$\therefore$ $$f = \mu m_A g = 5 \text{ N}$$

From free body diagram of block A,

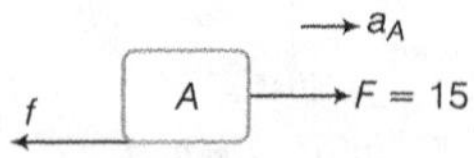

a_A in the frame of ground is found using,

$$F - f = m_A a_A \quad \text{or} \quad 15 - 5 = 1 \times a_A$$

$\therefore$ $$a_A = 10 \text{ m/s}^2$$

(c) For $F = 6$ N, common acceleration $= a_0 = \frac{F}{3} = \frac{6}{3} = 2 \text{ m/s}^2$

The free body diagram of block A in the frame of block B,

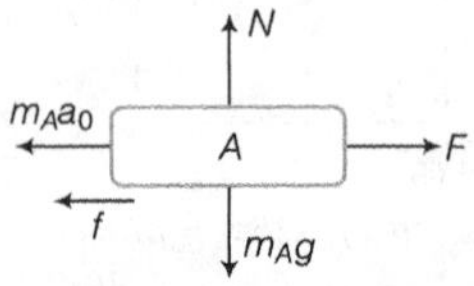

Here, $$F - m_A a_0 = 6 - 1 \times 2 = 4 \text{ N}$$

$$f_{max} = \mu m_A g = 0.5 \times 1 \times 10 = 5 \text{ N}$$

$\because$ $(F - m_A a_0) < f_{max}$, the blocks will move forward with common acceleration a_0.

$\therefore$ Friction on block $A = f = F - m_A a_0 = 4$ N

This option is also correct.

16. (a,b,d)

(a) Common acceleration $(a_0) = \frac{F}{m_A + m_B} = \frac{9}{3} = 3 \text{ m/s}^2$

Free body diagram of block A in the frame of block B is shown below,

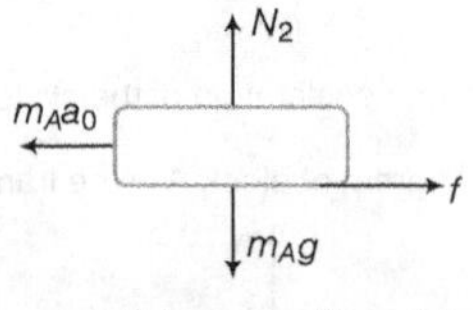

$$(f_s)_{max} = \mu_s N_2 = 0.5 \times 1 \times 10 = 5 \text{ N}$$

Also, $$m_A a_0 = 1 \times 3 = 3 \text{ N}$$

$\because$ $m_A a_0 < (f_s)_{max}$, the blocks will move forward with common acceleration and static friction exists between A and B.

$$f = m_A a_0 = 3 \text{ N}$$

(b) $$a_0 = \frac{16}{3} = 5.33 \text{ m/s}^2$$

$$m_A a_0 = 1 \times \frac{16}{3} = 5.33 \text{ N}$$

$\because$ $(m_A a_0) > (f_s)_{max}$, there will be relative motion between the blocks and kinetic friction acts between A and B.

Free body diagram of block B is shown,

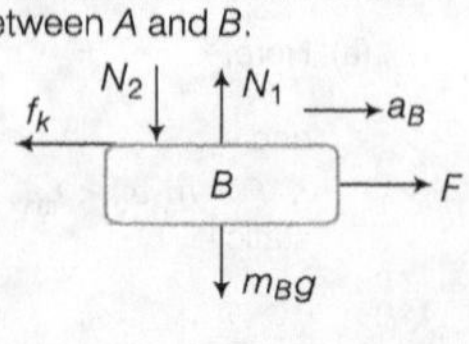

$$F - f_k = m_B a_B$$

or $$F - \mu_k N_2 = m_B a_B \quad \text{...(i)}$$

Also, $$N_2 = m_A g \quad \text{...(ii)}$$

From Eqs. (i) and (ii), we get

$$a_B = \frac{F - \mu_k m_A g}{m_B}$$

or $$a_B = \frac{16 - 0.4 \times 1 \times 10}{2} = 6 \text{ m/s}^2$$

Free body diagram of A in the frame of ground,

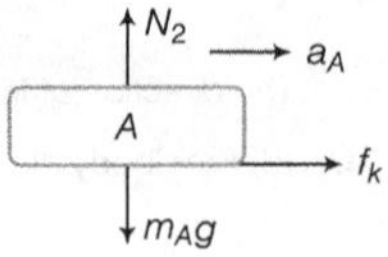

$$f_k = m_A a_A$$

or $$a_A = \frac{f_k}{m_A} = 4 \text{ m/s}^2$$

(d) $$\frac{F - \mu_k m_A g}{m_B} = a_B \quad \text{...(i)}$$

Also, $$f = \mu_k m_A g = m_A a_A$$

or $$a_A = \mu_k g \quad \text{...(ii)}$$

Combining Eqs. (i) and (ii),

Consider, $a_B > a_A \Rightarrow \frac{F - \mu_k m_A g}{m_B} > \mu_k g$

or $$F > \mu_k g\,(m_A + m_B)$$

Hence, $$a_B > a_A \text{ for } F > \mu_k g (m_A + m_B)$$

17. (a,b,c,d)

(a) Let blocks are moving with common acceleration upto $t = t_0$.

For $0 \le t \le t_0$, the common acceleration is $a_0 = \frac{3t}{1+2} = t$

The static friction on block A is

$$f = m_A a_0 = 1 \times t = t$$

$\because$ $$f \le f_{max}$$

or $$t \le \mu m_A g \quad \text{or} \quad t \le 5$$

$\therefore$ $$t_0 = 5 \text{ s}$$

If $0 \le t \le t_0$, static friction acts between blocks and they move with common acceleration.

If $t > t_0$, kinetic friction acts between blocks and acceleration of block B is greater than that of block A.

(b) $a_0 = t = 1 \text{ m/s}^2$ $\quad \left(\because \text{for } t \le 5\text{s, blocks move with common acceleration}\right)$

$$f = t = 1 \text{ N}$$

(c) For $t = 10$ s, relative acceleration takes place.

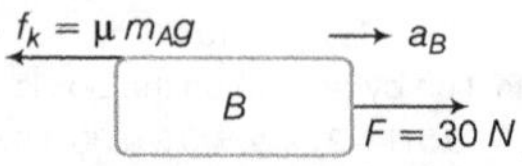

$\therefore$ From free body diagram of B,

$\because$ $$30 - \mu m_A g = m_B a_B$$

or $$30 - 0.5 \times 1 \times 10 = 2a_B$$

or $$30 - 5 = 2a_B$$

$\therefore$ $$a_B = \frac{25}{2} = 12.5 \text{ m/s}^2$$

(d) From free body diagram of A,

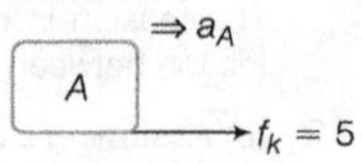

$$5 = 1 \times a_A$$

$\therefore$ $$a_A = 5 \text{ m/s}^2$$

18. (b) For $0 \le t \le t_0$, common acceleration $= \frac{dv}{dt} = a_0 = t$

$\Rightarrow$ $$v(t) = \frac{t^2}{2}$$

i.e. graph is parabolic opening upwards.

19. (a,b,c) Here, $$f_{1\,max} = \mu_s m_1 g = 10 \text{ N}$$

and $$f_{2\,max} = \mu_s m_2 g = 15 \text{ N}$$

(a) If $F = 5$ N

$\therefore F < f_{max}$, $\therefore$ $F = f_1$ and $f_1 = 5$ N

Also, $f_2 = 0$

Free body diagram of blocks are shown below,

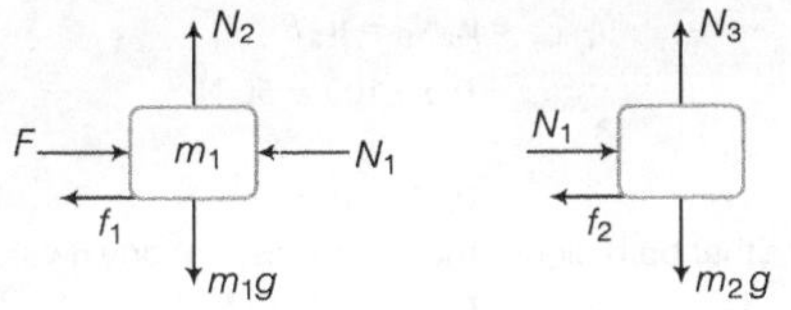

$$F = N_1 + f_1$$

$\because$ $F = 5\text{ N}$

$\Rightarrow$ $f_1 = 5\text{ N}$

Hence, $f_2 = N_1 = 0$

(b) If $F = 15$ N

Here $f_{1\max} < F < f_{1\max} + f_{2\max}$, static friction acts on both blocks.

$\therefore$ From free body diagram of block A

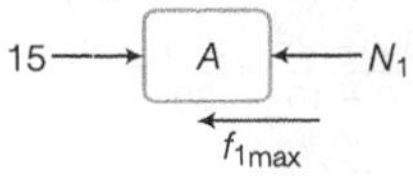

$\therefore$ $15 = N_1 + 10$

$\therefore$ $N_1 = 5\text{ N}$

From free body diagram of block B,

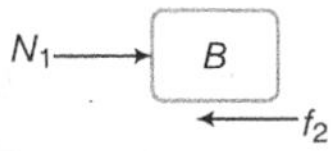

Here, $N_1 = f_2$

$\therefore$ $f_2 = N_1 = 5\text{ N}$

Here, N_1 is contact force between blocks.

(c) If $F = 40$ N

$\therefore$ $F > f_{1\max} + f_{2\max}$

Kinetic friction comes into play.

$\therefore$ $F - \mu_k m_1 g - \mu_k m_2 g = (m_1 + m_2)a$

$\therefore$ $a = \dfrac{40 - 8 - 12}{5} = 4\text{ m/s}^2$

For A,

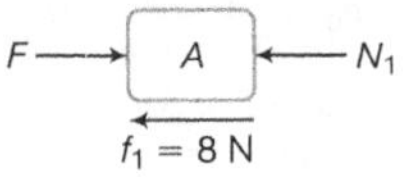

$\therefore$ $40 - N_1 - 8 = 2a$

$\because$ $N_1 = 40 - 8 - 8 = 24\text{ N}$

20. (c) $m_A g - T = m_A a$ ...(i)

$$T - \mu(m_B + m_C)g = (m_B + m_C)a \quad \text{...(ii)}$$

From Eqs. (i) and (ii),

$$m_A g - \mu(m_B + m_C)g = (m_A + m_B + m_C)a$$

$\therefore$ $a = \dfrac{m_A g - \mu(m_B + m_C)g}{m_A + m_B + m_C} = \dfrac{50 - 0.2 \times 100}{15} = 2\text{ m/s}^2$

21. (b) $m_A g - T = m_A a$

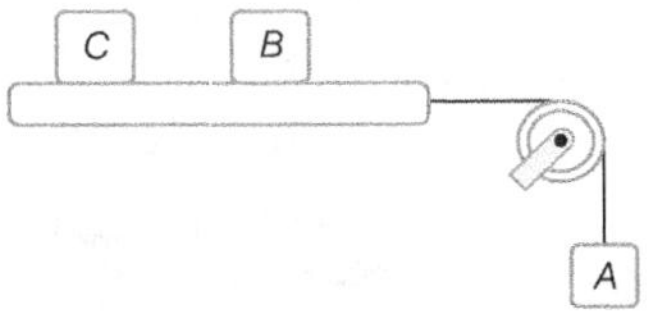

[**Note** : *The situation can be imagined like above arrangement*]

or $50 - T = 5 \times 2 = 10$

$\therefore$ $T = 50 - 10 = 40\text{ N}$

22. (d) $T_C - \mu m_C g = m_C a$

$\therefore$ $T_C = m_C a + \mu m_C g$

$= 7 \times 2 + 0.2 \times 7 \times 10$

$= 14 + 14 = 28\text{ N}$

23. (d) Here, $f_{1\max} = \mu\, m_A g = 5\text{ N}$

$$f_{2\max} = \mu m_B g = 10\text{ N}$$

The common acceleration is

$$a_0 = \frac{F_1 - F_2 + F_3}{m_A + m_B + m_C} = \frac{F_1 - F_2 + F_3}{6}$$

From free body diagram of A in the frame of C,

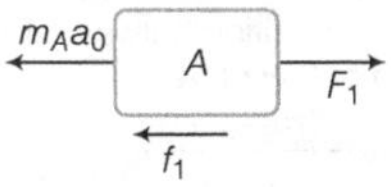

If $F_1 - m_A a_0 = f_1 \le f_{1\max}$

or $F_1 \le 5 + \dfrac{F_1 - F_2 + F_3}{6}$

or $6F_1 \le 30 + F_1 - F_2 + F_3$

or $5F_1 + F_2 - F_3 \le 30$

Static friction acts on A and A and C move together.

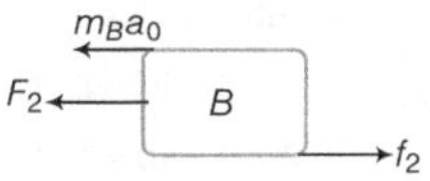

From free body diagram of B in the frame of C,

Here, $m_B a_0 + F_2 = f_2 \le f_{2\max}$

or $2a_0 + F_2 \le 10$

or $\dfrac{F_1 - F_2 + F_3}{3} + F_2 \le 10$

or $F_1 + 2F_2 + F_3 \le 30 \rightarrow$ static friction acts on B and B and C move together.

$$5F_1 + F_2 - F_3 = 4\text{ N} < 30\text{ N}$$

Here, $a_0 = \dfrac{F_1 - F_2 + F_3}{6} = \dfrac{1}{3}\text{ m/s}^2$

$f_1 = F_1 - m_A a_0 = 1 - \dfrac{1}{3} = \dfrac{2}{3}\text{ N}$ (leftwards)

(c) $f_2 = F_2 + m_B a_0 = 5 + 2 \times \dfrac{1}{3} = \dfrac{17}{3} = 5.67\text{ N}$

Think the solution of other option as above.

24. (a) Here, $5F_1 + F_2 - F_3 = 25 < 30\text{ N}$

and $F_1 + 2F_2 + F_3 = 53 > 30\text{ N}$

$\therefore$ Kinetic friction acts between B and C.

$\therefore$ $f_2 = \mu m_B g = 10\text{ N}$

The acceleration of B is

$$a_2 = \frac{F_2 - \mu m_B g}{m_B} = \frac{10}{2} = 5\text{ m/s}^2$$

25. (a) When time $\le t_0$, static friction comes into play.

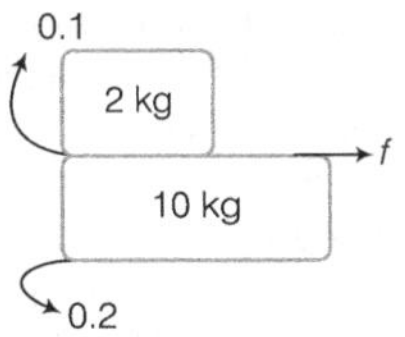

As, $t_1 < t_0$

$f = \mu mg = 0.1 \times 2 \times 10 = 2\text{ N}$ (leftwards)

26. (a) Free body diagram of block A is as shown,

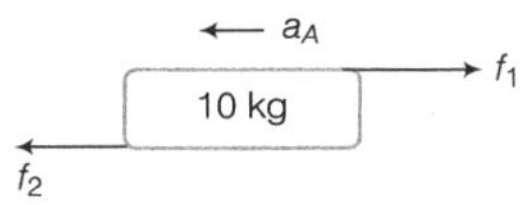

$f_1 = 0.1 \times 2 \times 10 = 2\text{ N}$

$f_2 = 0.2 \times 12 \times 10 = 24\text{N}$

$a_A = \dfrac{f_2 - f_1}{m_A} = \dfrac{24 - 2}{10}$

$= 2.2\text{ m/s}^2$

Free body diagram of B in the frame of A is

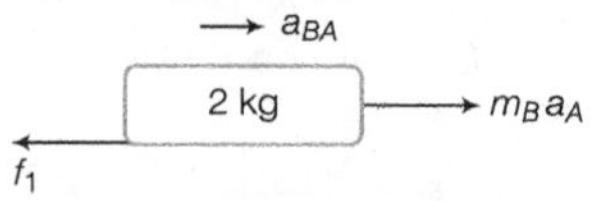

$$f_1 = m_B a_A = 2 \times 2.2$$
$$= 4.4 \text{ N} > (f_1)_{max} = 2 \text{ N}$$

Hence, there is relative motion between A and B, let a_{BA} be the relative acceleration of B w.r.t. A.

$$a_{BA} = \frac{m_B a_A - f_1}{m_B}$$
$$= \frac{2 \times 2.2 - 2}{2} = 1.2 \text{ m/s}^2$$

Using, $\quad s_{BA} = (u_{BA})t + \frac{1}{2}(a_{BA})t^2$

$\Rightarrow \quad 1.6 = \frac{1}{2} \times 1.2 \times t^2$

or $\quad t = \sqrt{2.66} = 1.63$ s

27. (6) $\quad f_{1\,max} = 0.3 \times 20 \times 10 = 60$ N

$$f_{2\,max} = 0.3 \times 5 \times 10 = 15 \text{ N}$$

For motion of m_2,

$$f_1 \geq f_2$$

But $\quad f_1 = 2f_2 \Rightarrow f_1 > f_{2\,max}$ or $f_1 > 15$ N

But $\quad f_1 = 2f_2 \Rightarrow f_1 = 30$ N

But $\quad f_{1\,max} = 60$ N

$\therefore \quad f_1 < f_{1\,max}$

It means static friction comes into play between m_1 and M. Hence, all blocks move together.

$\therefore \quad F - f_{2\,max} = (m_1 + m_2 + M)a$

or $\quad F - 15 = 75\, a_0$

For block M,

$$F - f_1 = 50\, a;\ F - 30 = 50\, a$$
$$F - 30 = 50\left(\frac{F - 15}{75}\right)$$

$\therefore \quad F = 60$ N

$\therefore \quad 10n = 60$

$\therefore \quad n = 6$

28. (4) The maximum static friction on block A is

$$f_{1\,max} = \mu_s m_1 g$$
$$= 0.6 \times 2 \times 10 = 12 \text{ N}$$

The maximum static friction between ground and the box B

$$f_{2\,max} = \mu_s(m_1 + m_2)g$$
$$= 0.6 \times 4 \times 10 = 24 \text{ N}$$

Assume both blocks move together.

$$F - \mu_K(m_1 + m_2)g = (m_1 + m_2)a$$

or $\quad F - 0.4 \times 4 \times 10 = 4a$

$\because \quad F - 16 = 4a$

For block A, $\quad f_1 = 2a = 2\left(\frac{F - 16}{4}\right)$

as $\quad f_1 = \frac{F - 16}{2}$

For just sliding, $\quad f_1 = f_{1\,max}$

$\Rightarrow \quad \frac{F - 16}{2} = 12$

$\therefore \quad F = 40 \text{ N} = 10x$

$\Rightarrow \quad x = 4$

29. (1) For B,

$$f_{2\,max} = \mu_2 N_B = \mu_2 F$$
$$f_{2\,max} = 0.5 \times 100 = 50 \text{ N}$$

For A,

$$f_{1\,max} = 0.25 \times 100 = 25 \text{ N}$$

Assume that both blocks move together net downward force is

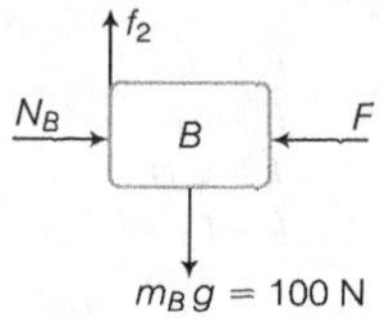

$$(30) \times 10 = 300 \text{ N}$$

$\therefore \quad 300 > f_{1\,max}$

$\therefore$ Dynamic friction comes into play at wall.

$\therefore \quad a = \frac{300 - 25}{20 + 10} = \frac{275}{30} = \frac{55}{6}$ m/s^2

For B,

$$100 - f_2 = 10\, a$$

$\Rightarrow \quad 100 - f_2 = 10 \times \frac{55}{6} = \frac{550}{6}$

$\therefore \quad f_2 = 100 - \frac{550}{6} = \frac{600 - 550}{6} = \frac{50}{6}$ N

$$f_2 = 8.33 < f_{max}$$

Hence, static friction comes into play.

$\therefore$ Both blocks move together.

$\therefore \quad$ Ratio $= 1$

$\therefore \quad n = 1$

30. (a)

(i) The common acceleration is

$$a_0 = \frac{F_1 - F_2}{m_A + m_B} = 0$$

From free body diagram of A in the frame of B.

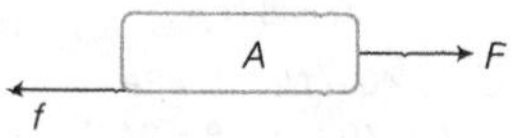

$$F = f \leq \mu m_A g$$

or $\quad F \leq 5$ N $\rightarrow$ static friction $\quad (\because F = 4$ N$)$

(ii) The common acceleration is

$$a_0 = \frac{10}{3} \text{ m/s}^2 \quad \text{(towards left)}$$

From the free body diagram of block A in the frame of block B,

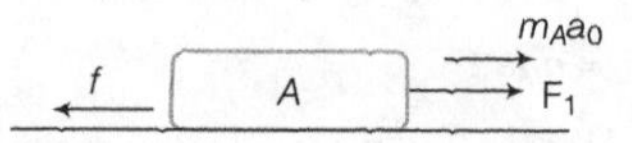

$$m_A a_0 + F_1 = f$$

$\Rightarrow \quad f = 1 \times \frac{10}{3} + 2$

$$= 5.33 \text{ N} > (f_{max} = 5 \text{ N})$$

Thus, kinetic friction is applicable.

(iii) The maximum friction on B due to A is $0.1 \times 10 = 1$ N, while the maximum friction by ground on block B is

$$0.3(1 + 2) \times 10 = 9 \text{ N}$$

Thus, B remains stationary. But $a_A = \frac{5 - 1}{1} = 4$ m/s^2

and $a_B = 0$

(iv) Common acceleration $= a_0 = \dfrac{F_2 - F_1}{(m_1 + m_2)} = \dfrac{12}{3} = 4\ m/s^2$

Free body diagram of A in the frame of B is,

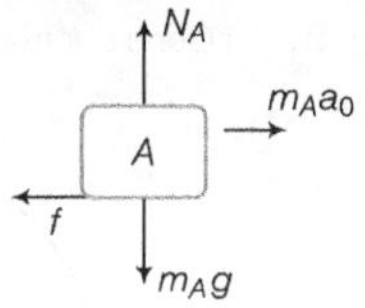

$$f = m_A a_0 = 4\ N$$

$$(f_{max}) = \mu_1 m_A g = 0.4 \times 1 \times 10 = 4\ N$$

Hence, static friction acts between blocks.

(i) → (p,s), (ii) → (q), (iii) → (q,r,s), (iv) → (p)

31. (a) (i) If $\mu_1 = \mu_2 = 0$, no frictional force exists between m_1, m_2 and plane.

Free body diagram of system

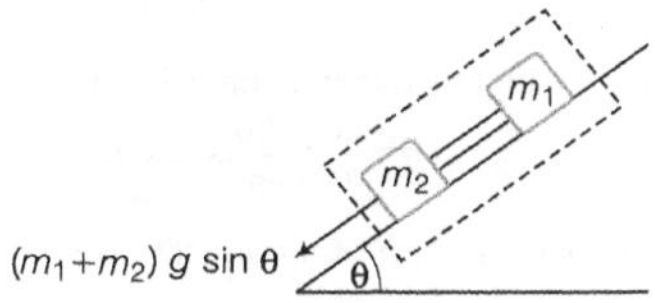

$$a = \frac{(m_1 + m_2)\ g \sin\theta}{(m_1 + m_2)} = g \sin\theta$$

(ii) and (iii)

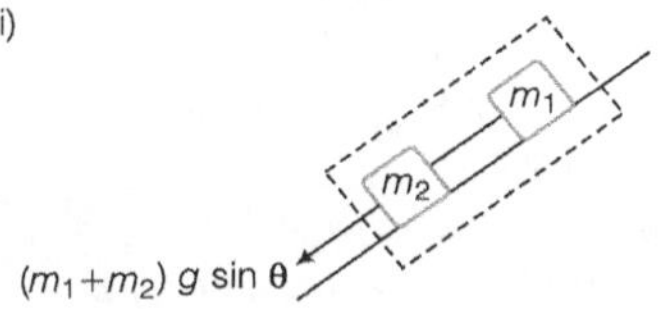

For along the inclined plane $= F$

$$= (m_1 + m_2)\ g \sin\theta = 2mg \sin\theta$$

Maximum value of friction acting on system $= (f)_{max}$

$$= (f_1)_{max} + (f_2)_{max} = (\mu_1 + \mu_2)\ mg \cos\theta$$

For (ii), given, $\sin\theta \le \dfrac{(\mu_1 + \mu_2)\cos\theta}{2}$

$$2 \sin\theta \le (\mu_1 + \mu_2) \cos\theta$$

or $\quad 2mg \sin\theta \le (\mu_1 + \mu_2)\ mg \cos\theta \quad$ or $\quad F \le (f)_{max}$

Hence, blocks remain stationary.

For (iii), given, $2mg \sin\theta > (\mu_1 + \mu_2)\ mg \cos\theta$

or $\quad F > (f)_{max}$

Hence, block will move down with acceleration less than $g \sin\theta$.

For (iv), $(f)_{max} = \mu_1 m_1\ g \cos\theta$

$$F = (m_1 + m_2)\ g \sin\theta$$

Given, $\quad (m_1 + m_2) \sin\theta \le (\mu_1 m_1 + \mu_2 m_2) \cos\theta$

$\Rightarrow \quad (m_1 + m_2)\ g \sin\theta \le \mu_1 m_1\ g \cos\theta \qquad (\because \mu_2 = 0)$

$\Rightarrow \quad F \le (f)_{max}$

Hence, blocks will remain stationary.

(i) → (q,r), (ii) → (p,q), (iii) → (s,r) (iv) → (p,r)

32. (a)

(i) Since, box B is connected with string to the wall. So, it always remains stationary.

For A, $\quad f_{1\,max} = \mu_1 m_1 g = 0.3 \times 10 \times 10 = 30\ N$

But $\quad F = 20\ N$

∴ Hence, static friction comes into play.

∴ $\quad f_1 = F = 20\ N$

Also, force on B in forward direction is $f_1 = 20$ N. But $f_{2\,max} = \mu_2 (m_1 + m_2) g = 30\ N$

∴ $\quad f_1 = f_2 = 20$

∴ $\quad T = 0$

(ii) $\quad F = f_1 \qquad (\therefore f_1 = 30\ N)$

on B, $\quad f_1 = f_2 = 30\ N$

$\quad T = 0$

(iii) $F > f_{1\,max}$

Hence, dynamic friction comes into play.

∴ $\quad F - f_{1\,max} = m_A a_1$

Free body diagram of blocks A and B are shown below.

$\Rightarrow a_1$

(i) → (p,q,r,s), (ii) → (p) (iii) → (q,r), (iv) → (p,q,s)

33. (b)

(i) Free body diagram of block B is as shown below.

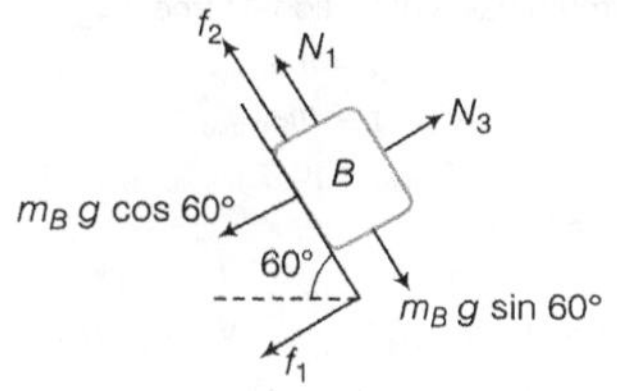

With respect to block A, block B has tendency to move up the plane, hence friction is down the plane.

(ii) Free body diagram of block A is shown below.

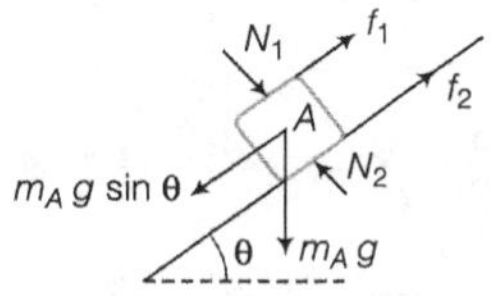

$f_1 + f_2 = m_A g \sin 30° \qquad$ (given, $\theta = 30°$)

$$= 3 \times 10 \times \frac{1}{2} = 15\ N$$

(iii) Free body diagram of incline is

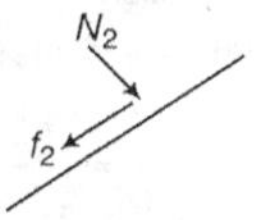

(iv) Block A is moving downward. So, friction due to in direction is up along inclined plane.

(i) → (r), (ii) → (q,s) (iii) → (r) (iv) → (q)

34. (d) Block will not slip, if

$$(m_1 + m_2)\ g \sin\theta \le \mu m_2 g \cos\theta$$

$\Rightarrow \quad 3 \sin\theta \le \left(\dfrac{3}{10}\right)(2) \cos\theta$

$\tan\theta \le \dfrac{1}{5} \quad \Rightarrow \quad \theta \le 11.5°$

(i) If $\theta = 5°$, friction is static

$$f = (m_1 + m_2)\ g \sin\theta$$

(ii) If $\theta = 10°$, friction is static

$$f = (m_1 + m_2)\ g \sin\theta$$

(iii) If $\theta = 15°$, friction is kinetic

$$f = \mu m_2 g \cos\theta$$

(iv) If $\theta = 20°$, friction is kinetic

$\Rightarrow \quad f = \mu m_2 g \cos\theta$

(i) → (q), (ii) → (q), (iii) → (r), (iv) → (r)

35. (b) Cloth can be pulled out without dislodging the dishes from the table due to inertia. The law of inertia is Newton's first law. It is not Newton's third law.

36. (b) During pushing, normal reaction increases, while during pulling normal reaction decreases. Hence during pushing, $(f_s)_{max} = \mu_s N$ increases for two surfaces in contact.

DPP-5 Friction in Daily Life

1. (i) (a) In case of walking or running static friction between the foot and ground prevent the slipping of foot in backward direction and as a result we are able to walk.

(ii) (b) $\because a = 0 \Rightarrow f = ma = 0$

(iii) (a) In case of walking or running the static friction acts in the direction of motion [as the static friction avoids the slipping of foot in backward direction while walking].

(iv) (a) In this case, the frictional force will provide the required acceleration

$$\Rightarrow \quad f = ma_0$$

(v) (a) Maximum value of frictional force, $f_{max} = \mu_s mg$

(vi) (b)

$$f_{max} = ma_{0\,max} \qquad (a_0 = a_{max})$$
$$\Rightarrow \quad \mu_s mg = ma_{0\,max}$$
$$\Rightarrow \quad a_{0\,max} = \mu_s g$$

(vii) (b) $a_{max} = \mu_s g$

Using, $v = u + at$

$$v = 0 + a_{max} t_{min} \Rightarrow v_0 = (\mu_s g) t_{min}$$
$$t_{min} = \frac{v_0}{\mu_s g}$$

(viii) (b) $a_{max} = \mu_s g = 0.8 \times 10 = 8 \text{ m/s}^2$

$$u = 0, s = 100 \text{ m}$$

Using, $s = ut + \frac{1}{2}at^2 = \frac{1}{2}a_{max}t_{min}^2$

$$100 = \frac{1}{2} \times 8 \times t_{min}^2 \Rightarrow 100 = 1 \times 4 \times t_{min}^2$$
$$t_{min} = 5 \text{ s}$$

(ix) (a) In this case, initial velocity $v_0 = u + a_{max} t_{min}$

$$v_0 = 8 \times 5 = 40 \text{ m/s}$$

and final velocity, $v = 0$

Retardation in this case is provided by static friction acting in opposite direction of motion.

Maximum retardation, $a_{max} = -\mu_s g = -0.8 \times 10 = -8 \text{ m/s}^2$

$$v = v_0 + a_{max} t_{min}$$
$$0 = 40 - 8 \times t_{min}$$
$$\Rightarrow \quad t_{min} = 5 \text{ s}$$

(x) (c) In this case, the static friction will act in the opposite direction of her velocity.

2. (c) Acceleration of the car, $a = \dfrac{mg \sin\theta - \mu_s mg \cos\theta}{m}$

$\therefore \quad a = g\sin\theta - \mu_s g \cos\theta$

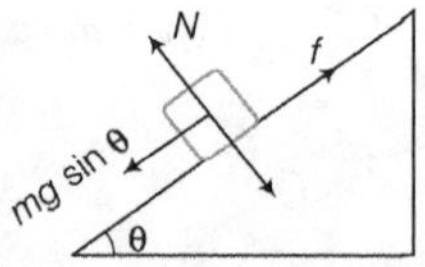

Using equation of motion, we have

$$v^2 = v_0^2 + 2as$$
$$= v_0^2 + 2gl(\sin\theta - \mu_s \cos\theta)$$
$$\therefore \quad v = \sqrt{v_0^2 + 2gl(\sin\theta - \mu_s\cos\theta)}$$

3. (a) As, $a_{max} = \dfrac{f_{max}}{m} = \dfrac{\mu\, mg}{m}$

and $\quad a = \dfrac{f}{m} \Rightarrow \int_{v_0}^{v} v.dv = \mu_0 g \int_0^x x\,dx \quad \left(\text{as, } a = \dfrac{dv}{dx}\cdot\dfrac{dx}{dt}\right)$

or $\quad \dfrac{v^2}{2} - \dfrac{v_0^2}{2} = \mu_0 g \dfrac{x^2}{2}$ or $v^2 = v_0^2 + \mu_0 g x^2$

$$\therefore \quad v = \sqrt{v_0^2 + \mu_0 g x^2}$$

4. (a) Retardation due to friction, $a = \dfrac{f}{m}$

or $\quad a = \dfrac{\mu mg}{m} \Rightarrow a = \mu g$

Now, using $v^2 = u^2 - 2as \Rightarrow 0 = 10^2 - 2\mu gs$

$$\therefore \quad s = \frac{100}{2 \times 0.5 \times 10} = 10 \text{ m}$$

5. (d) Consider free body diagram of insect,

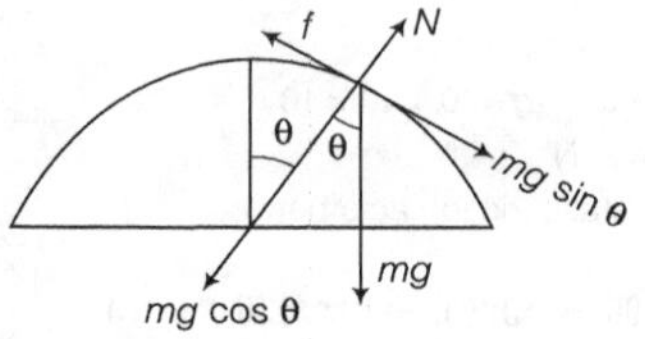

Here, $mg \sin\theta = f \le \mu_s N$

or $\quad mg\sin\theta \le \mu_s mg \cos\theta$ or $\tan\theta \le \mu_s$

$$\therefore \quad \theta_{max} = \frac{\pi}{6} \qquad \left(\because \mu_s = \frac{1}{\sqrt{3}}\right)$$

Distance travelled by the insect before sliping,

$$s = r\theta = \frac{\pi}{6} \times \frac{30}{\pi} = 5 \text{ cm}$$

6. (a) For the proper crawling of the insect up on a hemispherical surface,

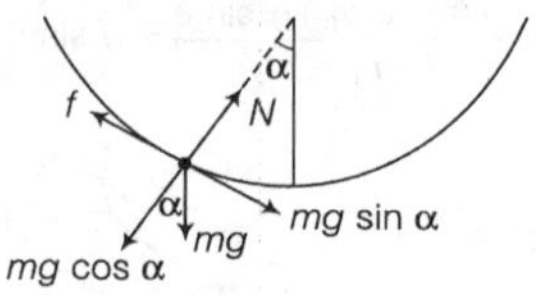

we can write

$$mg\sin\theta \le f \Rightarrow mg\sin\alpha = f_{max}$$
$$\Rightarrow \quad mg\sin\alpha_{max} = \mu mg\cos\alpha_{max}$$
$$\text{or} \quad \frac{\sin\alpha_{max}}{\cos\alpha_{max}} = \mu \quad \text{or} \quad \tan\alpha_{max} = \frac{1}{3}$$
$$\Rightarrow \quad \cot\alpha_{max} = 3$$

7. (d) For bug, $\quad f = ma$

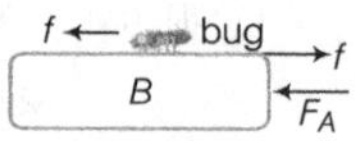

For plank, $f = F_A$ and for sugar cube,

$$F_A = f_A \le \mu m_A g \quad \text{or} \quad F_A \le \mu m_A g$$

or $\quad ma \le \mu m_A g$ or $a \le \dfrac{\mu m_A g}{m}$

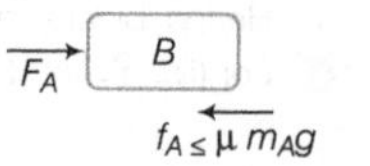

$$\therefore \quad a_{max} = \frac{0.001 \times 4 \times 10}{1}$$
$$= 0.04 \text{ m/s}^2 = 4 \text{ cm/s}^2$$

8. (c) For bug A, $\mathbf{f}_1 = (3 \times 1)\,\hat{\mathbf{i}} = (3 \text{ dyne})\,\hat{\mathbf{i}}$

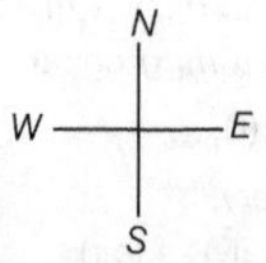

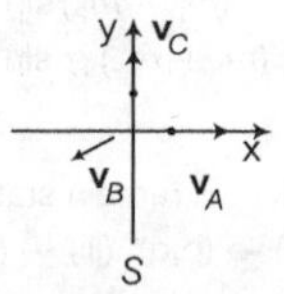

For bug C, $\mathbf{f}_3 = (2 \times 2)\,\hat{\mathbf{j}} = (4 \text{ dyne})\,\hat{\mathbf{j}}$

For cardboard to be stationary,

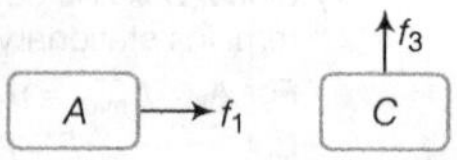

$$-\mathbf{f}_1 - \mathbf{f}_2 - \mathbf{f}_3 = 0$$

($\because$ cardboard is stationary)

$$\therefore \quad \mathbf{f}_2 = -(\mathbf{f}_1 + \mathbf{f}_3)$$

or $\quad 5\mathbf{a}_2 = -(3\hat{\mathbf{i}} + 4\hat{\mathbf{j}})$

$$\therefore \quad \mathbf{a}_2 = -\frac{3}{5}\hat{\mathbf{i}} - \frac{4}{5}\hat{\mathbf{j}}$$

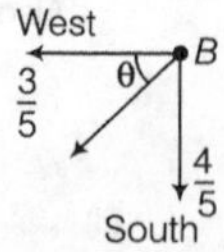

$$\therefore \quad a_2 = \sqrt{\left(-\frac{3}{5}\right)^2 + \left(-\frac{4}{5}\right)^2} = 1 \text{ cm/s}^2$$

and $\quad \tan\theta = \dfrac{4}{5 \times \dfrac{3}{5}} = \dfrac{4}{3} \Rightarrow \theta = 53°$

9. (a) From free body diagram of monkey,

$$mg = f \quad (\because a = 0)$$
$$\therefore \quad f = 10 \times 10 = 100 \text{ N}$$

10. (a) For winning of team A over team B, the frictional force between team A and ground should be greater than the frictional force between team B and the ground.

$$\Rightarrow \quad \mu_1 N_A > \mu_2 N_B \quad \Rightarrow \quad \mu_1 m_1 g > \mu_2 m_2 g$$
$$\Rightarrow \quad \mu_1 m_1 > \mu_2 m_2$$

11. (b) As man is standing in non-inertial frame, it will experience a Pseudo force (F_p). This force will act in the direction of velocity (as shown in the figure), because the moving train is decelerating.

Hence, the friction will act in opposite direction of motion of train to keep the man in equilibrium.

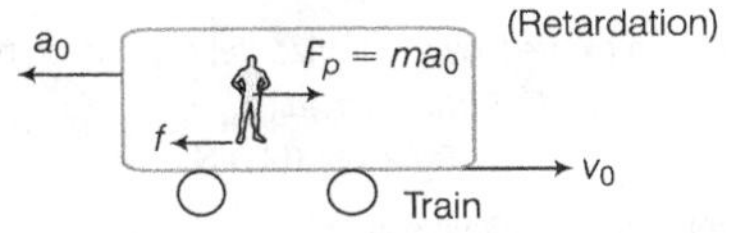

$F_p \rightarrow$ Pseudo force acting on the man.

12. (a) The wagon is certainly not pulled by the horse when the frictional force between the horse and ground is less than the frictional force between wagon and ground.

Hence, $f_{1\,max} < f_{2\,max}$ or $\mu_1 m_1 g < \mu_2 m_2 g$

$$\therefore \quad \mu_1 m_1 < \mu_2 m_2$$

or $\mu_1 < \mu_2$ and $m_1 < m_2$

13. (a)

14. (a,d)

(a) The friction between the legs of horse and horizontal ground is static in nature.

$$\because \quad v = u + at \quad (\because u = 0)$$
$$\text{or} \quad v = at = \frac{f}{m}t$$
$$\therefore \quad t = \frac{mv}{f}$$
$$\text{or} \quad t_{min} = \frac{mv}{f_{max}} = \frac{mv}{\mu_s mg} = \frac{v}{\mu_s g} = \frac{16}{8} = 2\text{ s}$$

(d) The distance travelled to achieve maximum speed in 2 s is

$$s_1 = \frac{1}{2} \times \mu_s g t^2 \quad \left(\because a = \frac{\mu_s mg}{m}\right)$$
$$= \frac{1}{2} \times 8 \times 2^2 = 16\text{ m}$$

The remaining distance (112 − 16) m is travelled by the horse with maximum speed of 16 m/s.

$$\therefore \quad 112 - 16 = 16t_2$$
$$\therefore \quad t_2 = \frac{96}{16} = 6\text{ s}$$

Thus, $t = t_1 + t_2 = 2 + 6 = 8$ s

15. (a,b,c) (a) When $a = 0$

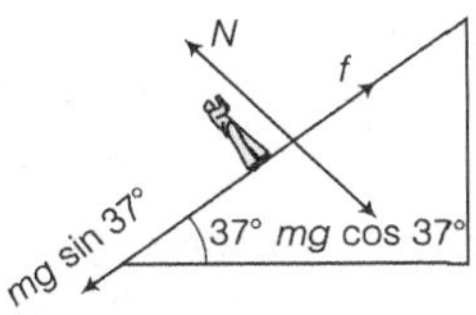

$$\therefore \quad f = mg \sin 37^\circ$$
$$f = 20 \times 10 \times \frac{3}{5} = 120\text{ N}$$

(b) $f - mg \sin 37^\circ = ma$

$$\therefore \quad f = 120 + 20 \times 1 = 140\text{ N}$$

(c) $f - mg \sin 37^\circ = ma$

$$\text{or} \quad f_{max} - mg \sin 37^\circ = ma_{max}$$
$$\text{or} \quad \mu_s mg \cos 37^\circ - mg \sin 37^\circ = ma_{max}$$
$$\text{or} \quad 0.9 \times 10 \times \frac{4}{5} - 10 \times \frac{3}{5} = a_{max}$$
$$\therefore \quad a_{max} = 1.2\text{ m/s}^2$$

16. (a,b,c) (a) The static friction exerted by plank on man provides the acceleration to man. Thus, $f = ma = 50 \times 2 = 100$ N

(b) The acceleration of plank is $a = \dfrac{f}{m} = \dfrac{100}{100} = 1\text{ m/s}^2$

(c) $f \le \mu_s N$ or $100 \le \mu \times 500$

$$\therefore \quad \mu \ge 0.2$$
$$\therefore \quad \mu_{min} = 0.2$$

17. (a,b) (a) From free body diagram of rope $T = f$

The static friction on the rope due monkey is in downward direction.

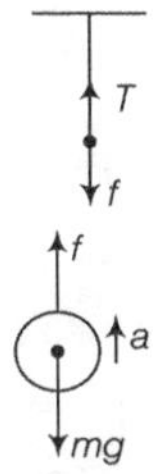

(b) From free body diagram of monkey,

$$f - mg = ma$$
$$\therefore \quad f = ma + mg = (10 \times 2) + (10 \times 10) = 120\text{ N}$$

18. (a) For cat A,

For the block to be stationary $\rightarrow f_1 \cos 60^\circ + N_2 \cos 30^\circ = f_2 \cos 60^\circ + N_1 \cos 30^\circ$

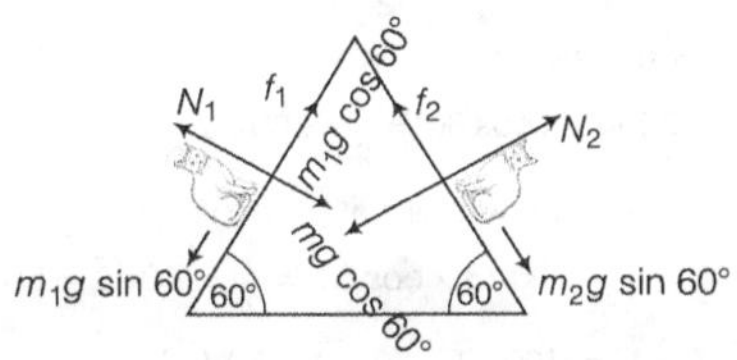

$$\text{or} \quad \frac{f_1}{2} + m_2 g \cos 60^\circ \cos 30^\circ = \frac{f_2}{2} + m_1 g \cos 60^\circ \cos 30^\circ$$

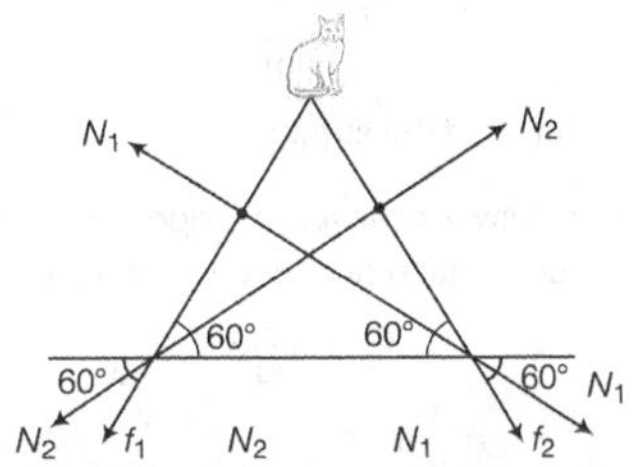

$$\text{or} \quad \frac{f_1}{2} + 100\sqrt{3} = \frac{f_2}{2} + 50\sqrt{3} \quad \text{...(i)}$$

But $\quad f_{1\,max} = \mu\, m_1 g \cos 60^\circ$

$$= 2 \times 20 \times 10 \times \frac{1}{2} = 200\text{ N} \quad \text{...(ii)}$$

and $\quad f_{2\,max} = 2 \times 40 \times 10 \times \dfrac{1}{2} = 400\text{ N} \quad$ ...(iii)

From analysis of Eqs. (i), (ii) and (iii), we get

$$\frac{f_{1\,max}}{2} + 100\sqrt{3} = \frac{f_2}{2} + 50\sqrt{3}$$
$$\text{or} \quad 100 + 100\sqrt{3} - 50\sqrt{3} = \frac{f_2}{2}$$
$$\text{or} \quad f_2 = 200 + 100\sqrt{3}$$

For A, $\quad f_{1\,max} - m_1 g \sin\theta = m_1 a_1$

$$\text{or} \quad 200 - 100\sqrt{3} = 20a_1$$
$$\therefore \quad a_1 = 10 - 5\sqrt{3}$$

For B,

$$f_{2\,max} - m_2 g \sin\theta = m_2 a_2$$
$$\text{or} \quad 200 + 100\sqrt{3} - 200\sqrt{3} = 40a_2$$
$$\text{or} \quad 200 - 100\sqrt{3} = 40a_2$$
$$\therefore \quad a_2 = 5 - 2.5\sqrt{3}$$
$$\therefore \quad a_1 > a_2$$

Thus, cat A will catch the rat first.

19. (d) $\frac{a_1}{a_2} = \frac{10 - 5\sqrt{3}}{5 - 2.5\sqrt{3}} = 2$

20. (d) At the top of the triangular block both the cats (A and B) are at rest. This means the component of weight of each cat along the inclined plane is balanced by static friction.

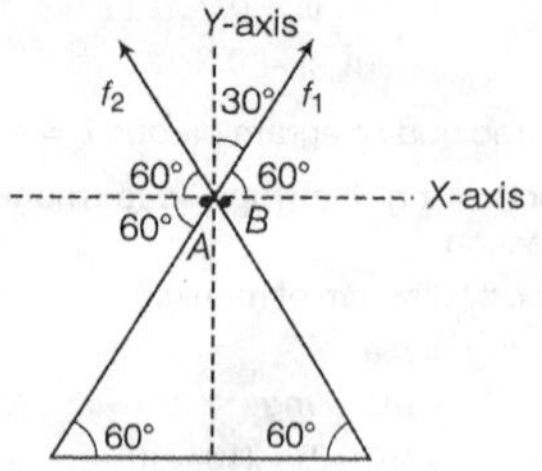

$\Rightarrow \quad f_1 = m_A g \sin 60° \text{ and } f_2 = m_B g \sin 60°$

or $\quad f_1 = 20 \times 10 \times \frac{\sqrt{3}}{2} = 100\sqrt{3}$ N

and $\quad f_2 = 40 \times 10 \times \frac{\sqrt{3}}{2} = 200\sqrt{3}$ N

Horizontal component of f_2 is

$$f_2' = 200\sqrt{3} \times \cos 60° = \frac{1}{2} \times 200\sqrt{3} = 100\sqrt{3} \text{ N}$$

and horizontal component of f_1 is

$$f_1' = 100\sqrt{3} \times \cos 60° = \frac{1}{2} \times 100\sqrt{3} = 50\sqrt{3} \text{ N}$$

$$f_2' - f_1' = (100\sqrt{3} - 50\sqrt{3}) = 50\sqrt{3} \text{ N}$$

21. (6) The acceleration of the bug is

$$a_1 = \frac{f}{m} = \frac{f}{m} = \frac{f}{3}$$

and the acceleration of the straw is $a_2 = \frac{f}{1}$

Since, bug and straw are moving in opposite direction.

$\therefore$ The relative acceleration of bug with respect to straw is

$$a_{rel} = a_1 + a_2 = \frac{f}{3} + f = \frac{4f}{3}$$

$\because \quad s_{rel} = \frac{1}{2} a_{rel} t^2$

$\therefore \quad t = \sqrt{\frac{2s_{rel}}{a_{rel}}} = \sqrt{\frac{8}{\frac{4f}{3}}}$

$\therefore \quad t_{min} = \sqrt{\frac{24}{4f_{max}}} = \sqrt{\frac{24}{4 \times \mu_s mg}}$

$\quad = \sqrt{\frac{24}{4 \times 0.05 \times 3 \times 1000}}$

$\Rightarrow \quad t_{min} = \sqrt{600} = \sqrt{100n}$

$\Rightarrow \quad n = 6$

22. (3) Let static fractional forces make a common angle θ with the vertical wall. For static equilibrium,

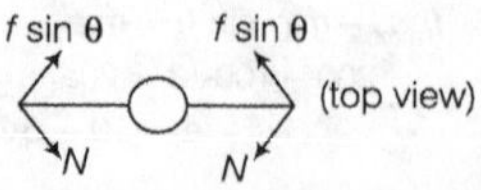

$$mg = 2 f \cos \theta \text{ and } N = f \sin \theta$$

$\therefore \quad N = \frac{1}{2} mg \tan\theta$

$\therefore$ Total force required is

$$F = \sqrt{N^2 + f^2} \quad \text{or} \quad F^2 = \left(\frac{mg}{2}\right)^2 \left(\frac{1 + \sin^2 \theta}{\cos^2 \theta}\right)$$

But $\quad f \le \mu N \Rightarrow \sin \theta \ge \frac{1}{\mu_0}$

$\therefore \quad (\sin \theta)_{min} = \frac{1}{\sqrt{2}} \Rightarrow \theta_{min} = 45°$

$\therefore \quad F^2 = \frac{100\sqrt{3}}{2}\left(\frac{1 + \sin^2 45}{\cos^2 45}\right)$

$\Rightarrow \quad F = 15$ N

But $\quad F = 3x \Rightarrow 3x = 15$

$\therefore \quad x = 3$

23. (a) 24. (b)

25. (i) Since, horse rider is moving with the acceleration of the horse. Pseudo force on horse rider is balanced by frictional force between the horse and horse rider.

$\therefore \quad f = m_1 a_0$ or $f_{max} = m_1 a_{0\,max}$

or $\quad \mu_s m_1 g = m_1 a_{0\,max}$

$\because \quad a_{0\,max} = \mu_s g = 0.5 \times 10 = 5 \text{ m/s}^2$

(ii) The friction exerted by the horizontal surface on horse,

$$f' = (m_1 + m_2) a_{0\,max}$$
$$= 300 \times 5 = 1500 \text{ N}$$

26. Here, acceleration of bicycle is

$$a = \frac{\frac{mg}{5} - \frac{mg}{20}}{m} = \frac{g}{5} - \frac{g}{20}$$
$$= 2 - \frac{1}{2} = 1.5 \text{ m/s}^2$$

(i) $v = at = 1.5 \times 10 = 15$ m/s $\quad (\because u = 0)$

(ii) $s = \frac{1}{2} at^2 = \frac{1}{2} \times 1.5 \times 4 \times 3600$

$= 10800 \text{ m} = 10.8$ km

27. (i) 3 m/s^2

(ii) 5 m/s^2

From free body diagram of boy,

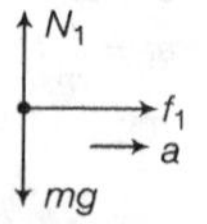

$f_1 = ma$ and $N_1 = mg = 100$ N

Free body diagram of plank is

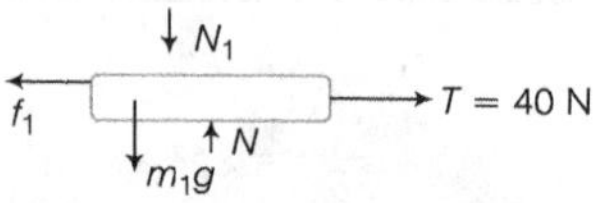

Here, $\quad N_1 + m_1 g = N$

$\therefore \quad N = 100 + 100 = 200$ N

$$f_{max} = \mu_s N = 200 \times 0.05 = 10 \text{ N}$$

If a is minimum, f_1 is minimum.

Thus, $T > f_1$. So, friction on the plank due to ground is leftward.

$\therefore \quad f + f_1 = T = 40$ or $f_1 = 40 - f$

or $\quad ma = 40 - f$

or $\quad ma_{min} = 40 - f_{max} = 40 - 10 = 30$

$\therefore \quad a_{min} = \frac{30}{10} = 3 \text{ m/s}^2$

If a is maximum, f_1 is maximum.

Thus, $f_1 > T_0$. So, friction on the plank due to ground acts rightward.

$\therefore \quad f_1 = T + f$

or $\quad ma_{max} = 40 + f_{max}$

$\quad = 40 + 10 = 50$

$\therefore \quad a_{max} = \frac{50}{10} = 5 \text{ m/s}^2$

28. (i) Consider free body diagram of rat as shown below

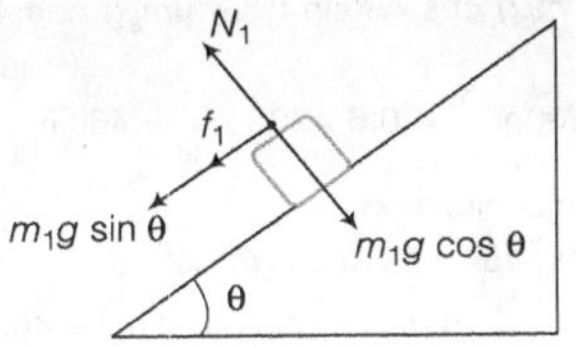

$$f_1 + m_1g \sin\theta = m_1a \Rightarrow f_1 = m_1a - m_1g \sin\theta$$

Free body diagram of plate,

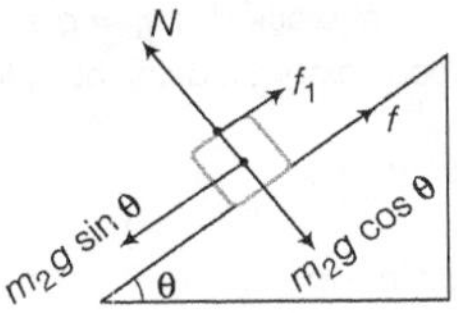

$$m_2g \sin\theta = f_1 + f$$

or $m_2g \sin\theta = m_1a - m_1g \sin\theta + f$

or $m_1 a_{min} = m_2g \sin\theta + m_1g \sin\theta - f_{max}$

$$= m_2g \sin\theta + m_1g \sin\theta - \mu(m_1 + m_2)g \cos\theta$$

$$= \frac{(m_2 + m_1)}{m_1} g (\sin\theta - \mu \cos\theta)$$

$\Rightarrow$ $a_{min} = 60\,(0.8 - 0.3) = 60 \times 0.5 = 30 \text{ m/s}^2$

(ii) For maximum acceleration,

$m_2g \sin\theta + f = f_1$ (for plank)

or $m_2g \sin\theta + f = m_1a - m_1g \sin\theta$

or $m_1a_{max} = m_2g \sin\theta + m_1g \sin\theta + \mu(m_1 + m_2)\, g \cos\theta$

or $$a_{max} = \left(\frac{m_2 + m_1}{m_1}\right) g (\sin\theta + \mu \cos\theta)$$

$$= 60\,(0.8 + 0.3) = 66 \text{ m/s}^2$$

Revisal Problems (JEE Main)

1. (c) The acceleration is zero, so net force on pillow is zero.

$\therefore$ $\mathbf{F} + \mathbf{f}_1 = 0$

2. (d) Free body diagram of the block, when it is moving towards right.

$$T + f = 10a \quad \text{...(i)}$$

or $T + 4 = 10a$ and $m_Bg - T = 2a$

$$20 - T = 2a \quad \text{...(ii)}$$

$\therefore$ $a = \frac{24}{12} = 2 \text{ m/s}^2$

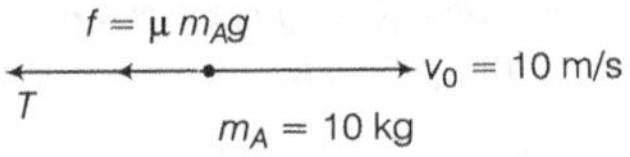

Using $v = u + at \Rightarrow 0 = 10 - 2 \times t$

$$t = \frac{10}{2} = 5 \text{ s}$$

When the velocity of the block becomes zero, it will move towards left as

$$T > f_{max}$$

Free body diagram of block,

$$T - f = 10a'$$

$T - 4 = 10a'$ and $m_Bg - T = m_Ba'$

or $20 - T = 2a'$

$\therefore$ $a' = 4/3 \text{ m/s}^2$

This acceleration will increase the velocity of the block in leftward direction.

3. (b) The maximum friction on block C is

$$f_{3\,max} = \mu_3 mg = 0.9 \times 50 \times 10 = 450 \text{ N}$$

But $T_2 = 500$ N

Thus, the friction $f_3 = 450$ N and tension in string connecting B and C is

$$T_3 = T_2 - f_3 = (50 \times 10) - 450 = 50 \text{ N}$$

$\therefore$ The block C is stationary.

$\therefore$ $T' = T_1 = 50$ N

$$f_{1max} = 50 \times 10 \times 0.3 = 150 \text{ N}$$

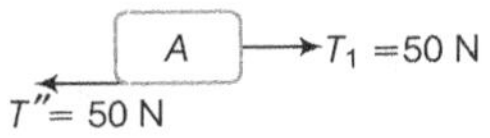

For A, $f_{1\,max} = 50 \times 10 \times 0.3 = 150$ N

$T'' = 50$ N

$\because$ $T = T''$

$\therefore$ $f_1 = 0$

4. (a) Free body diagram of block,

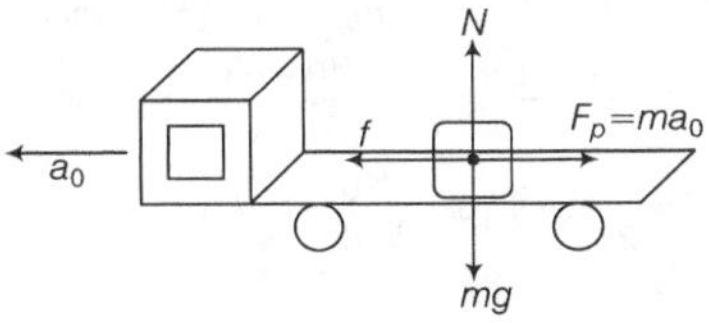

Pseudo force acting on the block $= F_P = ma_0 = 1 \times 4 = 4$ N

$\Rightarrow$ $f_{max} = \mu_s mg = 0.3 \times 1 \times 10 = 3$ N

As $F_P > f_{max}$, the block slides towards right as the truck will accelerate towards left (see figure) this means the kinetic friction will act on the block. Let a be the acceleration of the block in the frame of ground.

$\therefore$ $\mu_k mg = ma$

$\therefore$ $a = \mu_k g = 0.2 \times 10 = 2 \text{ m/s}^2$

Using, $s = ut + \frac{1}{2}at^2$ $(\because u = 0)$

$$s = \frac{1}{2} \times 2 \times (1)^2 = 1 \text{ m}$$

5. (b) At maximum speed, acceleration of the block, $a = 0$

Thus, $F - \mu mg = 0$

or $10 - x - 5 = 0$ or $5 = x$

$\therefore$ $x = 5$ m

The acceleration of the block is

$$a = \frac{10 - x - 5}{1} = 5 - x$$

or $\frac{vdv}{dx} = 5 - x$ or $\int_0^v v\,dv = 5\int_0^5 dx - \int_0^5 x\,dx$

$\Rightarrow$ $$\frac{v^2}{2} = 5[x]_0^5 - \left[\frac{x^2}{2}\right]_0^5$$

or $$\frac{v^2}{2} = 25 - \frac{25}{2} = \frac{25}{2}$$

or $v = 5$ m/s

6. (d) When length x of the rope is in rough surface, then the frictional force,

$$f = \frac{\mu m}{L} xg$$

$\because$ $$a = -\frac{\mu xg}{L} \Rightarrow \frac{vdv}{dx} = \frac{-\mu xg}{L}$$

or $$\int_{v_0}^{0} v\,dv = -\frac{\mu g}{L}\int_0^L x\,dx \quad \text{or} \quad -\frac{v_0^2}{2} = -\frac{\mu g}{L}\frac{L^2}{2}$$

$\therefore$ $$v_0 = \sqrt{\mu g L}$$

Thus, $$v_0 = \sqrt{0.5 \times 10 \times 5} = 5 \text{ m/s}$$

7. (a) Acceleration of the block, when it is in contact of spring, $a = -\left(\frac{kx + \mu mg}{m}\right)$

or $$\int_{\sqrt{2}}^{0} - v\,dv = \int_0^{x_{max}} \frac{k}{m} x\,dx + \mu g \int_0^{x_{max}} dx$$

or $$1 = \frac{x_{max}^2}{2} + \frac{x_{max}}{2} \Rightarrow x_{max} = 1 \text{ m}$$

8. (a) Normal reaction on each block is $N = ma_0 = 5$ N

FBD of block A FBD of block B

$$a_{rel} = a_A + a_B = \left(\frac{m_A g - \mu N}{m_A}\right) + \left(\frac{F - mg - \mu N}{m_B}\right)$$

$$= \left(\frac{10-5}{1}\right) + \left(\frac{20-10-5}{1}\right) = 10 \text{ m/s}^2$$

$\therefore$ $$s_{rel} = \frac{1}{2} a_{rel} t^2 \quad \text{or} \quad \sqrt{\frac{10}{10}} = t$$

$\therefore$ $$t = 1 \text{s}$$

9. (a) For equilibrium of block,

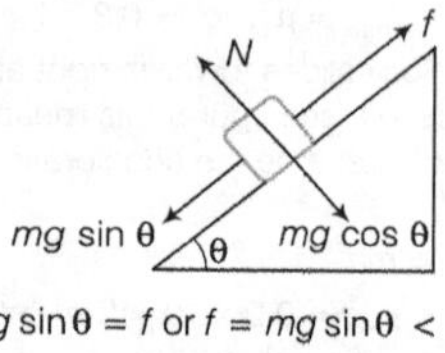

$$mg\sin\theta = f \text{ or } f = mg\sin\theta < mg$$

$\therefore$ $$f < mg \Rightarrow f < 10 \text{ N}$$

and $$N = mg\cos\theta < mg$$

$\therefore$ $$N < mg \Rightarrow N < 10 \text{N}$$

10. (c) As, $s = \frac{1}{2}at^2$ or $\frac{100}{100} = a$

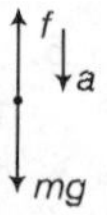

$\therefore$ $$a = 1 \text{ m/s}^2$$

Free body diagram of monkey gives

$\therefore$ $$mg - f = ma \quad \text{or} \quad 500 - f = 50 \times 1$$

$\therefore$ $$f = 450 \text{ N}$$

11. (c) From free body diagram of the wedge, we have

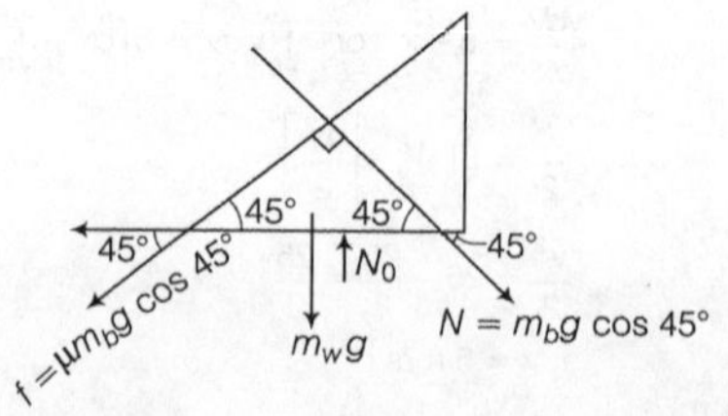

Normal reaction between wedge and ground

$$N_0 = m_w g + m_b g \cos 45° \sin 45° + \mu m_b g \cos 45° \sin 45°$$

($\because$ in vertical direction)

$$= 300 + 200 \times \frac{1}{2} + 0.6 \times 200 \times \frac{1}{2} \approx 460 \text{ N}$$

and in horizontal direction,

$$m_b g \cos^2 45° = f + \mu m_b g \cos^2 45° \quad \text{or} \quad f = 40,$$

But $$f \le \mu_s N_0 \quad \text{or} \quad 40 \le \mu_s \times 460$$

$\therefore$ $$\mu_{s\,min} \approx \frac{40}{460} = \frac{2}{23}$$

12. (d) The acceleration of block B is $a_0 = g \sin\theta$ along the inclined plane. From free body diagram of the baby in the frame triangular block B.

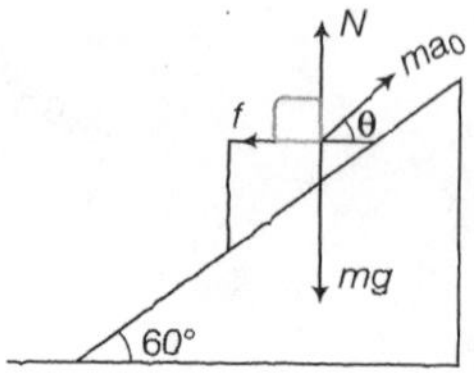

$$ma_0 \cos\theta = f$$

and $$N + ma_0 \sin\theta = mg$$

or $$N + mg\sin^2\theta = mg$$

or $$N = mg\cos^2\theta$$

Thus, $$ma_0\cos\theta = f \le \mu_s N$$

or $$ma_0\cos\theta \le \mu_s\, mg\cos^2\theta$$

or $$\sin\theta\cos\theta \le \mu_s \cos^2\theta$$

$$(a_0 = g\sin\theta)$$

$\therefore$ $$\mu_s \ge \tan\theta$$

$\therefore$ $$\mu_{s\,min} = \tan 60°$$

$$\mu_{s\,min} = \sqrt{3}$$

13. (d) For block C, $m_C g \sin 37° = 20 \times \frac{3}{5} = 12$ N

and $$f_{3\,max} = \mu_3 m_C g \cos 37° = 1.6 \text{ N}$$

$\because$ $$m_C g \sin 37 > f_{3\,max},$$

Kinetic friction comes into play. Thus,

$$m_C g \sin 37° - \mu_3 m_C g \cos 37° = m_C a_3$$

or $$12 - 1.6 = 2a_3 \quad \text{or} \quad a_3 = 5.2 \text{ m/s}^2$$

For block A, $$m_A g \sin\theta = 5 \times 10 \times \frac{3}{5} = 30 \text{ N}$$

$\Rightarrow$ $$f_{1\,max} = \mu_1 (m_B + m_C) g\cos\theta$$

$$= 0.5 \times 120 \times \frac{4}{5} = 48 \text{ N}$$

Here, $m_A g \sin\theta < f_{1\,max}$, static friction acts on block A.

Thus, $m_A g \sin\theta = f_1$

$\therefore$ $$f_1 = 30 \text{ N}$$

(c) For block B,

$\because$ The acceleration of block B is zero.

$$T = m_B g \sin\theta + f_1 + f_3$$

$$= 60 + 30 + 1.6 = 91.6 \text{ N}$$

14. (d) The maximum force exerted by the man on the lower book should be equal to the maximum friction on his legs.

$\therefore$ The maximum horizontal force on the books is

$$F_{max} = 0.5 \times 50 \times 10 = 250 \text{ N}$$

The maximum frictional force by table on the lower book is

$$f_{max} = 0.5 \times 60 \times 10 = 300 \text{ N}$$

$\therefore$ $$F_{max} \le f_{max}$$

Hence, option (d) is correct.

15. (d) The maximum tension for downward motion of hanging block is

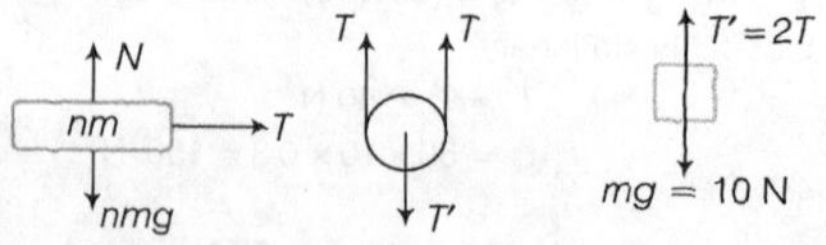

$$T' = 10 \Rightarrow 2T = 10$$

$\therefore$ $$T = 5 \text{ N}$$

The maximum friction on 1st block is

$$f_{1max} = \mu m_1 g = 0.8 \times 1 \times 10 = 8\text{ N}$$

$\therefore$ $T < f_{1max}$

Thus, hanging block remains stationary.

$\therefore$ $2T = mg = 10 \Rightarrow T = 5\text{ N}$

For 1st block,

$\therefore$ $T = f_1 \Rightarrow f_1 = T = 5\text{ N}$

So, no pulling force is on other numbered blocks except 1st block. Hence, option (d) is correct.

16. (d) The common acceleration is

$$a_0 = \frac{F}{m_A + m_B + m_C + m_D} = \frac{20}{20} = 1\text{ m/s}^2$$

The free body diagram of A in the frame of B,

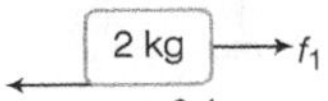

Here, $f_{1\max} = \mu_1 mg = 0.1 \times 2 \times 10 = 2\text{ N}$

So, block A does not move with respect to B. In similar way, blocks C and D are not moving with respect to B. Thus, all blocks move together.

$\therefore$ The acceleration of blocks is $a_0 = 1\text{ m/s}^2$

For block A, $f_1 = m_1 a_0 = 2 \times 1 = 2\text{ N}$

For block D, $f_3 = 10 \times 1 = 10\text{ N}$

For block C, $f_2 - f_3 = m_c a_0 = 4 \times 1 = 4$ or $f_2 = 14\text{ N}$

17. (d) Here, $F - 10 = 8a$ (where, a is the acceleration of the system)

or $30 - 10 = 8a$ or $a = \frac{20}{8} = \frac{5}{2}\text{ m/s}^2$

Since, blocks are moving together,

Free body diagram of A,

For block A, $F - f = 4a$

or $30 - 4 \times \frac{5}{2} = f \leq f_{max}$ or $20 \leq 30\mu$

or $30\mu \geq 20$

$\Rightarrow$ $\mu \geq \frac{2}{3} \Rightarrow \mu_{min} = \frac{2}{3}$

18. (c) Let coefficient of friction be μ, then retardation will be μg. From equation of motion, $v = u + at$

$\Rightarrow$ $0 = 6 - \mu g \times 10$

or $\mu = \frac{6}{100} = 0.06$

19. (a) A block of mass m is placed on a surface with a vertical cross-section, then

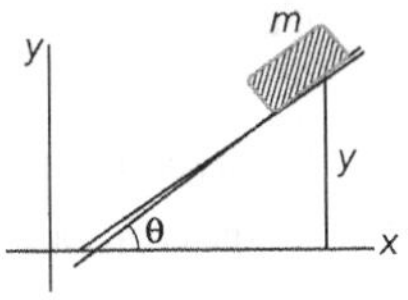

$$\tan\theta = \frac{dy}{dx} = \frac{d\left(\frac{x^3}{6}\right)}{dx} = \frac{x^2}{2}$$

At limiting equilibrium, we get

$$\mu = \tan\theta,\ 0.5 = x^2/2$$

$\Rightarrow$ $x^2 = 1 \Rightarrow x = \pm 1$

Now, putting the value of x in $y = x^3/6$, we get,

When $x = 1$

$\therefore y = \frac{(1)^3}{6} = \frac{1}{6}$

When $x = -1$

$y = \frac{(-1)^3}{6} = \frac{-1}{6}$

So, the maximum height above the ground at which the block can be placed without slipping is $\frac{1}{6}$ m.

20. (d) $F_1 = mg(\sin\theta + \mu\cos\theta)$

(as body just in position to move up, friction force downward)

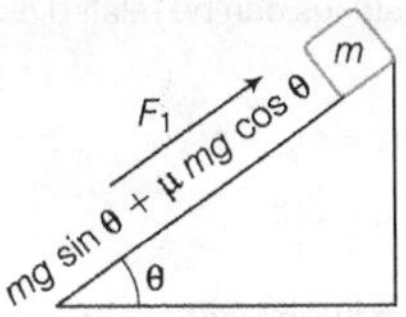

$F_2 = mg(\sin\theta - \mu\cos\theta)$ (as body just in position to slide down, friction force upward)

$$\therefore \frac{F_1}{F_2} = \frac{\sin\theta + \mu\cos\theta}{\sin\theta - \mu\cos\theta} = \frac{\tan\theta + \mu}{\tan\theta - \mu} = \frac{2\mu + \mu}{2\mu - \mu} = 3$$

21. (a) Frictional force comes into role when there is a tendency of relative motion between the surfaces in contact.

22. (c) Free body diagram of block A

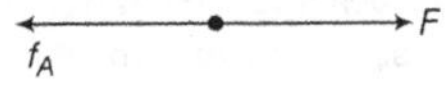

f_A-friction on block A due to block B.

If $F > f_A$, (block A)

then $a_A = \frac{F - f_A}{m_A}$

Free body diagram of block B

f_B–friction on block B due to ground

f_A–friciton on block B due to block A

and if $f_B > f_A$, (block B)

the acceleration (a_B) of block B will be zero

$$a_A > a_B$$

23. (a) Contact force is the resultant of normal reaction and the force of friction on block.

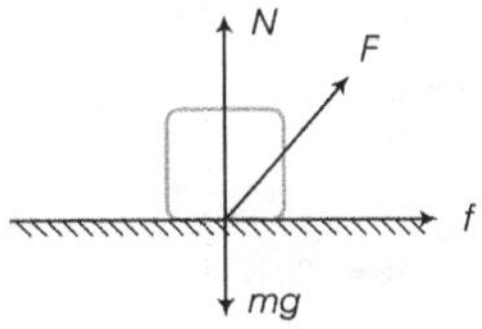

$$F = \sqrt{N^2 + f^2} \Rightarrow N = mg$$

and $0 \leq f \leq \mu mg$

Thus, $mg \leq F = \sqrt{(mg)^2 + (\mu mg)^2}$

24. (a)

25. (a) Free body diagram of man,

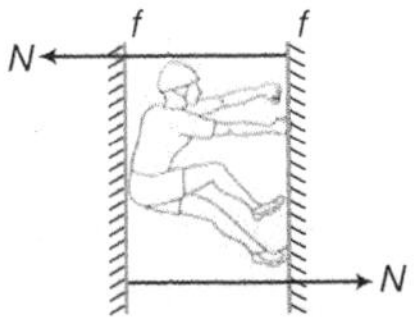

$$f + f = mg$$

$$2f = 50 \times 10 \Rightarrow f = 250\text{ N}$$

26. (d) Free body diagram of the block,

From free body diagram,

$N = F\cos\theta$

$\Rightarrow$ $f = \mu_k N = \mu_k F\cos\theta$

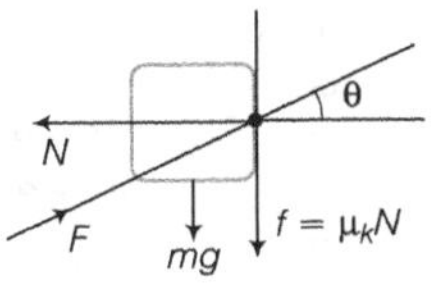

Revisal Problems (JEE Advanced)

1. (b) Tension in two strings can be related as

$$T_2 = T_1 e^{\mu\alpha}$$

or $\quad 2m_A = m_A e^{\pi\mu}$ or $\ln 2 = \pi\mu$

$\therefore \quad \mu = \dfrac{\ln 2}{\pi}$

2. (b) For sliding down the inclined plane,

$(m_1 + m_2 + m_3 + m_4)g \sin\alpha \ge \mu_1 m_1 g \cos\alpha + \mu_2 m_2 g \cos\alpha + \mu_3 m_3 g \cos\alpha + \mu_4 m_4 g \cos\alpha$

$\Rightarrow \quad \tan\alpha \ge \dfrac{\mu_1 m_1 + \mu_2 m_2 + \mu_3 m_3 + \mu_4 m_4}{m_1 + m_2 + m_3 + m_4}$

$\therefore \quad (\tan\alpha)_{min} = \left(\dfrac{\mu_1 m_1 + \mu_2 m_2 + \mu_3 m_3 + \mu_4 m_4}{m_1 + m_2 + m_3 + m_4}\right)$

3. (a) If block A has tendency to slip down.

$m_1 g \sin 60° - m_2 g \sin 30° \le \mu m_1 g \cos 60° + \mu m_2 g \cos 30°$

or $\quad \dfrac{m_1}{m_2} \le \dfrac{\frac{\sqrt{3}}{2}\mu + \frac{1}{2}}{\frac{\sqrt{3}}{2} - \frac{\mu}{2}} = \dfrac{\sqrt{3}\mu + 1}{\sqrt{3} - \mu} = \dfrac{\sqrt{3} + 2}{2\sqrt{3} - 1}$

If block B has tendency to slip down,

$m_2 g \sin 30° - m_1 g \sin 60° \le \mu m_2 g \cos 30° + \mu m_1 g \cos 60°$

or $\quad m_2 g \sin 30° - \mu m_2 g \cos 30° \le \mu m_1 g \cos 60° + m_1 g \sin 60°$

or $\quad \dfrac{m_2}{2} - \dfrac{\sqrt{3}\mu m_2}{2} \le \dfrac{\mu m_1}{2} + \dfrac{\sqrt{3}}{2} m_1$

or $\quad \dfrac{m_1}{m_2} \ge \dfrac{(1 - \sqrt{3}\mu)}{(\mu + \sqrt{3})}$ or $\dfrac{m_1}{m_2} \ge \dfrac{2 - \sqrt{3}}{1 + 2\sqrt{3}}$

4. (b) $F - \mu mg = ma$ or $10 - 2t - 5 = a$

or $5 - 2t = \dfrac{dv}{dt}$ or $\int_0^v dv = \int_0^{2.5} (5 - 2t)\, dt$

or $v = [5t - t^2]_0^{2.5} = 5 \times 2.5 - 6.25 = 12.5 - 6.25 = 6.25$ m/s

5. (a) Friction is external force on the bus which produces deceleration in the bus.

6. (b) $T_{max} = \mu Mg \quad$ ($\because$ block is stationary)

$= 0.5 \times 20 \times 10 = 100$ N

For monkey, $T - mg = ma$ or $T_{max} - mg = ma_{max}$

or $\quad 100 - 50 = 5a_{max}$

$\therefore \quad a_{max} = \dfrac{50}{5} = 10$ m/s^2

7. (a,b)

(a) As, $x = ut - \dfrac{1}{2}at^2$

$\Rightarrow \quad x = 10 \times 2 - \dfrac{1}{2} \times a \times 4$

$= 20 - 2a = 20 - 2\mu g = 20 - 20\mu$

$\because \quad 0 < x \le 10$ m or $0 < 20 - 20\mu \le 10$

or $\quad -20 < -20\mu \le -10$

$20 > 20\mu \ge 10$ or $1 > \mu \ge 0.5$

$\Rightarrow \quad 1 > \mu \ge 0.5$ is possible

(b) $10 < 20 - 20\mu < 20$

or $\quad 1 < 2 - 2\mu < 2$ or $2 - 2\mu < 2$

or $\quad 2 < 2 + 2\mu$ or $0 < 2\mu \Rightarrow \mu > 0$

Also, $\quad 1 < 2 - 2\mu$

or $\quad 2\mu < 1$ or $\mu < 0.5$

$\Rightarrow \quad 0 < \mu < 0.5$ is possible.

8. (a,b,d)

(a) For static friction, $F = f \le \mu mg$

$\therefore \quad F \le \mu mg$

$\therefore \quad F_0 = \mu mg$

(b) $\because \quad \dfrac{F_0}{2} < F_0$, static friction comes into play.

$\therefore \quad F = f$

$\therefore \quad \dfrac{F_0}{2} = f$

(d) For $F = F_0$, frictional force is static in nature.

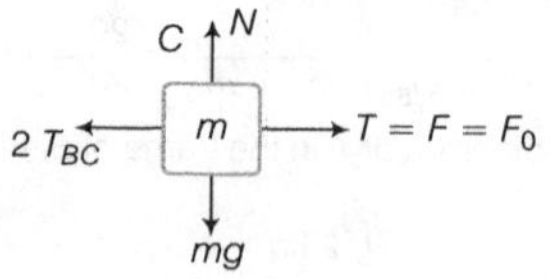

From free body diagram of block C,

$T = 2T_{BC}$ or $F_0 = 2T_{BC}$

$\therefore \quad T_{BC} = \dfrac{F_0}{2}$

9. (a,b,c) Here, $f_{max} = \mu mg = 10$ N

For sliding starts, $F > \mu mg$ or $2t > 10$

$\therefore \quad t_{min} = 5$ s

Hence, option (a) is correct.

$\because \quad a = \dfrac{2t - 10}{2} = t - 5$ or $\dfrac{dv}{dt} = t - 5$

or $\quad \int_0^v dv = \int_5^{10} (t - 5)dt = \left[\dfrac{t^2}{2} - 5t\right]_5^{10} = 12.5$ m/s

$\because \quad v = \dfrac{ds}{dt} = \dfrac{t^2}{2} - 5t$

$\therefore \quad s = \int_5^{10} \left(\dfrac{t^2}{2} - 5t\right) dt = \left[\dfrac{t^3}{6} - \dfrac{5}{2}t^2\right]_5^{10} = \dfrac{125}{6}$ m

10. (a,b) The maximum friction on man is

$f_{1\,max} = \mu m_1 g = 0.5 \times 50 \times 10 = 250$ N

The maximum friction on the block is

$f_{2\,max} = \mu m_2 g = 0.5 \times 100 \times 10 = 500$ N

If 500 N $< T <$ 1000 N, block does not move but man moves.

If $T >$ 1000 N, both start to move.

In this case, $a_1 = \dfrac{T - 500}{50}$

and $\quad a_2 = \dfrac{T - 1000}{100}$

$\therefore \quad a_1 > a_2$

Hence, options (a) and (b) are correct.

11. (c) The common acceleration is

$$a = \frac{F}{m + M}$$

For cup,

$f = Ma$

But $\quad f \le \mu mg$ or $Ma \le \mu mg$

or $\quad \dfrac{MF}{m + M} \le \mu mg$

or $\quad F \le \dfrac{\mu mg\,(m + M)}{M}$

or $\quad F \le \mu mg\left(1 + \dfrac{m}{M}\right)$

or $\quad F \le 0.06$ N

12. (a) For plate, $F = f + \mu mg$

where, f is frictional force between the plate and the coin.

But $f \le \mu mg$

$\Rightarrow$ $F - \mu mg \le \mu mg$

$\therefore$ $F \le 2\mu mg$

or $F \le 2 \times 0.5 \times 10 \times 10^{-3} \times 10$

$\therefore$ $F \le 0.1$ N

13. (a) Limiting frictional force on the block is

$$f_{max} = \mu_s N = (0.4)(10)(10) = 40 \text{ N}$$

As, $F < f_{max}$, block does not move

Thus, there is no relative motion between plank and the block and hence only static friction acts.

14. (a) Free body diagram of plank at $t = 0$

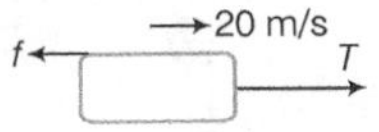

Since, plank is moving with constant velocity, $a = 0$

Free body diagram of block with respect to plank

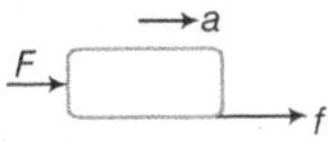

$$a = \frac{F + f}{m} = \frac{15 + 25}{10} = \frac{40}{10} = 4 \text{ m/s}^2$$

Time tile which this type of motion continuous

$$t = \frac{v - u}{a} = \frac{20 - 0}{4} = 5 \text{ s}$$

After $t = 5$ s, free body diagram of block is shown in figure,

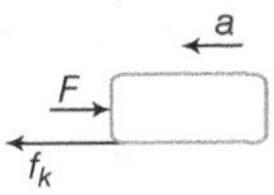

Now, there will be retardation relative to plank given by

$$a = \frac{f_k - F}{m} = \frac{25 - 15}{10} = 1 \text{ m/s}^2$$

15. (b) Free body diagram of block A,

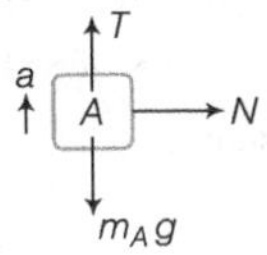

$$T - m_A g = m_A a \qquad \text{...(i)}$$

Free body diagram of block C,

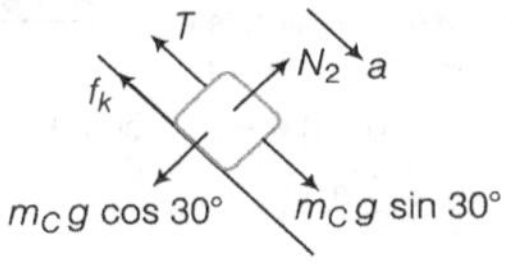

$$m_C g \sin 30° - (T + f_k) = m_C a$$

or $$\frac{m_C g}{2} - (T + \mu m_c g \cos 30°) = m_C a \qquad \text{...(ii)}$$

From adding Eqs. (i) and (ii), we get

$$\frac{m_C g}{2} - m_A g - \frac{\mu m_C g \sqrt{3}}{2} = a(m_A + m_C)$$

or $$\frac{3g}{2} - g - \frac{3g\sqrt{3}}{2} = 4a$$

or $$a = 0.233\, g \approx \frac{g}{4}$$

16. (b) Considering no motion of block B, acceleration of block A is same as that of C.

Acceleration of block $C = \dfrac{g}{4}$

17. (3) Balancing vertical forces, we have

$$f + T\cos\theta = mg$$

and $T \sin\theta = N$ or $f + Mg\cos\theta = mg$

or $$f = mg - Mg\cos\theta \qquad \text{...(i)}$$

Here, N is normal reaction on man due to wall.

or $T\sin\theta = N$ or $Mg \sin\theta = N$

$$f \le \mu N$$

or $mg - Mg\cos\theta \le \mu Mg \sin\theta$

or $m - M\cos\theta \le \mu M \sin\theta$ or $m \le M(\cos\theta + \mu\sin\theta)$

or $$\left(\frac{m}{M}\right) \le (\cos\theta + \mu \sin\theta)$$

or $$\left(\frac{m}{M}\right)_{max} = \frac{4}{5} + \frac{3}{10} = \frac{8 + 3}{10} = \frac{11}{10}$$

$\therefore$ $$\frac{11x}{30} = \frac{11}{10}$$

$\therefore$ $x = 3$

18. (4) For just sliding block B over A,

$$mg\sin\theta = T + \mu\, mg\cos\theta \qquad \text{...(i)}$$

and for sliding B over the inclined plane,

$$4T + \mu mg\cos\theta = mg\sin\theta + 2\mu\, mg\cos\theta$$

or $$4mg\sin\theta - 4\mu mg\cos\theta + \mu mg\cos\theta = mg\sin\theta + 2\mu mg\cos\theta$$

or $$3\, mg \sin\theta = 5\mu mg\cos\theta$$

or $$\tan\theta = \frac{5}{3}\mu = \frac{5}{3} \times 0.6 = 1$$

$\therefore$ $$\theta = \frac{\pi}{4} = \frac{\pi}{n}$$

$\therefore$ $n = 4$

19. (5) Equation of motion for block C is

$$m_C g - T = m_C a_C$$

or $$180 - T = 18 a_C \qquad \text{...(i)}$$

$$T - 45 = 9\, a_A \qquad \text{...(ii)}$$

and $$2T - 45 = 9 a_B = 9\left(\frac{a_C - a_A}{2}\right) \qquad \text{...(iii)}$$

or $$4T - 90 = 9a_C - 9a_A$$

or $$4T - 90 = 9\left(\frac{180 - T}{18}\right) - 9\left(\frac{T - 45}{9}\right)$$

or $$4T + \frac{T}{2} + T = 90 + 45 = 135$$

$\therefore$ $$T = \frac{135}{5.5} = \frac{1350}{55} = \frac{1350}{11n}$$

Comparing with the given value, we have $n = 5$

20. (1) The force of kinetic friction is always opposite to instantaneous velocity causes the magnitude of velocity to decrease, while $mg \sin\theta$ increases the component of velocity parallel to greatest slope.

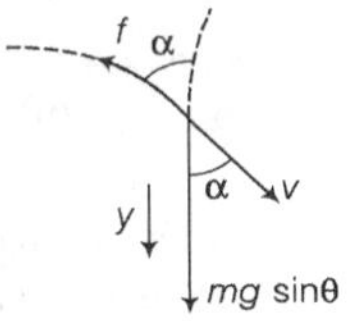

These two effects are cause of motion of the block on a curved path with a variable acceleration.

Since, initially block is stationary.

$\therefore$ $$mg\sin\theta = f = \mu\, mg\cos\theta$$

Considering the motion along the path of the particle,

$$mg\sin\theta\cos\alpha - f = m\frac{dv}{dt}$$

or $$f\cos\alpha - f = m\frac{dv}{dt} \qquad \text{...(i)}$$

Considering the motion parallel to the greatest slope,

$$mg\sin\theta - f\cos\alpha = m\frac{dv_y}{dt}$$

or $$f - f\cos\alpha = m\frac{dv_y}{dt} \qquad \text{...(ii)}$$

From Eqs. (i) and (ii),

$$-\frac{dv}{dt}=\frac{dv_y}{dt} \quad \text{or} \quad -\int_{v_0}^{v} dv=\int_0^{v\cos\alpha} dv_y$$

or $$v_0-v=v\cos\alpha \quad \text{or} \quad v=\frac{v_0}{1+\cos\alpha}$$

$\therefore$ $$v_{max}=\frac{v_0}{1+(\cos\alpha)_{max}}=\frac{2}{1+1}=1\text{ m/s}$$

21. (a) The maximum friction on block A is $f_{1\,max}=\mu(20+t)=10+\frac{t}{2}$

(i) When $t=4$ s, $f_{max}=12$ N

The horizontal force on block A is $F_A=t=4$ N, thus static friction acts on the block A.

(ii) When $t=20$ s

$\because$ $$f_{max}=20+\frac{20}{2}=30\text{ N}$$

$\because$ Horizontal component of F is 20 N and maximum friction is 30 N. Static friction acts on the block A.

(iii) When $t=40$ s

$$f_{max}=20+20=40\text{ N}$$

In this case, limiting friction acts on the block.

$\because$ $$f=40\text{ N}$$

(iv) When $t=60$ s

$$f_{max}=20+30=50\text{ N}$$

The horizontal component of F is

$$F\cos 45^\circ=t=60\text{ N}$$

Kinetic friction comes into play.

$\therefore$ $$f=50\text{ N}$$

and $$60-50=(m_A+m_B)a$$

$\therefore$ $$a=\frac{10}{5}=2\text{ m/s}^2$$

Free body diagram of block A is,

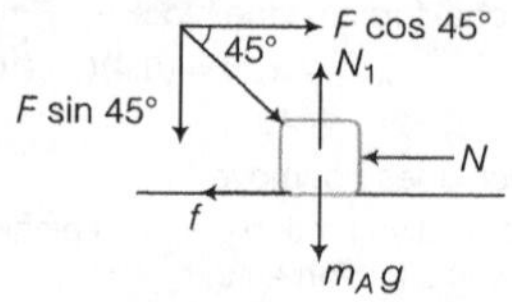

Free body diagram of block B is,

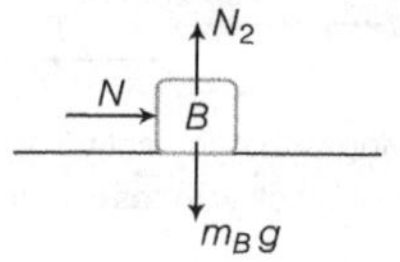

(i) $\rightarrow$ (p), (ii) $\rightarrow$ (p), (iii) $\rightarrow$ (p,q), (iv) $\rightarrow$ (q,r,s)

JEE Main & AIEEE Archive

1. (a) A block of mass m is placed on a surface with a vertical cross-section, then

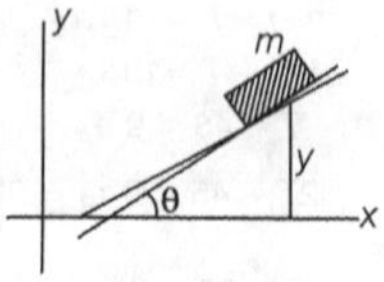

$$\tan\theta=\frac{dy}{dx}=\frac{d\left(\frac{x^3}{6}\right)}{dx}=\frac{x^2}{2}$$

At limiting equilibrium, we get

$$\mu=\tan\theta,\ 0.5=x^2/2$$

$\Rightarrow$ $$x^2=1 \Rightarrow x=\pm 1$$

Now, putting the value of x in $y=x^3/6$, we get

When $x=1$	When $x=-1$
$\therefore\ y=\frac{(1)^3}{6}=\frac{1}{6}$	$y=\frac{(-1)^3}{6}=\frac{-1}{6}$

So, the maximum height above the ground at which the block can be placed without slipping is $\frac{1}{6}$ m.

2. (d) $F_1=mg(\sin\theta+\mu\cos\theta)$ [as body just in position to move up, friction force downward]

$F_2=mg(\sin\theta-\mu\cos\theta)$ [as body just in position to slide down, friction upward]

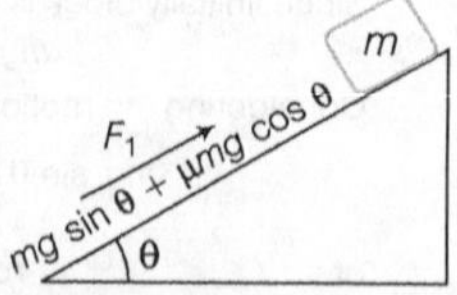

$\therefore$ $$\frac{F_1}{F_2}=\frac{\sin\theta+\mu\cos\theta}{\sin\theta-\mu\cos\theta}=\frac{\tan\theta+\mu}{\tan\theta-\mu}=\frac{2\mu+\mu}{2\mu-\mu}=3$$

3. (b) From Newton's equations, we have

$$v^2=u^2-2as$$

Given, $v=0$ [car is stopped]

As friction provide the retardation

$$a=\mu g,\ v=100\text{ ms}^{-1}$$

$\therefore$ $$(100)^2=2\mu gs$$

$\Rightarrow$ $$s=\frac{100\times 100}{2\times 0.5\times 10}=\frac{100\times 100}{5\times 2}=1000\text{ m}$$

4. (a) When friction is absent

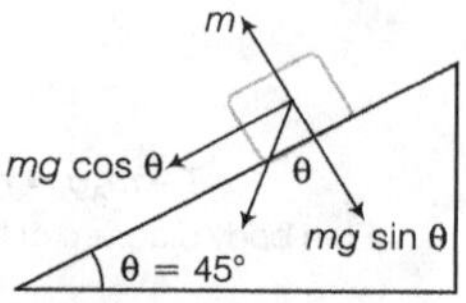

$$ma_1=mg\sin\theta$$

$$a_1=g\sin\theta$$

$\therefore$ $$s_1=\frac{1}{2}a_1t_1^2 \quad \ldots\text{(i)}$$

When friction is present, friction is in opposite to the direction of motion.

$$a_2=g\sin\theta-\mu_k g\cos\theta$$

$\therefore$ $$s_2=\frac{1}{2}a_2t_2^2 \quad \ldots\text{(ii)}$$

From Eqs. (i) and (ii), we get

$$\frac{1}{2}a_1t_1^2=\frac{1}{2}a_2t_2^2$$

$\Rightarrow$ $$a_1t_1^2=a_2(nt_1)^2 \quad [\because t_2=nt_1]$$

or $$a_1=n^2a_2$$

$\Rightarrow$ $$\frac{a_2}{a_1}=\frac{g\sin\theta-\mu_k g\cos\theta}{g\sin\theta}=\frac{1}{n^2}$$

or $$\frac{g\sin 45^\circ-\mu_k g\cos 45^\circ}{g\sin 45^\circ}=\frac{1}{n^2}$$

or $$1-\mu_k=\frac{1}{n^2} \text{ or } \mu_k=1-\frac{1}{n^2}$$

5. (c) According to work-energy theorem,

Work done = Change in kinetic energy

$$W = \Delta K = 0$$

$\Rightarrow$ Work done by friction + Work done by gravity = 0

$$\Rightarrow \quad -(\mu\, mg \cos\phi)\frac{l}{2} + mgl \sin\phi = 0$$

or $$\frac{\mu}{2}\cos\phi = \sin\phi \text{ or } \mu = 2\tan\phi$$

6. (a) Let mass of the block be m.

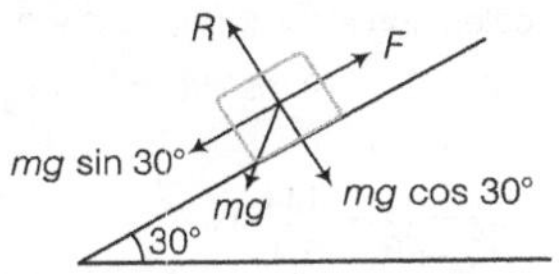

Frictional force in rest position

$$F = mg \sin 30^\circ$$

[this is static frictional force and may be less than the limiting frictional force]

$$\therefore \quad 10 = m \times 10 \times \frac{1}{2} \quad \text{or} \quad m = \frac{2\times 10}{10} = 2 \text{ kg}$$

7. (d) Let R be the normal contact force by wall on the block.

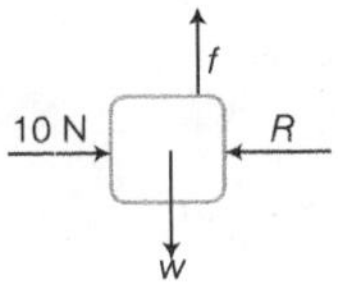

$$R = 10 \text{ N}$$

$$f_L = w \quad \text{and} \quad f_L = \mu R$$

$$\therefore \quad \mu R = w \quad \text{or} \quad w = 0.2 \times 10 = 2 \text{ N}$$

8. (c) Let coefficient of friction be μ, then retardation will be μg. From equation of motion, $v = u + at$

$$\Rightarrow \quad 0 = 6 - \mu g \times 10$$

or $$\mu = \frac{6}{100} = 0.06$$

9. (b) As centrifugal force is balanced by the centripetal force, i.e. frictional force.

Using the relation

$$\frac{mv^2}{r} = \mu R, \quad R = mg$$

$$\Rightarrow \quad \frac{mv^2}{r} = \mu mg$$

or $$v^2 = \mu rg$$

$$\Rightarrow \quad v^2 = 0.6 \times 150 \times 10$$

or $$v = 30 \text{ ms}^{-1}$$

10. (b) The system of masses is shown below.

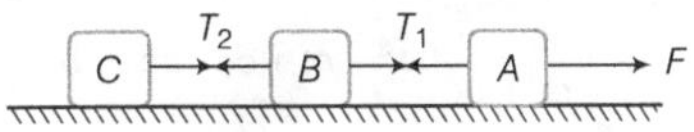

From the figure,

$$F - T_1 = ma \quad \ldots\text{(i)}$$

and $$T_1 - T_2 = ma \quad \ldots\text{(ii)}$$

Eq. (i) gives,

$$10.2 - T_1 = 2 \times 0.6$$

$$\Rightarrow \quad T_1 = 10.2 - 1.2 = 9 \text{ N}$$

Again, from Eq. (ii), we get

$$9 - T_2 = 2 \times 0.6$$

$$\Rightarrow \quad T_2 = 9 - 1.2 = 7.8 \text{ N}$$

JEE Advanced & IIT JEE Archive

1. (d) Block will not slip if $(m_1 + m_2)\, g \sin\theta \le \mu m_2 g \cos\theta$

$$\Rightarrow \quad 3\sin\theta \le \left(\frac{3}{10}\right)(2)\cos\theta$$

$$\tan\theta \le \frac{1}{5} \quad \Rightarrow \quad \theta \le 11.5^\circ$$

(i) If $\theta = 5^\circ$, friction is static

$$f = (m_1 + m_2)\, g \sin\theta$$

(ii) If $\theta = 10^\circ$, friction is static

$$f = (m_1 + m_2)\, g \sin\theta$$

(iii) If $\theta = 15^\circ$, friction is kinetic

$$f = \mu m_2 g \cos\theta$$

(iv) If $\theta = 20^\circ$, friction is kinetic $\Rightarrow$ $f = \mu m_2 g \cos\theta$

(i) $\rightarrow$ (q), (ii) $\rightarrow$ (q), (iii) $\rightarrow$ (r), (iv) $\rightarrow$ (r)

2. (5)

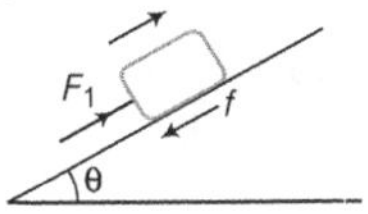

Moving upwards

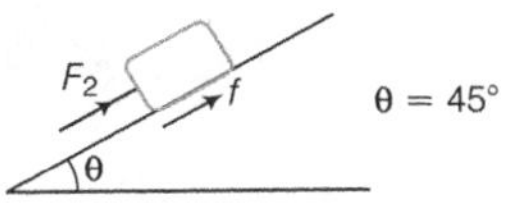

Just remains stationary

$$F_1 = mg \sin\theta + \mu mg \cos\theta$$

$$F_2 = mg \sin\theta - \mu mg \cos\theta$$

Given that $$F_1 = 3F_2 \text{ or } (\sin 45^\circ + \mu\cos 45^\circ) = 3(\sin 45^\circ - \mu \cos 45^\circ)$$

On solving, we get $\mu = 0.5$

$$\therefore \quad N = 10\mu = 5$$

3. (a) When

$$P = mg\,(\sin\theta - \mu\cos\theta)$$

$$f = \mu\, mg\cos\theta \quad \text{(upwards)}$$

When $$P = mg\sin\theta$$

$$f = 0$$

and when $$P = mg\,(\sin\theta + \mu\cos\theta)$$

$$f = \mu mg \cos\theta \quad \text{(downwards)}$$

Hence friction is first positive, then zero and then negative.

$\therefore$ Option (a) is correct.

4. (b) Both Statements are correct. But Statement II does not explain correctly Statement I.

Correct explanation is : There is increase in normal reaction when the object is pushed and there is decrease in normal reaction when the object is pulled (but strictly not horizontally).

5. (10) Normal reaction in vertical direction $N_1 = mg$

Normal reaction from side to the groove $N_2 = ma \sin 37^\circ$

Therefore, acceleration of block with respect to disc

$$a_r = \frac{ma\cos 37^\circ - \mu N_1 - \mu N_2}{m}$$

Substituting the values, we get

$$a_r = 10 \text{ m/s}^2$$

3. Circular Motion

DPP-1 Kinematics of Circular Motion

1. (i) axis (ii) vector
(iii) scalar (iv) commutative
(v) accelerated (vi) $\hat{\mathbf{k}}$
(vii) $\mathbf{v} = \omega \times \mathbf{r}$
(viii) directed towards the centre of circular path
(ix) not
(x) tangential acceleration

2. (i) (a) $\theta = at^2 - bt + c$

$$\theta = \theta(t = 0) = c$$
$$\theta(t) = at^2 - bt + c$$

(ii) (b) Angular displacement,
$$\Delta\theta = \theta(t) - \theta_i = at^2 - bt + c - c = at^2 - bt$$

(iii) (a) Angular velocity,
$$\omega = \frac{d\theta}{dt} = \frac{d}{dt}(at^2 - bt + c) = 2at - b$$

(iv) (b) Speed $|\mathbf{v}| = \omega \times \mathbf{r} = \omega r \sin\theta$
where, $\theta = 90°$ when centre of path is same as origin
So, speed $= |\mathbf{v}| = r\omega = r(2at - b)$

(v) (b) Angular acceleration $\propto = \frac{d\omega}{dt} = 2a$

(vi) (b) The angular acceleration is
$$\alpha = \frac{\Delta\omega}{\Delta t} = \frac{\text{rad}}{\text{s} \times \text{s}} = \text{rad/s}^2$$

(vii) (b) Tangential acceleration of a rotating particle is the rate of change of tangential velocity v.
$\therefore$ Tangential acceleration is
$$\frac{dv}{dt} = \frac{rd\omega}{dt} = r\alpha = 2ar$$

3. (b) $T = 24$ h

$\therefore$ $$\omega = \frac{2\pi}{T} = \frac{2\pi}{24} = \frac{\pi}{12} \text{ rad/h}$$

4. (a) a_r = centripetal acceleration

$$a_r = r\omega^2 = 1.5 \times 10^{11} \times \left(\frac{2\pi}{1 \text{ yr}}\right)^2 = 58.8 \text{ m/yr}^2$$

5. (b) $a = \sqrt{a_r^2 + a_t^2} = a_r = R\omega^2$

$$= 6400 \times 10^3 \times \left(\frac{2\pi}{24 \text{ h}}\right)^2 = 4.36 \times 10^5 \text{ m/h}^2$$

6. (c) The co-latitude is $(90° - 60°) = 30°$

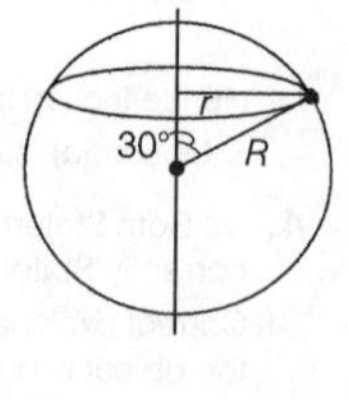

$\therefore$ $\sin 30° = \frac{r}{R}$

$\therefore$ $$a = \sqrt{a_r^2 + a_t^2}$$
$$= a_r = r\omega^2 = \frac{R}{2}\left(\frac{2\pi}{24 \text{ h}}\right)^2$$
$$= \frac{3200 \times 10^3 \times 4\pi^2}{576}$$
$$= 2.18 \times 10^5 \text{ m/h}^2$$

7. (b) $\theta = 1 - e^{-t/3}$, so, $\omega = \frac{d\theta}{dt} = \frac{e^{-t/3}}{3}$

or $$\alpha = \frac{d\omega}{dt} = -\frac{e^{-t/3}}{9}$$

At $t = 0$,
$$\alpha = -\frac{1}{9} \text{ rad/s}^2$$

8. (d) Given, $s = t^3 + 5$

$\therefore$ Speed, $v = \frac{ds}{dt} = 3t^2$

and rate of change of speed, $a_t = \frac{dv}{dt} = 6t$

$\therefore$ Tangential acceleration at $t = 2$ s, $a_t = 6 \times 2 = 12 \text{ ms}^{-2}$

and at $t = 2$ s, $v = 3(2)^2 = 12 \text{ ms}^{-1}$

$\therefore$ Centripetal acceleration
$$a_c = \frac{v^2}{R} = \frac{144}{20} \text{ ms}^{-2}$$

$\therefore$ Net acceleration $= \sqrt{a_t^2 + a_c^2} = \sqrt{(12)^2 + \left(\frac{144}{20}\right)^2} = 14 \text{ ms}^{-2}$

9. (b) The bead is moving on a circular path of radius a.

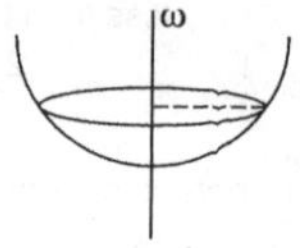

So, the acceleration is $a\omega^2$.

10. (c) As, $a_t = R\alpha$

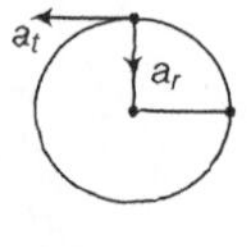

Here, $$\theta = \frac{1}{2}\alpha t^2$$

or, $$\theta = \frac{1}{2}\alpha \times \frac{\pi}{\alpha} = \frac{\pi}{2}$$

and $$\omega = \alpha t = \alpha\sqrt{\pi/\alpha} = \sqrt{\pi\alpha}$$

$\therefore$ The centripetal acceleration is $a_r = R\omega^2 = \pi R\alpha$

$\therefore$ $$\mathbf{a} = -a_t\hat{\mathbf{i}} - a_r\hat{\mathbf{j}} = -R\alpha\hat{\mathbf{i}} - \pi R\alpha\hat{\mathbf{j}}$$

11. (b) For a particle moving in a circle with constant angular speed, velocity vector is always along the tangent to the circle and the acceleration vector always points towards the centre of the circle or is always along radius of the circle.
Since, tangential vector is perpendicular to radial vector, therefore velocity vector will be perpendicular to the acceleration vector. But in no case, acceleration vector is tangent to the circle.

12. (b) $a_t = a_r$

or $r\alpha = r\omega^2$ or $\alpha = \omega^2$ or $0.25 = \omega^2$

$\therefore$ $$\omega = \sqrt{25 \times 10^{-2}} = 0.5 \text{ rad/s}$$

$\because$ $$\omega = \alpha t \quad \text{or} \quad 0.5 = 0.25\,t$$

$\therefore$ $$t = \frac{5 \times 10^{-1}}{25 \times 10^{-2}} = \frac{10}{5} = 2 \text{ s}$$

13. (c) For a particle in a uniform circular motion

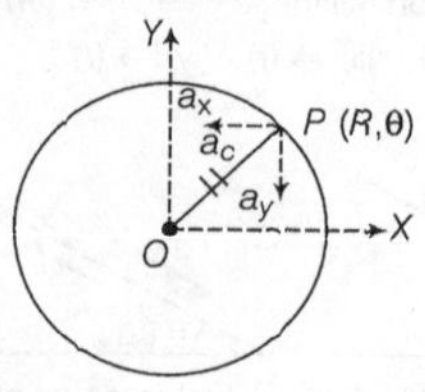

$$\mathbf{a} = \frac{v^2}{R} \text{ towards centre of circle} \quad \text{[centripetal acceleration]}$$

$\therefore$ $$\mathbf{a} = \frac{v^2}{R}(-\cos\theta\hat{\mathbf{i}} - \sin\theta\hat{\mathbf{j}})$$

or $$\mathbf{a} = -\frac{v^2}{R}\cos\theta\hat{\mathbf{i}} - \frac{v^2}{R}\sin\theta\hat{\mathbf{j}}$$

14. (a,c) (a) From triangle OAB,

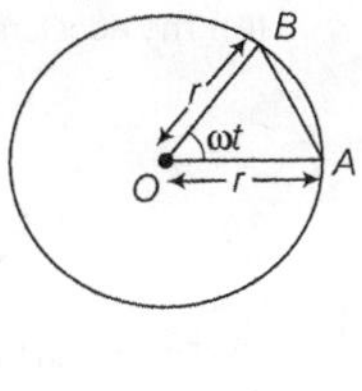

$$\cos \omega t = \frac{r^2 + r^2 - (AB)^2}{2r^2}$$

or $(AB)^2 = 2r^2(1-\cos\omega t) = 4r^2 \sin^2 \frac{\omega t}{2}$

$\therefore \quad AB = 2r\sin\frac{\omega t}{2} = 2r\sin\frac{\omega}{2}\times\frac{\pi}{3\omega}$

$= 2\times 2\times\frac{1}{2} = 2\,\text{m}$

(c) The distance travelled is the arc length $= r\theta = 2\times\frac{\pi}{3} = \frac{2\pi}{3}$ m

15. (a,b) (a) The top point of the wheel is moving on a circular path with respect to axis. Thus, the speed of top point with respect to cyclist is $v_{rel} = R\omega$

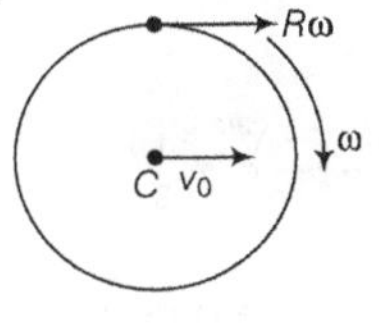

(b) $v_{rel} = v - v_0$

$\therefore \quad v = v_{rel} + v_0 = R\omega + v_0$

16. (a,b) As, $\omega = \frac{1}{\pi}\times\frac{2\pi}{60}$ rad/s $= \frac{1}{30}$ rad/s

(a) $\because \; v = R\omega = 30\times\frac{1}{30} = 1\,\text{cm/s}$

$\therefore \; v = \sqrt{v_0^2 + R^2\omega^2} = \sqrt{1^2+1^2} = \sqrt{2}\,\text{cm/s}$

(b) $a = R\omega^2 = 30\times\frac{1}{900} = \frac{1}{30}\,\text{cm/s}^2$

17. (a,b,d) $a_r = \frac{v^2}{r} = \frac{100}{25} = 4\,\text{m/s}^2$, $\qquad a_t = -3\,\text{m/s}^2$

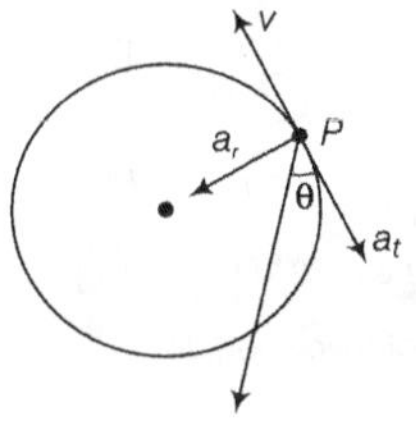

$\therefore \quad a = \sqrt{a_r^2 + a_t^2} = 5\,\text{m/s}^2$

Here, $\tan\theta = \frac{a_r}{a_t} = \frac{4}{3}$

$\therefore \quad \theta = 53°$

Thus, angle between acceleration and velocity is $180° - 53° = 127°$

18. (a,b,c)

(a) Here, $\theta = 3t + 4t^2$

$$\omega = \frac{d\theta}{dt} = 3 + 8t$$

The centripetal acceleration is

$a_r = r\omega^2 = 1\times(3+8t)^2$

At $t = 15 = 1\times(3+8\times 1)^2 = 1\times(3+8)^2 = (11)^2 = 121\,\text{m/s}^2$

$= 121\,\text{m/s}^2$

(b) The angular acceleration

$$\alpha = \frac{d\omega}{dt} = 8\,\text{rad/s}^2$$

The tangential acceleration is $a_t = r\alpha = 8\times 1 = 8\,\text{m/s}^2$

(c) The acceleration is $a = \sqrt{a_r^2 + a_t^2}$

$= \sqrt{(121)^2 + 8^2} = 121.26\,\text{m/s}^2$

19. (a,c,d)

(a) From figure, acceleration **a** is not perpendicular to velocity v.

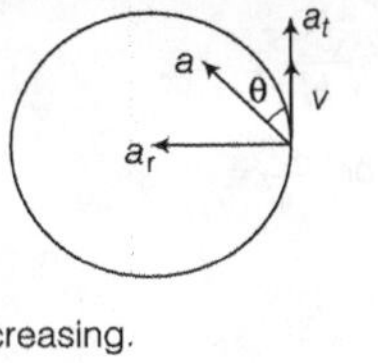

As speed is increasing.

(c) So, angular acceleration must points in same direction as that of angular velocity.

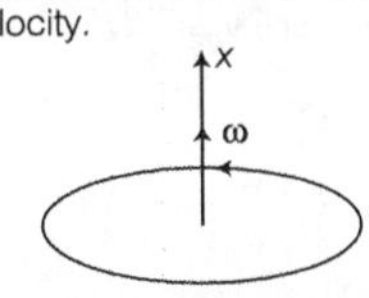

(d) Tangential acceleration is always along tangent to circular path while angular acceleration is along the axis of circular path.

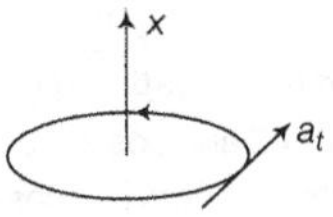

20. (d) The radius of circular path of the bead is

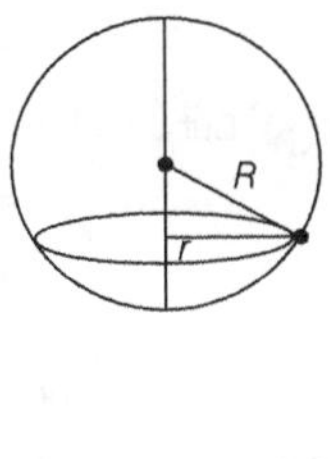

$r = R\sin 30° = \frac{R}{2}$

$\therefore \quad v = \frac{R}{2}\omega = \frac{R}{2}\alpha t$

Then, the acceleration is

$$a = \sqrt{a_r^2 + a_t^2}$$

Here, $\quad a_r = r\omega^2 = \frac{R}{2}\alpha^2 t^2$

and $\quad a_t = r\alpha = \frac{R}{2}\alpha$

Thus, $\quad a = \sqrt{\left(\frac{R}{2}\alpha^2 t^2\right)^2 + \left(\frac{R}{2}\alpha\right)^2}$

$= \frac{R}{2}\alpha\sqrt{1 + \alpha^2 t^4}$

21. (a) The bead rotated through

$$\theta = \frac{1}{2}\alpha t^2 = \frac{1}{2}\alpha\times\frac{\pi}{2\alpha} = \frac{\pi}{4}$$

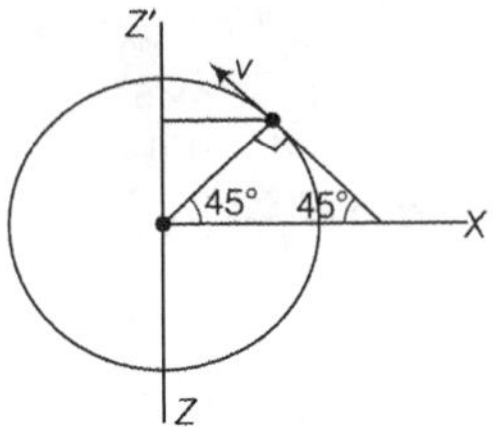

The speed of the bead is

$$v = \frac{R}{2}\omega = \frac{R}{2}\alpha\sqrt{\frac{\pi}{2\alpha}}$$

Thus, velocity of the bead is $\mathbf{v} = -v\cos 45°\,\hat{\mathbf{i}} - v\sin 45\,\hat{\mathbf{k}}$

$= -\frac{R\sqrt{\pi\alpha}}{4}\hat{\mathbf{i}} - \frac{R\sqrt{\pi\alpha}}{4}\hat{\mathbf{k}}$

$\therefore \quad \mathbf{v}_{rel} = \mathbf{v} - \mathbf{v}_0\hat{\mathbf{j}}$

$= -\frac{R}{4}\sqrt{\pi\alpha}\,\hat{\mathbf{i}} - \frac{R}{4}\sqrt{\pi\alpha}\,\hat{\mathbf{k}} - v_0\hat{\mathbf{j}}$

So, the momentum of bead in the frame of the bug is $\mathbf{p} = m\mathbf{v}_{rel}$

$= -\frac{mR}{4}\sqrt{\pi\alpha}\,\hat{\mathbf{i}} - \frac{mR}{4}\sqrt{\pi\alpha}\,\hat{\mathbf{k}} - mv_0\hat{\mathbf{j}}$

22. (2) From figure,

$a \cos 30° = a_r$ and $a \sin 30° = a_t$

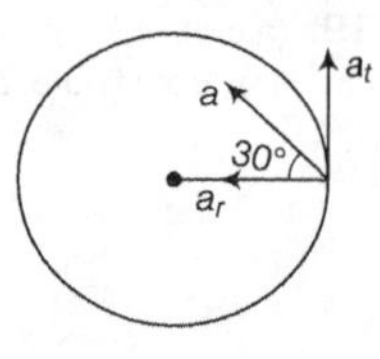

$\therefore \quad \tan 30° = \dfrac{a_t}{a_r} = \dfrac{\sqrt{3t}}{a_r}$

$a_r = 3t$ or $\dfrac{v^2}{1} = 3t$

$\therefore \quad v^2 = 3t$

But $\quad a_t = \dfrac{dv}{dt} = \sqrt{3t}$

or $\quad \int_0^v dv = \sqrt{3}\int_0^t t dt$ or $v = \dfrac{\sqrt{3}}{2}t^2$

$\therefore \quad \dfrac{3}{4}t^4 = 3t$ or $t^3 = 4$

Given $\quad t = 2^{2/3}$ s

$\therefore \quad n = 2$

23. (2) The equation of line is $\dfrac{x}{1} = \dfrac{y}{2} = \dfrac{z}{2}$

The standard equation of line is

$$\frac{x - x_1}{\cos\alpha} = \frac{y - y_1}{\cos\beta} = \frac{z - z_1}{\cos\gamma}$$

Thus, $\cos\alpha = 1, \cos\beta = 2$ and $\cos\gamma = 2$

$\therefore$ The unit vector along the line is $\hat{\mathbf{e}} = \cos\alpha\hat{\mathbf{i}} + \cos\beta\hat{\mathbf{j}} + \cos\gamma\hat{\mathbf{k}}$

$$= \frac{\hat{\mathbf{i}} + 2\hat{\mathbf{j}} + 2\hat{\mathbf{k}}}{\sqrt{9}} = \frac{1}{3}\hat{\mathbf{i}} + \frac{2}{3}\hat{\mathbf{j}} + \frac{2}{3}\hat{\mathbf{k}}$$

But $\quad \mathbf{v} = \omega \times \mathbf{R}$

$$= \left(\frac{\omega}{3}\hat{\mathbf{i}} + \frac{2\omega}{3}\hat{\mathbf{j}} + \frac{2\omega}{3}\hat{\mathbf{k}}\right) \times (2\hat{\mathbf{i}} + 3\hat{\mathbf{j}} + 5\hat{\mathbf{k}})$$

$$\mathbf{v} = \omega\left|\frac{4}{3}\hat{\mathbf{i}} - \frac{1}{3}\hat{\mathbf{j}} - \frac{1}{3}\hat{\mathbf{k}}\right|$$

$\Rightarrow \quad |\mathbf{v}| = \sqrt{2}\omega = \sqrt{2} \times \sqrt{2} = 2$ m/s

24. (5) As, $-\dfrac{dv}{dt} = \dfrac{v^2}{2R} = \dfrac{v^2}{2 \times 0.5}$

or $\quad -\dfrac{dv}{dt} = v^2$

or $\quad -\int_{20}^{v}\dfrac{dv}{v^2} = \int_0^5 dt$

or $\quad \left[\dfrac{1}{v}\right]_{20}^{v} = 5$

or $\quad \dfrac{1}{v} - \dfrac{1}{20} = 5$

or $\quad \dfrac{1}{v} = \dfrac{1}{20} + 5 = \dfrac{101}{20}$

$\therefore \quad v = \dfrac{20}{101} = \dfrac{4 \times 5}{101}$

According to the question, $\dfrac{4 \times 5}{101} = \dfrac{4x}{101}$

$\therefore \quad x = 5$

25. (a) (i) $v = r\omega$

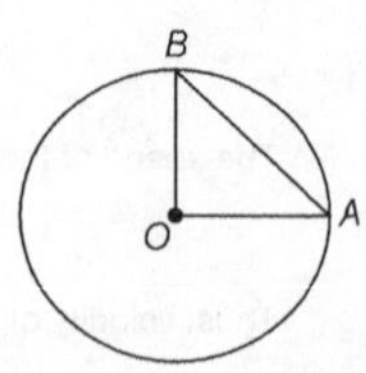

$\therefore \quad \omega = \dfrac{v}{r} = \dfrac{7}{7} = 1$ rad/s

$\therefore \quad \theta = \omega t$ or $\dfrac{\pi}{2} = 1 \times t$

$\therefore \quad t = \dfrac{\pi}{2}$ s

The magnitude of displacement is $|\mathbf{AB}|$

$$= \sqrt{r^2 + r^2} = \sqrt{2}r = 7\sqrt{2} \text{ m}$$

(ii) $\langle v \rangle = \dfrac{\text{Total displacement}}{\text{Total time}}$

$$= \frac{7\sqrt{2}}{\pi/2} = \frac{7\sqrt{2}}{22} \times 2 \times 7$$

$$= \frac{49\sqrt{2}}{11} = 6.3 \text{ m/s}$$

(iii) The acceleration is $a = \sqrt{a_r^2 + a_t^2} = \sqrt{a_r^2 + 0^2}$

$$= a_r = \frac{v^2}{r} = 7 \text{ m/s}^2$$

(iv) $\langle \mathbf{a} \rangle = \dfrac{\mathbf{v}_f - \mathbf{v}_i}{\Delta t} = -\dfrac{7\hat{\mathbf{i}} - 7\hat{\mathbf{j}}}{\dfrac{\pi}{2}}$

The magnitude of average acceleration is

$$|\langle \mathbf{a} \rangle| = \frac{7\sqrt{2} \times 2}{3.14} = 6.3 \text{ m/s}^2$$

(i) → (s), (ii) → (p), (iii) → (q), (iv) → (p)

26. (d) The acceleration $a = \sqrt{a_r^2 + a_t^2}$

$\because \quad a_t$ is non-zero.

$\therefore \quad a > a_r$

27. (a) $\because \quad T^2 \propto r^3$

$\therefore \quad \dfrac{T_1}{T_2} = \left(\dfrac{r_1}{r_2}\right)^{3/2} = \left(\dfrac{r}{4r}\right)^{3/2} = \dfrac{1}{8}$

28. (i) As, $\quad \omega = \omega_0 e^{-\frac{t}{t_0}}$

or $\quad \dfrac{\omega_0}{2} = \omega_0 e^{-t/t_0}$ or $\dfrac{1}{2} = e^{-t/t_0}$ or $-\ln 2 = -\dfrac{t}{t_0}$

$\therefore \quad t = t_0 \ln 2 = 10^4 \ln 2$ s

(ii) The angular acceleration is

$$\alpha = \frac{d\omega}{dt} = -\frac{\omega_0}{t_0}e^{-\frac{t}{t_0}}$$

At $t = 0$, $\quad \alpha = -\dfrac{\omega_0}{t_0} = \dfrac{3.6 \times 10^3}{10^4} \times \dfrac{2\pi}{60} - \dfrac{0.6 \times 10^3}{10^4} \times \dfrac{2\pi}{10}$

$$= -1.2\,\pi \times 10^{-2} \text{ rad/s}^2$$

(iii) $\theta = \int_0^t \omega dt = -\omega_0 t_0 = -\omega_0 t_0 [e^{-t/t_0}]_0^t$

$$= -\omega_0 t_0 [e^{-t/t_0} - 1]$$

$\therefore \quad \theta = \omega_0 t_0 [1 - e^{-t/t_0}]$

$\therefore \quad \theta = \dfrac{3.6}{60} \times 10^4 (1 - e^{-t/t_0})$

$$= 0.6 \times 10^3 [1 - e^{-t/t_0}]$$

Thus number of revolution, $N = \dfrac{\theta}{2\pi}$

$$= \frac{0.6 \times 10^3(1 - e^{-t/t_0})}{2\pi}$$

29. (i) As, $\quad \alpha = \dfrac{d\omega}{dt}$

$\therefore \quad \int_0^\omega d\omega = \int_0^t \alpha dt = \int_0^t (t^2 - t + 1)dt$

$\therefore \quad \omega = \dfrac{t^3}{3} - \dfrac{t^2}{2} + t = \dfrac{1}{3} - \dfrac{1}{2} + 1 = \dfrac{5}{6}$ rad/s

(ii) $\omega = \dfrac{d\theta}{dt}$

$\therefore \quad \int d\theta = \int_0^t \omega dt$

$\therefore \quad \theta = \dfrac{t^4}{12} - \dfrac{t^3}{6} + \dfrac{t^2}{2}$

At $t = 1$ s

$$\theta = \frac{1}{12} - \frac{1}{6} + \frac{1}{2} = \frac{5}{12} \text{ rad}$$

(iii) $\langle \alpha \rangle = \dfrac{\int_0^1 \alpha dt}{\int_0^1 dt} = \dfrac{\frac{5}{6}}{1} = \dfrac{5}{6}$ rad/s²

30. $-mv_2 \sin\dfrac{v_2}{R}t\hat{\mathbf{i}} + m\left(v_2\cos\dfrac{v_2 t}{R} - v_1\right)\hat{\mathbf{j}}$

DPP-2 Uniform Circular Motion

1. (i) (a,c,d) When a particle moves along a circular path with constant velocity, its tangential acceleration $a_t = \frac{dv}{dt} = 0$ and its radial (or centripetal) acceleration

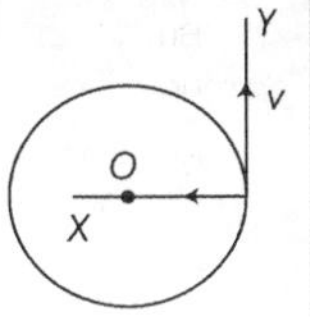

$$a_r = \frac{v^2}{r} = \omega^2 r.$$

So, for a particle rotating in XOY-plane,

$$\Sigma F_x = \frac{mv^2}{r}, \Sigma F_y = ma_t = 0$$

$$\Sigma F_z = 0$$

(ii) (a,b,c,d) Motion of a planet around the sun is an example of approximate a uniform circular motion. (As eccentricity of elliptical orbit is very less). Only force present is gravitational attraction which provides necessary centripetal acceleration for rotational motion.

$$F_g = \frac{mv^2}{r}$$

(iii) (a,b,c,d) In a Bohr model of H-atom, proton and electron attract each other with an electrostatic force,

$$F = \frac{kq_1q_2}{r^2} = \frac{9\times10^9\times(1.6\times10^{-19})^2}{r^2}$$

This electrostatic force provides necessary centripetal acceleration.

The electron moves in a circular orbit around the proton and its tangential acceleration is zero.

(iv) (a) Electrostatic force provides necessary centripetal force.

$$\therefore \quad F_e = \frac{9\times10^9(1.6\times10^{-19})^2}{r^2} = \frac{mv^2}{r}$$

Also, there is no tangential force.

So, $\Sigma F_y = 0$

(v) (b) $\frac{9\times10^9\times2.56\times10^{-38}}{r^2} = \frac{mv^2}{r}$

$$v = \sqrt{\frac{9\times10^9\times2.56\times10^{-38}}{5.3\times10^{-11}\times9.1\times10^{-31}}} = 2.2\times10^6 \text{ m/s}$$

2. (i) For circular motion of the block, $mg - N = \frac{mv^2}{R}$

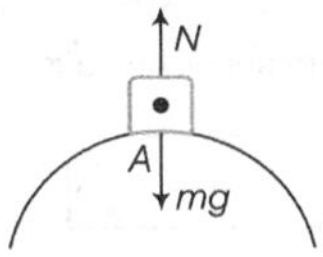

$$\therefore \quad N = mg - \frac{mv^2}{R}$$

(ii) Again for lowest point, $N - mg = \frac{mv^2}{R}$

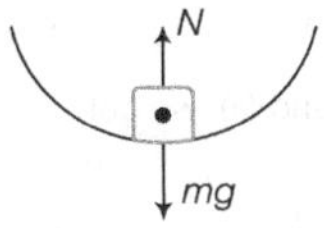

$$\therefore \quad N = mg + \frac{mv^2}{R}$$

(iii) $mg\cos\theta - N = \frac{mv^2}{R}$

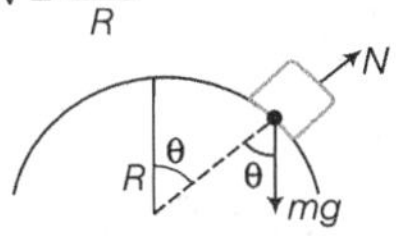

$$\therefore \quad N = mg\cos\theta - \frac{mv^2}{R}$$

(iv) $N - mg\cos\theta = \frac{mv^2}{R}$

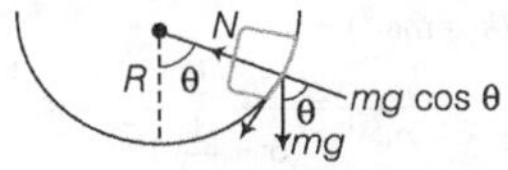

$$\therefore \quad N = mg\cos\theta + \frac{mv^2}{R}$$

(v) The elongation in spring $= \sqrt{2}R - R = (\sqrt{2}-1)R$

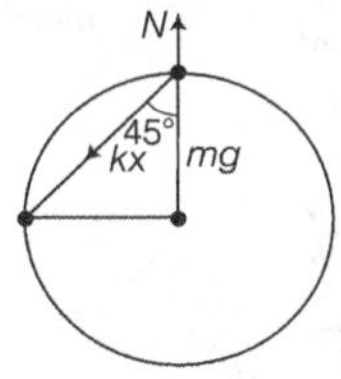

Now, $mg + kx\cos 45° - N = \frac{mv^2}{R} = 0$ $\quad (\because v = 0)$

$\Rightarrow \quad mg + \sqrt{2}\,mg(\sqrt{2}-1)\frac{R}{\sqrt{2}} - N = 0$ or

$mg + mg = N$

$\therefore \quad N = 2\,mg$

3. (i) $T - mg = \frac{mv_0^2}{l}$

$$\therefore \quad T = mg + \frac{mv_0^2}{l}$$

(ii) $T - mg\cos\theta = \frac{mv_0^2}{l}$

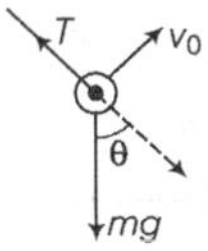

$$T = mg\cos\theta + \frac{mv_0^2}{l}$$

(iii) $T = \frac{mv_0^2}{l}$

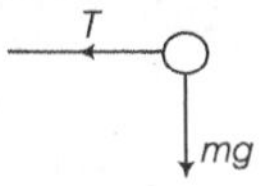

(iv) $T + mg\cos\alpha = \frac{mv_0^2}{l}$

$$\therefore \quad T = \frac{mv_0^2}{l} - mg\cos\alpha$$

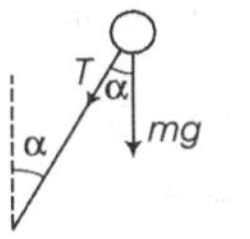

(v) $mg + T = \frac{mv_0^2}{l}$

$$\Rightarrow \quad T = \frac{mv_0^2}{l} - mg$$

(vi) $mg - T = \frac{mv_0^2}{l} = 0$

$\therefore \quad T = mg$

4. (d) Here, $kx = mr\omega^2$

$$k(r-1) = mr\omega^2$$

or $$kr - k = mr\omega^2$$

or $$r(k - m\omega^2) = k$$

$\therefore$ $$r = \frac{k}{k - m\omega^2} = \frac{5}{5 - \frac{1}{\pi^2}4\pi^2} = 5 \text{ m}$$

5. (b) Here, $N\cos\theta = mg$

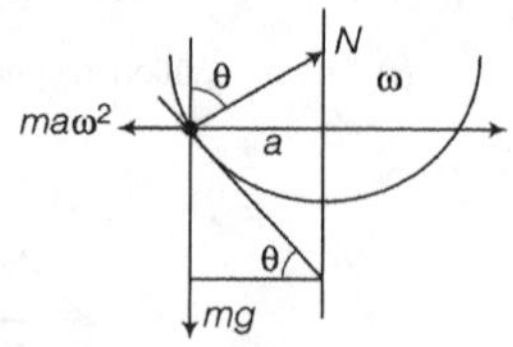

and $$N\sin\theta = ma\omega^2$$

or $$\tan\theta = \frac{a\omega^2}{g}$$

$$x^2 = cy$$

$\Rightarrow$ $$2x = c\frac{dy}{dx}$$

or $$\frac{dy}{dx} = \frac{2x}{c}$$

$$-\tan\theta = \frac{2x}{c} = \frac{2(-a)}{c}$$

$\therefore$ $$\tan\theta = \frac{2a}{c}$$

or $$\frac{a\omega^2}{g} = \frac{2a}{c}$$

$\therefore$ $$\omega = \sqrt{\frac{2g}{c}}$$

6. (b) Here, $T = m_1 r\omega^2 = 1 \times 0.1 \times \omega^2$

and $T = m_2 g = 50$ or $0.1 \times \omega^2 = 50$

or $\omega = \sqrt{500} = 10\sqrt{5}$ rad/s

7. (b) $T = mr\omega^2 = 1 \times 1 \times 100 = 100$ N

8. (c) From free body diagram,

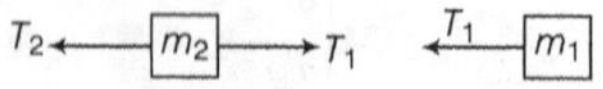

$$T_2 - T_1 = m_2 l\omega^2$$

and $$T_1 = m_1 2l\omega^2$$

or $$T_2 = (2m_1 + m_2)l\omega^2$$

$\therefore$ $$\frac{T_1}{T_2} = \frac{m_1 2l\omega^2}{(2m_1 + m_2)l\omega^2}$$

$$= \frac{2m_1}{2m_1 + m_2} = \frac{2 \times 10}{20 + 20} = \frac{20}{40} = 0.5$$

9. (a) Here, friction provides centripetal acceleration for the coin.

$\therefore$ $$f = mr\omega^2$$

But, $f \le \mu_s mg$ or $mr\omega^2 \le \mu_s mg$

or $$\mu_s \ge \frac{r\omega^2}{g} \quad \text{or} \quad \mu_s \ge \frac{0.5 \times 5}{10}$$

$\therefore$ $$\mu_{s\,(min)} = 0.25$$

10. (c) Here,

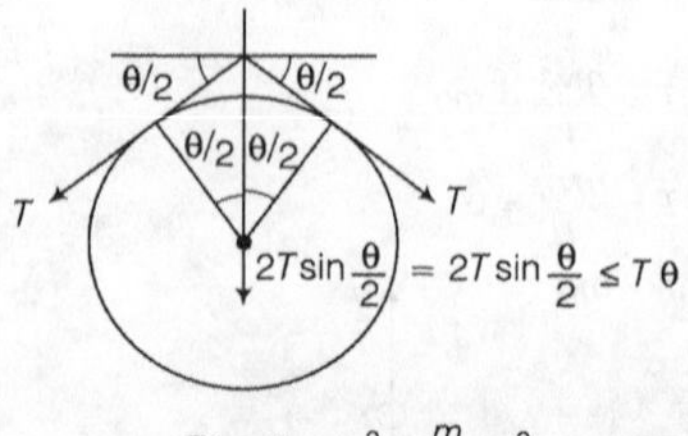

$$2T\sin\frac{\theta}{2} = 2T\sin\frac{\theta}{2} \le T\theta$$

$\Rightarrow$ $$T\theta = \Delta m r\omega^2 = \frac{m}{2\pi}r\omega^2$$

$\therefore$ $$T = \frac{m}{2\pi}r\omega^2 = \frac{3.14}{2\pi} \times 0.05 \times 100 = 2.5 \text{ N}$$

11. (d) Here, $N = ma_r = ml\omega^2$

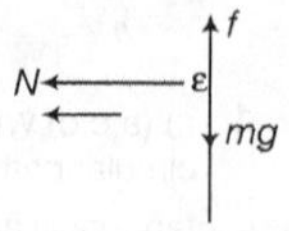

and $$f = mg$$

But $$f \le \mu_s N$$

or $$mg \le \mu_s ml\omega^2$$

or $$\mu_s \ge \frac{g}{l\omega^2}$$

$\therefore$ $$\mu_{s(min)} = \frac{g}{l\omega^2}$$

12. (c) The mass of part AB is assumed to be concentrated at its centre.

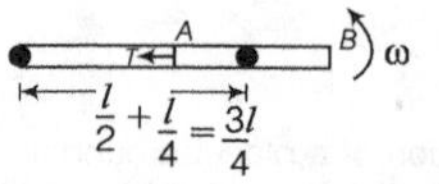

$\therefore$ $$T = \left(\frac{m}{2}\right)\frac{3l}{4}\omega^2 = \frac{3ml\omega^2}{8}$$

13. (b) Ratio $= \dfrac{mR\omega^2}{mg} = \dfrac{R\omega^2}{g} = \dfrac{6400 \times 10^3}{9.8}\left(\dfrac{2\pi}{24 \times 60 \times 60}\right)^2 = \dfrac{1}{303}$

14. (b) Here frictional force provides centripetal acceleration

$\therefore$ $$f = \frac{mv^2}{r} \Rightarrow f \le \mu mg$$

$\therefore$ $$\frac{mv^2}{r} \le \mu mg$$

$$v_{max} = \sqrt{\mu gr} = \sqrt{0.6 \times 10 \times 150}$$
$$= 30 \text{ m/s}$$

15. (b) The speed for side slipping is $v^2 = \mu gr$ and for safe turning

$$v^2 = \frac{grd}{2h} \Rightarrow \mu gr = \frac{grd}{2h}$$

$\therefore$ $$h = \frac{d}{2\mu}$$

16. (c) This will happen if speeds for side slipping and overturning are same.

$\therefore$ $$\left(\frac{\mu + \tan\alpha}{1 - \mu\tan\alpha}\right)gr = \left(\frac{d + 2h\tan\alpha}{2h - d\tan\alpha}\right)gr$$

By solving, $$h = \frac{d}{2\mu}$$

17. (d) From force diagram shown in figure,

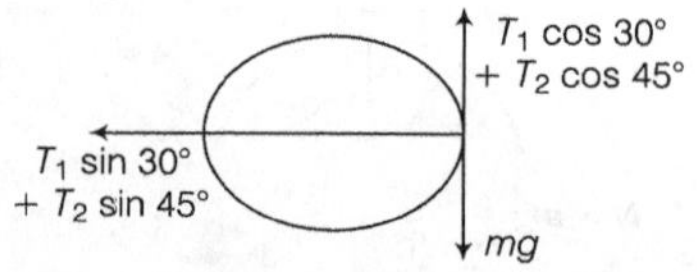

$$T_1\cos 30° + T_2\cos 45° = mg \quad \text{...(i)}$$

$$T_1\sin 30° + T_2\sin 45° = \frac{mv^2}{r} \quad \text{...(ii)}$$

After solving Eqs. (i) and (ii), we get

$$T_1 = \frac{mg - \frac{mv^2}{r}}{\frac{(\sqrt{3}-1)}{2}}$$

But $$T_1 > 0$$

$\therefore$ $$\frac{mg - \frac{mv^2}{r}}{\frac{\sqrt{3}-1}{2}} > 0$$

or $$mg > \frac{mv^2}{r} \text{ or } v < \sqrt{rg}$$

$\therefore$ $$v_{max} = \sqrt{rg} = \sqrt{1.6 \times 9.8} = 3.96 \text{ m/s}$$

18. (c) Here, $R = 0.2$ m

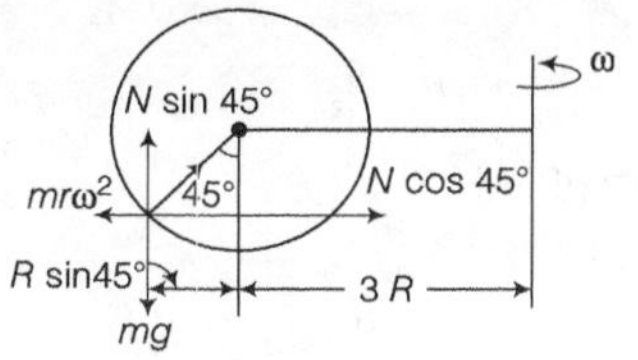

From figure, $r = 3R + R \sin 45°$

In the frame of rotating spherical chamber (non-inertial)

$$N \cos 45° = mr\omega^2 \quad \text{...(i)}$$

$$N \sin 45° = mg \quad \text{...(ii)}$$

$$\therefore \quad \tan 45° = \frac{mg}{mr\omega^2} = \frac{g}{r\omega^2}$$

$$\therefore \quad \omega = \sqrt{\frac{g}{r}} = \sqrt{\left(\frac{g}{3R + \frac{R}{\sqrt{2}}}\right)} = 3.64 \text{ rad/s}$$

19. (c,d) At highest point,

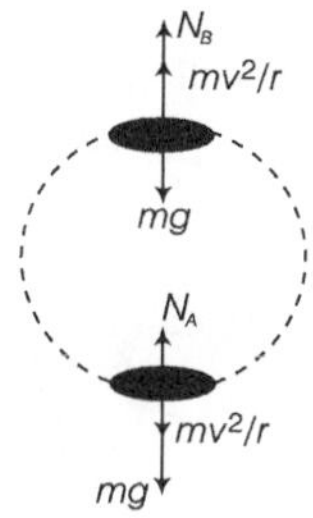

$$mg - N_B = \frac{mv^2}{R}$$

$$\therefore \quad N_B = 1000 - \frac{100 \times 100 \times 100}{1000}$$

$$N_B = 1000 - 1000 = 0$$

At lowest point, $N_A - mg = \dfrac{mv^2}{R}$

$$\therefore \quad N_A = mg + \frac{mv^2}{R}$$

$$= 1000 + 100 \times \frac{100 \times 100}{1000}$$

$$= 2000 \text{ N}$$

20. (a,b,c,d) $\tan \theta = \dfrac{v^2}{rg}$

Also, $v_{max} = \sqrt{\mu_s rg} = \sqrt{\dfrac{1}{\sqrt{3}} \times \sqrt{3} \times 10} = \sqrt{10}$ m/s

$$\therefore \quad \tan \theta_{max} = \frac{v_{max}^2}{rg} = \frac{\mu_s rg}{rg} = \frac{1}{\sqrt{3}}$$

$$\Rightarrow \quad \theta_{max} = 30°$$

21. (b) At the ball B,

$T_2 \sin 45° = mg + T_1 \sin 45°$ (in vertical direction)

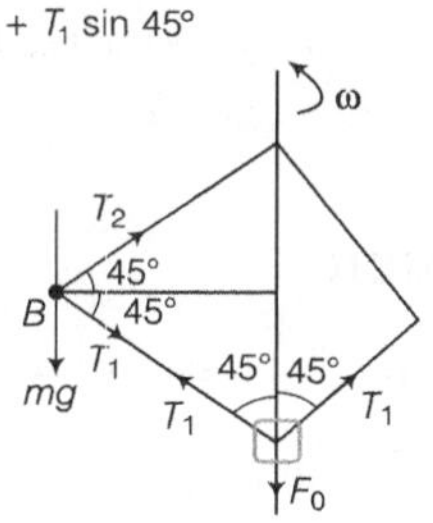

$$T_2 - T_1 = \frac{mg}{\sin 45°} = \sqrt{2}\, mg \quad \text{...(i)}$$

22. (a) $T_1 \cos 45° + T_2 \cos 45° = mr\omega^2$ (in horizontal direction)

$$(T_1 + T_2) \cos 45° = m\frac{l}{\sqrt{2}}\omega^2$$

or $\quad T_1 + T_2 = ml\omega^2 \quad$ ...(ii)

From Eqs. (i) and (ii), we get

$$T_1 = \frac{ml\omega^2}{2} - \frac{\sqrt{2}}{2}mg$$

For bead, $2T_1 \cos 45° = F_0$

$$\therefore \quad F_0 = \sqrt{2}ml\omega^2 - 2\, mg$$

23. (2) $N \cos 45° + T \cos 45° = mg$

(in vertical direction)

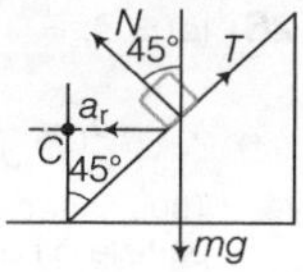

or $\quad \dfrac{N}{\sqrt{2}} + \dfrac{T}{\sqrt{2}} = 10$

$\therefore \quad N + T = 10\sqrt{2}$ N ...(i)

Also, in horizontal direction

$$N \sin 45° - T \sin 45° = mr\omega^2 = m \times \frac{1}{\sqrt{2}} \sin 45° \times 2^2$$

or $\quad \dfrac{N}{\sqrt{2}} - \dfrac{T}{\sqrt{2}} = 2$

or $\quad N - T = 2\sqrt{2} \quad$ ...(ii)

From Eqs. (i) and (ii), we get

$$T = 4\sqrt{2} \text{ N} = \sqrt{2 \times 16} = (32)^{1/2}$$

$$\therefore \quad x = 2$$

24. (3) As, in the block, A is in limiting equilibrium,

$$\therefore \quad m_B g = T_0 = \mu m_A g$$

$$\therefore \quad m_B = \mu m_A \quad \text{...(i)}$$

The initial motion of the ball in the frame of string is on circular path of radius l.

$$\therefore \quad T + m_B a_0 - m_B g = \frac{m_B v_0^2}{l} \quad \text{...(ii)}$$

For block A, $\quad T - \mu m_A g = m_A a_0 \quad$ ...(iii)

From Eqs. (i), (ii) and (iii), we get

$$\mu m_A g + m_A a_0 + m_B a_0 - m_B g = \frac{m_B v_0^2}{l}$$

or $\quad m_B g + m_A a_0 + m_B a_0 - m_B g = \dfrac{m_B v_0^2}{l}$

or $\quad \dfrac{m_B}{\mu} a_0 + m_B a_0 = \dfrac{m_B v_0^2}{l}$

or $\quad a_0\left(1 + \dfrac{1}{\mu}\right) = \dfrac{v_0^2}{l}$

$$\therefore \quad a_0 = \frac{\mu v_0^2}{(1 + \mu)\, l} = \frac{0.5 \times 1^2}{(1.5) \times 0.5} = \frac{10}{15} = \frac{2}{3} \text{ m/s}^2 = \frac{2}{n}$$

$$\therefore \quad n = 3$$

25. (a) (i) $N_B - mg = \dfrac{mv_0^2}{R}$

$$\therefore \quad N_B = mg + \frac{mv_0^2}{R} = 11 \text{ N}$$

(ii) At D,

Since, speed is constant.

$$\therefore \quad a_t = 0 \Rightarrow f_D = ma_t = 0$$

(iii) Just before C,

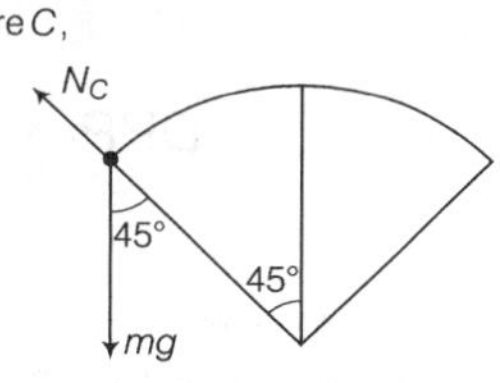

$$mg \cos 45° - N_C = \frac{mv_0^2}{R}$$

$$\therefore \quad N_C = \frac{mg}{\sqrt{2}} - \frac{mv_0^2}{R} = (5\sqrt{2} - 1) \text{ N}$$

(iv) The bug is about to slide just before point C,

$$f - mg \sin 45° = ma_t = 0$$

$$\therefore \quad f = mg \sin 45° \text{ or } \mu N_C = \frac{mg}{\sqrt{2}}$$

$$\therefore \quad \mu = \frac{mg}{\sqrt{2}\left(\frac{mg}{\sqrt{2}} - \frac{mv_0^2}{R}\right)} = \frac{g}{\left(g - \frac{\sqrt{2}v_0^2}{R}\right)} = \frac{10}{10 - \sqrt{2}}$$

(i) – (s), (ii) – (p), (iii) – (q), (iv) – (r)

26. (d) $\because a_r = \frac{v^2}{r} = 2t^2 \Rightarrow v = \sqrt{2r}\,t$

$\therefore \quad a_t = \frac{dv}{dt} = \sqrt{2r}$

Thus, tangential acceleration is non-zero. So, the motion of the particle is not a uniform circular motion.

The tangential acceleration in the case of a uniform circular motion is zero. So, acceleration of the particle in the case of a uniform circular motion is centripetal acceleration.

$\therefore F = ma_r$ directed towards centre.

27. (a) Stationary particle starts to move in the direction of acceleration.

28. (a)

29. (i) Here, $\sin\theta = \frac{r}{l} = \frac{0.5}{1.3} = 5/13$ and $\tan\theta = \frac{5}{12}$, $T\cos\theta = mg$

and $\quad T\sin\theta = mr\omega^2$

$\therefore \quad \tan\theta = \frac{r\omega^2}{g}$

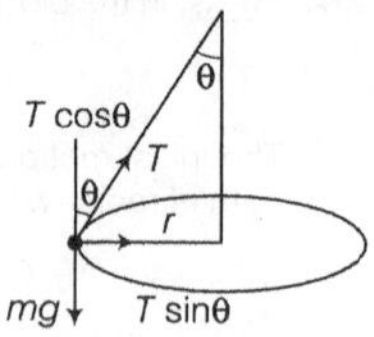

$\Rightarrow \quad \sqrt{\frac{5}{12} \times \frac{10}{0.5}} = \omega$

$\therefore \quad \omega = \sqrt{\frac{100}{12}} = \frac{10}{\sqrt{12}}$

$\therefore \quad t = \frac{2\pi}{\omega} = \frac{2\pi}{10}\sqrt{12} = \frac{2\pi \times 2\sqrt{3}}{10}$

$\therefore \quad = \frac{2\pi}{5}\sqrt{3}\text{ s}$

(ii) $T = \frac{mg}{\cos\theta} = \frac{0.2 \times 10}{\cos\theta} = \frac{2}{12} \times 13 = \frac{13}{6} = 2.17\text{ N}$

30. (i) From figure,

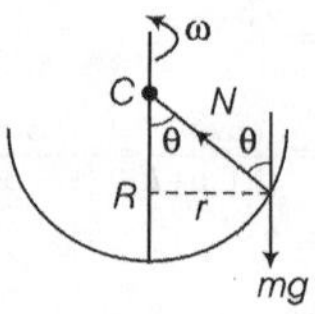

$\tan\theta = \frac{r}{R-h}$

and $\quad N\cos\theta = mg$ (in vertical direction)

Also, $\quad N\sin\theta = mr\omega^2$

$\therefore \quad \frac{N\sin\theta}{N\cos\theta} = \frac{mr\omega^2}{mg} = \frac{r\omega^2}{g}$

or $\quad \tan\theta = \frac{r\omega^2}{g}$ or $\frac{r}{R-h} = \frac{r\omega^2}{g}$

$\therefore \quad h \approx R - \frac{g}{\omega^2}$

For non-zero value of h,

$R > \frac{g}{\omega^2}$

$\therefore \quad \omega_{min} = \sqrt{\frac{g}{R}} = \sqrt{\frac{9.8}{0.1}} = 9.89\text{ rad/s}$

(ii) $\because \quad h = R - \frac{g}{\omega^2}$

If R and ω are known precisely, then

$\Delta h \approx \frac{-\Delta g}{\omega^2}$

$\therefore \quad (\Delta g)_{min} = \omega^2_{min}\,\Delta h = 9.8 \times 10^{-3}\text{ m/s}^2$

31. If there is no friction,

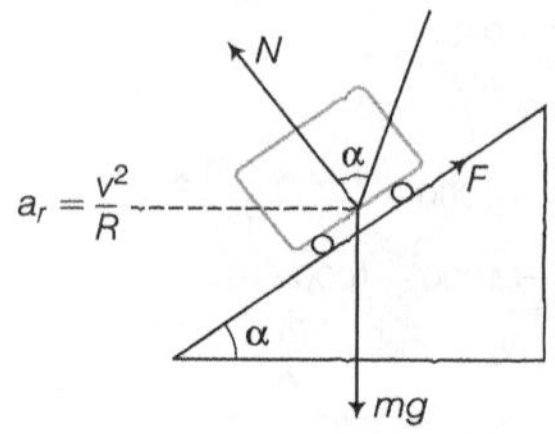

$N\cos\alpha = mg$ (in vertical direction)

and $\quad N\sin\alpha = \frac{mv_0^2}{R}$ (in horizontal direction)

$\therefore \quad \tan\alpha = \frac{v_0^2}{Rg}$

If friction is present, then

$N\cos\alpha + f\sin\alpha = mg$ (in vertical direction)

and $\quad N\sin\alpha - f\cos\alpha = \frac{mv^2}{R}$ (in horizontal direction)

$\therefore \quad f = \frac{mg\cos\alpha\,(v_0^2 - v^2)}{gR}$

(i) If $v < v_0$, f is positive, i.e. friction has the direction towards outer wheels.

(ii) If $v > v_0$, f is negative, i.e. friction has the direction towards inner wheels.

DPP-3 Non-Uniform Circular Motion

1. (b) $a_r = \frac{v^2}{r}$

Which looks like a parabola $y = 4ax^2$

So, option (b) is true. $a_r \propto v^2$, thus graph is parabolic.

2. (d) Since, ω is constant, so no net force or torque is acting on ring. The force experienced by any particle is only along radial direction or we can say the centripetal radial force.

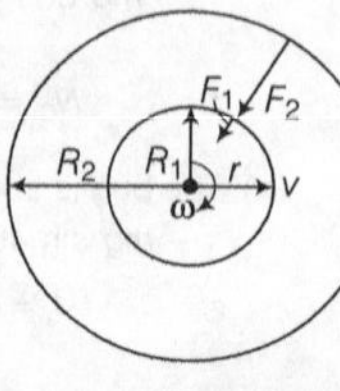

The force experienced by inner part $F_1 = m\omega^2 R_1$ and the force experienced by outer part,

$F_2 = m\omega^2 R_2 \Rightarrow \frac{F_1}{F_2} = \frac{R_1}{R_2}$

3. (d) $v = \mu g t = 0.2 \times 10 \times 2 = 4\text{ m/s}$

4. (b) $\quad N = \frac{mv^2}{R}$ and $\mu N = ma_t$

or $\quad \frac{\mu mv^2}{R} = ma_t$ or $a_t = \frac{\mu v^2}{R}$

or $\quad \frac{v\,dv}{ds} = \frac{\mu v^2}{R}$ or $\frac{v\,dv}{R\,d\theta} \approx \frac{\mu v^2}{R}$

or $\quad \int_{v_0}^{v} \frac{dv}{v} = \mu \int_0^{2\pi} d\theta$ or $\ln\frac{v}{v_0} = 2\pi\mu$

$\therefore \quad v = v_0 e^{2\pi\mu}$

5. (a) Since, block raises to some height in all cases speed of block v_0 is same at highest point. Let speed of block at highest point is v_0, then

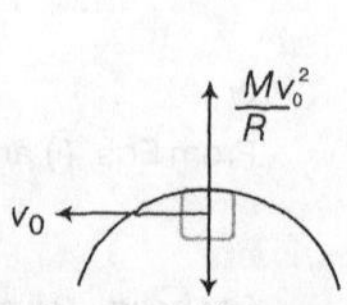

$N + mg = \frac{mv_0^2}{R} \Rightarrow N = \frac{mv_0^2}{R} - mg$

As R is minimum in first case, so N will be maximum.

6. (b) Different angular positions associated with the oscillation of pendulum is shown below.

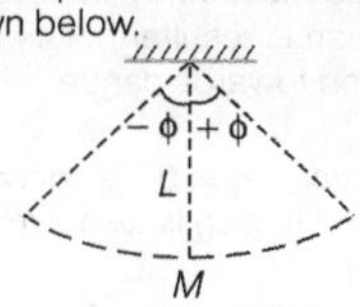

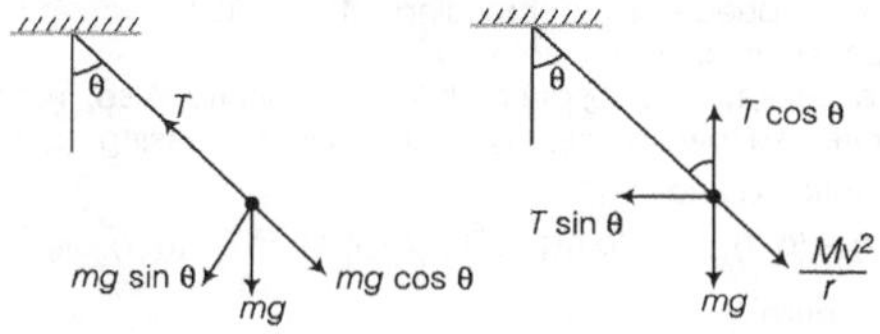

Thus, equations of motion are

$$T - Mg\cos\theta = \frac{Mv^2}{L} \quad \text{and} \quad a_T = g\sin\theta$$

7. (a) From figure, $N = ma_t = m < \alpha$

$$f = mL\omega^2 \quad \text{or} \quad \mu N = mL\omega^2$$

or $\mu L\alpha = mL\omega^2$ or $\mu L\alpha = mL(\alpha t)^2$

$\therefore$ $t = \sqrt{\frac{\mu}{\alpha}}$

8. (a,b,c,d) All options are true. When a particle moves along a circular path with a variable speed, it posseses a tangential acceleration along with a centripetal (radial) acceleration.

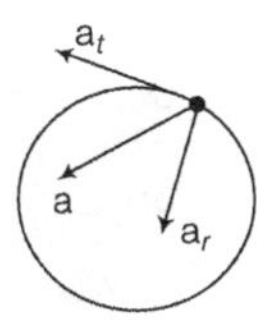

Clearly, net force is not directed towards the centre of path.

Also, a_t = tangential acceleration = $\frac{dv}{dt} = \frac{rd\omega}{dt} = r\alpha$

$\Rightarrow F_t$ = tangential force = $mr\alpha$ and a_r = radial acceleration = $\frac{v^2}{r}$

$\Rightarrow F_r$ = radial force = $\frac{mv^2}{r}$.

Kinematic equations given are also true.

9. (a,b,c,d) Here, $T - mg\cos\theta = \frac{mv_0^2}{l}$

All options are true.

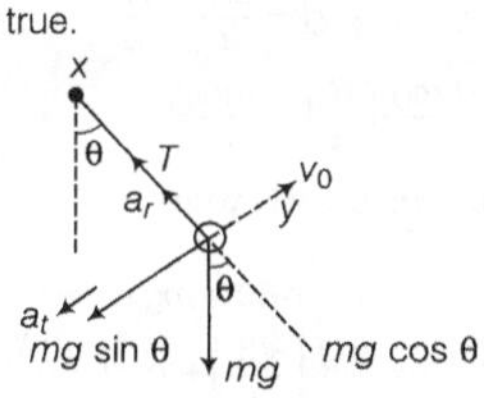

(a) $\therefore$ $T = mg\cos\theta + \frac{mv_0^2}{l}$

and $-mg\sin\theta = ma_t$

(b) $\therefore$ $a_t = -g\sin\theta$

(c) $\therefore$ $a_t = g\sin\theta$ (in magnitude)

Also, $a_t = l\alpha$

(d) $\alpha = \frac{g}{l}\sin\theta$

10. (a,c) Options (a) and (c) are true.

$$T - mg\cos 37^\circ = \frac{mv^2}{l}$$

$$= \frac{m}{l}(0)^2 = 0$$

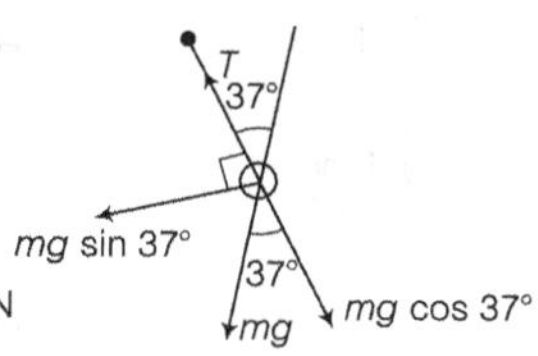

$\therefore T = mg\cos 37^\circ = 10 \times \frac{4}{5} = 8\text{ N}$

and $mg\sin 37^\circ = ma_t$

$\therefore$ $a_t = 10 \times \frac{3}{5} = 6\text{ m/s}^2$

11. (c,d) Free body diagram for cyclist is shown as below:

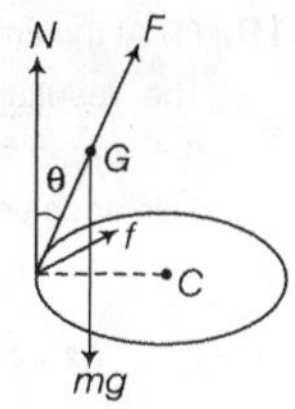

Here, $f = ma = m\sqrt{a_r^2 + a_t^2} = m\sqrt{\left(\frac{v^2}{r}\right)^2 + a_0^2}$

Thus, friction is not directed towards the centre of circular path.

Also, friction provides tangential as well as centripetal acceleration.

12. (d) Friction is zero and hence block starts to slip in infinitesimally small time.

13. (a) $\mu mg = ml\omega_{max}^2$ $\therefore$ $\omega_{max} = \sqrt{\frac{\mu g}{l}}$

14. (d) For slipping of bead to start, $f \le mr\omega^2$ but in given case, $f = 0$ as there is no gravity, so it take no time for sliding of bead.

15. (b) Here, $N = ma_t$ or $N = ml\alpha$

or $\mu N = ml\omega^2$ or $\mu ml\alpha = ml\omega^2$

$\therefore$ $\omega = \sqrt{\mu\alpha}$ or $\alpha t = \sqrt{\mu\alpha}$

$\therefore$ $t = \sqrt{\frac{\mu}{\alpha}}$

16. (a) Here, normal reactions are as given below:

$$N = \sqrt{(mg)^2 + (ml\alpha)^2}$$

$\because$ $\mu N = ml\omega^2$ or $\mu N = ml(\alpha t)^2$

or $\mu\sqrt{(mg)^2 + (ml\alpha)^2} = ml\alpha^2 t^2$

$\therefore$ $t^2 = \frac{\mu\sqrt{g^2 + l^2\alpha^2}}{l\alpha^2}$

$\therefore$ $t = \sqrt{\frac{\mu}{l\alpha^2}}(g^2 + l^2\alpha^2)^{\frac{1}{4}}$

17. (b) From figure, $mg\sin\theta = ma_t$

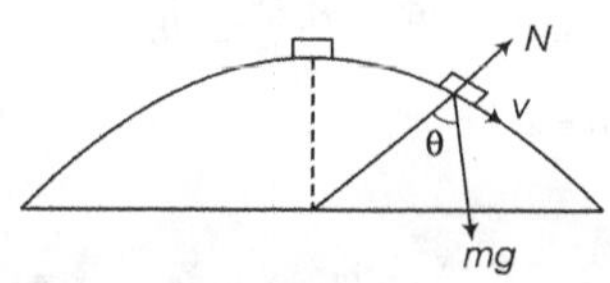

or $g\sin\theta = a_t = \frac{vdv}{ds} = \frac{vdv}{rd\theta}$ or $gr\int_0^\alpha \sin\theta d\theta = \int_0^v vdv$

or $gr[-\cos\theta]_0^\alpha = \frac{v^2}{2}$ or $gr[1 - \cos\alpha] = \frac{v^2}{2}$

or $\frac{v^2}{r} = 2g(1 - \cos\alpha)$

Also, at the time of breaking, $N = 0$

$\therefore$ $mg\cos\alpha = \frac{mv^2}{r}$

or $g\cos\alpha = 2g(1 - \cos\alpha)$

or $\cos\alpha = \frac{2}{3}$ $\Rightarrow$ $\sin\alpha = \frac{\sqrt{5}}{3}$

$\therefore$ $a_t = g\sin\alpha$

(at the time of breaking the surface)

$$= \frac{\sqrt{5}}{3}g$$

At the time of break off the surface,

$$mg = ma \Rightarrow a = g$$

18. (d) From figure, it is clear that acceleration is not directed towards centre of circular path.

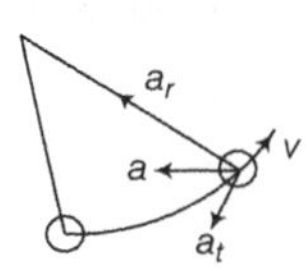

19. (4) At the time of skidding, friction is limiting friction (i.e. μmg).

The resultant acceleration of bug at the time of skidding is $a = \sqrt{a_r^2 + a_0^2}$

This acceleration is provided by the friction.

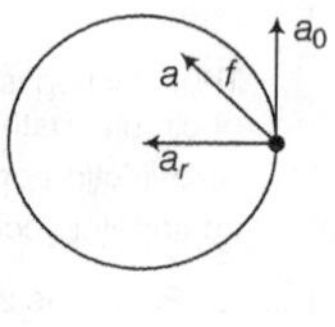

$\therefore \quad f = ma \quad$ or $\quad \mu mg = m\sqrt{a_r^2 + a_0^2}$

or $\quad \mu^2 g^2 - a_0^2 = \left(\frac{v^2}{R}\right)^2$

or $\quad v^4 = R^2(\mu^2 g^2 - a_0^2)$

$\therefore \quad v = \{R^2(\mu^2 g^2 - a_0^2)\}^{1/4}$

But $\quad v = a_0 t$

$\therefore \quad t = \frac{v}{a_0} = \frac{\{R^2(\mu^2 g^2 - a_0^2)\}^{1/4}}{a_0}$

$\therefore \quad n = 4$

20. (2) Since, board is accelerating upward, the normal reaction on the car is $N - mg = ma_1$

$\therefore \quad N = mg + ma_1$

At the time of skidding, limiting friction provides the required acceleration.

$\therefore \quad \mu N = ma$

or $\quad \mu m(g + a_1) = m\sqrt{a_r^2 + a_t^2}$

or $\quad \mu^2(g + a_1)^2 = \left(\frac{v^2}{R}\right)^2 + a_0^2$

$\therefore \quad v = [R^2\{\mu^2(g + a_1)^2 - a_0^2\}]^{1/4}$

$\therefore \quad n = 2$

21. (2) Here, frictional force provides the acceleration in the car.

$\therefore \quad f = m\sqrt{a_r^2 + a_t^2} = m\sqrt{(R\omega^2)^2 + (R\alpha)^2}$

Here, $\quad \omega^2 = \frac{b^2}{R}\theta \quad$ or $\quad 2\omega\frac{d\omega}{d\theta} = \frac{b^2}{R} \quad$ or $\quad \alpha = \frac{b^2}{2R}$

$\therefore \quad f = \sqrt{\frac{R^2 b^4 \theta^2}{R^2} + \frac{R^2 b^4}{4R^2}} = \sqrt{b^4\theta^2 + \frac{b^4}{4}}$

$\therefore \quad n = 2$

22. (1) $a_r = bt^2$

or $\quad \frac{v^2}{R} = bt^2 \quad$ or $\quad v = \sqrt{bRt^2}$

or $\quad v = \sqrt{bR}\, t \quad$ or $\quad a_t = \frac{dv}{dt} = \sqrt{bR}$

The required acceleration for vehicle is provided by friction

$\therefore \quad f = ma = m\sqrt{a_r^2 + a_t^2} = m\sqrt{b^2 t^4 + bR}$

$\therefore \quad n = 1$

23. (5) For block, $N = m_2 a_2 = 2a_2$

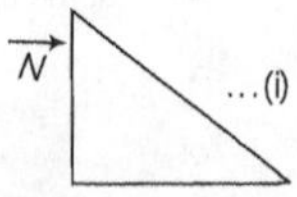

$\therefore \quad N = 2a_2 \quad$ ...(i)

For ball,

In tangential direction,

$mg \sin 45° - N \sin 45° = ma_t \quad$ ...(ii)

From the concept of wedge constraint, the horizontal component of tangential acceleration of ball is equal to the acceleration of the triangular block.

Thus, $\quad a_t \sin 45° = a_2 \quad$ or $\quad \frac{a_t}{\sqrt{2}} = a_2$

or $\quad a_t = \sqrt{2}a_2$

From Eq. (ii), $\quad \frac{mg}{\sqrt{2}} - \frac{N}{\sqrt{2}} = 1 \times \sqrt{2}a_2 \quad$ or $\quad 10 - N = 2a_2$

or $\quad 10 - 2a_2 = 2a_2 \quad$ [from Eq. (i)]

$\therefore \quad a_2 = \frac{10}{4} = 2.5 \text{ m/s}^2$

$\therefore \quad 2a_2 = 5 \text{ m/s}^2$

24. (a) (i) When a particle moves with increasing speed, net acceleration is resultant of a_T and a_r which is not directed towards centre.

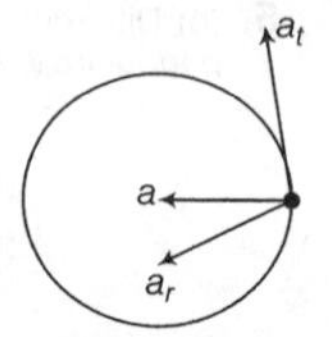

Also, $a_r \neq 0$, (q,r)

(ii) For a rotating car, $a_r \neq 0$, is directed towards centre and friction is also directed towards centre. (p, r)

(iii) As speed is decreasing tangential acceleration is present and acceleration is not directed towards centre. Also, friction is not directed towards centre as speed is decreasing. (q, r, s)

(iv) Option q and r.

(i) – (q ,r), (ii) – (p,r,q), (iii) – (q, r, s), (iv) – (q, r)

25. From figure,

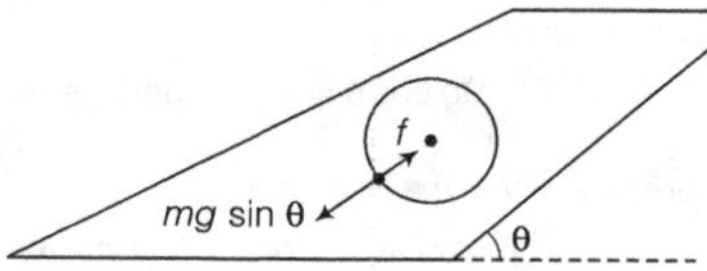

$$f - mg\sin\theta = \frac{mv^2}{r}$$

or $$\mu N - mg\sin\theta = \frac{mv_{max}^2}{r}$$

or $$\mu mg\cos\theta - mg\sin\theta = \frac{mv_{max}^2}{r}$$

or $$\mu g\cos\theta - g\sin\theta = \frac{v_{max}^2}{r} \quad \text{(here, } \theta = 30°)$$

$\therefore \quad v_{max} = \sqrt{gr(\mu\cos\theta - \sin\theta)} = \sqrt{gr(\mu\cos 30° - \sin 30°)}$

$$= \sqrt{10 \times 10\left(\frac{2}{\sqrt{3}} \times \frac{\sqrt{3}}{2} - \frac{1}{2}\right)} = \frac{10}{\sqrt{2}} = 5\sqrt{2} \text{ m/s}$$

26. From the diagram shown below, we have

$CP = CO \Rightarrow \angle CPO = \angle POC = 60°$

ΔOCP is an equilateral triangle.

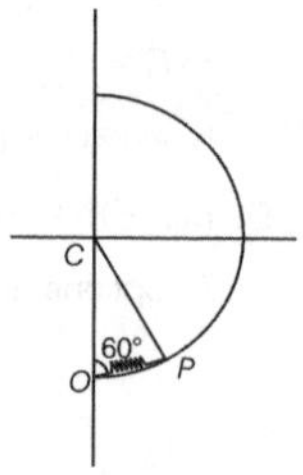

Natural length of spring is $\frac{3R}{4}$

Extension in spring, $\quad x = R - \frac{3R}{4} = \frac{R}{4}$

Spring force, $\quad kx = \frac{mg}{R}\left(\frac{R}{4}\right) = \frac{mg}{4}$.

Ring will move towards point O when released.

$\therefore \quad F_t = kx \sin 60° + mg \sin 60°$

$$= \frac{mg}{4}\left(\frac{\sqrt{3}}{2}\right) + mg\left(\frac{\sqrt{3}}{2}\right) = \frac{5\sqrt{3}}{8}mg$$

(i) Free body diagram of ring

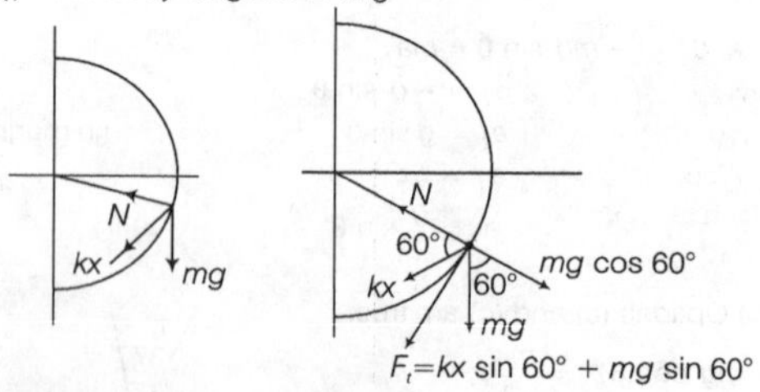

(ii) Tangential acceleration of the ring

$$a_T = \frac{F_t}{m} = \frac{5\sqrt{3}}{8}g$$

Net force along radial direction on ring just after release will be zero

$\Rightarrow \quad F_r = 0 \quad$ or $\quad N + kx\cos 60° = mg\cos 60°$

$\Rightarrow \quad N = \frac{3\,mg}{8}$.

DPP-4 Centrifugal Force

1. (a) Outward centrifugal force acts through the centre of mass and friction oppose the motion acting along surfaces of contact.

2. (b) Centrifugal force with respect to frame

$= mr\omega^2$, where ω = angular speed of the frame.

3. (c) Equation of motion can be written as,

$$mr\omega^2 - \mu mg = ma_{rel}$$

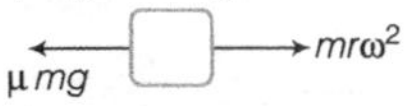

or $\quad r\omega^2 - \mu g = a_{rel}$

or $\quad 1 \times 25 - 0.5 \times 10 = a_{rel}$

$\therefore \quad a_{rel} = 20\ \text{m/s}^2$

4. (c) In the frame of platform,

$$m_1 r_1 \omega^2 - m_2 r_2 \omega^2 = (m_1 + m_2) a_{rel}$$

$$\therefore \quad a_{rel} = \left(\frac{m_1 r_1 - m_2 r_2}{m_1 + m_2}\right)\omega^2 = 1\ \text{m/s}^2$$

$$\therefore \quad a_1 = \sqrt{a_{rel}^2 + (r_1\omega^2)^2} = \sqrt{1^2 + 4} = \sqrt{5}\ \text{m/s}^2$$

5. (c) In the frame of platform,

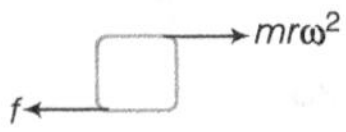

$f = mr\omega^2$

or $\quad \mu mg = mr\omega^2 \quad$ or $\quad \mu mg = mv_0 t\ \omega^2$

or $\quad t = \dfrac{\mu g}{v_0 \omega^2} = \dfrac{0.5 \times 10}{1 \times 1} = 5\ \text{s}$

6. (a) In the frame of rod,

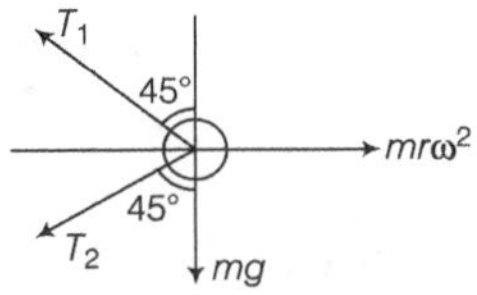

$T_1 \sin 45° + T_2 \sin 45° = mr\omega^2$

or $T_1 + T_2 = \dfrac{ml \sin 45°\, \omega^2}{\sin 45°}$

$= \sqrt{2} \times \sqrt{2} \times 10 = 20 \quad \ldots\text{(i)}$

In vertical direction, $\quad T_1 \cos 45° = T_2 \cos 45° + mg$

or $\quad T_1 - T_2 = \dfrac{mg}{\cos 45°} = \dfrac{\sqrt{2} \times 10}{\frac{1}{\sqrt{2}}} = 20 \quad \ldots\text{(ii)}$

From Eqs. (i) and (ii), we get $2T_1 = 40$

$\therefore \quad T_1 = 20\ \text{N}$

7. (b) $T \cos \alpha = mg$ and $T \sin \alpha = mr\omega^2$

So, when ω increases, α tends to increase and $\cos \alpha$ tends to zero. Hence, correct relation of $\cos \alpha$ and ω is graph b.

8. (a) Magnitude of relative speed is

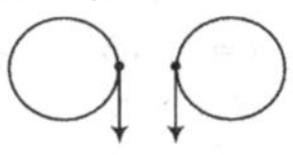

$t = 0, |v_r| = 0$

$t = T/4, |v_r| = 2v = 2$

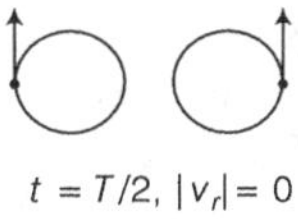

$t = T/2, |v_r| = 0$

So, correct graph is option a.

9. (c) At $t = \dfrac{1}{8} \times \dfrac{2\pi}{\omega} = \dfrac{\pi}{4\omega}$

x-coordinate of $P = R\omega\left(\dfrac{\pi}{4\omega}\right) = \dfrac{\pi R}{4} > R \cos 45°$

$\therefore$ Both particles P and Q land in unshaded region.

10. (c) In the frame of table,

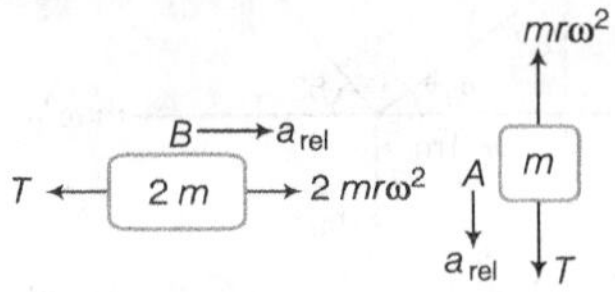

$2mr\omega^2 - T = 2ma_{rel} \quad \ldots\text{(i)}$ (for block B)

and $\quad T - mr\omega^2 = ma_{rel} \quad \ldots\text{(ii)}$ (for block A)

Adding Eqs. (i) and (ii), we get

$$mr\omega^2 = 3ma_{rel}$$

$$\therefore \quad a_{rel} = \frac{r\omega^2}{3}$$

From Eq. (ii), $\quad T = mr\omega^2 + \dfrac{mr\omega^2}{3} = \dfrac{4}{3} mr\omega^2$

$= \dfrac{4}{3} \times 1 \times 3 \times 4 = 16\ \text{N}$

11. (a,b) Here, $T \cos \theta = mg$

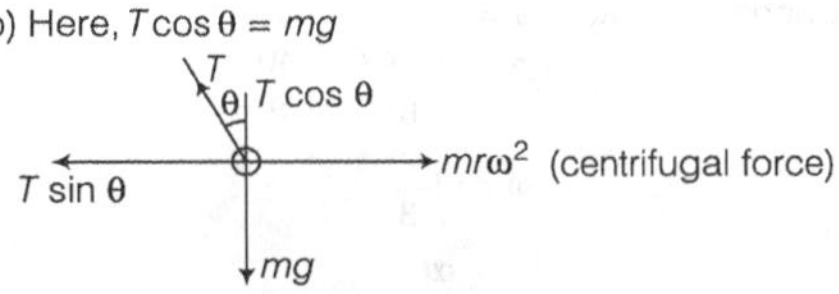

and $\quad T \sin \theta = mr\omega^2$

or $\quad \tan \theta = \dfrac{r\omega^2}{g} = \dfrac{\sqrt{2} \sin \theta\, \omega^2}{g}$ or $\dfrac{1}{\cos \theta} = \sqrt{2}$

$\therefore \quad \theta = 45°$

$\therefore \quad T = \dfrac{mg}{\cos \theta} = \dfrac{1 \times 10}{\frac{1}{\sqrt{2}}} = 10\sqrt{2}\ \text{N}$

12. (c,d)

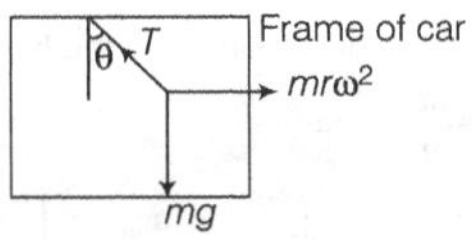

Resolving tension in the string,

$$T \cos \theta = mg,\ T \sin \theta = mr\omega^2$$

$$\tan \theta = \frac{r\omega^2}{g} = \frac{rv^2}{r^2 g} = \frac{v^2}{rg} = \frac{100}{10 \times 10} = 1$$

$$\Rightarrow \quad \theta = 45°$$

13. (a,d) To meet given condition,

$$v^2 = \left(\frac{\mu + \tan \theta}{1 - \mu \tan \theta}\right) gr$$

at $\theta = 45°, \mu = 1 \quad \Rightarrow \quad v = \infty$

Similarly for $\tan \theta = \mu$

Hence, any speed is possible.

14. (a) In the frame of turntable,

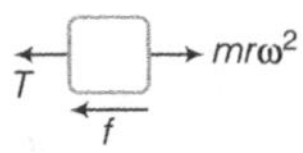

$T + f = mr\omega^2$

or $\quad 0 + \mu mg = mr\omega^2$

$\therefore \quad \omega = \sqrt{\dfrac{\mu g}{r}} = \sqrt{\dfrac{0.4 \times 10}{1}} = 2\ \text{rad/s}$

15. (d) $T + \mu mg = mr\omega^2$

$T + 0.4 \times 1 \times 10 = 1 \times 1 \times \omega^2$

$T + 4 = 16$

$\therefore \quad T = 16 - 4$

$= 12\ \text{N}$

16. (b) In the frame of triangular block,

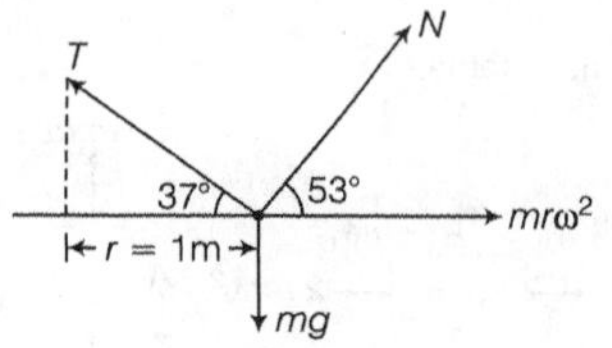

$$T\cos 37^\circ = N\cos 53^\circ + mr\omega^2 \quad \text{(in horizontal direction)}$$

or $\frac{4}{5}T - \frac{3}{5}N = \omega^2$

or $4T - 3N = 5\omega^2 \quad \ldots(i)$

and $T\sin 37^\circ + N\sin 53^\circ = mg$

or $\frac{3}{5}T + \frac{4}{5}N = 10$

or $3T + 4N = 50 \quad \ldots(ii)$

or $3\left(\frac{5\omega^2 + 3N}{4}\right) + 4N = 50$

or $\frac{15}{4}\omega^2 + \frac{9}{4}N + 4N = 50$

For leaving the surface, $N = 0$

$\Rightarrow \quad \omega^2 = \frac{50 \times 4}{15} = \frac{40}{3}$

$\therefore \quad \omega = \sqrt{\frac{40}{3}}$

$\because \quad \omega = \alpha t$

$\therefore \quad t = \frac{\sqrt{40/3}}{\sqrt{40/3}} = 1\text{s}$

17. (b) $\because 3T + 4N = 50$ or $3T = 50 \quad [\because N = 0]$

$\therefore \quad T = \frac{50}{3} = 16.67 \text{ N}$

18. (2) In the frame of rod, the force on sleeve along the rod is only centrifugal force.

Thus, the acceleration of the sleeve in the frame of rod is

$$a_{rel} = \frac{mr\omega^2}{m} = r\omega^2$$

or $v_{rel}\frac{dv_{rel}}{dr} = r\omega^2$ or $\int_0^{v_{rel}} v_{rel}\, dv_{rel} = \int_{r_0}^{l} r\omega^2 dr$

or $\frac{v_{rel}^2}{2} = \frac{\omega^2}{2}(l^2 - r_0^2)$

$\therefore \quad v_{rel} = \omega\sqrt{l^2 - r_0^2}$

The speed of the sleeve in the frame of ground is

$$v = \sqrt{v_{rel}^2 + (l\omega)^2} = \sqrt{\omega^2(l^2 - r_0^2) + l^2\omega^2}$$

$$= \sqrt{2l^2\omega^2 - \omega^2 r_0^2}$$

$\therefore \quad n = 2$

19. (5) Centripetal acceleration is provided by normal contact force

thus, we can write, $N = \frac{mv^2}{R}$

Also, $-\mu N = m\frac{dv}{dt}$ or $-\mu\frac{mv^2}{R} = m\frac{dv}{dt}$

or $-\frac{\mu}{R}\int_0^t dt = \int_{v_0}^{v}\frac{dv}{v^2}$ or $-\frac{\mu}{R}t = \left[\frac{v^{-2+1}}{-2+1}\right]_{v_0}^{v}$

or $-\frac{\mu t}{R} = \left[-\frac{1}{v}\right]_{v_0}^{v}$ or $-\frac{\mu t}{R} = \left[-\frac{1}{v} + \frac{1}{v_0}\right]$

or $\frac{1}{v} = \frac{1}{v_0} + \frac{\mu t}{R}$

or $v = \frac{v_0 R}{R + \mu t v_0} = \frac{10 \times 1}{1 + 0.1 \times 1 \times 10} = 5 \text{ m/s}$

20. (a) Centrifugal force is pseudo force, so it depends on angular velocity of the frame.

$\therefore$ Centrifugal force $= mr\omega_1^2$

21. (b) From free body diagram of block in the frame of disc,

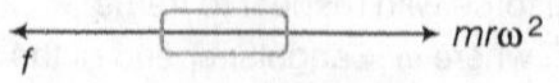

$$f = mr\omega^2$$

From free body diagram of block in the frame of ground,

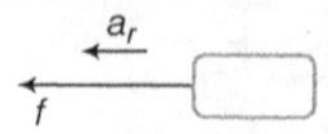

$$f = ma_r = mr\omega^2$$

22. (a) $R_4 < R_3 < R_1 < R_2 = R_5$

$\because \quad F = \frac{mv^2}{R} \quad$ (centrifugal force)

$\therefore \quad F_4 > F_3 > F_1 > F_2 = F_5$

(i) $\frac{|F_4|}{|F_3|} > 1$ (ii) $\frac{|F_2|}{|F_5|} = 1$

(iii) $\frac{|F_1|}{|F_4|} < 1$ (iv) $\frac{|F_2|}{|F_4|} < 1$

23. The situation can be illustrated as shown below.

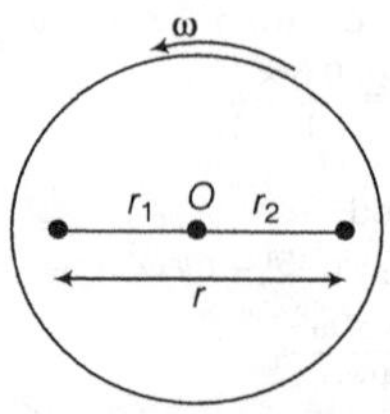

Given, $m_1 = 10 \text{ kg}, \ m_2 = 5 \text{ kg}$

$\omega = 10 \text{ rad/s}$

$r = 0.3 \text{ m}$

$r_1 = 0.124 \text{ m}$

$r_2 = r - r_1 = 0.176 \text{ m}$

(i) Free body diagrams for masses are

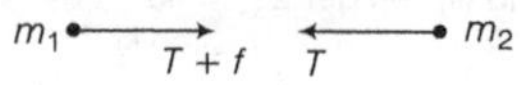

$T = m_2 r_2 \omega^2 \quad \ldots(i)$

As, $m_2 r_2 \omega^2 < m_1 r_1 \omega^2$

$\therefore \quad m_1 r_1 \omega^2 > T$

Now, $f + T = m_1 r_1 \omega^2 \quad \ldots(ii)$

$\Rightarrow \quad f = m_1 r_1 \omega^2 - m_2 r_2 \omega^2 = 36 \text{ N}$

$\therefore \quad f = 36 \text{ N}$

(ii) From above, $f = (m_1 r_1 - m_2 r_2)\omega^2$ masses start slipping when

$(m_1 r_1 - m_2 r_2)\omega^2 > f_{max} > \mu m_1 g$

$\therefore \quad \omega_{min} = \sqrt{\frac{\mu m_1 g}{m_1 r_1 - m_2 r_2}}$

$= 11.67 \text{ rad/s}$

(iii) From Eq. (ii), $f = 0$

When $m_1 r_1 = m_2 r_2$

or $\frac{r_1}{r_2} = \frac{m_2}{m_1} = \frac{5}{10} = \frac{1}{2}$

and $r_1 + r_2 = 0.3 \text{ m}$

$\therefore \quad r_1 = 0.1 \text{ m}$

and $r_2 = 0.2 \text{ m}$

DPP-5 Motion on General Curve

1. (i) (a) Acceleration can be resolved into two components $a\cos\theta$ along $\mathbf{v}$ and $a\sin\theta$ perpendicular to $\mathbf{v}$.

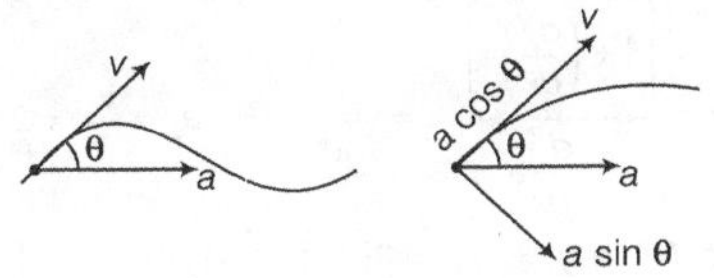

So, at point P, tangential acceleration is $a_t = a\cos\theta$.

(ii) (b) $\frac{d|\mathbf{v}|}{dt}$ = rate of change of magnitude of v = acceleration component along $v = a\cos\theta$

(iii) (c) Radial component of acceleration is perpendicular to the tangential component.

$\therefore \quad a_r = a\sin\theta$

(iv) (a) At point P, radius is perpendicular to velocity at point P

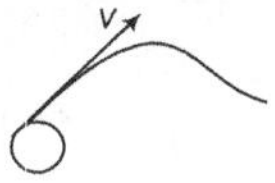

(v) (b) Radial acceleration = centripetal acceleration

$$= a_r = a\sin\theta = \frac{v^2}{r}$$

(vi) (a,c) As, $\quad a\sin\theta = \frac{v^2}{r} \quad \Rightarrow \quad \frac{v^2}{a\sin\theta} = r$

$$r = \frac{v^3}{av\sin\theta}$$

$$\Rightarrow \quad r = \frac{v^3}{|\mathbf{a}\times\mathbf{v}|}$$

2. (i) (b) Acceleration due to gravity, g can be resolved as shown below.

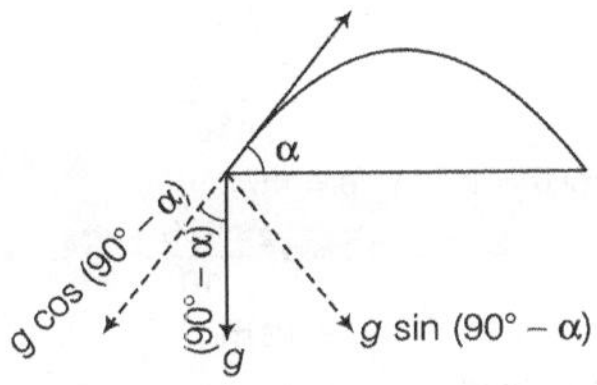

$\therefore \quad a_t = -g\cos(90°-\alpha) = -g\sin\alpha$

(ii) (c) $a_r = g\sin(90°-\alpha) = g\cos\alpha$

(iii) (a) Using, $\frac{v^2}{r} = a_{rad}$

$\Rightarrow \quad \frac{v^2}{r} = g\cos\alpha \quad$ or $\quad r = \frac{u^2}{g\cos\alpha}$

(iv) (b) At highest point, $u\cos\alpha$ is perpendicular to radius.

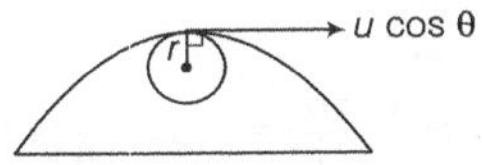

(v) (c) At maximum height, instantaneous velocity has only horizontal component and acceleration is perpendicular to it so at maximum height, $a_t = 0$

(vi) (a) At maximum height, $a_r = g$

(vii) (a) Using, $a_r = \frac{v^2}{r}$

$$a_r = g = \frac{u^2\cos^2\alpha}{r} \text{ at maximum height.}$$

(viii) (b) Also, at maximum height

$$g = \frac{u^2\cos^2\alpha}{r} \quad \text{or} \quad r = \frac{u^2\cos^2\alpha}{g}$$

3. (a) $\therefore$ Angular velocity, $\omega = \frac{10\sin 37°}{r} = \frac{10\times\frac{3}{5}}{5} = \frac{6}{5} = 1.2$ rad/s

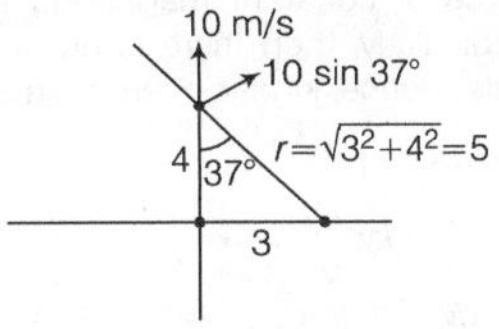

4. (c) $\mathbf{r} = \mathbf{PA}$ = Position vector of A – Position vector of P

$$= \hat{i} + 2\hat{j} + 2\hat{k} - (2\hat{i} + 4\hat{j} + 4\hat{k})$$

$$= -\hat{i} - 2\hat{j} - 2\hat{k}$$

$$\therefore \quad \omega = \frac{v\sin\theta}{r} = \frac{|\mathbf{v}\times\mathbf{r}|}{r^2} = 3.4 \text{ rad/s}$$

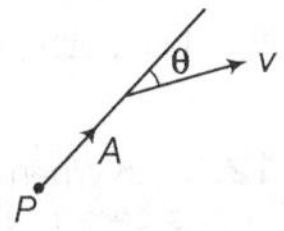

5. (a) Given, $\omega = \frac{d\theta}{dt} = 0.1$ rad/s

Also, $\tan\theta = \frac{OP}{OS} \quad \Rightarrow \quad OP = OS\tan\theta$

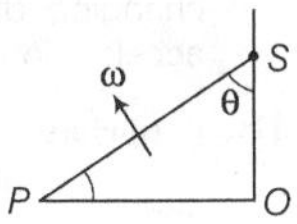

$\Rightarrow \quad OP = OS\tan\theta$

Differentiating w.r.t. t,

$$\frac{d}{dt}OP = \frac{d}{dt}OS\tan\theta$$

$$= OS\frac{d}{dt}\tan\theta = OS\sec\theta\cdot\frac{d\theta}{dt}$$

and velocity of sport S on wall $= \frac{d}{dt}OP = OS\sec^2\theta\cdot\frac{d\theta}{dt}$

$$= 3\times\sec^2 45°\times 0.1$$

$$= 3\times 2\times 0.1 = 0.6 \text{ m/s}$$

6. (d) $\omega_{rad} = \frac{8\sin 30° + 6\sin 30°}{10} = 0.7$ rad/s

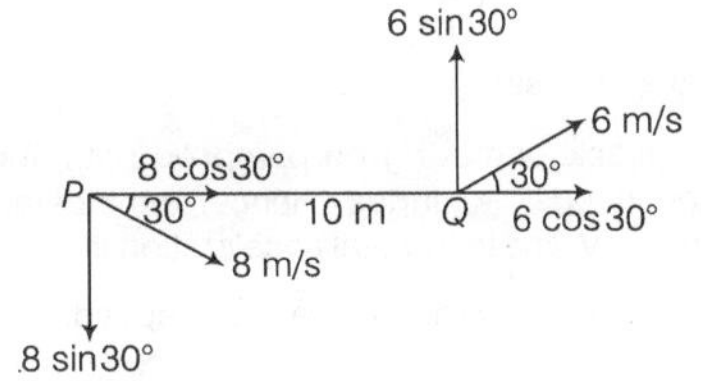

7. (b) Here, $N - mg = \frac{mv^2}{R}$

or $\quad 2mg - mg = \frac{mv^2}{R} \quad$ or $\quad mg = \frac{mv^2}{R}$

$\therefore \quad R = \frac{2gb}{g} = 2b$

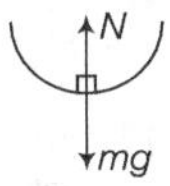

8. (b) $x = a\cos\omega t$ and $y = b\sin\omega t$

or $\quad v_x = -a\omega\sin\omega t$

and $\quad v_y = b\omega\cos\omega t$

and $\quad a_x = -a\omega^2\cos\omega t$

and $\quad a_y = -b\omega^2\sin\omega t$

Thus, acceleration at A is

$$a_0 = a_x = -a\omega^2 \quad (\because t = 0 \text{ at point } A)$$

and velocity $\quad v = v_y = b\omega$

$\therefore \quad |a_0| = \frac{v^2}{R} \quad$ or $\quad a\omega^2 = \frac{b^2\omega^2}{R}$

$\therefore \quad R = \frac{b^2}{a}$

9. (a) Here, $\omega = \dfrac{u}{R}$

$\therefore \quad a_r = r\omega^2 = \dfrac{ru^2}{R^2}$

10. (c) When a force of constant magnitude acts on velocity of particle perpendicularly, then there is no change in the kinetic energy of particle. Hence, kinetic energy remains constant.

11. (d) Given, $\mathbf{v} = ky\,\mathbf{i} + kx\,\mathbf{j}$

$$\Rightarrow \quad \frac{dx}{dt} = ky,\ \frac{dy}{dt} = kx$$

$$\frac{dy}{dx} = \frac{dy}{dt} \times \frac{dt}{dx} = \frac{kx}{ky}$$

$$ydy = xdx \quad \text{...(i)}$$

Integrating both sides of Eq. (i), we have

$$y^2 = x^2 + c$$

12. (c,d) When force is perpendicular to velocity, work done by force is zero or $\mathbf{F} \cdot \mathbf{v} = 0$. The force then produces an acceleration by changing direction of velocity and it produces a centripetal acceleration which rotates the particle is circular path.

13. (a,b) Here, $\dfrac{d\theta}{dt} = \omega$

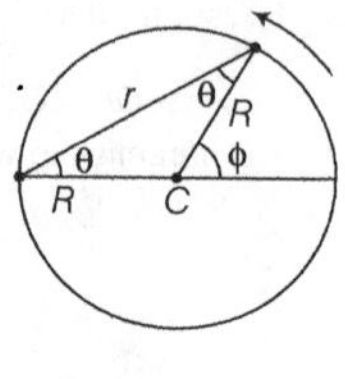

and $\quad \phi = 2\theta$

or $\quad \dfrac{d\phi}{dt} = \dfrac{2d\theta}{dt} = 2\omega$

Thus, acceleration is directed towards centre of circular path and is equal to

$$a = R(2\omega)^2 = 4R\omega^2$$

14. (a,b,c) $x = a\cos pt \Rightarrow \dfrac{x}{a} = \cos pt$

$$y = b\sin pt$$

$$\Rightarrow \quad \frac{y}{b} = \sin pt$$

$$\Rightarrow \quad \frac{x^2}{a^2} + \frac{y^2}{b^2} = 1$$

Hence, it is an ellipse.

15. (d) When particle is moving on parabolic path, the direction of motion continuously changes. Hence, centripetal acceleration comes into play. The tangential acceleration is

$$a_t = \frac{d|\mathbf{v}|}{dt} = \text{rate of change of speed.}$$

16. (a) $a = \sqrt{a_r^2 + a_t^2} \Rightarrow F = ma$ or $mg = ma$

$\therefore \quad a = g$

17. (3) If rod is rotated about end A, then vertical component of velocity v at end A will be zero.

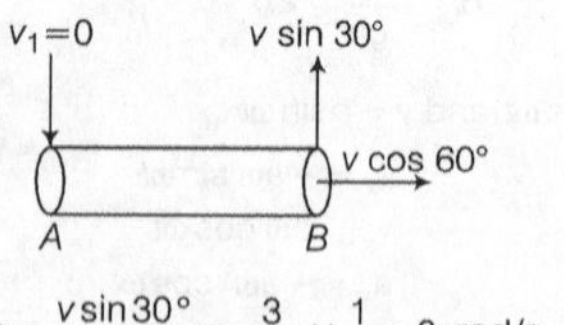

$$\omega = \frac{v\sin 30^\circ}{l} = \frac{3}{0.5} \times \frac{1}{2} = 3 \text{ rad/s}$$

18. (4) Here, $s = 2t^3$

$\therefore \quad v = \dfrac{ds}{dt} = 6t^2$

$\therefore \quad a_t = \dfrac{d|\mathbf{v}|}{dt} = 12t = 12 \text{ m/s}^2$ at $t = 1$ s

$\therefore \quad 3x = 12$

or $\quad x = 4$

19. (6) As, $\dfrac{dy}{dx} = -0.5 \times \dfrac{2\pi}{l} \sin\dfrac{2\pi x}{l} = 0$ at B

$$\Rightarrow \quad \frac{d^2y}{dx^2} = -0.5 \times \frac{4\pi^2}{l^2}\cos\frac{2\pi x}{l} = 2\pi^2 \text{ at } B$$

$$\therefore \quad r = \frac{\left[1 + \left(\frac{dy}{dx}\right)^2\right]^{3/2}}{\frac{d^2y}{dx^2}} = \frac{1}{2\pi^2} \Rightarrow a_r = \frac{v^2}{r} = 2\pi^2$$

At point B, $N - mg = ma_r = 2\pi^2 m$

$$\therefore \quad N = 2\pi^2 m + mg = mg\left(\frac{2\pi^2}{g} + 1\right)$$

$$= mg \times 3 = 30\text{ N} = 5n$$

$\therefore \quad n = 6$

20. (4) $x^2 = 200y$ or $2x = 200\dfrac{dy}{dx}$

$$\therefore \quad \frac{dy}{dx} = \frac{x}{100} \Rightarrow \frac{d^2y}{dx^2} = \frac{1}{100}$$

So, radius of curvature is

$$r = \frac{\left[1 + \left(\frac{dy}{dx}\right)^2\right]^{3/2}}{\frac{d^2y}{dx^2}} = \frac{\left[1 + \left(\frac{x}{100}\right)^2\right]^{3/2}}{\frac{1}{100}}$$

At deepest point, $x = 0$

$\therefore \quad r = 100$ m

$$\therefore \quad a_r = \frac{v^2}{r} = \frac{20 \times 20}{100} = 4 \text{ m/s}^2$$

21. (9) If wire is assumed as parabolic path, then

$$R = 10 = \frac{u^2 \sin 2\theta}{g}$$

and $\quad H = 2.5 = \dfrac{u^2 \sin^2\theta}{2g}$

$$\therefore \quad 4 = \frac{2\sin 2\theta}{\sin^2\theta}$$

$\therefore \quad \cot\theta = 1 \Rightarrow \theta = 45°$

and $\quad 10 = u^2 \dfrac{\sin 90°}{10}$

$\therefore \quad u = 10$ m/s

Thus, at highest point

$$g = \frac{(u\cos\theta)^2}{r}$$

The radius of curvature at highest point is

$$r = \frac{u^2\cos^2\theta}{g} = \frac{100}{10 \times 2} = 5 \text{ m}$$

$\therefore$ For bug at highest point,

$$mg - N = \frac{mv^2}{r}$$

$$\therefore \quad N = mg - \frac{mv^2}{r} = m\left(10 - \frac{5}{5}\right) = m \times 9$$

$$= 5 \times 10^{-3} \times 9 = 45 \times 10^{-3}\text{ N} = 5n \times 10^{-3}$$

$\therefore \quad n = 9$

22. (d) The position of the particle seen by the observer is

$$\mathbf{r}_{\text{rel}} = b\sin\omega t\ \hat{\mathbf{i}} + b\cos\omega t\ \hat{\mathbf{j}} + ct\,\hat{\mathbf{k}} - ct\,\hat{\mathbf{k}}$$

$$= b\sin\omega t\,\hat{\mathbf{i}} + b\cos\omega t\,\hat{\mathbf{j}}$$

Here, $\quad x_{r\mu} = b\sin\omega t$

$\quad y_{r\mu} = b\cos\omega t$

So, $x_{r\mu}^2 + y_{r\mu}^2 = b^2$ is a circle.

23. (a) Here, $\mathbf{v} = \frac{d\mathbf{r}}{dt} = b\omega\cos\omega t\,\hat{\mathbf{i}} - b\omega\sin\omega t\,\hat{\mathbf{j}}$

$\therefore \quad \mathbf{a} = \frac{d\mathbf{v}}{dt} = -b\omega^2\sin\omega t\,\hat{\mathbf{i}} - b\omega^2\cos\omega t\,\hat{\mathbf{j}}$

$a = \sqrt{(-b\omega^2\sin\omega t)^2 + (-b\omega^2\cos\omega t)^2} = b\omega^2$ directed in XY-plane.

24. (b) $\because \quad \mathbf{v}\cdot\mathbf{a} = 0$ or $va\cos\theta = 0$

$\therefore \quad \theta = \pi/2$

25. (c) $15\sin 53° = \frac{mv^2}{R}$

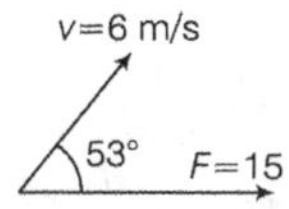

$\therefore \quad R = \frac{1\times 36}{15\times\frac{4}{5}} = \frac{36}{12} = 3\text{ m}$

26. (a) Here, $15\cos 53° = ma_t$ or $15\times\frac{3}{5} = 1\times a_t$

$\therefore \quad a_t = 9\text{ m/s}^2$

27. (b) The force is always perpendicular to velocity and is constant in magnitude. Thus, the path of the particle is circular.

28. (a) $F = v_0 c\sin 90° = \frac{mv_0^2}{r}$

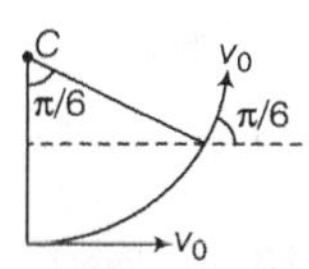

$\therefore \quad r = \frac{mv_0}{c} = \frac{mr\omega}{c}$

$\therefore \quad \omega = \frac{c}{m}$

$\therefore \quad \theta = \omega t = \frac{c}{m}\times\frac{m\pi}{6c} = \pi/6$

$\therefore \quad \mathbf{v} = \mathbf{v}_0\cos\frac{\pi}{6}\hat{\mathbf{i}} + v_0\sin\frac{\pi}{6}\hat{\mathbf{j}} = \frac{\sqrt{3}}{2}v_0\hat{\mathbf{i}} + \frac{v_0}{2}\hat{\mathbf{j}}$

29. (a) Path is a helical.

30. (a) We can break the velocity in two components $v_0\sin 30°$ parallel to **c** and $v_0\cos 30°$ perpendicular to **c**. The force $\mathbf{v}\times\mathbf{c}$ is directly perpendicular to **c**, thus $v_0\sin 30°$ parallel to **c** remains constant. In YZ-plane, the particle traces the circular path. Thus, the path is helical.

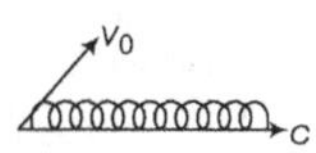

$\therefore \quad x = (v_0\sin 30°)t = 1\times\frac{1}{2}\times 2 = 1\text{ m}$

31. (b) In the case of three-dimensional motion discussion of a particle, the motion of the particle along axes are assumed as independent to each other. In this case, the particle's motion is divided in two parts.

(i) The motion along Z-axis,

Here, $a_z = \frac{b}{m}$ because, $\mathbf{v}\times\mathbf{c}$ is perpendicular to Z-axis.

$\therefore \quad z = \frac{1}{2}a_z t^2 = \frac{b}{2m}t^2.$

(ii) Motion in XY-plane. $\mathbf{v}\times\mathbf{c}$ is perpendicular to Z-axis. So, this force is acting in XY-plane and is always directed perpendicular to the velocity. Thus, in this plane, the motion of the particle is on a circular path with constant speed v_0.

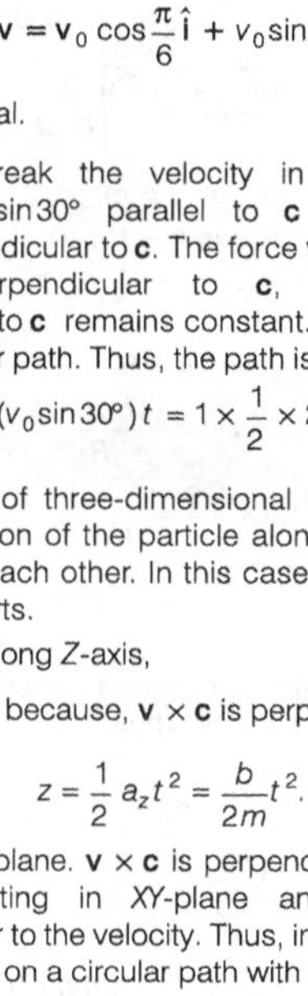

In this case, $v_0 c = \frac{mv_0^2}{r}$

$\therefore \quad r = \frac{mv_0}{c}$

But, $\quad v_0 = r\omega$

$\therefore \quad \omega = \frac{v_0}{\frac{mv_0}{c}} = \frac{c}{m}$

Thus, $\quad T = \frac{2\pi}{\omega} = \frac{2\pi m}{c}$

$\therefore \quad z = \frac{b}{2m}T^2 = \frac{b}{2m}\times\left(\frac{2\pi m}{c}\right)^2 = \frac{2\pi^2 mb}{c}$

32. (a) Here, $\quad \mathbf{v} = v_0\hat{\mathbf{i}} + v_z\hat{\mathbf{k}}$

$\therefore \quad \tan\theta = \frac{v_0}{v_z} = \frac{v_0}{a_z T} = \frac{v_0}{\frac{b}{m}\frac{2nm}{c}} = \frac{v_0 c}{2\pi b}$

$\therefore \quad \theta = \tan^{-1}\left(\frac{v_0 c}{2\pi b}\right)$

The resultant path is helical.

33. (c) Here, $z = \frac{b}{2m}t^2$

$\therefore \quad t = \sqrt{\frac{2mz}{b}} \Rightarrow x = r\sin\omega t$

$\therefore \quad l = r\sin\omega t$ and $y = -(r - r\cos\omega t)$

or $\quad y = -(r - r\cos\omega t) = -r\,2\sin^2\frac{\omega t}{2} = -2r\sin^2\frac{\omega t}{2}$

$\therefore \quad \frac{x}{y} = \frac{r2\sin\frac{\omega t}{2}\cos\omega t/2}{-2r\sin^2\frac{\omega t}{2}} = -\cot\frac{\omega t}{2}$

or $\quad \frac{y}{l} = -\tan\frac{\omega t}{2} \Rightarrow \frac{y}{l} = \frac{\omega t}{2} \qquad (\because y << l)$

or $\quad \frac{y}{l} = \frac{c}{2m}\sqrt{\frac{2mz}{b}}$

$\therefore \quad y^2 = \frac{cl^2}{4m^2}\frac{2mz}{b} \longrightarrow$ Parabola

34. (b)

(i) The velocity of the particle is $\mathbf{v} = \frac{d\mathbf{r}}{dt} = 3t^2\hat{\mathbf{i}} + 2t\hat{\mathbf{j}} + \hat{\mathbf{k}}$

At $\quad t = 1\text{ s},$

$\mathbf{v} = 3\hat{\mathbf{i}} + 2\hat{\mathbf{j}} + \hat{\mathbf{k}}$

$\therefore \quad |\mathbf{v}| = v = \sqrt{9+4+1} = \sqrt{14}$

Also, the acceleration of the particle is

$\mathbf{a} = \frac{d\mathbf{v}}{dt} = 6t\hat{\mathbf{i}} + 2\hat{\mathbf{j}}$

At $\quad t = 1\text{ s}, \mathbf{a} = 6\hat{\mathbf{i}} + 2\hat{\mathbf{j}}$

$\therefore \quad a = \sqrt{36+4} = \sqrt{40} = 6.32\text{ m/s}^2$

(ii) $\because$ The radius of curvature is

$R = \frac{v^3}{|\mathbf{a}\times\mathbf{v}|} = v^3$

$\therefore \quad R = \frac{v^3}{\sqrt{76}} = \frac{14\sqrt{14}}{\sqrt{76}}$

$= 14\sqrt{\frac{14}{76}} = 14\sqrt{\frac{7}{38}} = \sqrt{\frac{14\times 14\times 7}{38}}$

$\approx 6\text{ m}$

(iii) The centripetal acceleration is

$a_r = \frac{v^2}{R} = \frac{14}{6} = \frac{7}{3} = 2.33\text{ m/s}^2$

(iv) $\because a^2 = a_r^2 + a_t^2$

$\therefore$ The tangential acceleration is

$a_t = \sqrt{a^2 - a_r^2} = \sqrt{40 - \frac{49}{9}} = \sqrt{\frac{360-49}{9}} \approx 5.88\text{ m/s}^2$

(i) → (r), (ii) → (q), (iii) → (s), (iv) → (p)

Revisal Problems (JEE Main)

1. (c) Here, $x = b \sin \omega t$ and $y = b \cos \omega t$

$\therefore \quad \sin^2 \omega t + \cos^2 \omega t = \frac{x^2 + y^2}{b^2}$

$\therefore \quad x^2 + y^2 = b^2,$

Thus, this is the equation of circle.

2. (a) $bv = \frac{mv^2}{r} \Rightarrow v = \frac{rb}{m}$

3. (a) The bug is moving with constant speed. So, tangential acceleration should be zero. At the top and bottom,

$$f = ma_t = m \times 0 = 0$$

4. (a) Here, $N \cos \theta = mg$

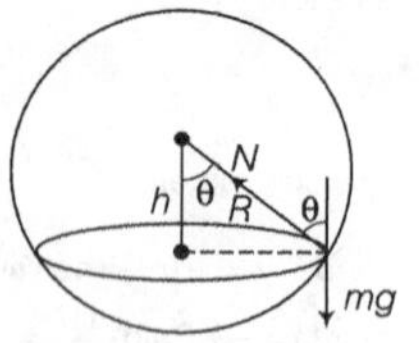

and $\quad N \sin \theta = \frac{mv^2}{r}$

or $\quad \tan\theta = \frac{v^2}{rg} \quad \ldots(i)$

Also, $\quad \cos \theta = \frac{h}{R} = \frac{1}{30 \times \frac{1}{15}} = \frac{1}{2} \Rightarrow \theta = 60°$

and $\quad \sin \theta = \frac{r}{R} \Rightarrow r = \frac{\sqrt{3}}{2}R$

From Eq. (i), we get

$$\tan 60° = \frac{v^2}{\frac{\sqrt{3}}{2} \times \frac{1}{15} \times 10}$$

$\therefore \quad v = 1\,\text{m/s}$

5. (b) Here, $\quad v^2 = 2a_t s = 2a_t 2\pi R = 4\pi a_t R$

and $\quad a_r = \frac{v^2}{R} = 4\pi a_t$

From the figure,

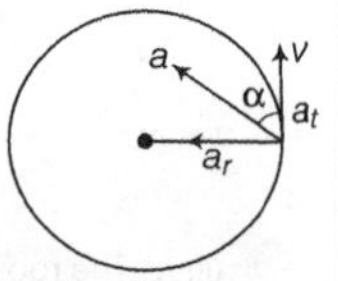

$$\tan \alpha = \frac{a_r}{a_t} = \frac{4\pi a_t}{a_t} = 4\pi$$

$\therefore \quad \frac{\tan \alpha}{2\pi} = \frac{4\pi}{2\pi} = 2$

6. (b) Here, $\quad N_1 - mg = \frac{mv^2}{R}$

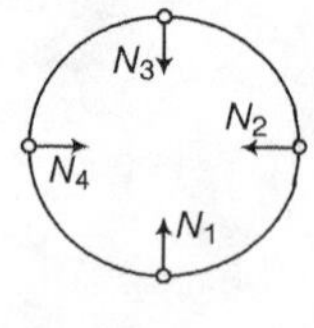

$$N_1 = mg + \frac{mv^2}{R}$$

$$N_2 = \frac{mv^2}{R} = N_4$$

and $\quad N_3 + mg = \frac{mv^2}{R}$

$\therefore \quad N_3 = \frac{mv^2}{R} - mg$

7. (c) If a path is more curve at a point, then radius of curvature is lesser.

Thus, $\quad mg - N_C = \frac{mv^2}{R}$

$\therefore \quad N_C = mg - \frac{mv^2}{R}$

8. (b) Normal reaction should be towards the centre to provide required centripetal acceleration. Hence, block will be in the contact of outer wall.

9. (b) When ball is at the lowest point,

$$T - mg = \frac{mv^2}{l}$$

$\therefore \quad T = mg + \frac{mv^2}{l}$

$\therefore \quad T > 10\,\text{N}$

10. (b) $a_r = \frac{v^2}{R} = \frac{16}{0.04} = 400\,\text{m/s}^2$

11. (c) $\quad a_r = r\omega^2 = 18 \times \omega_0^2 t^2$

$$= 18 \times \frac{1}{81} \times 9 = 2\,\text{m/s}^2$$

and $\quad \alpha = \frac{d\omega}{dt} = \omega_0 = \frac{1}{9}$

$\therefore \quad a_t = r\alpha = \frac{1}{9} \times 18 = 2\,\text{m/s}^2$

Thus, $\quad a = \sqrt{a_r^2 + a_t^2} = 2\sqrt{2}\,\text{m/s}^2$

12. (c) $T = \frac{mv_0^2}{r} = \frac{mv_0^2}{(L - \pi R)} = \frac{mv_0^2}{\pi R}$

13. (b) Here, $T = m_B g = 20\,\text{N}$

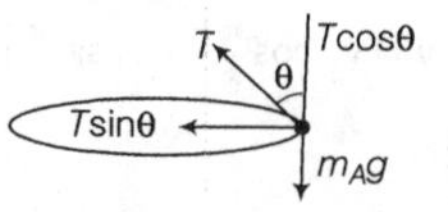

and $\quad T \cos \theta = m_A g$

or $\quad 20 \cos \theta = 1 \times 10$

or $\quad \cos \theta = \frac{1}{2}$

$\Rightarrow \quad \theta = 60°$

Also, $\quad T \sin \theta = \frac{m_A v^2}{R} = \frac{v^2}{R}$

or $\quad 20 \times \frac{\sqrt{3}}{2} = \frac{v^2}{R}$

$\therefore \quad \frac{v^2}{R} = 10\sqrt{3}\,\text{N}$

14. (d) The small ball starts to move on a circular path about point O, just after cutting the string A. Thus,

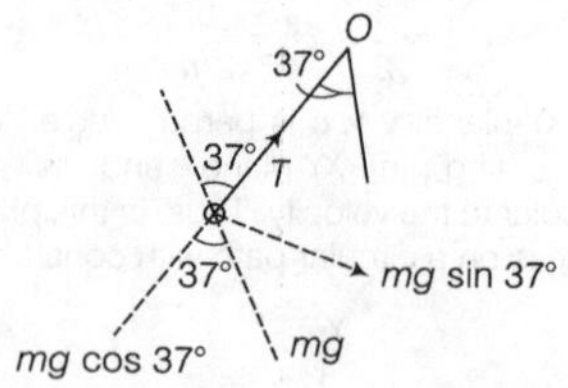

$$T - mg \cos 37° = \frac{mv^2}{l} = 0 \quad (\because v = 0)$$

$\therefore \quad T = mg \cos 37° = 10 \times \frac{4}{5} = 8\,\text{N}$

and $\quad mg \sin 37° = ma_t$

(perpendicular to the string)

$\therefore \quad a_t = 10 \times \frac{3}{5} = 6\,\text{m/s}^2$

15. (a) From the figure,

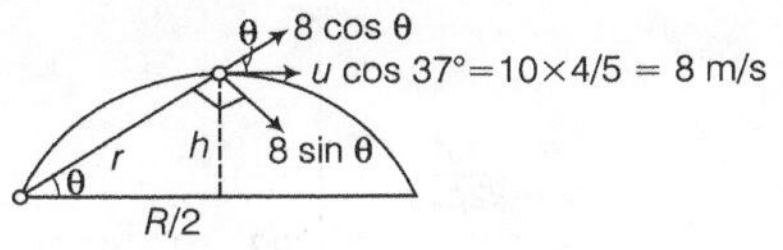

$$\omega = \frac{8 \sin \theta}{r} = \frac{8h}{r^2}$$

Here, $h = \dfrac{u^2 \sin^2 37°}{2g} = \dfrac{100}{2 \times 10} \times \dfrac{9}{25} = 1.8 \text{ m}$

and $R = \dfrac{u^2 \sin 2\alpha}{g} = \dfrac{100 \times 2 \times \frac{3}{5} \times \frac{4}{5}}{10} = \dfrac{48}{5} = 9.6 \text{ m}$

$\therefore \quad r^2 = (1.8)^2 + \left(\dfrac{9.6}{2}\right)^2 = (1.8)^2 + (4.8)^2 = 26.28$

$\therefore \quad \omega = \dfrac{8h}{r^2} = \dfrac{8 \times 1.8}{26.28} = \dfrac{14.4}{26.28} = 0.55 \text{ rad/s}$

16. (b) From the figure

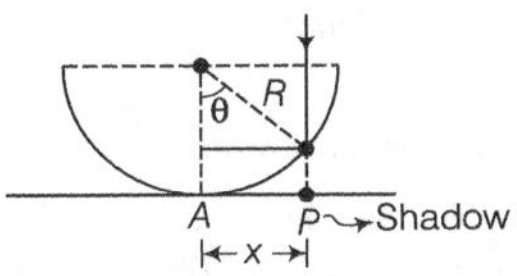

$$\theta = \omega t = \frac{v_0}{R} t = \frac{\pi}{2 \times 2} \times 1 = \frac{\pi}{4}$$

Also, $\sin \theta = \dfrac{x}{R}$

$\therefore \quad x = R \sin \theta \quad \text{or} \quad x = R \sin \dfrac{v_0}{R} t$

or $\dfrac{dx}{dt} = v_0 \cos \dfrac{v_0}{R} t$

Thus, the speed of shadow is

$$v = \frac{dx}{dt} = v_0 \cos \frac{v_0}{R} t$$

$$= \frac{\pi}{2} \cos \frac{\pi}{4} = \frac{\pi}{2\sqrt{2}} \text{ cm/s}$$

17. (b) The time to pass one spoke is

$$t = \frac{\theta}{\omega} = \frac{2\pi}{8 \times 5\pi} = \frac{1}{20} \text{ s}$$

$\therefore \quad s = \dfrac{1}{2} g t^2 = \dfrac{1}{2} \times 10 \times \dfrac{1}{400} = \dfrac{1}{80} \text{ m}$

18. (c) From free body diagram of ball in the frame of L-shaped rod,
$T \cos 45° = mg$

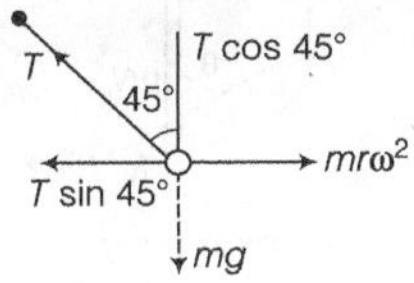

and $T \sin 45° = mr\omega^2$

$\therefore \quad \tan 45° = \dfrac{r\omega^2}{g} = r$

$\therefore \quad r = 1 \text{ m}$

But $r = 0.5 + l \sin 45°$

or $\dfrac{1}{2} = l \times \dfrac{1}{\sqrt{2}}$

$\therefore \quad l = \dfrac{1}{\sqrt{2}} \text{ m}$

19. (d) Centrifugal force is pseudo force.

20. (b) $\because$ The magnitude of pseudo force
= mass of body × acceleration of frame

21. (c) Fluid particles goes to outer edge due to centrifugal force

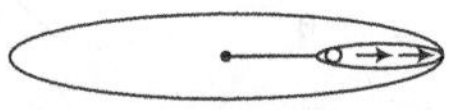

22. (d) Electrostatic force provides the necessary centripetal acceleration

$$\frac{mv^2}{r} = \frac{Ke^2}{r^2}$$

23. (a) Due to inertia, the velocity of car is along the tangent of the path.

24. (b) Net torque about centre of gravity should be zero.

25. (a) From the figure, $\theta < \dfrac{\pi}{2}$.

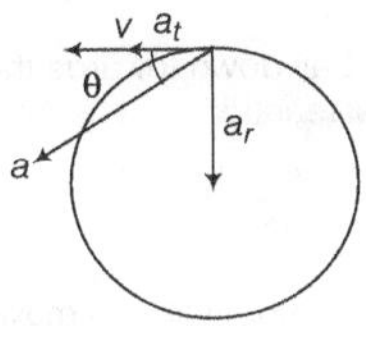

Revisal Problems (JEE Advanced)

1. (b) Here, $kx \cos 60° = mg$

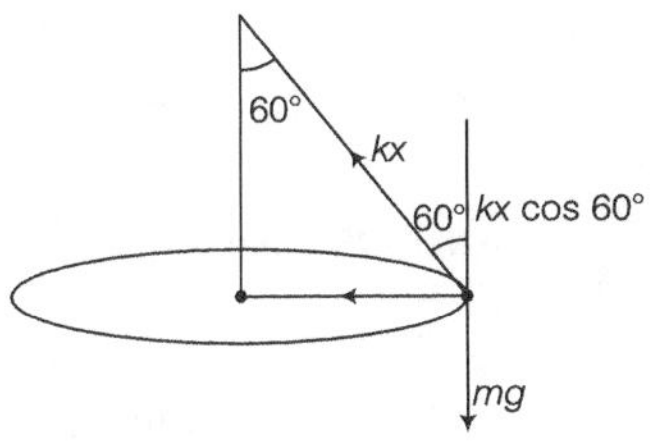

and $kx \sin 60° = \dfrac{mv^2}{r}$

$\therefore \quad x = \dfrac{mg}{k \cos 60°} = \dfrac{1 \times 10}{100 \times \frac{1}{2}} = \dfrac{20}{100} = 0.2 \text{ m}$

2. (b) Due to rotation in shaft, the centre of gravity of the rope moves on the circular path whose centre is lying on the rotation axis.

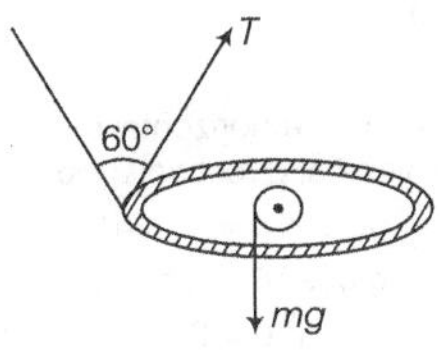

$\therefore \quad T \cos 60° = mg$

and $T \sin 60° = \dfrac{mv^2}{r}$

Thus, $T = \dfrac{mg}{\cos 60°} = \dfrac{1 \times 10}{\frac{1}{2}} = 20 \text{ N}$

3. (b) Let normal reaction makes an angle α with vertical.

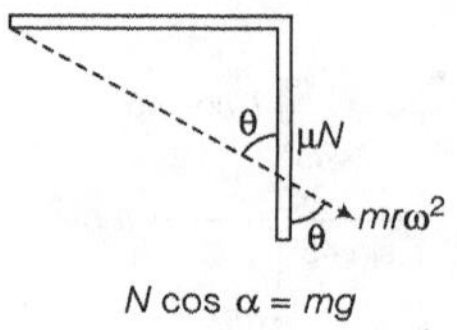

$$\therefore \quad N\cos\alpha = mg \quad \ldots(i)$$

and $\quad N\sin\alpha = mr\omega^2 \sin\theta$

Also, $\quad \mu N = mr\omega^2\cos\theta$

or $\quad \dfrac{\mu\, mr\omega^2 \sin\theta}{\sin\alpha} = mr\omega^2\cos\theta \quad (\because \theta = 45^\circ)$

or $\quad \mu\tan\theta = \sin\alpha \quad \ldots(ii)$

$\therefore \quad \sin\alpha = \mu = 0.5$

$\therefore \quad \alpha = 30^\circ$

$\therefore \quad N\cos\alpha = mg = 10 \quad$ [from Eq. (i)]

or $\quad N = \dfrac{20}{\sqrt{3}}$ N

4. (b) Here, $mg\sin\theta = ma_t$

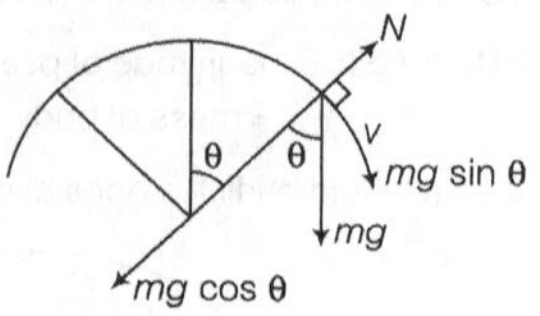

or $\quad a_t = g\sin\theta$

or $\quad v\dfrac{dv}{ds} = g\sin\theta$

or $\quad v\dfrac{dv}{Rd\theta} = g\sin\theta$

or $\quad \int_0^{v_0} v\,dv = gR\int_0^{\theta_0}\sin\theta\,d\theta \quad$ or $\quad \dfrac{v_0^2}{2} = gR\,[-\cos\theta]_0^{\theta_0}$

or $\quad v_0^2 = 2gR(1-\cos\theta_0)$

But, at $\theta = \theta_0$, $N = 0$

$\therefore \quad mg\cos\theta_0 = \dfrac{mv_0^2}{R} \quad$ or $\quad v_0^2 = gR\cos\theta_0$

or $\quad 2gR(1-\cos\theta) = gR\cos\theta_0$

or $\quad 2gR = 3gR\cos\theta_0$

$\quad \cos\theta_0 = 2/3$

$\therefore \quad \theta_0 = \cos^{-1}\left(\dfrac{2}{3}\right)$

5. (b) Gravity pulls the ball down towards the table and table holds the ball up by normal reaction.

6. (b) $r = \dfrac{v^3}{|\mathbf{a}\times\mathbf{v}|} = \dfrac{v^3}{gv\sin 90^\circ}$

$\therefore \quad r = \dfrac{(u\cos\alpha)^2}{g}$ = minimum value at maximum height.

7. (b) Here, $mg\sin\theta = ma_t$

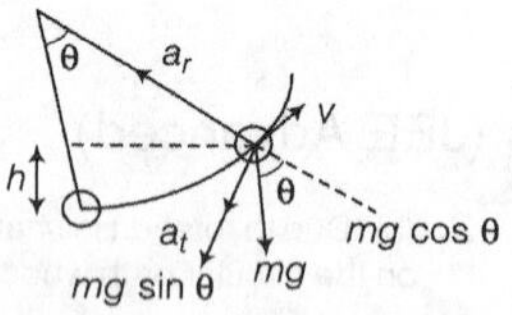

$\therefore \quad a_t = g\sin\theta$

and $\quad a_r = \dfrac{v^2}{l} = \dfrac{u^2 - 2gh}{l}$

$\quad = \dfrac{2gl - 2gl(1-\cos\theta)}{l}$

or $\quad a_r = 2g\cos\theta$

and $\quad a_t = g\sin\theta$

Since, acceleration is in the horizontal direction, so the sum of components in the vertical direction should be zero.

$\therefore \quad a_r\cos\theta = a_t\sin\theta$

or $\quad 2g\cos^2\theta = g\sin^2\theta$

$\therefore \quad \tan^2\theta = 2 \Rightarrow \tan\theta = \sqrt{2}$

8. (a,b,c) $T + mg = \dfrac{mv_0^2}{l}$ or $T = \dfrac{mv_0^2}{l} - mg$

For $v_0 = \sqrt{gl}$, $T = 0$

For $v_0 < \sqrt{gl}$, T is negative

For $v_0 > \sqrt{gl}$, T is positive

9. (a,b,c,d) (a) At lowest point,

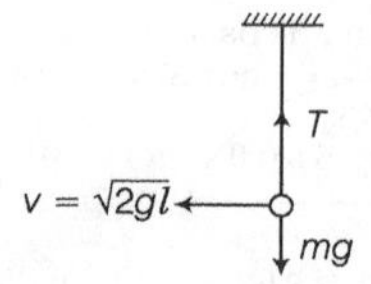

$$T - mg = \frac{mv^2}{l}$$

$$\therefore \quad T = mg + \frac{mv^2}{l} = mg + \frac{m}{l}2gl$$

$\Rightarrow T = 3mg$, hence tension is maximum at lowest point.

(b) The acceleration of ball is resultant of radial and tangential acceleration. From figure, it is clear that the resultant of radial and tangential acceleration may be horizontal. (c) and (d) initially velocity is zero. So, $a_r = \dfrac{v^2}{l} = 0$

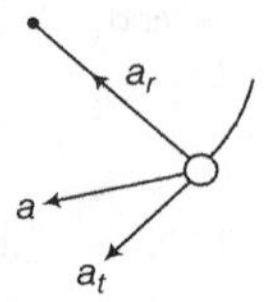

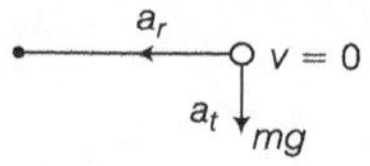

and $\quad mg = ma_t$

$\therefore \quad a_t = g$ directed downward.

10. (a,b,d) In vertical direction, $T\cos 30^\circ + T\cos 60^\circ = mg$

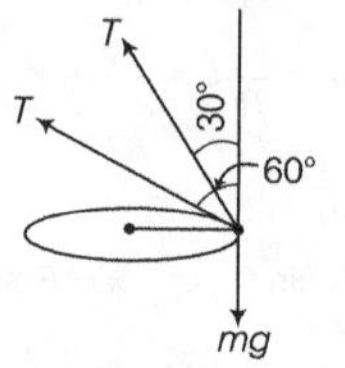

$\therefore \quad T = \dfrac{mg}{\cos 30^\circ + \cos 60^\circ} = \dfrac{2mg}{(\sqrt{3}+1)}$

and $\quad T\sin 30^\circ + T\sin 60^\circ = mr\omega^2 \quad$ (towards centre)

or $\quad \dfrac{(\sqrt{3}+1)}{2}T = mr\omega^2$

or $\quad \left(\dfrac{\sqrt{3}+1}{2}\right)\dfrac{2mg}{(\sqrt{3}+1)} = mr\omega^2$

$\therefore \quad \omega = \sqrt{\dfrac{g}{r}}$

Hence, options (a,b) and (d) are correct.

11. (a,b,c) If friction is absent.

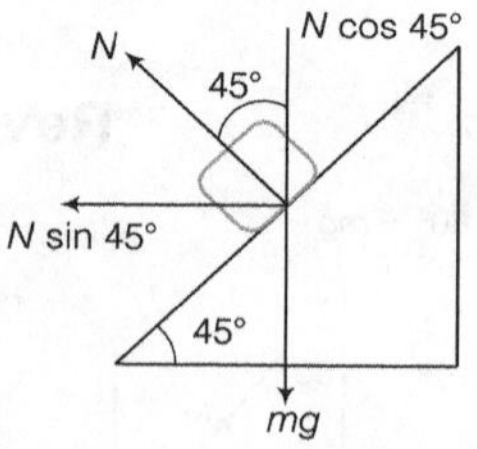

$N\cos 45^\circ = mg \quad$ and $\quad N\sin 45^\circ = \dfrac{mv^2}{r}$

$\therefore \quad \tan 45^\circ = \dfrac{v^2}{rg}$

$\Rightarrow \quad v = \sqrt{rg} = 10$ m/s

In the presence of friction velocity is either greater than 10 m/s or less than 10 m/s depending on the direction of friction.

Hence, options a, b and c are correct.

Use concept of centripetal force

12. (a,b,c,d)

13. (a,b) From free body diagram of 10 kg ball in the frame of turntable,

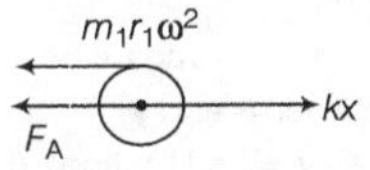

$$F_A + m_1 r_1 \omega^2 = kx = 500$$

$$\therefore \quad F_A = 500 - m_1 r_1 \omega^2 = 500 - 10 \times 0.2 \times (2\pi)^2$$

$$= 500 - 80 = 420 \text{ N}$$

From free body diagram of 15 kg ball in the frame of turntable,

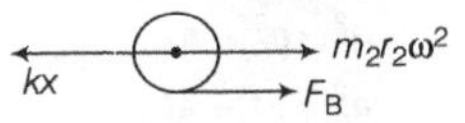

$$kx = F_B + m_2 r_2 \omega^2$$

$$\therefore \quad F_B = kx - m_2 r_2 \omega^2 = 500 - 15 \times 0.3 \times (2\pi)^2$$

$$= 500 - 180 = 320 \text{ N}$$

14. (a,c) Before the string AB is cut, the ball A is in equilibrium,

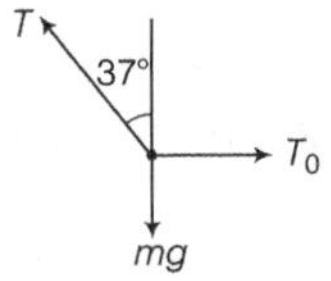

From free body diagram of the ball A,

$$T \cos 37° = mg$$

and $\quad T \sin 37° = T_0$

$$\therefore \quad T = \frac{mg}{\cos 37°} = \frac{10 \times 5}{4} = 12.5 \text{ N}$$

When string AB is cut, the forces on the ball A are its weight mg, tension in string OA. At that instant, the velocity of ball A is zero.

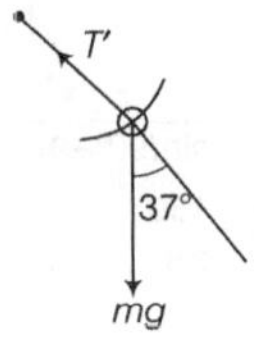

Here $\quad T' - mg \cos 37° = ma_r = \dfrac{mv^2}{l} = 0 \qquad (\because v = 0)$

$$\therefore \quad T' = mg \cos 37° = 10 \times \frac{4}{5} = 8 \text{ N}$$

15. (b) From free body diagram of the block,

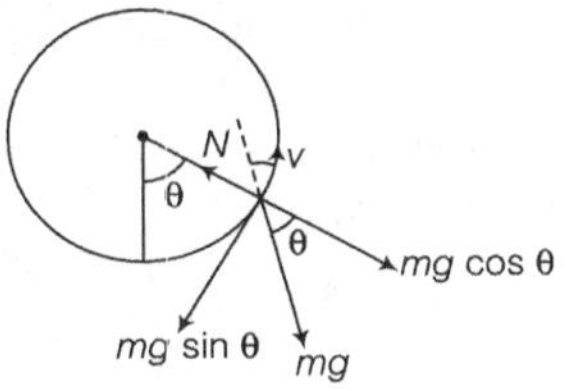

$$-mg \sin \theta = ma_t$$

$$\therefore \quad a_t = -g \sin \theta \quad \text{or} \quad v\frac{dv}{ds} = -g \sin \theta$$

or $$\frac{v\,dv}{R d\theta} = -g \sin \theta$$

or $$\frac{1}{R}\int_{\sqrt{2gR}}^{v} v\,dv = -g \int_0^\theta \sin \theta \, d\theta$$

$$v^2 - 2gR = -2gR\,[\cos \theta]_0^\theta$$

or $$v^2 = 2gR - 2gR(1 - \cos \theta)$$

or $$v = \sqrt{2gR \cos \theta}$$

16. (c) From the figure, vertical component of velocity is

$$v_\perp = v \sin \theta = \sqrt{2gR \cos \theta} \sin \theta$$

For maximum, $\dfrac{dv_\perp}{d\theta} = 0$, after solving, $\cos \theta = \dfrac{1}{\sqrt{3}}$

Also, $\quad N - mg \cos \theta = \dfrac{mv^2}{R}$

$$\therefore \quad N = mg \cos \theta + \frac{mv^2}{R}$$

$$= mg \times \frac{1}{\sqrt{3}} + \frac{m}{R} 2gR \cos \theta$$

$$= \frac{mg}{\sqrt{3}} + \frac{2mg}{\sqrt{3}} = \sqrt{3}\, mg$$

17. (c) From the figure, the centripetal acceleration is $\dfrac{v^2}{R} = a \sin \theta$

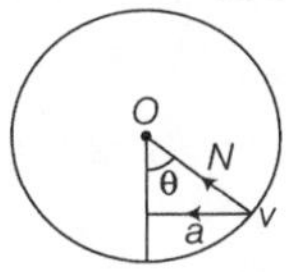

and a tangential acceleration is $a_t = a \cos \theta$

or $$g \sin \theta = a \cos \theta$$

$$\therefore \quad a = g \tan \theta$$

$$\therefore \quad \frac{v^2}{R} = a \sin \theta = g \tan \theta \sin \theta$$

or $$v^2 = gR \tan \theta \sin \theta$$

or $$2gR \cos \theta = gR \tan \theta \sin \theta$$

$$\therefore \quad 2 \cos \theta = \frac{\sin^2 \theta}{\cos \theta} \quad \text{or} \quad \tan \theta = \sqrt{2}$$

18. (b) From the figure, $u = v_A \cos 60°$

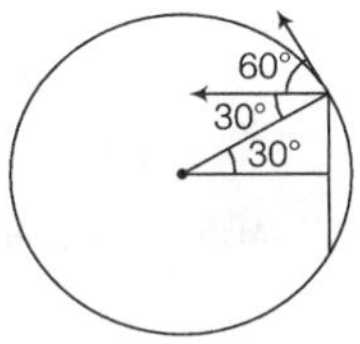

$$\therefore \quad v_A = \frac{u}{\cos 60°}$$

$$= \frac{10}{\frac{1}{2}} = 20 \text{ cm/s}$$

19. (d) The centripetal acceleration is

$$a_r = \frac{v_A^2}{R} = \frac{400}{1} = 400 \text{ cm/s}^2 = 4 \text{ m/s}^2$$

20. (b) $a_r = a \cos 60°$, and $a_t = a \sin 60°$

$$a = \frac{a_r}{\cos 60°} = \frac{4}{1/2} = 8 \text{ m/s}^2$$

$$\therefore \quad a_t = a \sin 60° = 8 \times \frac{\sqrt{3}}{2} = 4\sqrt{3} \text{ m/s}^2$$

21. $\left(\dfrac{1}{2}\right)$ At extreme point, $v = 0$

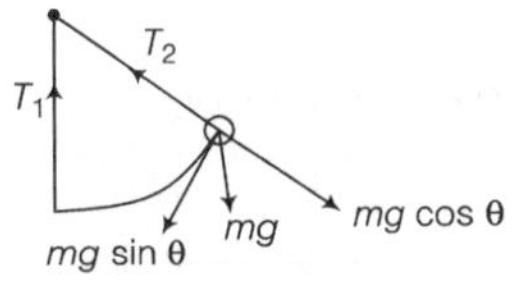

$$\therefore \quad T_2 - mg \cos \theta = \frac{m_2 v^2}{l} = 0$$

$$\therefore \quad T_2 = mg \cos \theta$$

and $$mg \sin \theta = ma_t$$

$$\therefore \quad a_t = g \sin \theta$$

At lowest point, $a_t = 0$ and $a = a_r = \dfrac{u^2}{l}$

$\because$ Acceleration at lowest point and extreme points are same.

$$\therefore \quad \frac{u^2}{l} = g \sin \theta$$

$$\because \quad u^2 = gl \sin \theta$$

But $$a_t = -g \sin \theta \quad \text{or} \quad v\frac{dv}{ds} = -gs \sin \theta$$

or $$\int_u^0 v\,dv = -gl \int_0^\theta \sin \theta \, d\theta$$

or $$-\frac{u^2}{2} = -2gl\,[-\cos \theta]_0^\theta$$

or $$u^2 = 2gl\,(1-\cos\theta)$$

or $$gl\sin\theta = 2gl\,(1-\cos\theta)$$

or $$\sin\theta = 2\times 2\sin^2\frac{\theta}{2}$$

or $$2\sin\frac{\theta}{2}\cos\frac{\theta}{2} = 4\sin^2\frac{\theta}{2}$$

or $$\tan\frac{\theta}{2} = \frac{1}{2}$$

22. (3) From free body diagram of particle in the frame of plank,

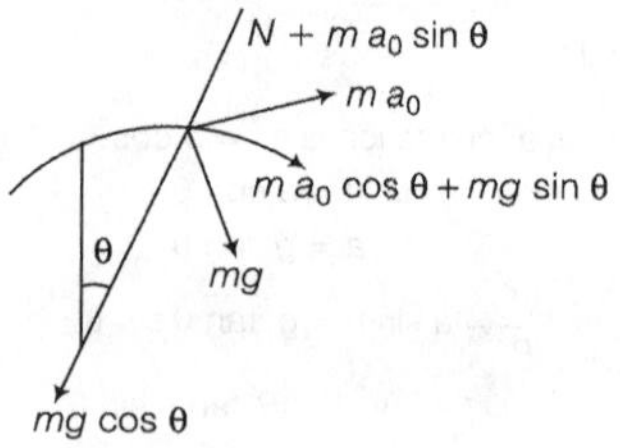

$$mg\sin\theta + ma_0\cos\theta = ma_t$$

or $$g\sin\theta + a_0\cos\theta = v\frac{dv}{ds} = \frac{v\,dv}{R\,d\theta}$$

or $$\int_0^v v\,dv = gR\int_0^{\theta_0}\sin\theta + a_0R\int_0^{\theta_0}\cos\theta\,d\theta$$

or $$\frac{v_0^2}{2} = gR\,(1-\cos\theta_0) + a_0R\sin\theta_0$$

Also, $mg\cos\theta - N - ma_0\sin\theta = \dfrac{mv^2}{R}$

For $$\theta = \theta_0,\ N = 0,\ v = v_0$$

or $$mg\cos\theta_0 - ma_0\sin a_0 = \frac{mv_0^2}{R}$$

or $$g\cos\theta_0 - a_0\sin\theta_0 = 2g\,(1-\cos\theta_0) + 2a_0\sin\theta_0$$

or $$3\times 10\times\frac{\sqrt{3}}{2} = 2\times 10 + 3a_0\times\frac{1}{2}$$

or $$15\sqrt{3} = 20 + \frac{3}{2}a_0$$

or $$15\sqrt{3} - 20 = \frac{3}{2}a_0$$

$$a_0 = \frac{30\sqrt{3}}{3} - \frac{40}{3} = 10\sqrt{3} - \frac{40}{3} = \frac{30\sqrt{3}-40}{3}$$

$\Rightarrow$ $$n = 3$$

23. (1) In the frame shaft, here $T = mL\omega^2$

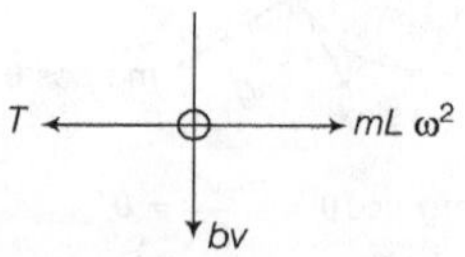

and $$-bv = ma_t = mL\,\alpha$$

or $$-bL\omega = mL\alpha$$

or $$\alpha = -\frac{b}{m}\omega$$

or $$\frac{\omega\,d\omega}{d\theta} = -\frac{b}{m}\omega$$

or $$\int_{\omega_0}^{\omega} d\omega = -\frac{b}{m}\int_0^{\frac{s}{L}} d\theta$$

$$\omega - \omega_0 = -\frac{b}{m}\frac{s}{L} = -\frac{bs}{mL}$$

$\Rightarrow$ $$\omega = \omega_0 - \frac{bs}{mL}$$

$\therefore$ $$T = mL\omega^2 = mL\left(\omega_0 - \frac{bs}{mL}\right)^2$$

$\therefore$ $$n = 1$$

24. (1) Here, $\mathbf{a} = b\cdot\dfrac{dy}{dt}\hat{\mathbf{i}} = bv_y\,\hat{\mathbf{i}} = bc\,\hat{\mathbf{i}}$

The component of acceleration in the direction of velocity is tangential acceleration.

$\therefore$ $$a_t = a\cos\theta$$

where, θ is angle between $\mathbf{a}$ and $\mathbf{v}$.

$\because$ $$\mathbf{a}\cdot\mathbf{v} = |\mathbf{a}||\mathbf{v}|\cos\theta$$

$\therefore$ $$\cos\theta = \frac{\mathbf{a}\cdot\mathbf{v}}{|\mathbf{a}||\mathbf{v}|}$$

$\therefore$ $$a_t = a\frac{(\mathbf{a}\cdot\mathbf{v})}{|\mathbf{a}||\mathbf{v}|} = \frac{(\mathbf{a}\cdot\mathbf{v})}{|\mathbf{v}|} = \frac{b^2cy}{\sqrt{b^2y^2+c^2}}$$

But, $$a^2 = a_r^2 + a_t^2$$

$\therefore$ $$a_r^2 = a^2 - a_t^2$$

$$= b^2c^2 - \frac{b^4c^2y^2}{b^2y^2+c^2} = \frac{b^2c^4}{b^2y^2+c^2}$$

$\therefore$ $$a_r = \frac{bc^2}{\sqrt{b^2y^2+c^2}}$$

$\therefore$ $$n = 1$$

25. (1) From the figure,

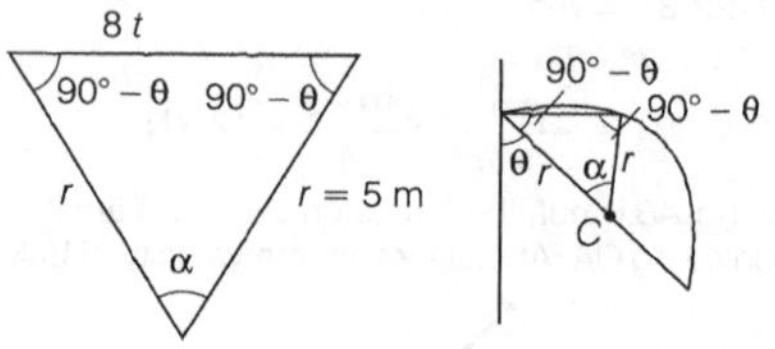

$$\frac{8t}{\sin\alpha} = \frac{r}{\sin(90°-\theta)}$$

But $90°-\theta + 90°-\theta + \alpha = 180°$

$\therefore$ $$\alpha = 2\theta$$

$\therefore$ $$\frac{8t}{\sin 2\theta} = \frac{5}{\cos\theta} \quad\text{or}\quad \frac{8t}{2\sin\theta\cos\theta} = \frac{5}{\cos\theta}$$

or $$\frac{10}{8}\sin\theta = t \quad\Rightarrow\quad \frac{10}{8}\cos\theta\,d\theta = dt$$

$\therefore$ $$\frac{d\theta}{dt} = \frac{8}{10\cos\theta}$$

$\therefore$ $$\omega = \frac{d\theta}{dt} = \frac{8}{10\cos\theta} = \frac{8}{10\times\frac{4}{5}} = 1\ \text{rad/s}$$

26. (3) Here, $v_A = r_1\omega_1$ and $v_B = r_2\omega_2$

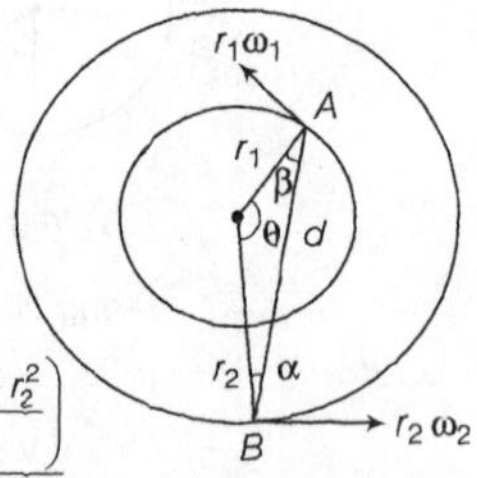

The relative angular velocity of B with respect to A is

$$\omega_{rel} = \frac{r_2\omega_2\cos\alpha + r_1\omega_1\cos\beta}{AB}$$

$$\omega_{rel} = \frac{r_2\omega_2\cos\alpha + r_1\omega_1\cos\beta}{d}$$

$$= \frac{r_2\omega_2\left(\dfrac{r_2^2+d^2-r_1^2}{2r_2d}\right) + r_1\omega_1\left(\dfrac{r_1^2+d^2-r_2^2}{2r_1d}\right)}{d}$$

$\therefore$ $$\omega_{rel} = \left(\frac{\omega_1+\omega_2}{2}\right) + \frac{(\omega_2-\omega_1)(r_2^2-r_1^2)}{2(r_1^2+r_2^2-2r_1r_2\cos\theta)}$$

But, $\omega_{rel} = 0$

$\therefore$ $$\cos\theta = \frac{r_2^2\omega_2 + r_1^2\omega_1}{2r_1r_2(\omega_1+\omega_2)}$$

Here, $r_1 = 1\ \text{m}$, $r_2 = 2\ \text{m}$, $\omega_1 = 1\ \text{rad/s}$ and $\omega_2 = 0.5\ \text{rad/s}$

$\therefore$ $$\cos\theta = \frac{1}{2} \Rightarrow \theta = \frac{\pi}{3} = \frac{\pi}{n}$$

$\Rightarrow$ $$n = 3$$

27. (8) $\because \quad \mathbf{v} = \mathbf{v}_0 + \mathbf{a}t$

$\therefore \quad \mathbf{v} = 24\hat{\mathbf{i}} + 10\hat{\mathbf{j}}$ (at $t = 2$ s)

The radius of curvature,

$$r = \frac{v^3}{|\mathbf{a} \times \mathbf{v}|} = \frac{(\sqrt{24^2 + 10^2})^3}{|(-3\hat{\mathbf{i}} - 15\hat{\mathbf{j}}) \times (24\hat{\mathbf{i}} + 10\hat{\mathbf{j}})|}$$

$$= \frac{26 \times 26 \times 26}{|-30\,\hat{\mathbf{k}} + 360\,\hat{\mathbf{k}}|} = \frac{26 \times 26 \times 26}{330} = \frac{2197x}{330}$$

$\therefore \quad x = 8$

28. (c)

(i) Unit vector along angular velocity is $\hat{\mathbf{k}}$ or particle is rotating in XY-plane in anti-clockwise sense.

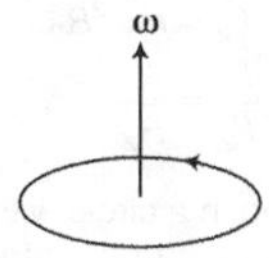

(ii) As, speed is increasing, angular acceleration must points in the direction of angular velocity.

So, α must be directed along $\hat{\mathbf{k}}$.

(iii) At point B, velocity is parallel to X-axis.

So, unit vector is $-\hat{\mathbf{i}}$.

(iv) At B, acceleration points along the vector $\hat{\mathbf{e}}$ or along $-\left(\frac{\hat{\mathbf{i}} + \hat{\mathbf{j}}}{\sqrt{2}}\right)$

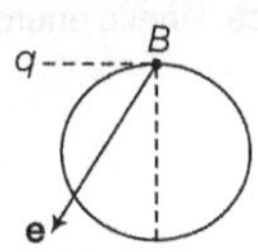

(i) – (p), (ii) – (p), (iii) – (r), (iv) – (q,s)

29. (d)

(i) If $\mathbf{F}_0 + \mathbf{v} \times \mathbf{c} = 0$, then velocity of particle may be constant. If $\mathbf{F}_0$ is parallel to $\mathbf{v} \times \mathbf{c}$, then path of particle is helical.

(ii) As gravity is also present, then

$$F_{net} = F_g + F_0 + \mathbf{v} \times \mathbf{c}$$

(iii) $F_{net} = \mathbf{v} \times \mathbf{c} + F_g$

(iv) When only gravity is present only option q is possible.

(i) → (p,q,s), (ii) → (p,q,r,s), (iii) → (p,q,r,s), (iv) → (q)

30. (a) (i) In position A and C centripetal force acts parallel to floor and force of friction is maximum at positions A and C.

(i) → (p,r)

(ii) At position D, both mg and $mr\omega^2$ acts downwards, reading on scale reads maximum at D.

(ii) → (s)

(iii) Contact force is resultant of mg and $mr\omega^2$, hence $\quad mg - mr\omega^2 \le F_C \le mg + mr\omega^2$

(iii) → (p,q,r,s)

(iv) At positions A and C, mg is perpendicular to $mr\omega^2$

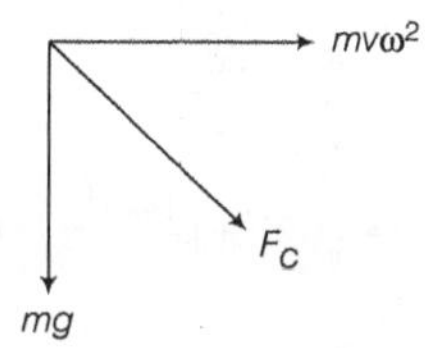

$$\therefore \quad F_C = \sqrt{m^2g^2 + m^2r^2\omega^4}$$

(iv) → (p,r)

31. (b) (i) → (p), (ii) → (r, s), (iii) → (r), (iv) → (q, s)

JEE Main & AIEEE Archive

1. (c) As, the force is exponentially decreasing, so its acceleration, i.e. rate of increase of velocity will decrease with time. Thus, the graph of velocity will be an increasing curl with decreasing slope with time.

$$\therefore \quad a = \frac{F}{M} = \frac{F_0}{m} e^{-bt} = \frac{dv}{dt}$$

$$\Rightarrow \quad \int_0^v dv = \int_0^t \frac{F_0}{m} e^{-bt} dt$$

$$\Rightarrow \quad v = \frac{F_0}{m}\left[\left(\frac{1}{-b}\right) e^{-bt}\right]_0^t = \frac{F_0}{mb}[e^{-bt}]_t^0$$

$$= \frac{F_0}{mb}(e^0 - e^{-bt}) = \frac{F_0}{mb}(1 - e^{-bt})$$

$$v_{max} = \frac{F_0}{mb}$$

2. (b) For the motion of the block,

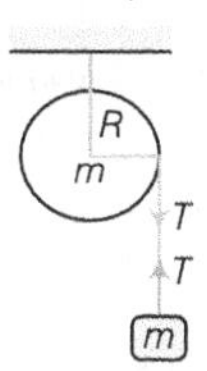

$$mg - T = ma \quad \ldots(i)$$

For the rotation of the pulley,

$$\tau = TR = I\alpha$$

$$\Rightarrow \quad T = \frac{1}{2} mR^2 \alpha \quad \ldots(ii)$$

As, string does not slip on the pulley.

$$a = R\alpha \quad \ldots(iii)$$

On solving Eqs. (i), (ii) and (iii), we get

$$a = 2g/3$$

3. (c) For a particle in a uniform circular motion,

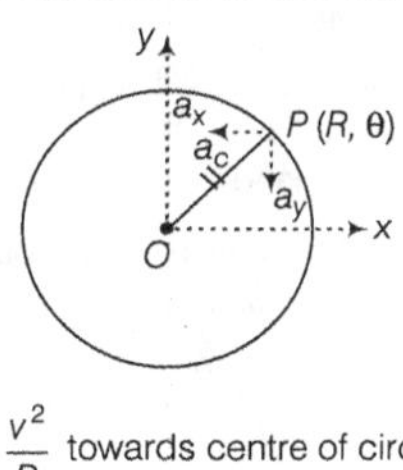

$$\mathbf{a} = \frac{v^2}{R} \text{ towards centre of circle}$$

[centripetal acceleration]

$$\therefore \quad \mathbf{a} = \frac{v^2}{R}(-\cos\theta\,\hat{\mathbf{i}} - \sin\theta\,\hat{\mathbf{j}})$$

$$\text{or} \quad \mathbf{a} = -\frac{v^2}{R}\cos\theta\,\hat{\mathbf{i}} - \frac{v^2}{R}\sin\theta\,\hat{\mathbf{j}}$$

4. (d) Given, $\mathbf{v} = ky\hat{\mathbf{i}} + kx\hat{\mathbf{j}}$

$$\Rightarrow \quad \frac{dx}{dt} = ky, \frac{dy}{dt} = kx$$

$$\Rightarrow \quad \frac{dy}{dx} = \frac{dy}{dt} \times \frac{dt}{dx} = \frac{kx}{ky}$$

or $\quad ydy = xdx \quad \ldots(i)$

Integrating both sides Eq. (i), we get

$$y^2 = x^2 + c$$

5. (d) Given, $s = t^3 + 5$

$\therefore$ Speed, $v = \frac{ds}{dt} = 3t^2$

and rate of change of speed, $a_t = \frac{dv}{dt} = 6t$

$\therefore$ Tangential acceleration at $t = 2$ s,

$$a_t = 6 \times 2 = 12 \text{ ms}^{-2}$$

and at $t = 2$ s,

$$v = 3(2)^2 = 12 \text{ ms}^{-1}$$

$\therefore$ Centripetal acceleration

$$a_c = \frac{v^2}{R} = \frac{144}{20} \text{ ms}^{-2}$$

$\therefore$ Net acceleration $= \sqrt{a_t^2 + a_c^2}$

$$= \sqrt{(12)^2 + \left(\frac{144}{20}\right)^2} = 14 \text{ ms}^{-2}$$

6. (d) Since, ω is constant, so no net force or torque is acting on ring. The force experienced by any particle is only along radial direction or we can say the centripetal radial force.

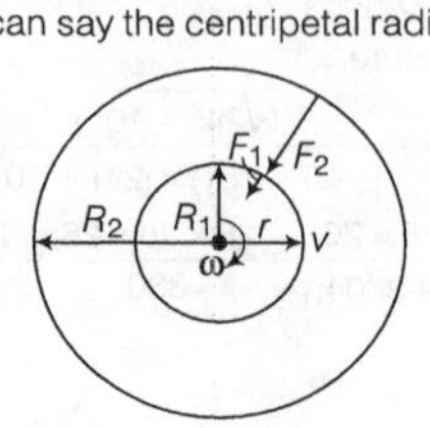

The force experienced by inner part $F_1 = m\omega^2 R_1$ and the force experienced by outer part,

$$F_2 = m\omega^2 R_2$$

$$\Rightarrow \quad \frac{F_1}{F_2} = \frac{R_1}{R_2}$$

7. (b) For a particle moving in a circle with constant angular speed, velocity vector is always along the tangent to the circle and the acceleration vector always points towards the centre of the circle or is always along radius of the circle.

Since, tangential vector is perpendicular to radial vector, therefore velocity vector will be perpendicular to the acceleration vector. But in no case, acceleration vector is tangent to the circle.

8. (c) When a force of constant magnitude acts on velocity of particle perpendicularly, then there is no change in the kinetic energy of particle. Hence, kinetic energy remains constant.

JEE Advanced & IIT JEE Archive

1. (a) Magnitude of relative speed is

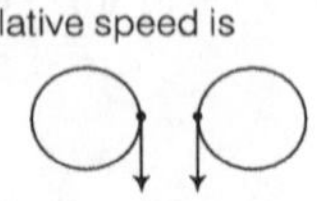

$t = 0, |v_r| = 0$

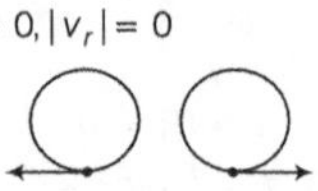

$t = T/4, |v_r| = 2v = 2$ Rul

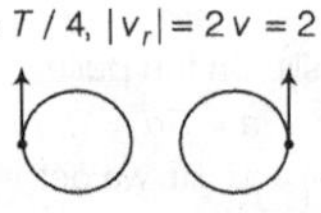

$t = T/2, |v_r| = 0$

2. (c) At $t = \frac{1}{8} \times \frac{2\pi}{\omega} = \frac{\pi}{4\omega}$

x-coordinate of $P = R\omega\left(\frac{\pi}{4\omega}\right) = \frac{\pi R}{4} > R\cos 45°$

$\therefore$ Both particles P and Q land in the unshaded region.

3. (d) Consider motion of the ball as shown below

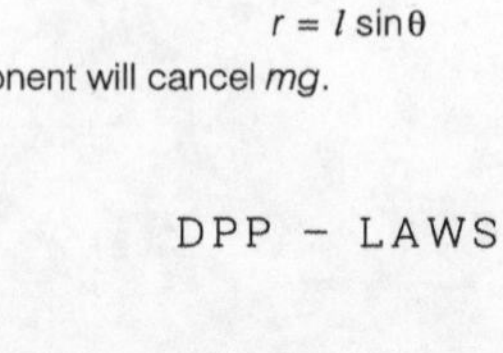

$r = l\sin\theta$

$T\cos\theta$ component will cancel mg.

$T\sin\theta$ component will provide necessary centripetal force to the ball towards centre C.

$$\therefore \quad T\sin\theta = mr\,\omega^2 = m\,(l\sin\theta)\,\omega^2$$

$$\text{or} \quad T = ml\omega^2$$

$$\therefore \quad \omega = \sqrt{\frac{T}{ml}}$$

$$\text{or} \quad \omega_{max} = \sqrt{\frac{T_{max}}{ml}}$$

$$= \sqrt{\frac{324}{0.5 \times 0.5}}$$

$$= 36 \text{ rad/s}$$

So, correct option is d.

4. (a) Since, block raises to same height in all cases, speed of block v_0 is same at highest point. Let speed of block at highest point is v_0, then

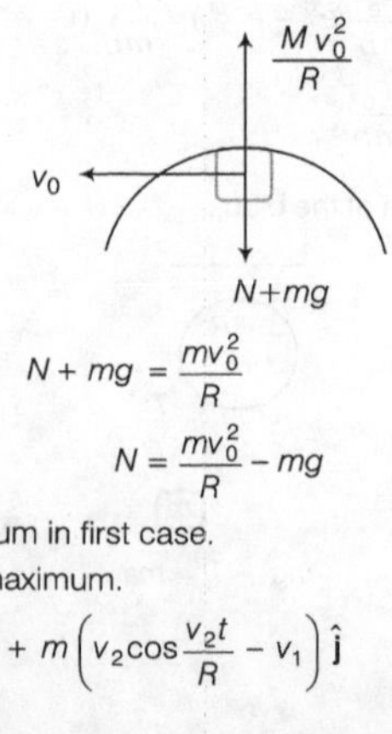

$$N + mg = \frac{mv_0^2}{R}$$

$$\Rightarrow \quad N = \frac{mv_0^2}{R} - mg$$

As, R is minimum in first case.

So, N will be maximum.

5. $-mv_2\sin\frac{v_2 t}{R}\hat{\mathbf{i}} + m\left(v_2\cos\frac{v_2 t}{R} - v_1\right)\hat{\mathbf{j}}$

4. Work, Power and Energy

DPP-1 Kinetic Energy

1. (i) (b) KE w.r.t. bus $= \frac{1}{2}m(v_{rel})^2 = 0$ $\quad (\because v_{rel} = 0)$

(ii) (b) KE measured by friend $= \frac{1}{2}mv^2 = \frac{1}{2} \times 50 \times (5)^2 = 625$ J

(iii) (a) Magnitude of velocity of bus w.r.t. driver in the car

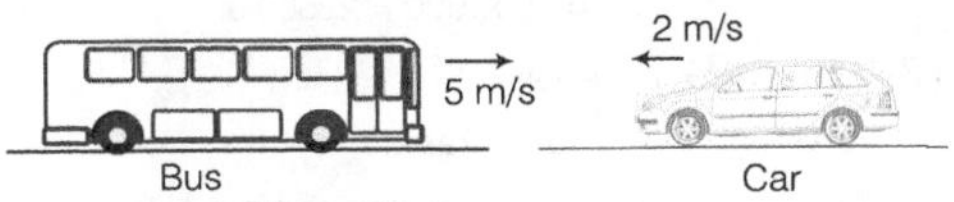

$$|\mathbf{v}_b - \mathbf{v}_c| = |5 - (-2)| = 7 \text{ m/s}$$

$$KE = \frac{1}{2} \times 50 \times 49 = 1225 \text{ J}$$

(iv) (d) The speed and the kinetic energy of a single object is frame dependent (relative), it can take any non-negative value.

2. (a) $KE = \frac{1}{2} \times 100 \times 5^2 + \frac{1}{2} \times 1000 \times 10^2$

$$= 1250 + 50000 = 51250 \text{ J}$$

3. (d) The given question is somewhat based on approximation. Let mass of athlete be 65 kg. Average velocity from the given data is 10 m/s.

So, $\quad KE = \frac{65 \times 100}{2} = 3250$ J

So, option (d) is the most probable answer.

4. (c) Area under $F(t)$ curve

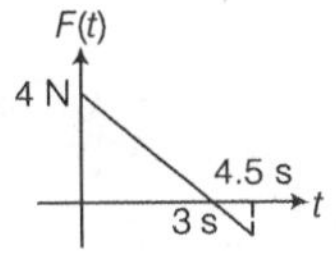

$$= \int_0^{4.5} F(t)\,dt = \int_0^v m\,dv$$

$$= \int_0^{4.5} F(t)\,dt = mv$$

$$= \frac{1}{2} \times 3 \times 4 - \frac{1}{2} \times 2 \times 1.5$$

$\Rightarrow \quad mv = 4.5$ kg-m/s

or $\quad v = \left(\frac{4.5}{2}\right)$ m/s

$\Rightarrow \quad KE = \frac{1}{2}mv^2 = \frac{1}{2} \times 2 \times \left(\frac{4.5}{2}\right)^2 = 5.06$ J

5. (b) Initial velocity of stone is equal to that of the balloon. If v is the velocity just before reaching at the ground.

Then, $\quad v^2 = u^2 - 2gh$

or $\quad v^2 = 10^2 - 2 \times 10(-15) = 100 + 300 = 400$

or $\quad v = 20$ m/s

The kinetic energy of stone is

$$KE = \frac{1}{2}mv^2 = \frac{1}{2} \times 1 \times 400 = 200 \text{ J}$$

6. (a) Let the time of descent of each ball is t_0.

$\therefore \quad h = \frac{1}{2}gt_0^2$ or $\frac{20 \times 2}{10} = t_0^2$

$\therefore \quad t_0 = 2$ s

$\therefore$ Each ball remains in air for four second. Thus, the time interval between balls is 1 s.

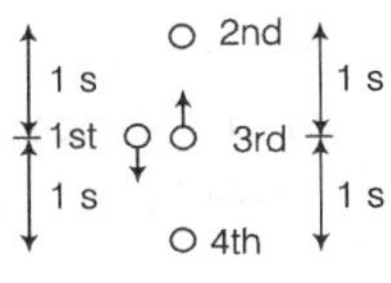

The speed of first ball is $v = gt = 10 \times 1 = 10$ m/s

$\therefore$ The kinetic energy of the first ball is $KE = \frac{1}{2}mv_0^2$

$$= \frac{1}{2} \times 10 \times 10^{-3} \times 100 = 0.5 \text{ J}$$

7. (b) From figure,

$$BD = \sqrt{(6\sqrt{5})^2 - 6^2} = 12 \text{ m}$$

and $\quad DC = \sqrt{10^2 - 6^2} = 8$ m

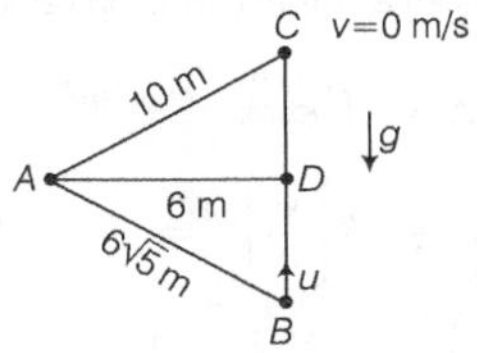

$\therefore$ The maximum height attained by the ball is

$$h = BD + DC = 12 + 8 = 20 \text{ m}$$

$\therefore \quad v^2 = u^2 - 2gh$

or $\quad 0^2 = u^2 - 2g \times 20$

$\therefore \quad u = \sqrt{2 \times 10 \times 20} = 20$ m/s

$\therefore$ Initial kinetic energy of the ball is

$$KE = \frac{1}{2}mu^2 = \frac{1}{2} \times 10 \times 10^{-3} \times 400 = 2 \text{ J}$$

8. (a) $KE = \frac{1}{2}mv^2$

$\because \quad v^2 = u^2 - 2gh$

$$KE = \frac{1}{2}mu^2 - \frac{1}{2}m \cdot 2gh$$

$KE = \lambda - kh \quad (\lambda = \text{constant}, k = \text{constant})$

Hence, it represents a straight line with negative slope.

9. (c) The motion is a projectile motion.

Given, Kinetic energy at point of projection $= \frac{1}{2}mu^2 = K$

At highest point,

$$v_y = 0$$

$\Rightarrow \quad v_x = u\cos 60° = \frac{u}{2}$

$\therefore$ KE (at highest point) $= \frac{1}{2}\frac{mu^2}{4} = \frac{K}{4}$

10. (a) In the frame of river, the float is at rest. The swimmer moves with constant speed v_{rel} in the frame of river. Thus, the time taken by float to move a distance 6 km is $2h$.

$\therefore$ Speed of river flow

$$v_r = \frac{l}{2\text{ h}} = \frac{6}{2} = 3 \text{ km/h}$$

$\therefore$ Kinetic energy per unit volume

$$= \frac{1}{2}\left(\frac{m}{V}\right)v_r^2 = \frac{1}{2} \times \rho \times v_r^2$$

$$= \frac{1}{2} \times 1000 \times \left(3 \times \frac{5}{18}\right)^2$$

or $\quad KE = \frac{1}{2} \times 1000 \times \frac{25}{36}$

$$= 347.22 \text{ J/m}^3$$

$$\approx 347.2 \text{ J/m}^3$$

11. (d) $\because v = b\sqrt{s}$

or $\frac{ds}{dt} = b\sqrt{s}$ or $\int_0^s \frac{ds}{\sqrt{s}} = b\int_0^t dt$

or $\left[\frac{s^{-\frac{1}{2}+1}}{-\frac{1}{2}+1}\right]_0^s = bt$ or $2\sqrt{s} = bt$ or $s = \frac{b^2t^2}{4}$

$\therefore$ $KE = \frac{1}{2}mv^2 = \frac{1}{2}mb^2s = \frac{1}{2}\times 1\times 1\times \frac{1\times 1}{4} = \frac{1}{8}$ J

12. (a) From given information $a = -kx$, where a is acceleration, x is displacement and k is a proportionality constant.

$$\frac{vdv}{dx} = -kx \Rightarrow vdv = -kx\,dx$$

Let for any displacement from 0 to x, the velocity changes from v_0 to v.

$$\int_{v_0}^{v} vdv = -\int_0^x kx\,dx$$

$\Rightarrow$ $\left[\frac{v^2}{2}\right]_{v_0}^{v} = -k\left[\frac{x^2}{2}\right]_0^x \Rightarrow \frac{v^2 - v_0^2}{2} = -\frac{kx^2}{2}$

or $m\left(\frac{v^2 - v_0^2}{2}\right) = -\frac{mkx^2}{2} \Rightarrow \frac{1}{2}mv^2 - \frac{1}{2}mv_0^2 = -\frac{mkx^2}{2}$

$\Delta K \propto x^2$ (ΔK is loss in KE)

13. (b) Here, kinetic energy

$$KE = \frac{1}{2}mv^2 = \frac{1}{2}\times 1\times k^2t^2 = \frac{t^2}{2}$$

$\therefore$Mean KE, $\langle KE \rangle = \frac{\int_1^2 KE\,dt}{\int_1^2 dt} = \frac{\frac{1}{2}\left[\frac{t^3}{3}\right]_1^2}{[t]_1^2} = \frac{\frac{1}{6}[8-1]}{1} = \frac{7}{6}$ J

14. (d) $\frac{1}{2}mv^2 \propto t$, $v \propto \sqrt{t}$, $F = ma = m\frac{dv}{dt}$

As, $a = \frac{dv}{dt} = k\frac{1}{2}t^{-1/2}$

$a \propto \frac{1}{\sqrt{t}}$, so $F \propto \frac{1}{\sqrt{t}}$

15. (b) The speed of turns of spring varies linearly from one end to another end.
The mass of considered element is

$$dm = \frac{m_0}{l}dx$$

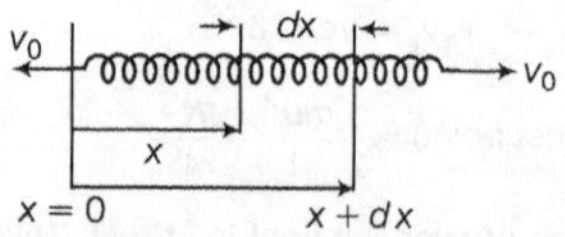

where, l is the length of the spring.

Here, $v = bx + c$

For x = 0, $v = -v_0$

$\therefore$ $c = -v_0$

$\therefore$ $v = bx - v_0$

At $x = l$, $v = v_0$ or $2v_0 = bl$

$\therefore$ $b = \frac{2v_0}{l}$

$\therefore$ $v = \frac{2v_0}{l}x - v_0$

$\therefore$ Kinetic energy of the element is

$$dt = \frac{1}{2}dm\,v^2$$

$\therefore$ $T = \int_0^l \frac{1}{2}\left(\frac{m_0}{l}dx\right)\left(\frac{2v_0}{l}x - v_0\right)^2 = \frac{1}{6}m_0v_0^2$

16. (a,b) Since speed is constant, then Kinetic energy is constant given by

$$KE = \frac{1}{2}mv^2 = \frac{1}{2}\times 1\times (10)^2 = 50 \text{ J}$$

Chagne in KE = 0 ($\because$ speed is constant)

17. (a, b) (a) $KE = \frac{1}{2}mu^2 = 0$

(b) $KE = \frac{1}{2}mu_{rel}^2 = \frac{1}{2}\times 50\times\left(108\times\frac{5}{18}\right)^2$

$= 25 \times 900 = 22500$ J

18. (a,d) $\because v^2 = u^2 + 2as$

$\therefore$ $v^2 - u^2 = 2as$

or $v^2 - u^2 = 2\times$ area of a-s graph

or $\frac{1}{2}m(v^2 - u^2) = \frac{1}{2}\times 1\times 2\times\frac{1}{2}\times 10\times 10$

$= 50$ J

Alternate Method

From graph,

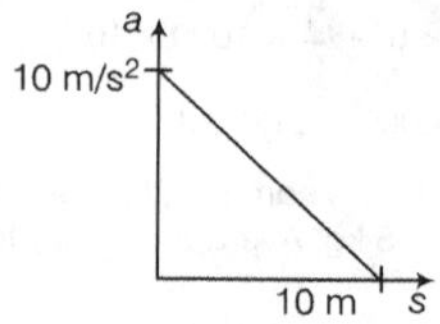

Equation of graph can be written as

$a(s) = 10 - s$

or $v\frac{dv}{ds} = (10 - s)$

$\int_{v_1}^{v_2} vdv = \int_0^{10}(10 - s)\,ds$

or $\left[\frac{v^2}{2}\right]_{v_1}^{v_2} = \left[10s - \frac{s^2}{2}\right]_0^{10}$

or $\frac{1}{2}(v_2^2 - v_1^2) = (100 - 50)$

or $\frac{1}{2}m(v_2^2 - v_1^2) = 50\times 1 = 50$ J

19. (c,d) $\mathbf{v} = \frac{dr}{dt} = 6\cos 2t\hat{i} + 6\sin 2t\hat{j}$

$\therefore$ The speed of the particle at an instant$t$ is

$$v = |\mathbf{v}| = \sqrt{(6\cos 2t)^2 + (6\sin 2t)^2} = 6 \text{ m/s}$$

$\therefore$ $KE = \frac{1}{2}mv^2 = \frac{1}{2}\times 1\times 36 = 18$ J

20. (i) (5) As, $v = u + at = at = 10$ m/s ($\because u = 0$)

$\therefore$ Change in $KE = \frac{1}{2}mv^2 - \frac{1}{2}mu^2$

$= \frac{1}{2}\times 1000\times 100$

$= 5\times 10^4$ J

$\therefore$ $n = 5$

(ii) (2) At $t = 1$ s, $v_1 = at = 5\times 1 = 5$ m/s^2

and at$t = 2$ s, $v_2 = 10$ m/s

$\therefore$ Change in $KE = \frac{1}{2}mv_2^2 - \frac{1}{2}mv_1^2$

$= 5\times 10^4 - \frac{1}{2}\times 1000\times 25$

$= 5\times 10^4 - 12500$

$= 37500$ J

$\therefore$ $n = 2$

(iii) (4) $v^2 = u^2 + 2as$

or $\quad v^2 = 0 + 2 \times 5 \times 10 = 100$

$\therefore$ Change in KE $= \frac{1}{2}mv^2 - \frac{1}{2}mu^2$

$= \frac{1}{2} \times 1000 \times 100 - 0 = 50000 \text{ J} = 5 \times 10^4$

$\therefore \quad x = 4$

21. (i) (2) The area of (a-t) graph gives change in velocity.

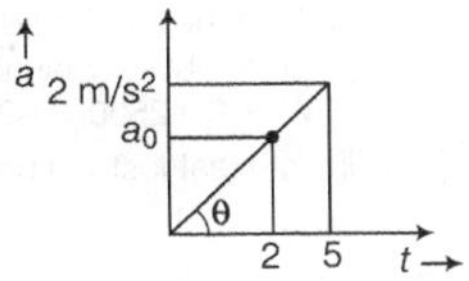

At $\quad t = 2 \text{ s}$

Here, $\quad \tan\theta = \frac{2}{5} = \frac{a_0}{2}$

$\therefore \quad a_0 = \frac{4}{5}$

$\therefore \quad v - u =$ area of (a-t) graph

$= \frac{1}{2} \times a_0 \times 2$

$v - 5 = a_0 = \frac{4}{5} = 0.8$

$\therefore \quad v = 5.8 \text{ m/s}$

$\therefore \quad \text{KE} = \frac{1}{2}mv^2 = 16.82 \text{ J}$

$\therefore \quad n = 2$

(ii) (2) $v - 5 = \frac{1}{2} \times 2 \times 5$

$\therefore \quad v = 5 + 5 = 10 \text{ m/s}$

$\therefore$ Change in KE $= \frac{1}{2}mv^2 - \frac{1}{2}mu^2$

$= 50 - 12.5 = 37.5 \text{ J} = \frac{75}{2} \text{ J}$

$\therefore \quad x = 2$

(iii) (0) Think solution same as above.

From $t = 5$ s to $t = 15$ s, area under curve gives change in velocity.

$|v| - |u| = \frac{1}{2} \times 5 \times 2 - \frac{1}{2} \times 5 \times 2 = 0$

or $\quad |v| = |u|$

Hence, change in KE = 0

22. (6) Here, $\frac{1}{2}mv^2 = bt^4$ or $mv^2 = 2bt^4$

The centripetal force on the body is

$$F_c = \frac{mv^2}{R} = \frac{2bt^4}{R} \text{ and } a_t = \frac{dv}{dt} = \sqrt{\frac{2b}{m}}\frac{d(t)^2}{dt} = \sqrt{\frac{8b}{m}}\,t$$

$\therefore$ Tangential force is

$$F_t = ma_t = m\sqrt{\frac{8b}{m}}\,t$$

The force on the particle is

$$F = \sqrt{\left(\frac{2bt^4}{R}\right)^2 + 8mbt^2} = \sqrt{4+8} = \sqrt{12} = 2\sqrt{3} \text{ N}$$

$\therefore 2\sqrt{3} = (2x)^{1/2} \Rightarrow 4 \times 3 = 2x \Rightarrow x = 6$

23. (2) Here, $\mathbf{v}_1 = 3\hat{\mathbf{i}} - gt_0\hat{\mathbf{j}}$

and $\quad \mathbf{v}_2 = -3\hat{\mathbf{i}} - gt_0\hat{\mathbf{j}}$

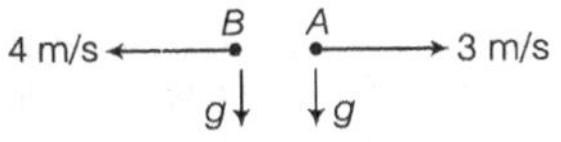

But $\mathbf{v}_1$ and $\mathbf{v}_2$ are perpendicular to each other.

$\therefore \quad \mathbf{v}_1 \cdot \mathbf{v}_2 = 0$ or $-9 + g^2t_0^2 = 0 \Rightarrow t_0 = \frac{\sqrt{9}}{\sqrt{100}}$

$\therefore \quad t_0 = 0.3 \text{ s}$

$\therefore \quad \mathbf{v}_1 = 3\hat{\mathbf{i}} - gt_0\hat{\mathbf{j}} = 3\hat{\mathbf{i}} - 10 \times 0.3\,\hat{\mathbf{j}}$

$\therefore \quad v_1 = 3\sqrt{2} \text{ m/s}$

and $\quad \mathbf{v}_2 = -3\hat{\mathbf{i}} - 10 \times 0.3\,\hat{\mathbf{j}} = -3\hat{\mathbf{i}} - 3\hat{\mathbf{j}}$

$\therefore \quad v_2 = 3\sqrt{2} \text{ m/s}$

$\therefore \quad v_{rel} = \sqrt{v_1^2 + v_2^2} = \sqrt{36} = 6 \text{ m/s}$

$\therefore \quad \text{KE} = \frac{1}{2}mv_{rel}^2 = \frac{1}{2} \times 1 \times 36 = 18 \text{ J}$

$\therefore \quad 9N = 18 \Rightarrow N = 2$

24. (a) The acceleration of A with respect to B is zero. So, velocity of A with respect to B remains constant. Hence, kinetic energy of A with respect to B remains constant.

25. (b) As, $v = r\omega = 1 \times 10\sqrt{2} = 10\sqrt{2}$ m/s

$\therefore$ The kinetic energy of insect is $= \frac{1}{2}mv^2$

$= \frac{1}{2} \times 10^{-3} \times 100 \times 2 = 0.1 \text{ J}$

26. (c) $\frac{1}{2}mv^2 = 2$ or $v^2 = 4$

$\therefore \quad v = 2 \text{m/s}$

Here, $v_{rel} = 2$ m/s

$\therefore$ The maximum speed in the frame S_1 is

$v_2 = v_{rel} + v = 4 \text{ m/s}$

$\therefore$ Kinetic energy in the frame S_2 is $\frac{1}{2}mv_2^2$

$= \frac{1}{2} \times 1 \times 16 = 8 \text{ J}$

27. (a) We have, $\quad 15 = v_0 + 6a \quad$ ...(i)

Here, v_0 = velocity at the 1st point

Also, $\quad 60 = 6v_0 + \frac{1}{2}a \times 6^2$

or $\quad 60 = 6v_0 + 18a \quad$ ...(ii)

From Eqs. (i) and (ii), we get

$a = \frac{5}{3} \text{ m/s}^2$ and $v_0 = 5$ m/s

$\therefore$ Kinetic energy at first point is

$\frac{1}{2}mv_0^2 = \frac{1}{2} \times 1000 \times 25 = 12500 \text{ J}$

28. (a) As, $v_0^2 = u^2 + 2as$ or $25 = u^2 + 2 \times \frac{5}{3} \times 7.5$

$\therefore \quad u = 0$

$\therefore$ Kinetic energy is $\frac{1}{2}mu^2 = 0$

29. (a)

(i) Initially, the acceleration of the particle is in opposite direction of velocity. So, speed decreases up to zero, then the speed increases in the direction of acceleration.

(ii) From figure, speed decreases up to maximum height, then starts to increase.

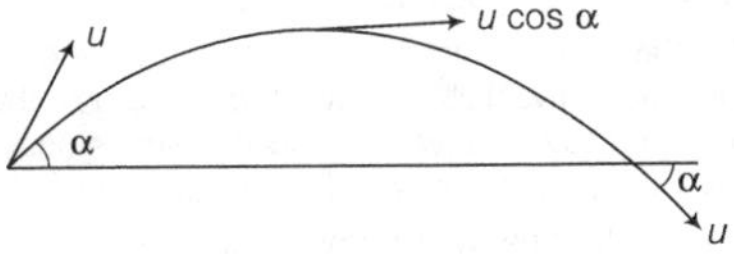

(iii) In the case of a uniform circular motion, speed remains constant.

(iv) The speed of a car decreases continuously and finally the car comes in the rest.

(i) → p, (ii) → p, (iii) → q, (iv) → r

DPP-2 **Work Done by Constant Force**

1. Maximum frictional force acting on the block,

$$f_{max} = \mu mg = 0.5 \times 10 \times 10 = 50\ \text{N}$$

As, $f_{max} > F$

$\Rightarrow$ $f = F = 20$ N

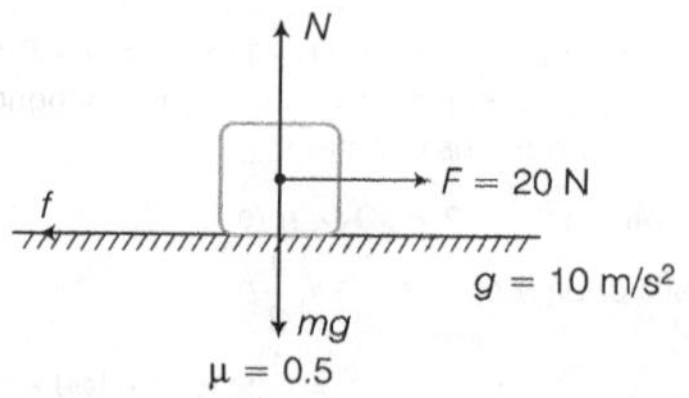

Hence, the block will not move with respect to ground.

(i) (a) Work done by friction on the block in time $t = 5$ s is zero, as the displacement of the block is zero w.r.t. ground.

(ii) (a) The frame S is moving with constant velocity of 10 m/s in rightward direction that means the block will appear to move leftward with velocity 10 m/s with respect to frame S. The displacement of the block w.r.t. reference frame $S = 10 \times 5 = 50$ m in leftward direction.

(iii) (c) Work done by friction on the block as seen from the frame S.

W_B = frictional force on the block × displacement of the block w.r.t. frame $S \times \cos 0° = 20 \times 50 = +1000$ J

(iv) (c) With respect to frame S, the ground will appear to move in leftward direction and the frictional force on the ground is acting in rightward direction.

$$W_g = 20 \times 50 \cos 180° = -1000\ \text{J}$$

(v) (a) $W_n = W_b + W_g = 1000 - 1000 = 0$ J

(vi) (d) conclusion can be made through previous discussions.

2. As, the block is moving with constant velocity

$\Rightarrow$ $f = F = 50$ N

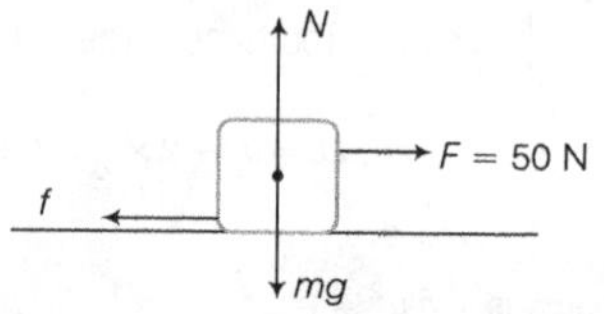

(i) (b) Distance moved by the block in 5 s, $s = 5 \times 10 = 50$ m

Work done by dynamic friction $= fs \cos 180°$

$$W_f = -50 \times 50 = -2500\ \text{J}$$

(ii) (b) Net work done by friction = work done by friction on (block + ground)

$= -2500$ J

(iii) (b) $W_f = -\Delta H = -(-2500) = 2500$ J

(iv) (a) As, the frame is moving with constant velocity of 10 m/s in the direction of velocity (righward) of the block, the ground will appear to move in leftward direction with velocity 10 m/s and the block will appear in rest. Option (a) is correct.

(v) (c) The work done by dynamic friction on the block in $t = 5$ s with respect to the frame of reference will be zero as the block is in rest with respect to the frame.

(vi) (b) Work done by dynamic friction exerted by the block on the ground $W = fs \cos 180°$

Friction on the ground acting in rightward direction and displacement of the ground w.r.t. frame is in leftward direction.

$$W = -50 \times 50 \approx -2500\ \text{J}$$

(vii) (b) If the block and ground are taken as system, then the work done by dynamic friction between contact surfaces is $W = 0 - 2500 = -2500$ J

(viii) (c) Heat lost will be non-zero, it is equal to 2500 J.

(ix) (a,b,c,d)

3. (c) Free body diagram of block

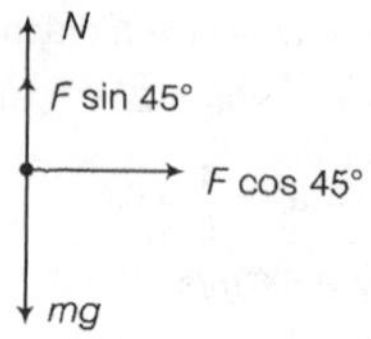

$$W = Fs \cos 45° = 10\sqrt{2} \times 2 \times \frac{1}{\sqrt{2}} = 20\ \text{J}$$

4. (a) $W = \mathbf{F} \cdot \Delta \mathbf{s}$ $\Rightarrow$ $\mathbf{r}_i = \hat{i} + \hat{j}$ and $\mathbf{r}_f = 2\hat{i} + 3\hat{j}$

$\Delta \mathbf{s} = \mathbf{r}_f - \mathbf{r}_i$

Thus, work done, $W = (3\hat{i} + 4\hat{j}) \cdot (2\hat{i} + 3\hat{j} - \hat{i} - \hat{j})$

$= (3\hat{i} + 4\hat{j}) \cdot (\hat{i} + 2\hat{j}) = 3 + 8 = 11$ J

5. (b) Work done in displacing the particle,

$W = \mathbf{F} \cdot \mathbf{r} = (5\hat{i} + 3\hat{j} + 2\hat{k}) \cdot (2\hat{i} - \hat{j})$

$= 5 \times 2 + 3 \times (-1) + 2 \times 0 = 10 - 3 = 7$ J

6. (c) The work done by constant force depends on the initial and final position of an object.

$\therefore$ $\mathbf{s} = \mathbf{r}_f - \mathbf{r}_i = 4\hat{i} + 5\hat{j} - (\hat{i} + \hat{j}) = 3\hat{i} + 4\hat{j}$

$\therefore$ $W = \mathbf{F} \cdot \mathbf{s} = (3\hat{i} + 4\hat{j}) \cdot (3\hat{i} + 4\hat{j}) = 9 + 16 = 25$ J

7. (b) In this case, point of application of force does not move. Thus, work done by the force is zero.

8. (d) The force of friction on the block is $f = mg \sin \theta$ (as block is in rest with respect to triangular block) and displacement of the block is $s = v_0 t$ (with respect to ground)

$\therefore$ $W = fs \cos \alpha$

where, α is angle between **f** and **s**.

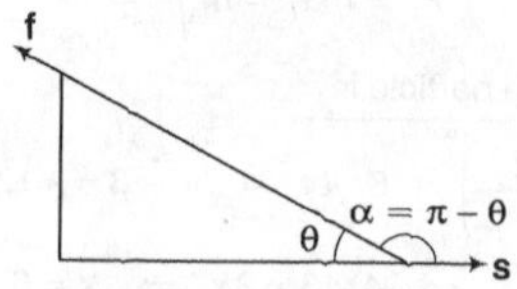

$\therefore$ $W = fs \cos(\pi - \theta) = -fs \cos \theta$

$$= -mg \sin \theta \cos \theta\,(v_0 t) = -\frac{mgv_0 t}{2} \sin 2\theta$$

9. (a) Work done by constant force does not depend on the path. It only depends on initial and final positions. Here, initial and final positions are same. So, work done is zero.

10. (b) In ΔABC,

$$\sin 30^\circ = \frac{BC}{AC}$$

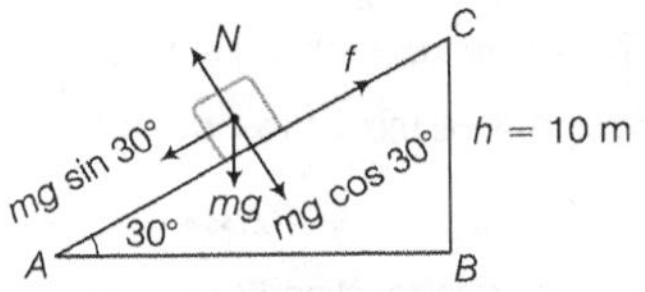

$\Rightarrow \quad \frac{1}{2} = \frac{10}{AC}$ or $AC = 20\text{ m}$

Frictional force acting on the block $f = \mu mg \cos 30^\circ$

$$= 0.5 \times 5 \times 10 \times \frac{\sqrt{3}}{2} = \frac{50\sqrt{3}}{4}$$

Work done by frictional force on the block,

$$W_f = f \times AC \times \cos 180^\circ$$
$$= \frac{-50\sqrt{3}}{4} \times 20 = -250\sqrt{3}\text{ J}$$

11. (c) FBD of the block

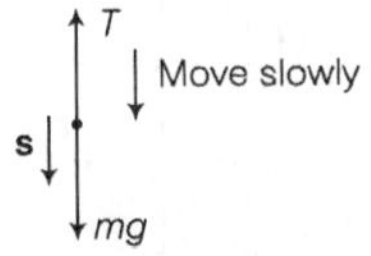

$$T = mg \qquad (\because a = 0)$$

$\therefore \quad W = -Ts = -mg\,s$

$$= -10 \times 10 \times 2 = -200\text{ J}$$

12. (b) Here, $\quad \mathbf{s} = \mathbf{r}_f - \mathbf{r}_i$

$$= 3\hat{i} - 4\hat{j} + 8\hat{k} - (2\hat{i} + 3\hat{j} + 4\hat{k})$$
$$= \hat{i} - 7\hat{j} + 4\hat{k}$$

and $\quad \mathbf{F} = 2\hat{i} + 4\hat{j} - 10\hat{k}$

$\therefore \quad W = \mathbf{F} \cdot \mathbf{s} = (2\hat{i} + 4\hat{j} - 10\hat{k}) \cdot (\hat{i} - 7\hat{j} + 4\hat{k}) = -66\text{ J}$

13. (c) Force acting on the block $= ma$

$$= 20 \times 0.5 = 10\text{ N}$$

Work done by the coolie on the box during the motion

$$W = Fs\cos 0^\circ = 10 \times 10 = 100\text{ J}$$

14. (a) If the coolie moves with constant velocity of 10 m/s, that means no acceleration of the box.

Work done by the coolie on the box is zero.

15. (a,b,c,d) (a) Since, no relative motion is present between blocks A and B. So, the nature of friction will be static.

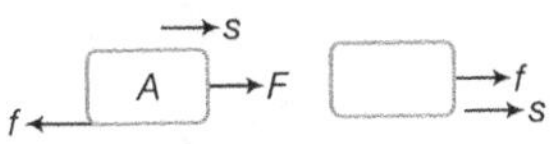

(b) Frictional force on block A is in opposite direction of displacement.

(c) The friction force on block B is in the direction of displacement.

(d) If static friction is internal force of a system, the work done by static friction on the system is zero in all frames of reference.

All options are correct.

16. (a,b,c,d) Work done depends on the frame of reference because displacement is different for different frames of reference.

Options (a) and (b) are correct.

As, there is no relative motion between blocks A and B, no heat loss takes place.

Work done by internal forces does not depend on the frame of reference. Hence, the work done by the friction on the system (block A + block B) does not depend on the frame of reference.

17. (a,b,c,d) Since, relative motion is present between the blocks, then nature of friction will be kinetic friction.

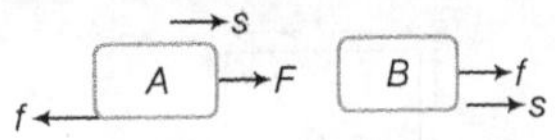

(b) The displacement of the block B is in the direction of frictional force.

(c) The displacement of the block A is in opposite direction of frictional force on the block A.

(d) The work done by kinetic friction on the system is the cause of heat loss.

18. (a,c)

19. (a,b,c,d) As the block A always remains stationary with respect to triangular block.

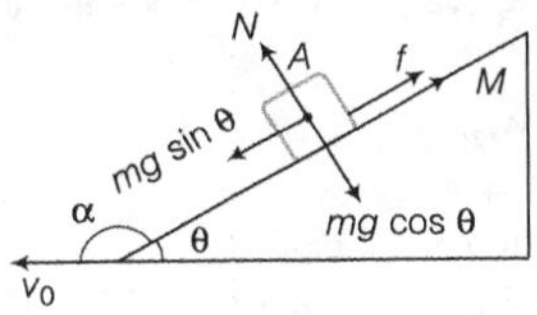

$$f = mg\sin\theta$$
$$N = mg\cos\theta$$

Displacement of the triangular block w.r.t. ground $= v_0 t$

Work done by normal reaction $W_1 = -Nv_0 t\cos(90^\circ - \theta)$

$$W_1 = -mg\cos\theta\, v_0 t\cos(\pi/2 - \theta)$$
$$= -mg\cos\theta\sin\theta\, v_0 t$$
$$= \frac{-mg\, v_0 t\sin 2\theta}{2}$$

(b) If normal reaction is internal force of a system, work done by normal reaction on the system is zero in all frames of reference.

(c) $W_1 + W_2 = 0$

$\therefore \quad W_2 = -W_1 = \frac{mgv_0 t\sin 2\theta}{2}$

(d) $W_1 + W_2 = 0$

20. (a,b) As, the velocity of the block is constant, this means that

$$v = \text{constant}$$

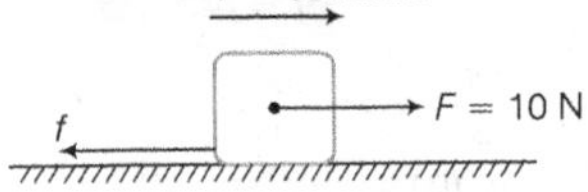

$$f = F = 10\text{ N}$$

(a) Work done by friction on the block $W_f = -f_s$

$$= -10 \times 10 = -100\text{ J}$$

(b) Work done by applied force on the block

$$W_F = Fs = 10 \times 10 = 100\text{ J}$$

Net work done on the block $W_N = W_f + W_F$

$$= -100 + 100 = 0$$

21. (a) Maximum friction acting on the block $f_{max} = \mu mg$

$$f_{max} = 0.5 \times 10 \times 10 = 50\text{ N}$$
$$f_{max} > F \qquad (\because F = 40\text{ N})$$

Hence, the applied force is balanced by static friction.

$$f_s = F$$

The block will not move work done by friction on the block is zero.

22. (a,b,c) $s_{rel} = v_{rel}t = 10 \times 2 = 20\text{ m}$

(a) $W_f = -40 \times 20 = -800\text{ J}$

(b) $W_f = 40 \times 20 = 800\text{ J}$

(c) $W_f = 40 \times 20 = 800\text{ J}$

23. (b) $W_{f(\text{total})} = W_{f(AB)} + W_{f(BC)}$

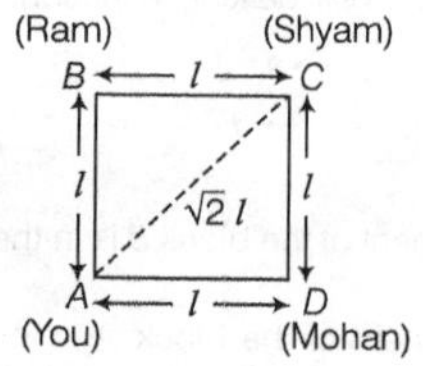

From $A \to B$ and from $B \to C$

$$= -\mu mgl - \mu mgl = -2\,\mu mgl$$
$$= -2 \times 0.3 \times 2.5 \times 9.8 \times 8$$
$$W_f = -118\text{ J}$$

24. (b) From $A \to C$ directly

$$W_f = -\mu\, mg\,(\sqrt{2}l) = -0.3 \times 2.5 \times 9.8 \times \sqrt{2} \times 8$$
$$W_f = -83\text{ J}$$

25. (a) $W_f = W_{f1} + W_{f2}$

$A \to D$, then $D \to A$

$$= -\mu\, mgl - \mu\, mgl$$
$$= -2\,\mu\, mgl = -2 \times 0.3 \times 2.5 \times 9.8 \times 8$$
$$W_f = -118\text{ J}$$

26. (b) Here, $s_A = 2s_B$

$v_A = 2v_B$ and $a_A = 2a_B$

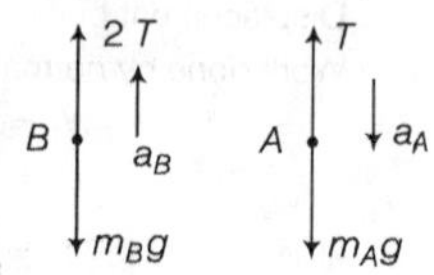

Also, $100 - T = 10a_A = 20a_B$ ...(i)

and $2T - 50 = 5a_B$ ...(ii)

From Eqs. (i) and (ii), we get

$$200 - 2T = 40a_B$$
$$-50 + 2T = 5a_B$$
$$\Rightarrow 150 = 45\,a_B$$

$\therefore$ $a_B = \dfrac{150}{45} = \dfrac{10}{3}\text{ m/s}^2$

$\therefore$ $a_A = \dfrac{20}{3}\text{ m/s}^2$

$\because$ $2T - 50 = 5a_B = \dfrac{50}{3}$ or $2T = \dfrac{200}{3}$

or $T = \dfrac{100}{3}$

$$W_B = 2T \times s_B = \frac{200}{3} \times 2.5 = \frac{500}{3}\text{ J} = 116.67\text{ J}$$

27. (a) $W_A = -Ts_A = -\dfrac{100}{3} \times 5 = -\dfrac{500}{3}\text{ J} = -166.67\text{ J}$

$$W_T = W_A + W_B = 0$$

28. (c) $W_g = 100 \times 5 - 50 \times 2.5 = 375\text{ J}$

$$W = W_T + W_g = 0 + 375 = 375\text{ J}$$

29. (a) As, $F - (mg\sin\theta + \mu mg\cos\theta) = ma$

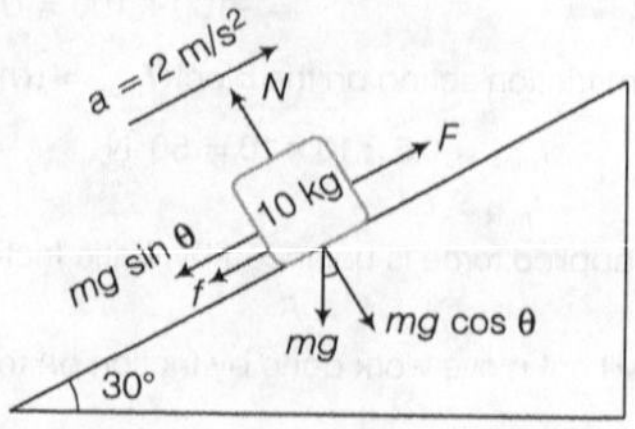

$$\mu = 1/\sqrt{3}$$

$\Rightarrow$ $F = ma + mg\sin\theta + \mu mg\cos\theta$

or $F = (10 \times 2) + 10 \times 10 \times \dfrac{1}{2} + \dfrac{1}{\sqrt{3}} \times 10 \times 10 \times \dfrac{\sqrt{3}}{2}$

or $F = 20 + 50 + 50 = 120\text{ N}$

Displacement of block, $s = ut + \dfrac{1}{2}at^2$

$$s = \frac{1}{2} \times 2 \times 2 \times 2 = 4\text{ m}$$

Work done, $W_1 = Fs\cos 0^\circ = 120 \times 4 = 480\text{ J}$

30. (c) $W_g = (mg\sin\theta)s\cos 180^\circ = \left(10 \times 10 \times \dfrac{1}{2}\right) \times 4 \times (-1)$

$$= -50 \times 4 = -200\text{ J}$$

31. (c) Work done by normal reaction, $W_N = Ns\cos 90^\circ = 0$

$$W_f = (\mu mg\cos\theta)s\cos 180^\circ$$
$$= -\frac{1}{\sqrt{3}} \times 10 \times 10 \times \frac{\sqrt{3}}{2} \times 4 = -200\text{ J}$$

Net work done on the block $= W_1 + W_g + W_N + W_f$

$$= 480 - 200 + 0 - 200\text{ J} = 80\text{ J}$$

32. (d) All the options are correct.

33. (8) $m_A = 0.36$ kg and $m_B = 0.72$ kg

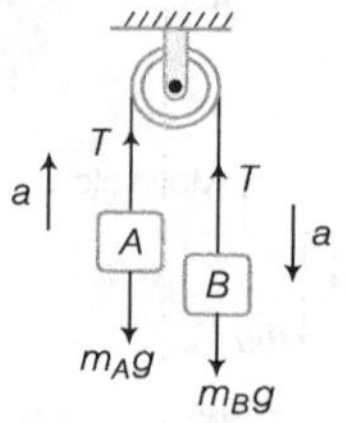

$$m_Bg - T = m_Ba \quad \text{...(i)}$$
$$T - m_Ag = m_Aa \quad \text{...(ii)}$$

$\Rightarrow$ $(m_B - m_A)g = (m_B + m_A)a$

or $(0.72 - 0.36) \times 10 = (0.72 + 0.36)a$

or $a = \dfrac{0.36}{1.08} \times 10 = 3.3\text{ m/s}^2$

$\Rightarrow$ $0.72 \times 10 - T = 0.72 \times 3.3$

or $7.2 - 2.4 = T$

$\Rightarrow$ $T = 4.8\text{ N}$

Displacement of the block A in 1 s

$$s_A = ut + \frac{1}{2}at^2$$
$$= \frac{1}{2} \times 3.3 \times (1)^2 = 1.65\text{ m}$$

Work done by string on the block $A = T \times s_A \times \cos 0^\circ$

$$= 4.8 \times 1.65 = 7.92\text{ J} \approx 8\text{ J}$$

34. (i) (3) $s = \dfrac{1}{2}at^2 = \dfrac{1}{2} \times 2 \times 5^2 = 25\text{ m}$

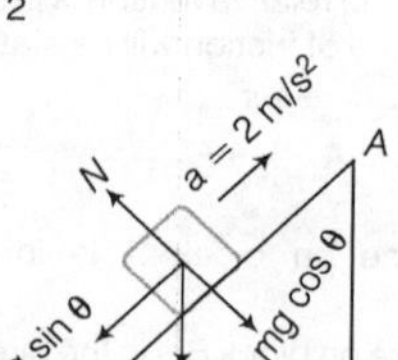

$$W_f = (-\mu\, mg\cos\theta)\, s$$
$$= -\frac{1}{\sqrt{2}} \times 10 \times 10 \times \frac{1}{\sqrt{2}} \times 25$$
$$= -\frac{100}{2} \times 25$$
$$= -1250\text{ J}$$

$\therefore$ $n = 3$

(ii) (0) $N = mg\cos 45^\circ$

$\therefore W_N = Ns\cos 90^\circ = 0$

(iii) (2) $F - mg \sin \theta - \mu mg \cos \theta = ma$

$\therefore \quad F = mg \sin\theta + \mu mg \cos \theta + ma$

$= \dfrac{100}{\sqrt{2}} + 50 + 20$

Work done by applied force,

$\therefore \quad W_F = Fs = \left(70 + \dfrac{100}{\sqrt{2}}\right) 25$

$\therefore \quad n = 2$

(iv) (2) $W_g = -(mg \sin \theta)\, s = \dfrac{100}{\sqrt{2}} \times 25 \qquad (\theta = 45°)$

$= \dfrac{-2500}{\sqrt{2}}$

$\therefore \quad n = 2$

(v) (2) $W = W_f + W_N + W_F + W_g = 500$ J

$\therefore \quad n = 2$

35. For triangular block, $N \cos 60° = 10\, a_0$

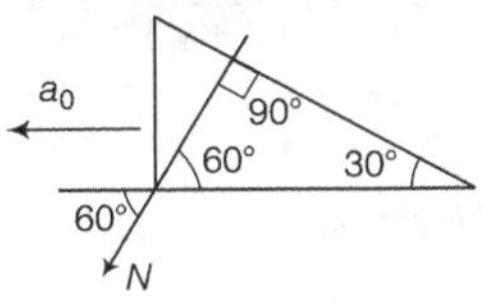

From free body diagram of smaller block in the frame of triangular block, $N + ma_0 \sin 30° = mg \cos 30°$

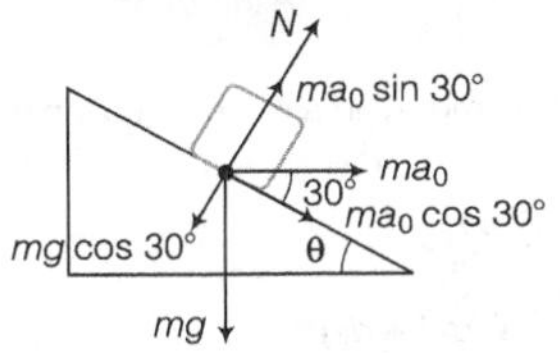

$\therefore \quad N = mg \cos 30 - ma_0 \sin 30$

or $\quad N = 5\sqrt{3} - \dfrac{N}{40}$

or $\quad N = \dfrac{5\sqrt{3} \times 40}{41} = \dfrac{120\sqrt{3}}{41}$

(i) (3) $W_1 = N \cos 60° \times 5 = \dfrac{300\sqrt{3}}{41}$

$\therefore \quad n = 3$

(ii) (2) $W_1 + W_2 = 0, W_2 = -W_1$

$\therefore \quad x = 2$

36. Acceleration of the system,

$$a_0 = \frac{F}{m_A + m_B} = \frac{60}{10 + 20} = \frac{60}{30} = 2 \text{ m/s}^2$$

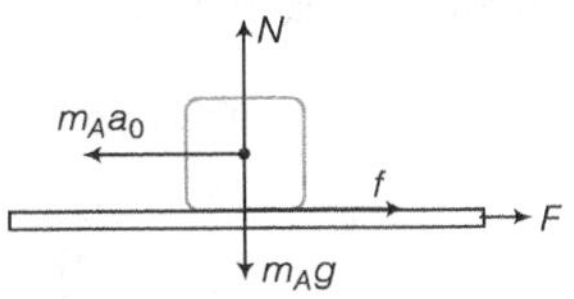

$f_{A(\max)} = \mu m_A g = 0.5 \times 10 \times 10 = 50$ N

Pseudo force (F_p) acting on the block $A = m_A a_0 = 10 \times 2 = 20$ N

As $f_{\max} > F_P$, the block and the plank will move together.

$f = F_P$

Displacement of the block in 2 s

$$s = ut + \frac{1}{2} at^2$$

$$s_A = \frac{1}{2} \times 2 \times (2)^2 = 4 \text{ m}$$

Work done by friction on block A, $W_f = f \times s_A \times \cos 0°$

$20 \times 4 = 80$ J

Work done by friction on plank B, $W_f = F \times s_B \times \cos 180°$

$s_A = s_B = 4$ m

$W_f = -20 \times 4 = -80$ J

(i) (0) Work done by friction on the system

$W_T = \underset{\text{(Block)}}{W_f} + \underset{\text{(Plank)}}{W_f} = 80 - 80 = 0$

(ii) (3) If block A and plank B are taken as system, the net work on the system will be due to the applied force.

$W_N = F \times s \times \cos 0° = 60 \times 4 = 240 \text{ J} = 80 \times 3$

$\therefore \quad x = 3$

37. (4) Mass of water lifted from swimming pool is

$$m = \frac{1}{2} \times \text{volume of vessel} \times \text{density of water}$$

$$= \frac{1}{2} \times \pi \times \left(\frac{1}{\sqrt{\pi}}\right)^2 \times 200 \times 1000 = 10^5 \text{ kg}$$

Decrease in depth of water in swimming pool is $\Delta y = \dfrac{m}{\rho A}$

$$= \frac{10^5}{10^3 \times 200} = \frac{100}{200} = 0.5 \text{ m}$$

The normal section of water in vessel is semicircle.

$\therefore$ The centre of gravity of semicircle is $\dfrac{4r}{3\pi} = \dfrac{4}{3\pi\sqrt{\pi}}$

$\therefore$ The vertical distance through which centre of gravity lifted is

$$h = \frac{\Delta y}{2} + 10 - \frac{4}{3\pi\sqrt{\pi}}$$

$$= 0.25 + 10 - \frac{4}{3\pi\sqrt{\pi}}$$

$$= 10.25 - \frac{4}{3\pi\sqrt{\pi}}$$

$\therefore \quad W = mgh = 10^5 \times 10\left(10.25 - \dfrac{4}{3\pi\sqrt{\pi}}\right)$

$= 10^6\left(10.25 - \dfrac{4}{3\pi\sqrt{\pi}}\right)$

$\Rightarrow \quad n = 4$

38. (2) To slide the block, the boy has to overcome the force of friction

$\mu mg = 300$ N

$\therefore \quad W_1 = 300 \times 2 = 600$ J

To displace the block by tipping about corner A, the change in the height of centre of gravity is

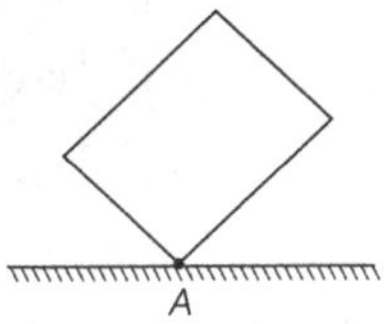

$h = \sqrt{2} - 1 = 1.414 - 1 = 0.414$

$\therefore \quad W_2 = mgh = 414 = 207 \times 2 \qquad$ (least)

$\therefore \quad n = 2$

39. (a) $W = \mathbf{F} \cdot \mathbf{s}$

Here, displacement of paper is zero. So, work is zero.

40. (a) $\because W = \mathbf{F} \cdot \mathbf{s}$

Here, displacement only depends on initial and final position of the particle.

41. (a) (i) Centripetal force is perpendicular to displacement.

$\therefore$ Work done is zero.

(ii) The friction on block B is in rightward direction. So, frictional force and displacement are in same direction. Hence, work done is non-zero positive.

(iii) Static friction comes into play. If static friction is internal force of a system, work done by static friction on the system is zero.

(iv) Due to presence of tangential acceleration, a component of force is required either in the direction of velocity or in the opposite direction of velocity.

(i) → (p), (ii) → (q), (iii) → (p), (iv) → (s)

DPP-3 **Work Done by Variable Force**

1. (a) $W = \int F dx = \int_0^d (ax^2 + bx + c) dx$

$$= a\left[\frac{x^3}{3}\right]_0^d + b\left[\frac{x^2}{2}\right]_0^d + c\,[x]_0^d$$

$$= \frac{ad^3}{3} + \frac{bd^2}{2} + cd$$

2. (c) We know that change in potential energy of a system corresponding to a conservative internal force as

$$U_f - U_i = -W = -\int_i^f \mathbf{F}\cdot d\mathbf{r}$$

Given, $F = ax + bx^2$

According to work-energy theorem, we know that work done in stretching the rubber band by dx is

$$|dW| = fdx$$

$$\Rightarrow \quad |W| = \int_0^L (ax + bx^2)\, dx$$

$$= \left[\frac{ax^2}{2}\right]_0^L + \left[\frac{bx^3}{3}\right]_0^L$$

$$= \left[\frac{aL^2}{2} - \frac{a\times(0)^2}{2}\right] + \left[\frac{b\times L^3}{3} - \frac{b\times(0)^3}{3}\right]$$

$$\therefore \quad |W| = \frac{aL^2}{2} + \frac{bL^3}{3}$$

3. (c) $W = \int_{0,0,0}^{1,1,1} \mathbf{F}\cdot d\mathbf{r}$

$$= \int_{0,0,0}^{1,1,1} (x^2\hat{\mathbf{i}} + y^2\hat{\mathbf{j}} + z^2\hat{\mathbf{k}})\cdot(dx\hat{\mathbf{i}} + dy\hat{\mathbf{j}} + dz\hat{\mathbf{k}})$$

$$= \int_0^1 x^2 dx + \int_0^1 y^2 dy + \int_0^1 z^2 dz$$

$$= \frac{1}{3} + \frac{1}{3} + \frac{1}{3} = 1\text{ J}$$

4. (d) $W = \int \mathbf{F}\cdot d\mathbf{r}$

$$= \int_{1,1}^{2,-1} (3x^2y\hat{\mathbf{i}} + x^3\hat{\mathbf{j}})\cdot(dx\hat{\mathbf{i}} + dy\hat{\mathbf{j}} + dz\hat{\mathbf{k}})$$

$$= \int_{1,1}^{2,-1} (3x^2y dx + x^3 dy)$$

$$= \int_{1,1}^{2,-1} d(x^3y) = [x^3y]_{1,1}^{2,-1}$$

$$= [-8 - 1] = -9\text{ J}$$

5. (c)

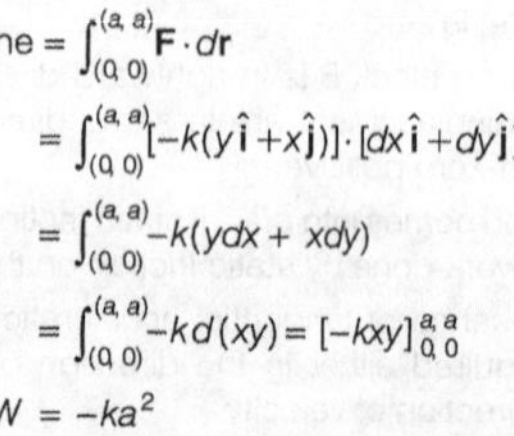

Path of particle

$$\text{Total work done} = \int_{(0,0)}^{(a,a)} \mathbf{F}\cdot d\mathbf{r}$$

$$= \int_{(0,0)}^{(a,a)} [-k(y\hat{\mathbf{i}} + x\hat{\mathbf{j}})]\cdot[dx\hat{\mathbf{i}} + dy\hat{\mathbf{j}}]$$

$$= \int_{(0,0)}^{(a,a)} -k(ydx + xdy)$$

$$= \int_{(0,0)}^{(a,a)} -kd(xy) = [-kxy]_{0,0}^{a,a}$$

or $\quad W = -ka^2$

6. (a) $\tan\theta = \frac{F_y}{F_x} = -\frac{x}{y}$

or $\quad \frac{dy}{dx} = -\frac{x}{y}$

or $\quad \int ydy = -\int xdx$

or $\quad x^2 + y^2 = c$ is a circular curve.

Thus, the force is along tangent to the path of the block.

The magnitude of the force is

$$F = \sqrt{F_x^2 + F_y^2} = k$$

$$\therefore \quad W = \int_C Fds = k\int_C ds$$

$$= k2\pi R = 2\pi Rk$$

7. (d) As, $\mathbf{F} = k\left[\frac{x}{(x^2 + y^2)^{3/2}}\hat{\mathbf{i}} + \frac{y}{(x^2 + y^2)^{3/2}}\hat{\mathbf{j}}\right] = F_x\hat{\mathbf{i}} + F_y\hat{\mathbf{j}}$

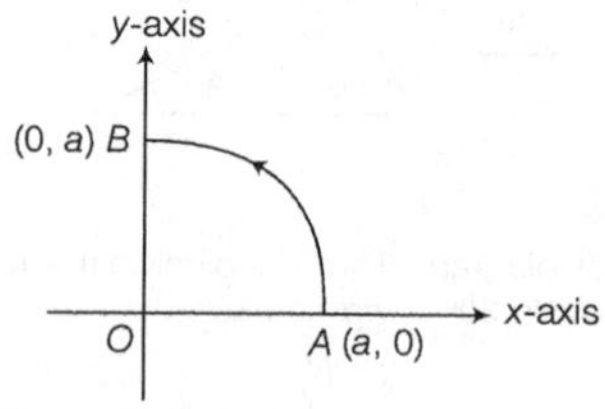

Since, the particle moves in a circle, equaiton of its path is

$$x^2 + y^2 = a^2 \qquad \text{...(i)}$$

$$\text{Work done} = \int \mathbf{F}\cdot d\mathbf{r}$$

$$= \int_{(a,0)}^{(0,a)} \mathbf{F}(dx\hat{\mathbf{i}} + dy\hat{\mathbf{j}})$$

$$= \int_{(a,0)}^{(0,a)} F_x dx + \int_{(a,0)}^{(0,a)} F_y dy$$

$$= k\int_{(a,0)}^{(0,a)} \frac{xdx}{(x^2 + y^2)^{3/2}} + k\int \frac{ydy}{(x^2 + y^2)^{3/2}}$$

$$= \frac{k}{a^3}\left[\int_a^0 xdx + \int_0^a ydy\right] \qquad \text{[from Eq. (i)]}$$

$$= \frac{k}{a^3}\left[\left.\frac{x^2}{2}\right|_a^0 + \left.\frac{y^2}{2}\right|_0^a\right]$$

$$= (k/a^3)\left[\frac{-a^2}{2} + \frac{a^2}{2}\right] = 0$$

8. (a) $W_{ext} = \int Fdr = \int_a^{2a} \frac{k}{r} dr = k[\ln r]_a^{2a} = k\ln 2$

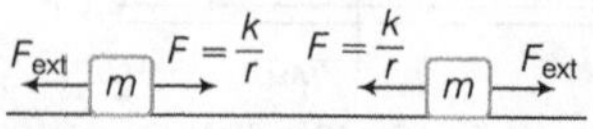

9. (b) $F_r = \frac{b\cos\theta}{r^3}$ and $F_\theta = \frac{b\sin\theta}{r^3}$

$$\therefore \quad W_1 = \int_a^r F_r dr = 0 \text{ and } W_2 = \int_0^{\pi/2} F_\theta ds = \int_0^{\pi/2} F_\theta r d\theta$$

or $\quad W_2 = \int_0^{\pi/2} \frac{b\sin\theta}{r^2} d\theta = \int_0^{\pi/2} \frac{b}{a^2}\sin\theta\, d\theta$

$$= \frac{b}{a^2}[-\cos\theta]_0^{\pi/2} = \frac{b}{a^2}[+1] = +\frac{b}{a^2}$$

$$\therefore \quad W = W_1 + W_2 = +\frac{b}{a^2}$$

10. (c) $\int_{r_0}^{2r_0} mr\omega^2 dr = \frac{m\omega^2}{2}[4r_0^2 - r_0^2]$

$= \frac{3}{2} m\omega^2 r_0^2 = \frac{3}{2} \times \frac{1}{2} \times 100 = 75$ J

11. (b) $W = \int F_x dx = \int_0^2 x dx = \left[\frac{x^2}{2}\right]_0^2 = 2$ J

Note *Since, particle moves in the xy-plane, hence work done by the z-component of the force is zero.*

12. (c) Work done by the spring, $W_s = \frac{1}{2}kx_i^2 - \frac{1}{2}kx_f^2$

$= \frac{1}{2} \times 100 \times 0.04 - \frac{1}{2} \times 100 \times 0.01$

$= 2 - 0.5 = 1.5$ J

13. (b) Work done to stretch the spring by 5 cm from mean position,

$W_1 = \frac{1}{2}kx^2$

$= \frac{1}{2} \times 5 \times 10^3 \times (5 \times 10^{-2})^2$

$= 6.25$ J

Work done to stretch the spring by 10 cm from mean position,

$W_2 = \frac{1}{2}k(x_1 + x_2)^2$

$= \frac{1}{2} \times 5 \times 10^3 (5 \times 10^{-2} + 5 \times 10^{-2})^2$

$= 25$ J

Net work done to stretch the spring from 5 cm to 10 cm.

$= W_2 - W_1$

$= 25 - 6.25 = 18.75$ J

$= 18.75$ N-m

14. (b) The work done on the spring is stored as the potential energy of the body and is given by

$U = \int_{x_1}^{x_2} F_{ext} dx$ or $U = \int_{x_1}^{x_2} kx dx$

$= \frac{1}{2}k(x_2^2 - x_1^2)$

$= \frac{800}{2}[(0.15)^2 - (0.05)^2]$

$= 400 (0.2 \times 0.1)$

$= 8$ J

15. (a) Work done on a rubber in stretching gets stored as elastic potential energy, hence positive.

16. (d) $dW = -(\mu mg \cos\theta) ds = -\mu mg (ds \cos\theta)$

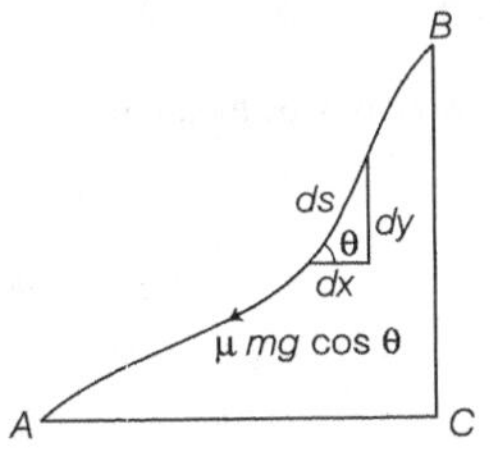

$= -\mu mg\, dx = -\mu mg \int_0^1 dx$ $(\because ds \cos\theta = dx)$

$= -0.5 \times 1 \times 10 \times 1 = -5$ J

17. (b,d) $W = \int F \cdot dy = \int_{-1}^{+1} (y^2 + 2y + 5) dy$

$= \left[\frac{y^3}{3} + y^2 + 5y\right]_{-1}^{+1} = \left[\frac{y^3}{3}\right]_{-1}^{+1} + [y^2]_{-1}^{1} + 5[y]_{-1}^{1}$

$= \frac{2}{3} + 0 + 10 = \frac{32}{3} = 10.67$ J

18. (a,b,c) Total work done is

$W = W_A + W_B = \frac{1}{2}kx_i^2 - \frac{1}{2}kx_f^2 = 2 - 0 = 2$ J

$\therefore \quad W_A = W_B = 1$ J

19. (a,d) The work done by spring does not depend on the frame of reference.

$\therefore \quad W_{ext} = -$(work done by spring)

$= -\left(-\frac{1}{2}kx_0^2\right) = \frac{1}{2}kx_0^2$

20. (a,b,c) The area of F-x graph gives work done.

(a) $W_1 = \frac{1}{2} \times 100 \times 5 = 250$ J

(b) $W_2 = \frac{1}{2} \times 100 \times 5 - \frac{1}{2} \times (10 - 5) \times 100 = 0$

(c) $W_3 = -\frac{1}{2} \times (100)(10 - 5) = -250$ J

21. (c) Work done by a variable force

$= \int \mathbf{F}_{ext} \cdot d\mathbf{x}$

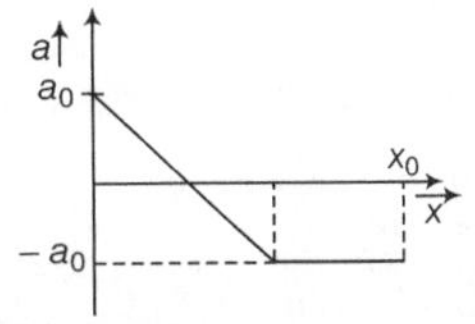

Since, $\mathbf{F}_{ext} = m\mathbf{a}$

$W = m\int a(x)\, dx$

$= m$ (Area under a-x curve)

$=$ non-zero negative

22. (i) (6) Work done is obtained by finding the area under F-x graph.

Work done from $x_1 = -10$ m to $x_2 = 5$ m

$= F(x_2 - x_1) = (40\text{ N}) \times [5 - (-10)]\text{ m} = 600$ J

$\therefore \quad n = 6$

(ii) (2) Work done from $x_1 = 0$ to $x_2 = 10$ m

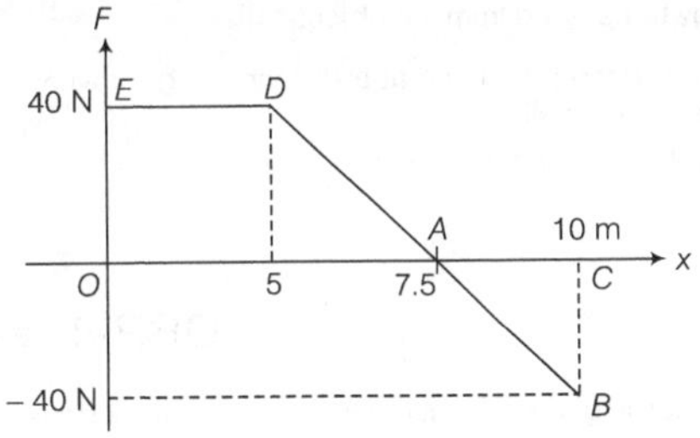

= Area under curve taking area below x-axis as negative.

= Area of trapezium $ADEO$ + Area of ΔABC

$= \left[\frac{1}{2} \times (7.5 + 5) \times 40\right] - \left[\frac{1}{2} \times 2.5 \times 40\right]$

$= \frac{1}{2} \times 12.5 \times 40 - \frac{1}{2} \times 2.5 \times 40 = 200$ J

$\therefore n = 2$

(iii) (3) Work done from $x = -10$ m to $x = 20$ m is

= Area under complete graph

$= \left[15 \times 40 + \frac{1}{2} \times 40 \times 2.5\right] - \left[\frac{1}{2}(12.5 + 5) \times 40\right]$

$= [650 - 350]$ J $= 300$ J

$\therefore \quad n = 3$

23. (2) $W = \int \mathbf{F} d\mathbf{r} = \int_0^a br dr + \int_a^\infty \frac{b}{r^2} dr = \frac{a^2 b}{2} + \frac{b}{a}$

$\therefore \quad n = 2$

24. (8) $W = \int \mathbf{F} \cdot d\mathbf{r}$

$= \int \{2(x + 4y)\hat{i} + 8x\hat{j}\} \cdot (dx\hat{i} + dy\hat{j})$

$= \int (2xdx + 8ydx + 8xdy) = \int_0^2 8dy + 8\int_{0,0}^{4,2} d(xy)$

$= 8\lfloor y \rfloor_0^2 + 8\lfloor xy \rfloor_{0,0}^{4,2}$

$= 16 + 8 \times 8 = 80$ J

$\therefore \quad n = 8$

25. (8) $W = \int_{OABC} \mathbf{F} \cdot d\mathbf{r} = \int_{OABC} \mathbf{F} \cdot d\mathbf{r} + \int_{AB} \mathbf{F} \cdot d\mathbf{r} + \int_{BC} \mathbf{F} \cdot d\mathbf{r}$

$\int_{OA} \mathbf{F} \cdot d\mathbf{r} = \int_0^2 2xyzdx = 0 \qquad (\because y = 0 \text{ and } z = 0)$

$\int_{AB} \mathbf{F} \cdot d\mathbf{r} = \int_0^2 x^2 zdy = 0 \qquad (\because z = 0)$

$\int_{BC} \mathbf{F} \cdot d\mathbf{r} = \int_0^2 x^2 yaz = (2^2)(1)\int_0^2 dz = 4\lfloor z \rfloor_0^2 = 8$ J

26. (2) Here, $\frac{\pi}{2} - \alpha + \frac{\pi}{2} - \alpha = \theta$

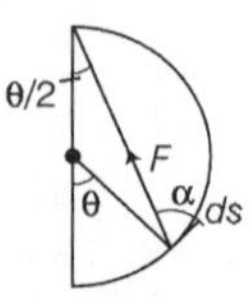

or $\quad \pi - 2\alpha = \theta \quad$ or $\quad 2\alpha = \pi - \theta$

or $\quad \alpha = \frac{\pi}{2} - \frac{\theta}{2}$

$dW = Fds\cos\alpha = F(rd\theta)\cos\left(\frac{\pi}{2} - \frac{\theta}{2}\right)$

Total work done, $W = \int dW = Fr\int_0^\pi \sin\frac{\theta}{2} d\theta$

$= 2Fr\left[-\cos\frac{\theta}{2}\right]_0^\pi = 2Fr\,[0 + 1]$

$= 2Fr = 2 \times 10 \times 0.1 = 2$ J

27. (b) A force is said to do work if when acting on a body, there is a displacement of the point of application in the direction of the force. When work is done on the object, it implies energy transferred to it and when work is done by the object. it implies energy transferred from the object.

28. (b) Considering the force at a distance, e.g. gravitational pull of the earth on a ball.

Action-force by the earth on ball.

Reaction-force by ball on the earth.

The force is given by $F = \frac{Gm_1m_2}{d^2}$, but ball and the earth have different acceleration and hence distance travelled is different in a particular time interval.

29. (b) Gravitational force is a conservative force and hence work done by the gravitational force on the block is independent of the path taken. The force of gravity always acts vertically downwards.

30. (b) $W_1 = \int_A^B (20\cos\theta)\, ds$

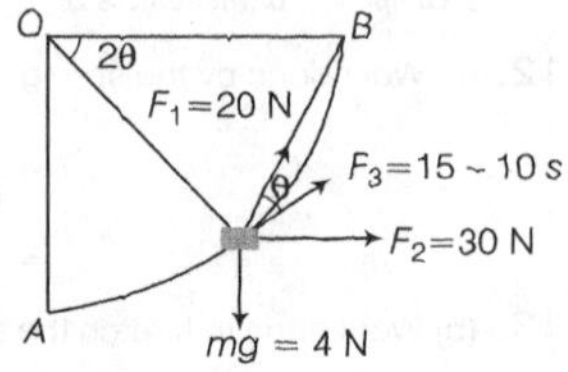

From figure, $s = R\left(\frac{\pi}{2} - 2\theta\right)$

$ds = -2Rd\theta = -12d\theta$

$\therefore \quad W_1 = \int_A^B (20\cos\theta)(-12d\theta)$

$= -240\int_{\frac{\pi}{4}}^0 \cos d\theta = 240\lfloor \sin\theta \rfloor_{\pi/4}^0$

$= -240 \times \left(-\frac{1}{\sqrt{2}}\right) = 120\sqrt{2}$ J

31. (c) $W_2 = F_2(OB) = 30 \times 6 = 180$ J

$W_3 = \int F_3 ds = \int_0^{6\left(\frac{\pi}{2}\right)} (15 - 10s)\, ds = -302.8$ J

Work done by weight, $\quad W_4 = -mg \times 6 = -4 \times 6 = -24$ J

$\therefore$ Total work done, $\quad W = W_1 + W_2 + W_3 + W_4 = 23$ J

32. (a) (i) $W_1 = \frac{-F^2}{2k_1} = \frac{-50 \times 50}{2 \times 100} = -12.5$ J

(ii) $W_2 = \frac{F^2}{2k_1} = 12.5$ J

(iii) $W_3 = \frac{-F^2}{2k_2} = -\frac{50 \times 50}{2 \times 200} = -6.25$ J

(iv) $W_4 = \frac{F^2}{2}\left(\frac{1}{k_1} + \frac{1}{k_2}\right)$

$= \frac{2500}{2}\left(\frac{1}{100} + \frac{1}{200}\right)$

$= \frac{1250}{200} \times 3 = 18.75$ J

(i) → (s), (ii) → (r), (iii) → (q), (iv) → (p)

DPP-4 **Work-Energy Theorem**

1. (b) According to work-energy theorem, net work done is equal to the change in kinetic energy.

i.e. $\quad W = \Delta KE$

or $\quad W_{you} + W_{gravity} = 0$

or $\quad W_{you} + mgh = 0$

$\therefore \quad W_{you} = -mgh$

2. (d) According to work-energy theorem,

$W = \Delta T = T_f - T_i$

Here, ΔT = change in kinetic energy

$= \frac{1}{2}mv^2 - \frac{1}{2}mu^2$

$v_{(t=2s)} = 3t^2 = 3(2)^2 = 12$ m/s

$u_{(t=1s)} = 3t^2 = 3(1)^2 = 3$ m/s

$= \frac{1}{2} \times 1 \times 144 - \frac{1}{2} \times 1 \times 9 = 67.5$ J

3. (b) According to work-energy theorem,

$W = \Delta T = T_f - T_i = 0 - \frac{1}{2}mv_0^2$

or $\quad W_f = -\frac{1}{2}mv_0^2 \quad$ or $-\mu mg\, s = \frac{-1}{2}mv_0^2$

$\therefore \quad s = \frac{v_0^2}{2\mu g}$

4. (b) In a round trip, work done by gravity is zero because gravitational force is a conservative force.

According to work-energy theorem,

$W = \Delta T$

$W_f + W_N + W_g = T_f - T_i = 0 - \frac{1}{2}mv_0^2$

where, $\quad W_f$ = work done by frictional force

W_N = work done by normal reaction

W_g = work done by gravity

$$W_f + 0 + 0 = -\frac{1}{2}mv_0^2$$

$\therefore$ $$W_f = -\frac{1}{2}mv_0^2 = -\frac{1}{2} \times 1 \times 100 = -50\,J$$

5. (c) In a round trip, work done by gravity is zero.
According to work-energy theorem,

$$W = \Delta T$$

or $$W_g + W_{air} = T_f - T_i$$

$$= \frac{1}{2} \times 10 \times 10^{-3} \times 5^2 - \frac{1}{2} \times 10 \times 10^{-3} \times 10^2$$

or $$W_{air} = -0.375\,J$$

6. (a) At maximum elongation, the velocity of the block is zero.
According to work-energy theorem, $W = \Delta T$

or $$W_S + W_F = T_f - T_i = 0 - 0 = 0$$

or $$-\frac{1}{2}kx_{max}^2 + Fx_{max} = 0$$

$\therefore$ $$x_{max} = \frac{2F}{k} = \frac{2 \times 50}{100} = 1\,m$$

7. (c) At maximum speed, acceleration of the block should be zero.

$\therefore$ $$F = kx \Rightarrow x = \frac{F}{k}$$

According to work-energy theorem, $W = \Delta T = \frac{1}{2}mv_{max}^2 - 0$

or $$W_S + W_F = \frac{1}{2}mv_{max}^2$$

or $$-\frac{1}{2}kx^2 + Fx = \frac{1}{2}mv_{max}^2$$

or $$-\frac{1}{2}k\left(\frac{F}{k}\right)^2 + F \times \left(\frac{F}{k}\right)^2 = \frac{1}{2}mv_{max}^2$$

or $$-\frac{F^2}{2k} + \frac{F^2}{k} = \frac{1}{2}mv_{max}^2 \quad \text{or} \quad \frac{F^2}{2k} = \frac{1}{2}mv_{max}^2$$

$\therefore$ $$v_{max} = \sqrt{\frac{F^2}{mk}} = \frac{F}{\sqrt{mk}} = \frac{50}{\sqrt{1 \times 100}} = \frac{50}{10} = 5\,m/s$$

8. (a) According to work-energy theorem,
Loss in kinetic energy = Work done against friction + Potential energy of spring

$$\Rightarrow \quad \frac{1}{2}mv^2 = fx + \frac{1}{2}kx^2$$

where, x is the compression of the spring.

$$\Rightarrow \quad \frac{1}{2} \times 2 \times (4)^2 = 15x + \frac{1}{2} \times 10000\,x^2$$

$$\Rightarrow \quad 5000x^2 + 15x - 16 = 0$$

$\therefore$ $$x = 0.055\,m = 5.5\,cm$$

9. (b) Here, $\mathbf{F} = \mathbf{F}_1 + \mathbf{F}_2$

$\therefore$ $$\mathbf{F}_1 = F_0\hat{\mathbf{k}} \quad \text{and} \quad \mathbf{F}_2 = (bv)\hat{\mathbf{j}}$$

The work done by $\mathbf{F}_1$ is $W_1 = F_0 z$

$\mathbf{F}_2$ is always perpendicular to the velocity, so the work done by $\mathbf{F}_2$ is zero.

$\therefore$ Net work done $W = W_1 + W_2 = F_0 z$

According to work-energy theorem,

$$W = \Delta T = T_f - T_i = \frac{1}{2}mv^2 - 0 \quad \text{or} \quad F_0 z = \frac{1}{2}mv_0^2$$

$\therefore$ $$v_0 = \sqrt{\frac{2F_0 z}{m}}$$

10. (b,c) $\mathbf{a} = \mathbf{b} \times \mathbf{v} \Rightarrow \mathbf{a}$ (acceleration) is always perpendicular to velocity or we can say that force $\mathbf{F}$ is perpendicular to velocity. Hence, the work done on the particle by the force is zero, its change in kinetic energy is also zero and the speed of the particle remains constant.

Options (b) and (c) are correct.

11. (a,c) Block moves in the direction of tension, hence work done by the string (tension) on the block is positive. As, the woman is in rest while pulling the string, the work done by string (tension) on the woman is zero.

Also, the net work done on the block may be zero, if it is moving with constant velocity in this case sum of work done by the tension and work done by gravity on the block will be zero.

Options (a) and (c) are correct.

12. (b,c) If $F = 0$, $x = \sqrt{2}$ m

When $x > \sqrt{2}$ m, force is directed away from the origin.

When $0 \le x < \sqrt{2}$ m, force is directed towards origin.

If the particle just cross $x = \sqrt{2}$ m, then it starts to accelerate. For minimum value of initial speed, the speed of particle should be zero at $x = \sqrt{2}$ m. This means that KE at $x = \sqrt{2}$ must be zero and it is non-zero at origin.

13. (a,b,c) $W = \int F dx = \int_2^{\sqrt{2}} (2x^2 - 4)\,dx$

$$= \left[\frac{2x^3}{3} - 4x\right]_2^{\sqrt{2}} = \left(\frac{4\sqrt{2}}{3} - 4\sqrt{2}\right) - \left(\frac{16}{3} - 8\right)$$

$$= \frac{-8\sqrt{2}}{3} - \left(\frac{-8}{3}\right) = \frac{-8}{3}(\sqrt{2} - 1)$$

(a) $\because W = \Delta T$ or $\frac{-8}{3}(\sqrt{2} - 1) = 0 - T_i$

$\therefore$ $$T_i = \frac{8}{3}(\sqrt{2} - 1)\,J$$

(b) Also, $$W = \left[\frac{2x^3}{3} - 4x\right]_{\sqrt{2}}^{0} = -\left(\frac{4\sqrt{2}}{3} - 4\sqrt{2}\right) = \frac{8\sqrt{2}}{3}\,J$$

But $$W = \Delta T \quad \text{or} \quad \frac{8\sqrt{2}}{3} = T_f - T_i = T_f - 0$$

$\therefore$ $$T_f = \frac{8\sqrt{2}}{3}\,J$$

(c) $T(x = \sqrt{2}) \ge 0$

14. (a,b,c,d) The area of F_x-x graph gives work done.

According to work-energy theorem,

$$W = \Delta T = T_f - T_i = T_f$$

(a) For $x = 15$ m area of F_x-X graph is maximum. So, kinetic energy is maximum.

(b) For $x = 25$ m, area of F_x-X graph

$$= \frac{1}{2} \times 2 \times (15 + 5) - \frac{1}{2} \times 4 \times 10 = 0$$

$\therefore$ $T_f = 0$

(c) At $x = 5$ m, $W = \frac{1}{2} \times 2 \times 5 = 5\,J = T_f$

(d) At $x = 10$ m, $W = \frac{1}{2} \times 2 \times 5 + 2 \times 5 = 15\,J = T_f$

15. (a) The net work done on the particle is always positive because one of the components of resultant force is in the direction of displacement (or velocity) of the particle and hence its kinetic energy continuously increases (as work done = ΔKE, work done is positive, this means kinetic energy always increases).

It is not necessary that the particle moves in a straight line.

16. (a) Force is always perpendicular to the velocity. So, tangential acceleration is zero while centripetal acceleration is non-zero. Thus, speed is constant. But direction of velocity continuously changes. According to work-energy theorem, net work done is zero.

17. (c) For maximum value of x, the velocity of block will be zero.
According to work-energy theorem,

$$W = \Delta T = T_f - T_i = 0 - 0 = 0$$

or $$\int_0^{x_{max}} F\,dx = 0 \quad \text{or} \quad \int_0^{x_{max}} (b - cx)\,dx = 0$$

or $\left[bx - \dfrac{cx^2}{2}\right]_0^{x_{max}} = 0$ or $b\,x_{max} - \dfrac{cx_{max}^2}{2} = 0$

$\therefore \quad x_{max} = \dfrac{2b}{c}$

18. (a) The speed of the block is maximum at the point, where acceleration of the particle is zero.

$\therefore \quad F = 0$ or $b - cx_0 = 0 \Rightarrow x_0 = \dfrac{b}{c}$

According to work-energy theorem,

$W = \Delta T = T_f - T_i = \dfrac{1}{2}mv_{max}^2 - 0$ or $\int_0^{x_0} F\,dx = \dfrac{1}{2}mv_{max}^2$

or $\left[bx - \dfrac{cx^2}{2}\right]_0^{x_0} = \dfrac{1}{2}mv_{max}^2$ or $bx_0 - \dfrac{cx_0^2}{2} = \dfrac{1}{2}mv_{max}^2$

or $\dfrac{b^2}{c} - \dfrac{cb^2}{2c^2} = \dfrac{1}{2}mv_{max}^2$ or $\dfrac{b^2}{2c} = \dfrac{1}{2}mv_{max}^2$

$\therefore \quad v_{max} = \dfrac{b}{\sqrt{mc}}$

Explanations for Q. Nos. (19-20)

In both cases, the work done by friction is $-\mu mgx$.

According to work-energy theorem,

$mgy - \mu mgx = \dfrac{1}{2}mv_C^2 = \dfrac{1}{2}mv_F^2 \Rightarrow v_C = \sqrt{2gy - \mu gx}$

Also, $v_C = v_F$

19. (a) **20.** (c)

21. (1) For $0 \le x \le l$, only $\mathbf{F} = c(\hat{\mathbf{j}} \times \hat{\mathbf{k}})$ is present in

$\therefore$ The value of $\mathbf{F} = c\hat{\mathbf{i}}$

$\therefore \quad W_1 = \mathbf{F} \cdot l\hat{\mathbf{i}} = cl$

For $x > l$, only friction is present,

$\therefore \quad W_2 = -\mu mg\, l_2$

According to work-energy theorem,

$W = \Delta T = T_f - T_i = 0$ or $W_1 + W_2 = 0$

or $cl - \mu mg\, l_2 = 0$

$\therefore \quad l_2 = \dfrac{cl}{\mu mg}$

Thus, total distance travelled is

$$s = l + l_2 = l + \frac{cl}{\mu mg} = l\left(1 + \frac{c}{\mu mg}\right)$$

$\therefore \quad n = 1$

22. (2) At the maximum compression, velocity of the block is zero.

According to work-energy theorem,

$W = \Delta T = T_f - T_i = 0$

or $W_s + W_f + W_g = 0$

or $-\dfrac{1}{2}kx_{max}^2 - (\mu mg\cos\theta)(4.8 + x_{max}) + mg\sin\theta\,(4.8 + x_{max}) = 0$

or $-\dfrac{1}{2} \times 1000 x_{max}^2 + 0.5 \times 2 \times 9.8 \cos 37°(4.8 + x_{max}) + 2 \times 9.8 \times \sin 37°(4.8 + x_{max})$

or $-\dfrac{1}{2} \times 1000\, x_{max}^2 - 8(4.8 + x_{max}) + 12\,(4.8 + x_{max}) = 0$

After solving, we get $x_{max} = 0.2$ m

$\therefore \quad 10x_{max} = 2$

23. (3) If the block leaves the surface at point B, then

$\therefore \quad kx\cos\theta = mg \quad (\because N = 0)$

or $\quad 40x\cos\theta = 10$

or $\quad x\cos\theta = \dfrac{1}{4} \quad \ldots(i)$

From the figure,

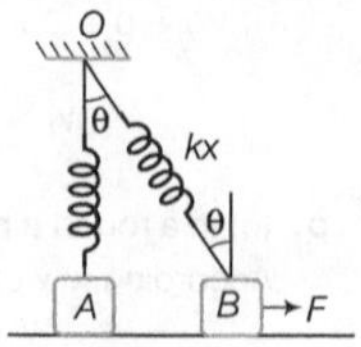

$\cos\theta = \dfrac{OA}{OB} = \dfrac{0.5}{0.5 + x}$

or $\dfrac{1}{4x} = \dfrac{0.5}{0.5 + x}$

or $0.5 + x = 2x$ or $\Rightarrow 0.5 = x$

$\therefore \quad \cos\theta = \dfrac{1}{4x} = \dfrac{1}{4 \times 0.5} = \dfrac{1}{2}$

$\therefore \quad \theta = 60°$

$\therefore \quad AB = OA\tan\theta = 0.5 \times \sqrt{3} = \dfrac{\sqrt{3}}{2}$

Applying work-energy theorem,

$W = \Delta T$ or $W_S + W_F = \dfrac{1}{2}mv^2 - 0$

$\Rightarrow \quad -\dfrac{1}{2}k(AB)^2 + F(AB) = \dfrac{1}{2}mv^2$

or $-\dfrac{1}{2} \times 40 \times \dfrac{3}{4} + 20\sqrt{3} \times \dfrac{\sqrt{3}}{2} = \dfrac{1}{2} \times 1 \times v^2$

or $-15 + 30 = \dfrac{v^2}{2}$ or $v = \sqrt{30}$ m/s

$\therefore \quad n = 3$

24. (5) If x portion is outside of frictional region, then frictional force on the chain is $f = \mu\dfrac{m}{l}(l - x)g$

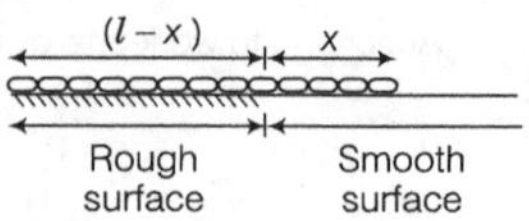

$\therefore$ The work done by friction is

$W_f = -\int f\,dx = -\dfrac{\mu mg}{l}\int_0^l (l - x)\,dx$

$= -\dfrac{\mu mg}{l}\left[lx - \dfrac{x^2}{2}\right]_0^l = \dfrac{-\mu mg}{l}\left[l^2 - \dfrac{l^2}{2}\right] = \dfrac{-\mu mg\,l}{2}$

According to work-energy theorem,

$W = \Delta T = T_f - T_i = 0 - \dfrac{1}{2}mv_0^2$

or $\dfrac{-\mu mg\,l}{2} = -\dfrac{1}{2}mv_0^2$

$v_0 = \sqrt{\mu g l} = \sqrt{0.5 \times 10 \times 5} = 5$ m/s

25. (3) $df = \mu\dfrac{M}{L}g\,dx$

$\Rightarrow \quad f = \int_y^L kx\dfrac{M}{L}g\,dx$ or, $f = \dfrac{M}{L}kg\int_y^L x\,dx$

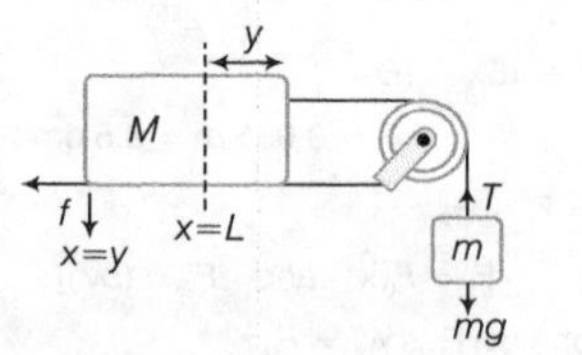

or, $f = \dfrac{M}{L}kg\left[\dfrac{x^2}{2}\right]_y^L$ or, $f = \dfrac{Mkg}{2L}[L^2 - y^2]$

$\therefore \quad dW = -f\,dy$

$\int dW = -\dfrac{Mkg}{2L}\int_0^L (L^2 - y^2)\,dy$

$= -\dfrac{Mkg}{2L}\left[L^2 y - \dfrac{y^3}{3}\right]_0^L$

$\Rightarrow \quad W = -\dfrac{Mkg}{2L}\left[L^2 - \dfrac{L^3}{3}\right]$

$= -\dfrac{Mkg}{2L} \times \dfrac{2L^3}{3} = -\dfrac{MkgL^2}{3}$

$\therefore \quad n = 3$

26. (2) For breaking $F_0 + mg = k(y_1 + y_0)$, where k is force constant.

For initial condition, $mg = ky_0$

$\therefore \quad F_0 = ky_1$

Thus, for breaking the thread, elongation in thread should be $y_0 + y_1$.

If a constant force F_1 is applied. Then, at lower extreme position (where, $v = 0$), the thread breaks and F_1 is minimum.

$\therefore$ According to work-energy theorem,

$$W = \Delta T = T_f - T_i = 0$$

$$F_1 y_1 + mgy_1 + \frac{1}{2}ky_0^2 - \frac{1}{2}k(y_1 + y_0)^2 = 0$$

or $$F_1 \times \frac{F_0}{k} + mg\frac{F_0}{k} + \frac{1}{2}k\left(\frac{mg}{k}\right)^2 = \frac{1}{2}k\left(\frac{F_0}{k} + \frac{mg}{k}\right)^2$$

or $$\frac{F_1F_0}{k} + \frac{mg\,F_0}{k} + \frac{(mg)^2}{2k} = \frac{(F_0 + mg)^2}{2k}$$

or $$\frac{F_1F_0}{k} + \frac{mgF_0}{k} + \frac{(mg)^2}{2k} = \frac{F_0^2}{2k} + \frac{(mg)^2}{2k} + \frac{2F_0mg}{2k}$$

or $$\frac{F_1F_0}{k} = \frac{F_0^2}{2k}$$

or $$F_1F_0 = \frac{F_0^2}{2}$$

$\therefore$ $$F_1 = \frac{F_0}{2} = \frac{300}{2} = 150\text{ N} \Rightarrow n = 2$$

27. (5) Work done by gravity, $W_g = mg(OQ)$

$$= -1 \times 10 \times 4 = -40\text{ J}$$

Work done by the applied force, $W_F = F(PQ)\cos 0^\circ$

$$W_F = 18 \times 5 \times 1 = 90\text{ J}$$

According to work-energy theorem,

$$W_g + W_F = \Delta\text{KE}$$

or $$-40 + 90 = \Delta\text{KE}$$

or $$\Delta\text{KE} = 50\text{ J}$$

$\therefore$ $$n = 5$$

28. As, $t = \sqrt{x} + 3$

or $(\sqrt{x})^2 = (t-3)^2 \Rightarrow x = t^2 - 6t + 9$

Velocity of the particle, $v = \dfrac{dx}{dt} = \dfrac{d}{dt}(t^2 - 6t + 9)$

$\Rightarrow$ $$v = 2t - 6$$

(a) $v = 0 \Rightarrow 2t - 6 = 0 \Rightarrow t = 3\text{ s}$

Position of the particle at $t = 3$ s (or $v = 0$ m/s)

$$x = (3)^2 - 6 \times 3 + 9$$

$x = 0$ m (particle will be at origin)

Initial position of the particle (at $t = 0$ s)

$$x = (0)^2 - 6 \times 0 + 9 = 9\text{ m}$$

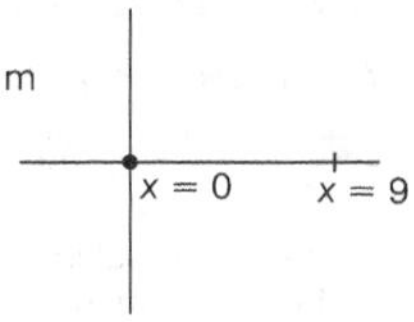

Displacement of the particle in 3 s

$$s = 9 - 0 = 9\text{ m}$$

(b) Position of the particle at $t = 6$ s

$$x = t^2 - 6t + 9$$

or $$x = (6)^2 - (6 \times 6) + 9 = 9\text{ m}$$

$$x_{t=0\text{ s}} = x_{t=6\text{ s}} = 9\text{ m}$$

As, the initial and final position is same, displacement of the particle after 6 s will be zero, hence work done by the constant force on the particle is zero.

29. For shifting of B, $kx_0 > mg\sin\theta + \mu mg\cos\theta$

or $$kx_0 > \sqrt{2} \times 10 \times \frac{1}{\sqrt{2}} + 0.5 \times \sqrt{2} \times 10 \times \frac{1}{\sqrt{2}}$$

or $$kx_0 > 15$$

$\therefore$ $$x_{0\text{ min}} = \frac{15}{k}$$

For minimum value of F, the block B will be in the condition of just shift, when the block A is at its extreme position ($v = 0$). Applying work-energy theorem for the block A,

$$W = \Delta T = T_f - T_i = 0 - 0 = 0$$

or $W_F + W_S + W_f = 0$

or $$Fx_{0\text{ min}} - \frac{1}{2}kx_{0\text{ min}}^2 - \mu mg\,x_{0\text{ min}} = 0$$

or $$F - \frac{k}{2}x_{0\text{ min}} - \mu mg = 0$$

or $$F - \frac{k}{2} \times \frac{15}{k} - \mu mg = 0$$

or $$F - 7.5 - 0.5 \times \sqrt{2} \times 10 = 0$$

$\therefore$ $$F = 7.5 + 5\sqrt{2} = 14.57\text{ N}$$

30. The work done by friction is $W_f = -\mu mgs$

$$= -\mu mgvt = -0.2 \times 2 \times 9.8 \times 2 \times 5$$

$$= -39.2\text{ J} = \frac{-39.2}{4.2}\text{ cal} = -9.33\text{ cal}$$

$\therefore$ $$\Delta H = -W_f = 9.33\text{ cal}$$

31. Total heat required to just melt the bullet is

$\Delta H = ms\Delta\theta + mL$

$$= m \times 10^3 \times 0.03 \times 4.2(327 - 27) + m \times 10^3 \times 6 \times 4.2$$

$$= 6.3 \times 10^4\text{ m}$$

$\therefore$ $$\frac{1}{2}mv^2 \times 75\% = 6.3 \times 10^4\text{ m}$$

or $$\frac{1}{2}v^2 \times \frac{75}{100} = 6.3 \times 10^4$$

$\therefore$ $$v = 409.8\text{ m/s}$$

32. Let the block comes in rest for a moment when spring compresses for a compression x.

According to work-energy theorem,

$$-\mu_k mg(BD + x) - \frac{1}{2}kx^2 = 0 - \frac{1}{2}mv^2$$

Putting the values, we get

$$x = 0.1\text{ m}$$

Total distance covered, $d = AB + BD + x$

$$= 2 + 2.14 + 0.1 = 4.24\text{ m}$$

33. (a) (i) The normal reaction and displacement of block in the frame of car are perpendicular to each other, so work done by normal reaction on the block in the frame of car is zero.

$\therefore$ $$W_1 = 0$$

(ii) The free body diagram of the block in the frame of car is shown in figure below

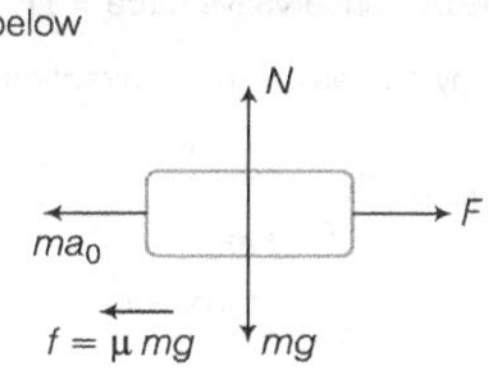

From free body diagram of block in the frame of car,

$\therefore$ $$F - ma_0 - \mu mg = ma_{\text{rel}}$$

or $$20 - 1 \times 5 - 0.5 \times 1 \times 10 = a_{\text{rel}}$$

or $$a_{\text{rel}} = 10\text{ m/s}^2$$

$\therefore$ $$s_{\text{rel}} = \frac{1}{2}a_{\text{rel}}t^2 = \frac{1}{2} \times 10 \times 1^2 = 5\text{ m}$$

$\therefore$ $$W_3 = Fs_{\text{rel}} = 20 \times 5 = 100\text{ J}$$

(iii) From free body diagram of block,

$$W_5 = -ma_0s_{\text{rel}} = -1 \times 5 \times 5 = -25\text{ J}$$

(iv) According to work-energy theorem,

$$W_1 + W_2 + W_3 + W_4 + W_5 = \Delta T = 50\text{ J}$$

(i) $\rightarrow$ (q), (ii) $\rightarrow$ (s), (iii) $\rightarrow$ (p), (iv) $\rightarrow$ (r)

DPP-5 Potential Energy

1. (i) True, the work done by conservative force is independent of the path taken and depends only on the position. Hence, it is a state function.

(ii) True, work done by conservative force does not depend on path followed. Hence, it is not a path function. It depends only on initial and final position.

(iii) True, work done by a conservative force in a closed path is zero as it is a state function.

(iv) True, work done by a conservative force depends only on the position.

(v) True, work done by a conservative forces are state functions.

(vi) True, conservative forces are position dependent and path independent.

(vii) True, frictional force is non-conservative and path dependent hence the work done in a closed path is non-zero.

(viii) True, gravitational force, electrostatic force and spring force are conservative forces.

(ix) True, push or pull is not a conservative force.

2. (i) (a) $\mathbf{F} = -\frac{\partial U}{\partial x}\hat{\mathbf{i}} - \frac{\partial U}{\partial y}\hat{\mathbf{j}} - \frac{\partial U}{\partial z}\hat{\mathbf{k}} = -2ax\hat{\mathbf{i}} - 2by\hat{\mathbf{j}}$

(ii) (c) $F_x = -2ax$

$F_y = -2by$

$\therefore \tan\theta = \frac{dy}{dx} = \frac{F_y}{F_x} = \frac{yb}{ax}$ or $\int \frac{dy}{y} = \frac{b}{a}\int \frac{dx}{x}$

$\therefore \quad \ln x = \frac{a}{b}\ln y + \ln c$

$\therefore \quad x = cy^{a/b}$

(iii) (b) For $a = b$, $x = c\,y$ or $y = \frac{1}{c}x$

The graph is a straight line.

(iv) (c) $W = -\Delta U = U_i - U_f = -3(a+b)d^2$.

(v) (a) Potential energy is a constant at equipotential surface. Hence, $U = ax^2 + by^2$ = constant represents the locus of all the points where potential energy is constant.

(vi) (a) Equipotential surface is represented by

$$ax^2 + by^2 = \text{constant}$$

or $$\frac{x^2}{(a')^2} + \frac{y^2}{(b')^2} = 1$$

which is an equation of ellipse.

(vii) (d) Work done by conservative force $= \int \mathbf{F}\cdot d\mathbf{s} = W = -\Delta U$.

Potential energy decreases in the direction of conservative force.

3. (i) (a) Here, $F_x = ax$

$$F_y = ay$$

$\therefore$ $$\frac{dy}{dx} = \tan\theta = \frac{y}{x}$$

$\therefore$ $$\int \frac{dy}{y} = \int \frac{dx}{x}$$

$$\ln y = \ln x + \ln k$$

$\therefore$ $$y = kx$$

(ii) (a) Force $\mathbf{F}$ is conservative in nature.

$$\mathbf{F} = -\frac{\partial U}{\partial x}\hat{\mathbf{i}} - \frac{\partial U}{\partial y}\hat{\mathbf{j}} - \frac{\partial U}{\partial z}\hat{\mathbf{k}}$$

(iii) (c) $dU = -\mathbf{F}\cdot d\mathbf{r}$

or $$dU = -(a x\hat{\mathbf{i}} + ay\hat{\mathbf{j}})\cdot(dx\hat{\mathbf{i}} + dy\hat{\mathbf{j}})$$

or $$U = -a\int x dx - a\int y dy$$

$\therefore$ $$U = -\frac{ax^2}{2} - \frac{ay^2}{2} + k$$

(iv) (c) The shape of equipotential surface is circular since, U = constant for equipotential surface

$\Rightarrow$ $$-\frac{ax^2}{2} - \frac{ay^2}{2} + k = \text{constant}$$

or $$x^2 + y^2 = \text{constant}$$

which represents a circle.

(v) (b) Work done by force $= -\Delta U = U_i - U_f$

Potential enrgy at (0, 0) $= U_i = -\frac{ax^2}{2} - \frac{ay^2}{2} + k = k$

Potential energy at $(b, b) = U_f = -\frac{ax^2}{2} - \frac{ay^2}{2} + k$

$$= -\frac{ab^2}{2} - \frac{ab^2}{2} + k$$

$$= -ab^2 + k$$

or $$W = -\Delta U = (k) - (-ab^2 + k) = ab^2$$

(vi) (a) $U(x, y) = -\frac{ax^2}{2} - \frac{ay^2}{2} + k$

$U(0, 0) = k = 0$ (given)

So, $U(x, y) = -\frac{ax^2}{2} - \frac{ay^2}{2}$

4. (i) (a) Potential energy is minimum at stable equilibrium. Hence, at $(x = x_0)$ potential energy is minimum and equilibrium is stable.

(ii) (b) $(x = x_3)$ is a point of maximum potential energy. Hence, the equilibrium is unstable.

In the vicinity of point $(x = x_5)$, the potential energy is constant and hence is a point of neutral equilibrium.

5. (a) $s = v_0 t - \frac{1}{2}(g\sin\alpha)t^2$

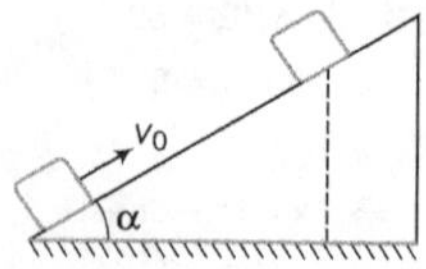

$\therefore$ $$U = mgh = mgs\sin\theta$$

Hence, option (a) is correct.

6. (a) $dU = -Fdx = \frac{-b}{x^2}dx$

$\therefore$ $$U = -b\left[\frac{x^{-2+1}}{-2+1}\right] + c \quad \text{or} \quad U = \frac{b}{x} + c$$

which represents a hyperbola.

7. (b) As, $F = kx_0$

$$U = \frac{1}{2}kx_0^2 \quad \text{or} \quad 200 = \frac{F^2}{2k}$$

If spring is cut into two parts, spring constant of each part becomes twice.

$\therefore$ $$F = 2kx_0$$

$\therefore$ $$x_0 = \frac{F}{2k}$$

$\therefore$ $$U' = \frac{1}{2}(2k)x_0^2 = k\frac{F^2}{4k^2}$$

$$= \frac{F^2}{4k} = \frac{200}{2} = 100\,\text{J}$$

8. (c) $W_g = -\Delta U = U_i - U_f$

$$= Mg\frac{H}{2} + mgH - 0 = \left(\frac{M}{2} + m\right)gH$$

9. (a) Given, $F = kx$

From $F = -\frac{dU}{dx}$, $\quad dU = -Fdx$ or $\int_0^{U(x)} dU = -\int_0^x Fdx$

or $\quad U(x) = -\frac{kx^2}{2}$ as $U(0) = 0$

10. (d) Given, $F(x) = -kx + ax^3$ ...(i)

$$F = -\frac{dU}{dx}$$

$\therefore \quad dU = -Fdx$ or $U(x) = -\int_0^x (-kx + ax^3)\,dx$

or $\quad U(x) = \frac{kx^2}{2} - \frac{ax^4}{4}$ ...(ii)

From Eq. (ii),

$$U(x) = 0$$

$\Rightarrow \quad \frac{kx^2}{2}\left(1 - \frac{ax^2}{2k}\right) = 0$ or $x = 0$ and $x = \pm\sqrt{\frac{2k}{a}}$

Also, $\quad U(x) =$ negative for $x > \sqrt{\frac{2k}{a}}$

From Eq. (i), we get

$$F(x) = -\frac{dU}{dx} = 0 \Rightarrow x = (-k + ax^2) = 0$$

or $\quad x = 0$ and $x = \pm\sqrt{\frac{k}{a}}$

The slope of U-x graph is zero at these points. Therefore, the most appropriate option is (d).

11. (d) Here, $m = \rho a^3$

Here, $\quad W_1 = mg\frac{a}{2}$ (for first cube) and $W_2 = mg\left(a + \frac{a}{2}\right)$

$W_3 = mg\left(2a + \frac{a}{2}\right)$ and so on.

$\therefore \quad W = W_1 + W_2 + W_3 + \cdots$

$$W = \frac{mga}{2}[1 + 3 + 5 + \cdots]$$

$$W = \frac{mga}{2}n^2 = \frac{\rho a^4 gn^2}{2}$$

No work is required to arrange first block.

$\therefore$ Required work done

$$W_r = W - \frac{mgd}{2} = \frac{\rho a^4 gn^2}{2} - \frac{mga}{2} = \frac{\rho a^4 g}{2}(n^2 - 1)$$

Hence, option (d) is correct.

12. (b) From given figure,

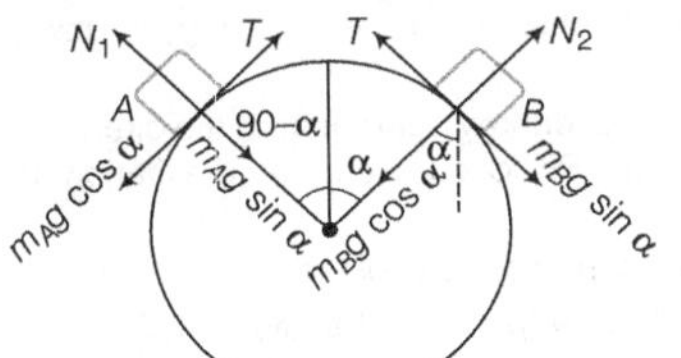

$$T = m_A g\cos\alpha \text{ and } T = m_B g\sin\alpha$$

$\therefore \quad m_A g\cos\alpha = m_B g\sin\alpha$ or $\tan\alpha = \frac{m_A}{m_B} = \sqrt{3}$

$\therefore \quad \alpha = \frac{\pi}{3}$

When block B is slightly displaced downward, α increases. Due to this $m_B g\sin\alpha$ becomes greater than $m_A g\cos\alpha$.

So, net force is found in the direction of displacement. Thus, equilibrium is unstable.

13. (c) Take the horizontal surface as reference line

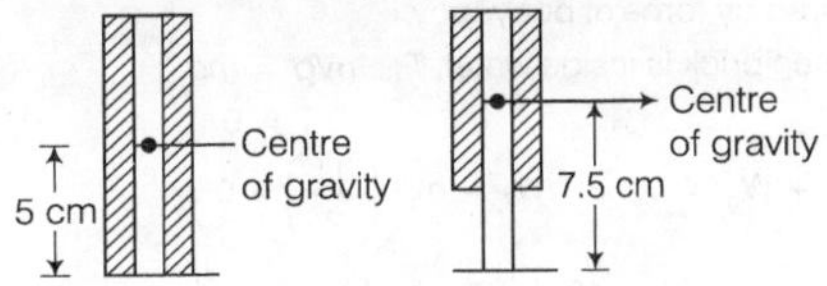

$$W_{ext} = \Delta U$$

In first case, $\quad W_1 = U_f - U_i$

$$= mg \times (5\text{ cm}) - 0 = mg\,(5\text{ cm})$$

In second case,

$$W_2 = mg \times (7.5\text{ cm}) - 0$$
$$= mg \times (7.5\text{ cm})$$

Thus, $\quad \frac{W_1}{W_2} = \frac{mg \times (5\text{ cm})}{mg \times (7.5\text{ cm})} = \frac{50}{75} = \frac{2}{3}$

14. (a) $mg = kx_0 + ma_0 \Rightarrow x_0 = 0$ $\quad [\because a = g]$

or $\quad U = \frac{1}{2}kx_0^2 = 0$

15. (a) From given figures, the acceleration of each ball is in opposite direction of displacement of each ball from equilibrium position. Hence, equilibrium is stable.

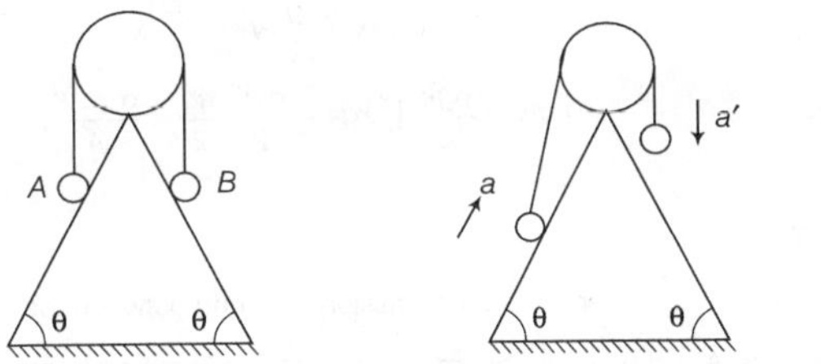

At equilibrium $\qquad$ Displaced positions of balls

16. (i) (a,b) For conservative force, work done by the force is a state function and is path independent.

Hence, if $W_1 = W_2 = W_3$

F is conservative. (b) Here, force is path function, so the force is non-conservative.

(ii) (b) Gravitational field is a conservative force field. In a conservative force field, work done is independent of path taken.

$\therefore \quad W_1 = W_2 = W_3$

17. (a,b,c,d) $W_{ext} = \Delta U = mgh$

(b) $W_F + W_g = \Delta T$ or $W_F = \Delta T + mgh$ $\quad$ (here, $\Delta T > 0$)

(c) $W_F = \Delta T + mgh$ $\quad$ (here, $\Delta T < 0$)

(d) Pushing force is non-conservative in nature.

18. (a) $W_g = -\Delta U = U_i - U_f = 0 - mgh = -mgh$

19. (b) Path dependent force is non-conservative in nature.

20. (a,b) If the springs are compressed to same amount, then

$$W_A = \frac{1}{2}k_A x^2, W_B = \frac{1}{2}k_B x^2$$

$\therefore \quad k_A > k_B \Rightarrow W_A > W_B$

If the springs are compressed by same force, then

$$F = k_A x_A = k_B x_B$$

$$x_A = \frac{F}{k_A}, x_B = \frac{F}{k_B}$$

$$\frac{W_A}{W_B} = \frac{\frac{1}{2}k_A \cdot \frac{F^2}{k_A^2}}{\frac{1}{2}k_B \cdot \frac{F^2}{k_B^2}} = \frac{k_B}{k_A}$$

Hence, $\quad W_A < W_B$

21. (a,b) The density of block is same as that of water. Thus, mg is balanced by force of buoyancy.

(a) When brick is inside water, $T + \rho Vg = mg$

$\therefore \quad T = 0$

(b) $W_T + W_g = 0 \Rightarrow W_T - mg\left(\frac{0.5}{2}\right) = 0$

$\therefore \quad W_T = 2 \times 10 \times \frac{0.5}{2} = 5\text{ J}$

22. (a,c) The contact point does not move.

(c) $W = \Delta T$

$\therefore \quad W_T = W_{int} + W_g = \Delta T = 0$

Here, W_T = work done by tension = 0

W_g = work done by gravity = $-mgh$

$\therefore \quad 0 + W_{int} - mgh = 0$

$\therefore \quad W_{int} = mgh \neq 0.$

Hence, options (a) and (c) are correct.

23. (d) The change in potential energy measured in any frame are same.

24. (4) The mass of considered element is

$$dm = \sigma y dx = \sigma\sqrt{kx}\, dx$$

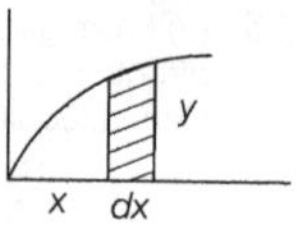

The potential energy of considered element is

$$dU = (dm)\, g\frac{y}{2} = \sigma\sqrt{kx}\, dx\, \frac{g}{2}\sqrt{kx}$$

or $$U = \int dU = \frac{\sigma g k}{2}\int_0^a x dx = \frac{\sigma g k}{2}\frac{a^2}{2} = \frac{\sigma g a^2 k}{4}$$

$\therefore \quad n = 4$

25. (6) $m_A g = kx_1$

$\therefore x_1 = \frac{1}{10}$ m, where x_1 is compression in spring between the blocks A and B at equilibrium. If compression in spring between blocks B and C is x_2. Then, $(m_A + m_B)g = kx_2$

$\therefore \quad x_2 = \frac{20}{100} = \frac{1}{5}$ m

Similarly, $x_3 = \frac{(m_A + m_B + m_C)g}{k} = \frac{30}{100} = \frac{3}{10}$ m

When lower spring comes in natural length,

$$ky_2 = m_C g$$

$\therefore \quad y_2 = \frac{10}{k} = \frac{1}{10}$ m

i.e. elongation in the spring between blocks B and C.

But, block B is in equilibrium.

$\therefore \quad ky_1 = m_B g + ky_2$

$\therefore \quad y_1 = \frac{m_B g}{k} + y_2 = \frac{1}{10} + \frac{1}{10} = \frac{1}{5}$ m

Here, y_1 and y_2 are elongations in springs between A and B and between B and C. Thus, upward shifting in block C is

$$x_3 = \frac{3}{10}\text{ m} = 0.3\text{ m}$$

The upward shifting in block B is $x_3 + x_2 + y_2$,

$$= \frac{3}{10} + \frac{2}{10} + \frac{1}{10} = 0.6\text{ m}$$

Similarly, the upward shifting in block A is

$$x_3 + x_2 + y_2 + x_1 + y_1 = \frac{3}{10} + \frac{2}{10} + \frac{1}{10} + \frac{1}{10} + \frac{2}{10} = \frac{9}{10} = 0.9\text{ m}$$

$\therefore \quad \Delta U = m_C g x_3 + m_B g \times 0.6 + m_A g \times 0.9$

$= 3 + 6 + 9 = 18\text{ J} = 3n$

$\therefore \quad n = 6$

26. (1) $W_{Rishu} - W_g = 0 \Rightarrow W_{Rishu} - mgh = 0$

$\therefore \quad W_{Rishu} = mgh = 10 \times 10 \times 10 = 1000\text{ J} = 1 \times 10^3\text{ J}$

$\therefore \quad n = 1$

27. (1) Here, $\mathbf{F} = -\frac{\partial U}{\partial x}\hat{\mathbf{i}} - \frac{\partial U}{\partial y}\hat{\mathbf{j}}$

or $\mathbf{F} = -\sqrt{2}\frac{\partial}{\partial x}(x^2y)\,\hat{\mathbf{i}} - \sqrt{2}\frac{\partial}{\partial y}(x^2y)\,\hat{\mathbf{j}}$

or $\mathbf{F} = -2\sqrt{2}\, xy\hat{\mathbf{i}} - \sqrt{2}\, x^2\hat{\mathbf{j}}$

$\because \quad \mathbf{r} = \hat{\mathbf{i}} + \hat{\mathbf{j}}$

$\therefore \quad x = 1$ and $y = 1$

$\therefore \quad \mathbf{F} = -2\sqrt{2}\,\hat{\mathbf{i}} - \sqrt{2}\,\hat{\mathbf{j}}$

The component of $\mathbf{F}$ along $\mathbf{r}$ gives radial acceleration. Thus, $F_r = \mathbf{F}\cdot\hat{\mathbf{e}}_r$

$\therefore \quad F_r = [-2\sqrt{2}\,\hat{\mathbf{i}} - \sqrt{2}\,\hat{\mathbf{j}}]\cdot\left(\frac{\hat{\mathbf{i}} + \hat{\mathbf{j}}}{\sqrt{2}}\right) = -2 - 1 = -3$

$\therefore \quad ma_r = -3 \Rightarrow a_r = 3\text{ m/s}^2$ (in magnitude)

If $\mathbf{a}$ is net acceleration,

$\therefore \quad \mathbf{F} = m\mathbf{a}$ or $-2\sqrt{2}\,\hat{\mathbf{i}} - \sqrt{2}\,\hat{\mathbf{j}} = 1 \times \mathbf{a}$

$\therefore \quad \mathbf{a} = -2\sqrt{2}\,\hat{\mathbf{i}} - \sqrt{2}\,\hat{\mathbf{j}}$

$\therefore \quad a = \sqrt{8 + 2} = \sqrt{10}$

But, $a^2 = a_r^2 + a_t^2 \Rightarrow a_t^2 = 10 - 9$

$\therefore \quad a_t = 1\text{ m/s}^2$

28. (b) (i) Work done by conservative force is a state function dependent only on position and is zero for work done in closed path.

(ii) Static friction is a non-conservative force, hence work done is dependent on path taken.

(iii) Displacement is the shortest distance travelled.

(iv) Work done by non-conservative force is path function.

(i) → (q,r,s), (ii) → (p,r,s) (iii) → (r,s) (iv) → (p,r,s)

29. (a) (i) $mg = \rho_1 Vg$, F_B = buoyancy force = $\rho_2 vg$

Here, $\rho_1 = \rho_2$ (both air)

$\therefore$ PE remains constant.

(ii) Here, $\rho_1 > \rho_2$, hence, net conservative force is downward. So, balloon moves downward.

$\therefore$ PE and gravitation PE decrease.

(iii) $\rho_2 > \rho_1$

$\therefore$ Net conservative force (i.e. $F_B - mg$) is upward. So, balloon rises upward.

Since, potential energy decreases in the direction of conservative force. Hence, PE decreases but gravitation potential energy increases.

(iv) Here, $\rho_2 > \rho_1$, hence explanation is same as (r).

(i) → (q,r,s) (ii) → (p) (iii) → (q), (iv) → (r,s)

30. (c) As, $F = -\frac{dU}{dx}$

Given, $U = \frac{a}{x^{12}} - \frac{b}{x^6}$

$\therefore \quad F = -\frac{dU}{dx} = -\left[\frac{-12a}{x^{13}} + \frac{6b}{x^7}\right]$ or $F = \frac{12a}{x^{13}} - \frac{6b}{x^7}$

31. (c) $U(x) = \frac{a}{x^{12}} - \frac{b}{x^6}$

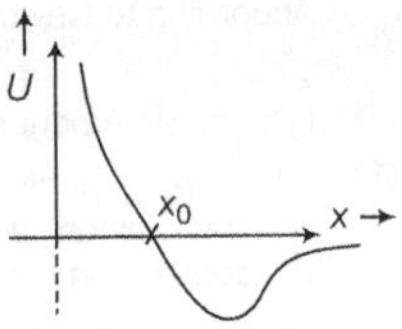

For $x \to \infty$, $U(x) = 0$

For $x \to 0$, $U(x) \to \infty$

Also, $U(x) = 0 \Rightarrow \frac{a}{x^{12}} - \frac{b}{x^6} = 0$

or $x^6 = \frac{a}{b}$ or $x = x_0 = \left(\frac{a}{b}\right)^{1/6}$

32. (a) $F(x) = -\frac{dU}{dx}$ or $F = \frac{12a}{x^{13}} - \frac{6b}{x^7}$

As $x \to \infty$, $F \to 0$

As $x \to 0$, $F \to \infty$

Graph (a) represents the variation of force as a function of x.

33. (a) Given, $U(x) = \frac{a}{x^{12}} - \frac{b}{x^6}$; At $x = x_0 = \left(\frac{2a}{b}\right)^{1/6}$

$$U(x_0) = \frac{a}{\left[\left(\frac{2a}{b}\right)^{1/6}\right]^{12}} - \frac{b}{\left[\left(\frac{2a}{b}\right)^{1/6}\right]^{6}} = \frac{ab^2}{4a^2} - \frac{b^2}{2a} = \frac{-ab^2}{4a^2}$$

or $U(x_0) = -\frac{b^2}{4a}$

Since, $a > 0$, $b > 0, U(x_0) < 0$

Also, $\left.\frac{dU(x)}{dx}\right]_{x_0 = \left(\frac{2a}{b}\right)^{1/6}} = 0$

$\because \quad \frac{Du(x)}{dx} = -F = \frac{b}{x^6} - \frac{a}{x^{12}}$

$\therefore \quad \frac{d^2U(x)}{dx^2} = \frac{-6b}{x^7} + \frac{12a}{x^{13}}$

$\Rightarrow \quad \left[\frac{d^2U(x)}{dx^2}\right]_{x_0} = \left(\frac{2a}{b}\right)^{1/6} > 0$

Hence, potential energy is minimum at $x = \left(\frac{2a}{b}\right)^{1/6}$ and equilibrium is stable.

34. (a) $W_{ext} = \Delta U = U_f - U_i$

The work done by pulling agent is W_{ext} which is positive.

$\therefore \quad W_{ext} > 0$ or $U_f - U_i > 0$

$\therefore \quad U_f > U_i$

35. (d) $F = -\frac{dU}{dx} = -[(x-3)(7-3x)]$

For equilibrium, $F = 0$ or $(x-3)(7-3x)$

$\therefore \quad x = 3$ or $7/3$

$\because \quad \frac{d^2U}{dx^2} = 26x - 16$

At $x = 3$, $\frac{d^2U}{dx^2} = +2 > 0$

$\therefore$ At $x = 3$, potential energy is minimum.

Hence, equilibrium is stable.

36. (d) In one-dimensional motion, Statement I is correct but in two-dimensional or three-dimensional motion, it is not the correct.

DPP-6 Power and Other Forms of Energy

1. (i) (b) Mean power $= \frac{1}{b-a}\int_a^b P dt = \frac{1}{2}\int_0^2 (3t^2 - 2t + 1) dt$

$$= \frac{1}{2}\left[\frac{3t^3}{3} - \frac{2t^2}{2} + t\right]_0^2 = \frac{1}{2}\{(2^3 - 2^2 + 2) - (0)\} = 3 \text{ W}$$

(ii) (b) Work done $= \frac{dW}{dt} = P \Rightarrow dW = Pdt$

$$W(t = 0 \text{ to } t = 2) = \int_0^2 P dt = \int_0^2 (3t^2 - 2t + 1) dt = [t^3 - t^2 + t]_0^2$$

$$= (8 - 4 + 2) - (0) = 6 \text{ W/s} = 6 \text{ J}$$

(iii) (c) Change in KE = Work done on the particle

$$= \int_2^4 P dt = [t^3 - t^2 + t]_2^4 = (4^3 - 4^2 + 4) - (2^3 - 2^2 + 2)$$

$$= (64 - 16 + 4) - (8 - 4 + 2) = 52 - 6 = 46 \text{ J}$$

(iv) (c) Work done on the particle $= \int_2^4 P dt = 46$ J

2. (i) (b) $W = Pt$

By work-energy theorem,

$$\frac{1}{2}mv^2 = Pt \text{ or } v \propto t^{1/2}$$

$$\frac{ds}{dt} = kt^{1/2} \Rightarrow s = k\int t^{1/2} dt$$

$$s = kt^{3/2} \Rightarrow s \propto t^{3/2}$$

(ii) (a) A body of mass m with a uniform acceleration, then force

$$F = ma = \frac{mv}{T} \qquad \left[\because a = \frac{v - 0}{T}\right]$$

Instantaneous power $= Fv = mav$

$$= \frac{mv}{T} \cdot at = \frac{mv}{T} \cdot \frac{v}{T} \cdot t = \frac{mv^2}{T^2} t$$

3. (i) (b) When truck strikes a tree, part of KE is converted into heat, temperature and internal energy of truck increases.

(ii) (a,b,c) Coil uses electrical energy and gets heated up. So, internal energy of coil increases. Apart of heat is carried by air molecules and internal energy of air increases. Coil radiates infrared and visible part of spectrum when hot.

(iii) (c) Both (a) and (b) happen simultaneously.

(iv) (b,c) While transmitting, electrons oscillate in antenna creating electromagnetic wave. During the process of oscillation, heat is created due to interatomic collisions and lattice vibration results in humming sound.

(v) (a,b,c) During breaking, heat is generated due to friction, hot bike parts radiates heat in the surrounding atmosphere.

(vi) (a,b,c) In all (a), (b), (c) cases, friction is present, resulting in dissipation of energy in the form of heat.

4. (a) The rate of loss of mechanical energy

$= -$ (power developed by friction) $= (\mu mg \cos \alpha)v$

Here, $a = g \sin \alpha - \mu g \cos \alpha = 10 \times \frac{3}{5} - 4 = 2$

$\therefore \quad v = at = 2$

$\therefore$ The rate of loss of mechanical energy $= (\mu\, mg \cos \alpha)\, v$

$$= 0.5 \times 1 \times 10 \times \frac{4}{5} \times 2 = 8 \text{ J}$$

5. (c) $P = mg \sin \theta \times v = 1 \times 10 \times \frac{3}{5} \times 2 = 12$ J

6. (a) Average power $= \langle P \rangle = \frac{\text{Total work done}}{\text{Total time}}$

$$= \frac{\text{Total increase in energy}}{\text{Time}}$$

$$= \frac{mg\frac{l}{2} + \frac{1}{2}mv^2}{\frac{l}{v_0}} = \frac{1 \times 10 \times \frac{1}{2} + \frac{1}{2} \times 1 \times 1^2}{\frac{1}{1}} = 5.5 \text{ W}$$

7. (d) $W_A = Fs \cos \theta$ and $W_B = Fs \cos \theta$

But, the acceleration of block A is greater than that of block B.

$\because \quad a_A > a_B$

$\therefore \quad t_A < t_B$

$\therefore \quad P_A = \frac{W_A}{t_A} = \frac{Fs \cos \theta}{t_A}$ and $P_B = \frac{Fs \cos \theta}{t_B} \Rightarrow P_A > P_B$

8. (a) Initial $KE = \frac{1}{2}mv_0^2$, Also, $v = v_0 - gt$

$\therefore$ $KE(t) = \frac{1}{2}m(v_0 - gt)^2$

KE becomes zero at some time and then again increases. Also, KE remains always positive.

9. (c) Total power, P = rate of change of kinetic energy

$\therefore$ P = slope of kinetic energy *versus* time graph.

10. (b) Statement is always correct.

11. (b) $P = \frac{mgh}{t} = \frac{80 \times 10 \times 4}{20} = 160$ W

12. (a) $P = \mathbf{F} \cdot \mathbf{v} = (10\hat{\mathbf{i}} + 10\hat{\mathbf{j}} + 20\hat{\mathbf{k}}) \cdot (5\hat{\mathbf{i}} - 3\hat{\mathbf{j}} + 6\hat{\mathbf{k}})$

$= 50 - 30 + 120 = 140$ W

13. (b) $a_c = k^2rt^2 \Rightarrow \frac{v^2}{r} = k^2rt^2$ or $v = krt$

$\therefore$ Tangential acceleration, $a_t = \frac{dv}{dt} = kr$

Work done by tangential force per sec $= F_t v = (mkr)(krt)$

Power input $= mk^2r^2t$

14. (d) The power, $P = Fv = v^2\frac{dm}{dt} = v^2\rho s\frac{dx}{dt} = \rho s v^3$

where, ρ = density of water, s = cross-sectional area of the pipe.

v = speed of stream of water

$\therefore$ $P \propto v^3$

$\therefore$ $\frac{P'}{P} = \frac{v'^3}{v^3} = \frac{(2v)^3}{v^3} = 8$

15. (c) $P = Fv = \left(v\frac{dm}{dt}\right) \cdot v = v\left(\frac{d}{dt}\rho \times \text{volume}\right) \cdot v = \rho v^2 \cdot \frac{d}{dt}(Avt)$

$= A\rho v^3 \Rightarrow P \propto v^3$

16. (c) For block A, $\Delta E_1 = \frac{1}{2}mv_0^2 - 0 = \frac{1}{2}mv_0^2$

$\Delta E_2 = \frac{1}{2}mv_0^2$ − gain in gravitational potential energy

$\Delta E_3 = \frac{1}{2}mv_0^2$ + loss in gravitational potential energy

$\therefore$ $\Delta E_3 > \Delta E_1 > \Delta E_2$

17. (a,d) The displacement of swimmer with respect to float is same in both cases.

18. (b,c) Work done against gravity by the labourer is stored in the form of PE of brick.

19. (b,d) Work done = Change of energy of the system

$\therefore$ $200 = 100 + 30 + \Delta E$

$\therefore$ $\Delta E = 200 - 100 - 30 = 70$ J

20. (a,d) Net work done on system = Total increase in the energy of system

$\therefore$ $400 - 80 = 250 - 30 + \Delta E$

$\therefore$ $\Delta E = 150 + 30 - 80 = 100$ J

21. (c) $\Delta U_1 = \frac{1}{2}mv^2 \Rightarrow \Delta U_2 = \frac{1}{2}mv^2 - mgh$

$\therefore$ $\Delta U_2 < \Delta U_1$

Coefficient of friction depends on the nature of surfaces and it is a constant.

$\therefore$ Statement I is correct and II is incorrect.

22. (a) $P = \mathbf{F} \cdot \mathbf{v}$

$\because$ **F** and **v** are perpendicular. So, power is zero.

23. (b) According to work-energy theorem,

$W_f + W_{int} = \Delta T$ or $0 + W_{int} = \Delta T$

$\therefore$ $W_{int} = \Delta T > 0$

So, Statement I is correct.

According to Newton's second law of motion,

$f = ma$

Hence, Statement II is correct.

24. (c) During positive acceleration, KE increases, it remains constant when acceleration is zero and it reduces due to deacceleration.

25. (a) Power is positive initially and then it is negative during deacceleration.

26. (c) (i) According to work-energy theorem,

$W = \Delta T$

$\therefore$ $W_{max} = \Delta T_{max}$

This is found in time interval

$t = 3$ s to $t = 5$ s

(ii) Power = Rate of change of kinetic energy

$= \frac{d}{dt}\left(\frac{1}{2}mv^2\right) = mv\frac{dv}{dt}$

$= (mv)$ slope of (v-t) graph

$\therefore$ P_{max} is found in time interval.

$t = 2$ s to $t = 3$ s

(iii) From $t = 1$ s to $t = 2$ s,

Slope of (v-t) graph is zero.

$\therefore$ $F = 0$

$\therefore$ $W = 0$

(iv) $F = ma = m \times$ slope of (v-t) graph

So, F_{max} is found in time intervals $t = 0$ to $t = 1$ and $t = 2$ to $t = 3$

(i) → (q), (ii) → (r), (iii) → (p), (iv) → (r,s)

27. (2) $P = \frac{dT}{dt}$ = rate of change of kinetic energy.

or $\int_0^T dT = \int_0^t P dt$ or $T = Pt$

or $\frac{1}{2}mv^2 = Pt$

or $v = \sqrt{\frac{2Pt}{m}} = \sqrt{\frac{2 \times 1000 \times 4}{2000}} = 2$ m/s

28. (5) Power = constant

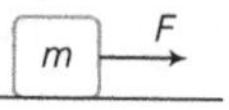

$\therefore$ $v^2 = \frac{2P}{m} \cdot t = 2 \times \frac{0.5}{0.2} \times 5$

$v = 5$ m/s

29. (2) $P = \frac{mgh}{t} = \frac{m}{t}gh = \frac{7500 \times 1.05 \times 10^3 \times 10 \times 1.63}{24 \times 60 \times 60} = 1.48$ W

$\therefore$ $n = 2$

30. $\langle P \rangle = \frac{mgh}{t} = mgv = (1000 \times 10 \times 5)$ W

$= \frac{1000 \times 10 \times 5}{746} \approx 67$ HP

31. (i) $\langle P_s \rangle = \frac{\text{Work done by spring}}{\text{Time}}$

$= \frac{-\frac{1}{2}kx_0^2}{t_0} = \frac{-\frac{1}{2} \times 10^4 \times (0.055)^2}{15.17}$

Here, $\langle P \rangle = \frac{15.17}{15.17} = -1$ W

(ii) $\langle P \rangle = \frac{\text{Work done by friction}}{\text{Time}}$

$= -\frac{fx_0}{t_0} = -\frac{15 \times 0.055}{15.17} = -54.38 \times 10^{-3}$ W

(iii) $\langle P \rangle = \frac{\text{Total work done}}{\text{Time}} = \frac{\frac{1}{2}mv_0^2}{t_0}$

$= \frac{\frac{1}{2} \times 2 \times 16}{15.17} = 1.05$ W

DPP-7 **Mechanical Energy**

1. (d) According to mechanical energy conservation principle,

$$U_i + T_i = U_f + T_f$$

$$mgh + \frac{1}{2}mv_0^2 = \frac{1}{2}mv^2$$

$$\therefore \quad v^2 = v_0^2 + 2gh$$

$$= 44 + 2 \times 10 \times 5$$

$$\therefore \quad v = 12 \text{ m/s}$$

2. (b) According to conservation principle of mechanical energy,

$$U_i + T_i = U_f + T_f$$

or $$mgh + 0 = 0 + \frac{1}{2}mv^2$$

$$\therefore \quad v = \sqrt{2gh}$$

$$= \sqrt{2 \times 10 \times 5} = 10 \text{ m/s}$$

So, shape of track is not required.

3. (b) According to conservation principle of mechanical energy,

$$U_i + T_i = U_f + T_f$$

or $$mg(5) + \frac{1}{2}mv^2 = mg(10) + \frac{1}{2}mv^2$$

or $$5\,mg + \frac{1}{2}mv_{\min}^2 = mg(10) + \frac{1}{2}mv_{\min}^2$$

or $$5mg + \frac{1}{2}mv_{\min}^2 = mg\,(10) + 0$$

$$\therefore \quad v_{\min} = \sqrt{2g(10-5)} = 10 \text{ m/s}$$

4. (d) At the maximum compression, velocity of each blocks are zero.

$\therefore$ Gain in PE = ΔU = Loss in KE

$$= \frac{1}{2}mv_0^2 + \frac{1}{2}mv_0^2 - 0 = mv_0^2$$

$\because$ Spring force is conservative in nature.

$\therefore$ The work done by spring is

$$W_s = -\Delta U = -mv_0^2$$

$\therefore$ The work done on each block by spring is $W = \frac{W_s}{2} = -\frac{1}{2}mv_0^2$

$$= -\frac{1}{2} \times 1 \times 100 = -50 \text{ J}$$

5. (c) $W_s = -\Delta U \Rightarrow \Delta U = -W_s$

6. (a) When gravity is present, then

$$mg = kx_0$$

$$\therefore \quad x_0 = \frac{mg}{k} = \frac{1 \times 10}{100} = \frac{1}{10} \text{ m}$$

When gravity is switched off, then total loss in spring potential energy appears is kinetic energy of the block.

$$\therefore \quad \frac{1}{2}kx_0^2 = \frac{1}{2}mv_{\max}^2$$

$$v_{\max} = \sqrt{\frac{k}{m}}x_0 = \sqrt{\frac{100}{1}} \times \frac{1}{10} = 1 \text{ m/s}$$

7. (a) Momentum would be maximum, when KE would be maximum and this is the case when total elastic PE is converted into KE.

According to conservation of energy,

$$\frac{1}{2}kL^2 = \frac{1}{2}Mv_{\max}^2$$

$$p_{\max} = \sqrt{(kL^2M)} = L\sqrt{Mk}$$

8. (c) $$\frac{1}{2}k_1x_1^2 = \frac{1}{2}k_2x_2^2$$

$$\Rightarrow \quad \frac{x_1}{x_2} = \sqrt{\frac{k_2}{k_1}}$$

9. (c) As friction is absent, energy is conserved.

$$\therefore \quad \frac{1}{2}kx^2 = \frac{1}{2}(4k)y^2$$

$$\frac{y^2}{x^2} = \frac{1}{4} \Rightarrow \frac{y}{x} = \frac{1}{2}$$

10. (b) In the frame of point A, the initial speed of the block is $2v_0$. At the time of maximum elongation, block comes at rest in the frame of point A.

By applying conservation principle of mechanical energy in the frame of A,

$$U_i + T_i = U_f + T_f$$

or $$0 + \frac{1}{2}m(2v_0)^2 = \frac{1}{2}kx_{\max}^2 + 0$$

$$\therefore \quad x_{\max} = \sqrt{\frac{4m}{k}}v_0$$

11. (c) In the frame of car,

$$\therefore \quad ma_0 = kx_0$$

$$\therefore \quad x_0 = \frac{ma_0}{k}$$

When car stops, the maximum speed is achieved by the block when spring comes in natural length. Thus, loss in spring potential energy appears as kinetic energy of the block

$$\frac{1}{2}kx_0^2 = \frac{1}{2}mv_{\max}^2 \Rightarrow v_{\max} = \sqrt{\frac{m}{k}}a_0$$

12. (b) Decrease in gravitational PE = Increase in elastic PE

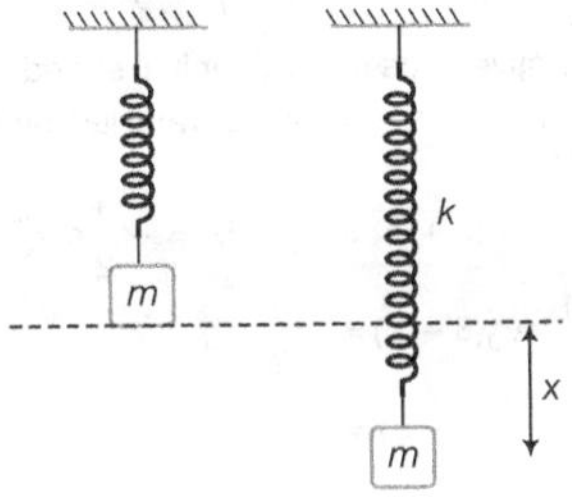

$$Mgx = \frac{1}{2}kx^2$$

$$\Rightarrow \quad x = \frac{2Mg}{k}$$

13. (b) Mass per unit length $= \frac{M}{L} = \frac{4}{2} = 2$ kg/m

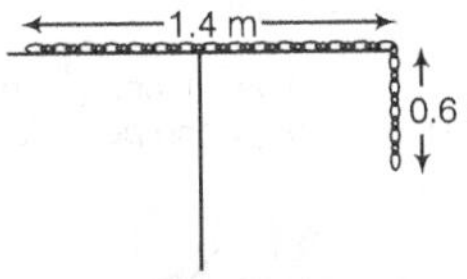

The mass of 0.6 m of chain

$$= 0.6 \times 2 = 1.2 \text{ kg}$$

The height of centre of mass of hanging part

$$h = \frac{0.6 + 0}{2} = 0.3 \text{ m}$$

Hence, work done in pulling the chain on the table = work done against gravity force

i.e. $W = mgh = 1.2 \times 10 \times 0.3 = 3.6$ J

14. (d) Hanging part $= \frac{L}{3}$

$\therefore$ Height of centre of mass $= \frac{L}{6}$

Mass of hanging part $= \frac{M}{L} \times \frac{L}{3} = \frac{M}{3}$

Work done against gravity

$$= \frac{M}{3} \times g \times \frac{L}{6} = \frac{MgL}{18}$$

15. (b) Work done by external force = Work done against friction + change in KE of block.

= Heat gained by block + Change in KE = Change in total energy

16. (a) According to work-energy theorem,

net work done = change in KE

17. (b) The speed at point B is $v_B = \sqrt{2gH}$

For BC,

$$W = \Delta T \quad \text{or} \quad W_f = 0 - \frac{1}{2}mv_B^2$$

or $\quad -\mu mg\, d = -\frac{1}{2}mv_B^2$

(here, μ = coefficient of friction)

or $\quad -\mu gd = -\frac{1}{2} \times 2gH$

or $\quad \mu d = H$

$\therefore \quad H = \mu d$

$\therefore \quad H \propto d$

18. (d) According to work-energy theorem,

Total work done = Change in kinetic energy

$$W = \Delta T$$

Case I $\quad -F \times 3 = \frac{1}{2}m\left(\frac{v_0}{2}\right)^2 - \frac{1}{2}mv_0^2$

Where F is resistive force and v_0 is initial speed.

Case II Let, the further distance travelled by the bullet before coming to rest be s.

$\therefore \quad -F(3 + s) = K_f - K_i = -\frac{1}{2}mv_0^2$

$\Rightarrow \quad -\frac{1}{8}mv_0^2(3 + s) = -\frac{1}{2}mv_0^2$

or $\quad \frac{1}{4}(3 + s) = 1$

or $\quad \frac{3}{4} + \frac{s}{4} = 1$ or $s = 1$ cm

19. (b) The loss in mechanical energy in curved path appears as thermal energy.

$\therefore \quad \Delta H = 20 - 16 = 4$ J

or $\quad -W_f = 4$ J

$\therefore \quad W_f = -4$ J

20. (a) According to conservation of energy, potential energy at height H is sum of kinetic energy and potential energy at h_2.

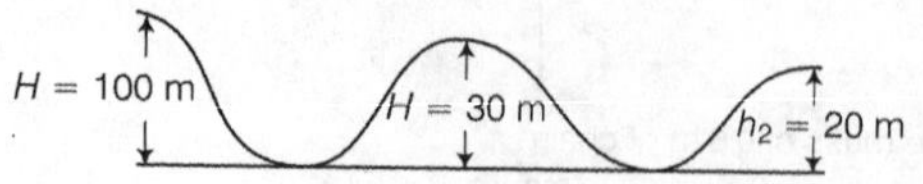

$\therefore \quad mgH = \frac{1}{2}mv^2 + mgh_2$

$\Rightarrow \quad mg(H - h_2) = \frac{1}{2}mv^2$

or $\quad v = \sqrt{2g(100 - 20)}$

or $\quad v = \sqrt{2 \times 10 \times 80} = 40$ m/s

21. (a) Given, $v(x) = \left(\frac{x^4}{4} - \frac{x^2}{2}\right)$

For minimum value of v, $\frac{dv}{dx} = 0$

$\Rightarrow \quad \frac{4x^3}{4} - \frac{2x}{2} = 0$

$\Rightarrow \quad x = 0, x = \pm 1$

So, $\quad v_{max}(x = \pm 1) = \frac{1}{4} - \frac{1}{2} = \frac{-1}{4}$ J

Now, $\quad K_{max} + v_{min}$ = Total mechanical energy

$\Rightarrow \quad K_{max} = \left(\frac{1}{4}\right) + 2$ or $K_{max} = \frac{9}{4}$

or $\quad \frac{mv^2}{2} = \frac{9}{4}$ or $v = \frac{3}{\sqrt{2}}$ ms^{-1}

22. (b,c) Total mechanical energy = Potential energy + Kinetic energy = constant

When potential energy is maximum, kinetic energy is minimum.

At $x = 1$ m, potential energy is maximum. So, kinetic energy is minimum at $x = 1$ m.

23. (a) Here, $\quad U_i + T_i = U_f + T_f$

or $\quad 0 + 19 = U_f + 0$

$\therefore \quad U_f = 19$ J

$0 < 19$ J < 20 J

$\therefore \quad 0 < x_0 < 1$ m

24. (a,b,d) When spring unstretches, block gains KE.

Work done by spring on the block is positive.

If initially block spring, block losses its KE and work done by spring on the block is negative.

25. (c,d) In the frame of block, block is stationary. So, the work done by spring on the block is zero.

26. (a,b,c,d) Maximum loss in PE $= mgh = \frac{1}{2}mv_{max}^2$

or $\quad v_{max} = \sqrt{2gh} = \sqrt{2 \times 10 \times 10}$

$= 10\sqrt{2}$ m/s

For reaching at point B, minimum mechanical energy required is $mg \times 11 >$ initial mechanical energy. That is why, block never reach at the point B and repeats its path continuously.

27. (a,b,c) To reach B, the initial mechanical energy should be greater than 10 J. So, particle will never reach in well B.

At A, PE = 4 J

$\therefore$ Loss of PE = Gain in KE = 8 − 4 = 4 J

28. (c) $\frac{1}{2}kx^2 = mgh = h \propto x^2$

$\therefore \quad \frac{h_2}{h_1} = \frac{x_2^2}{x_1^2}$

$h_1 = \left(\frac{11}{10}\right)^2 3\text{ m} = 3.63$ m

29. (d) $\frac{1}{2}kx^2 = mgh \Rightarrow h = \frac{kx^2}{2mg} \Rightarrow h \propto \frac{1}{m}$

$$\frac{h_2}{h_1} = \frac{m_1}{m_2} = \frac{m}{m + \frac{m}{10}} = \frac{10}{11}$$

$\therefore \quad h_2 = \frac{10}{11} \times 3.3 = 3$ m

30. (c) $U = \frac{1}{2}kx^2 = \frac{1}{2} \times 5 \times 10^3 \times 2^2 = 10$ kJ

31. (d) Required gravitational potential energy

$= mgh = 1000 \times 10 \times 10 = 100$ kJ

$\therefore$ Number of bounce $= \frac{100 \text{ kJ}}{10 \text{ kJ}} = 10$

32. (b) Let, the maximum compression in spring is x_0.

$\therefore \quad \frac{1}{2}k \cdot x_0^2 = mgh$

or $\quad x_0 = \sqrt{\frac{2mgh}{k}} = \sqrt{\frac{2\times10^3\times10\times10}{5\times10^3}} = \sqrt{40}$

$\because \quad ma_{max} = kx_0$

$\therefore \quad a_{max} = \frac{kx_0}{m} = \frac{5\times10^3\times\sqrt{40}}{1000} = 31.6\text{ m/s}^2$

33. (b) Loss in potential energy = Gain in kinetic energy

$\therefore \quad mgh = \frac{1}{2}mv^2$

$\therefore \quad v = \sqrt{2gh}$ is independent of mass

Kinetic energy at the bottom is mgh.

$\therefore \quad m_A > m_B$

$\therefore$ Kinetic energy of $A >$ kinetic energy of B.

34. (a) Since, acceleration of the block is variable. So, kinematic equations with constant acceleration are not applicable.

According to mechanical energy construction principle

$$U_i + T_i = U_f + T_f$$

or $$mgh + \frac{1}{2}mu^2 = \frac{1}{2}mv^2$$

$\therefore$ $$v^2 = u^2 + 2gh$$

35. (a) As, $\quad U_i + T_i = U_f + T_f$

$\Rightarrow \quad mgh + 0 = 0 + T_f$

$\therefore \quad T_f = mgh$

36. (a) Potential energy does not depend on the frame of reference. Thus, mechanical energy (PE+KE) is same in both reference frame.

37. (1) The equivalent spring constant is $k_{eq} = k_1 + k_2 + k_3$

$= 40 + 15 + 45 = 100$ N/m

When block will achieve maximum speed, then springs come in relaxed positions.

According to mechanical energy conservation principle,

$$\frac{1}{2}k_{eq}\,x_0^2 = \frac{1}{2}mv_0^2$$

or $$100 \times (0.1)^2 = 1 \times v_0^2$$

$\therefore$ $$v_0 = 1\text{ m/s}$$

38. (i) (2) At the time of maximum compression, velocity of the block is zero. According to conservation principle of mechanical energy,

$$U_i + T_i = U_f + T_f$$

$\therefore$ $$x_{max} = (2 + 2\sqrt{3})$$

or $$mgh + 0 = -\,mgh_{max} + \frac{1}{2}kx_{max}^2$$

When speed of the block is maximum, then acceleration of the block is zero.

For this, $\quad mg = kx_0$

$\therefore \quad x_0 = \frac{mg}{k} = \frac{2\times10}{10} = 2\text{ m}$

(ii) (2) According to conservation principle of mechanical energy,

$U_i + T_i = U_f + T_f$

or $\quad mgh + 0 = -\,mg\,x_0 + \frac{1}{2}mv_{max}^2 + \frac{1}{2}kx_0^2$

or $\quad 2\times10\times2 = -\,2\times10\times2 + \frac{1}{2}\times2\times v_{max}^2 + \frac{1}{2}\times10\times4$

$\therefore \quad v_{max} = \sqrt{60} = 2\sqrt{15}$ m/s

$(60)^{\frac{1}{2}} = (60)^{\frac{1}{n}}$

$\therefore \quad n = 2$

39. (5) When the block A just starts to leave the surface, then

$$kx_0 = Mg$$

For minimum value of m, x_0 should be maximum elongation in the spring. At the instant of maximum elongation, the velocity of block B will be zero.

$\therefore$ Loss in gravitational potential energy of block B = Gain in elastic potential energy by spring

$\therefore \quad mg\,x_0 = \frac{1}{2}k\,x_0^2$

$\therefore \quad x_0 = \frac{2mg}{k}$ or $\frac{Mg}{k} = \frac{2mg}{k} \Rightarrow m = \frac{M}{2} = 5$ kg

40. (5) Let for bouncing of the block A, the elongation in the spring is x_1.

$$kx_1 > m_A g \quad\text{or}\quad kx_1 > 100$$

$\therefore \quad x_1 > \frac{100}{150}$ or $x_1 > \frac{2}{3}$

$\therefore \quad x_{1\,min} = \frac{2}{3}$

For $x_{1\,min}$, the velocity of block B will be zero.

According to conservation principle of mechanical energy,

$$U_i + T_i = U_f + T_f$$

$$\frac{1}{2}kx_0^2 + \frac{1}{2}mv_{0\,min}^2 = mg(x_0 + x_{1\,min}) + \frac{1}{2}kx_{1\,min}^2$$

Here, $\quad x_0 = \frac{mg}{k}$

By putting the value,

$$v_{0_{min}} = \frac{20}{\sqrt{15}}\text{ m/s}$$

$\therefore \quad n = 5$

41. (5) According to conservation principle of mechanical energy,

$$U_i + T_i = U_f + T_f$$

$$\frac{1}{2}k(1\times\sec 60 - 1)^2 + 0 = \frac{1}{2}k(1\times\sec 37 - 1)^2 + \frac{1}{2}mv^2$$

or $$\frac{1}{2}\times240\times1 = \frac{1}{2}\times240\left(\frac{1}{4}\right)^2 + \frac{1}{2}\times1\times v^2$$

or $$120 = \frac{120}{16} + \frac{v^2}{2} \quad\text{or}\quad 120 - \frac{120}{16} = \frac{v^2}{2}$$

or $$120\times\frac{15}{16}\times2 = v^2$$

or $$15\times15 = v^2$$

$\therefore \quad v = 15$ m/s

$\therefore \quad n = 5$

42. (5) Net loss in gravitational potential energy = Gain in kinetic energy

$\therefore \quad m_B g\times1 - m_A g\times\sin 30° = \frac{1}{2}m_A v_0^2 + \frac{1}{2}m_B v_0^2$

or $\quad 100 - 50 = \frac{1}{2}\times20\times v_0^2 = 10\,v_0^2$

or $\quad \frac{50}{10} = v_0^2$

$\therefore \quad v_0 = \sqrt{5}$ m/s

$\therefore \quad v_0^2 = 5$ m/s

43. (2) When block A travels distance of 2 m, block B moves by 1 m.

$\therefore$ Net loss in PE of system = Loss in PE of A – Gain in PE of B

$= mg(2) - mg(1) = mg$

Now, this loss in PE appears or KE of system (no friction)

$\therefore$ KE gained by blocks A and $B = \frac{1}{2}mv_A^2 + \frac{1}{2}m\left(\frac{v_A}{2}\right)^2 = mg$

or $\quad \frac{1}{2}mv_A^2 \times \frac{5}{4} = mg$

$\therefore \quad v_A = 4\text{ m/s}$

and $\quad v_B = \frac{v_A}{2} = 2\text{ m/s}$

$\therefore \quad \frac{v_A}{v_B} = 2$

44. (2) Take the level of pulley as reference line

$\because \quad U_i + T_i = U_f + T_f$

or $\quad \frac{-mg}{3}\frac{l}{6} - \frac{2mg}{3}\frac{l}{3} = -mg\frac{l}{2} + \frac{1}{2}mv^2$

$\therefore \quad v = \frac{2}{3}\sqrt{gl} = \frac{2}{3}\sqrt{10 \times 0.9} = 2\text{ m/s}$

45. (1) Let for just shifting of the block of mass m_1, the elongation in spring is x_1

$\therefore \quad kx_1 > \mu m_1 g$

$\because \quad x_{1\,min} = \frac{\mu m_1 g}{k}$

For minimum value of m_2, the speed of block of mass m_2 is zero when the elongation in spring is $x_{1\,min}$.

According to the conservation principle of mechanical energy, Loss in gravitational potential energy of the block mass m_2 is equal to gain in elastic potential energy in the spring.

$\therefore \quad m_2 g\, x_{1\,min} = \frac{1}{2}kx_{min}^2$

$\therefore \quad x_{1\,min} = \frac{2m_2 g}{k}$

or $\quad \frac{\mu m_1 g}{k} = \frac{2m_2 g}{k}$

$\therefore \quad m_2 = \frac{\mu m_1}{2} = \frac{0.5 \times 4}{2} = 1\text{ kg}$

Using conservation of energy, we have

46. (i) $\quad \frac{1}{2}mv^2 = \frac{1}{2}k_1x_1^2 = \frac{1}{2}k_2x_2^2$

or $\quad mv^2 = k_1x_1^2 = k_2x_2^2$

or $\quad 1 \times 100 = 100 \times x_1^2 = 50 \times x_2^2$

$\because \quad x_1 = 1\text{ m}$

and $\quad x_2 = \sqrt{2}\text{ m/s}^2$

$\therefore \quad \frac{x_1}{x_2} = \frac{1}{\sqrt{2}}$

(ii) The magnitude of maximum deceleration is

$\frac{k_1x_1}{m} = \frac{100 \times 1}{1} = 100\text{ m/s}^2$

47. (i) $\because \quad \frac{1}{2}kx^2 = \frac{1}{2}mv_{max}^2$

or $\quad \frac{1}{2} \times 100 \times (0.1)^2 = \frac{1}{2} \times 1 \times v_{max}^2$

$\therefore \quad v_{max} = 1\text{ m/s}$

(ii) $\quad U_i + T_i = U_f + T_f$

or $\quad \frac{1}{2} \times 100 \times (0.1)^2 = \frac{1}{2}kx^2 + \frac{1}{2}m\left(\frac{v_{max}}{2}\right)^2$

or $\quad \frac{1}{2} \times 100 \times 0.01 = \frac{1}{2} \times 100 \times r^2 + \frac{1}{2} \times 1 \times \frac{1}{4}$

$\therefore \quad x = 8.66\text{ cm}$

48. (i) At the time of maximum elongation, velocity of the block will be zero.

Here, loss in gravitational potential energy is equal to gain in elastic potential energy of the spring.

Thus, $\quad (mgx_{max}\sin 30°) = \frac{1}{2}kx_{max}^2$

$\therefore \quad x_{max} = \frac{mg}{k}$

(ii) At the instant of maximum speed of the block, acceleration of the block is zero.

$\therefore \quad mg\sin 30° = kx_0$

$\therefore \quad x_0 = \frac{mg}{2k}$

Here, loss in gravitational potential energy = Gain in elastic potential energy of spring + Gain in kinetic energy.

$\therefore \quad mg\,x_0\sin 30° = v \cdot \frac{1}{2}kx_0^2 + \frac{1}{2}mv_{max}^2$

$v_{max} = g/2\sqrt{m/k}$

49. Let M strikes with wall with speed v. Then, speed of m at this instant is $v\cos\theta = \frac{2v}{\sqrt{5}}$. When M will fall a distance 1 m, then m will rise a distance $(\sqrt{5}-1)$ m. Thus, decrease in PE of M = Increase in PE of m + Increase in KE of both blocks

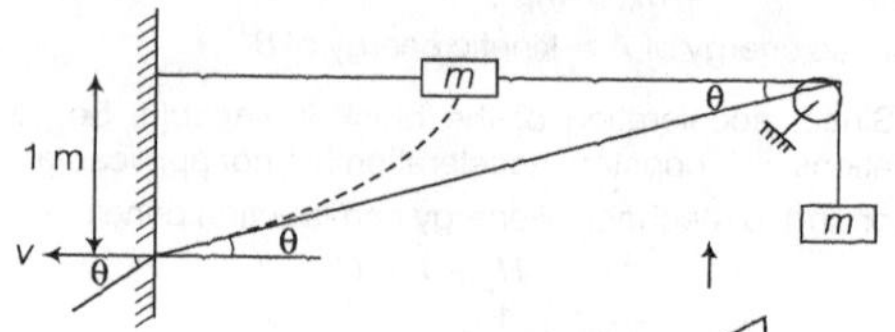

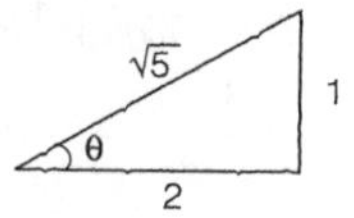

$\therefore \quad 2 \times 9.8 \times 1 = 0.5 \times 9.8(\sqrt{5}-1) + \frac{1}{2} \times 2 \times v^2 + \frac{1}{2} \times 0.5\left(\frac{2v}{\sqrt{5}}\right)^2$

$\therefore \quad v = 3.29\text{ m/s}$

50. Let velocity at the top is v_0.

$\therefore \quad \frac{1}{2}m \times 30^2 = mgh + \frac{1}{2}mv_0^2$

or $\quad v_0^2 = 30^2 - 2gh$

or $\quad v_0^2 = 900 - 20 \times 33.75 = 225\text{ m}$

$\therefore \quad v_0 = 15\text{ m/s}$

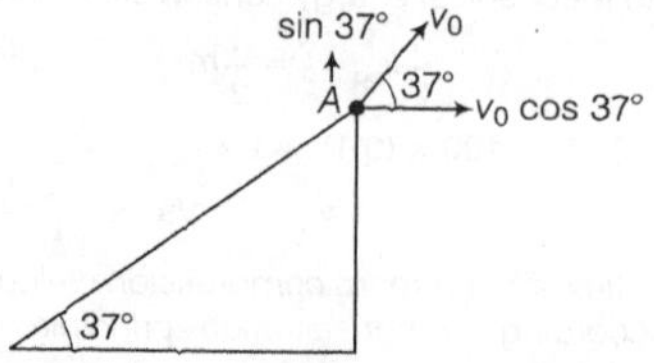

For vertical motion after point A,

$-33.75 = (v_0\sin 37°)T - \frac{1}{2}gT^2$

or $\quad -33.75 = 9T - \frac{1}{2}gT^2$

or $\quad 5T^2 - 9T - 33.75 = 0$

$\therefore \quad T = \frac{9 \pm \sqrt{81 + 4 + 5 \times 33.75}}{2 \times 5} = 3.65\text{ s}$

$\therefore \quad x = (v_0\cos 37°)T = 15 \times \frac{4}{5} \times 3.65 = 43.8\text{ m}$

51. (a) For any path,

$w = \mathbf{F}\cdot\mathbf{s} = (4\hat{i} + 6\hat{j})\cdot(8\hat{i} + 8\hat{j}) = 32 + 48 = 80\text{ J}$

Since, work does not depend on paths so force is conservative in nature.

Hence, mechanical energy remains constant in all paths. The work done is positive, so kinetic energy increases.

(i) → (p,q,r,s) (ii) → (p,q,r,s) (iii) → (p,q,r,s) (iv) → (p,q,r,s)

DPP-8 Vertical Circular Motion

1. (i) Consider motion of the ball as shown below

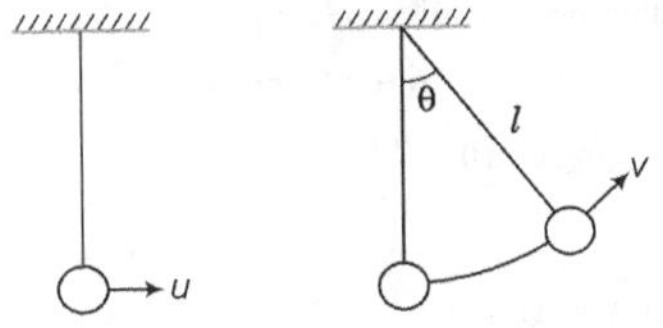

Here, $T = mg\cos\theta + \dfrac{mu^2}{l}$

At lowest point, $u_{min} = \sqrt{5gl}$

$\therefore \quad T = mg\cos\theta + \dfrac{mu_{min}^2}{l}$

$= mg + 5\,mg = 6\,mg$

(ii) Here $u < \sqrt{2gl}$, the ball will oscillate about lowest point with angular amplitude less than $\dfrac{\pi}{2}$.

Here $\quad v^2 = u^2 - 2gh$

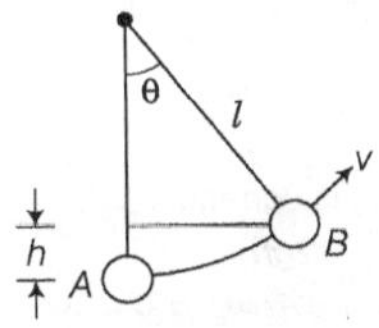

Here, $v = 0$ at the extreme position.

$\therefore \quad 0 = 20 - 20h$

$\therefore \quad h = 1\,\text{m}$

But $\quad h = l - l\cos\theta$

$1 = 2 - 2\cos\theta$

or $\quad \cos\theta = \dfrac{1}{2}$

$\therefore \quad \theta = \dfrac{\pi}{3}$

The distance travelled $= s = 2l\theta$

$= 2 \times 2 \times \dfrac{\pi}{3} = \dfrac{4\pi}{3}\,\text{m}$

(iii) Here, $\quad T = mg\cos\theta + \dfrac{mv^2}{l}$

or $\quad T = mg\cos\theta + \dfrac{m}{l}(u^2 - 2gh)$

For horizontal position,

$\theta = \dfrac{\pi}{2}$ and $h = l = 1\,\text{m}$

$\therefore \quad T_1 = \dfrac{1}{1} \times (100 - 2 \times 10 \times 1)$

$= 80\,\text{N}$

For highest point, $\theta = \pi$ and $h = 2l = 2\,\text{m}$

$\therefore \quad T_2 = 1 \times 10 \times \cos\theta + \dfrac{1}{1}(100 - 2 \times 10 \times 2)$

$= -10 + 60 = 50\,\text{N}$

$\therefore \quad \dfrac{T_1}{T_2} = \dfrac{80}{50} = \dfrac{8}{5}$

2. (i) (b,d) To complete vertical circle, velocity at lowest point must be greater than or equals to $\sqrt{5gl}$.

As, $u > \sqrt{2gl}$, ball completes half circle and it leaves vertical circle above the lower half circle.

Also, energy is conserved even when ball leaves the vertical circle.

(ii) (b) $\theta + \alpha = \pi,\ \theta = \pi - \alpha$

(iii) (c) $\sin(90° - \alpha) = \dfrac{x}{l} \Rightarrow x = l\cos\alpha$

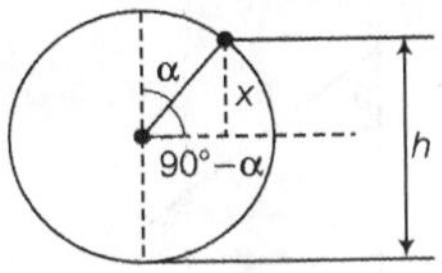

$\therefore \quad h = l + l\cos\alpha$

(iv) (b) Velocity of moving mass at any angle α is

$v = \sqrt{u^2 - 2gl(1 - \cos(\pi - \alpha))}$

$= \sqrt{u^2 - 2gl(1 + \cos\alpha)}$

$= \sqrt{u^2 - 2gh}$

(v) (a,b,c,d) Using, $T = 3mg\cos\theta - 2mg + \dfrac{mu^2}{l}$

$\theta = \pi - \alpha,\ \cos(\pi - \alpha) = -\cos\alpha,\ u^2 = 4gl$

We have,

$T = -3mg\cos\alpha - 2mg + 4mg$

$= 2mg - 3\,mg\cos\alpha$

(vi) (a) $\sqrt{5gl} > u > \sqrt{2gl}$, the ball leaves the vertical circle above the half circle. But point of leaving the vertical circle is not specified. Hence, option (a) is correct.

(vii) (c) $\because T = 2mg - 3mg\cos\alpha$

At $T = 0,\ \cos\alpha = \dfrac{2}{3}$

(viii) (a) At point P, $T = 0$, $\cos\alpha = \dfrac{2}{3}$

As, $\quad v = \sqrt{u^2 - 2gl(1 + \cos\alpha)}$

At point $P = \sqrt{4gl - 2gl\left(1 + \dfrac{2}{3}\right)} = \sqrt{\dfrac{2}{3}gl}$

(ix) (b) In this case, acceleration is constant and initial velocity and acceleration are not parallel so, path is parabolic.

(x) (a,b) $H = \dfrac{v_0^2\sin^2\alpha}{2g} = \dfrac{\frac{2}{3}gl\left(1 - \frac{4}{9}\right)}{2g} = \dfrac{5l}{27}$

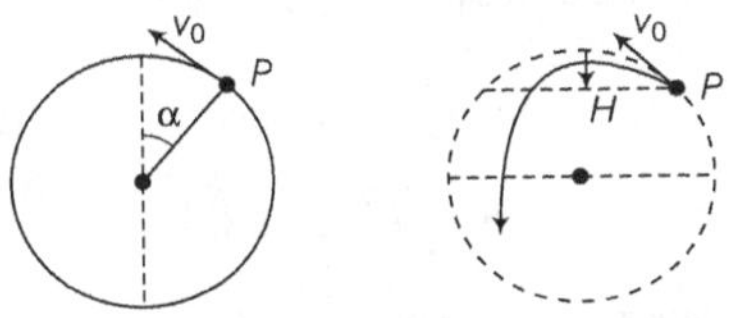

(xi) (a,c) Height w.r.t. lower point

$= l + l\cos\alpha + \dfrac{v_0^2\sin^2\alpha}{2g}$

$= l + l\dfrac{2}{3} + \dfrac{5l}{27} = \dfrac{50l}{27}$

3. Similar to Q. No. 2.

(i) (b)	(ii) (a)
(iii) (a)	(iv) (a,b)
(v) (a)	(vi) (b)
(vii) (a,b,c)	(viii) (b)

4. (i) (a) Tension is acting towards centre of path and weight acts downwards.

(ii) (a) $T - mg = \dfrac{mu^2}{l}$

$\therefore \quad T = mg + \dfrac{mu^2}{l}$

(iii) (b) $v^2 = u^2 - 2gl(1-\cos\theta)$

or $v^2 = u^2 - 2gh,\ h = l(1-\cos\theta)$

(iv) (a,b,c) $T - mg\cos\theta = \frac{mv^2}{l}$

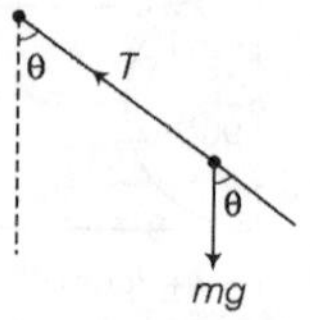

$$\therefore \quad T = mg\cos\theta + \frac{mv^2}{l}$$

Here, $v^2 = u^2 - 2gh$

or $v^2 = u^2 - 2gl(1-\cos\theta)$

(v) (a,c) $v = \sqrt{u^2 - 2gl(1-\cos\theta)}$

At top point $\theta = \pi$, $\cos\pi = -1$

$$\therefore \quad v = \sqrt{u^2 - 2gl \times 2} = \sqrt{u^2 - 4gl}$$

(vi) (a,c) $v_{top} = 0$

$\Rightarrow \quad u^2 - 4gl = 0$

$\therefore \quad u = \sqrt{4gl}$

$\because \quad T = mg\cos\theta + \frac{mv^2}{l}$

Here, $\theta = \pi$ and $v = 0$

$\therefore \quad T = mg\cos\pi + 0$

$\therefore \quad T = -mg$

(vii) (a) $u_{min} = \sqrt{4gl}$

(viii) (c) For $u = u_{min} = \sqrt{4gl}$

$T = 0$

(ix) (a) $v^2 = u^2 - 2gh$

or $0 = u^2 - 2gh$

or $u^2 - 2gh = 0$

or $u^2 = 2gh$

or $u^2 < 4gl$

$\therefore \quad h < 2l$

Hence, (a) is correct.

(x) (b) For $u = \sqrt{3gl}$

$$KE = \frac{1}{2}mu^2 = mgh \qquad \text{(at light point)}$$

$$\therefore \quad h = \frac{3}{2}l$$

5. (i) (b,c) Radius of circular path is $r = R + \frac{d}{2}$

For $R >> d$, $R = r$

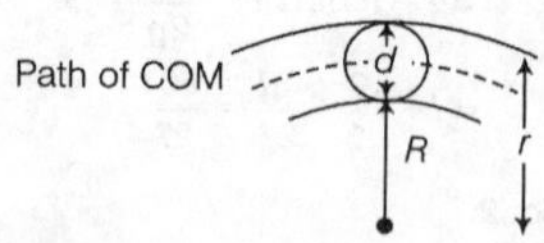

(ii) (c) At lowest point,

$$N - mg = \frac{mu^2}{r}$$

$$\Rightarrow \quad N = mg + \frac{mu^2}{r}$$

(iii) (b) For u_{min}, $v = 0$

$$\therefore \quad v^2 = u^2 - 2g \times 2R \qquad [\because R >> d]$$

$$\Rightarrow \quad 0 = u^2 - 4gR$$

$$\therefore \quad u_{min} = \sqrt{4gR}$$

(iv) (a,b,d) At top point, $v = 0$

$$\therefore \quad Mg - N = \frac{mv^2}{R} = 0 \Rightarrow N = mg$$

(v) (a,b) At an angle θ, $v^2 = u^2 - 2gh$

where, $h = r(1-\cos\theta)$

Also, $N = mg\cos\theta + \frac{mv^2}{r}$

6. (a)

(i) Option (a) is correct.

(ii) (c) Here, $\theta = \pi - \alpha$

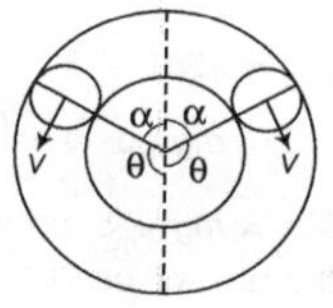

$\because \quad N = mg\cos\theta + \frac{mv^2}{R}$

or $N = mg\cos(\pi - \alpha) + \frac{mv^2}{R}$

$\therefore \quad N = -mg\cos\alpha + \frac{mv^2}{R}$

Hence, option (c) is correct.

(iii) (a) Here, $v^2 = u^2 + 2gh$

or $v^2 = 0^2 + 2g(R - R\cos\alpha)$

$$(\because h = R - R\cos\alpha = R - R\cos 60^\circ = \frac{R}{2})$$

or $v^2 = 2g\frac{R}{2} = gR$

$\therefore \quad v = \sqrt{gR}$

(iv) (a) For $\alpha = 60^\circ$

Normal reaction on each ball is

$$N = -mg\cos 60^\circ + \frac{mv^2}{R} = -\frac{mg}{2} + \frac{m}{R}gR = \frac{mg}{2}$$

Here, normal reaction on each ball is positive means it is directed radially inward. So, normal reaction on pipe by each ball is directed radially outward.

Hence, option (a) is correct.

(v) (a) For bouncing, upward force due to balls should be greater than or equals to weight of the pipe.

$$\therefore \quad 2N\cos 60^\circ \geq Mg$$

Hence, option (a) is correct.

(vi) (a) $\because 2N\cos 60^\circ \geq Mg$

or $N \geq Mg$ or $\frac{mg}{2} \geq Mg$

$\therefore \quad \frac{m}{M} \geq 2$

Hence, option (a) is correct.

7. (d) Here, the constant horizontal force required to take the body from position 1 to position 2 can be calculated by using work-energy theorem. Let us assume that body be taken slowly. So that its speed does not change, then

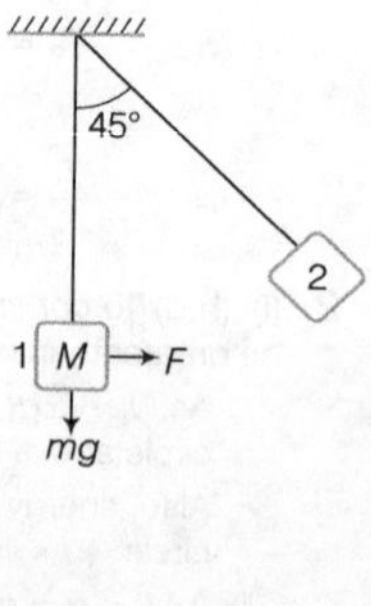

$$\Delta K = 0 = W_F + W_{Mg} + W_{tension}$$

(symbols have their usual meanings)

$$W_F = F \times l\sin 45^\circ = \frac{Fl}{\sqrt{2}}$$

$$W_{Mg} = Mg\,(l - l\cos 45^\circ),$$

$$W_{tension} = 0$$

$$\therefore \quad F = Mg\,(\sqrt{2} - 1)$$

8. (c) Velocity at point A, $v_A = \sqrt{5gl}$, as bob reaches at B

At same angle θ,

$$v = \sqrt{u^2 - 2gl(1-\cos\theta)}$$

$\therefore$ $$\frac{v_A}{2} = \sqrt{v_A^2 - 2gl(1-\cos\theta)}$$

or $$\frac{v_A^2}{4} - v_A^2 = -2gl(1-\cos\theta)$$

or $$\frac{3}{4}v_A^2 = 2gl(1-\cos\theta)$$

or $$\frac{3}{4} \times 5gl = 2gl(1-\cos\theta)$$

$\therefore$ $$\cos\theta = -\frac{7}{8} \Rightarrow \theta < \frac{3\pi}{4}$$

9. (d) As, $|\Delta \mathbf{v}| = \sqrt{u^2 + v^2}$

$$= \sqrt{u^2 + u^2 - 2gl} = \sqrt{2(u^2 - gl)}$$

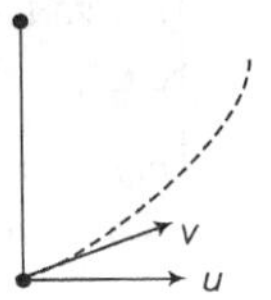

10. (d)

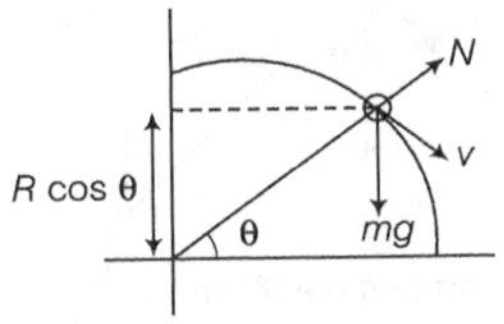

Bead applies an inward force till $\cos\theta > \frac{2}{3}$ and outward force when $\cos\theta < \frac{2}{3}$

$$N = mg\cos\theta - \frac{mv^2}{R}$$

$$= mg(3\cos\theta - 2)$$

$\Rightarrow$ $N = 0$, when $\cos\theta = \frac{2}{3}$

11. (b) In this case,

$$mg\cos\theta - T = \frac{mv^2}{l}$$

Here, $$v^2 = 2gh = 2g(l - l\sin\theta)$$

$$T = mg\cos\theta - 2mg(1 - \cos\theta)$$

$$= 3mg\cos\theta - 2mg$$

For $T = 0$, $\cos\theta = \frac{2}{3}$

12. (b) In the case of particle A, loss in gravitational potential energy appears as spring potential and kinetic energy.

But in the case of particle B, loss in gravitational potential energy appears as only kinetic energy.

13. (c) Loss in potential energy = Gain in kinetic energy

or $$mg(2) = \frac{1}{2}mv_B^2$$

or $$4mg = mv_B^2$$

$\therefore$ $$N_B + mg = \frac{mv_B^2}{R}$$

$\therefore$ $$N_B = 3mg$$

$\therefore$ $$\frac{N_B}{mg} = 3$$

14. (c) Applying conservation principle of mechanical energy at points A and B,

$$U_i + T_i = U_f + T_f$$

or $$mgR + \frac{1}{2}k(2R - \sqrt{2}R)^2 = \frac{1}{2}mv^2$$

or $$2mgR + k(2 - \sqrt{2})R^2 = mv^2$$

or $$2mg + k(2 - \sqrt{2})R = \frac{mv^2}{R}$$

Also, at point B,

$$N = \frac{mv^2}{R} = 2mg + k(2 - \sqrt{2})R$$

$$= 2mg + (2 + \sqrt{2})(2 - \sqrt{2})R$$

$$= 2mg + (4 - 2)R$$

$$= 20 + 2 \times 5 = 30 \text{ N}$$

15. (b,c) Here, $T = mg\cos\theta + \frac{mv^2}{l}$

$$= mg\cos\theta + \frac{m}{l}(v_0^2 + 2gh)$$

$$= mg\cos\theta + \frac{m}{l}(v_0^2 + 2gl\cos\theta)$$

or $$20 = mg\cos\theta + \frac{mv_0^2}{l} + 2mg\cos\theta$$

(at the time of breaking)

or $$20 = 10\cos\theta + \frac{25}{2} + 20\cos\theta$$

or $$30\cos\theta = \frac{15}{2}$$

$\therefore$ $$\theta = \cos^{-1}\left(\frac{1}{4}\right)$$

16. (a,c) According to work-energy theorem,

$$W = \Delta T = \frac{1}{2}mv^2 - \frac{1}{2}mu^2$$

or $$W_F + W_g + W_N = \frac{1}{2}mv^2$$

or $$FR\sin 60° - mgR\cos 60° = \frac{1}{2}mv^2$$

or $$200 \times 10 \times \frac{\sqrt{3}}{2} - 100 \times 10 \times \frac{1}{2} = \frac{1}{2} \times 10 \times v^2$$

or $$200\sqrt{3} - 100 = v^2$$

or $$100(2\sqrt{3} - 1) = v^2$$

$\therefore$ $$v = 10\sqrt{2\sqrt{3} - 1} = 15.7 \text{ m/s}$$

17. (a,b) $U_i + T_i = U_f + T_f$

or $$mgh + \frac{1}{2}mv_0^2 = \frac{1}{2}mv^2$$

$\therefore$ $$v_B^2 = v_0^2 + 2gh$$

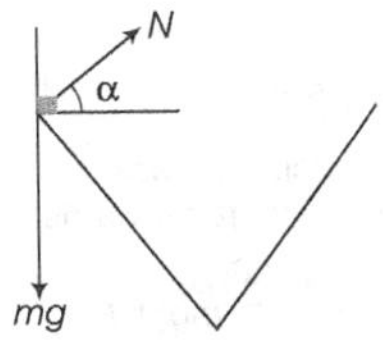

Here, $$N\cos\alpha = \frac{mv_0^2}{r_0}$$

and $$N\sin\alpha = mg$$

$\therefore$ $$\cot\alpha = \frac{v_0^2}{r_0 g}$$

$\therefore$ $$v_0 = \sqrt{r_0 g \cot\alpha}$$

18. (b,c,d) $\frac{k}{r^2} = \frac{mv^2}{r}$

$\therefore \quad \frac{1}{2}mv^2 = KE = \frac{k}{2r} \quad \Rightarrow \quad F = -\frac{dU}{dr}$

$\therefore \quad dU = \frac{k}{r^2}dr$

or $\quad U = \int_0^U dU = \int_\infty^r \frac{k}{r^2}dr$

$\therefore \quad U = k\left[\frac{r^{-2+1}}{-2+1}\right]_\infty^r = k\left[-\frac{1}{r}\right]_\infty^r = -\frac{k}{r}$

$\therefore$ Total mechanical energy = PE + KE

$$= -\frac{k}{r} + \frac{k}{2r} = -\frac{k}{2r}$$

19. (b,c) According to work-energy theorem,

$$W = \Delta T$$

$$W_g + W_T + W_F = 0 \quad \text{or} \quad -mgh + 0 + W_F = 0$$

$\therefore \quad W_F = mgh = mgl(1-\cos\theta)$

$$= 1 \times 10 \times 1\left(1 - \frac{1}{2}\right) = 5\text{ J}$$

20. (a) $\frac{1}{2}kx^2 = \frac{1}{2}mv^2 + mg \times 1.4 \times \sin 15°$

or $\frac{1}{2} \times 200 \times 20 \times 20 \times 10 = \frac{1}{2}v^2 + 0.1 \times 9.79 \times 1.9 \times \frac{15}{180} \times \frac{22}{7}$

or $\quad 4 = \left(\frac{1}{2}0.1\,v^2 + 0.35\right) \times 10^{-2}$

or $\quad 0.1\,v^2 = 2 \times 3.65 \times 10^{-2}$

or $\quad v^2 = 0.73 \times 10^{-2}$

$\Rightarrow \quad v = 0.85$ m/s

21. (c) $\frac{1}{2}kx^2 = mg\,(1.4 \sin 15°)$

or $\frac{1}{2} \times 200 \times x^2 = 0.1 \times 9.79 \times 1.4 \times 1.4 \times \frac{15}{180} \times \frac{22}{7}$

or $\quad x^2 = 0.35 \quad \Rightarrow \quad x^2 = 0.35 \times 10^{-2}$

$\Rightarrow \quad x = 0.59 \times 10^{-1}$m

$\therefore \quad x = 5.9$ cm

22. (a) As, N will not become zero as $v = 0$

So, ball will strike on bumper A.

23. (b) Height fallen to Q is $R\sin 30° = \frac{R}{2} = 20$cm

Using conservation of energy,

$$0 - \left(-mgh + \frac{1}{2}mv^2\right) = 150$$

or $\quad v = 10$ m/s

24. (a) $N - mg\cos 60° = \frac{mv^2}{R}$

$\therefore \quad N = 7.5$ N

25. (a) (i) For minimum velocity at lowest point, the tension in string is zero where net force is directed towards centre.

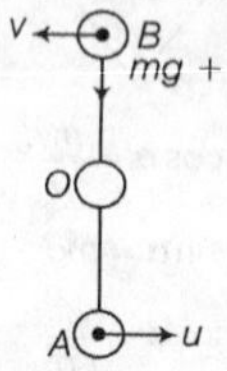

At point B, $PT = 0$

$\therefore \quad mg + F = \frac{mv^2}{l}$

or $\quad 10 + 20 = v^2 \quad$ According to work-energy theorem,

$$W_g + W_F + W_T = \frac{1}{2}mv^2 - \frac{1}{2}mv^2$$

or, $\quad -mg \times 2l - F \times 2l = \frac{1}{2}mv^2 - \frac{1}{2}mu^2 \quad \ldots$(i)

and $\quad mg + F = \frac{mu^2}{l} \quad \Rightarrow \quad mu^2 = 30$

$\therefore \quad -20 - 40 = 15 - \frac{1}{2} \times 1 \times u^2$

$$u^2 = 150$$

$\therefore \quad u = \sqrt{150}$ m/s

(ii) $\therefore \; 20 - 10 = \frac{mu^2}{l}$

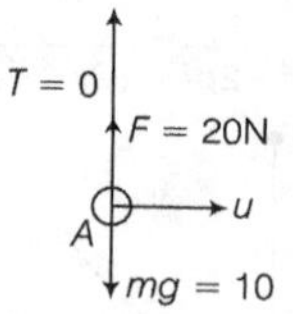

$\therefore \quad 10 = u^2 \quad \Rightarrow \quad u = \sqrt{10}$ m/s

(iii) From given figure,

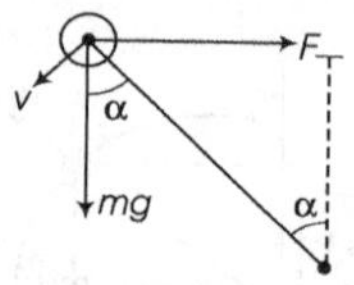

$$mg\cos\alpha + F\sin\alpha = \frac{mv^2}{l}$$

or $\quad 10\cos\alpha + 20\sin\alpha = v^2$

and $\quad mg\sin\alpha = F\cos\alpha$

$\therefore \quad \tan\alpha = \frac{20}{10} = 2$

$\therefore \quad \sin\alpha = \frac{2}{\sqrt{5}}$ and $\cos\alpha = \frac{1}{\sqrt{5}}$

$\therefore \quad v^2 = 10 \times \frac{1}{\sqrt{5}} + 20 \times \frac{2}{\sqrt{5}} = \frac{50}{\sqrt{5}}$

According to work-energy theorem,

$$W_g + W_F + W_T = \frac{1}{2}mv^2 - \frac{1}{2}mu^2$$

or $\quad -mg\,(l + l\cos\alpha) - F\,l\sin\alpha = \frac{1}{2}mv^2 - \frac{1}{2}mu^2$

or $-1 \times 10\left(1 + \frac{1}{\sqrt{5}}\right) - 20 \times 1 \times \frac{2}{\sqrt{5}} = \frac{1}{2} \times 1 \times \frac{50}{\sqrt{5}} - \frac{1}{2}mu^2$

$\therefore \quad u = \sqrt{30\sqrt{5} + 20}$ m/s

(iv) $u = \sqrt{5gl} = \sqrt{50}$ m/s

(i) $\to$ (q) (ii) $\to$ (r) (iii) $\to$ (s) (iv) $\to$ (p)

26. (b) At lowest point speed is maximum

$\therefore \quad T - mg = \frac{mv^2}{l}$

$T_{max} = mg + \frac{mv^2}{l}$ at lowest point.

27. (d) When the component of mg provides centripetal force, then tension in rod becomes zero.

28. (b) $W_f + W_g + W_N = 0 \quad$ or $\quad W_f + mgR + 0 = 0$

$\therefore \quad W_f = -mgR$

$\Rightarrow \quad \Delta H = -W_f = mgR$

29. (9) According to conservation principle of mechanical energy,

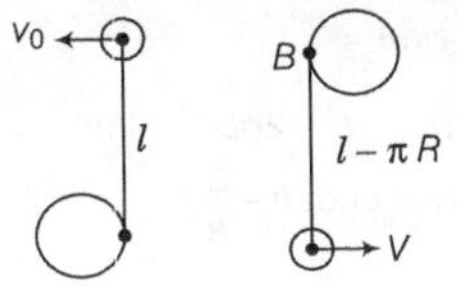

$$U_i + T_i = U_f + T_f$$

or $$mgl + \frac{1}{2}mv_0^2 = -mg(l-\pi R) + \frac{1}{2}mv^2$$

Thus, $mgl + mg(l-\pi R) + \frac{1}{2}mv_0^2 = \frac{1}{2}mv^2$

or $$2mgl - mg\pi R + \frac{1}{2}mv_0^2 = \frac{1}{2}mv^2$$

Also, $$T - mg = \frac{mv^2}{l-\pi R}$$

$$\therefore T = mg + \frac{mv^2}{l-\pi R} = mg + \left(\frac{1}{l-\pi R}\right)\{4\,mgl - 2\,mg\pi R + mv_0^2\}$$

$$= mg + \frac{2}{l}\{4mgl - 2\,mg\frac{l}{2} + mv_0^2\}$$

$$= mg + 8mg - 2mg + 2mv_0^2$$

$$= 7mg + 2mv_0^2$$

$$= 70 + 2\times 1\times 100$$

$$= 270\text{ N} = 30x$$

$\therefore x = 9$

30. (3) Here, $\sqrt{2gl} < u < \sqrt{5gl}$

So, the ball will not complete the vertical circle.

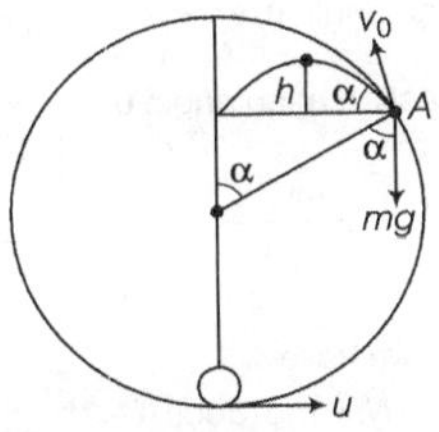

Let at an angle α with upper vertical, it leaves the vertical circle, then moves on the projectile path.

$\therefore$ $$v_0^2 = u^2 - 2gh$$

$$v_0^2 = 200 - 20h$$

But $$mg\cos\alpha = \frac{mv_0^2}{l} \quad (\because T = 0 \text{ at point } A)$$

or $$10\cos\alpha = \frac{v_0^2}{5}$$

$\therefore$ $$v_0^2 = 50\cos\alpha$$

$\therefore$ $$50\cos\alpha = 200 - 20h$$

or $$50\cos\alpha = 200 - 20(l + l\cos\alpha)$$

or $$50\cos\alpha = 200 - 100 - 100\cos\alpha \quad (\because l = 5\text{ cm})$$

or $$150\cos\alpha = 100$$

$\Rightarrow$ $$\cos\alpha = \frac{100}{150} = \frac{2}{3}$$

$\therefore$ $$h = l + l\cos\alpha = 5 + 5\times\frac{2}{3} = \frac{25}{3}$$

Also $$v_0^2 = 50\cos\alpha$$

$$= 50\times\frac{2}{3} = \frac{100}{3}$$

$\therefore$ $$v_0 = \frac{10}{\sqrt{3}}\text{ m/s} = \frac{10}{\sqrt{n}}\text{ m/s}$$

$\therefore$ $n = 3$

31. (2) As shown in figure,

The time of flight in air is

$$T = \frac{2v\sin 60°}{g} = \frac{\sqrt{3}v}{g}$$

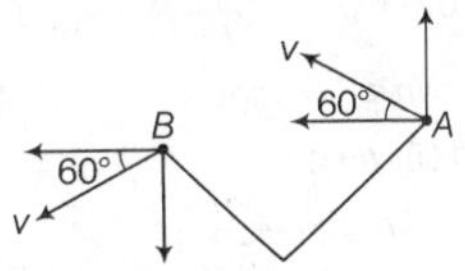

The block must travel the horizontal distance $AB = 2R\sin 60°$ in the time T with constant horizontal velocity $v\cos 60°$

$\therefore$ $$(v\cos 60°)\,T = \sqrt{3}R \quad (\text{here, } R = 0.2)$$

$\therefore$ $$v = 2\text{ m/s}$$

According to mechanical energy consecutive principle,

$$U_C + T_C = U_A + T_A$$

or $$0 + \frac{1}{2}mv_0^2 = mg(R + R\sin 30°) + \frac{1}{2}mv^2$$

or $$\frac{v_0^2}{2} = 10\times 0.2\times\frac{3}{2} + \frac{1}{2}\times 4$$

$\therefore$ $$v_0 = \sqrt{10}\text{ m/s} = (10)^{\frac{1}{x}}$$

$\therefore$ $x = 2$

32. (4) At point B,

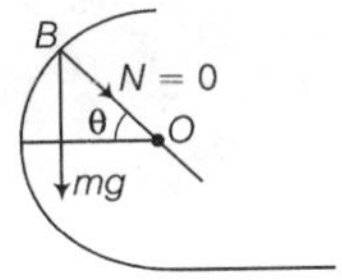

$$N = 0$$

$\therefore$ $$mg\sin\theta = \frac{mv^2}{R}$$

or $$g\sin\theta = \frac{v^2}{R}$$

$\therefore$ $$v^2 = gR\sin\theta$$

Also, $$v^2 = v_0^2 - 2gh$$

or $$v^2 = \frac{22Rg}{5} - 2g(R + R\sin\theta)$$

or $$5v^2 = 22gR - 10\,gR - 10gR\sin\theta$$

or $$5gR\sin\theta + 10gR\sin\theta = 12\,gR$$

or $$15gR\sin\theta = 12\,gR$$

$\therefore$ $$\sin\theta = \frac{12}{15} = \frac{4}{5} = \frac{x}{5}$$

$\therefore$ $x = 4$

33. (5) If the block completes its turn on vertical track, at the lowest point of track, then speed of block will be

$$u_{min} = \sqrt{5g(R-r)} \quad \ldots\text{(i)}$$

According to conservation principle of mechanical energy,

$$mg(r_{min}\cos\theta + R - r_{min}) = \frac{1}{2}mu^2$$

$\therefore$ $$r_{min} = \frac{3R}{2\cos\theta + 3} = \frac{3\times 23}{2\times\frac{4}{5} + 3}$$

$$= \frac{3\times 5\times 23}{23} = 15\text{ m} = 3n$$

$\therefore$ $n = 5$

34. (1) If the car brakes with deceleration a_0, then in the frame of car, a pseudo force of magnitude ma_0 in the direction of car's motions will appear to act on the ball. If pseudo force acted continuously, then ball could certainly not reach at the highest point of path because work done by pseudo force is zero. Thus, the string will be horizontal when the car stop.

In the frame of car, let the velocity of ball is v.

$\therefore$ Work done by pseudo force + Work done by gravity = Change in kinetic energy of ball.

or $\quad ma_0 l - mg\,(2l) = \frac{1}{2}mv^2 - 0 \qquad \ldots(i)$

But at top point, $T + mg = \frac{mv^2}{l}$

$$0 + mg = \frac{mv_{min}^2}{l}$$

$\Rightarrow \quad mv_{min}^2 = mgl \qquad \ldots(ii)$

From Eqs. (i) and (ii), we get

$$ma_{0\,min}l - 2mgl = \frac{mgl}{2}$$

$\Rightarrow \quad a_{0\,min} = \frac{5}{2}g$

$\therefore$ Maximum stopping distance of car is

$$s_{max} = \frac{u^2}{2a_{0\,min}} = \frac{50}{2 \times \frac{5}{2}g} = 1\text{ m}$$

35. (i) Here, $v^2 = u^2 + 2gh$

or $\quad v^2 = 0 + 2g\,(R - R\cos 37°)$

or $\quad v^2 = 2 \times 10 \times R\left(1 - \frac{4}{5}\right) = 20R \times \frac{1}{5}$

$$= 20 \times 0.12 \times \frac{1}{5} = 0.48$$

$\therefore \quad mg\cos 37° - N = \frac{mv^2}{R}$

$\therefore \quad N = mg\cos 37° - \frac{mv^2}{R}$

$$= 1 \times 10 \times \frac{4}{5} - 1 \times \frac{0.48 \times 0.48}{0.12}$$

$$= 8 - 1.92 = 6.1\text{ N}$$

(ii) Here, $mg\cos\theta - N = \frac{mv^2}{R}$

For $\quad N = 0, \theta = \alpha \quad$ or $\quad mg\cos\alpha = \frac{mv^2}{R}$

or $\quad g\cos\alpha = \frac{2g\,(R - R\cos\alpha)}{R}$

$\therefore \quad \cos\alpha = 2/3$

36. Let the string slacks at Q.

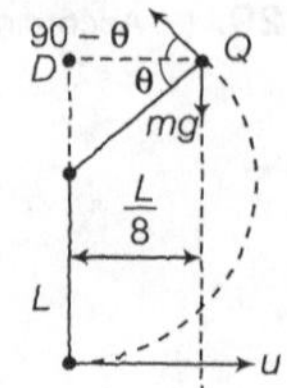

$$T_Q = 0,\ mg\sin\theta = \frac{mv^2}{L}$$

$$v^2 = u^2 - 2gh = u^2 - 2gL(1 + \sin\theta)$$

$$QD = \frac{1}{2}\text{ range} = L\cos\theta - \frac{L}{8}$$

$$= \frac{v^2}{gL}\sin\theta\cos\theta$$

As, $\quad u^2 = v^2 + 2gh(1 + \cos\theta)$

$$= \frac{\sqrt{3}}{2}gL + 2gL\left(1 + \frac{\sqrt{3}}{2}\right)$$

$\Rightarrow \quad u^2 = gL\left(2 + \frac{3\sqrt{3}}{2}\right)$

$\Rightarrow \quad u = \sqrt{gL\left(2 + \frac{3\sqrt{3}}{2}\right)}$

37. (c) From the figure shown below we have,

$$h = \left(R + \frac{d}{2}\right)(1 - \cos\theta)$$

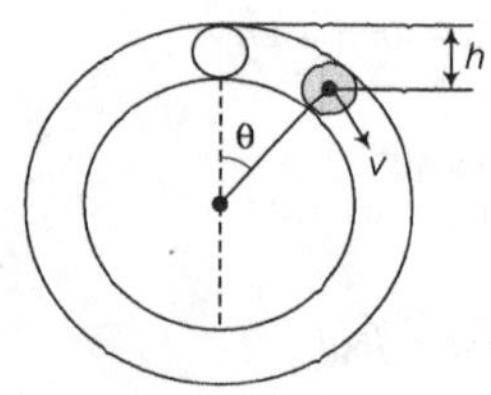

Velocity of ball at an angle θ, $v^2 = 2gh = 2\left(R + \frac{d}{2}\right)(1 - \cos\theta)g$

Let N = normal reaction at an angle θ

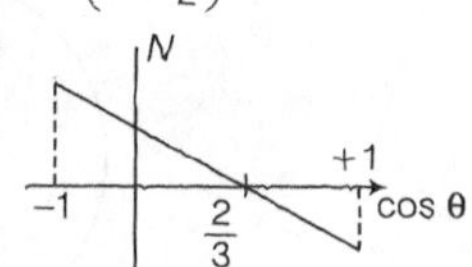

Then, $mg\cos\theta - N = \frac{mv^2}{\left(R + \frac{d}{2}\right)}$

Substituting for v^2, we have

$$N = mg(3\cos\theta - 2)$$

Revisal Problems (JEE Main)

1. (d) If only conservative forces operate on a system. Then,

$$U_f - U_i = -W$$

and $W = T_f - T_i$ = work-energy theorem

$$U_f - U_i = -(T_f - T_i)$$

$\Rightarrow \quad U_f + T_f = U_i + T_i$

Total mechanical energy is constant.

2. (b) According to mechanical energy conservation principle

$$U_i + T_i = U_f + T_f$$

or $\quad \frac{1}{2}kx_0^2 + 0 = mgh + 0$

or $\quad \frac{1}{2} \times 4000 \times x_0^2 = 20 \times 4 = 80$

or $\quad x_0^2 = \frac{160}{4000} = 4 \times 10^{-2}$

$\therefore \quad x_0 = 0.2\text{ m} = 20\text{ cm}$

3. (b) The work done by the force is

$$W = \int_0^x F dx = \int_0^x (F_0 - bx)\,dx = F_0x - \frac{bx^2}{2}$$

According to work-energy theorem,

$$W = \Delta T \quad \text{or} \quad W = 0$$

or $\quad F_0x - \frac{bx^2}{2} = 0 \Rightarrow x = \frac{2F_0}{b}$

4. (c) Here, $\mathbf{F} = 10\sqrt{2}\,(\cos 45°\,\hat{\mathbf{i}} + \sin 45°\,\hat{\mathbf{j}}) = 10\hat{\mathbf{i}} + 10\hat{\mathbf{j}}$

and $\quad \mathbf{s} = \hat{\mathbf{i}} + \hat{\mathbf{j}}$

$\therefore \quad W = \mathbf{F} \cdot \mathbf{s} = 10 + 10 = 20\text{ J}$

5. (c) Applying work-energy theorem, in the frame of car,

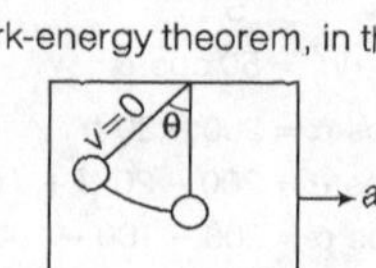

$$W_g + W_T + W_{pseudo} = \Delta T = T_f - T_i = 0$$

or $\quad -mg\,(l - l\cos\theta) + 0 + ma_0 l\sin\theta = 0$

or $\quad -mgl + mgl\cos\theta + \frac{mg}{\sqrt{3}}l\sin\theta = 0$

or $\quad \frac{\sin\theta}{\sqrt{3}} = 1 - \cos\theta = 2\sin^2\frac{\theta}{2}$

or $\quad \frac{2}{\sqrt{3}}\sin\frac{\theta}{2}\cos\frac{\theta}{2} = 2\sin^2\frac{\theta}{2}$

or $\quad \tan\frac{\theta}{2} = \frac{1}{\sqrt{3}} = \tan 30°$

$\therefore \quad \theta = 60°$

6. (b) According to work-energy theorem,

$$W = \Delta T = T_f - T_i = 0$$

or $\quad W_f + W_s + W_g = 0$

or $\quad W_f - \frac{1}{2}kx_0^2 + mg \times 4 = 0$

or $\quad W_f = \frac{1}{2} \times 100 \times (2)^2 - 10 \times 10 \times 4$

$= 200 - 400$

$= -200$ J

7. (a) $\because y = 3 + x$

$\therefore \tan\theta = 1 \Rightarrow \theta = 45°$

$\therefore \quad \mathbf{F} = 100 \cos 45° \,\hat{\mathbf{i}} + 100 \sin 45° \,\hat{\mathbf{j}}$

$= \frac{100}{\sqrt{2}}\hat{\mathbf{i}} + \frac{100}{\sqrt{2}}\hat{\mathbf{j}}$

The displacement of the particle is

$\mathbf{s} = \hat{\mathbf{i}} + 3\hat{\mathbf{j}} - (3\hat{\mathbf{i}} + \hat{\mathbf{j}})$

$= -2\hat{\mathbf{i}} + 2\hat{\mathbf{j}}$

$\therefore \quad W = \mathbf{F}\cdot\mathbf{s} = -\frac{200}{\sqrt{2}} + \frac{200}{\sqrt{2}} = 0$

8. (d) As, $y = 2x$

$\therefore \quad m_1 = \frac{dy}{dx} = 2$

The conservative force should be directed perpendicular to the line $y = 2x$.

$\therefore \quad m_1 m_2 = -1$

$\therefore \quad m_2 = -\frac{1}{2}$

$\therefore \quad \tan\theta = -\frac{1}{2}$

In option (d), $\tan\theta = \frac{4}{-8} = -\frac{1}{2}$

9. (b) As, $dU = -\mathbf{F}\cdot d\mathbf{r}$

$= \{-(3x^2 + yz)\hat{\mathbf{i}} - (3y^2 + xz)\hat{\mathbf{j}} - (3z^2 + xy)\hat{\mathbf{k}}\}\cdot(dx\,\hat{\mathbf{i}} + dy\,\hat{\mathbf{j}} + dz\,\hat{\mathbf{k}})$

$= -(3x^2 + yz)\,dx - (3y^2 + xz)\,dy - (3z^2 + xy)\,dz$

$\Rightarrow \int dU = \int -d(x^3 + y^3 + z^3 + xyz)$

$\therefore \quad U = -(x^3 + y^3 + z^3 + xyz) + \text{constant}$

10. (b) Let $F = \alpha v$

$\therefore \quad P = Fv = \alpha v^2$

or $\quad 3P = \alpha v'^2$

or $\quad \frac{3P}{P} = \frac{\alpha v'^2}{\alpha v'^2}$

or $\quad v' = \sqrt{3}\,v$

11. (c) $P = -\mu\, mg \times 1 \cos 45°$

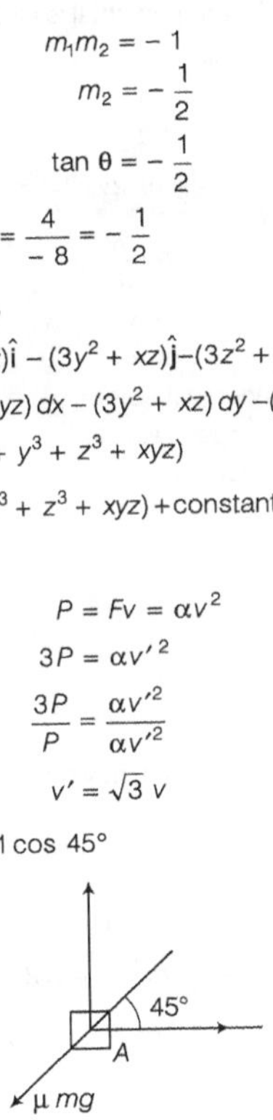

$-\mu\, mg \times 1 \cos 45° = 0.1 \times 1 \times 10 \times 2 \times \frac{1}{\sqrt{2}}$

$= \sqrt{2}$ J/s

12. (b) $P_{max} = mg\, v_{max}$

For v_{max},

$mg = kx_0$

$\therefore \quad x_0 = \frac{mg}{k}$

$\because \quad U_i + T_i = U_f + T_f$

$0 = -mgx_0 + \frac{1}{2}kx_0^2 + \frac{1}{2}mv_{max}^2$

$\therefore \quad v_{max} = \frac{mg}{k}\sqrt{\frac{k}{m}} = g\sqrt{\frac{m}{k}}$

$\therefore \quad P_{max} = mg \times g\sqrt{\frac{m}{k}} = mg^2\sqrt{\frac{m}{k}}$

$= 1 \times 100\sqrt{\frac{1}{100}}$

$= \frac{100}{10} = 10$ W

13. (a) $W_{AB} = \int_0^a F_x\,dx = \frac{ba^2}{2}$

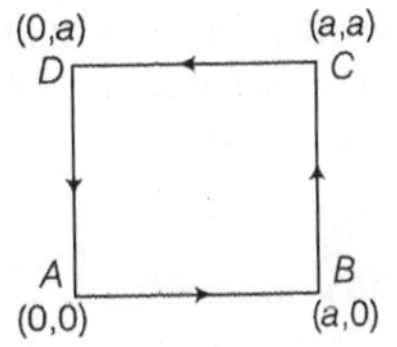

$W_{BC} = \int_0^a F_y\,dy = \frac{ba^2}{2}$

$W_{CA} = \int_a^0 F_x\,dx = -\frac{ba^2}{2}$

$W_{DA} = \int_a^0 F_y\,dy = -\frac{ba^2}{2}$

$\therefore \quad W_{AB} + W_{BC} + W_{CA} + W_{DA} = 0$

14. (d) Here, $F_x = -\frac{\partial U}{\partial x}$

The negative slope of U-x graph gives x-component of conservative force.

The slope at C is in negative at point C in the given graph. Hence, F_x is positive at C.

15. (c) Applying work-energy theorem, in the frame of cylinder,

$ma_0R - mgR = \Delta T = 0$

$ma_0R = mgR$

$\therefore \quad a_0 = g$

16. (b) At the maximum speed, the tangential acceleration of the particle with respect to the block should be zero.

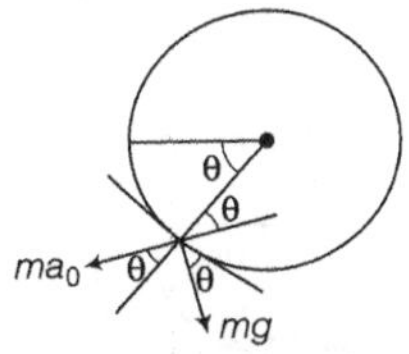

From figure, $\quad ma_0 \sin\theta = mg\cos\theta$

$\therefore \quad \tan\theta = \frac{g}{\cos\theta} = \frac{g}{g} = 1$

$\therefore \quad \theta = 45°$

17. (c) Loss in potential energy = Gain in kinetic energy

$$\therefore \quad mg\frac{l}{2} = \frac{1}{2}mv^2$$

$$\therefore \quad v = \sqrt{gl} = \sqrt{10 \times 10} = 10 \text{ m/s}$$

18. (b) $F_x = \frac{U_A - U_D}{0.2} = \frac{2-5}{0.2} = -15 \text{ N}$

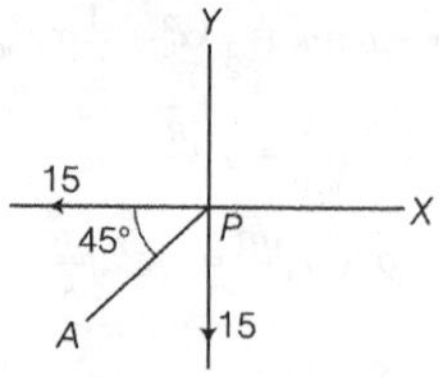

$$F_y = \frac{U_A - U_B}{AB} = \frac{2-5}{0.2} = -15 \text{ N}$$

$$\therefore \quad \mathbf{F} = -15\hat{\mathbf{i}} - 15\hat{\mathbf{j}}$$

$$\Rightarrow \quad F = \sqrt{15^2 + 15^2} = 15\sqrt{2} \text{ N along } PA.$$

19. (b) When B falls down 10 m, A will be lifted upward by 5 m.
The speed of A is half of that of B.
Hence, loss in PE = Gain in KE

$$\Rightarrow \quad m_B gh - m_A g\frac{h}{2} = \frac{1}{2}m_B v_B^2 + \frac{1}{2}m_A v_A^2$$

$$\Rightarrow \quad 10 \times 10 \times 10 - 10 \times 10 \times 5 = \frac{1}{2} \times 10 \times v_B^2 + \frac{1}{2} \times 10\left(\frac{v_B}{2}\right)^2$$

$$\Rightarrow \quad 500 = 5v_B^2 + \frac{10}{8}v_B^2$$

$$\Rightarrow \quad 500 = \left(\frac{50}{8}\right)v_B^2$$

$$\therefore \quad v_B^2 = \frac{500 \times 8}{50} = 80$$

$$\Rightarrow \quad v_B = \sqrt{80} \text{ m/s}$$

20. (d) The loss of potential energy by m_1 is

$$m_1 gH = 100 \times 0.04 = 4 \text{ J}$$

when m_1 has been pulled down by distance H each portion of the string passing round pulley above m_2 is decreased in length by $\frac{H}{2}$ and hence, m_2 rises up by distance $\frac{H}{2}$.

$\therefore$ Increase in potential energy of m_2

$$= m_2 g\frac{H}{2} = 2 \text{ J}$$

The system is initially in equilibrium.

$$\therefore \quad 2m_1 g = m_2 g + kx_i$$

$$\therefore \quad x_i = \frac{2m_1 g - m_2 g}{k} = \frac{100}{k} = \frac{100}{2000} = \frac{1}{20} \text{ m}$$

$$\text{and} \quad x_f = x_i + \frac{H}{2} = \frac{100}{k} + 0.02 = \frac{1}{20} + \frac{1}{200}$$

Thus, increase in spring energy is

$$\frac{1}{2}kx_f^2 - \frac{1}{2}kx_i^2 = 2.4 \text{ J}$$

Thus, W_{ext} = increase in potential energy of system
$= 2 + 2.4 - 4 = 0.4$ J

21. (d) In the perpendicular direction of conservative force potential energy remains constant. In Statement I conservative force may be perpendicular to line along which potential energy is constant so, Statement I is incorrect.
Hence, option (d) is correct.

22. (a) Statements I and II both are correct, also Statement II is correct explanation of Statement I.
As in projectile motion, only force present in gravitation, which is a conservative force and mechanical energy is conserved.

23. (b) Statement I is correct as, in a non-inertial frame, work done by all forces (including the pseudo forces) = Change in KE in non-inertial frame. But Statement II is also correct.

24. (a) Rate of change of KE

$$= \frac{d}{dt}\frac{1}{2}mv^2$$

$$= \frac{1}{2}m \cdot 2v \cdot \frac{dv}{dt}$$

$$= mv \cdot \frac{dv}{dt}$$

$$= ma \cdot v = F \cdot v$$

$$= \text{Power}$$

According to work-energy theorem,
Net work done = Change in kinetic energy
So, option (a) is correct.

25. (b)

Revisal Problems (JEE Advanced)

1. (c) In path (I) the total work done is 120 J. But in path (II) the work done is

$$(60 - 60) = 0.$$

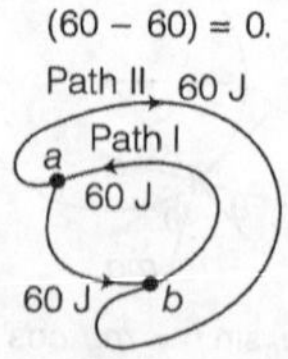

$\therefore$ As the work done in a closed loop in different paths is different.
$\therefore$ The force must be non-conservative.

2. (b, c) Work done on the book = mgh
It is independent of time taken and also independent of path of the book.

3. (c) Actual efficiency $= \frac{12}{96} \times 100 = \frac{100}{8} = \frac{25}{2} = 12.5\%$

4. (a) The mass of viscous liquid moving upward with velocity 0.01 m/s is
m = density of viscous liquid × volume of the ball
$= 2 \times 1 = 2$ g
$\therefore$ Kinetic energy of viscous liquid

$$= \frac{1}{2}mv^2 = \frac{1}{2} \times 2 \times (1 \text{ cm/s})^2$$

$$= 1 \text{ erg}$$

5. (d) $W = -\Delta U = U_i - U_f = 0 - U_f = -U_f$

$$\text{Here,} \quad \frac{4}{3}\pi R^3 = \pi R^2 h$$

$$\therefore \quad h = \frac{4R}{3}$$

$$\therefore \quad U_f = -mg(H - h) = -mg\left(H - \frac{4}{3}R\right)$$

$$\therefore \quad W = -U_f = mg\left(H - \frac{4}{3}R\right)$$
$$= 3 \times 10\left(4R - \frac{4}{3}R\right) = 3 \times 10\left(4 - \frac{4}{3}\right)$$
$$= 3 \times 10 \times \frac{8}{3} = 80 \text{ J}$$

6. (a,b,c) At the instant when upper ball leaves the vertical wall, tension in the rod is zero.

(a) $\therefore\ mg = ma$

$\therefore \quad a = g$

(b) and (c) $N = mg$ and $a = 0$ for lower ball.

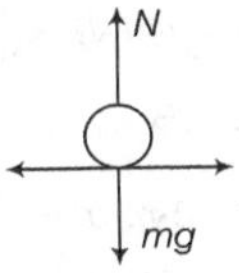

7. (a,b,c,d) $\mathbf{F} = \frac{-\partial U}{\partial x}\hat{\mathbf{i}} - \frac{\partial U}{\partial y}\hat{\mathbf{j}} = 3\hat{\mathbf{i}} - 4\hat{\mathbf{j}}$

$$\therefore \quad \mathbf{a} = \frac{\mathbf{F}}{m} = 3\hat{\mathbf{i}} - 4\hat{\mathbf{j}}, |\mathbf{a}| = \sqrt{3^2 + 4^2} = 5 \text{ m/s}^2$$
$$\therefore \quad \mathbf{v} = \mathbf{u} + \mathbf{a}t$$
$$= -3\hat{\mathbf{i}} + 4\hat{\mathbf{j}} + (3\hat{\mathbf{i}} - 4\hat{\mathbf{j}}) \times 1 = 0$$

Displacement, $\mathbf{s} = \mathbf{r}_f - \mathbf{r}_i = \mathbf{v}_0 t + \frac{1}{2}\mathbf{a}t^2$

$$= -3\hat{\mathbf{i}} + 4\hat{\mathbf{j}} + \frac{1}{2}(3\hat{\mathbf{i}} - 4\hat{\mathbf{j}})$$
$$= -\frac{3}{2}\hat{\mathbf{i}} + 2\hat{\mathbf{j}}$$
$$\mathbf{r}_f = -\frac{3}{2}\hat{\mathbf{i}} + 2\hat{\mathbf{j}} - \frac{3}{2}\hat{\mathbf{i}} + 2\hat{\mathbf{j}}$$
$$= -3\hat{\mathbf{i}} + 4\hat{\mathbf{j}}$$

8. (a,b,c,d) (if a then c, if b then d) If block is moving upward,

$$(2F - mg) = m\frac{dv}{dt}$$

or $\quad 2Ft - mgt = mv$

or $\quad 2F \times 2 - 2 \times 10 \times 2 = 2v$

or $\quad 2F - 20 = v$

Hence, $\quad \frac{1}{2}mv^2 = 16$

$\therefore \quad v = 4 \text{ m/s}$

$\therefore \quad F = 12 \text{ N}$

If block is moving downward,

$$mgt - 2Ft = mv$$

or $\quad 2 \times 10 \times 2 - 2 \times 2F = 8$

$\therefore \quad F = 8 \text{ N}$

9. (a,b,d) Work done by spring force on the block is

$$W = \frac{1}{2}kx_i^2 - \frac{1}{2}kx_f^2$$

10. (b,d) If particle is just able to complete the vertical circle, then

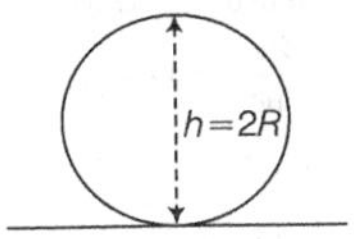

at the highest point, $N + mg = \frac{mv^2}{R}$

or $\quad 0 + mg = \frac{mv^2}{R}$

$\therefore \quad v = \sqrt{gR}$

$\because \quad v^2 = v_0^2 - 2gh$

or $\quad gR = v_0^2 - 2gR$

$\therefore \quad v_0 = \sqrt{3gR}$

At $v_0 = \sqrt{4gR}$, block is just able to complete the circle and normal reaction and velocity at the highest point is just zero.

11. (a,b,c) At the time of maximum speed, the acceleration of block A is zero.

$$m_A g \sin 37° = T = kx_0$$
$$\Rightarrow \quad 5 \times 10 \times \frac{3}{5} \times \frac{1}{100} = x_0$$
$$x_0 = 0.3 \text{ m}$$

Loss in gravitational potential energy

= Gain in spring energy + Gain in Kinetic energy

$$\Rightarrow \quad (mg \sin 37°)\, x_0 = \frac{1}{2}kx_0^2 + \frac{1}{2}mv^2 + \frac{1}{2}mv^2$$
$$\Rightarrow \quad 30 \times 0.3 = \frac{1}{2} \times 100 \times (0.3)^2 + 5v^2$$
$$\Rightarrow \quad 9 - 4.5 = 5v^2 \quad \text{or} \quad \frac{9}{2} = 5v^2$$
$$\therefore \quad v^2 = \frac{9}{10}$$
$$\therefore \quad v = \frac{3}{\sqrt{10}} \text{ m/s}$$

At the time of maximum elongation, velocity is zero.

$$\therefore \quad (mg \sin 37°)\, x_{max} = \frac{1}{2}kx_{max}^2$$
$$\therefore \quad x_{max} = \frac{30 \times 2}{100} = 0.6 \text{ m}$$

12. (b) $W_f = \Delta T = 0 - \frac{1}{2}mv_0^2 = -\frac{1}{2}mv_0^2 = -250 \text{ J}$

13. (b) From this frame, the block starts from rest and attains a backward velocity $(= -v_0)$. Plank continuous with $-v_0$.

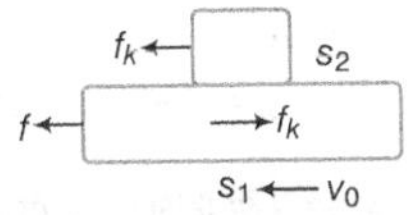

Let f_k be frictional force on the block. Then, attained velocity

$$v_0 = \left(\frac{f_k}{m}\right)t \quad \ldots\text{(i)}$$

Suffered displacement, $s_1 = \frac{1}{2}\left(\frac{f_k}{m}\right)t^2 \quad \ldots\text{(ii)}$

The work done on the block is given by

$W_1 = f_k s_1$ (as f_k and displacement s_1 have the same direction)

$$W_1 = \frac{1}{2}\frac{f_k^2}{m}t^2 = \frac{1}{2}\frac{f_k^2}{m}\cdot\left(\frac{mv_0}{f_k}\right) = \frac{1}{2}mv_0^2$$

In the same time the base surface moves by $S_2 = v_0 t$. Work done by friction on the base is

$W_2 = -f_k(v_0 t)$ ($v_0 t$ and force f_k are oppositely directed)

or $\quad W_2 = -f_k \cdot v_0 \cdot \left(\frac{mv_0}{f_k}\right)$ [using Eq. (i)]

$$= -mv_0^2$$

Total work done by frictional force on contact surfaces

$$W_f = W_1 + W_2$$
$$= \frac{1}{2}mv_0^2 + (-mv_0^2)$$
$$= -\frac{1}{2}mv_0^2$$
$$= -250 \text{ J}$$

14. (c) Heat loss does not depend upon frame of reference.

15. (a) We can see that $\angle ACB = 90°$

From constrained relations we can understand that speeds of both the particles will be

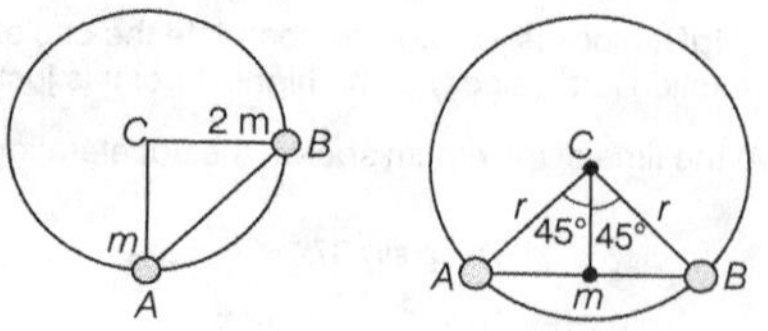

(a) (b)

From conservation of mechanical energy. (refer figures a and b).
Decrease in potential energy of mass $2m$ = Increase in potential energy of mass m + Increase in kinetic energies of both.

Hence, $2mg\,(h_B) = mg\,(h_A) + \frac{1}{2}(3m)(v^2)$

or $2g\,(h_B) = g\,(h_A) + \frac{3}{2}v^2$ ...(i)

Here, $h_B = r\cos 45° = 0.707\,r$

$h_A = r(1 - \cos 45°) = 0.293\,r$

Substituting the values, we have

$(2)(0.707)\,gr = 0.293\,gr + 1.5\,v^2$

or $v = 0.864\sqrt{gr} = 864$ cm/s

16. (c) At any angle θ

$h_B = r\sin\theta$ and $h_A = r(1 - \cos\theta)$ ...(i)

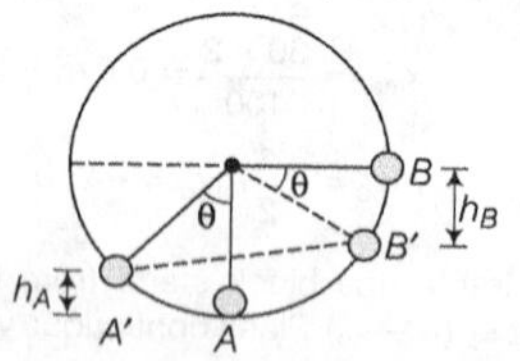

Hence, Eq. (i) can be written as

$2\,gr\sin\theta = gr(1 - \cos\theta) + \frac{3}{2}v^2$

or $v^2 = \frac{2}{3}gr(2\sin\theta + \cos\theta - 1) = k$ (say)...(ii)

For v to be maximum,

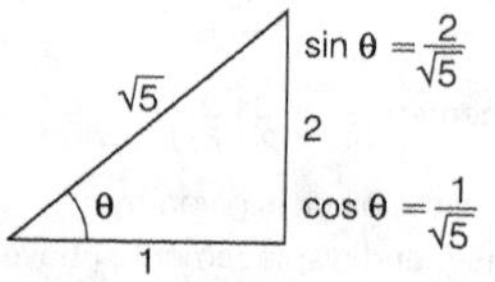

$\frac{dk}{d\theta} = 0$ or $2\cos\theta - \sin\theta = 0$

or $\tan\theta = 2$

$\therefore$ $v_{max} = \sqrt{\frac{2}{3}gr(2\sin\theta + \cos\theta - 1)}$

Substituting the values of θ, we get

$v_{max} = 0.908\sqrt{gr}$

$= 908$ cm/s

17. (c) At $\theta = \theta_{max}$, velocity of both the particles will become zero for a moment.

Hence, substituting $v^2 = 0$ in Eq. (ii), of previous problem we get

$2\sin\theta + \cos\theta - 1 = 0$

18. (2) Loss in gravitational potential energy = Gain in kinetic energy.

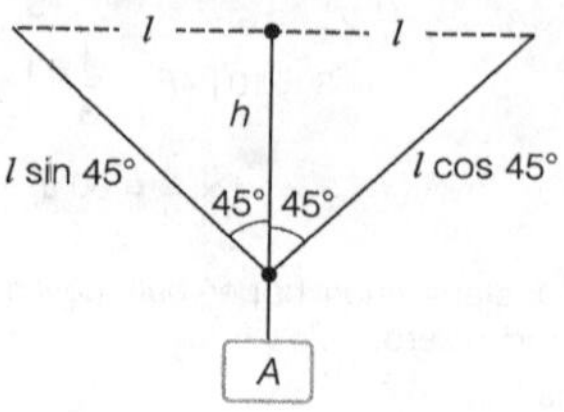

or $mgh - 2mg\,(l\,\text{cosec}\,45° - l) = \frac{1}{2}mv_A^2 + 2 \times \frac{1}{2}mv^2$

or $mgl + 2mgl - 2mgl\sqrt{2} = \frac{1}{2}mv_A^2 + m\,(v_A\cos 45°)^2$

or $3mgl - 2\sqrt{2}\,mgl = \frac{1}{2}mv_A^2 + \frac{m}{2}v_A^2$

or $3gl - 2\sqrt{2}\,gl = v_A^2$

$\therefore$ $v_A = \sqrt{(3 - 2\sqrt{2})\,gl} = 10$ m/s

$\therefore$ $x = 2$

19. (5) Applying work-energy theorem, when $v = 4$ m/s at $n = 0$,

$W_f = 0 - \frac{1}{2}mv^2$

$= -\frac{1}{2}m \times 4^2 = -8$ m

If $v = 5$ m/s, at $x = 0$

$W_f' = \frac{1}{2}m \times 4^2 - \frac{1}{2}m \times 5^2$ (in travelling at distance 2 m)

$= 8\text{ m} - \frac{25}{2}\text{ m}$

and $W_f'' = \frac{1}{2}m \times 3^2 - \frac{1}{2}m \times 4^2$ (In travelling next 3 m)

$= \frac{9}{2}\text{ m} - 8\text{ m}$

Here, $W'_f + W''_f = -8\text{ m} = W_f$

Hence, distance covered by the block before stop is $(2 + 3) = 5$ m.

20. (4) For maintaining constant velocity, tension in string should be equal to frictional force between block and plank (i.e. μmg).

For the block, $2v_0 = \mu g t_0$

$\therefore$ $t_0 = \frac{2v_0}{\mu g}$

$\therefore$ $s_0 = \left(\frac{0 + 2v_0}{2}\right)t_0 = v_0 t_0$

The distance travelled by the plank is

$s = 2v_0 t_0$

According to work-energy theorem,

$W_f + W_{\text{naughty boy}} = \frac{1}{2}m(2v_0)^2$

$\Rightarrow \mu mg\,s_0 - \mu mg\,(2v_0 t_0) - W_{\text{naughty boy}} = 2\,mv_0^2$

$\Rightarrow \mu mg v_0 t_0 - 2\,\mu mg v_0 t_0 + W_{\text{naughty boy}} = 2\,mv_0^2$

$\Rightarrow -\mu mg\,v_0 t_0 + W_{\text{naughty boy}} = 2\,mv_0^2$

$\Rightarrow -\mu mg v_0 \times \frac{2v_0}{\mu g} + W_{\text{naughty boy}} = 2\,mv_0^2$

$\therefore$ $W_{\text{naughty boy}} = 2\,mv_0^2 + 2mv_0^2 = 4 \times 5 \times (0.75)^2$

$= 11.25\text{ J} = \frac{45}{4}\text{ J}$

$\therefore$ $n = 4$

21. (3) In the frame of turn table,

$$r\omega^2 = v_{rel}\frac{dv_{rel}}{dr} \qquad \text{(for particle)}$$

$$\Rightarrow \quad v_{rel}dv_{rel} = \omega^2\int_{r_0}^{2r_0} r\,dr$$

$$\Rightarrow \quad v_{rel}^2 = \omega^2(4r_0^2 - r_0^2) = 3r_0^2\omega^2$$

In the frame of ground, initial velocity of particle is $v_i = r_0\omega$

and $\quad v_f = \sqrt{(2r_0\omega)^2 + v_{rel}^2}$

According to work-energy theorem, $W = \Delta T$

$$\therefore \quad W = \frac{1}{2}mv_f^2 - \frac{1}{2}mv_i^2 = \frac{1}{2}m(v_f^2 - v_i^2)$$

$$= \frac{1}{2}m(4r_0^2\omega^2 + v_{rel}^2 - r_0^2\omega^2)$$

$$= \frac{1}{2}m(3r_0^2\omega^2 + 3r_0^2\omega^2)$$

$$= \frac{1}{2}m \times 6r_0^2\omega^2 = 3mr_0^2\omega^2$$

$$\therefore \quad n = 3$$

22. (4) Work done against friction = Decrease in mechanical energy

$$\Rightarrow \quad \mu mgx = \frac{1}{2}mv^2 - \frac{1}{2}kx^2$$

$$\Rightarrow \quad v = \sqrt{\frac{2\mu mgx + kx^2}{m}}$$

Substituting the values, we get

$$v = 0.4 \text{ m/s} = \frac{4}{10} \text{ m/s}$$

$$\therefore \quad N = 4$$

23. (a) (i) Heat is generated due to friction.

(ii) Spring force is conservative,so mechanical energy remains conserved .

(iii) In case of pendulum, mechanical energy remains conserved.

(iv) In absence of air, no heat loss occurs.

(i) → (q,r,s), (ii) → (p,q), (iii) → (p,q), (iv) → (p,q,r)

24. (b) $W_1 = 3(4-1) = 9\text{J}$ and $W_2 = -3(1-4) = 9\text{J}$

$$\therefore \quad W = W_1 + W_2 = 18\text{ J}$$

(ii) $W_1 = 5(4-1) = 15\text{ J}$ and $W_2 = 5(1-4) = -15\text{ J}$

$$\therefore \quad W = W_1 + W_2 = 0$$

(iii) $W_1 = \int_1^4 2x\,dx = 2\left[\frac{x^2}{2}\right]_1^4 = 15\text{ J}$

and $\quad W_2 = \int_4^1 -2x\,dx = 15\text{ J}$

$$\therefore \quad W = W_1 + W_2 = 30\text{ J}$$

(iv) $W_1 = \int_4^1 3x^2dx = 63\text{ J}$ and $W_2 = \int_4^1 3x^2dx = -63\text{ J}$

$$\therefore \quad W = W_1 + W_2 = 0$$

(i) → (q), (ii) → (p,s), (iii) → (r), (iv) → (p,s)

JEE Main & AIEEE Archive

1. (c) Given, $F = ax + bx^2$

According to work-energy theorem, we know that work done in stretching the rubber band by L is

$$|dW| = Fdx$$

$$\Rightarrow \quad |W| = \int_0^L (ax + bx^2)\,dx = \left[\frac{ax^2}{2}\right]_0^L + \left[\frac{bx^3}{3}\right]_0^L$$

$$= \left[\frac{aL^2}{2} - \frac{a \times (0)^2}{2}\right] + \left[\frac{b \times L^3}{3} - \frac{b \times (0)^3}{3}\right]$$

$$\therefore \quad |W| = \frac{aL^2}{2} + \frac{bL^3}{3}$$

2. (d) Given, $k \propto t \Rightarrow \frac{dk}{dt} = \text{constant}$

$$\Rightarrow \quad k \propto t \Rightarrow \frac{1}{2}mv^2 \propto t \Rightarrow v \propto \sqrt{t}$$

Also, $P = Fv = \frac{dk}{dt} = \text{constant}$

$$\Rightarrow \quad F \propto \frac{1}{v} \Rightarrow F \propto \frac{1}{\sqrt{t}}$$

Alternate Method $k \propto t$, $v \propto \sqrt{t}$

$$F = ma = m\frac{dv}{dt}$$

As, $\quad a = \frac{dv}{dt} = k\frac{1}{2}t^{-1/2}$

$$a \propto \frac{1}{\sqrt{t}}, \text{ so } F \propto \frac{1}{\sqrt{t}}$$

3. (d) The given question is somewhat based on approximations. Let mass of athlete be 65 kg.

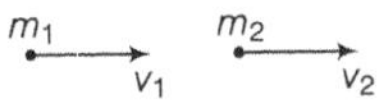

Approx velocity from the given data is 10 m/s.

So, $\quad \text{KE} = \frac{65 \times 100}{2} = 3250 \text{ J}$

So, option (d) is the most probable answer.

4. (a) According to work-energy theorem,

Loss in kinetic energy = Work done against friction + Potential energy of spring

$$\frac{1}{2}mv^2 = fx + \frac{1}{2}kx^2 \Rightarrow \frac{1}{2} \times 2(4)^4 = 15x + \frac{1}{2} \times 10000x^2$$

$$\Rightarrow \quad 5000x^2 + 15x - 16 = 0 \Rightarrow x = 0.055 \text{ m} = 5.5 \text{ cm}$$

5. (a) Given, $V(x) = \left(\frac{x^4}{4} - \frac{x^2}{2}\right)\text{J}$

For minimum value of V, $\frac{dV}{dx} = 0 \Rightarrow \frac{4x^3}{4} - \frac{2x}{2} = 0$

$$\Rightarrow \quad x = 0, \; x = \pm 1$$

So, $\quad V_{min}(x = \pm 1) = \frac{1}{4} - \frac{1}{2} = \frac{-1}{4}\text{ J}$

Now, $K_{max} + V_{min}$ = Total mechanical energy

$$\Rightarrow \quad K_{max} = \left(\frac{1}{4}\right) + 2 \quad \text{or} \quad K_{max} = \frac{9}{4}$$

or $\quad \frac{mv^2}{2} = \frac{9}{4} \quad \text{or} \quad v = \frac{3}{\sqrt{2}}\text{ ms}^{-1}$

6. (d) According to work-energy theorem,

Total work done = Change in kinetic energy

$$W = \Delta K$$

Case I $-F \times 3 = \frac{1}{2}m\left(\frac{v_0}{2}\right)^2 - \frac{1}{2}mv_0^2$

where, F is resistive force and v_0 is initial speed.

Case II Let the further distance travelled by the bullet before coming to rest be s.

$$\therefore \quad -F(3 + s) = K_f - K_i = -\frac{1}{2}mv_0^2$$

$$\Rightarrow \quad -\frac{1}{8}mv_0^2(3 + s) = -\frac{1}{2}mv_0^2$$

or $\quad \frac{1}{4}(3 + s) = 1 \quad \text{or} \quad \frac{3}{4} + \frac{s}{4} = 1$

or $\quad s = 1\text{ cm}$

7. (a) A body of mass m with uniform acceleration, then force

$$F = ma = \frac{mv}{T} \qquad \left(\because a = \frac{v-0}{T}\right)$$

Instantaneous power $= Fv = mav$

$$= \frac{mv}{T} \cdot at = \frac{mv}{T} \cdot \frac{v}{T} \cdot t = \frac{mv^2}{T^2} t$$

8. (a) According to conservation of energy, potential energy at height H is sum of kinetic energy and potential energy at h_2.

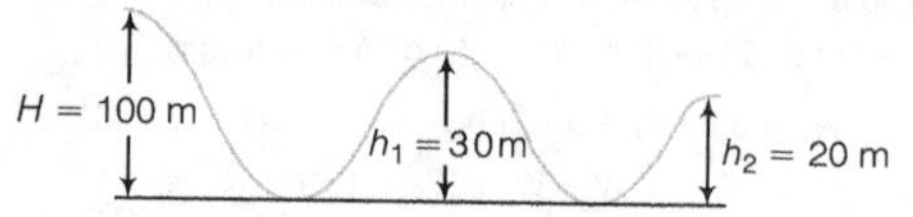

$$\therefore \quad mgH = \frac{1}{2}mv^2 + mgh_2$$

$$\Rightarrow \quad mg(H - h_2) = \frac{1}{2}mv^2$$

or $\quad v = \sqrt{2g(100-20)} = \sqrt{2 \times 10 \times 80} = 40 \text{ m/s}$

9. (a) Momentum would be maximum when KE would be maximum and this is the case when total elastic PE is converted into KE.

According to conservation of energy,

$$\frac{1}{2}kL^2 = \frac{1}{2}Mv^2$$

$$\Rightarrow \quad kL^2 = \frac{(Mv)^2}{M}$$

or $\quad MkL^2 = p^2 \qquad (\because p = Mv)$

$$\Rightarrow \quad p = L\sqrt{Mk}$$

10. (a) From given information $a = -kx$, where a is acceleration, x is displacement and k is a proportionality constant.

$$\frac{vdv}{dx} = -kx$$

$$\Rightarrow \quad v\,dv = -kx\,dx$$

Let for any displacement from 0 to x, the velocity changes from v_0 to v.

$$\Rightarrow \quad \int_{v_0}^{v} vdv = -\int_0^x kx\,dx \Rightarrow \left[\frac{v^2}{2}\right]_{v_0}^{v} = -k\left[\frac{x^2}{2}\right]_0^x$$

$$\Rightarrow \quad \frac{v^2 - v_0^2}{2} = -\frac{kx^2}{2}$$

$$\Rightarrow \quad m\left(\frac{v^2 - v_0^2}{2}\right) = -\frac{mkx^2}{2}$$

$$\frac{1}{2}mv^2 - \frac{1}{2}mv_0^2 = -\frac{mkx^2}{2}$$

$$\Rightarrow \quad \Delta K \propto x^2 \qquad (\Delta K \text{ is loss in KE})$$

11. (b) Mass per unit length $= \frac{M}{L} = \frac{4}{2} = 2$ kg/m

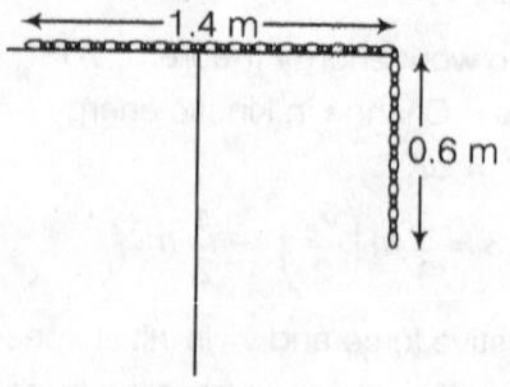

The mass of 0.6 m of chain $= 0.6 \times 2 = 1.2$ kg

The height of centre of mass of hanging part

$$h = \frac{0.6 + 0}{2} = 0.3 \text{ m}$$

Hence, work done in pulling the chain on the table

= Work done against gravity force

i.e. $\quad W = mgh = 1.2 \times 10 \times 0.3 = 3.6$ J

12. (b) Work done in displacing the particle

$$W = \mathbf{F} \cdot \mathbf{r} = (5\hat{i} + 3\hat{j} + 2\hat{k}) \cdot (2\hat{i} - \hat{j})$$

$$= 5 \times 2 + 3 \times (-1) + 2 \times 0 = 10 - 3 = 7\text{J}$$

13. (b) Work done to stretch the spring by 5 cm from mean position,

$$W_1 = \frac{1}{2}kx_1^2 = \frac{1}{2} \times 5 \times 10^3 \times (5 \times 10^{-2})^2 = 6.25 \text{ J}$$

Work done to stretch the spring by 10 cm from mean position,

$$W_2 = \frac{1}{2}k(x_1 + x_2)^2$$

$$= \frac{1}{2} \times 5 \times 10^3 (5 \times 10^{-2} + 5 \times 10^{-2})^2 = 25 \text{ J}$$

Net work done to stretch the spring from 5 cm to 10 cm

$$= W_2 - W_1$$

$$= 25 - 6.25 = 18.75 \text{ J} = 18.75 \text{ N-m}$$

14. (b) Delivering power of a machine P = constant

$$P = F \cdot v \qquad \left(\because \text{Power} = \frac{W}{t} = F \cdot \frac{s}{t} = F \cdot v\right)$$

$$= mav = m\frac{dv}{dt}v$$

$$P = mv\frac{dv}{dt}$$

$$vdv = \frac{P}{m}dt$$

Integrating on both sides, we get

$$\int_0^v v\,dv = \int_0^t \frac{P}{m}dt$$

$$\Rightarrow \quad \frac{v^2}{2} = \frac{Pt}{m}$$

$$v = \left(\frac{2Pt}{m}\right)^{1/2}$$

$$\Rightarrow \quad v = \frac{ds}{dt} = \left(\frac{2Pt}{m}\right)^{1/2}$$

or $\quad ds = \left(\frac{2Pt}{m}\right)^{1/2} dt$

$$\Rightarrow \quad \int_0^s ds = \sqrt{\frac{2P}{m}} \int_0^t t^{1/2}\,dt$$

or $\quad s = \sqrt{\frac{2P}{m}} \frac{t^{3/2}}{3/2}$

$$\Rightarrow \quad s \propto t^{3/2}$$

15. (b) The work done on the spring is stored as the PE of the body and is given by

$$U = \int_{x_1}^{x_2} F_{ext}\,dx \text{ or } U = \int_{x_1}^{x_2} kx\,dx = \frac{1}{2}k(x_2^2 - x_1^2)$$

$$= \frac{800}{2}[(0.15)^2 - (0.05)^2]$$

$$= 400(0.2 \times 0.1) = 8 \text{ J}$$

JEE Advanced & IIT JEE Archive

1. (b)

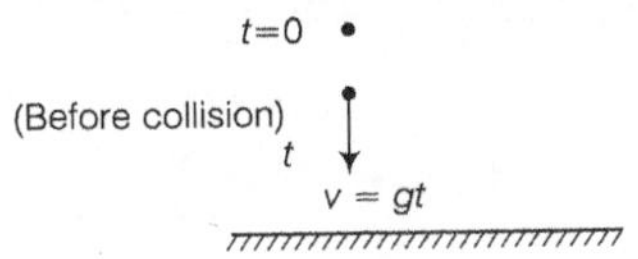

$$K = \frac{1}{2} mg^2t^2$$

$K \propto t^2$. Therefore, K-t graph is parabola.

During collision retarding force is just like the spring force ($F \propto x$), therefore kinetic energy first decreases to elastic potential energy and then increases.

Most appropriate graph is therefore (b).

2. (5) From work-energy theorem,

Work done by all forces = Change in kinetic energy

or $\quad W_F + W_{mg} = K_f - K_i$

or $\quad 18 \times 5 + (1 \times 10)(-4) = K_f$

or $\quad 90 - 40 = K_f \Rightarrow K_f = 50 \text{ J} = 5 \times 10 \text{ J}$

3. (b) Final momentum of object $= \dfrac{\text{Power} \times \text{time}}{\text{Speed of light}}$

$$= \frac{30 \times 10^{-3} \times 100 \times 10^{-9}}{3 \times 10^8}$$

$$= 1.0 \times 10^{-17} \text{ kg-m/s}$$

4. (d) $\mathbf{r} = OP = x\hat{\mathbf{i}} + y\hat{\mathbf{j}}$

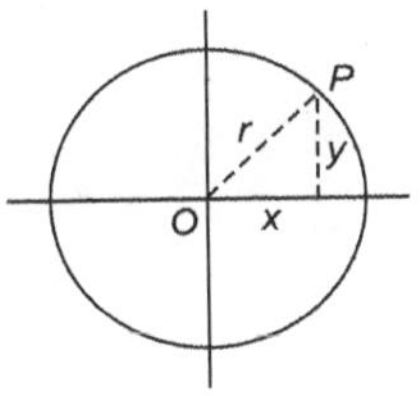

$$F = \frac{K}{(x^2 + y^2)^{3/2}}(x\hat{\mathbf{i}} + y\hat{\mathbf{j}}) = \frac{K}{r^3}(\mathbf{r})$$

Since, **F** is along **r** or in radial direction. Therefore, work done is zero.

5. (b) Height fallen upto Q is $R \sin 30° = \dfrac{R}{2} = 20$ m

$E_i - E_f$ = work done against friction.

$\therefore \quad 0 - \left[-mgh + \frac{1}{2}mv^2\right] = 150$

$\therefore \quad (1)(10)(20) - \frac{1}{2} \times 1 \times v^2 = 150 \quad$ or $\quad v = 10$ m/s

6. (a) $N = mg \cos 60° + \dfrac{mv^2}{R}$

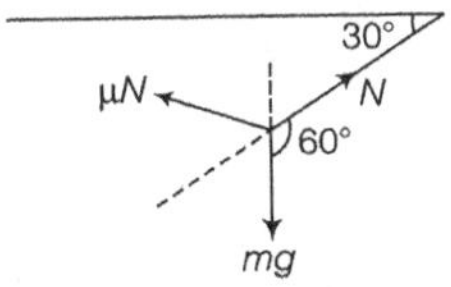

$\therefore N = \dfrac{mv^2}{R} + \dfrac{mg}{2} = \dfrac{(1)(10)^2}{40} + \dfrac{(1)(10)}{2} = 7.5$ N

7. (4) Decrease in mechanical energy = Work done against friction

$\therefore \quad \frac{1}{2}mv^2 - \frac{1}{2}kx^2 = \mu mgx$

or $\quad v = \sqrt{\dfrac{2\mu mgx + kx^2}{m}}$

Substituting the values, we get

$$v = 0.4 \text{ m/s} = \left(\frac{4}{10}\right) \text{m/s}$$

$\therefore \quad N = 4$

8. (c) Area under F-t graph = Momentum

$$= p = \sqrt{2km}$$

$\therefore \quad k = \dfrac{A^2}{2m} \quad$ (A = net area of F-t graph)

$$= \frac{\left\{\left(\frac{4 \times 3}{2}\right) - \left(\frac{1.5 \times 2}{2}\right)\right\}^2}{2 \times 2} = 5.0625 \text{ J}$$

9. (8) $a = \dfrac{\text{Net pulling force}}{\text{Total mass}}$

$$= \frac{0.72g - 0.36g}{0.72 + 0.36} = \frac{g}{3}$$

$$s = \frac{1}{2}at^2 = \frac{1}{2}\left(\frac{g}{3}\right)(1)^2 = \frac{g}{6}$$

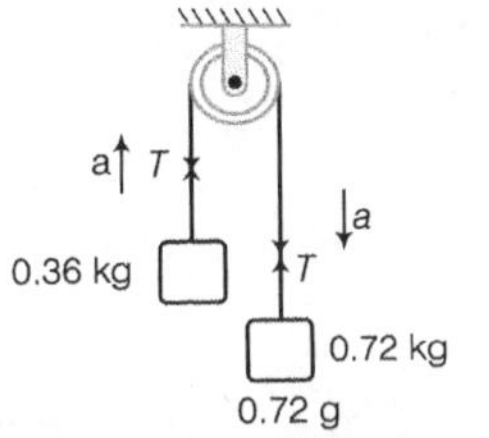

$\Rightarrow \quad T - 0.36g = 0.36a = 0.36\dfrac{g}{3}$

$\therefore \quad T = 0.48g$

Now, $\quad W_T = TS \cos 0° \quad$ (on 0.36 kg mass)

$$= (0.48g)\left(\frac{g}{6}\right)(1)$$

$$= 0.08(g^2)$$

$$= 0.08(10)^2 = 8 \text{ J}$$

10. (c) From energy conservation,

$$\frac{1}{2}kx^2 = \frac{1}{2}(4k)y^2$$

$\therefore \quad \dfrac{y}{x} = \dfrac{1}{2}$

11. (d) As, $\quad v = \sqrt{5gL} \quad$ ... (i)

$$\left(\frac{v}{2}\right)^2 = v^2 - 2gh \quad \text{... (ii)}$$

and $\quad h = L(1 - \cos\theta) \quad$ ... (iii)

Solving Eqs. (i), (ii) and (iii), we get

$$\cos\theta = -\frac{7}{8}$$

or $\quad \theta = \cos^{-1}\left(-\dfrac{7}{8}\right) = 151°$

12. (c) **Statement I** Decrease in mechanical energy in case I will be

$$\Delta U_1 = \frac{1}{2}mv^2$$

But decrease in mechanical energy in case II will be

$$\Delta U_2 = \frac{1}{2}mv^2 - mgh$$

$\therefore \quad \Delta U_2 < \Delta U_1$

or Statement I is correct.

Statement II Coefficient of friction will not change or this Statement II is wrong.

$\therefore$ Option (c) is correct.

Last 3 Years' Questions
JEE Main & Advanced
(2016-2018)

JEE Main

2016

1. A point particle of mass m, moves along the uniformly rough track PQR as shown in the figure. The coefficient of friction, between the particle and the rough track equals μ. The particle is released from rest, from the point P and it comes to rest at a point R. The energies lost by the ball, over the parts PQ and QR of the track, are equal to each other, and no energy is lost when particle changes direction from PQ to QR. The values of the coefficient of friction μ and the distance $x (= QR)$, are respectively close to **[Work, Power and Energy]**

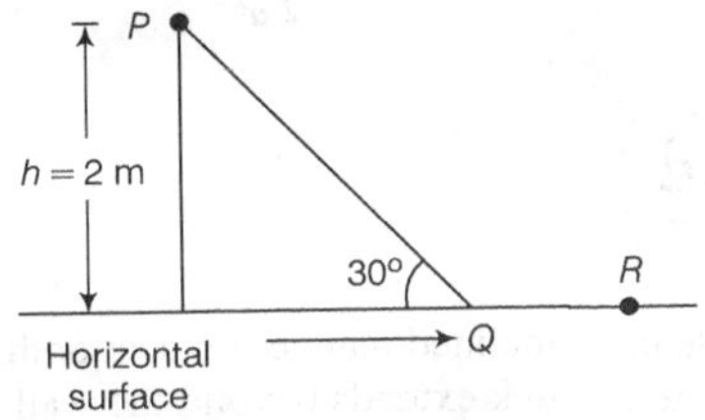

(a) 0.2 and 6.5 m
(b) 0.2 and 3.5 m
(c) 0.29 and 3.5 m
(d) 0.29 and 6.5 m

2. A person trying to lose weight by burning fat lifts a mass of 10 kg upto a height of 1 m 1000 times. Assume that the potential energy lost each time he lowers the mass is dissipated. How much fat will he use up considering the work done only when the weight is lifted up? Fat supplies 3.8×10^7 J of energy per kg which is converted to mechanical energy with a 20% efficiency rate. (Take, $g = 9.8 \text{ ms}^{-2}$) **[Work, Power and Energy]**

(a) 2.45×10^{-3} kg
(b) 6.45×10^{-3} kg
(c) 9.89×10^{-3} kg
(d) 12.89×10^{-3} kg

2017

3. A body of mass $m = 10^{-2}$ kg is moving in a medium and experiences a frictional force $F = -kv^2$. Its initial speed is $v_0 = 10 \text{ ms}^{-1}$. If, after 10 s, its energy is $\frac{1}{8} mv_0^2$, the value of k will be **[Work, Power and Energy]**

(a) 10^{-3}kgs^{-1}
(b) 10^{-4}kgm^{-1}
(c) $10^{-1} \text{kgm}^{-1}\text{s}^{-1}$
(d) 10^{-3}kgm^{-1}

4. A time dependent force $F = 6t$ acts on a particle of mass 1 kg. If the particle starts from rest, the work done by the force during the first 1 s will be **[Work, Power and Energy]**

(a) 22 J
(b) 9 J
(c) 18 J
(d) 4.5 J

2018

5. Two masses $m_1 = 5\,\text{kg}$ and $m_2 = 10\,\text{kg}$ connected by an inextensible string over a frictionless pulley, are moving as shown in the figure. The coefficient of friction of horizontal surface is 0.15. The minimum weight m that should be put on top of m_2 to stop the motion is **[Friction]**

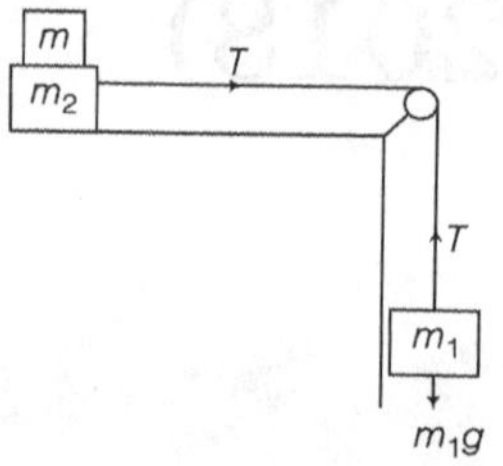

(a) 18.3 kg (b) 27.3 kg
(c) 43.3 kg (d) 10.3 kg

6. A particle is moving in a circular path of radius a under the action of an attractive potential $U = -\frac{k}{2r^2}$. Its total energy is **[Work, Power and Energy]**

(a) $-\frac{k}{4a^2}$ (b) $\frac{k}{2a^2}$ (c) zero (d) $-\frac{3}{2}\frac{k}{a^2}$

JEE Advanced

2016

1. A uniform wooden stick of mass 1.6 kg of length l rests in an inclined manner on a smooth, vertical wall of height $h(<l)$ such that a small portion of the stick extends beyond the wall. The reaction force of the wall on the stick is perpendicular to the stick. The stick makes an angle of 30° with the wall and the bottom of the stick is on a rough floor. The reaction of the wall on the stick is equal in magnitude to the reaction of the floor on the stick. The ratio $\frac{h}{l}$ and the frictional force f at the bottom of the stick are $(g = 10\,\text{ms}^{-2})$ **[Friction, Type 1]**

(a) $\frac{h}{l} = \frac{\sqrt{3}}{16}$, $f = \frac{16\sqrt{3}}{3}$N (b) $\frac{h}{l} = \frac{3}{16}$, $f = \frac{16\sqrt{3}}{3}$N
(c) $\frac{h}{l} = \frac{3\sqrt{3}}{16}$, $f = \frac{8\sqrt{3}}{3}$N (d) $\frac{h}{l} = \frac{3\sqrt{3}}{16}$, $f = \frac{16\sqrt{3}}{3}$N

2018

2. A particle of mass m is initially at rest at the origin. It is subjected to a force and starts moving along the X-axis. Its kinetic energy K changes with time as $dK/dt = \gamma t$, where γ is a positive constant of appropriate dimensions. Which of the following statements is (are) true? **[Work, Power and Energy Type 2]**

(a) The force applied on the particle is constant
(b) The speed of the particle is proportional to time
(c) The distance of the particle from the origin increases linearly with time
(d) The force is conservative

3. A solid horizontal surface is covered with a thin layer of oil. A rectangular block of mass $m = 0.4$ kg is at rest on this surface. An impulse of 1.0 N s is applied to the block at time $t = 0$, so that it starts moving along the X-axis with a velocity $v(t) = v_0 e^{-t/\tau}$, where v_0 is a constant and $\tau = 4$ s. The displacement of the block, in metres, at $t = \tau$ is .... (Take, $e^{-1} = 0.37$). **[Laws of Motion, Type 4]**

Note *Type-1 Only one correct option, Type-2 More then one correct option, Type-3 Single digit Answer, Type-4 Numerical value based*

Answer *with* Explanations

JEE Main

1. (c) As energy loss is same, thus

$$\mu mg\cos\theta \cdot (PQ) = \mu mg \cdot (QR)$$

$$\therefore \quad QR = (PQ)\cos\theta$$

$$\Rightarrow \quad QR = 4 \times \frac{\sqrt{3}}{2} = 2\sqrt{3} \approx 3.5 \text{ m}$$

Further, decrease in potential energy = loss due to friction

$$\therefore \quad mgh = (\mu mg\cos\theta)d_1 + (\mu mg)d_2$$

$$m \times 10 \times 2 = \mu \times m \times 10 \times \frac{\sqrt{3}}{2} \times 4 + \mu \times m \times 10 \times 2\sqrt{3}$$

$$\Rightarrow \quad 4\sqrt{3}\,\mu = 2$$

$$\Rightarrow \quad \mu = \frac{1}{2\sqrt{3}} = 0.288 = 0.29$$

2. (d) Work done in lifting mass

$$= (10 \times 9.8 \times 1) \times 1000$$

If m is mass of fat burnt, then energy

$$= m \times 3.8 \times 10^7 \times \frac{20}{100}$$

Equating the two, we get

$$\therefore \quad m = \frac{49}{3.8} \approx 12.89 \times 10^{-3}\text{kg}$$

3. (b) Given, force,

$$F = -kv^2$$

$$\therefore \text{ Acceleration, } a = \frac{-k}{m}v^2$$

or

$$\frac{dv}{dt} = \frac{-k}{m}v^2$$

$$\Rightarrow \quad \frac{dv}{v^2} = -\frac{k}{m}\,.\,dt$$

Now, with limits, we have

$$\int_{10}^{v} \frac{dv}{v^2} = -\frac{k}{m}\int_0^t dt$$

$$\Rightarrow \quad \left(-\frac{1}{v}\right)_{10}^{v} = -\frac{k}{m}t$$

$$\Rightarrow \quad \frac{1}{v} = 0.1 + \frac{kt}{m}$$

$$\Rightarrow \quad v = \frac{1}{0.1 + \frac{kt}{m}} = \frac{1}{0.1 + 1000k}$$

$$\Rightarrow \quad \frac{1}{2} \times m \times v^2 = \frac{1}{8} m v_0^2$$

$$\Rightarrow \quad v = \frac{v_0}{2} = 5$$

$$\Rightarrow \quad \frac{1}{0.1 + 1000\,k} = 5$$

$$\Rightarrow \quad 1 = 0.5 + 5000k$$

$$\Rightarrow \quad k = \frac{0.5}{5000} \Rightarrow k = 10^{-4}\text{kg/m}$$

4. (d) From Newton's second law, $\frac{\Delta p}{\Delta t} = F$

$$\Rightarrow \quad \Delta p = F\Delta t$$

$$\therefore \quad p = \int dp = \int_0^1 F\,dt$$

$$\Rightarrow \quad p = \int_0^1 6t\,dt = 3 \text{ kg}\left(\frac{\text{m}}{\text{s}}\right)$$

Also, change in kinetic energy

$$\Delta k = \frac{\Delta p^2}{2m} = \frac{3^2}{2 \times 1} = 4.5$$

From work-energy theorem, work done

= change in kinetic energy.

So, work done = Δk = 4.5 J

5. (b) Motion stops when pull due to $m_1 \leq$ force of friction between m and m_2 and surface.

$$\Rightarrow \quad m_1 g \leq \mu(m_2 + m)g$$

$$\Rightarrow \quad 5 \times 10 \leq 0.15\,(10 + m) \times 10$$

$$\Rightarrow \quad m \geq 23.33 \text{ kg}$$

Here, nearest value is 27.3 kg

So, $\quad m_{\text{min}} = 27.3$ kg

6. (c) $\therefore$ Force $= -\frac{dU}{dr}$

$$\Rightarrow \quad F = -\frac{d}{dr}\left(\frac{-k}{2r^2}\right) = -\frac{k}{r^3}$$

As particle is on circular path, this force must be centripetal force.

$$\Rightarrow \quad |F| = \frac{mv^2}{r}$$

So, $\quad \frac{k}{r^3} = \frac{mv^2}{r}$

$$\Rightarrow \quad \frac{1}{2}mv^2 = \frac{k}{2r^2}$$

$\therefore$ Total energy of particle = KE + PE $= \frac{k}{2r^2} - \frac{k}{2r^2} = 0$

Total energy = 0

JEE Advanced

1. (d)

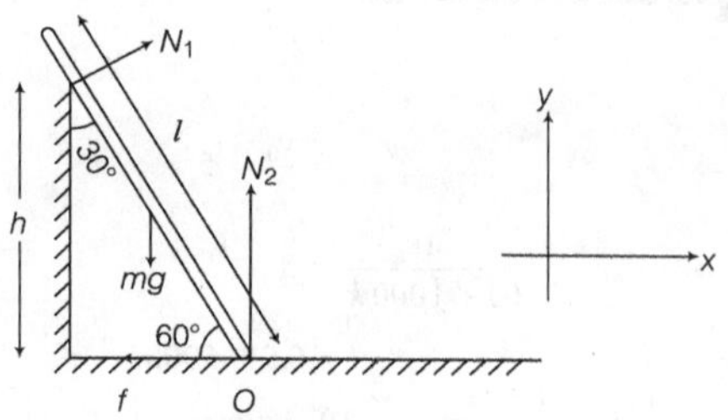

$$\Sigma F_x = 0, \quad N_1 \cos 30° - f = 0 \quad \text{...(i)}$$

$$\Sigma F_y = 0, \quad N_1 \sin 30° + N_2 - mg = 0 \quad \text{...(ii)}$$

$$\Sigma \tau_0 = 0$$

$$mg \frac{l}{2} \cos 60° - N_1 \frac{h}{\cos 30°} = 0 \quad \text{...(iii)}$$

Also, given $N_1 = N_2$ ...(iv)

Solving Eqs. (i), (ii), (iii) and (iv) we have

$$\frac{h}{l} = \frac{3\sqrt{3}}{16} \text{ and } f = \frac{16\sqrt{3}}{3}$$

2. (a, b) $K = \frac{1}{2} mv^2 \Rightarrow \frac{dK}{dt} = mv \frac{dv}{dt}$

Given $\frac{dK}{dt} = \gamma t \quad \Rightarrow \quad mv \frac{dv}{dt} = \gamma t$

$$\Rightarrow \int_0^v v\,dv = \int_0^t \frac{\gamma}{m} t\,dt \quad \Rightarrow \quad \frac{v^2}{2} = \frac{\gamma}{m} \frac{t^2}{2}$$

$$\Rightarrow \quad v = \sqrt{\frac{\gamma}{m}}\, t \quad \Rightarrow \quad a = \frac{dv}{dt} = \sqrt{\frac{\gamma}{m}}$$

$$\therefore \quad F = ma = \sqrt{\gamma m} = \text{constant}$$

$$\therefore \quad V = \frac{ds}{dt} = \sqrt{\frac{\gamma}{m}}\, t \quad \Rightarrow \quad s = \sqrt{\frac{\gamma}{m}} \frac{t^2}{2}$$

Note *Force is constant. In the website of IIT, option (d) is given correct. In the opinion of author all constant forces are not necessarily conservative. For example : viscous force at terminal velocity is a constant force but it is not conservative.*

3. (6.30) Linear impulse, $J = mv_0$

$$\therefore \quad v_0 = \frac{J}{m} = 2.5 \text{ m/s}$$

$$\therefore \quad v = v_0 e^{-t/\tau}$$

$$\frac{dx}{dt} = v_0 e^{-t/\tau}$$

$$\int_0^x dx = v_0 \int_0^\tau e^{-t/\tau} dt$$

$$x = v_0 \left[\frac{e^{-t/\tau}}{-1/\tau} \right]_0^\tau$$

$$x = 2.5\,(-4)\,(e^{-1} - e^0)$$

$$= 2.5(-4)\,(0.37 - 1)$$

$$x = 6.30 \text{ m}$$